NONNEGATIVE MATRIX AND TENSOR FACTORIZATIONS

NONNEGATIVE MATRIX AND TENSOR FACTORIZATIONS

APPLICATIONS TO EXPLORATORY MULTI-WAY DATA ANALYSIS AND BLIND SOURCE SEPARATION

Andrzej Cichocki

Laboratory for Advanced Brain Signal Processing, Riken Brain Science Institute, Japan; and Warsaw University of Technology and Systems Research Institute, PAN, Poland

Rafal Zdunek

Institute of Telecommunications, Teleinformatics and Acoustics, Wroclaw University of Technology, Poland; and Riken Brain Science Institute, Japan

Anh Huy Phan

Laboratory for Advanced Brain Signal Processing, Riken Brain Science Institute, Japan

Shun-ichi Amari

Research Unit for Mathematical Neuroscience, Riken Brain Science Institute, Japan

WILEY

A John Wiley and Sons, Ltd, Publication

This edition first published 2009

Registered office
John Wiley & Sons Ltd, The Atrium, Southern Gate, Chichester, West Sussex, PO19 8SQ, United Kingdom

For details of our global editorial offices, for customer services and for information about how to apply for permission to reuse the copyright material in this book please see our website at www.wiley.com.

Library of Congress Cataloging-in-Publication Data

Nonnegative matrix and tensor factorizations : applications to exploratory multiway data analysis and blind source separation / Andrzej Cichocki ... [et al.].
p. cm.
Includes bibliographical references and index.
ISBN 978-0-470-74666-0 (cloth)
1. Computer algorithms. 2. Data mining. 3. Machine learning 4. Data structures (Computer science)
I. Cichocki, Andrzej.
QA76.9.A43N65 2009
005.1–dc22 2009016049

A catalogue record for this book is available from the British Library.

ISBN: 978-0-470-74666-0

Set in 9/11 pt Times by Thomson Digital, Noida, India

Contents

Preface

Signal processing, data analysis and data mining are pervasive throughout science and engineering. Extracting an interesting knowledge from experimental raw datasets, measurements, observations and understanding complex data has become an important challenge and objective. Often datasets collected from complex phenomena represent the integrated result of several inter-related variables or they are combinations of underlying latent components or factors. Such datasets can be first decomposed or separated into the components that underlie them in order to discover structures and extract hidden information. In many situations, the measurements are gathered and stored as data matrices or multi-way arrays (tensors), and described by linear or multi-linear models.

Approximative low-rank matrix and tensor factorizations or decompositions play a fundamental role in enhancing the data and extracting latent components. A common thread in various approaches for noise removal, model reduction, feasibility reconstruction, and Blind Source Separation (BSS) is to replace the original data by a lower dimensional approximate representation obtained via a matrix or multi-way array factorization or decomposition. The notion of a matrix factorization arises in a wide range of important applications and each matrix factorization makes a different assumption regarding component (factor) matrices and their underlying structures, so choosing the appropriate one is critical in each application domain. Very often the data, signals or images to be analyzed are nonnegative (or partially nonnegative), and sometimes they also have sparse or smooth representation. For such data, it is preferable to take these constraints into account in the analysis to extract nonnegative and sparse/smooth components or factors with physical meaning or reasonable interpretation, and thereby avoid absurd or unpredictable results. Classical tools cannot guarantee to maintain nonnegativity.

In this book, we provide a wide survey of models and algorithmic aspects of Nonnegative Matrix Factorization (NMF), and its various extensions and modifications, especially Nonnegative Tensor Factorization (NTF) and Nonnegative Tucker Decomposition (NTD). In NTF and NTD approaches, high-dimensional data, such as hyper-spectral or medical images, are factored or decomposed directly and approximated by a sum of rank-one nonnegative tensors. The motivation behind NMF, NTF and NTD is that besides the dimensionality reduction sought in many applications, the underlying data ensemble is nonnegative and can be better modeled and interpreted by means of nonnegative and, ideally, sparse or smooth components.

The notions of NMF, NTF and NTD play a major role in a wide range of important applications, including bioinformatics, micro-array analysis, neuroscience, text mining, image understanding, air pollution research, chemometrics, and spectral data analysis. Other applications include linear sparse coding, image classification, clustering, neural learning process, sound recognition, remote sensing, and object characterization. For example, NMF/NTF processing permits the detection of alternative or context-dependent patterns of gene expression in complex biological systems and especially the recovery of meaningful biological information from cancer-related microarray data. We believe that a potential impact of NMF and its extensions on scientific advancements might be as great as the Independent Component Analysis (ICA), or the Singular Value Decomposition (SVD) and Principal Component Analysis (PCA). In contrast to ICA or SVD/PCA approaches, NMF/NTF and NTD techniques, if successively realized, may improve interpretability and visualization of large-scale data while maintaining the physical feasibility more closely.

Researchers from various research fields are interested in different, usually very diverse, aspects of NMF and NTF. For example, neuroscientists and biologists need reliable methods and techniques which can extract or separate useful information from superimposed biomedical data corrupted by a large level of noise and interference; for example, by using non-invasive recordings of human brain activities targeted at understanding the ability of the brain to sense, recognize, store and recall patterns, and comprehending crucial elements of learning: association, abstraction and generalization. A second group of researchers, engineers and computer scientists, are fundamentally interested in developing and implementing flexible and efficient algorithms for specific practical engineering and scientific applications. A third group of researchers, mathematicians and physicists, have an interest in the development of a fundamental theory to understand mechanisms, properties and abilities of the developed algorithms, and their generalizations to more complex and sophisticated models. Interactions between such groups permit real progress to be made in the very interdisciplinary research devoted to NMF/NTF and NTD, and each group benefits from the others.

The theory built up around NMF, NTF and NTD is so extensive and the applications so numerous that we are, of course, not able to cover all of them. Our selection and treatment of material reflects our background, and our own research interests and results in this fascinating area over the last five years.

The book provides wide coverage of the models and algorithms for nonnegative matrix factorizations and tensor decompositions both from a theoretical and practical point of view. The main objective is to derive and implement, in MATLAB, efficient and relatively simple iterative algorithms that work well in practice for real-world data. In fact, almost all of the algorithms presented in the book have been implemented in MATLAB and extensively tested. We have attempted to present the concepts, models and algorithms in general or flexible forms to stimulate the reader to be creative in visualizing new approaches and adopting methods or algorithms for their specific needs and applications.

In *Chapter 1* we describe the basic NMF models and their extensions, and formulate the fundamental problems related to the calculation of component (factor) matrices. Special emphasis is given to basic properties and mathematical operations for multi-way arrays, also called multi-dimensional matrices or tensors. Chapter 1 introduces the basic linear and multi-linear models for matrix factorizations and tensor decompositions, and formulates the fundamental analytical framework for the solution of the problems posed in this book. The workhorse is the NMF model for sparse representations of data, and its extensions, including the multi-layer NMF, semi-NMF, sparse NMF, tri-NMF, symmetric NMF, orthogonal NMF, non-smooth NMF (nsNMF), overlapping NMF, convolutive NMF (CNMF), and large-scale NMF. Our particular emphasis is on the NMF and semi-NMF models and their extensions to multi-way models (i.e., multi-linear models which perform multi-way array (tensor) decompositions) with nonnegativity and sparsity constraints, especially a family of Tucker and PARAFAC models.

In *Chapter 2*, we give an overview and discuss properties of a large family of generalized and flexible divergences or similarity distances between two nonnegative sequences or patterns. They are formulated for probability distributions used in the development of novel algorithms for NMF and NTF. Information theory, convex analysis, and information geometry play key roles in the formulation of the divergences. The scope of these results is vast since the generalized divergence functions and their variants include quite a large number of useful loss functions, including those based on relative entropies, generalized Kullback-Leibler or I-divergence, Hellinger distance, Jensen-Shannon divergence, J-divergence, Pearson and Neyman Chi-squared divergences, triangular discrimination, and arithmetic-geometric Taneya divergence. Many of these measures belong to the class of Alpha-divergences and Beta-divergences, and have been applied successfully in disciplines such as signal processing, pattern recognition, probability distributions, information theory, finance and economics. In the following chapters we apply such divergences as cost functions (possibly with additional constraints and regularization terms) to derive novel multiplicative and additive projected gradient and fixed-point NMF/NTF algorithms. They provide working solutions for the problems where nonnegative latent (hidden) components can be generally statistically dependent, and satisfy some other conditions or additional constraints such as sparsity or smoothness.

In *Chapter 3*, we introduce a wide family of iterative multiplicative algorithms for NMF and related problems, subject to additional constraints such as sparsity and/or smoothness. Although a standard multiplicative update rule for NMF achieves a sparse representation of its components, we can impose a control over the sparsity of the components by designing a suitable cost function with additional penalty terms. In this chapter we consider a wide class of cost functions or divergences, leading to generalized multiplicative algorithms with regularization and/or penalty terms. Such relaxed forms of the multiplicative NMF algorithms usually provide better performance and convergence speed, and allow us to extract the desired components uniquely up to the scale and permutation ambiguities. As special cases we introduce the multiplicative algorithms for Alpha and Beta divergences, squared Hellinger, Pearson's Chi-squared, and Itakura-Saito distances.

In *Chapter 4*, we derive and give an overview of the Alternating Least Squares algorithms referred to as the ALS algorithms for NMF, semi-NMF, and multi-layer NMF. This is important as many existing NMF techniques are prohibitively slow and inefficient, especially for large-scale problems. For such problems, a promising approach is to apply the ALS algorithms with the regularization and/or extended line search techniques. Special emphasis in this chapter is put on various regularization and penalty terms together with local learning rules. By incorporating the regularization and penalty terms into the weighted Frobenius norm, we show that it is possible to achieve sparse, orthogonal, or smooth representations, thus helping to obtain a desired global solution. The main objective of this chapter is to develop efficient and robust Regularized ALS (RALS) algorithms. For this purpose, we use several approaches from constrained optimization and regularization theory, and in addition introduce several heuristic algorithms. The proposed algorithms are characterized by improved efficiency and often very good convergence properties, especially for large-scale problems.

In *Chapter 5*, we present a wide class of Projected Gradient (PG) algorithms and compare their performance, especially for large-scale problems. In contrast to the multiplicative NMF algorithms discussed in Chapter 3, the PG algorithms have additive updates, and provide an approximate solution to the Non-Negative Least Squares (NNLS) problems. Chapter 5 focuses on the following PG algorithms: Oblique Projected Landweber (OPL), projected gradients with Armijo rule, Barzilai-Borwein gradient projection, Projected Sequential Subspace Optimization (PSESOP), Interior-Point Newton (IPN), Minimal Residual Norm Steepest Descent (MRNSD), and Sequential Coordinate-Wise Algorithm (SCWA).

In *Chapter 6*, we introduce learning algorithms for NMF, using second-order approximations, that is, the information about the Hessian and gradient of a cost function. Using the information about the curvature of the cost function, which is intimately related to second-order derivatives, the convergence can be considerably accelerated. This, however, also introduces many related practical problems that must be addressed prior to applying learning algorithms. For example, the Hessian must be positive-definite to ensure the convergence of approximations to a local minimum of a specific cost function. Unfortunately, this is not guaranteed using the NMF alternating minimization rule, and we need to resort to some suitable Hessian approximation techniques. In addition, the Hessian matrix may be very large, and of a severely ill-conditioned nature (in particular for large-scale problems), which gives rise to many difficult problems related to its inversion. Moreover, the nonlinear projections may be performed in many ways, similarly to the PG and steepest descent algorithms. This chapter provides a comprehensive study of the solutions to the above-mentioned problems. We also give some heuristics on the selection of a cost function and the related regularization terms which restrict the area of feasible solutions, and help the algorithms to converge to the global minimum of a specific cost function. In particular, we discuss the simplest approach to the projected quasi-Newton optimization using the Levenberg-Marquardt regularization of the Hessian. We then extend the discussion to more sophisticated second-order algorithms that iteratively update only the strictly positive (inactive) variables. The example includes the Gradient Projection Conjugate Gradient (GPCG). Furthermore, as a special case of the second-order method, we present one Quadratic Programming (QP) method that solves a QP problem using the trust-region subproblem algorithm. The QP problem is formulated from the Tikhonov regularized squared Euclidean cost function extended with a logarithmic barrier function to satisfy nonnegativity constraints. The BSS experiments demonstrate the high efficiency of the proposed algorithms.

In *Chapter 7*, we attempt to extend and generalize the results and algorithms from the previous chapters for the NTF, BOD and NTD models. In fact, almost all the NMF algorithms described in the earlier chapters can be extended or generalized to the various nonnegative tensor factorizations and decompositions formulated in Chapter 1. However, in this chapter we mainly focus on NTF, that is, PARAFAC with nonnegativity constraints and NTD. In order to make this chapter as self-contained as possible, we re-introduce some concepts and derive several novel and efficient algorithms for nonnegative and semi-nonnegative tensor (multi-way arrays) factorizations and decompositions. Our particular emphasis is on a detailed treatment of the generalized cost functions, including Alpha- and Beta-divergences. Based on these cost functions, several classes of algorithms are introduced, including: (1) multiplicative updating; (2) ALS; and (3) Hierarchical ALS (HALS). These algorithms are then incorporated into multi-layer hierarchical networks in order to improve their performance. Special emphasis is given to the ways to impose nonnegativity or semi-nonnegativity, together with optional constraints such as orthogonality, sparsity and/or smoothness. The developed algorithms are tested for several applications such as denoising, compression, feature extraction, clustering, EEG data analysis, brain computer interface and video tracking. To understand the material in this chapter it would be helpful to be familiar with the previous chapters, especially Chapters 1, 3 and 4.

Finally, in *Chapter 8*, we briefly discuss the selected applications of NMF and multi-dimensional array decompositions, with a special emphasis on these applications to which the algorithms described in the previous chapters are applicable. We review the following applications: data clustering, text mining, email surveillance, musical instrument classification, face recognition, handwritten digit recognition, texture classification, Raman spectroscopy, fluorescence spectroscopy, hyper-spectral imaging, chemical shift imaging, and gene expression classification.

The book is partly a textbook and partly a research monograph. It is a textbook because it gives the detailed introduction to the basic models and algorithms of nonnegative and sparse matrix and tensor decompositions. It is simultaneously a monograph because it presents many new results, methods, ideas, models, further developments and implementation of efficient algorithms which are brought together and published in this book for the first time. As a result of its twofold character, the book is likely to be of interest to graduate and postgraduate students, engineers and scientists working in the fields of biomedical engineering, data analysis, data mining, multidimensional data visualization, signal/image processing, mathematics, computer science, finance, economics, optimization, geophysics, and neural computing. Furthermore, the book may also be of interest to researchers working in different areas of science, because a number of the results and concepts included may be advantageous for their further research. One can read this book through sequentially but it is not necessary, since each chapter is essentially self-contained, with as few cross references as possible. So, browsing is encouraged.

Acknowledgments

The authors would like to express their appreciation and gratitude to a number of researchers who helped them in a variety of ways, directly and also indirectly, in the development of this book.

First of all, we would like to express our sincere gratitude to Professor Susumu Tonegawa and Professor Keiji Tanaka directors of the RIKEN Brain Science Institute for creating a great scientific environment for multidisciplinary research and the promotion of international scientific collaboration.

Although this book is derived from the research activities of the authors over the past five years on this subject, many influential results and approaches are developed in collaboration with our colleagues and researchers from the RIKEN Brain Science Institute, Japan, and several worldwide universities.

Many of them have made important and influential contributions. Special thanks and gratitude go to Professors Jonathon Chambers from Loughborough University, Robert Plemmons from Wake Forest University, Seungjin Choi, Hyekyoung Lee and Yong-Deok Kim from the Department of Computer Science and Engineering at Pohang University of Science and Technology (POSTECH), Danilo Mandic from Imperial College London, Sergio A. Cruces-Alvarez from E.S. Ingenieros, University of Seville, Spain and Cesar Caiafa from Universidad de Buenos Aires, Argentina.

Over various phases of writing this book, several people have kindly agreed to read and comment on parts or all of the manuscript. We are very grateful for the insightful comments and suggestions of Danilo Mandic, Jonathon Chambers, Cesar Caifa, Saied Sanei, Nicolas Gillis, Yoshikazu Washizawa, Noboru Murata, Shinto Eguchi, Ivica Kopriva, Abd-Krim Seghouane, Scott Douglas, Sergio Cruces, Wenwu Wang, Zhaoshui He, and Paul Sajda.

Finally, we must acknowledge the help and understanding of our families and friends during the past two years while we carried out this challenging project.

A. Cichocki, R. Zdunek, A.H. Phan, and S. Amari

January 25, 2009, Tokyo, JAPAN

Glossary of Symbols and Abbreviations

Principal Symbols

$\mathbb{R}_+$	nonnegative real number
$\mathbb{R}^n$	n-dimensional real vector space
$\mathbf{A} = [a_{ij}] \in \mathbb{R}^{I \times J}$	I by J matrix (mixing, basis or dictionary matrix)
a_{ij}	ij-th element of the matrix $\mathbf{A} \in \mathbb{R}^{I \times J}$
I	number of observations, sensors or available time series
J	number of components or common factors
$\mathbf{A}^{-1}$	inverse of a nonsingular matrix $\mathbf{A}$
$\mathbf{A}^{\dagger}$	Moore-Penrose pseudo-inverse of a matrix $\mathbf{A}$
$\mathbf{A} \geq \mathbf{0}$	nonnegative matrix $\mathbf{A}$ with all entries $a_{ij} \geq 0$
$\mathbf{A} \geq \boldsymbol{\varepsilon}$	positive matrix $\mathbf{A}$ with entries $a_{ij} \geq \varepsilon$, where ε is small positive constant
$\mathbf{X} \in \mathbb{R}^{J \times T}$	matrix representing hidden components
$\mathbf{Y} \in \mathbb{R}^{I \times T}$	matrix representing input data, e.g., measurements
$\mathbf{E} \in \mathbb{R}^{I \times T}$	matrix representing error or noise
$\arg\min_{\theta} J(\theta)$	denotes the value of θ that minimizes $J(\theta)$
x_i	i-th element of vector $\boldsymbol{x}$
$y_{it} = y_i(t)$	element of time series $y_i(t)$ for a time instant t or it-th entry of $\mathbf{Y}$
$[\mathbf{AX}]_j$	j-th column vector of the matrix $\hat{\mathbf{Y}} = \mathbf{AX}$
$[\mathbf{AX}]_{it}$	it-th element of the matrix $\hat{\mathbf{Y}} = \mathbf{AX}$
$\underline{\mathbf{A}} \in \mathbb{R}^{I \times J \times Q}$	three-way array (third-order tensor) representing basis matrices (slices)
a_{ijq}	ijq-th element of a three-way array $\underline{\mathbf{A}} \in \mathbb{R}^{I \times J \times Q}$
$\underline{\mathbf{G}} \in \mathbb{R}^{J \times R \times P}$	third-order core tensor representing links between factor matrices
g_{jrp}	jrp-th element of a three-way array $\underline{\mathbf{G}} \in \mathbb{R}^{J \times R \times P}$

$\underline{\mathbf{X}} \in \mathbb{R}^{J \times T \times Q}$	three-way array representing hidden components
x_{jtq}	jtq-th element of a three-way array $\underline{\mathbf{X}} \in \mathbb{R}^{J \times T \times Q}$
$\underline{\mathbf{Y}} \in \mathbb{R}^{I \times T \times Q}$	three-way array (a third-order tensor) representing input (observed) data
y_{itq}	itq-th element of the three-way array $\underline{\mathbf{Y}} \in \mathbb{R}^{I \times T \times Q}$
$\underline{\mathbf{E}} \in \mathbb{R}^{I \times T \times Q}$	three-way array (third-order tensor) representing error or noise
e_{itq}	the itq-th element of the three-way array $\underline{\mathbf{E}} \in \mathbb{R}^{I \times T \times Q}$
$\underline{\mathbf{Y}} \in \mathbb{R}^{I_1 \times I_2 \cdots \times I_N}$	Nth order tensor representing usually input (observed) data
$\mathbf{Y}_{(n)}$	n-mode matricized (unfolded) version of $\underline{\mathbf{Y}}$
$\underline{\mathbf{Y}}_{i_n=p}$	subtensor of the tensor $\underline{\mathbf{Y}} \in \mathbb{R}^{I_1 \times I_2 \cdots \times I_N}$ obtained by fixing the n-th index to some value p
$\underline{\mathbf{G}} \in \mathbb{R}^{J_1 \times J_2 \cdots \times J_N}$	N-th order core tensor
$\mathbf{A}^{(n)}$	the n-th factor (loading matrix) of N-th order tensor
$\boldsymbol{a}_j^{(n)}$	j-th column vector of factor $\mathbf{A}^{(n)}$
$\{\mathbf{A}\}$	set of factors (loading matrices) $\mathbf{A}^{(1)}, \mathbf{A}^{(2)}, \ldots, \mathbf{A}^{(N)}$
$\{\boldsymbol{a}_j\}$	set of vectors $\boldsymbol{a}_j^{(1)}, \boldsymbol{a}_j^{(2)}, \ldots, \boldsymbol{a}_j^{(N)}$
$D(\boldsymbol{p}\|\|\boldsymbol{q})$	divergence between two nonnegative 1D sequences: $\boldsymbol{p} = \{p_t\}$ and $\boldsymbol{q} = \{q_t\}$
$D(\mathbf{P}\|\|\mathbf{Q})$	divergence between two nonnegative 2D sequences: $\mathbf{P} = \{p_{it}\}$ and $\mathbf{Q} = \{q_{it}\}$
$D(\underline{\mathbf{P}}\|\|\underline{\mathbf{Q}})$	divergence between two nonnegative three-way arrays: $\underline{\mathbf{P}} = \{p_{itq}\}$ and $\underline{\mathbf{Q}} = \{q_{itq}\}$
$\mathbf{D}$	diagonal scaling matrix
$\det(\mathbf{A})$	determinant of matrix $\mathbf{A}$
$\mathbf{1}$	vector or matrix of suitable dimension with all ones
$\mathbf{1}_{I \times J} \in \mathbb{R}^{I \times J}$	I by J matrix with all ones
$\exp\{\cdot\}$	exponential function
$E\{\cdot\}$	expectation operator
$\mathbf{I}$	identity matrix
$\mathbf{I}_n$	identity matrix of dimension $n \times n$
$J(\mathbf{X})$	penalty or regularization term of a cost function
ln	natural logarithm (equivalent to log)
k	k-th alternating step
K	total number of alternating steps

Q	number of frontal slices, trials, experiments or subjects
I, T, Q	dimensions of different modes in three-way array $\underline{\mathbf{Y}}$
$p(\boldsymbol{x})$ or $p_x(\boldsymbol{x})$	probability density function (p.d.f.) of $\boldsymbol{x}$
$p_y(\boldsymbol{y})$	probability density function (p.d.f.) of $\boldsymbol{y}(t)$
$\mathbf{P}$	permutation matrix
$\text{sign}\,(x)$	sign function (= 1 for $x > 0$ and $= -1$ for $x < 0$)
t	continuous or discrete time
$\text{tr}\,(\mathbf{A})$	trace of matrix $\mathbf{A}$
$\|x\|$	absolute value (magnitude) of x
$\\|\boldsymbol{x}\\|_p$	p-norm (length) of the vector $\boldsymbol{x}$, where $p = 0, 1, 2, \ldots, \infty$
δ_{ij}	Kronecker delta
η	learning rate for discrete-time algorithms
$\lambda_{max}(\mathbf{A})$	maximal eigenvalue of matrix $\mathbf{A}$
$\lambda_{min}(\mathbf{A})$	minimal eigenvalue of matrix $\mathbf{A}$
σ^2	variance
ω	over-relaxation parameter $0 < \omega \leq 2$
∇	gradient operator
$\nabla_{x_i} D$	gradient of D with respect to variable x_i
$\nabla_{\mathbf{X}} D$	gradient of cost function D with respect to matrix $\mathbf{X}$
$[\boldsymbol{x}]_+ = \max\{0, \boldsymbol{x}\}$	nonlinear "half-wave rectifying" projection replacing negative values of $\boldsymbol{x}$ (element-wise) by zero or by a small positive value ε
$[\cdot]^{\dagger}$	superscript symbol for Moore-Penrose pseudo-inversion of a matrix
$[\cdot]^T$	transpose of matrix or vector
$\langle\cdot\rangle$	inner product or average operator
$:=$ or $\triangleq$	left hand variable is defined by the right hand variable
$\circledast$ or $.*$	Hadamard product
$\oslash$ or $./$	element-wise division
$\mathbf{X}^{[.\alpha]}$	rise to the power α each element of matrix $\mathbf{X}$
$\circ$	outer product
$\odot$	Khatri-Rao product
$\otimes$	Kronecker product ($\mathbf{A} \otimes \mathbf{B} \triangleq [a_{ij}\mathbf{B}]$)
$\times_n$	$n - mode$ product of a tensor by matrix

$\bar{\times}_n$	$n - mode$ product of a tensor by vector
$\mathbf{A} \sqcup_n \mathbf{B}$	concatenation of two tensors along the n-th mode
$\mathbf{A}^{\odot}$	$\mathbf{A}^{(N)} \odot \mathbf{A}^{(N-1)} \odot \cdots \odot \mathbf{A}^{(1)}$
$\mathbf{A}^{\odot_{-n}}$	$\mathbf{A}^{(N)} \odot \cdots \odot \mathbf{A}^{(n+1)} \odot \mathbf{A}^{(n-1)} \odot \cdots \odot \mathbf{A}^{(1)}$
$\mathbf{A}^{\otimes}$	$\mathbf{A}^{(N)} \otimes \mathbf{A}^{(N-1)} \otimes \cdots \otimes \mathbf{A}^{(1)}$
$\mathbf{A}^{\otimes_{-n}}$	$\mathbf{A}^{(N)} \otimes \cdots \times \mathbf{A}^{(n+1)} \otimes \mathbf{A}^{(n-1)} \otimes \cdots \otimes \mathbf{A}^{(1)}$
$\underline{\mathbf{G}} \times \{\mathbf{A}\}$	$\underline{\mathbf{G}} \times_1 \mathbf{A}^{(1)} \times_2 \mathbf{A}^{(2)} \cdots \times_N \mathbf{A}^{(N)}$
$\underline{\mathbf{G}} \times_{-n} \{\mathbf{A}\}$	$\underline{\mathbf{G}} \times_1 \mathbf{A}^{(1)} \cdots \times_{n-1} \mathbf{A}^{(n-1)} \times_{n+1} \mathbf{A}^{(n+1)} \cdots \times_N \mathbf{A}^{(N)}$

Abbreviations

ALS	Alternating Least Squares
HALS	Hierarchical Alternating Least Squares
RALS	Regularized Alternating Least Squares
i.i.d.	independent identically distributed
BOD	Block Oriented Decomposition
BSE	Blind Signal Extraction
BSS	Blind Signal Separation
BSD	Blind Signal Deconvolution
EVD	Eigenvalue Decomposition
ICA	Independent Component Analysis
IIR	Infinite Impulse Response
SIR	Signal-to-Interference-Ratio
PSNR	Peak Signal-to-Noise-Ratio
JAD	Joint Approximative Diagonalization
LMS	Least Mean Squares
KL	Kullback Leibler divergence
MCA	Minor Component Analysis
MBD	Multichannel Blind Deconvolution
MED	Maximum Entropy Distribution
MoCA	Morphological Component Analysis
MIMO	Multiple-Input, Multiple-Output
NMF	Nonnegative Matrix Factorization

aNMF	affine NMF
CNMF	Convolutive NMF
nsNMF	non-smooth NMF
NTD	Nonnegative Tucker Decomposition
NTF	Nonnegative Tensor Factorization
CP	CANDECOMP/PARAFAC
PARAFAC	Parallel Factor Analysis (equivalent to CP)
CANNDECOPM	Canonical Decomposition (equivalent to PARAFAC or CP)
PCA	Principal Component Analysis
SCA	Sparse Component Analysis
SOD	Slice Oriented Decomposition
SVD	Singular Value Decomposition

1

Introduction – Problem Statements and Models

Matrix factorization is an important and unifying topic in signal processing and linear algebra, which has found numerous applications in many other areas. This chapter introduces basic linear and multi-linear[1] models for matrix and tensor factorizations and decompositions, and formulates the analysis framework for the solution of problems posed in this book. The workhorse in this book is Nonnegative Matrix Factorization (NMF) for sparse representation of data and its extensions including the multi-layer NMF, semi-NMF, sparse NMF, tri-NMF, symmetric NMF, orthogonal NMF, non-smooth NMF (nsNMF), overlapping NMF, convolutive NMF (CNMF), and large-scale NMF. Our particular emphasis is on NMF and semi-NMF models and their extensions to multi-way models (i.e., multi-linear models which perform multi-way array (tensor) decompositions) with nonnegativity and sparsity constraints, including, Nonnegative Tucker Decompositions (NTD), Constrained Tucker Decompositions, Nonnegative and semi-nonnegative Tensor Factorizations (NTF) that are mostly based on a family of the TUCKER, PARAFAC and PARATUCK models.

As the theory and applications of NMF, NTF and NTD are still being developed, our aim is to produce a unified, state-of-the-art framework for the analysis and development of efficient and robust algorithms. In doing so, our main goals are to:

1. Develop various working tools and algorithms for data decomposition and feature extraction based on nonnegative matrix factorization (NMF) and sparse component analysis (SCA) approaches. We thus integrate several emerging techniques in order to estimate physically, physiologically, and neuroanatomically meaningful sources or latent (hidden) components with morphological constraints. These constraints include nonnegativity, sparsity, orthogonality, smoothness, and semi-orthogonality.
2. Extend NMF models to multi-way array (tensor) decompositions, factorizations, and filtering, and to derive efficient learning algorithms for these models.
3. Develop a class of advanced blind source separation (BSS), unsupervised feature extraction and clustering algorithms, and to evaluate their performance using *a priori* knowledge and morphological constraints.
4. Develop computational methods to efficiently solve the bi-linear system $\mathbf{Y} = \mathbf{AX} + \mathbf{E}$ for noisy data, where $\mathbf{Y}$ is an input data matrix, $\mathbf{A}$ and $\mathbf{X}$ represent unknown matrix factors to be estimated, and the matrix $\mathbf{E}$ represents error or noise (which should be minimized using suitably designed cost function).

[1] A function in two or more variables is said to be multi-linear if it is linear in each variable separately.

Nonnegative Matrix and Tensor Factorizations: Applications to Exploratory Multi-way Data Analysis and Blind Source Separation Andrzej Cichocki, Rafal Zdunek, Anh Huy Phan and Shun-ichi Amari

5. Describe and analyze various cost functions (also referred to as (dis)similarity measures or divergences) and apply optimization criteria to ensure robustness with respect to uncertainty, ill-conditioning, interference and noise distribution.
6. Present various optimization techniques and statistical methods to derive efficient and robust learning (update) rules.
7. Study what kind of prior information and constraints can be used to render the problem solvable, and illustrate how to use this information in practice.
8. Combine information from different imaging modalities (e.g., electroencephalography (EEG), magnetoencephalography (MEG), electromyography (EMG), electrooculography (EOG), functional magnetic resonance imaging (fMRI), positron emission tomography (PET)), in order to provide data integration and assimilation.
9. Implement and optimize algorithms for NMF, NTF and NTD together with providing pseudo-source codes and/or efficient source codes in MATLAB, suitable for parallel computing and large-scale-problems.
10. Develop user-friendly toolboxes which supplement this book: NMFLAB and MULTI-WAY-LAB for potential applications to data analysis, data mining, and blind source separation.

Probably the most useful and best understood matrix factorizations are the Singular Value Decomposition (SVD), Principal Component Analysis (PCA), and LU, QR, and Cholesky decompositions (see Appendix). In this book we mainly focus on nonnegativity and sparsity constraints for factor matrices. We shall therefore attempt to illustrate why nonnegativity and sparsity constraints play a key role in our investigations.

1.1 Blind Source Separation and Linear Generalized Component Analysis

Blind source separation (BSS) and related methods, e.g., independent component analysis (ICA), employ a wide class of unsupervised learning algorithms and have found important applications across several areas from engineering to neuroscience [26]. The recent trends in blind source separation and generalized (flexible) component analysis (GCA) are to consider problems in the framework of matrix factorization or more general multi-dimensional data or signal decomposition with probabilistic generative models and exploit *a priori* knowledge about true nature, morphology or structure of latent (hidden) variables or sources such as nonnegativity, sparseness, spatio-temporal decorrelation, statistical independence, smoothness or lowest possible complexity. The goal of BSS can be considered as estimation of true physical sources and parameters of a mixing system, while the objective of GCA is to find a reduced or hierarchical and structured component representation for the observed (sensor) data that can be interpreted as physically or physiologically meaningful coding or blind signal decomposition. The key issue is to find such a transformation or coding which has true physical meaning and interpretation.

Throughout this book we discuss some promising applications of BSS/GCA in analyzing multi-modal, multi-sensory data, especially brain data. Furthermore, we derive some efficient unsupervised learning algorithms for linear blind source separation, and generalized component analysis using various criteria, constraints and assumptions.

Figure 1.1 illustrates a fairly general BSS problem also referred to as blind signal decomposition or blind source extraction (BSE). We observe records of I sensor signals $\boldsymbol{y}(t) = [y_1(t), y_2(t), \ldots, y_I(t)]^T$ coming from a MIMO (multiple-input/multiple-output) mixing and filtering system, where t is usually a discrete time sample,[2] and $(\cdot)^T$ denotes transpose of a vector. These signals are usually a superposition (mixture) of J unknown source signals $\boldsymbol{x}(t) = [x_1(t), x_2(t), \ldots, x_J(t)]^T$ and noises $\boldsymbol{e}(t) = [e_1(t), e_2(t), \ldots, e_I(t)]^T$. The primary

[2] Data are often represented not in the time domain but in the complex frequency or time-frequency domain, so, the index t may have a different meaning and can be multi-dimensional.

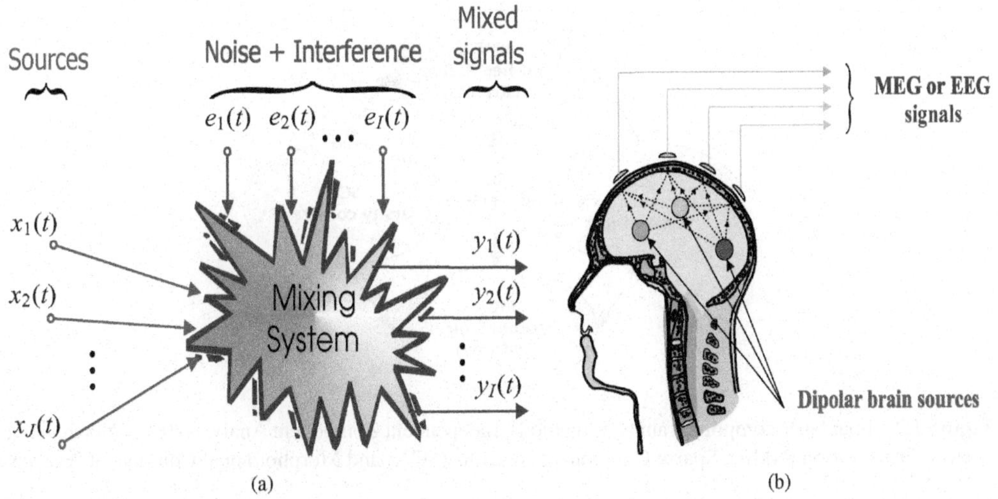

Figure 1.1 (a) General model illustrating blind source separation (BSS), (b) Such models are exploited, for example, in noninvasive multi-sensor recording of brain activity using EEG (electroencephalography) or MEG (magnetoencephalography). It is assumed that the scalp sensors (e.g., electrodes, magnetic or optical sensors) pick up a superposition of neuronal brain sources and non-neuronal sources (noise or physiological artifacts) related, for example, to movements of eyes and muscles. Our objective is to identify the individual signals coming from different areas of the brain.

objective is to estimate all the primary source signals $x_j(t) = x_{jt}$ or only some of them with specific properties. This estimation is usually performed based only on the output (sensor, observed) signals $y_{it} = y_i(t)$.

In order to estimate sources, sometimes we try first to identify the mixing system or its inverse (unmixing) system and then estimate the sources. Usually, the inverse (unmixing) system should be adaptive in such a way that it has some tracking capability in a nonstationary environment. Instead of estimating the source signals directly by projecting observed signals using the unmixing system, it is often more convenient to identify an unknown mixing and filtering system (e.g., when the unmixing system does not exist, especially when the system is underdetermined, i.e., the number of observations is lower than the number of source signals with $I < J$) and simultaneously estimate the source signals by exploiting some *a priori* information about the source signals and applying a suitable optimization procedure.

There appears to be something magical about blind source separation since we are estimating the original source signals without knowing the parameters of the mixing and/or filtering processes. It is difficult to imagine that one can estimate this at all. In fact, without some *a priori* knowledge, it is not possible to *uniquely* estimate the original source signals. However, one can usually estimate them up to certain indeterminacies. In mathematical terms these indeterminacies and ambiguities can be expressed as arbitrary scaling and permutation of the estimated source signals. These indeterminacies preserve, however, the waveforms of original sources. Although these indeterminacies seem to be rather severe limitations, in a great number of applications these limitations are not crucial, since the most relevant information about the source signals is contained in the temporal waveforms or time-frequency patterns of the source signals and usually not in their amplitudes or the order in which they are arranged in the system output.[3]

[3]For some models, however, there is no guarantee that the estimated or extracted signals have exactly the same waveforms as the source signals, and then the requirements must be sometimes further relaxed to the extent that the extracted waveforms are distorted (i.e., time delayed, filtered or convolved) versions of the primary source signals.

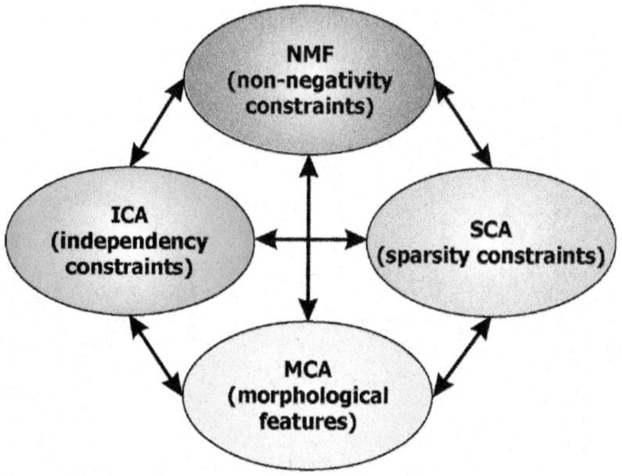

Figure 1.2 Four basic component analysis methods: Independent Component Analysis (ICA), Nonnegative Matrix Factorization (NMF), Sparse Component Analysis (SCA) and Morphological Component Analysis (MCA).

The problem of separating or extracting source signals from a sensor array, without knowing the transmission channel characteristics and the sources, can be briefly expressed as a number of related BSS or GCA methods such as ICA and its extensions: Topographic ICA, Multi-way ICA, Kernel ICA, Tree-dependent Component Analysis, Multi-resolution Subband Decomposition - ICA [28,29,41,77], Non-negative Matrix Factorization (NMF) [35,94,121], Sparse Component Analysis (SCA) [70,72,96,97,142], and Multi-channel Morphological Component Analysis (MCA) [13] (see Figure 1.2).

The mixing and filtering processes of the unknown input sources $x_j(t)$, $(j = 1, 2, \ldots, J)$ may have different mathematical or physical models, depending on the specific applications [4,77]. Most linear BSS models in their simplest forms can be expressed algebraically as some specific forms of matrix factorization: Given observation (often called sensor or input data matrix) $\mathbf{Y} = [y_{it}] = [\boldsymbol{y}(1), \ldots, \boldsymbol{y}(T)] \in \mathbb{R}^{I \times T}$ perform the matrix factorization (see Figure 1.3(a)):

$$\boxed{\mathbf{Y} = \mathbf{A}\mathbf{X} + \mathbf{E},} \tag{1.1}$$

where $\mathbf{A} \in \mathbb{R}^{I \times J}$ represents the unknown basis matrix or mixing matrix (depending on the application), $\mathbf{E} \in \mathbb{R}^{I \times T}$ is an unknown matrix representing errors or noises, $\mathbf{X} = [x_{jt}] = [\boldsymbol{x}(1), \boldsymbol{x}(2), \ldots, \boldsymbol{x}(T)] \in \mathbb{R}^{J \times T}$ contains the corresponding latent (hidden) components that give the contribution of each basis vector, T is the number of available samples, I is the number of observations and J is the number of sources or components. In general, the number of source signals J is unknown and can be larger, equal or smaller than the number of observations. The above model can be written in an equivalent scalar (element-wise) form (see Figure 1.3(b)):

$$y_{it} = \sum_{j=1}^{J} a_{ij}\, x_{jt} + e_{it} \qquad \text{or} \qquad y_i(t) = \sum_{j=1}^{J} a_{ij}\, x_j(t) + e_i(t). \tag{1.2}$$

Usually, the latent components represent unknown source signals with specific statistical properties or temporal structures. The matrices usually have clear statistical properties and meanings. For example, the rows of the matrix $\mathbf{X}$ that represent sources or components should be statistically independent for ICA or sparse for SCA [69,70,72,96,97], nonnegative for NMF, or have other specific and additional morphological properties such as sparsity, smoothness, continuity, or orthogonality in GCA [13,26,29].

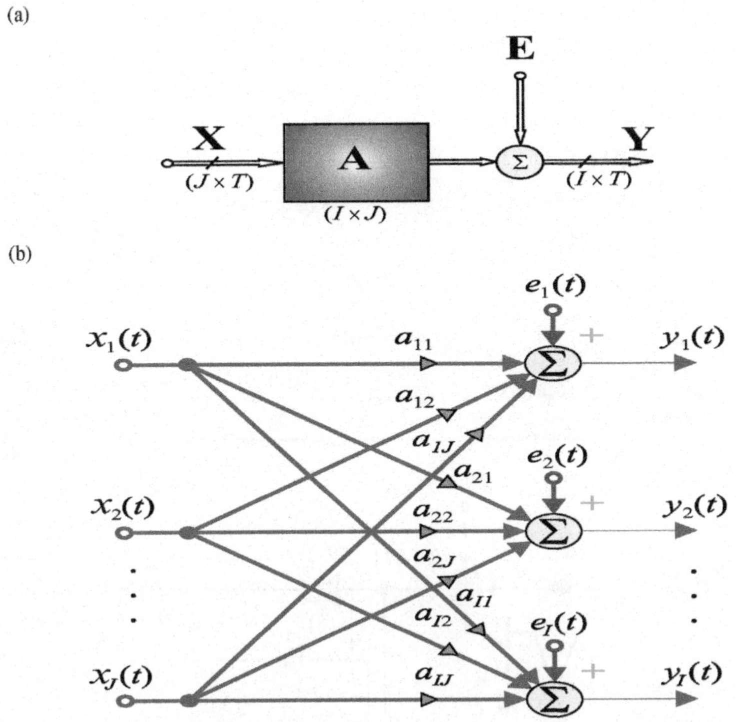

Figure 1.3 Basic linear instantaneous BSS model: (a) Block diagram, (b) detailed model.

In some applications the mixing matrix $\mathbf{A}$ is ill-conditioned or even singular. In such cases, some special models and algorithms should be applied. Although some decompositions or matrix factorizations provide an exact reconstruction of the data (i.e., $\mathbf{Y} = \mathbf{AX}$), we shall consider here factorizations which are approximative in nature. In fact, many problems in signal and image processing can be solved in terms of matrix factorization. However, different cost functions and imposed constraints may lead to different types of matrix factorization. In many signal processing applications the data matrix $\mathbf{Y} = [\boldsymbol{y}(1), \boldsymbol{y}(2) \ldots, \boldsymbol{y}(T)] \in \mathbb{R}^{I \times T}$ is represented by vectors $\boldsymbol{y}(t) \in \mathbb{R}^I$ $(t = 1, 2, \ldots, T)$ for a set of discrete time instants t as multiple measurements or recordings. As mentioned above, the compact aggregated matrix equation (1.1) can be written in a vector form as a system of linear equations (see Figure 1.4(a)), that is,

$$\boldsymbol{y}(t) = \mathbf{A}\,\boldsymbol{x}(t) + \boldsymbol{e}(t), \qquad (t = 1, 2, \ldots, T), \tag{1.3}$$

where $\boldsymbol{y}(t) = [y_1(t), y_2(t), \ldots, y_I(t)]^T$ is a vector of the observed signals at the discrete time instant t whereas $\boldsymbol{x}(t) = [x_1(t), x_2(t), \ldots, x_J(t)]^T$ is a vector of unknown sources at the same time instant. The problems formulated above are closely related to the concept of linear inverse problems or more generally, to solving a large ill-conditioned system of linear equations (overdetermined or underdetermined), where it is required to estimate vectors $\boldsymbol{x}(t)$ (also in some cases to identify a matrix $\mathbf{A}$) from noisy data [26,32,88]. Physical systems are often contaminated by noise, thus, our task is generally to find an optimal and robust solution in a noisy environment. Wide classes of extrapolation, reconstruction, estimation, approximation, interpolation and inverse problems can be converted into minimum norm problems of solving underdetermined systems

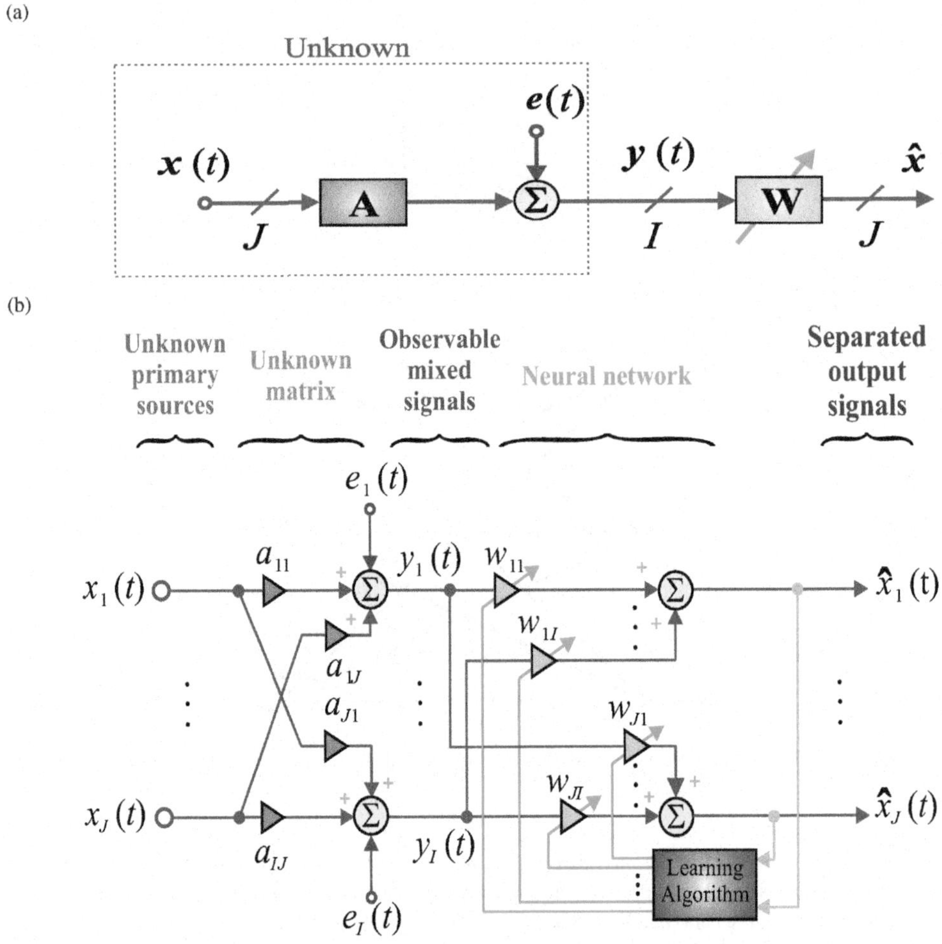

Figure 1.4 Blind source separation using unmixing (inverse) model: (a) block diagram, and(b) detailed model.

of linear equations (1.3) for $J > I$ [26,88].[4] It is often assumed that only the sensor vectors $\mathbf{y}(t)$ are available and we need to estimate parameters of the unmixing system online. This enables us to perform indirect identification of the mixing matrix $\mathbf{A}$ (for $I \geq J$) by estimating the separating matrix $\mathbf{W} = \hat{\mathbf{A}}^{\dagger}$, where the symbol $(\cdot)^{\dagger}$ denotes the Moore-Penrose pseudo-inverse and simultaneously estimate the sources. In other words, for $I \geq J$ the original sources can be estimated by the linear transformation

$$\hat{\mathbf{x}}(t) = \mathbf{W}\,\mathbf{y}(t), \qquad (t = 1, 2, \ldots, T). \tag{1.4}$$

[4]Generally speaking, in signal processing applications, an overdetermined ($I > J$) system of linear equations (1.3) describes filtering, enhancement, deconvolution and identification problems, while the underdetermined case describes inverse and extrapolation problems [26,32].

Although many different BSS criteria and algorithms are available, most of them exploit various diversities[5] or constraints imposed for estimated components and/or mixing matrices such as mutual independence, nonnegativity, sparsity, smoothness, predictability or lowest complexity. More sophisticated or advanced approaches use combinations or integration of various diversities, in order to separate or extract sources with various constraints, morphology, structures or statistical properties and to reduce the influence of noise and undesirable interferences [26].

All the above-mentioned BSS methods belong to a wide class of unsupervised learning algorithms. Unsupervised learning algorithms try to discover a structure underlying a data set, extract of meaningful features, and find useful representations of the given data. Since data can always be interpreted in many different ways, some knowledge is needed to determine which features or properties best represent our true latent (hidden) components. For example, PCA finds a low-dimensional representation of the data that captures most of its variance. On the other hand, SCA tries to explain data as a mixture of sparse components (usually, in the time-frequency domain), and NMF seeks to explain data by parts-based localized additive representations (with nonnegativity constraints).

Generalized component analysis algorithms, i.e., a combination of ICA, SCA, NMF, and MCA, are often considered as pure mathematical formulas, powerful, but rather mechanical procedures. There is an illusion that there is not very much left for the user to do after the machinery has been optimally implemented. However, the successful and efficient use of such tools strongly depends on *a priori* knowledge, common sense, and appropriate use of the preprocessing and postprocessing tools. In other words, it is the preprocessing of data and postprocessing of models where expertise is truly needed in order to extract and identify physically significant and meaningful hidden components.

1.2 Matrix Factorization Models with Nonnegativity and Sparsity Constraints

1.2.1 Why Nonnegativity and Sparsity Constraints?

Many real-world data are nonnegative and the corresponding hidden components have a physical meaning only when nonnegative. In practice, both nonnegative and sparse decompositions of data are often either desirable or necessary when the underlying components have physical interpretation. For example, in image processing and computer vision, involved variables and parameters may correspond to pixels, and nonnegative sparse decomposition is related to the extraction of relevant parts from the images [94,95]. In computer vision and graphics, we often encounter multi-dimensional data, such as images, video, and medical data, one type of which is MRI (magnetic resonance imaging). A color image can be considered as 3D nonnegative data, two of the dimensions (rows and columns) being spatial, and the third one being a color plane (channel) depending on its color space, while a color video sequence can be considered as 4D nonnegative data, time being the fourth dimension. A sparse representation of the data by a limited number of components is an important research problem. In machine learning, sparseness is closely related to feature selection and certain generalizations in learning algorithms, while nonnegativity relates to probability distributions. In economics, variables and data such as volume, price and many other factors are nonnegative and sparse. Sparseness constraints may increase the efficiency of a portfolio, while nonnegativity both increases efficiency and reduces risk [123,144]. In microeconomics, household expenditure in different commodity/service groups are recorded as a relative proportion. In information retrieval, documents are usually represented as relative frequencies of words in a prescribed vocabulary. In environmental science, scientists investigate a relative proportion of different pollutants in water or air [11]. In biology, each coordinate axis may correspond to a specific gene and the sparseness is necessary for finding local patterns hidden in data, whereas the nonnegativity is required to give physical or physiological meaning. This is also important for

[5]By diversities we mean usually different morphological characteristics or features of the signals.

the robustness of biological systems, where any observed change in the expression level of a specific gene emerges from either positive or negative influence, rather than a combination of both, which partly cancel each other [95,144].

It is clear, however, that with constraints such as sparsity and nonnegativity some of the explained variance (FIT) may decrease. In other words, it is natural to seek a trade-off between the two goals of interpretability (making sure that the estimated components have physical or physiological sense and meaning) and statistical fidelity (explaining most of the variance of the data, if the data are consistent and do not contain too much noise). Generally, compositional data (i.e., positive sum of components or real vectors) are natural representations when the variables (features) are essentially the probabilities of complementary and mutually exclusive events. Furthermore, note that NMF is an additive model which does not allow subtraction; therefore it often quantitatively describes the parts that comprise the entire entity. In other words, NMF can be considered as a part-based representation in which a zero-value represents the absence and a positive number represents the presence of some event or component. Specifically, in the case of facial image data, the additive or part-based nature of NMF has been shown to result in a basis of facial features, such as eyes, nose, and lips [94]. Furthermore, matrix factorization methods that exploit nonnegativity and sparsity constraints usually lead to estimation of the hidden components with specific structures and physical interpretations, in contrast to other blind source separation methods.

1.2.2 Basic NMF Model

NMF has been investigated by many researchers, e.g. Paatero and Tapper [114], but it has gained popularity through the works of Lee and Seung published in Nature and NIPS [94,95]. Based on the argument that the nonnegativity is important in human perception they proposed simple algorithms (often called the Lee-Seung algorithms) for finding nonnegative representations of nonnegative data and images.

The basic NMF problem can be stated as follows: Given a nonnegative data matrix $\mathbf{Y} \in \mathbb{R}_+^{I \times T}$ (with $y_{it} \geq 0$ or equivalently $\mathbf{Y} \geq \mathbf{0}$) and a reduced rank J ($J \leq \min(I, T)$), find two nonnegative matrices $\mathbf{A} = [\boldsymbol{a}_1, \boldsymbol{a}_2, \ldots, \boldsymbol{a}_J] \in \mathbb{R}_+^{I \times J}$ and $\mathbf{X} = \mathbf{B}^T = [\boldsymbol{b}_1, \boldsymbol{b}_2, \ldots, \boldsymbol{b}_J]^T \in \mathbb{R}_+^{J \times T}$ which factorize $\mathbf{Y}$ as well as possible, that is (see Figure 1.3):

$$\boxed{\mathbf{Y} = \mathbf{A}\mathbf{X} + \mathbf{E} = \mathbf{A}\mathbf{B}^T + \mathbf{E},} \tag{1.5}$$

where the matrix $\mathbf{E} \in \mathbb{R}^{I \times T}$ represents approximation error.[6] The factors $\mathbf{A}$ and $\mathbf{X}$ may have different physical meanings in different applications. In a BSS problem, $\mathbf{A}$ plays the role of mixing matrix, while $\mathbf{X}$ expresses source signals. In clustering problems, $\mathbf{A}$ is the basis matrix, while $\mathbf{X}$ denotes the weight matrix. In acoustic analysis, $\mathbf{A}$ represents the basis patterns, while each row of $\mathbf{X}$ expresses time points (positions) when sound patterns are activated.

In standard NMF we only assume nonnegativity of factor matrices $\mathbf{A}$ and $\mathbf{X}$. Unlike blind source separation methods based on independent component analysis (ICA), here we do not assume that the sources are independent, although we will introduce other assumptions or constraints on $\mathbf{A}$ and/or $\mathbf{X}$ later. Notice that this symmetry of assumptions leads to a symmetry in the factorization: we could just as easily write $\mathbf{Y}^T \approx \mathbf{X}^T\mathbf{A}^T$, so the meaning of "source" and "mixture" in NMF are often somewhat arbitrary.

The NMF model can also be represented as a special form of the bilinear model (see Figure 1.5):

$$\boxed{\mathbf{Y} = \sum_{j=1}^{J} \boldsymbol{a}_j \circ \boldsymbol{b}_j + \mathbf{E} = \sum_{j=1}^{J} \boldsymbol{a}_j \boldsymbol{b}_j^T + \mathbf{E},} \tag{1.6}$$

[6] Since we usually operate on column vectors of matrices (in order to avoid a complex or confused notation) it is often convenient to use the matrix $\mathbf{B} = \mathbf{X}^T$ instead of the matrix $\mathbf{X}$.

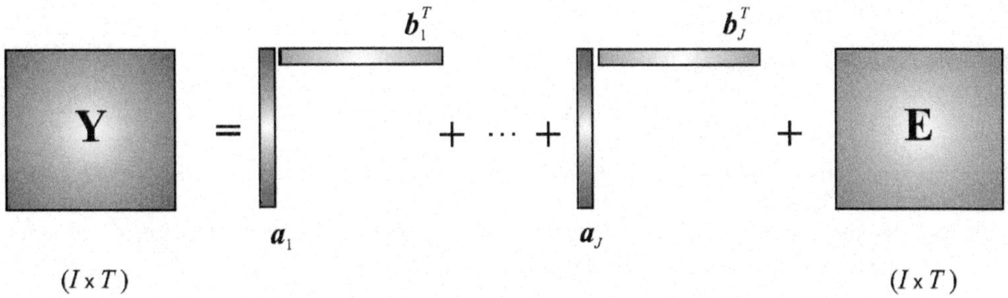

Figure 1.5 Bilinear NMF model. The nonnegative data matrix $\mathbf{Y} \in \mathbb{R}_+^{I \times T}$ is approximately represented by a sum or linear combination of rank-one nonnegative matrices $\mathbf{Y}^{(j)} = \boldsymbol{a}_j \circ \boldsymbol{b}_j = \boldsymbol{a}_j \boldsymbol{b}_j^T \in \mathbb{R}_+^{I \times T}$.

where the symbol $\circ$ denotes the outer product of two vectors. Thus, we can build an approximate representation of the nonnegative data matrix $\mathbf{Y}$ as a sum of rank-one nonnegative matrices $\boldsymbol{a}_j \boldsymbol{b}_j^T$. If such decomposition is exact (i.e., $\mathbf{E} = \mathbf{0}$) then it is called the Nonnegative Rank Factorization (NRF) [53]. Among the many possible series representations of data matrix $\mathbf{Y}$ by nonnegative rank-one matrices, the smallest integer J for which such a nonnegative rank-one series representation is attained is called the nonnegative rank of the nonnegative matrix $\mathbf{Y}$ and it is denoted by $\text{rank}_+(\mathbf{Y})$. The nonnegative rank satisfies the following bounds [53]:

$$\text{rank}(\mathbf{Y}) \leq \text{rank}_+(\mathbf{Y}) \leq \min\{I, T\}. \tag{1.7}$$

It should be noted that an NMF is not necessarily an NRF in the sense that the latter demands the exact factorization whereas the former is usually only an approximation.

Although the NMF can be applied to BSS problems for nonnegative sources and nonnegative mixing matrices, its application is not limited to BSS and it can be used in various and diverse applications far beyond BSS (see Chapter 8). In many applications we require additional constraints on the elements of matrices $\mathbf{A}$ and/or $\mathbf{X}$, such as smoothness, sparsity, symmetry, and orthogonality.

1.2.3 Symmetric NMF

In the special case when $\mathbf{A} = \mathbf{B} \in \mathbb{R}_+^{I \times J}$ the NMF is called a symmetric NMF, given by

$$\mathbf{Y} = \mathbf{A}\mathbf{A}^T + \mathbf{E}. \tag{1.8}$$

This model is also considered equivalent to Kernel K-means clustering and Laplace spectral clustering [50].

If the exact symmetric NMF ($\mathbf{E} = \mathbf{0}$) exists then a nonnegative matrix $\mathbf{Y} \in \mathbb{R}_+^{I \times I}$ is said to be completely positive (CP) and the smallest number of columns of $\mathbf{A} \in \mathbb{R}_+^{I \times J}$ satisfying the exact factorization $\mathbf{Y} = \mathbf{A}\mathbf{A}^T$ is called the cp-rank of the matrix $\mathbf{Y}$, denoted by $\text{rank}_{cp}(\mathbf{Y})$. If $\mathbf{Y}$ is CP, then the upper bound estimate of the cp-rank is given by[53]:

$$\text{rank}_{cp}(\mathbf{Y}) \leq \frac{\text{rank}(\mathbf{Y})(\text{rank}(\mathbf{Y}) + 1)}{2} - 1, \tag{1.9}$$

provided $\text{rank}(\mathbf{Y}) > 1$.

1.2.4 Semi-Orthogonal NMF

The semi-orthogonal NMF can be defined as

$$\mathbf{Y} = \mathbf{A}\mathbf{X} + \mathbf{E} = \mathbf{A}\mathbf{B}^T + \mathbf{E}, \tag{1.10}$$

subject to nonnegativity constraints $\mathbf{A} \geq \mathbf{0}$ and $\mathbf{X} \geq \mathbf{0}$ (component-wise) and an additional orthogonality constraint: $\mathbf{A}^T\mathbf{A} = \mathbf{I}_J$ or $\mathbf{X}\mathbf{X}^T = \mathbf{I}_J$.

Probably the simplest and most efficient way to impose orthogonality onto the matrix $\mathbf{A}$ or $\mathbf{X}$ is to perform the following transformation after each iteration

$$\boxed{\mathbf{A} \leftarrow \mathbf{A}\left[\mathbf{A}^T\mathbf{A}\right]^{-1/2}, \qquad \text{or} \qquad \mathbf{X} \leftarrow \left[\mathbf{X}\mathbf{X}^T\right]^{-1/2}\mathbf{X}.} \tag{1.11}$$

1.2.5 Semi-NMF and Nonnegative Factorization of Arbitrary Matrix

In some applications the observed input data are unsigned (unconstrained or bipolar) as indicated by $\mathbf{Y} = \mathbf{Y}_{\pm} \in \mathbb{R}^{I\times T}$ which allows us to relax the constraints regarding nonnegativity of one factor (or only specific vectors of a matrix). This leads to approximative semi-NMF which can take the following form

$$\mathbf{Y}_{\pm} = \mathbf{A}_{\pm}\mathbf{X}_{+} + \mathbf{E}, \qquad \text{or} \qquad \mathbf{Y}_{\pm} = \mathbf{A}_{+}\mathbf{X}_{\pm} + \mathbf{E}, \tag{1.12}$$

where the subscript in $\mathbf{A}_{+}$ indicates that a matrix is forced to be nonnegative.

In Chapter 4 we discuss models and algorithms for approximative factorizations in which the matrices $\mathbf{A}$ and/or $\mathbf{X}$ are restricted to contain nonnegative entries, but the data matrix $\mathbf{Y}$ may have entries with mixed signs, thus extending the range of applications of NMF. Such a model is often referred to as Nonnegative Factorization (NF) [58,59].

1.2.6 Three-factor NMF

Three-factor NMF (also called the tri-NMF) can be considered as a special case of the multi-layer NMF and can take the following general form [51,52]

$$\boxed{\mathbf{Y} = \mathbf{A}\mathbf{S}\mathbf{X} + \mathbf{E},} \tag{1.13}$$

where nonnegativity constraints are imposed to all or only to the selected factor matrices: $\mathbf{A} \in \mathbb{R}^{I\times J}$, $\mathbf{S} \in \mathbb{R}^{J\times R}$, and/or $\mathbf{X} \in \mathbb{R}^{R\times T}$.

It should be noted that if we do not impose any additional constraints to the factors (besides nonnegativity), the three-factor NMF can be reduced to the standard (two-factor) NMF by the transformation $\mathbf{A} \leftarrow \mathbf{A}\mathbf{S}$ or $\mathbf{X} \leftarrow \mathbf{S}\mathbf{X}$. However, the three-factor NMF is not equivalent to the standard NMF if we apply special constraints or conditions as illustrated by the following special cases.

1.2.6.1 Orthogonal Three-Factor NMF

Orthogonal three-factor NMF imposes additional constraints upon the two matrices $\mathbf{A}^T\mathbf{A} = \mathbf{I}_J$ and $\mathbf{X}\mathbf{X}^T = \mathbf{I}_J$ while the matrix $\mathbf{S}$ can be an arbitrary unconstrained matrix (i.e., it has both positive and negative entries) [51,52].

For uni-orthogonal three-factor NMF only one matrix $\mathbf{A}$ or $\mathbf{X}$ is orthogonal and all three matrices are usually nonnegative.

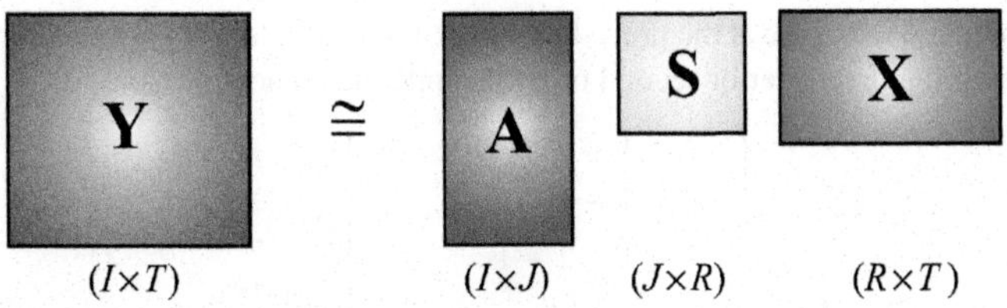

Figure 1.6 Illustration of three factor NMF (tri-NMF). The goal is to estimate two matrices $\mathbf{A} \in \mathbb{R}_+^{I\times J}$ and $\mathbf{X} \in \mathbb{R}_+^{R\times T}$, assuming that the matrix $\mathbf{S} \in \mathbb{R}^{J\times R}$ is given, or to estimate all three factor matrices $\mathbf{A}, \mathbf{S}, \mathbf{X}$ subject to additional constraints such as orthogonality or sparsity.

1.2.6.2 Non-Smooth NMF

Non-smooth NMF (nsNMF) was proposed by Pascual-Montano *et al.* [115] and is a special case of the three-factor NMF model in which the matrix **S** is fixed and known, and is used for controlling the sparsity or smoothness of the matrix **X** and/or **A**. Typically, the smoothing matrix $\mathbf{S} \in \mathbb{R}^{J\times J}$ takes the form:

$$\mathbf{S} = (1-\Theta)\,\mathbf{I}_J + \frac{\Theta}{J}\,\mathbf{1}_{J\times J}, \tag{1.14}$$

where $\mathbf{I}_J$ is $J \times J$ identity matrix and $\mathbf{1}_{J\times J}$ is the matrix of all ones. The scalar parameter $0 \le \Theta \le 1$ controls the smoothness of the matrix operator **S**. For $\Theta = 0$, $\mathbf{S} = \mathbf{I}_J$, the model reduces to the standard NMF and for $\Theta \to 1$ strong smoothing is imposed on **S**, causing increased sparseness of both **A** and **X** in order to maintain the faithfulness of the model.

1.2.6.3 Filtering NMF

In many applications it is necessary to impose some kind of filtering upon the rows of the matrix **X** (representing source signals), e.g., lowpass filtering to perform smoothing or highpass filtering in order to remove slowly changing trends from the estimated components (source signals). In such cases we can define the filtering NMF as

$$\boxed{\mathbf{Y} = \mathbf{A}\mathbf{X}\mathbf{F} + \mathbf{E},} \tag{1.15}$$

where **F** is a suitably designed (prescribed) filtering matrix. In the case of lowpass filtering, we usually perform some kind of averaging in the sense that every sample value x_{jt} is replaced by a weighted average of that value and the neighboring value, so that in the simplest scenario the smoothing lowpass filtering matrix **F** can take the following form:

$$\mathbf{F} = \begin{bmatrix} 1/2 & 1/3 & 0 & & & 0 \\ 1/2 & 1/3 & 1/3 & & & 0 \\ & 1/3 & 1/3 & 1/3 & & \\ & & \ddots & \ddots & \ddots & \\ 0 & & & 1/3 & 1/3 & 1/2 \\ 0 & & & 0 & 1/3 & 1/2 \end{bmatrix} \in \mathbb{R}^{T\times T}. \tag{1.16}$$

A standard way of performing highpass filtering is equivalent to an application of a first-order differential operator, which means (in the simplest scenario) just replacing each sample value by the difference between the value at that point and the value at the preceding point. For example, a highpass filtering matrix can take following form (using the first order or second order discrete difference forms):

$$\mathbf{F} = \begin{bmatrix} 1 & -1 & 0 & & & 0 \\ -1 & 2 & -1 & & & 0 \\ & -1 & 2 & -1 & & \\ & & \ddots & \ddots & \ddots & \\ 0 & & & -1 & 2 & -1 \\ 0 & & & & -1 & 1 \end{bmatrix} \in \mathbb{R}^{T \times T}. \tag{1.17}$$

Note that since the matrix $\mathbf{S}$ in the nsNMF and the matrix $\mathbf{F}$ in filtering NMF are known or designed in advance, almost all the algorithms known for the standard NMF can be straightforwardly extended to the nsNMF and Filtering NMF, for example, by defining new matrices $\mathbf{A} \triangleq \mathbf{AS}$, $\mathbf{X} \triangleq \mathbf{SX}$, or $\mathbf{X} \triangleq \mathbf{XF}$, respectively.

1.2.6.4 CGR/CUR Decomposition

In the CGR, also recently called CUR decomposition, a given data matrix $\mathbf{Y} \in \mathbb{R}^{I \times T}$ is decomposed as follows [55,60,61,101,102]:

$$\boxed{\mathbf{Y} = \mathbf{CUR} + \mathbf{E},} \tag{1.18}$$

where $\mathbf{C} \in \mathbb{R}^{I \times C}$ is a matrix constructed from C selected columns of $\mathbf{Y}$, $\mathbf{R} \in \mathbb{R}^{R \times T}$ consists of R rows of $\mathbf{Y}$ and matrix $\mathbf{U} \in \mathbb{R}^{C \times R}$ is chosen to minimize the error $\mathbf{E} \in \mathbb{R}^{I \times T}$. The matrix $\mathbf{U}$ is often the pseudo-inverse of a matrix $\mathbf{Z} \in \mathbb{R}^{R \times C}$, i.e., $\mathbf{U} = \mathbf{Z}^{\dagger}$, which is defined by the intersections of the selected rows and columns (see Figure 1.7). Alternatively, we can compute a core matrix $\mathbf{U}$ as $\mathbf{U} = \mathbf{C}^{\dagger}\mathbf{Y}\mathbf{R}^{\dagger}$, but in this case knowledge of the whole data matrix $\mathbf{Y}$ is necessary.

Since typically, $C << T$ and $R << I$, our challenge is to find a matrix $\mathbf{U}$ and select rows and columns of $\mathbf{Y}$ so that for the fixed number of columns and rows the error cost function $||\mathbf{E}||_F^2$ is minimized. It was proved by Goreinov *et al.* [60] that for $R = C$ the following bounds can be theoretically achieved:

$$||\mathbf{Y} - \mathbf{CUR}||_{\max} \leq (R+1)\, \sigma_{R+1}, \tag{1.19}$$

$$||\mathbf{Y} - \mathbf{CUR}||_F \leq \sqrt{1 + R(T-R)}\, \sigma_{R+1}, \tag{1.20}$$

where $||\mathbf{Y}||_{\max} = \max_{it}\{|y_{it}|\}$ denotes max norm and σ_r is the r-th singular value of $\mathbf{Y}$.

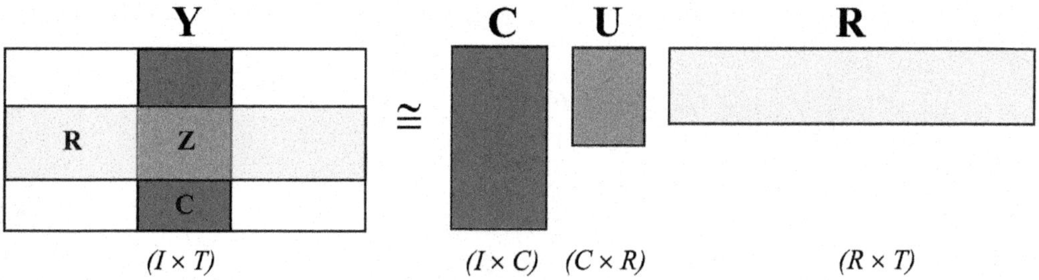

Figure 1.7 Illustration of CUR decomposition. The objective is to select such rows and columns of data matrix $\mathbf{Y} \in \mathbb{R}^{I \times T}$ which provide the best approximation. The matrix $\mathbf{U}$ is usually the pseudo-inverse of the matrix $\mathbf{Z} \in \mathbb{R}^{R \times C}$, i.e., $\mathbf{U} = \mathbf{Z}^{\dagger}$. For simplicity of graphical illustration, we have assumed that the joint R rows and the joint C columns of the matrix $\mathbf{Y}$ are selected.

Without loss of generality, let us assume for simplicity, that the first C columns and the first R rows of the matrix $\mathbf{Y}$ are selected so the matrix is partitioned as follows:

$$\mathbf{Y} = \begin{bmatrix} \mathbf{Y}_{11} & \mathbf{Y}_{12} \\ \mathbf{Y}_{21} & \mathbf{Y}_{22} \end{bmatrix} \in \mathbb{R}^{I\times T}, \quad \text{and} \quad \mathbf{C} = \begin{bmatrix} \mathbf{Y}_{11} \\ \mathbf{Y}_{21} \end{bmatrix} \in \mathbb{R}^{I\times C}, \quad \mathbf{R} = \begin{bmatrix} \mathbf{Y}_{11} & \mathbf{Y}_{12} \end{bmatrix} \in \mathbb{R}^{R\times T}, \tag{1.21}$$

then the following bound is obtained [61]

$$||\mathbf{Y} - \mathbf{C}\mathbf{Y}_{11}^{\dagger}\mathbf{R}||_F \leq \gamma_R \, \sigma_{R+1}, \tag{1.22}$$

where

$$\gamma_R = \min \left\{ \sqrt{(1 + ||\mathbf{Y}_{21}\mathbf{Y}_{11}^{\dagger}||_F^2)}, \sqrt{(1 + ||\mathbf{Y}_{11}^{\dagger}\mathbf{Y}_{12}||_F^2)} \right\}. \tag{1.23}$$

This formula allows us to identify optimal columns and rows in sequential manner [22]. In fact, there are several strategies for the selection of suitable columns and rows. The main principle is to select columns and rows that exhibit high "statistical leverage" and provide the best low-rank fit of the data matrix [55,60].

In the special case, assuming that $\mathbf{U}\mathbf{R} = \mathbf{X}$, we have CX decomposition:

$$\boxed{\mathbf{Y} = \mathbf{C}\mathbf{X} + \mathbf{E}.} \tag{1.24}$$

The CX and CUR (CGR) decompositions are low-rank matrix decompositions that are explicitly expressed in terms of a small number of actual columns and/or actual rows of the data matrix and they have recently received increasing attention in the data analysis community, especially for nonnegative data due to many potential applications [55,60,101,102]. The CUR decomposition has an advantage that components (factor matrices $\mathbf{C}$ and $\mathbf{R}$) are directly obtained from rows and columns of data matrix $\mathbf{Y}$, preserving desired properties such as nonnegativity or sparsity. Because they are constructed from actual data elements, CUR decomposition is often more easily interpretable by practitioners of the field from which the data are drawn (to the extent that the original data points and/or features are interpretable) [101].

1.2.7 NMF with Offset (Affine NMF)

In NMF with offset (also called affine NMF, aNMF), our goal is to remove the base line or DC bias from the matrix $\mathbf{Y}$ by using a slightly modified NMF model:

$$\boxed{\mathbf{Y} = \mathbf{A}\mathbf{X} + \boldsymbol{a}_0\mathbf{1}^T + \mathbf{E},} \tag{1.25}$$

where $\mathbf{1} \in \mathbb{R}^T$ is a vector of all ones and $\boldsymbol{a}_0 \in \mathbb{R}_+^I$ is a vector which is selected in such a way that a suitable error cost function is minimized and/or the matrix $\mathbf{X}$ is zero-grounded, that is, with a possibly large number of zero entries in each row (or for noisy data close to zero entries). The term $\mathbf{Y}_0 = \boldsymbol{a}_0\mathbf{1}^T$ denotes offset, which together with nonnegativity constraint often ensures the sparseness of factored matrices. The main role of the offset is to absorb the constant values of a data matrix, thereby making the factorization sparser and therefore improving (relaxing) conditions for the uniqueness of NMF (see next sections). Chapter 3 will demonstrates the affine NMF with multiplicative algorithms. However, in practice, the offsets are not the same and perfectly constant in all data sources. For image data, due to illumination flicker, the intensities of offset regions vary between images. Affine NMF with the model (1.25) fails to decompose such data. The Block-Oriented Decomposition (BOD1) model presented in section (1.5.9) will help us resolving this problem.

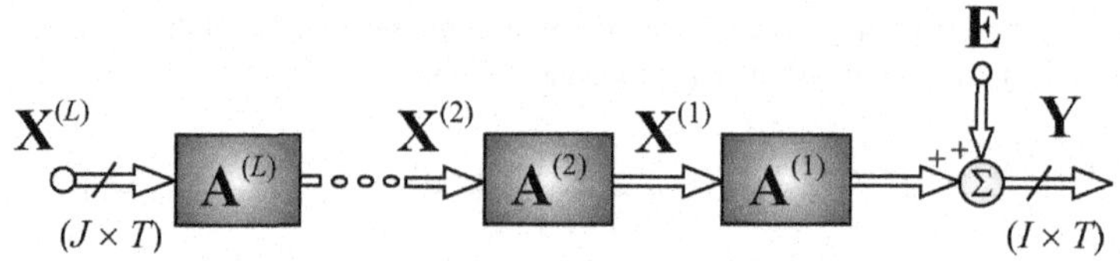

Figure 1.8 Multilayer NMF model. In this model the global factor matrix $\mathbf{A} = \mathbf{A}^{(1)}\mathbf{A}^{(2)} \cdots \mathbf{A}^{(L)}$ has distributed representation in which each matrix $\mathbf{A}^{(l)}$ can be sparse.

1.2.8 Multi-layer NMF

In multi-layer NMF the basic matrix $\mathbf{A}$ is replaced by a set of cascaded (factor) matrices. Thus, the model can be described as (see Figure 1.8)

$$\boxed{\mathbf{Y} = \mathbf{A}^{(1)}\mathbf{A}^{(2)} \cdots \mathbf{A}^{(L)}\mathbf{X} + \mathbf{E}.} \tag{1.26}$$

Since the model is linear, all the matrices can be merged into a single matrix $\mathbf{A}$ if no special constraints are imposed upon the individual matrices $\mathbf{A}^{(l)}$, $(l = 1, 2, \ldots, L)$. However, multi-layer NMF can be used to considerably improve the performance of standard NMF algorithms due to distributed structure and alleviating the problem of local minima.

To improve the performance of the NMF algorithms (especially for ill-conditioned and badly-scaled data) and to reduce the risk of converging to local minima of a cost function due to nonconvex alternating minimization, we have developed a simple hierarchical multi-stage procedure [27,37–39] combined with a multi-start initialization, in which we perform a sequential decomposition of nonnegative matrices as follows. In the first step, we perform the basic approximate decomposition $\mathbf{Y} \cong \mathbf{A}^{(1)}\mathbf{X}^{(1)} \in \mathbb{R}^{I\times T}$ using any available NMF algorithm. In the second stage, the results obtained from the first stage are used to build up a new input data matrix $\mathbf{Y} \leftarrow \mathbf{X}^{(1)}$, that is, in the next step, we perform a similar decomposition $\mathbf{X}^{(1)} \cong \mathbf{A}^{(2)}\mathbf{X}^{(2)} \in \mathbb{R}^{J\times T}$, using the same or different update rules. We continue our decomposition taking into account only the last obtained components. The process can be repeated for an arbitrary number of times until some stopping criteria are satisfied. Thus, our multi-layer NMF model has the form:

$$\mathbf{Y} \cong \mathbf{A}^{(1)}\mathbf{A}^{(2)} \cdots \mathbf{A}^{(L)}\mathbf{X}^{(L)}, \tag{1.27}$$

with the final results $\mathbf{A} = \mathbf{A}^{(1)}\mathbf{A}^{(2)} \cdots \mathbf{A}^{(L)}$ and $\mathbf{X} = \mathbf{X}^{(L)}$. Physically, this means that we build up a distributed system that has many layers or cascade connections of L mixing subsystems. The key point in this approach is that the learning (update) process to find parameters of matrices $\mathbf{X}^{(l)}$ and $\mathbf{A}^{(l)}$, $(l = 1, 2, \ldots, L)$ is performed sequentially, layer-by-layer, where each layer is randomly initialized with different initial conditions. We have found that the hierarchical multi-layer approach can improve performance of most NMF algorithms discussed in this book [27,33,36].

1.2.9 Simultaneous NMF

In simultaneous NMF (siNMF) we have available two or more linked input data matrices (say, $\mathbf{Y}_1$ and $\mathbf{Y}_2$) and the objective is to decompose them into nonnegative factor matrices in such a way that one of a factor matrix is common, for example, (which is a special form of the Nonnegative Tensor Factorization NTF2 model presented in Section 1.5.4),

$$\begin{aligned} \mathbf{Y}_1 &= \mathbf{A}_1\mathbf{X} + \mathbf{E}_1, \\ \mathbf{Y}_2 &= \mathbf{A}_2\mathbf{X} + \mathbf{E}_2. \end{aligned} \tag{1.28}$$

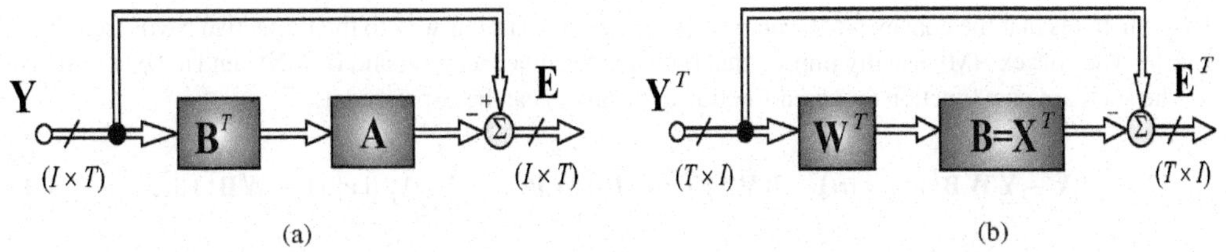

Figure 1.9 (a) Illustration of Projective NMF (typically, $\mathbf{A} = \mathbf{B} = \mathbf{W}$) and (b) Convex NMF.

Such a problem arises, for example, in bio-informatics if we combine gene expression and transcription factor regulation [8]. In this application the data matrix $\mathbf{Y}_1 \in \mathbb{R}^{I_1 \times T}$ is the expression level of gene t in a data sample i_1 (i.e., the index i_1 denotes samples, while t stands for genes) and $\mathbf{Y}_2 \in \mathbb{R}^{I_2 \times T}$ is a transcription matrix (which is 1 whenever transcription factor i_2 regulates gene t).

1.2.10 Projective and Convex NMF

A projective NMF model can be formulated as the estimation of sparse and nonnegative matrix $\mathbf{W} \in \mathbb{R}_+^{I \times J}$, $I > J$, which satisfies the matrix equation

$$\mathbf{Y} = \mathbf{W}\mathbf{W}^T\mathbf{Y} + \mathbf{E}. \tag{1.29}$$

In a more general nonsymmetric form the projective NMF involves estimation of two nonnegative matrices: $\mathbf{A} \in \mathbb{R}_+^{I \times J}$ and $\mathbf{B} \in \mathbb{R}_+^{I \times J}$ in the model (see Figure 1.9(a)):

$$\mathbf{Y} = \mathbf{A}\mathbf{B}^T\mathbf{Y} + \mathbf{E}. \tag{1.30}$$

This may lead to the following optimization problem:

$$\min_{\mathbf{A},\mathbf{B}} ||\mathbf{Y} - \mathbf{A}\mathbf{B}^T\mathbf{Y}||_F^2, \qquad \text{s.t. } \mathbf{A} \geq \mathbf{0}, \ \ \mathbf{B} \geq \mathbf{0}. \tag{1.31}$$

The projective NMF is similar to the subspace PCA. However, it involves nonnegativity constraints.

In the convex NMF proposed by Ding, Li and Jordan [51], we assume that the basis vectors $\mathbf{A} = [\boldsymbol{a}_1, \boldsymbol{a}_2, \ldots, \boldsymbol{a}_J]$ are constrained to be convex combinations of the data input matrix $\mathbf{Y} = [\boldsymbol{y}_1, \boldsymbol{y}_2, \ldots, \boldsymbol{y}_T]$. In other words, we require that the vectors $\boldsymbol{a}_j$ lie within the column space of the data matrix $\mathbf{Y}$, i.e.:

$$\boldsymbol{a}_j = \sum_{t=1}^{T} w_{tj}\boldsymbol{y}_t = \mathbf{Y}\boldsymbol{w}_j \qquad \text{or} \qquad \mathbf{A} = \mathbf{Y}\mathbf{W}, \tag{1.32}$$

where $\mathbf{W} \in \mathbb{R}_+^{T \times J}$ and $\mathbf{X} = \mathbf{B}^T \in \mathbb{R}_+^{J \times T}$. Usually each column in $\mathbf{W}$ satisfies the sum-to-one constraint, i.e., they are unit length in terms of the ℓ_1-norm. We restrict ourselves to convex combinations of the columns of $\mathbf{Y}$. The convex NMF model can be written in the matrix form as[7]

$$\mathbf{Y} = \mathbf{Y}\mathbf{W}\mathbf{X} + \mathbf{E} \tag{1.33}$$

and we can apply the transpose operator to give

$$\mathbf{Y}^T = \mathbf{X}^T\mathbf{W}^T\mathbf{Y}^T + \mathbf{E}^T. \tag{1.34}$$

[7] In general, the convex NMF applies to both nonnegative data and mixed sign data which can be written symbolically as $\mathbf{Y}_\pm = \mathbf{Y}_\pm\mathbf{W}_+\mathbf{X}_+ + \mathbf{E}$.

This illustrates that the convex NMF can be represented in a similar way to the projective NMF (see Figure 1.9(b)). The convex NMF usually implies that both nonnegative factors $\mathbf{A}$ and $\mathbf{B} = \mathbf{X}^T$ tend to be very sparse.

The standard cost function (squared Euclidean distance) can be expressed as

$$||\mathbf{Y} - \mathbf{Y}\,\mathbf{W}\,\mathbf{B}^T||_F^2 = \text{tr}(\mathbf{I} - \mathbf{B}\,\mathbf{W}^T)\,\mathbf{Y}^T\,\mathbf{Y}\,(\mathbf{I} - \mathbf{W}\,\mathbf{B}^T) = \sum_{j=1}^{J} \lambda_j \,||\, \boldsymbol{v}_j^T\,(\mathbf{I} - \mathbf{W}\,\mathbf{B}^T)\,||_2^2, \tag{1.35}$$

where λ_j is the positive j-th eigenvalue (a diagonal entry of diagonal matrix $\mathbf{\Lambda}$) and $\boldsymbol{v}_j$ is the corresponding eigenvector for the eigenvalue decomposition: $\mathbf{Y}^T\mathbf{Y} = \mathbf{V}\mathbf{\Lambda}\mathbf{V}^T = \sum_{j=1}^{J} \lambda_j \boldsymbol{v}_j \boldsymbol{v}_j^T$. This form of NMF can also be considered as a special form of the kernel NMF with a linear kernel defined as $\mathbf{K} = \mathbf{Y}^T\mathbf{Y}$.

1.2.11 Kernel NMF

The convex NMF leads to a natural extension of the kernel NMF [92,98,120]. Consider a mapping $\boldsymbol{y}_t \rightarrow \phi(\boldsymbol{y}_t)$ or $\mathbf{Y} \rightarrow \phi(\mathbf{Y}) = [\phi(\boldsymbol{y}_1), \phi(\boldsymbol{y}_2), \ldots, \phi(\boldsymbol{y}_T)]$, then the kernel NMF can be defined as

$$\phi(\mathbf{Y}) \cong \phi(\mathbf{Y})\,\mathbf{W}\,\mathbf{B}^T. \tag{1.36}$$

This leads to the minimization of the cost function:

$$||\phi(\mathbf{Y}) - \phi(\mathbf{Y})\mathbf{W}\,\mathbf{B}^T||_F^2 = \text{tr}(\mathbf{K}) - 2\,\text{tr}(\mathbf{B}^T\,\mathbf{K}\,\mathbf{W}) + \text{tr}(\mathbf{W}^T\,\mathbf{K}\,\mathbf{W}\,\mathbf{B}^T\mathbf{B}), \tag{1.37}$$

which depends only on the kernel $\mathbf{K} = \phi^T(\mathbf{Y})\phi(\mathbf{Y})$.

1.2.12 Convolutive NMF

The Convolutive NMF (CNMF) is a natural extension and generalization of the standard NMF. In the Convolutive NMF, we process a set of nonnegative matrices or patterns which are horizontally shifted (or time delayed) versions of the primary matrix $\mathbf{X}$ [126]. In the simplest form the CNMF can be described as (see Figure 1.10)

$$\boxed{\mathbf{Y} = \sum_{p=0}^{P-1} \mathbf{A}_p \overset{p\rightarrow}{\mathbf{X}} + \mathbf{E},} \tag{1.38}$$

where $\mathbf{Y} \in \mathbb{R}_+^{I\times T}$ is a given input data matrix, $\mathbf{A}_p \in \mathbb{R}_+^{I\times J}$ is a set of unknown nonnegative basis matrices, $\mathbf{X} = \overset{0\rightarrow}{\mathbf{X}} \in \mathbb{R}_+^{J\times T}$ is a matrix representing primary sources or patterns, $\overset{p\rightarrow}{\mathbf{X}}$ is a shifted by p columns version of $\mathbf{X}$. In other words, $\overset{p\rightarrow}{\mathbf{X}}$ means that the columns of $\mathbf{X}$ are shifted to the right p spots (columns), while the entries in the columns shifted into the matrix from the outside are set to zero. This shift (time-delay) is performed by a basic operator illustrated in Figure 1.10 as $\mathbf{S}_p = \mathbf{T}_1$. Analogously, $\overset{\leftarrow p}{\mathbf{Y}}$ means that the columns of $\mathbf{Y}$ are shifted to the left p spots. These notations will also be used for the shift operations of other matrices throughout this book (see Chapter 3 for more detail). Note that, $\overset{0\rightarrow}{\mathbf{X}} = \overset{\leftarrow 0}{\mathbf{X}} = \mathbf{X}$.
The shift operator is illustrated by the following example:

$$\mathbf{X} = \begin{bmatrix} 1 & 2 & 3 \\ 4 & 5 & 6 \end{bmatrix}, \quad \overset{1\rightarrow}{\mathbf{X}} = \begin{bmatrix} 0 & 1 & 2 \\ 0 & 4 & 5 \end{bmatrix}, \quad \overset{2\rightarrow}{\mathbf{X}} = \begin{bmatrix} 0 & 0 & 1 \\ 0 & 0 & 4 \end{bmatrix}, \quad \overset{\leftarrow 1}{\mathbf{X}} = \begin{bmatrix} 2 & 3 & 0 \\ 5 & 6 & 0 \end{bmatrix}.$$

In the Convolutive NMF model, temporal continuity exhibited by many audio signals can be expressed more efficiently in the time-frequency domain, especially for signals whose frequencies vary with time. We will present several efficient and extensively tested algorithms for the CNMF model in Chapter 3.

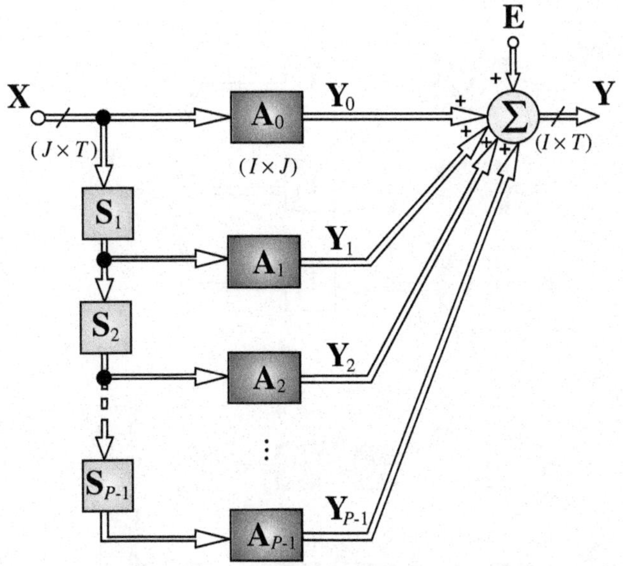

Figure 1.10 Illustration of Convolutive NMF. The goal is to estimate the input sources represented by nonnegative matrix $\mathbf{X} \in \mathbb{R}_+^{J \times T}$ (typically, $T >> I$) and to identify the convoluting system, i.e., to estimate a set of nonnegative matrices $\{\mathbf{A}_0, \mathbf{A}_1, \ldots, \mathbf{A}_{P-1}\}$ ($\mathbf{A}_p \in \mathbb{R}_+^{I \times J}$, $p = 0, 1, \ldots, P-1$) knowing only the input data matrix $\mathbf{Y} \in \mathbb{R}^{I \times T}$. Each operator $\mathbf{S}_p = \mathbf{T}_1$ ($p = 1, 2, \ldots, P-1$) performs a horizontal shift of the columns in $\mathbf{X}$ by one spot.

1.2.13 Overlapping NMF

In Convolutive NMF we perform horizontal shift of the columns of the matrix $\mathbf{X}$. In some applications, such as in spectrogram decomposition, we need to perform different transformations by shifting vertically the rows of the matrix $\mathbf{X}$. For example, the observed data may be represented by a linear combination of horizontal bars or features and modeled by transposing the CNMF model (1.38) as

$$\mathbf{Y} \approx \sum_{p=0}^{P} (\overset{\rightarrow p}{\mathbf{X}})^T \mathbf{A}_p^T = \sum_{p=0}^{P} (\mathbf{X}\underset{\rightarrow}{\mathbf{T}}_p)^T \mathbf{A}_p^T = \sum_{p=0}^{P} \underset{\rightarrow}{\mathbf{T}}_p^T \mathbf{X}^T \mathbf{A}_p^T, \tag{1.39}$$

where $\underset{\rightarrow}{\mathbf{T}}_p \triangleq \mathbf{T}_p$ is the horizontal-shift matrix operator such that $\overset{\rightarrow p}{\mathbf{X}} = \mathbf{X}\underset{\rightarrow}{\mathbf{T}}_p$ and $\overset{\leftarrow p}{\mathbf{X}} = \mathbf{X}\underset{\leftarrow}{\mathbf{T}}_p$. For example, for the fourth-order identity matrix this operator can take the following form

$$\underset{\rightarrow}{\mathbf{T}}_1 = \begin{bmatrix} 0&1&0&0 \\ 0&0&1&0 \\ 0&0&0&1 \\ 0&0&0&0 \end{bmatrix}, \quad \underset{\rightarrow}{\mathbf{T}}_2 = \underset{\rightarrow}{\mathbf{T}}_1 \underset{\rightarrow}{\mathbf{T}}_1 = \begin{bmatrix} 0&0&1&0 \\ 0&0&0&1 \\ 0&0&0&0 \\ 0&0&0&0 \end{bmatrix}, \quad \underset{\leftarrow}{\mathbf{T}}_1 = \begin{bmatrix} 0&0&0&0 \\ 1&0&0&0 \\ 0&1&0&0 \\ 0&0&1&0 \end{bmatrix}.$$

Transposing the horizontal shift operator $\underset{\rightarrow}{\mathbf{T}}_p := \mathbf{T}_p$ gives us the vertical shift operator $\mathbf{T}_{\uparrow p} = \underset{\leftarrow}{\mathbf{T}}_p^T$ and $\mathbf{T}_{\downarrow p} = \underset{\rightarrow}{\mathbf{T}}_p^T$, in fact, we have $\mathbf{T}_{\uparrow p} = \underset{\rightarrow}{\mathbf{T}}_p$ and $\mathbf{T}_{\downarrow p} = \underset{\leftarrow}{\mathbf{T}}_p$.

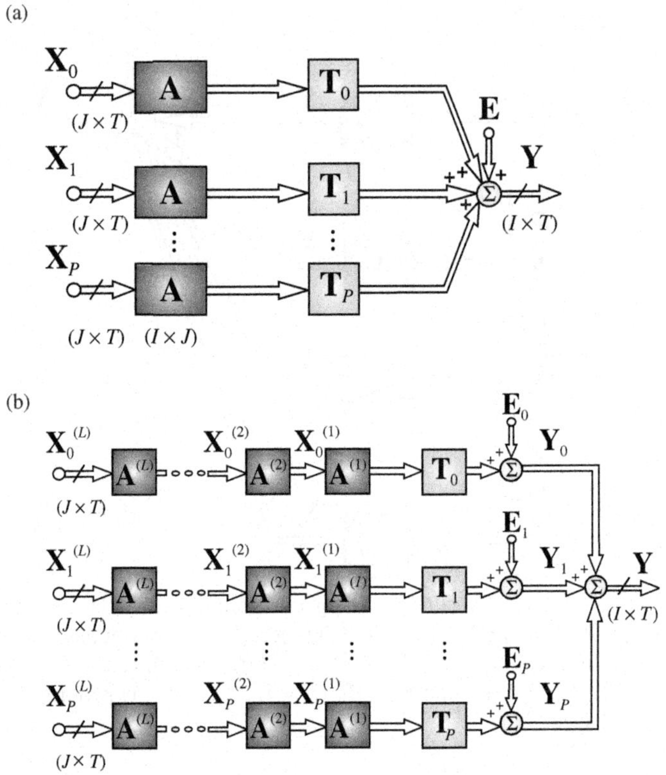

Figure 1.11 (a) Block diagram schema for overlapping NMF, (b) Extended Multi-layer NMF model.

It is interesting to note that by interchanging the role of matrices **A** and **X**, that is, $\mathbf{A} \stackrel{\Delta}{=} \mathbf{X}$ and $\mathbf{X}_p \stackrel{\Delta}{=} \mathbf{A}_p$, we obtain the overlapping NMF introduced by Eggert *et al.* [56] and investigated by Choi *et al.* [81], which can be described as (see Fig. 1.11(a))

$$\mathbf{Y} \cong \sum_{p=0}^{P} \mathbf{T}_{\uparrow p} \mathbf{A} \mathbf{X}_p. \tag{1.40}$$

Figure 1.11(b) illustrates the extended multi-layer overlapping NMF (by analogy to the standard multi-layer NMF in order to improve the performance of the overlapping NMF). The overlapping NMF model can be considered as a modification or variation of the CNMF model, where transform-invariant representations and sparseness constraints are incorporated [56,81].

1.3 Basic Approaches to Estimate Parameters of Standard NMF

In order to estimate factor matrices **A** and **X** in the standard NMF, we need to consider the similarity measure to quantify a difference between the data matrix **Y** and the approximative NMF model matrix $\widehat{\mathbf{Y}} = \mathbf{A}\mathbf{X}$. The choice of the similarity measure (also referred to as distance, divergence or measure of dissimilarity) mostly depends on the probability distribution of the estimated signals or components and on the structure of data or a distribution of noise. The simplest and most often used measure is based on Frobenius norm:

$$D_F(\mathbf{Y}||\mathbf{A}\mathbf{X}) = \frac{1}{2}||\mathbf{Y} - \mathbf{A}\mathbf{X}||_F^2, \tag{1.41}$$

which is also referred to as the squared Euclidean distance. It should be noted that the above cost function is convex with respect to either the elements of the matrix $\mathbf{A}$ or the matrix $\mathbf{X}$, but not both.[8] Alternating minimization of such a cost leads to the ALS (Alternating Least Squares) algorithm which can be described as follows:

1. Initialize $\mathbf{A}$ randomly or by using a specific deterministic strategy.
2. Estimate $\mathbf{X}$ from the matrix equation $\mathbf{A}^T\mathbf{A}\mathbf{X} = \mathbf{A}^T\mathbf{Y}$ by solving

$$\min_{\mathbf{X}} D_F(\mathbf{Y}||\mathbf{A}\mathbf{X}) = \frac{1}{2}||\mathbf{Y} - \mathbf{A}\mathbf{X}||_F^2, \qquad \text{with fixed} \qquad \mathbf{A}. \tag{1.42}$$

3. Set all the negative elements of $\mathbf{X}$ to zero or some small positive value.
4. Estimate $\mathbf{A}$ from the matrix equation $\mathbf{X}\mathbf{X}^T\mathbf{A}^T = \mathbf{X}\mathbf{Y}^T$ by solving

$$\min_{\mathbf{A}} D_F(\mathbf{Y}||\mathbf{A}\mathbf{X}) = \frac{1}{2}||\mathbf{Y}^T - \mathbf{X}^T\mathbf{A}^T||_F^2, \qquad \text{with fixed} \qquad \mathbf{X}. \tag{1.43}$$

5. Set all negative elements of $\mathbf{A}$ to zero or some small positive value ε.

The above ALS algorithm can be written in the following form:[9]

$$\mathbf{X} \leftarrow \max\left\{\varepsilon,\ (\mathbf{A}^T\mathbf{A})^{-1}\mathbf{A}^T\mathbf{Y}\right\} = [\mathbf{A}^{\dagger}\mathbf{Y}]_+, \tag{1.44}$$

$$\mathbf{A} \leftarrow \max\left\{\varepsilon,\ \mathbf{Y}\mathbf{X}^T(\mathbf{X}\mathbf{X}^T)^{-1}\right\} = [\mathbf{Y}\mathbf{X}^{\dagger}]_+, \tag{1.45}$$

where $\mathbf{A}^{\dagger}$ is the Moore-Penrose inverse of $\mathbf{A}$, ε is a small constant (typically 10^{-16}) to enforce positive entries. Various additional constraints on $\mathbf{A}$ and $\mathbf{X}$ can be imposed.

Today the ALS method is considered as a basic "workhorse" approach, however it is not guaranteed to converge to a global minimum nor even a stationary point, but only to a solution where the cost functions cease to decrease [11,85]. Moreover, it is often not sufficiently accurate. The ALS method can be dramatically improved and its computational complexity reduced as it will be shown in Chapter 4.

It is interesting to note that the NMF problem can be considered as a natural extension of a Nonnegative Least Squares (NLS) problem formulated as the following optimization problem: given a matrix $\mathbf{A} \in \mathbb{R}^{I\times J}$ and a set of observed values given by the vector $\boldsymbol{y} \in \mathbb{R}^I$, find a nonnegative vector $\boldsymbol{x} \in \mathbb{R}^J$ which minimizes the cost function $J(\boldsymbol{x}) = \frac{1}{2}||\boldsymbol{y} - \mathbf{A}\boldsymbol{x}||_2^2$, i.e.,

$$\min_{\boldsymbol{x}\geq 0} \frac{1}{2}||\boldsymbol{y} - \mathbf{A}\,\boldsymbol{x}||_2^2, \tag{1.46}$$

subject to $\boldsymbol{x} \geq \mathbf{0}$. There is a large volume of literature devoted to the NLS problems which will be exploited and adopted in this book.

Another frequently used cost function for NMF is the generalized Kullback-Leibler divergence (also called the I-divergence) [95]:

$$D_{KL}(\mathbf{Y}||\mathbf{A}\mathbf{X}) = \sum_{it}\left(y_{it}\ \ln\frac{y_{it}}{[\mathbf{A}\mathbf{X}]_{it}} - y_{it} + [\mathbf{A}\mathbf{X}]_{it}\right). \tag{1.47}$$

[8] Although the NMF optimization problem is not convex, the objective functions are separately convex in each of the two factors $\mathbf{A}$ and $\mathbf{X}$, which implies that finding the optimal factor matrix $\mathbf{A}$ corresponding to a fixed matrix $\mathbf{X}$ reduces to a convex optimization problem and vice versa. However, the convexity is lost as soon as we try to optimize factor matrices simultaneously [59].

[9] Note that the max operator is applied element-wise, that is, each element of a matrix is compared with scalar parameter ε.

Most existing approaches minimize only one kind of cost function by alternately switching between sets of parameters. In this book we adopt a more general and flexible approach in which instead of one cost function we rather exploit two or more cost functions (with the same global minima); one of them is minimized with respect to **A** and the other one with respect to **X**. Such an approach is fully justified as **A** and **X** may have different distributions or different statistical properties and therefore different cost functions can be optimal for them.

Algorithm 1.1: Multi-layer NMF using alternating minimization of two cost functions

Input: $\mathbf{Y} \in \mathbb{R}_+^{I \times T}$: input data, J: rank of approximation
Output: $\mathbf{A} \in \mathbb{R}_+^{I \times J}$ and $\mathbf{X} \in \mathbb{R}_+^{J \times T}$ such that some given cost functions are minimized.

```
begin
    X = Y, A = I
    for l = 1 to L do
        Initialize randomly A_(l) and X_(l)   ^a
        repeat
            A_(l) = arg min_{A_(l) ≥ 0} { D_1 (X || A_(l) X_(l)) }   for fixed X_(l)
            X_(l) = arg min_{X_(l) ≥ 0} { D_2 (X || A_(l) X_(l)) }   for fixed A_(l)
        until a stopping criterion is met          /* convergence condition */
        X = X_(l)
        A ← A A_(l)
    end
end
```

[a] Instead of random initialization, we can use ALS or SVD based initialization, see Section 1.3.3.

Algorithm 1.1 illustrates such a case, where the cost functions $D_1(\mathbf{Y}||\mathbf{AX})$ and $D_2(\mathbf{Y}||\mathbf{AX})$ can take various forms, e.g.: I-divergence and Euclidean distance [35,49] (see Chapter 2).

We can generalize this concept by using not one or two cost functions but rather a set of cost functions to be minimized sequentially or simultaneously. For $\mathbf{A} = [\boldsymbol{a}_1, \boldsymbol{a}_1, \ldots, \boldsymbol{a}_J]$ and $\mathbf{B} = \mathbf{X}^T = [\boldsymbol{b}_1, \boldsymbol{b}_2, \ldots, \boldsymbol{b}_J]$, we can express the squared Euclidean cost function as

$$J(\boldsymbol{a}_1, \boldsymbol{a}_1, \ldots, \boldsymbol{a}_J, \boldsymbol{b}_1, \boldsymbol{b}_2, \ldots, \boldsymbol{b}_J) = \frac{1}{2}||\mathbf{Y} - \mathbf{A}\mathbf{B}^T||_F^2 = \frac{1}{2}||\mathbf{Y} - \sum_{j=1}^{J} \boldsymbol{a}_j \boldsymbol{b}_j^T||_F^2. \tag{1.48}$$

An underlying idea is to define a residual (rank-one approximated) matrix (see Chapter 4 for more detail and explanation)

$$\mathbf{Y}^{(j)} \triangleq \mathbf{Y} - \sum_{p \neq j} \boldsymbol{a}_p \boldsymbol{b}_p^T \tag{1.49}$$

and alternately minimize the set of cost functions with respect to the unknown variables $\boldsymbol{a}_j$, $\boldsymbol{b}_j$:

$$D_A^{(j)}(\boldsymbol{a}) = \frac{1}{2}||\mathbf{Y}^{(j)} - \boldsymbol{a}\,\boldsymbol{b}_j^T||_F^2, \qquad \text{for a fixed } \boldsymbol{b}_j, \tag{1.50a}$$

$$D_B^{(j)}(\boldsymbol{b}) = \frac{1}{2}||\mathbf{Y}^{(j)} - \boldsymbol{a}_j\,\boldsymbol{b}^T||_F^2, \qquad \text{for a fixed } \boldsymbol{a}_j, \tag{1.50b}$$

for $j = 1, 2, \ldots, J$ subject to $\boldsymbol{a} \geq \mathbf{0}$ and $\boldsymbol{b} \geq \mathbf{0}$, respectively.

1.3.1 Large-scale NMF

In many applications, especially in dimension reduction applications the data matrix $\mathbf{Y} \in \mathbb{R}^{I\times T}$ can be very large (with millions of entries), but it can be approximately factorized using a rather smaller number of nonnegative components (J), that is, $J << I$ and $J << T$. Then the problem $\mathbf{Y} \approx \mathbf{AX}$ becomes highly redundant and we do not need to use information about all entries of $\mathbf{Y}$ in order to estimate precisely the factor matrices $\mathbf{A} \in \mathbb{R}^{I\times J}$ and $\mathbf{X} \in \mathbb{R}^{J\times T}$. In other words, to solve the large-scale NMF problem we do not need to know the whole data matrix but only a small random part of it. As we will show later, such an approach can considerably outperform the standard NMF methods, especially for extremely over determined systems.

In this approach, instead of performing large-scale factorization

$$\mathbf{Y} = \mathbf{AX} + \mathbf{E},$$

we can consider a two set of linked factorizations using much smaller matrices, given by (see Figure 1.12)

$$\mathbf{Y}_r = \mathbf{A}_r\mathbf{X} + \mathbf{E}_r, \qquad \text{for fixed (known)} \qquad \mathbf{A}_r, \tag{1.51}$$

$$\mathbf{Y}_c = \mathbf{A}\mathbf{X}_c + \mathbf{E}_c, \qquad \text{for fixed (known)} \qquad \mathbf{X}_c, \tag{1.52}$$

where $\mathbf{Y}_r \in \mathbb{R}_+^{R\times T}$ and $\mathbf{Y}_c \in \mathbb{R}_+^{I\times C}$ are the matrices constructed from the selected rows and columns of the matrix $\mathbf{Y}$, respectively. Analogously, we can construct the reduced matrices: $\mathbf{A}_r \in \mathbb{R}^{R\times J}$ and $\mathbf{X}_c \in \mathbb{R}^{J\times C}$ by using the same indices for the columns and rows as those used for the construction of the data sub-matrices $\mathbf{Y}_c$ and $\mathbf{Y}_r$. In practice, it is usually sufficient to choose: $J < R \leq 4J$ and $J < C \leq 4J$.

In the special case, for the squared Euclidean distance (Frobenius norm), instead of alternately minimizing the cost function

$$D_F(\mathbf{Y}\;||\;\mathbf{AX}) = \frac{1}{2}\;||\mathbf{Y} - \mathbf{AX}||_F^2\,, \tag{1.53}$$

we can minimize sequentially the two cost functions:

$$D_F(\mathbf{Y}_r\;||\;\mathbf{A}_r\mathbf{X}) = \frac{1}{2}\;||\mathbf{Y}_r - \mathbf{A}_r\mathbf{X}||_F^2\,, \qquad \text{for fixed} \qquad \mathbf{A}_r, \tag{1.54}$$

$$D_F(\mathbf{Y}_c\;||\;\mathbf{A}\mathbf{X}_c) = \frac{1}{2}\;||\mathbf{Y}_c - \mathbf{A}\mathbf{X}_c||_F^2\,, \qquad \text{for fixed} \qquad \mathbf{X}_c. \tag{1.55}$$

The minimization of these cost functions with respect to $\mathbf{X}$ and $\mathbf{A}$, subject to nonnegativity constraints, leads to the simple ALS update formulas for the large-scale NMF:

$$\boxed{\mathbf{X} \leftarrow \left[\mathbf{A}_r^{\dagger}\mathbf{Y}_r\right]_+ = \left[(\mathbf{A}_r^T\mathbf{A}_r)^{-1}\mathbf{A}_r\mathbf{Y}_r\right]_+, \qquad \mathbf{A} \leftarrow \left[\mathbf{Y}_c\mathbf{X}_c^{\dagger}\right]_+ = \left[\mathbf{Y}_c\mathbf{X}_c^T(\mathbf{X}_c\mathbf{X}_c^T)^{-1}\right]_+.} \tag{1.56}$$

A similar strategy can be applied for other cost functions and details will be given in Chapter 3 and Chapter 4.

There are several strategies to choose the columns and rows of the input data matrix [15,22,66,67,101]. The simplest scenario is to choose the first R rows and the first C columns of the data matrix $\mathbf{Y}$ (see Figure 1.12) or randomly select them using a uniform distribution. An optional strategy is to select randomly

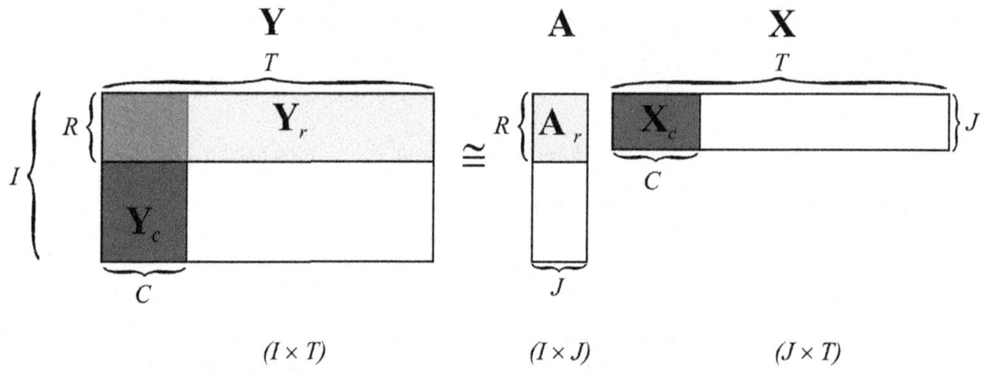

Figure 1.12 Conceptual illustration of block-wise data processing for large-scale NMF. Instead of processing the whole matrix $\mathbf{Y} \in \mathbb{R}^{I \times T}$, we can process much smaller block matrices $\mathbf{Y}_c \in \mathbb{R}^{I \times C}$ and $\mathbf{Y}_r \in \mathbb{R}^{R \times T}$ and corresponding factor matrices $\mathbf{X}_c \in \mathbb{R}^{J \times C}$ and $\mathbf{A}_r \in \mathbb{R}^{R \times J}$ with $C << T$ and $R << I$. For simplicity of graphical illustration, we have assumed that the first R rows and the first C columns of the matrices $\mathbf{Y}$, $\mathbf{A}$ and $\mathbf{X}$ are selected.

rows and columns from the set of all rows and columns with probability proportional to their relevance, e.g., with probability proportional to square of Euclidean ℓ_2-norm of rows and columns, i.e., $||\underline{\boldsymbol{y}}_i||_2^2$ and $||\boldsymbol{y}_t||_2^2$, respectively. Another heuristic option is to choose those rows and columns that provide the largest ℓ_p-norm. For noisy data with uncorrelated noise, we can construct new columns and rows as a local average (mean values) of some specific numbers of the columns and rows of raw data. For example, the first selected column is created as an average of the first M columns, the second column is an average of the next M columns, and so on; the same procedure applies for rows. Another strategy is to select optimal rows and columns using optimal CUR decomposition [22].

1.3.2 Non-uniqueness of NMF and Techniques to Alleviate the Ambiguity Problem

Usually, we perform NMF using the alternating minimization scheme (see Algorithm 1.1) of a set given objective functions. However, in general, such minimization does not guarantee a unique solution (neglecting unavoidable scaling and permutation ambiguities). Even the quadratic function with respect to both sets of arguments $\{\mathbf{A}\}$ and $\{\mathbf{X}\}$ may have many local minima, which makes NMF algorithms suffer from rotational indeterminacy (ambiguity). For example, consider the quadratic function:

$$D_F(\mathbf{Y}||\mathbf{AX}) = ||\mathbf{Y} - \mathbf{AX}||_F^2 = ||\mathbf{Y} - \mathbf{AR}^{-1}\mathbf{RX}||_F^2 = ||\mathbf{Y} - \tilde{\mathbf{A}}\tilde{\mathbf{X}}||_F^2. \tag{1.57}$$

There are many ways to select a rotational matrix $\mathbf{R}$ which is not necessarily nonnegative or not necessarily a generalized permutation matrix,[10] so that the transformed (rotated) $\tilde{\mathbf{A}} \neq \mathbf{A}$ and $\tilde{\mathbf{X}} \neq \mathbf{X}$ are nonnegative. Here, it is important to note that the inverse of a nonnegative matrix is nonnegative if and only if it is a generalized permutation matrix [119]. If we assume that $\mathbf{R} \geq 0$ and $\mathbf{R}^{-1} \geq 0$ (element-wise) which are sufficient (but not necessary) conditions for the nonnegativity of the transform matrices $\mathbf{AR}^{-1}$ and $\mathbf{RX}$, then $\mathbf{R}$ must be a generalized permutation (also called monomial) matrix, i.e., $\mathbf{R}$ can be expressed as a product of a nonsingular positive definite diagonal matrix and a permutation matrix. It is intuitively easy to understand

[10]Generalized permutation matrix is a matrix with only one nonzero positive element in each row and each column.

that if the original matrices **X** and **A** are sufficiently sparse only a generalized permutation matrix $\mathbf{P} = \mathbf{R}$ can satisfy the nonnegativity constraints of any transform matrices and NMF is unique.

To illustrate rotational indeterminacy consider the following mixing and source matrices:

$$\mathbf{A} = \begin{bmatrix} 3 & 2 \\ 7 & 2 \end{bmatrix}, \quad \mathbf{X} = \begin{bmatrix} x_1(t) \\ x_2(t) \end{bmatrix}, \tag{1.58}$$

which give the output

$$\mathbf{Y} = \begin{bmatrix} y_1(t) \\ y_2(t) \end{bmatrix} = \mathbf{A}\mathbf{X} = \begin{bmatrix} 3x_1(t) + 2x_2(t) \\ 7x_1(t) + 2x_2(t) \end{bmatrix}. \tag{1.59}$$

It is clear that there exists another nonnegative decomposition which gives us the following components:

$$\mathbf{Y} = \begin{bmatrix} 3x_1(t) + 2x_2(t) \\ 7x_1(t) + 2x_2(t) \end{bmatrix} = \tilde{\mathbf{A}}\tilde{\mathbf{X}} = \begin{bmatrix} 0 & 1 \\ 4 & 1 \end{bmatrix} \begin{bmatrix} x_1(t) \\ 3x_1(t) + 2x_2(t) \end{bmatrix}, \tag{1.60}$$

where

$$\tilde{\mathbf{A}} = \begin{bmatrix} 0 & 1 \\ 4 & 1 \end{bmatrix}, \quad \tilde{\mathbf{X}} = \begin{bmatrix} x_1(t) \\ 3x_1(t) + 2x_2(t) \end{bmatrix} \tag{1.61}$$

are new nonnegative components which do not come from the permutation or scaling indeterminacies.

However, incorporating some sparsity or smoothness measures to the objective function is sufficient to solve the NMF problem uniquely (up to unavoidable scale and permutation indeterminacies). The issues related to sparsity measures for NMF have been widely discussed [36,39,54,73,76,145], and are addressed in almost all chapters in this book.

When no prior information is available, we should perform normalization of the columns in **A** and/or the rows in **X** to help mitigate the effects of rotation indeterminacies. Such normalization is usually performed by scaling the columns $\boldsymbol{a}_j$ of $\mathbf{A} = [\boldsymbol{a}_1, \ldots, \boldsymbol{a}_J]$ as follows:

$$\mathbf{A} \leftarrow \mathbf{A}\mathbf{D}_A, \quad \text{where} \quad \mathbf{D}_A = \operatorname{diag}(||\boldsymbol{a}_1||_p^{-1}, ||\boldsymbol{a}_2||_p^{-1}, \ldots, ||\boldsymbol{a}_J||_p^{-1}), \qquad p \in [0, \infty). \tag{1.62}$$

Heuristics based on extensive experimentations show that best results can be obtained for $p = 1$, i.e., when the columns of **A** are normalized to unit ℓ_1-norm. This may be justified by the fact that the mixing matrix should contain only a few dominant entries in each column, which is emphasized by the normalization to the unit ℓ_1-norms.[11] The normalization (1.62) for the alternating minimization scheme (Algorithm 1.1) helps to alleviate many numerical difficulties, like numerical instabilities or ill-conditioning, however, it makes searching for the global minimum more complicated.

Moreover, to avoid rotational ambiguity of NMF, the rows of **X** should be sparse or zero-grounded. To achieve this we may apply some preprocessing, sparsification, or filtering of the input data. For example,

[11] In the case when the columns of **A** and rows of **X** are simultaneously normalized, the standard NMF model $\mathbf{Y} \approx \mathbf{A}\mathbf{X}$ is converted to a three-factor NMF model $\mathbf{Y} \approx \mathbf{A}\mathbf{D}\mathbf{X}$, where $\mathbf{D} = \mathbf{D}_A\mathbf{D}_X$ is a diagonal scaling matrix.

we may remove the baseline from the input data $\mathbf{Y}$ by applying the affine NMF instead of the regular NMF, that is,

$$\mathbf{Y} = \mathbf{A}\mathbf{X} + \boldsymbol{a}_0 \mathbf{1}_T^T + \mathbf{E}, \tag{1.63}$$

where $\boldsymbol{a}_0 \in \mathbb{R}_+^I$ is a vector selected in such a way that the unbiased matrix $\hat{\mathbf{Y}} = \mathbf{Y} - \boldsymbol{a}_0 \mathbf{1}_T^T \in \mathbb{R}_+^{I \times T}$ has many zeros or close to zero entries (see Chapter 3 for algorithms).

In summary, in order to obtain a unique NMF solution (neglecting unavoidable permutation and scaling indeterminacies), we need to enforce at least one of the following techniques:

1. Normalize or filter the input data $\mathbf{Y}$, especially by applying the affine NMF model (1.63), in order to make the factorized matrices zero-grounded.
2. Normalize the columns of $\mathbf{A}$ and/or the rows of $\mathbf{X}$ to unit length.
3. Impose sparsity and/or smoothness constraints to the factorized matrices.

1.3.3 Initialization of NMF

The solution and convergence provided by NMF algorithms usually highly depend on initial conditions, i.e., its starting guess values, especially in a multivariate context. Thus, it is important to have efficient and consistent ways for initializing matrices $\mathbf{A}$ and/or $\mathbf{X}$. In other words, the efficiency of many NMF strategies is affected by the selection of the starting matrices. Poor initializations often result in slow convergence, and in certain instances may lead even to an incorrect, or irrelevant solution. The problem of selecting appropriate starting initialization matrices becomes even more complicated for large-scale NMF problems and when certain structures or constraints are imposed on the factorized matrices involved. As a good initialization for one data set may be poor for another data set, to evaluate the efficiency of an initialization strategy and the algorithm we should perform uncertainty analysis such as Monte Carlo simulations. Initialization in NMF plays a key role since the objective function to be minimized may have many local minima, and the intrinsic alternating minimization in NMF is nonconvex, even though the objective function is strictly convex with respect to one set of variables. For example, the quadratic function:

$$D_F(\mathbf{Y}||\mathbf{A}\mathbf{X}) = ||\mathbf{Y} - \mathbf{A}\mathbf{X}||_F^2$$

is strictly convex in one set of variables, either $\mathbf{A}$ or $\mathbf{X}$, but not in both. The issues of initialization in NMF have been widely discussed in the literature [3,14,82,93].

As a rule of thumb, we can obtain a robust initialization using the following steps:

1. First, we built up a search method for generating R initial matrices $\mathbf{A}$ and $\mathbf{X}$. This could be based on random starts or the output from a simple ALS NMF algorithm. The parameter R depends on the number of required iterations (typically 10-20 is sufficient).
2. Run a specific NMF algorithm for each set of initial matrices and with a fixed but small number of iterations (typically 10-20). As a result, the NMF algorithm provides R initial estimates of the matrices $\mathbf{A}^{(r)}$ and $\mathbf{X}^{(r)}$.
3. Select the estimates (denoted by $\mathbf{A}^{(r_{\min})}$ and $\mathbf{X}^{(r_{\min})}$) corresponding to the lowest value of the cost function (the best likelihood) among the R trials as initial values for the final factorization.

In other words, the main idea is to find good initial estimates ("candidates") with the following multi-start initialization algorithm:

Algorithm 1.2: Multi-start initialization

Input: $\mathbf{Y} \in \mathbb{R}_+^{I \times T}$: input data,
J: rank of approximation , R: number of restarts,
K_{init}, K_{fin}: number of alternating steps for initialization and completion
Output: $\mathbf{A} \in \mathbb{R}_+^{I \times J}$ and $\mathbf{X} \in \mathbb{R}_+^{J \times T}$ such that a given cost function is minimized.

begin
  parfor $r = 1$ *to* R **do** /* process in parallel mode */
    Initialize randomly $\mathbf{A}^{(0)}$ or $\mathbf{X}^{(0)}$
    $\{\mathbf{A}^{(r)}, \mathbf{X}^{(r)}\} \leftarrow$ nmf_algorithm($\mathbf{Y}, \mathbf{A}^{(0)}, \mathbf{X}^{(0)}, K_{init}$)
    $d_r = D(\mathbf{Y}||\mathbf{A}^{(r)}\mathbf{X}^{(r)})$ /* compute the cost value */
  endfor
  $r_{min} = \arg\min_{1 \le r \le R} d_r$
  $\{\mathbf{A}, \mathbf{X}\} \leftarrow$ nmf_algorithm($\mathbf{Y}, \mathbf{A}^{(r_{min})}, \mathbf{X}^{(r_{min})}, K_{fin}$)
end

Thus, the multi-start initialization selects the initial estimates for **A** and **X** which give the steepest decrease in the assumed objective function $D(\mathbf{Y}||\mathbf{AX})$ via alternating steps. Usually, we choose the generalized Kullback-Leibler divergence $D_{KL}(\mathbf{Y}||\mathbf{AX})$ for checking the convergence results after K_{init} initial alternating steps. The initial estimates $\mathbf{A}^{(0)}$ and $\mathbf{X}^{(0)}$ which give the lowest values of $D_{KL}(\mathbf{Y}||\mathbf{AX})$ after K_{init} alternating steps are expected to be the most suitable candidates for continuing the alternating minimization. In practice, for $K_{init} \ge 10$, the algorithm works quite efficiently.

Throughout this book, we shall explore various alternative methods for the efficient initialization of the iterative NMF algorithms and provide supporting pseudo-source codes and MATLAB codes; for example, we use extensively the ALS-based initialization technique as illustrated by the following MATLAB code:

Listing 1.1 Basic initializations for NMF algorithms.

```
function [Ainit,Xinit] = NMFinitialization(Y,J,inittype)
% Y        :      nonnegative matrix
% J        :      number of components
% inittype        1 {random}, 2 {ALS}, 3 {SVD}
[I,T] = size(Y);
Ainit = rand(I,J);
Xinit = rand(J,T);

switch inittype
    case 2 % ALS
        Ainit = max(eps,(Y*Xinit')*pinv(Xinit*Xinit'));
        Xinit = max(eps,pinv(Ainit'*Ainit)*(Ainit'*Y));
    case 3 %SVD
        [Ainit,Xinit] = lsvNMF(Y,J);
end
Ainit = bsxfun(@rdivide,Ainit,sum(Ainit));
end
```

1.3.4 Stopping Criteria

There are several possible stopping criteria for the iterative algorithms used in NMF:

- The cost function achieves a zero-value or a value below a given threshold ε, for example,

$$D_F^{(k)}(\mathbf{Y} \,||\, \hat{\mathbf{Y}}^{(k)}) = \left\|\mathbf{Y} - \hat{\mathbf{Y}}^{(k)}\right\|_F^2 \le \varepsilon \,. \tag{1.64}$$

- There is little or no improvement between successive iterations in the minimization of a cost function, for example,

$$D_F^{(k+1)}(\hat{\mathbf{Y}}^{(k+1)} \; || \; \hat{\mathbf{Y}}^{(k)}) = \left\|\hat{\mathbf{Y}}^{(k)} - \hat{\mathbf{Y}}^{(k+1)}\right\|_F^2 \leq \varepsilon, \tag{1.65}$$

or

$$\frac{|D_F^{(k)} - D_F^{(k-1)}|}{D_F^{(k)}} \leq \varepsilon \, . \tag{1.66}$$

- There is little or no change in the updates for factor matrices **A** and **X**.
- The number of iterations achieves or exceeds a predefined maximum number of iterations.

In practice, the iterations usually continue until some combinations of stopping conditions are satisfied. Some more advanced stopping criteria are discussed in Chapter 5.

1.4 Tensor Properties and Basis of Tensor Algebra

Matrix factorization models discussed in the previous sections can be naturally extended and generalized to multi-way arrays, also called multi-dimensional matrices or simply tensor decompositions.[12]

1.4.1 Tensors (Multi-way Arrays) – Preliminaries

A tensor is a multi-way array or multi-dimensional matrix. The order of a tensor is the number of dimensions, also known as ways or modes. Tensor can be formally defined as

Definition 1.1 **(Tensor)** *Let $I_1, I_2, \ldots, I_N \in \mathbb{N}$ denote index upper bounds. A tensor $\underline{\mathbf{Y}} \in \mathbb{R}^{I_1 \times I_2 \times \cdots \times I_N}$ of order N is an N-way array where elements $y_{i_1 i_2 \cdots i_n}$ are indexed by $i_n \in \{1, 2, \ldots, I_n\}$ for $1 \leq n \leq N$.*

Tensors are obviously generalizations of vectors and matrixes, for example, a third-order tensor (or three-way array) has three modes (or indices or dimensions) as shown in Figure 1.13. A zero-order tensor is a

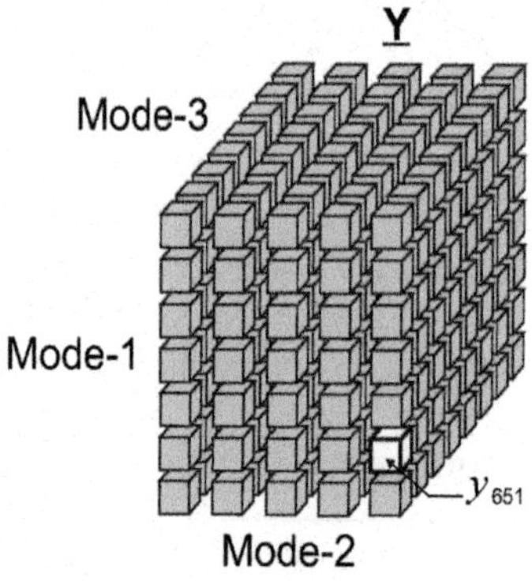

Figure 1.13 A three-way array (third-order tensor) $\underline{\mathbf{Y}} \in \mathbb{R}^{7 \times 5 \times 8}$ with elements y_{itq}.

[12]The notion of tensors used in this book should not be confused with field tensors used in physics and differential geometry, which are generally referred to as tensor fields (i.e., tensor-valued functions on manifolds) in mathematics [85]. Examples include, stress tensor, moment-of inertia tensor, Einstein tensor, metric tensor, curvature tensor, Ricci tensor.

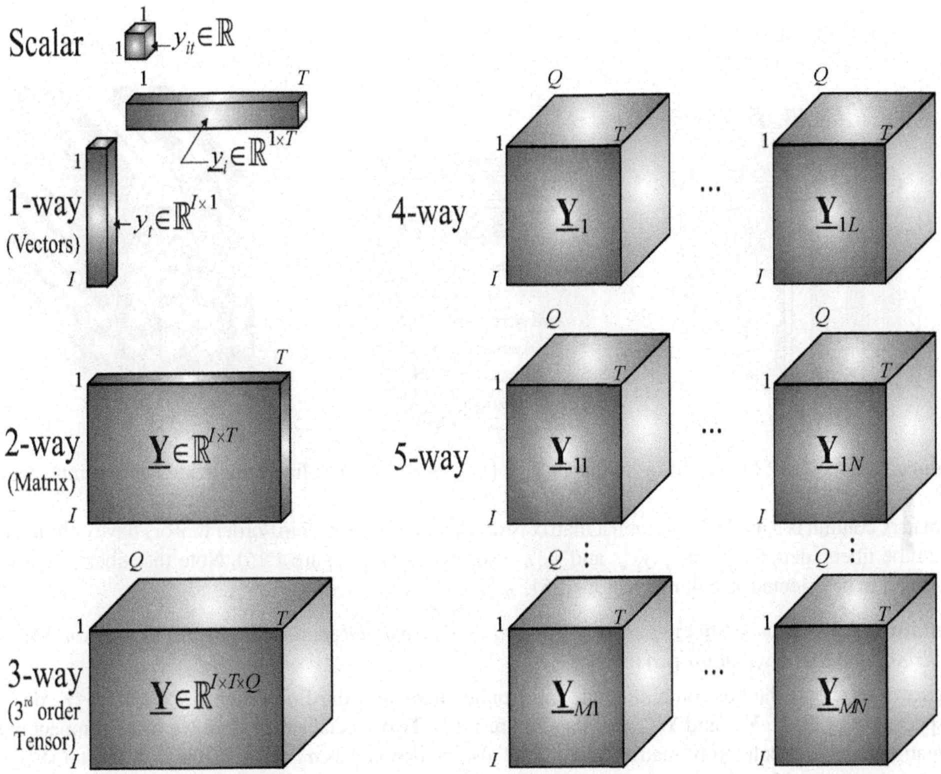

Figure 1.14 Illustration of multi-way data: zero-way tensor = scalar, 1-way tensor = row or column vector, 2-way tensor = matrix, N-way tensor = higher-order tensors. The 4-way and 5-way tensors are represented here as a set of the three-way tensors.

scalar, a first-order tensor is a vector, a second-order tensor is a matrix, and tensors of order three and higher are called higher-order tensors (see Figure 1.14).

Generally, N-order tensors are denoted by underlined capital boldface letters, e.g., $\underline{\mathbf{Y}} \in \mathbb{R}^{I_1 \times I_2 \times \cdots \times I_N}$. In contrast, matrices are denoted by boldface capital letters, e.g., $\mathbf{Y}$; vectors are denoted by boldface lowercase letters, e.g., columns of the matrix $\mathbf{A}$ by $\boldsymbol{a}_j$ and scalars are denoted by lowercase letters, e.g., a_{ij}. The i-th entry of a vector $\boldsymbol{a}$ is denoted by a_i, and the (i, j)-th element of a matrix $\mathbf{A}$ by a_{ij}. Analogously, the element (i, t, q) of a third-order tensor $\underline{\mathbf{Y}} \in \mathbb{R}^{I \times T \times Q}$ is denoted by y_{itq}. The values of indices are typically ranging from 1 to their capital version, e.g., $i = 1, 2, \ldots, I$; $t = 1, 2, \ldots, T; q = 1, 2, \ldots, Q$.

1.4.2 Subarrays, Tubes and Slices

Subtensors or subarrays are formed when a subset of the indices is fixed. For matrices, these are the rows and columns. A colon is used to indicate all elements of a mode in the style of MATLAB. Thus, the j-th column of a matrix $\mathbf{A} = [\boldsymbol{a}_1, \boldsymbol{a}_2, \ldots, \boldsymbol{a}_J]$ is formally denoted by $\boldsymbol{a}_{:j}$; likewise, the j-th row of $\mathbf{X}$ is denoted by $\underline{\boldsymbol{x}}_j = \boldsymbol{x}_{j:}$.

Definition 1.2 (Tensor Fiber) *A tensor fiber is a one-dimensional fragment of a tensor, obtained by fixing all indices except for one.*

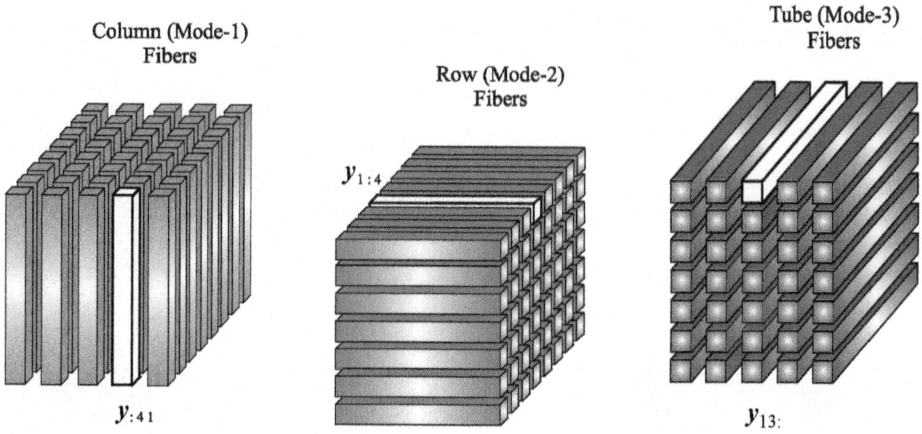

Figure 1.15 Fibers: for a third-order tensor $\underline{\mathbf{Y}} = [y_{itq}] \in \mathbb{R}^{I \times T \times Q}$ (all fibers are treated as column vectors).

A matrix column is a mode-1 fiber and a matrix row is a mode-2 fiber. Third-order tensors have column, row, and tube fibers, denoted by $\boldsymbol{y}_{:tq}$, $\boldsymbol{y}_{i:q}$, and $\boldsymbol{y}_{it:}$, respectively (see Figure 1.15). Note that fibers are always assumed to be oriented as column vectors [85].

Definition 1.3 **(Tensor Slice)** *A tensor slice is a two-dimensional section (fragment) of a tensor, obtained by fixing all indices except for two indices.*

Figure 1.16 shows the horizontal, lateral, and frontal slices of a third-order tensor $\underline{\mathbf{Y}} \in \mathbb{R}^{I \times T \times Q}$, denoted respectively by $\mathbf{Y}_{i::}$, $\mathbf{Y}_{:t:}$ and $\mathbf{Y}_{::q}$ (see also Figure 1.17). Two special subarrays have more compact representations: the j-th column of matrix $\mathbf{A}$, $\boldsymbol{a}_{:j}$, may also be denoted as $\boldsymbol{a}_j$, whereas the q-th frontal slice of a third-order tensor, $\mathbf{Y}_{::q}$ may also be denoted as $\mathbf{Y}_q$, $(q = 1, 2, \ldots, Q)$.

1.4.3 Unfolding – Matricization

It is often very convenient to represent tensors as matrices or to represent multi-way relationships and a tensor decomposition in their matrix forms. Unfolding, also known as matricization or flattening, is a process of reordering the elements of an N-th order tensor into a matrix. There are various ways to order the fibers of tensors, therefore, the unfolding process is not unique. Since the concept is easy to understand

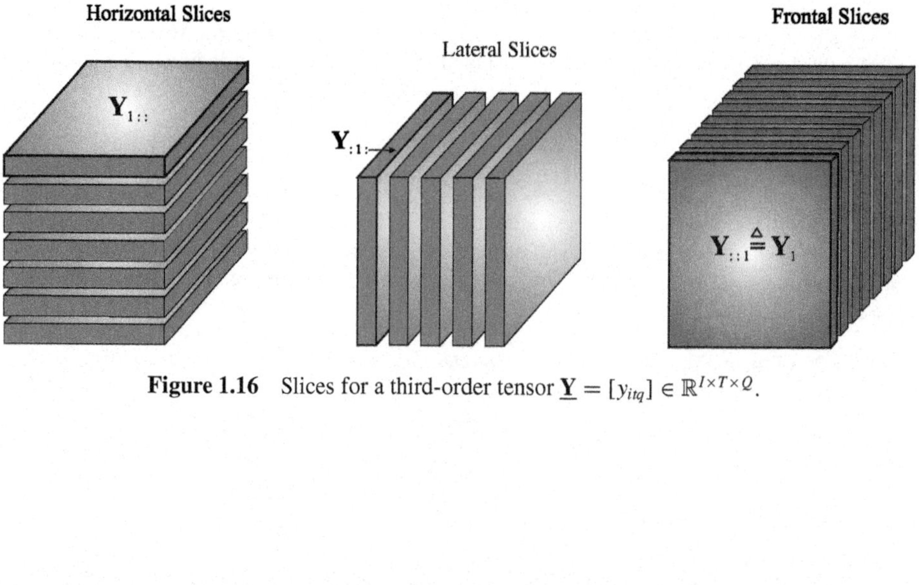

Figure 1.16 Slices for a third-order tensor $\underline{\mathbf{Y}} = [y_{itq}] \in \mathbb{R}^{I \times T \times Q}$.

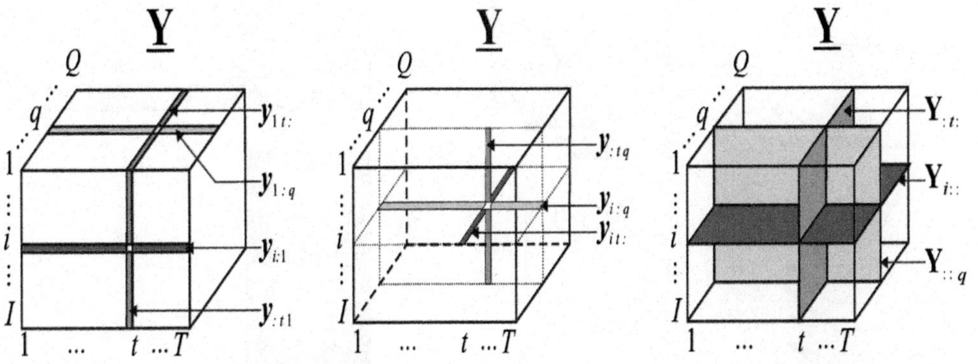

Figure 1.17 Illustration of subsets (subarrays) of a three-way tensor and basic tensor notations of tubes and slices.

by examples, Figures 1.18, 1.19 and 1.20 illustrate the various unfolding processes of a three-way array. For example, for a third-order tensor we can arrange frontal, horizontal and lateral slices in row-wise and column-wise ways. Generally speaking, the unfolding of an N-th order tensor can be understood as the process of the construction of a matrix containing all the mode-n vectors of the tensor. The order of the columns is not unique and in this book it is chosen in accordance with [85] and based on the following definition:

Definition 1.4 (Unfolding) *The mode-n unfolding of tensor* $\underline{\mathbf{Y}} \in \mathbb{R}^{I_1 \times I_2 \times \cdots \times I_N}$ *is denoted by*[13] $\mathbf{Y}_{(n)}$ *and arranges the mode-n fibers into columns of a matrix. More specifically, a tensor element* $(i_1, i_2, \ldots, i_N)$ *maps onto a matrix element* (i_n, j)*, where*

$$j = 1 + \sum_{p \neq n} (i_p - 1) J_p, \quad \textit{with} \quad J_p = \begin{cases} 1, & \textit{if } p = 1 \textit{ or if } p = 2 \textit{ and } n = 1, \\ \prod_{m \neq n}^{p-1} I_m, & \textit{otherwise.} \end{cases} \tag{1.67}$$

Observe that in the mode-n unfolding the mode-n fibers are rearranged to be the columns of the matrix $\mathbf{Y}_{(n)}$.

More generally, a subtensor of the tensor $\underline{\mathbf{Y}} \in \mathbb{R}^{I_1 \times I_2 \times \cdots \times I_N}$, denoted by $\mathbf{Y}_{(i_n = j)}$, is obtained by fixing the n-th index to some value j. For example, a third-order tensor $\underline{\mathbf{Y}} \in \mathbb{R}^{I_1 \times I_2 \times I_3}$ with entries y_{i_1, i_2, i_3} and indices (i_1, i_2, i_3) has a corresponding position (i_n, j) in the mode-n unfolded matrix $\mathbf{Y}_{(n)}$ $(n = 1, 2, 3)$ as follows

- mode-1: $j = i_2 + (i_3 - 1) I_2$,
- mode-2: $j = i_1 + (i_3 - 1) I_1$,
- mode-3: $j = i_1 + (i_2 - 1) I_1$.

Note that mode-n unfolding of a tensor $\underline{\mathbf{Y}} \in \mathbb{R}^{I_1 \times I_2 \cdots \times I_N}$ also represents mode-1 unfolding of its permuted tensor $\underline{\tilde{\mathbf{Y}}} \in \mathbb{R}^{I_n \times I_1 \cdots \times I_{n-1} \times I_{n+1} \cdots \times I_N}$ obtained by permuting its modes to obtain the mode-1 be I_n.

[13] We use the Kolda - Bader notations [85].

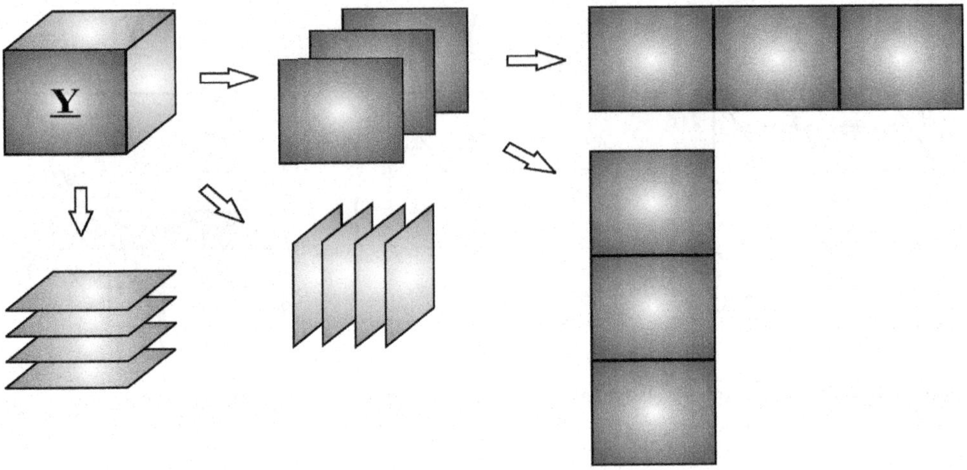

Figure 1.18 Illustration of row-wise and column-wise unfolding (flattening, matricizing) of a third-order tensor.

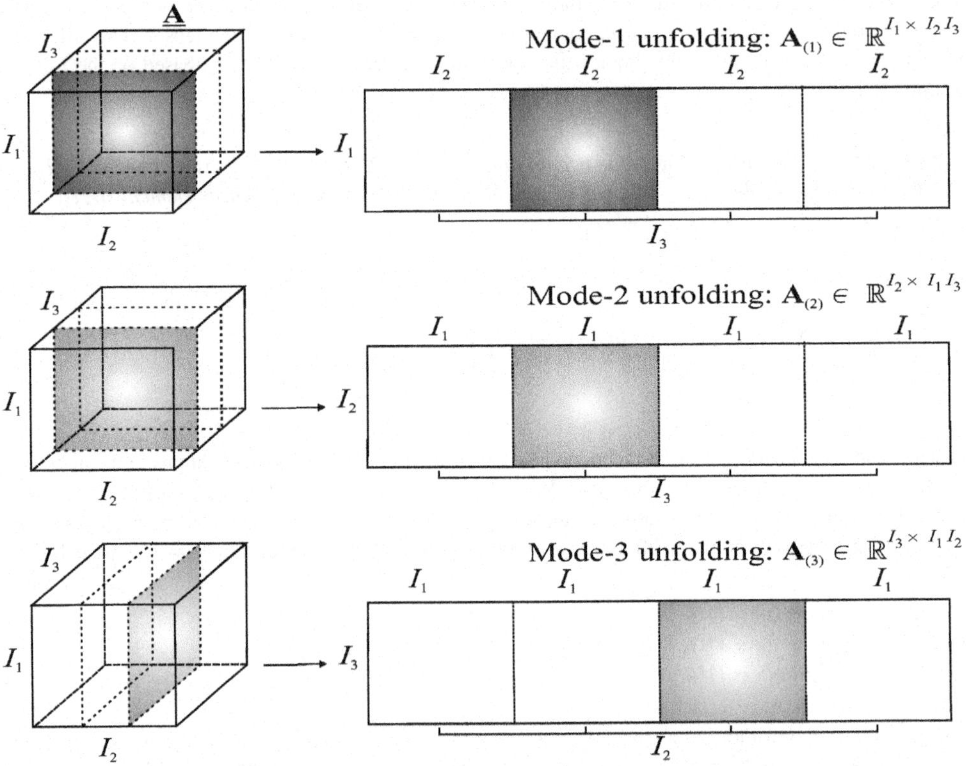

Figure 1.19 Unfolding (matricizing) of a third-order tensor. The tensor can be unfolded in three ways to obtain matrices comprising its mode-1, mode-2 and mode-3 vectors.

$$\mathbf{A}_{(1)} = \begin{bmatrix} a_{111} & a_{121} & a_{131} & a_{141} & a_{112} & a_{122} & a_{132} & a_{142} \\ a_{211} & a_{221} & a_{231} & a_{241} & a_{212} & a_{222} & a_{232} & a_{242} \\ a_{311} & a_{321} & a_{331} & a_{341} & a_{312} & a_{322} & a_{332} & a_{342} \end{bmatrix}$$

$$\mathbf{A}_{(2)} = \begin{bmatrix} a_{111} & a_{211} & a_{311} & a_{112} & a_{212} & a_{312} \\ a_{121} & a_{221} & a_{321} & a_{122} & a_{222} & a_{322} \\ a_{131} & a_{231} & a_{331} & a_{132} & a_{232} & a_{332} \\ a_{141} & a_{241} & a_{341} & a_{142} & a_{242} & a_{342} \end{bmatrix}$$

$$\mathbf{A}_{(3)} = \begin{bmatrix} a_{111} & a_{211} & a_{311} & a_{121} & a_{221} & a_{321} & a_{131} & a_{231} & a_{331} & a_{141} & a_{241} & a_{341} \\ a_{112} & a_{212} & a_{312} & a_{122} & a_{222} & a_{322} & a_{132} & a_{232} & a_{332} & a_{142} & a_{242} & a_{342} \end{bmatrix}$$

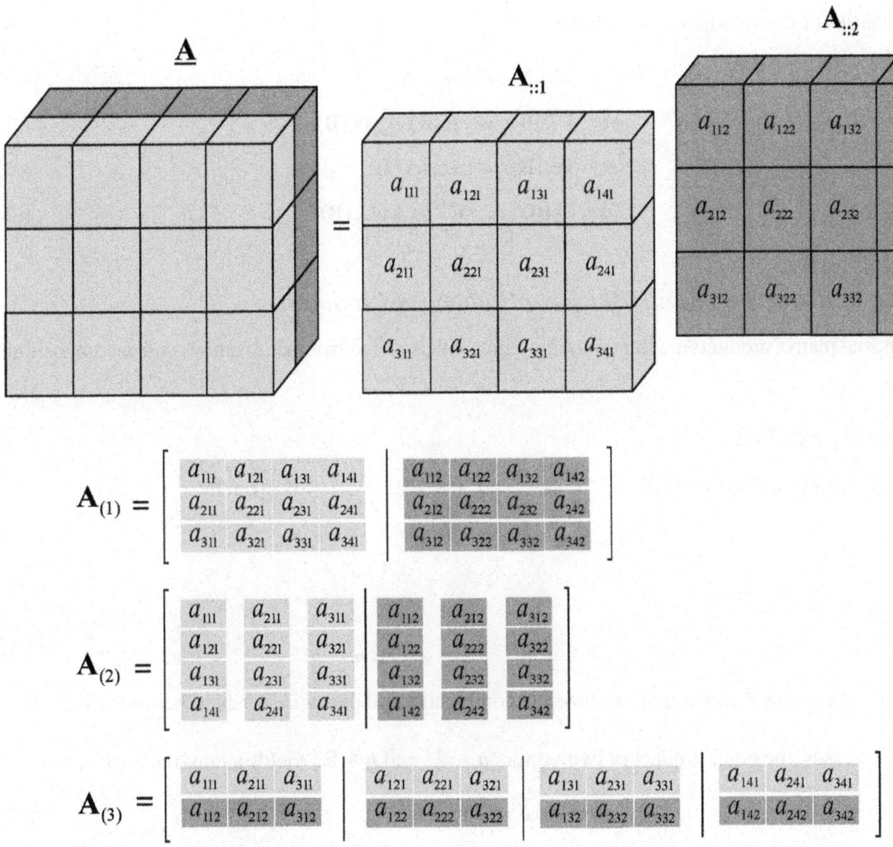

Figure 1.20 Example of unfolding the third-order tensor in mode-1, mode-2 and mode-3.

1.4.4 Vectorization

It is often convenient to represent tensors and matrices as vectors, whereby vectorization of matrix $\mathbf{Y} = [\mathbf{y}_1, \mathbf{y}_2, \dots, \mathbf{y}_T] \in \mathbb{R}^{I \times T}$ is defined as

$$\mathbf{y} = \text{vec}(\mathbf{Y}) = \left[\mathbf{y}_1^T, \mathbf{y}_2^T, \dots, \mathbf{y}_T^T\right]^T \in \mathbb{R}^{IT}. \tag{1.68}$$

The vec-operator applied on a matrix $\mathbf{Y}$ stacks its columns into a vector. The reshape is a reverse function to vectorization which converts a vector to a matrix. For example, reshape$(\mathbf{y}, I, T) \in \mathbb{R}^{I \times T}$ is defined as (using MATLAB notations and similar to the reshape MATLAB function):

$$\text{reshape}(\mathbf{y}, I, T) = \left[\mathbf{y}(1:I), \mathbf{y}(I+1:2I), \dots, \mathbf{y}((T-1)I:IT)\right] \in \mathbb{R}^{I \times T}. \tag{1.69}$$

Analogously, we define the vectorization of a tensor $\underline{\mathbf{Y}}$ as a vectorization of the associated mode-1 unfolded matrix $\mathbf{Y}_{(1)}$. For example, the vectorization of the third-order tensor $\underline{\mathbf{Y}} \in \mathbb{R}^{I \times T \times Q}$ can be written in the following form

$$\text{vec}(\underline{\mathbf{Y}}) = \text{vec}(\mathbf{Y}_{(1)}) = \left[\text{vec}(\mathbf{Y}_{::1})^T, \text{vec}(\mathbf{Y}_{::2})^T, \dots, \text{vec}(\mathbf{Y}_{::Q})^T\right]^T \in \mathbb{R}^{ITQ}. \tag{1.70}$$

Basic properties of the vec-operators include:

$$\text{vec}(c\,\mathbf{A}) = c\,\text{vec}(\mathbf{A}), \tag{1.71}$$

$$\text{vec}(\mathbf{A}+\mathbf{B}) = \text{vec}(\mathbf{A}) + \text{vec}(\mathbf{B}), \tag{1.72}$$

$$\text{vec}(\mathbf{A})^T\,\text{vec}(\mathbf{B}) = \text{trace}(\mathbf{A}^T\mathbf{B}), \tag{1.73}$$

$$\text{vec}(\mathbf{ABC}) = (\mathbf{C}^T \otimes \mathbf{A})\text{vec}(\mathbf{B}). \tag{1.74}$$

1.4.5 Outer, Kronecker, Khatri-Rao and Hadamard Products

Several special matrix products are important for representation of tensor factorizations and decompositions.

1.4.5.1 Outer Product

The outer product of the tensors $\underline{\mathbf{Y}} \in \mathbb{R}^{I_1 \times I_2 \times \cdots \times I_N}$ and $\underline{\mathbf{X}} \in \mathbb{R}^{J_1 \times J_2 \times \cdots \times J_M}$ is given by

$$\underline{\mathbf{Z}} = \mathbf{Y} \circ \mathbf{X} \in \mathbb{R}^{I_1 \times I_2 \times \cdots \times I_N \times J_1 \times J_2 \times \cdots \times J_M}, \tag{1.75}$$

where

$$z_{i_1,i_2,\ldots,i_N,j_1,j_2,\ldots,j_M} = y_{i_1,i_2,\ldots,i_N}\; x_{j_1,j_2,\ldots,j_M}. \tag{1.76}$$

Observe that, the tensor $\underline{\mathbf{Z}}$ contains all the possible combinations of pair-wise products between the elements of $\underline{\mathbf{Y}}$ and $\underline{\mathbf{X}}$.

As special cases, the outer product of two vectors $\boldsymbol{a} \in \mathbb{R}^I$ and $\boldsymbol{b} \in \mathbb{R}^J$ yields a rank-one matrix

$$\mathbf{A} = \boldsymbol{a} \circ \boldsymbol{b} = \boldsymbol{a}\boldsymbol{b}^T \in \mathbb{R}^{I \times J} \tag{1.77}$$

and the outer product of three vectors: $\boldsymbol{a} \in \mathbb{R}^I$, $\boldsymbol{b} \in \mathbb{R}^J$ and $\boldsymbol{c} \in \mathbb{R}^Q$ yields a third-order rank-one tensor:

$$\underline{\mathbf{Z}} = \boldsymbol{a} \circ \boldsymbol{b} \circ \boldsymbol{c} \in \mathbb{R}^{I \times J \times Q}, \tag{1.78}$$

where

$$z_{ijq} = a_i\, b_j\, c_q. \tag{1.79}$$

1.4.5.2 Kronecker Product

The Kronecker product of two matrices $\mathbf{A} \in \mathbb{R}^{I \times J}$ and $\mathbf{B} \in \mathbb{R}^{T \times R}$ is a matrix denoted as $\mathbf{A} \otimes \mathbf{B} \in \mathbb{R}^{IT \times JR}$ and defined as (see the MATLAB function `kron`):

$$\mathbf{A} \otimes \mathbf{B} = \begin{bmatrix} a_{11}\,\mathbf{B} & a_{12}\,\mathbf{B} & \cdots & a_{1J}\,\mathbf{B} \\ a_{21}\,\mathbf{B} & a_{22}\,\mathbf{B} & \cdots & a_{2J}\,\mathbf{B} \\ \vdots & \vdots & \ddots & \vdots \\ a_{I1}\,\mathbf{B} & a_{I2}\,\mathbf{B} & \cdots & a_{IJ}\,\mathbf{B} \end{bmatrix} \tag{1.80}$$

$$= \begin{bmatrix} \boldsymbol{a}_1 \otimes \boldsymbol{b}_1 & \boldsymbol{a}_1 \otimes \boldsymbol{b}_2 & \boldsymbol{a}_1 \otimes \boldsymbol{b}_3 & \cdots & \boldsymbol{a}_J \otimes \boldsymbol{b}_{R-1} & \boldsymbol{a}_J \otimes \boldsymbol{b}_R \end{bmatrix}. \tag{1.81}$$

For any given three matrices **A**, **B**, and **C**, where **B** and **C** have the same size, the following properties hold:

$$(\mathbf{A} \otimes \mathbf{B})^T = \mathbf{A}^T \otimes \mathbf{B}^T, \tag{1.82}$$

$$(\mathbf{A} \otimes \mathbf{B})^{\dagger} = \mathbf{A}^{\dagger} \otimes \mathbf{B}^{\dagger}, \tag{1.83}$$

$$\mathbf{A} \otimes (\mathbf{B} + \mathbf{C}) = (\mathbf{A} \otimes \mathbf{B}) + (\mathbf{A} \otimes \mathbf{C}), \tag{1.84}$$

$$(\mathbf{B} + \mathbf{C}) \otimes \mathbf{A} = (\mathbf{B} \otimes \mathbf{A}) + (\mathbf{C} \otimes \mathbf{A}), \tag{1.85}$$

$$(\mathbf{A} + \mathbf{B}) \otimes (\mathbf{C} + \mathbf{D}) = \mathbf{AC} \otimes \mathbf{BD}, \tag{1.86}$$

$$c\,(\mathbf{A} \otimes \mathbf{B}) = (c\,\mathbf{A}) \otimes \mathbf{B} = \mathbf{A} \otimes (c\,\mathbf{B}). \tag{1.87}$$

It should be mentioned that, in general, the outer product of vectors yields a tensor whereas the Kronecker product gives a vector. For example, for the three vectors $\boldsymbol{a} \in \mathbb{R}^J$, $\boldsymbol{b} \in \mathbb{R}^T$, $\boldsymbol{c} \in \mathbb{R}^Q$ their three-way outer product $\underline{\mathbf{Y}} = \boldsymbol{a} \circ \boldsymbol{b} \circ \boldsymbol{c} \in \mathbb{R}^{I \times T \times Q}$ is a third-order tensor with the entries $y_{itq} = a_j b_t c_q$, while the three-way Kronecker product of the same vectors is a vector $\text{vec}(\underline{\mathbf{Y}}) = \boldsymbol{c} \otimes \boldsymbol{b} \otimes \boldsymbol{a} \in \mathbb{R}^{ITQ}$.

1.4.5.3 Hadamard Product

The Hadamard product of two equal-size matrices is the element-wise product denoted by $\circledast$ (or .∗ for MATLAB notation) and defined as

$$\mathbf{A} \circledast \mathbf{B} = \begin{bmatrix} a_{11}\, b_{11} & a_{12}\, b_{12} & \cdots & a_{1J}\, b_{1J} \\ a_{21}\, b_{21} & a_{22}\, b_{22} & \cdots & a_{2J}\, b_{2J} \\ \vdots & \vdots & \ddots & \vdots \\ a_{I1}\, b_{I1} & a_{I2}\, b_{I2} & \cdots & a_{IJ}\, b_{IJ} \end{bmatrix}. \tag{1.88}$$

1.4.5.4 Khatri-Rao Product

For two matrices $\mathbf{A} = [\boldsymbol{a}_1, \boldsymbol{a}_2, \ldots, \boldsymbol{a}_J] \in \mathbb{R}^{I \times J}$ and $\mathbf{B} = [\boldsymbol{b}_1, \boldsymbol{b}_2, \ldots, \boldsymbol{b}_J] \in \mathbb{R}^{T \times J}$ with the same number of columns J, their Khatri-Rao product, denoted by $\odot$, performs the following operation:

$$\mathbf{A} \odot \mathbf{B} = [\boldsymbol{a}_1 \otimes \boldsymbol{b}_1 \;\; \boldsymbol{a}_2 \otimes \boldsymbol{b}_2 \;\; \cdots \;\; \boldsymbol{a}_J \otimes \boldsymbol{b}_J] \tag{1.89}$$

$$= \left[\text{vec}(\boldsymbol{b}_1 \boldsymbol{a}_1^T) \;\; \text{vec}(\boldsymbol{b}_2 \boldsymbol{a}_2^T) \;\; \cdots \;\; \text{vec}(\boldsymbol{b}_J \boldsymbol{a}_J^T)\right] \in \mathbb{R}^{IT \times J}. \tag{1.90}$$

The Khatri-Rao product is:

- associative

$$\mathbf{A} \odot (\mathbf{B} \odot \mathbf{C}) = (\mathbf{A} \odot \mathbf{B}) \odot \mathbf{C}, \tag{1.91}$$

- distributive

$$(\mathbf{A} + \mathbf{B}) \odot \mathbf{C} = \mathbf{A} \odot \mathbf{C} + \mathbf{B} \odot \mathbf{C}, \tag{1.92}$$

- non-commutative

$$\mathbf{A} \odot \mathbf{B} \neq \mathbf{B} \odot \mathbf{A}, \tag{1.93}$$

- its cross-product simplifies into

$$(\mathbf{A} \odot \mathbf{B})^T\,(\mathbf{A} \odot \mathbf{B}) = \mathbf{A}^T\mathbf{A} \;\circledast\; \mathbf{B}^T\mathbf{B}, \tag{1.94}$$

- and the Moore-Penrose pseudo-inverse can be expressed as

$$(\mathbf{A} \odot \mathbf{B})^{\dagger} = [(\mathbf{A} \odot \mathbf{B})^T(\mathbf{A} \odot \mathbf{B})]^{-1}(\mathbf{A} \odot \mathbf{B})^T = [(\mathbf{A}^T\mathbf{A}) \circledast (\mathbf{B}^T\mathbf{B})]^{-1}(\mathbf{A} \odot \mathbf{B})^T, \tag{1.95}$$

$$((\mathbf{A} \odot \mathbf{B})^T)^{\dagger} = (\mathbf{A} \odot \mathbf{B})[(\mathbf{A}^T\mathbf{A}) \circledast (\mathbf{B}^T\mathbf{B})]^{-1}. \tag{1.96}$$

1.4.6 Mode-n Multiplication of Tensor by Matrix and Tensor by Vector, Contracted Tensor Product

To multiply a tensor by a matrix, we need to specify which mode of the tensor is multiplied by the columns (or rows) of a matrix (see Figure 1.21 and Table 1.1).

Definition 1.5 (mode-n tensor matrix product) *The mode-n product $\underline{\mathbf{Y}} = \underline{\mathbf{G}} \times_n \mathbf{A}$ of a tensor $\underline{\mathbf{G}} \in \mathbb{R}^{J_1 \times J_2 \times \cdots \times J_N}$ and a matrix $\mathbf{A} \in \mathbb{R}^{I_n \times J_n}$ is a tensor $\underline{\mathbf{Y}} \in \mathbb{R}^{J_1 \times \cdots \times J_{n-1} \times I_n \times J_{n+1} \times \cdots \times J_N}$, with elements*

$$y_{j_1, j_2, \ldots, j_{n-1}, i_n, j_{n+1}, \ldots, j_N} = \sum_{j_n=1}^{J_n} g_{j_1, j_2, \ldots, J_N}\, a_{i_n, j_n}. \tag{1.97}$$

The tensor-matrix product can be applied successively along several modes, and it is commutative, that is

$$(\underline{\mathbf{G}} \times_n \mathbf{A}) \times_m \mathbf{B} = (\underline{\mathbf{G}} \times_m \mathbf{B}) \times_n \mathbf{A} = \underline{\mathbf{G}} \times_n \mathbf{A} \times_m \mathbf{B}, \qquad (m \neq n). \tag{1.98}$$

The repeated (iterated) mode-n tensor-matrix product for matrices $\mathbf{A}$ and $\mathbf{B}$ of appropriate dimensions can be simplified as

$$(\underline{\mathbf{G}} \times_n \mathbf{A}) \times_n \mathbf{B} = \underline{\mathbf{G}} \times_n (\mathbf{BA}). \tag{1.99}$$

For $\underline{\mathbf{G}} \in \mathbb{R}^{J_1 \times J_2 \times \cdots \times J_N}$ and a set of matrices $\mathbf{A}^{(n)} \in \mathbb{R}^{I_n \times J_n}$, their multiplication in all possible modes ($n = 1, 2, \ldots, N$) is denoted as

$$\underline{\mathbf{G}} \times \{\mathbf{A}\} = \underline{\mathbf{G}} \times_1 \mathbf{A}^{(1)} \times_2 \mathbf{A}^{(2)} \cdots \times_N \mathbf{A}^{(N)}, \tag{1.100}$$

and the resulting tensor has dimension $I_1 \times I_2 \times \cdots \times I_N$. Multiplication of a tensor with all but one mode is denoted as

$$\underline{\mathbf{G}} \times_{-n} \{\mathbf{A}\} = \underline{\mathbf{G}} \times_1 \mathbf{A}^{(1)} \cdots \times_{n-1} \mathbf{A}^{(n-1)} \times_{n+1} \mathbf{A}^{(n+1)} \cdots \times_N \mathbf{A}^{(N)} \tag{1.101}$$

giving a tensor of dimension $I_1 \times \cdots \times I_{n-1} \times J_n \times I_{n+1} \times \cdots \times I_N$. The above notation is adopted from [85].

It is not difficult to verify that these operations satisfy the following properties

$$\left[\underline{\mathbf{G}} \times \{\mathbf{A}\}\right]_{(n)} = \mathbf{A}^{(n)}\mathbf{G}_{(n)} \left[\mathbf{A}^{(N)} \otimes \mathbf{A}^{(N-1)} \cdots \otimes \mathbf{A}^{(n+1)} \otimes \mathbf{A}^{(n-1)} \cdots \otimes \mathbf{A}^{(1)}\right]^T. \tag{1.102}$$

Definition 1.6 (mode-n tensor-vector product) *The mode-n multiplication of a tensor $\underline{\mathbf{Y}} \in \mathbb{R}^{I_1 \times I_2 \times \cdots \times I_N}$ by a vector $\boldsymbol{a} \in \mathbb{R}^{I_n}$ is denoted by*[14]

$$\underline{\mathbf{Y}} \,\bar{\times}_n\, \boldsymbol{a} \tag{1.103}$$

and has dimension $I_1 \times \cdots \times I_{n-1} \times I_{n+1} \times \cdots \times I_N$, that is,

$$\underline{\mathbf{Z}} = \underline{\mathbf{Y}} \,\bar{\times}_n\, \boldsymbol{a} \in \mathbb{R}^{I_1 \times \cdots \times I_{n-1} \times I_{n+1} \times \cdots \times I_N}, \tag{1.104}$$

Element-wise, we have

$$z_{i_1, i_2, \ldots, i_{n-1}, i_{n+1}, \ldots, i_N} = \sum_{i_n=1}^{I_n} y_{i_1, i_2, \ldots, i_N}\, a_{i_n}. \tag{1.105}$$

[14] A bar over the operator $\times$ indicates a contracted product.

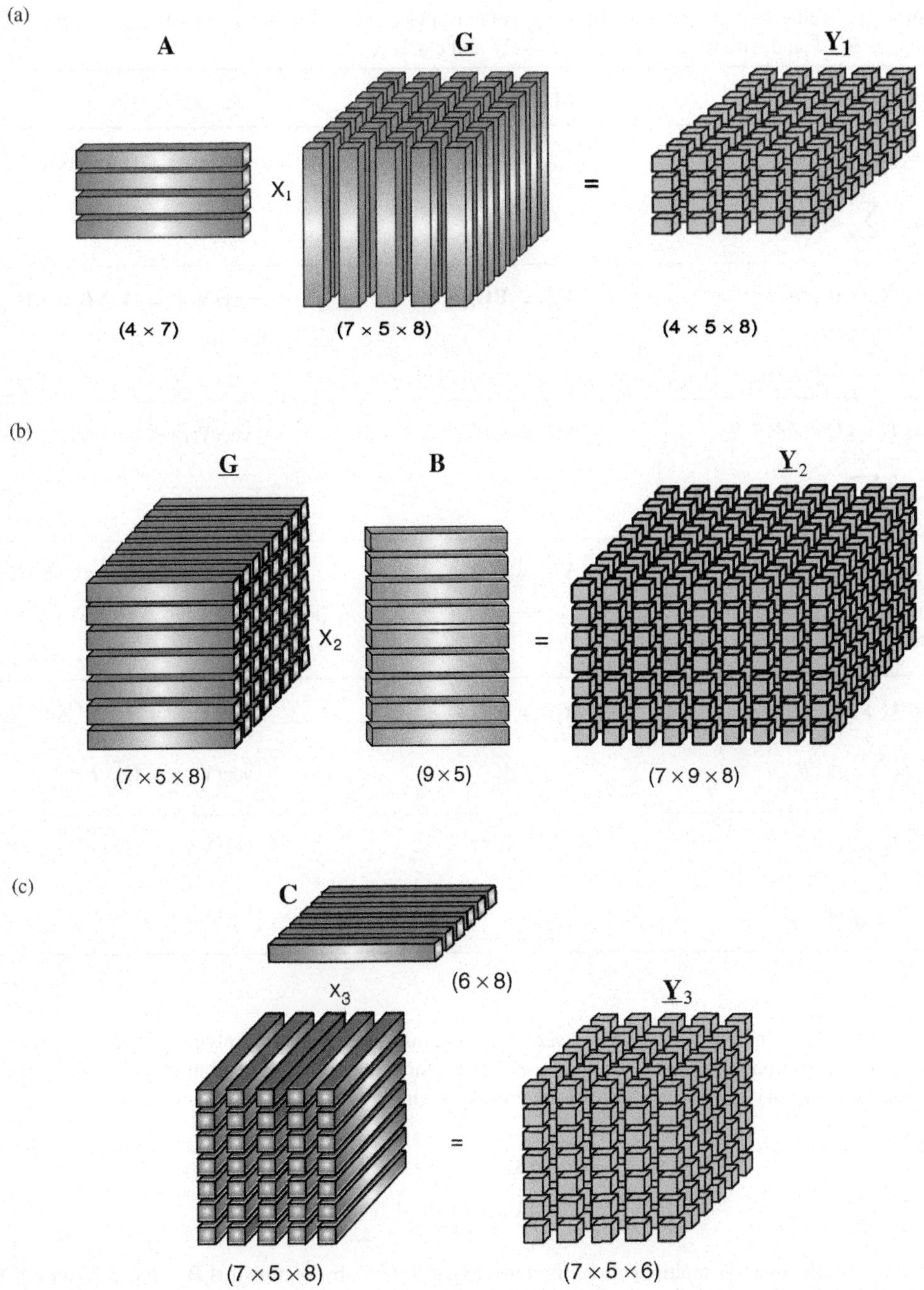

Figure 1.21 Illustration of the mode-n multiplications of a third-order tensor by matrices. (a) mode-1 multiplication $\underline{\mathbf{Y}}_1 = \underline{\mathbf{G}} \times_1 \mathbf{A}$, (b) mode-2 multiplication $\underline{\mathbf{Y}}_2 = \underline{\mathbf{G}} \times_2 \mathbf{B}$, (c) mode-3 multiplication $\underline{\mathbf{Y}}_3 = \underline{\mathbf{G}} \times_3 \mathbf{C}$.

Table 1.1 Rules for the mode-n multiplication of tensor $\underline{\mathbf{G}} \in \mathbb{R}^{J\times R\times P}$ with matrices $\mathbf{A} \in \mathbb{R}^{I\times J}$, $\mathbf{B} \in \mathbb{R}^{T\times R}$, and $\mathbf{C} \in \mathbb{R}^{Q\times P}$ and with vectors: $\boldsymbol{a} \in \mathbb{R}^{J}$, $\boldsymbol{b} \in \mathbb{R}^{R}$ and $\boldsymbol{c} \in \mathbb{R}^{P}$.

mode-n product	Matricized version	Vectorized version
$\underline{\mathbf{Y}} = \underline{\mathbf{G}} \times_1 \mathbf{A} \in \mathbb{R}^{I\times R\times P}$ $y_{irp} = \sum_{j=1}^{J} g_{jrp}\, a_{ij}$	$\mathbf{Y}_{(1)} = \mathbf{A}\mathbf{G}_{(1)}$	$\text{vec}(\mathbf{Y}_{(1)}) = (\mathbf{I} \otimes \mathbf{A})\text{vec}(\mathbf{G}_{(1)})$
$\underline{\mathbf{Y}} = \underline{\mathbf{G}} \times_2 \mathbf{B} \in \mathbb{R}^{J\times T\times P}$ $y_{jtp} = \sum_{r=1}^{R} g_{jrp}\, b_{tr}$	$\mathbf{Y}_{(2)} = \mathbf{B}\mathbf{G}_{(2)}$	$\text{vec}(\mathbf{Y}_{(2)}) = (\mathbf{I} \otimes \mathbf{B})\text{vec}(\mathbf{G}_{(2)})$
$\underline{\mathbf{Y}} = \underline{\mathbf{G}} \times_3 \mathbf{C} \in \mathbb{R}^{J\times R\times Q}$ $y_{jrq} = \sum_{p=1}^{P} g_{jrp}\, c_{qp}$	$\mathbf{Y}_{(3)} = \mathbf{C}\mathbf{G}_{(3)}$	$\text{vec}(\mathbf{Y}_{(3)}) = (\mathbf{I} \otimes \mathbf{C})\text{vec}(\mathbf{G}_{(3)})$
$\underline{\mathbf{Y}} = \underline{\mathbf{G}} \bar{\times}_1 \boldsymbol{a} \in \mathbb{R}^{R\times P}$ $y_{rp} = \sum_{j=1}^{J} g_{jrp}\, a_{j}$	$\mathbf{Y}_{(1)} = \boldsymbol{a}^T\mathbf{G}_{(1)}$	$\text{vec}(\mathbf{Y}_{(1)}) = (\mathbf{I} \otimes \boldsymbol{a}^T)\text{vec}(\mathbf{G}_{(1)})$ $\text{vec}(\mathbf{Y}_{(1)}) = \mathbf{G}_{(1)}^T\, \boldsymbol{a}$
$\underline{\mathbf{Y}} = \underline{\mathbf{G}} \bar{\times}_2 \boldsymbol{b} \in \mathbb{R}^{J\times P}$ $y_{jp} = \sum_{r=1}^{R} g_{jrp}\, b_{r}$	$\mathbf{Y}_{(2)} = \boldsymbol{b}^T\mathbf{G}_{(2)}$	$\text{vec}(\mathbf{Y}_{(2)}) = (\mathbf{I} \otimes \boldsymbol{b}^T)\text{vec}(\mathbf{G}_{(2)})$ $\text{vec}(\mathbf{Y}_{(2)}) = \mathbf{G}_{(2)}^T\, \boldsymbol{b}$
$\underline{\mathbf{Y}} = \underline{\mathbf{G}} \bar{\times}_3 \boldsymbol{c} \in \mathbb{R}^{J\times R}$ $y_{jp} = \sum_{p=1}^{P} g_{jrp}\, c_{p}$	$\mathbf{Y}_{(3)} = \boldsymbol{c}^T\mathbf{G}_{(3)}$	$\text{vec}(\mathbf{Y}_{(3)}) = (\mathbf{I} \otimes \boldsymbol{c}^T)\text{vec}(\mathbf{G}_{(3)})$ $\text{vec}(\mathbf{Y}_{(3)}) = \mathbf{G}_{(3)}^T\, \boldsymbol{c}$

It is also possible to multiply a tensor by a vector in more than one mode. Multiplying a three-way tensor by vectors in the two modes results in a 1-way tensor (a vector); multiplying it in all modes results in a scalar. We can exchange the order of multiplication by the following rule:

$$\begin{aligned}\underline{\mathbf{Y}} \,\bar{\times}_m\, \boldsymbol{a} \,\bar{\times}_n\, \boldsymbol{b} &= (\underline{\mathbf{Y}} \,\bar{\times}_m\, \boldsymbol{a}) \,\bar{\times}_n\, \boldsymbol{b} \\ &= (\underline{\mathbf{Y}} \,\bar{\times}_n\, \boldsymbol{b}) \,\bar{\times}_m\, \boldsymbol{a}, \qquad \text{for } m < n. \end{aligned} \tag{1.106}$$

For example, the mode-n multiplication of a tensor $\underline{\mathbf{G}} \in \mathbb{R}^{J\times R\times P}$ by vectors $\boldsymbol{a} \in \mathbb{R}^{J}$, $\boldsymbol{b} \in \mathbb{R}^{R}$ and $\boldsymbol{c} \in \mathbb{R}^{P}$ can be expressed as (see Figure 1.22 and Table 1.1)

$$z = \underline{\mathbf{G}} \,\bar{\times}_1\, \boldsymbol{a} \,\bar{\times}_2\, \boldsymbol{b} \,\bar{\times}_3\, \boldsymbol{c} = \sum_{j=1}^{J}\sum_{r=1}^{R}\sum_{p=1}^{P} g_{jrp}\, a_j\, b_r\, c_p.$$

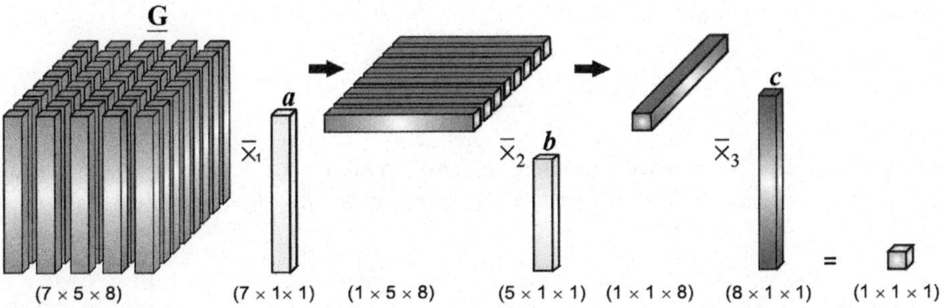

Figure 1.22 Illustration of mode-n multiplication of a third-order tensor $\underline{\mathbf{G}}$ by vectors, yielding scalar $y = \underline{\mathbf{G}} \,\bar{\times}_1\, \boldsymbol{a} \,\bar{\times}_2\, \boldsymbol{b} \,\bar{\times}_3\, \boldsymbol{c}$. Note that the dimension of the result is reduced by one. For example, multiplying a three-way (a third-order) tensor by a vector in mode-1 results in a 2-way tensor (a matrix).

More generally, for $\underline{\mathbf{G}} \in \mathbb{R}^{J_1 \times J_2 \times \cdots \times J_N}$ and $\boldsymbol{a}^{(n)} \in \mathbb{R}^{J_n}$, the multiplication by all vectors in all modes ($n = 1, 2, \ldots, N$) gives a scalar:

$$y = \underline{\mathbf{G}} \,\bar{\times}_1\, \boldsymbol{a}^{(1)} \,\bar{\times}_2\, \boldsymbol{a}^{(2)} \cdots \bar{\times}_N\, \boldsymbol{a}^{(N)} = \underline{\mathbf{G}} \,\bar{\times}\, \{\boldsymbol{a}\} \in \mathbb{R} \tag{1.107}$$

whereas multiplication in every mode except mode-n results in a vector $\boldsymbol{x}$ of length J_n:

$$\begin{aligned} \boldsymbol{x} &= \underline{\mathbf{G}} \,\bar{\times}_1\, \boldsymbol{a}^{(1)} \cdots \bar{\times}_{n-1}\, \boldsymbol{a}^{(n-1)} \,\bar{\times}_{n+1}\, \boldsymbol{a}^{(n+1)} \cdots \bar{\times}_N\, \boldsymbol{a}^{(N)} \\ &= \mathbf{G}_{(n)} \left(\boldsymbol{a}^{(N)} \otimes \cdots \otimes \boldsymbol{a}^{(n+1)} \otimes \boldsymbol{a}^{(n-1)} \otimes \cdots \otimes \boldsymbol{a}^{(1)}\right) = \underline{\mathbf{G}} \,\bar{\times}_{-n}\, \{\boldsymbol{a}\} \in \mathbb{R}^{J_n}. \end{aligned} \tag{1.108}$$

Also note that multiplication in every mode except mode-n and mode-m, results in a matrix of size $J_n \times J_m$.

A matrix $\mathbf{G}$ of dimension $(I \times J)$ can be considered as a third-order tensor $\underline{\mathbf{G}}$ in which the 3rd dimension is 1 that is of dimension $(I \times J \times 1)$, and its matricized versions in each mode are given by

$$\left[\underline{\mathbf{G}}\right]_{(1)} = \left[\underline{\mathbf{G}}\right]_{(2)}^T = \mathbf{G}, \tag{1.109}$$

$$\left[\underline{\mathbf{G}}\right]_{(3)} = \text{vec}\,(\mathbf{G})^T. \tag{1.110}$$

The mode-3 product of the tensor $\underline{\mathbf{G}}$ with a vector $\boldsymbol{a}$ is exactly the outer product of $\mathbf{G}$ and $\boldsymbol{a}$.

$$\underline{\mathbf{G}} \times_3 \boldsymbol{a} = \underline{\mathbf{G}} \circ \boldsymbol{a}. \tag{1.111}$$

Definition 1.7 **The scalar product** *(or inner product) of two tensors* $\underline{\mathbf{A}}, \underline{\mathbf{B}} \in \mathbb{R}^{I_1 \times I_2, \times \cdots \times I_N}$ *of the same order is denoted by* $\langle \underline{\mathbf{A}}, \underline{\mathbf{B}} \rangle$ *and is computed as a sum of element-wise products over all the indices, that is,*

$$c = \langle \underline{\mathbf{A}}, \underline{\mathbf{B}} \rangle = \sum_{i_1}^{I_1} \sum_{i_2}^{I_2} \cdots \sum_{i_N}^{I_N} b_{i_1, i_2, \ldots, i_N} a_{i_1, i_2, \ldots, i_N} \in \mathbb{R}. \tag{1.112}$$

The scalar product allows us to define the higher-order Frobenius norm of a tensor $\underline{\mathbf{A}}$ as

$$||\underline{\mathbf{A}}||_F = \sqrt{\langle \underline{\mathbf{A}}, \underline{\mathbf{A}} \rangle} = \sqrt{\sum_{i_1}^{I_1} \sum_{i_2}^{I_2} \cdots \sum_{i_N}^{I_N} a_{i_1, i_2, \ldots, i_N}^2}, \tag{1.113}$$

whereas the ℓ_1-norm of a tensor is defined as

$$||\underline{\mathbf{A}}||_1 = \sum_{i_1}^{I_1} \sum_{i_2}^{I_2} \cdots \sum_{i_N}^{I_N} |a_{i_1,i_2,\ldots,i_N}|. \tag{1.114}$$

Definition 1.8 The contracted product *of two tensors* $\underline{\mathbf{A}} \in \mathbb{R}^{I_1\times\cdots\times I_M\times J_1\times\cdots\times J_N}$ *and* $\underline{\mathbf{B}} \in \mathbb{R}^{I_1\times\cdots\times I_M\times K_1\times\cdots\times K_P}$ *along the first* M *modes is a tensor of size* $J_1 \times \cdots \times J_N \times K_1 \times \cdots \times K_P$, *given by*

$$\langle \underline{\mathbf{A}}, \underline{\mathbf{B}} \rangle_{1,\ldots,M;1,\ldots,M}(j_1, \ldots, j_N, k_1, \ldots, k_P) = \sum_{i_1=1}^{I_1} \cdots \sum_{i_M=1}^{I_M} a_{i_1,\ldots,i_M,j_1,\ldots,j_N}\, b_{i_1,\ldots,i_M,k_1,\ldots,k_P}. \tag{1.115}$$

The remaining modes are ordered such that those from $\underline{\mathbf{A}}$ come before $\underline{\mathbf{B}}$. The arguments specifying the modes of $\underline{\mathbf{A}}$ and those of $\underline{\mathbf{B}}$ for contraction need not be consecutive. However, the sizes of the corresponding dimensions must be equal. For example, the contracted tensor product along the mode-2 of a tensor $\underline{\mathbf{A}} \in \mathbb{R}^{3\times4\times5}$, and the mode-3 of a tensor $\underline{\mathbf{B}} \in \mathbb{R}^{7\times8\times4}$ returns a tensor $\underline{\mathbf{C}} = \langle \underline{\mathbf{A}}, \underline{\mathbf{B}} \rangle_{2;3} \in \mathbb{R}^{3\times5\times7\times8}$.

The contracted tensor product of $\underline{\mathbf{A}}$ and $\underline{\mathbf{B}}$ along the same M modes simplifies to

$$\langle \underline{\mathbf{A}}, \underline{\mathbf{B}} \rangle_{1,\ldots,M;1,\ldots,M} = \langle \underline{\mathbf{A}}, \underline{\mathbf{B}} \rangle_{1,\ldots,M}, \tag{1.116}$$

whereas the contracted product of tensors $\underline{\mathbf{A}} \in \mathbb{R}^{I_1\times\cdots\times I_N}$ and $\underline{\mathbf{B}} \in \mathbb{R}^{J_1\times\cdots\times J_N}$ along all modes except the mode-n is denoted as

$$\langle \underline{\mathbf{A}}, \underline{\mathbf{B}} \rangle_{-n} = \mathbf{A}_{(n)}\, \mathbf{B}_{(n)}^T \in \mathbb{R}^{I_n\times J_n}, \qquad (I_k = J_k, \;\; \forall k \neq n). \tag{1.117}$$

The tensor-vector, tensor-matrix and scalar multiplications can be expressed in a form of contracted product. For example, the contracted product along the mode-n of the tensor $\underline{\mathbf{A}}$ and the mode-2 of matrix $\mathbf{C} \in \mathbb{R}^{J\times I_n}$ can be obtained by permuting the dimensions of the mode-n product of $\underline{\mathbf{A}}$ and $\mathbf{C}$

$$\langle \underline{\mathbf{A}}, \mathbf{C} \rangle_{n;2} = \langle \underline{\mathbf{A}}, \mathbf{C}^T \rangle_{n;1} = \texttt{permute}(\underline{\mathbf{A}} \times_n \mathbf{C}, [1, \ldots, n-1, n+1, \ldots, N, n]). \tag{1.118}$$

We also have

$$\langle \mathbf{C}, \underline{\mathbf{A}} \rangle_{2;n} = \langle \mathbf{C}^T, \underline{\mathbf{A}} \rangle_{1;n} = \texttt{permute}(\underline{\mathbf{A}} \times_n \mathbf{C}, [n, 1, \ldots, n-1, n+1, \ldots, N]). \tag{1.119}$$

For two tensors of the same dimension, their contracted product along all their modes is their inner product

$$\langle \underline{\mathbf{A}}, \underline{\mathbf{B}} \rangle_{1,\ldots,N} = \langle \underline{\mathbf{A}}, \underline{\mathbf{B}} \rangle. \tag{1.120}$$

In a special case of $M = 0$, the contracted product becomes the outer product of two tensors.

1.4.7 Special Forms of Tensors

Tensors can take special forms or structures. For instance, often a tensor is sparse or symmetric.

1.4.7.1 Rank-One Tensor

Using the outer product, the rank of tensor can be defined as follows (see Figure 1.23)

Definition 1.9 (Rank-one tensor) *A tensor* $\underline{\mathbf{Y}} \in \mathbb{R}^{I_1\times I_2\times\cdots\times I_N}$ *of order* N *has rank-one if it can be written as an outer product of* N *vectors i.e.,*

$$\underline{\mathbf{Y}} = \boldsymbol{a}^{(1)} \circ \boldsymbol{a}^{(2)} \circ \cdots \circ \boldsymbol{a}^{(N)}, \tag{1.121}$$

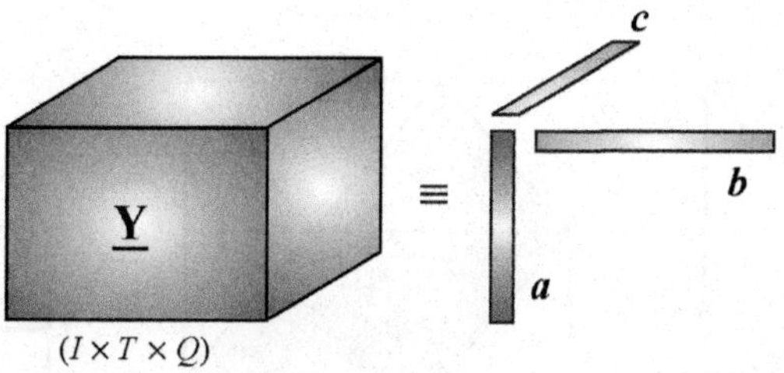

Figure 1.23 Rank-one third-order tensor: $\underline{\mathbf{Y}} = \boldsymbol{a} \circ \boldsymbol{b} \circ \boldsymbol{c} \in \mathbb{R}^{I \times T \times Q}$, $\boldsymbol{a} \in \mathbb{R}^{I}$, $\boldsymbol{b} \in \mathbb{R}^{T}$, $\boldsymbol{c} \in \mathbb{R}^{Q}$.

where $\boldsymbol{a}^{(n)} \in \mathbb{R}^{I_n}$ *and* $y_{i_1,i_2,\ldots,i_N} = a^{(1)}_{i_1} a^{(2)}_{i_2} \cdots a^{(N)}_{i_N}$. *The rank of a tensor* $\underline{\mathbf{Y}} \in \mathbb{R}^{I_1 \times I_2 \times \cdots \times I_N}$ *is defined as the minimal number of rank-one tensors* $\underline{\mathbf{Y}}_1, \ldots, \underline{\mathbf{Y}}_R$ *such that* $\underline{\mathbf{Y}} = \sum_{r=1}^{R} \underline{\mathbf{Y}}_r$.

This outer product is often computed via the Khatri-Rao product or the Kronecker product based on the following relation

$$\mathrm{vec}(\underline{\mathbf{Y}}) = \mathrm{vec}(\mathbf{Y}_{(1)}) = \mathrm{vec}(\boldsymbol{a}^{(1)}(\boldsymbol{a}^{(N)} \odot \cdots \odot \boldsymbol{a}^{(2)})^T) = \boldsymbol{a}^{(N)} \odot \cdots \odot \boldsymbol{a}^{(2)} \odot \boldsymbol{a}^{(1)}. \quad (1.122)$$

Rank-one tensors have many interesting properties and play an important role in multi-way analysis [30,31,68,99,116,139,140,146]. In general, rank of a higher-order tensor is defined as the minimal number of rank-one tensors whose linear combination yields $\underline{\mathbf{Y}}$. Such a representation of a tensor by a linear combination of rank-one tensors is just a CANonical DECOMPposition (CANDECOMP) or PARAFAC (PARAllel FACtor decomposition) which preserves the uniqueness under some mild conditions [91].

1.4.7.2 Symmetric and Super-Symmetric Tensors

For the particular case when all the N vectors $\boldsymbol{a}^{(j)}$ are equal to a vector $\boldsymbol{g}$, their outer product is called a supersymmetric rank-one tensor.[15] A super-symmetric tensor has the same dimension in every mode.

Tensors can also only be (partially) symmetric in two or more modes. For example, a three-way tensor $\underline{\mathbf{Y}} \in \mathbb{R}^{I \times I \times Q}$ is symmetric in modes one and two if all its frontal slices are symmetric, i.e., $\mathbf{Y}_q = \mathbf{Y}_q^T, \forall q = 1, 2, \ldots, Q$.

1.4.7.3 Diagonal Tensors

An N-th order cubical tensor $\underline{\mathbf{Y}} \in \mathbb{R}^{I_1 \times I_2 \times \cdots \times I_N}$ is diagonal if its elements $y_{i_1,i_2,\ldots,i_N} \neq 0$ only if $i_1 = i_2 = \cdots = i_N$. We use $\underline{\mathbf{I}}$ to denote the cubical identity tensor with ones on the superdiagonal and zeros elsewhere. This concept can be generalized or extended as illustrated in Figure 1.24(c).

1.5 Tensor Decompositions and Factorizations

Many modern applications generate large amounts of data with multiple aspects and high dimensionality for which tensors (i.e., multi-way arrays) provide a natural representation. These include text mining, clustering, Internet traffic, telecommunication records, and large-scale social networks.

[15]In general, by analogy to symmetric matrices a higher-order tensor is called supersymmetric if its entries are invariant under any permutation of their indices.

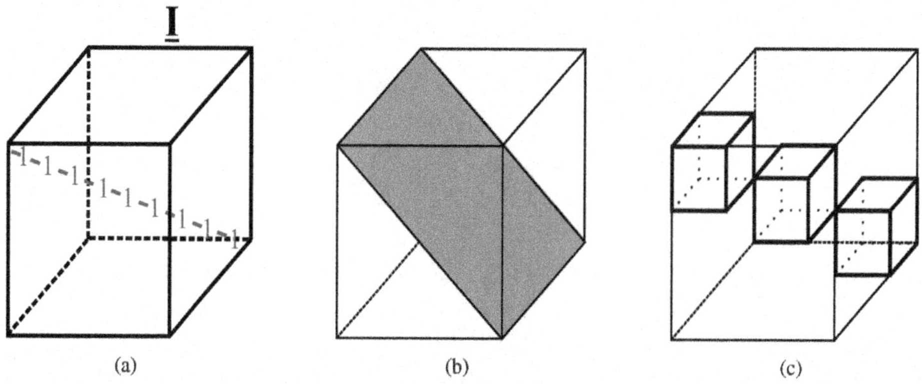

Figure 1.24 Special forms of third-order tensors: (a) Super-Identity cube tensor, (b) diagonal tensor with diagonally positioned nonzero matrix and (c) block diagonal tensor.

Tensor decompositions and factorizations were initiated by Hitchcock in 1927 [74], and later developed by Cattelin in 1944 [85] and by Tucker in 1966 [136,137]. These concepts and approaches received more attention after Carroll and Chang [24,25] proposed the Canonical Decomposition (CANDECOMP) and independently Harshman [62–64] proposed an equivalent model called the PARAFAC (Parallel Factor Analysis) in 1970.

Möck rediscovered the PARAFAC when tackling a neuroscience problem of event related potentials (ERP) in the context of brain imaging [105]. These foundations for tensor factorizations and decompositions also include results on the uniqueness of tensor factorizations and some recommendations on how to choose the number of components. The subsequent contributions put Möck's results in the framework proposed by Harshman, Kruskal and Carroll and Chang [24].

Most of the early results devoted to tensor factorizations and decompositions appeared in the psychometrics literature. Appellof, Davidson and Bro are credited as being the first to use tensor decompositions (in 1981–1998) in chemometrics, which have since become extremely popular in that field (see, e.g., [2,7,16,17,19,85]). In parallel with the developments in psychometrics and chemometrics, there was a great deal of interest in decompositions of bilinear forms in the field of algebraic complexity [80,85].

Although some tensor decomposition models have been proposed long time ago they have recently attracted the interest of researchers working in mathematics, signal processing, data mining, and neuroscience. This probably explains why available mathematical theory seldom deals with the computational and algorithmic aspects of tensor decompositions, together with many still unsolved fundamental problems.

Higher-order tensor decompositions are nowadays frequently used in a variety of fields including psychometrics, chemometrics, image analysis, graph analysis, and signal processing. Two of the most commonly used decompositions are the Tucker decomposition and PARAFAC (also known as CANDECOMP or simply CP) which are often considered (thought of) as higher-order generalizations of the matrix singular value decomposition (SVD) or principal component analysis (PCA). In this book, we superimpose different constraints such as nonnegativity, sparsity or smoothness, and generally such an analogy is no longer valid.

In this chapter we formulate the models and problems for three-way arrays. Extension for arbitrary N-th order tensors will be given in Chapter 7.

1.5.1 Why Multi-way Array Decompositions and Factorizations?

Standard matrix factorizations, such as PCA/SVD, ICA, NMF, and their variants, are invaluable tools for feature selection, dimensionality reduction, noise reduction, and data mining [26]. However, they have only two modes or 2-way representations (say, space and time), and their use is therefore limited. In many applications the data structures often contain higher-order ways (modes) such as trials, task conditions,

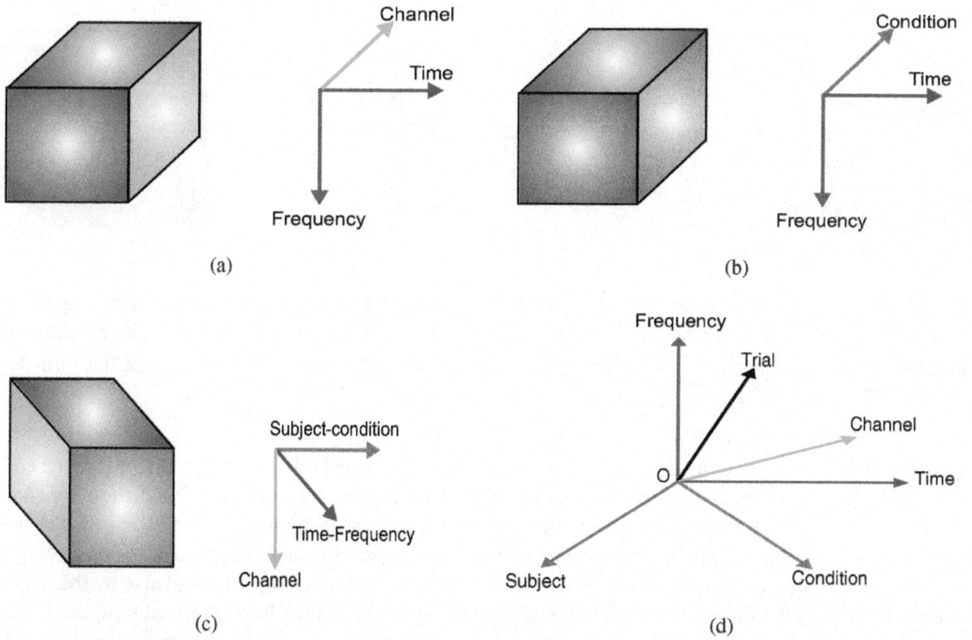

Figure 1.25 Illustration of various possible arrangements (organization) of three-way and multi-way multi-channel EEG/MEG data.

subjects, and groups together with the intrinsic dimensions of space, time, and frequency. For instance, a sequence of trials may lead to a large stream of data encompassing many dimensions: space, time-frequency, subjects, trials, and conditions [5,86,87].

Clearly the "flat-world view" provided by 2-way matrix factorizations (ICA, NMF, SCA) may be insufficient and it is natural to use tensor decomposition approaches. This way all dimensions or modes are retained by virtue of multi-linear models which often produce unique and physically meaningful components. For example, studies in neuroscience often involve multiple subjects (people or animals) and trials leading to experimental data structures conveniently represented by multiway arrays or blocks of three-way data. If the data for every subject were analyzed separately by extracting a matrix or slice from a data block we would lose the covariance information among subjects. To discover hidden components within the data and retain the integrative information, the analysis tools should reflect the multi-dimensional structure of the data.

The multi-way analysis (tensor factorizations and decompositions) is a natural choice, for instance, in EEG studies as it provides convenient multi-channel and multi-subject time-frequency-space sparse representations, artifacts rejection in the time-frequency domain, feature extraction, multi-way clustering and coherence tracking. Our main objective here is to decompose the multichannel time-varying EEG signals into multiple components with distinct modalities in the space, time, and frequency domains in order to identify among them the components common across these different domains, which at the same time are discriminative across different conditions (see Figure 1.25). The two most popular decomposition/factorization models for N-th order tensors are the Tucker model and the more restricted PARAFAC model. Especially, NMF and NTF in conjunction with sparse coding, have recently been given much attention due to their easy interpretation and meaningful representation. NTF has been used in numerous applications in environmental analysis, food studies, pharmaceutical analysis and in chemistry in general (see [2,18,85] for review).

As a result of such tensor decompositions, the inherent structures of the recorded brain signals usually become enhanced and better exposed. Further operations performed on these components can remove redundancy and achieve compact sparse representations. There are at least two possible operations we can perform.

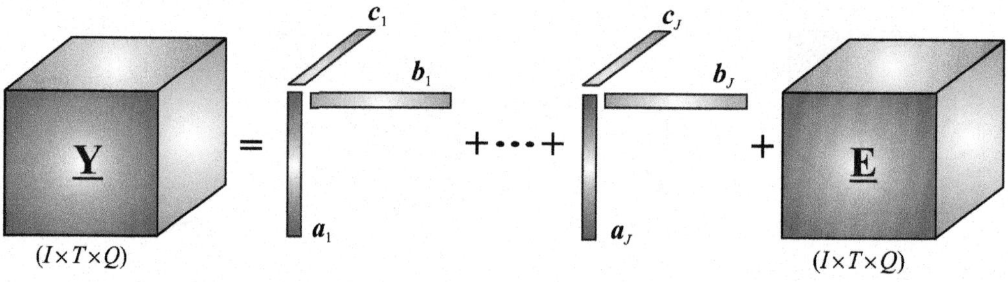

Figure 1.26 A graphical representation of the third-order PARAFAC as a sum of rank-one tensors. All the vectors $\{\boldsymbol{a}_j, \boldsymbol{b}_j, \boldsymbol{c}_j\}$ are treated as column vectors of factor matrices and are linked for each index j via the outer product operator, that is, $\underline{\mathbf{Y}} = \sum_{j=1}^{J} \boldsymbol{a}_j \circ \boldsymbol{b}_j \circ \boldsymbol{c}_j + \mathbf{E}$ or equivalently in a compact form $\underline{\mathbf{Y}} = \underline{\mathbf{I}} \times_1 \mathbf{A} \times_2 \mathbf{B} \times_3 \mathbf{C} + \mathbf{E}$. (In this model not all vectors are normalized to unit length).

First, the extracted factors or hidden latent components can be grouped (clustered) together and represented collectively in a lower dimensional space to extract features and remove redundancy. Second, the components can be simply pruned if they are correlated with a specific mental task. With the addition of extra dimensions it is possible to investigate topography and time and frequency patterns in one analysis. The resulting components can be described not only by the topography and the time-frequency signature but also by the relative contribution from different subjects or conditions. Regarding an application to brain signal analysis, various oscillatory activities within the EEG may overlap, however, the sparse and nonnegative tensor representation by means of the time-frequency-space transformation makes it possible in many cases to isolate each oscillatory behavior well, even when these activities are not well-separated in the space-time (2-way) domain.

Recent development in high spatial density arrays of EEG signals involve multi-dimensional signal processing techniques (referred to as multi-way analysis (MWA), multi-way-array (tensor) factorization/decomposition, dynamic tensor analysis (DTA), or window-based tensor analysis (WTA)). These can be employed to analyze multi-modal and multichannel experimental EEG/MEG and fMRI data [5,103,141].

1.5.2 PARAFAC and Nonnegative Tensor Factorization

The PARAFAC[16] can be formulated as follows (see Figures 1.26 and 1.27 for graphical representations). Given a data tensor $\underline{\mathbf{Y}} \in \mathbb{R}^{I \times T \times Q}$ and the positive index J, find three-component matrices, also called loading matrices or factors, $\mathbf{A} = [\boldsymbol{a}_1, \boldsymbol{a}_2, \ldots, \boldsymbol{a}_J] \in \mathbb{R}^{I \times J}$, $\mathbf{B} = [\boldsymbol{b}_1, \boldsymbol{b}_2, \ldots, \boldsymbol{b}_J] \in \mathbb{R}^{T \times J}$ and $\mathbf{C} = [\boldsymbol{c}_1, \boldsymbol{c}_2, \ldots, \boldsymbol{c}_J] \in \mathbb{R}^{Q \times J}$ which perform the following approximate factorization:

$$\underline{\mathbf{Y}} = \sum_{j=1}^{J} \boldsymbol{a}_j \circ \boldsymbol{b}_j \circ \boldsymbol{c}_j + \underline{\mathbf{E}} = [\![\mathbf{A}, \mathbf{B}, \mathbf{C}]\!] + \underline{\mathbf{E}}, \tag{1.123}$$

or equivalently in the element-wise form (see Table 1.2 for various representations of PARAFAC)

$$y_{itq} = \sum_{j=1}^{J} a_{ij} b_{tj} c_{qj} + e_{itq}. \tag{1.124}$$

[16] Also called the CANDECOMP (Canonical Decomposition) or simply CP decomposition (factorization).

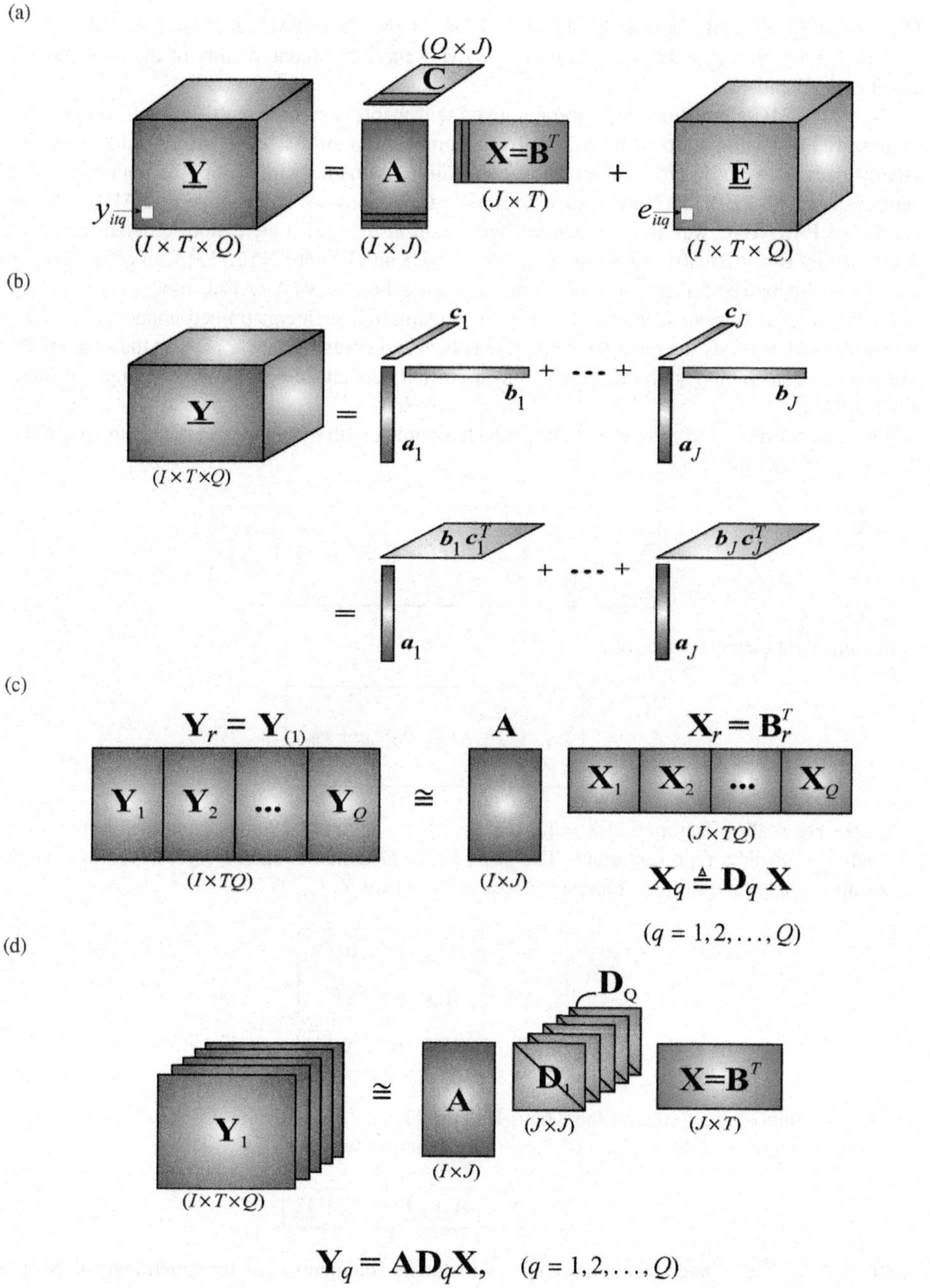

Figure 1.27 The alternative representations of the third-order PARAFAC model: (a) as a set of three matrices using a scalar representation (see Eq. (1.124)), (b) as a set of vectors using summation of rank-one tensors expressed by the outer products of the vectors (see Eq. (1.123)), (c) decomposition into two matrices using row-wise unfolding, and (d) representation by frontal slices (see Eq. (1.135)).

The symbol $\widehat{\underline{\mathbf{Y}}} = [\![\mathbf{A}, \mathbf{B}, \mathbf{C}]\!]$ is a shorthand notation for the PARAFAC factorization, and $\boldsymbol{a}_j = [a_{ij}] \in \mathbb{R}^I$, $\boldsymbol{b}_j = [b_{tj}] \in \mathbb{R}^T$, and $\boldsymbol{c}_j = [c_{qj}] \in \mathbb{R}^Q$ are respectively the constituent vectors of the corresponding factor matrices $\mathbf{A}, \mathbf{B}$ and $\mathbf{C}$.

The PARAFAC algorithms decompose a given tensor into a sum of multi-linear terms (in this case tri-linear), in a way analogous to the bilinear matrix decomposition. As discussed before, unlike SVD, PARAFAC usually does not impose any orthogonality constraints. A model which imposes nonnegativity on factor matrices is called the NTF (Nonnegative Tensor Factorization) or Nonnegative PARAFAC. A nonnegative version of PARAFAC was first introduced by Carroll *et al.* [25]. Later, more efficient approaches were developed by Bro [6,16,80], based on the modified NLS and Paatero [112,113] who generalized his earlier 2-way positive matrix factorization (PMF) method to the three-way PARAFAC model, referring to the result as PMF3 (three-way positive matrix factorization). Although such constrained nonnegativity based model may not match perfectly the input data (i.e., it may have larger residual errors $\mathbf{E}$ than the standard PARAFAC without any constraints) such decompositions are often very meaningful and have physical interpretation [30,116,118].

It is often convenient to assume that all vectors have unit length so that we can use the modified Harshman's PARAFAC model given by

$$\underline{\mathbf{Y}} = \sum_{j=1}^{J} \lambda_j \, \boldsymbol{a}_j \circ \boldsymbol{b}_j \circ \boldsymbol{c}_j + \underline{\mathbf{E}} \cong [\![\boldsymbol{\lambda}, \mathbf{A}, \mathbf{B}, \mathbf{C}]\!], \tag{1.125}$$

or in equivalent element-wise form

$$y_{itq} = \sum_{j=1}^{J} \lambda_j \, a_{ij} \, b_{tj} \, c_{qj} + e_{itq}, \tag{1.126}$$

where λ_j are scaling factors and $\boldsymbol{\lambda} = [\lambda_1, \lambda_2, \ldots, \lambda_J]^T$. Figure 1.28 illustrates the above model and its alternative equivalent representations. The basic PARAFAC model can be represented in compact matrix forms upon applying unfolding representations of the tensor $\underline{\mathbf{Y}}$:

$$\mathbf{Y}_{(1)} \cong \mathbf{A}\, \boldsymbol{\Lambda}\, (\mathbf{C} \odot \mathbf{B})^T, \tag{1.127}$$

$$\mathbf{Y}_{(2)} \cong \mathbf{B}\, \boldsymbol{\Lambda}\, (\mathbf{C} \odot \mathbf{A})^T, \tag{1.128}$$

$$\mathbf{Y}_{(3)} \cong \mathbf{C}\, \boldsymbol{\Lambda}\, (\mathbf{B} \odot \mathbf{A})^T, \tag{1.129}$$

where $\boldsymbol{\Lambda} = \text{diag}(\boldsymbol{\lambda})$ and $\odot$ means the Khatri-Rao product.

Using the mode-n multiplication of a tensor by a matrix, we have

$$\underline{\mathbf{Y}} = \underline{\boldsymbol{\Lambda}} \times_1 \mathbf{A} \times_2 \mathbf{B} \times_3 \mathbf{C} + \underline{\mathbf{E}}, \tag{1.130}$$

where $\underline{\boldsymbol{\Lambda}} \in \mathbb{R}^{J \times J \times J}$ is diagonal cubical tensor with nonzero elements λ_j on the superdiagonal. In other words, within Harshman's model for the core tensor all but the superdiagonal elements vanish (see Figure 1.28). This also means that PARAFAC can be considered as a special case of the Tucker3 model in which the core tensor is a cubical superdiagonal or super-identity tensor, i.e., $\underline{\mathbf{G}} = \underline{\boldsymbol{\Lambda}} \in \mathbb{R}^{J \times J \times J}$ with $g_{jjj} \neq 0$.

Another form of the PARAFAC model is the vectorized form given by

$$\text{vec}\,(\underline{\mathbf{Y}}) \cong (\mathbf{C} \odot \mathbf{B} \odot \mathbf{A})\boldsymbol{\lambda}. \tag{1.131}$$

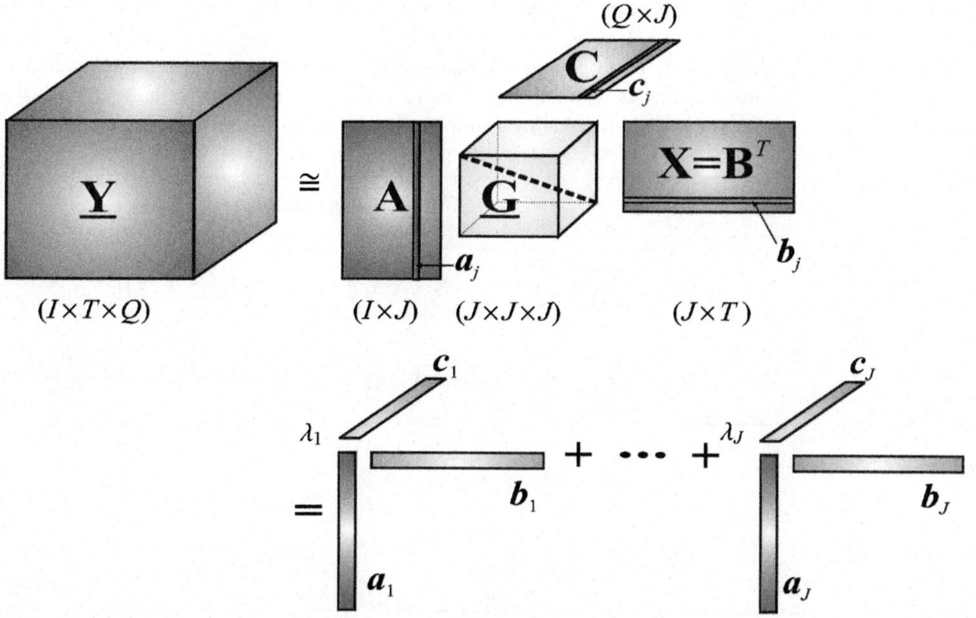

Figure 1.28 Harshman's PARAFAC model with a superdiagonal core tensor $\underline{\mathbf{G}} = \underline{\mathbf{\Lambda}} = \in \mathbb{R}^{J\times J\times J}$ for the third-order tensor $\underline{\mathbf{Y}} \cong \underline{\mathbf{\Lambda}} \times_1 \mathbf{A} \times_2 \mathbf{B} \times_3 \mathbf{C} = \sum_{j=1}^{J} \lambda_j \, \boldsymbol{a}_j \circ \boldsymbol{b}_j \circ \boldsymbol{c}_j$. (In this model all vectors are normalized to unit length).

The three-way PARAFAC model can also be described by using frontal, lateral and horizontal slices as follows[17]

$$\mathbf{Y}_{::q} \cong \mathbf{A}\, \mathbf{D}_q(\boldsymbol{c}_{q:})\, \mathbf{B}^T, \tag{1.132}$$

$$\mathbf{Y}_{:t:} \cong \mathbf{A}\, \mathbf{D}_t(\boldsymbol{b}_{t:})\, \mathbf{C}^T, \tag{1.133}$$

$$\mathbf{Y}_{i::} \cong \mathbf{B}\, \mathbf{D}_i(\boldsymbol{a}_{i:})\, \mathbf{C}^T, \tag{1.134}$$

where $\mathbf{D}_i(\boldsymbol{a}_{i:})$, $\mathbf{D}_t(\boldsymbol{b}_{t:})$ and $\mathbf{D}_q(\boldsymbol{c}_{q:})$ are diagonal matrices which take the i-th, t-th and q-th row of the matrices $\mathbf{A}$, $\mathbf{B}$, and $\mathbf{C}$, respectively, and produce diagonal matrices by placing the corresponding row on the main diagonal.

In particular, it is convenient to represent the three-way PARAFAC model in terms of the frontal slices, as $\mathbf{Y}_q = \mathbf{Y}_{::q}$ of $\underline{\mathbf{Y}}$

$$\boxed{\mathbf{Y}_q \cong \mathbf{A}\mathbf{D}_q\mathbf{B}^T, \; (q = 1, 2, \ldots, Q)} \tag{1.135}$$

where a matrix $\mathbf{D}_q := \mathbf{D}_q(\boldsymbol{c}_{q:})$ is the diagonal matrix based on the q-th row of $\mathbf{C}$.

The above representation of the PARAFAC model has striking similarity to the Approximative Joint Diagonalization method, thus having a clear interpretation for the BSS problem, where the matrix $\mathbf{A}$ represents a mixing matrix, $\mathbf{X} = \mathbf{B}^T$ represents unknown sources, and $\mathbf{C}$ represents a scaling matrix [44,45]. In fact, the PARAFAC factorization can be reformulated as simultaneous diagonalization of a set of matrices, which often leads to fast and reliable way to compute this factorization.

[17] Such a representation does not exist for higher-order tensors where $N > 3$.

Table 1.2 Mathematical formulations of the standard PARAFAC model for a third-order tensor $\underline{\mathbf{Y}} \in \mathbb{R}^{I \times T \times Q}$ with factor matrices: $\mathbf{A} = [\boldsymbol{a}_1, \boldsymbol{a}_2, \ldots, \boldsymbol{a}_J] \in \mathbb{R}^{I \times J}$, $\mathbf{B} = [\boldsymbol{b}_1, \boldsymbol{b}_2, \ldots, \boldsymbol{b}_J] \in \mathbb{R}^{T \times J}$ and $\mathbf{C} = [\boldsymbol{c}_1, \boldsymbol{c}_2, \ldots, \boldsymbol{c}_J] \in \mathbb{R}^{Q \times J}$. There are two optional and equivalent representations: one in which we have a linear combination of rank-one tensors with unit length vectors and a second in which the vectors $\boldsymbol{a}_j \in \mathbb{R}^I$ and $\boldsymbol{b}_j \in \mathbb{R}^T$ have a unit ℓ_2-norm and the scaling factors λ_j are absorbed by the non-normalized vectors $\boldsymbol{c}_j \in \mathbb{R}^Q$.

Operator	Formulation
Outer products	$\underline{\mathbf{Y}} = \sum_{j=1}^{J} \boldsymbol{a}_j \circ \boldsymbol{b}_j \circ \boldsymbol{c}_j + \underline{\mathbf{E}}, \quad (\text{s.t. } \lvert\lvert \boldsymbol{a}_j \rvert\rvert_2 = \lvert\lvert \boldsymbol{b}_j \rvert\rvert_2 = 1,\ \forall j)$ $\underline{\mathbf{Y}} = \sum_{j=1}^{J} \lambda_j\, \boldsymbol{a}_j \circ \boldsymbol{b}_j \circ \boldsymbol{c}_j + \underline{\mathbf{E}}, \quad (\text{s.t. } \lvert\lvert \boldsymbol{a}_j \rvert\rvert_2 = \lvert\lvert \boldsymbol{b}_j \rvert\rvert_2 = \lvert\lvert \boldsymbol{c}_j \rvert\rvert_2 = 1,\ \forall j)$
Scalar	$y_{itq} = \sum_{j=1}^{J} a_{ij}\, b_{tj}\, c_{qj} + e_{itq}$ $y_{itq} = \sum_{j=1}^{J} \lambda_j\, a_{ij}\, b_{tj}\, c_{qj} + e_{itq}$
Mode-n multiplications	$\underline{\mathbf{Y}} = \underline{\mathbf{I}} \times_1 \mathbf{A} \times_2 \mathbf{B} \times_3 \mathbf{C} + \underline{\mathbf{E}}$ $\underline{\mathbf{Y}} = \underline{\mathbf{\Lambda}} \times_1 \mathbf{A} \times_2 \mathbf{B} \times_3 \mathbf{C} + \underline{\mathbf{E}}$
Slice representations	$\mathbf{Y}_{::q} = \mathbf{A}\, \mathbf{D}_q\, (\boldsymbol{c}_{q:})\, \mathbf{B}^T + \mathbf{E}_{::q}$ $\mathbf{Y}_{i::} = \mathbf{B}\, \mathbf{D}_i\, (\boldsymbol{a}_{i:})\, \mathbf{C}^T + \mathbf{E}_{i::}$ $\mathbf{Y}_{:t:} = \mathbf{A}\, \mathbf{D}_t\, (\boldsymbol{b}_{t:})\, \mathbf{C}^T + \mathbf{E}_{:t:}$
Vectors	$\text{vec}(\underline{\mathbf{Y}}) = \text{vec}(\mathbf{Y}_{(1)}) = (\mathbf{C} \odot \mathbf{B} \odot \mathbf{A})\, \boldsymbol{\lambda} + \text{vec}(\mathbf{E}_{(1)}$
Kronecker products	$\mathbf{Y}_{(1)} = \mathbf{A}\, \mathbf{D}_{(1)}(\boldsymbol{c}_q)\, (\mathbf{I} \otimes \mathbf{B})^T + \mathbf{E}_{(1)}$
Khatri-Rao products	$\mathbf{Y}_{(1)} \cong \mathbf{A}\, (\mathbf{C} \odot \mathbf{B})^T + \mathbf{E}_{(1)} \qquad \mathbf{Y}_{(1)} \cong \mathbf{A}\, \mathbf{\Lambda}\, (\mathbf{C} \odot \mathbf{B})^T,$ $\mathbf{Y}_{(2)} \cong \mathbf{B}\, (\mathbf{C} \odot \mathbf{A})^T + \mathbf{E}_{(2)} \qquad \mathbf{Y}_{(2)} \cong \mathbf{B}\, \mathbf{\Lambda}\, (\mathbf{C} \odot \mathbf{A})^T,$ $\mathbf{Y}_{(3)} \cong \mathbf{C}\, (\mathbf{B} \odot \mathbf{A})^T + \mathbf{E}_{(3)} \qquad \mathbf{Y}_{(3)} \cong \mathbf{C}\, \mathbf{\Lambda}\, (\mathbf{B} \odot \mathbf{A})^T,$

The PARAFAC model has some severe limitations as it represents observed data by common factors utilizing the same number of components (columns). In other words, we do not have enough degrees of freedom as compared to other models. Moreover, the PARAFAC approximation may be ill-posed and may lead to unstable estimation of its components. The next sections discuss more flexible models. Summary notations are given in Table 1.2.

1.5.2.1 Basic Approaches to Solve NTF Problem

In order to compute the nonnegative component matrices $\{\mathbf{A}, \mathbf{B}, \mathbf{C}\}$ we usually apply constrained optimization approach as by minimizing a suitable design cost function. Typically, we minimize (with respect the component matrices) the following global cost function

$$D_F(\underline{\mathbf{Y}} \,||\, [\![\mathbf{A}, \mathbf{B}, \mathbf{C}]\!]) = ||\underline{\mathbf{Y}} - [\![\mathbf{A}, \mathbf{B}, \mathbf{C}]\!]||_F^2 + \alpha_{\mathbf{A}}\, ||\mathbf{A}||_F^2 + \alpha_{\mathbf{B}}\, ||\mathbf{B}||_F^2 + \alpha_{\mathbf{C}}\, ||\mathbf{C}||_F^2, \tag{1.136}$$

subject to nonnegativity constraints, where $\alpha_{\mathbf{A}}, \alpha_{\mathbf{B}}, \alpha_{\mathbf{C}}$ are nonnegative regularization parameters.

There are at least three different approaches for solving this optimization problem. The first approach is to use a vectorized form of the above cost function in the form $J(\boldsymbol{x}) = \text{vec}(\underline{\mathbf{Y}} - [\![\mathbf{A}, \mathbf{B}, \mathbf{C}]\!]) = \mathbf{0}$, and employ the Nonnegative Least Squares (NLS) approach. Such a method was first applied for NTF by Paatero [112] and also Tomasi and Bro [134,135]. The Jacobian of such function can be of large size $(I + T + Q)J \times ITQ$, yielding very high computation cost.

In the second approach, Acar, Kolda and Dunlavy propose to optimize the cost function simultaneously with respect to all variables using a modern nonlinear conjugate gradient optimization technique [1]. However, such a cost function is generally not convex and is not guaranteed to obtain the optimal solution although results are very promising.

The most popular approach is to apply the ALS technique (see Chapter 4 for more detail). In this approach we compute the gradient of the cost function with respect to each individual component matrix (assuming that the others are fixed and independent):

$$\nabla_{\mathbf{A}} D_F = -\mathbf{Y}_{(1)} (\mathbf{C} \odot \mathbf{B}) + \mathbf{A} [(\mathbf{C}^T\mathbf{C}) \circledast (\mathbf{B}^T\mathbf{B}) + \alpha_{\mathbf{A}} \mathbf{I}], \tag{1.137}$$

$$\nabla_{\mathbf{B}} D_F = -\mathbf{Y}_{(2)} (\mathbf{C} \odot \mathbf{A}) + \mathbf{B} [(\mathbf{C}^T\mathbf{C}) \circledast (\mathbf{A}^T\mathbf{A}) + \alpha_{\mathbf{B}} \mathbf{I}], \tag{1.138}$$

$$\nabla_{\mathbf{C}} D_F = -\mathbf{Y}_{(3)} (\mathbf{B} \odot \mathbf{A}) + \mathbf{C} [(\mathbf{B}^T\mathbf{B}) \circledast (\mathbf{A}^T\mathbf{A}) + \alpha_{\mathbf{C}} \mathbf{I}]. \tag{1.139}$$

By equating the gradient components to zero and applying the nonlinear projection to maintain nonnegativity of components we obtain efficient and relatively simple nonnegative ALS update rules for the NTF:

$$\mathbf{A} \leftarrow \left[\mathbf{Y}_{(1)} (\mathbf{C} \odot \mathbf{B}) [(\mathbf{C}^T\mathbf{C}) \circledast (\mathbf{B}^T\mathbf{B}) + \alpha_{\mathbf{A}} \mathbf{I}]^{-1}\right]_+, \tag{1.140}$$

$$\mathbf{B} \leftarrow \left[\mathbf{Y}_{(2)} (\mathbf{C} \odot \mathbf{A}) [(\mathbf{C}^T\mathbf{C}) \circledast (\mathbf{A}^T\mathbf{A}) + \alpha_{\mathbf{B}} \mathbf{I}]^{-1}\right]_+, \tag{1.141}$$

$$\mathbf{C} \leftarrow \left[\mathbf{Y}_{(3)} (\mathbf{B} \odot \mathbf{A}) [(\mathbf{B}^T\mathbf{B}) \circledast (\mathbf{A}^T\mathbf{A}) + \alpha_{\mathbf{C}} \mathbf{I}]^{-1}\right]_+. \tag{1.142}$$

In Chapter 4 and Chapter 6 we prove that the ALS algorithms are special cases of a quasi-Newton method that implicitly employ information about the gradient and Hessian of a cost function. The main advantage of ALS algorithms is high convergence speed and its scalability for large-scale problems.

1.5.3 NTF1 Model

Figure 1.29 illustrates the basic 3D NTF1 model, which is an extension of the NTF model [34]. A given data (observed) tensor $\underline{\mathbf{Y}} \in \mathbb{R}_+^{I\times T\times Q}$ is decomposed into a set of matrices $\mathbf{A} \in \mathbb{R}_+^{I\times J}$ and $\mathbf{C} \in \mathbb{R}_+^{Q\times J}$, as well as a third-order tensor with reduced dimension $(J < I)$, for which the frontal slices $\{\mathbf{X}_1, \mathbf{X}_2, ..., \mathbf{X}_Q\}$ have nonnegative entries. This three-way NTF1 model is given by

$$\mathbf{Y}_q = \mathbf{A}\mathbf{D}_q\mathbf{X}_q + \mathbf{E}_q, \qquad (q = 1, 2, \ldots, Q), \tag{1.143}$$

where $\mathbf{Y}_q = \mathbf{Y}_{::q} \in \mathbb{R}_+^{I\times T}$ are the frontal slices of $\underline{\mathbf{Y}} \in \mathbb{R}_+^{I\times T\times Q}$, Q is the number of the frontal slices, $\mathbf{A} = [a_{ij}] \in \mathbb{R}_+^{I\times J}$ is the basis (mixing matrix) representing common factors, $\mathbf{D}_q \in \mathbb{R}_+^{J\times J}$ is a diagonal matrix that holds the q-th row of matrix $\mathbf{C} \in \mathbb{R}_+^{Q\times J}$ in its main diagonal, $\mathbf{X}_q = [x_{jtq}] \in \mathbb{R}_+^{J\times T}$ is matrix representing the sources (or hidden components), and $\mathbf{E}_q = \mathbf{E}_{::q} \in \mathbb{R}^{I\times T}$ is the q-th vertical slice of the tensor $\underline{\mathbf{E}} \in \mathbb{R}^{I\times T\times Q}$ representing the errors or noise depending on the application. Typically, for BSS problems $T >> I \geq Q > J$.

We wish to estimate the set of matrices $\mathbf{A}$, $\mathbf{C}$, and $\{\mathbf{X}_1, \mathbf{X}_2, \ldots, \mathbf{X}_Q\}$ subject to nonnegativity constraints (and other constraints such as sparseness and/or smoothness), given only the observed data $\underline{\mathbf{Y}}$. Since the diagonal matrices $\mathbf{D}_q$ are scaling matrices, they can be absorbed into the matrices $\mathbf{X}_q$ by introducing the row-normalized matrices $\mathbf{X}_q := \mathbf{D}_q\mathbf{X}_q$, thus giving $\mathbf{Y}_q = \mathbf{A}\mathbf{X}_q + \mathbf{E}_q$. Therefore, in the multi-way BSS applications only the matrix $\mathbf{A}$ and the set of scaled source matrices $\mathbf{X}_1, \mathbf{X}_2, \ldots, \mathbf{X}_Q$ need be estimated.

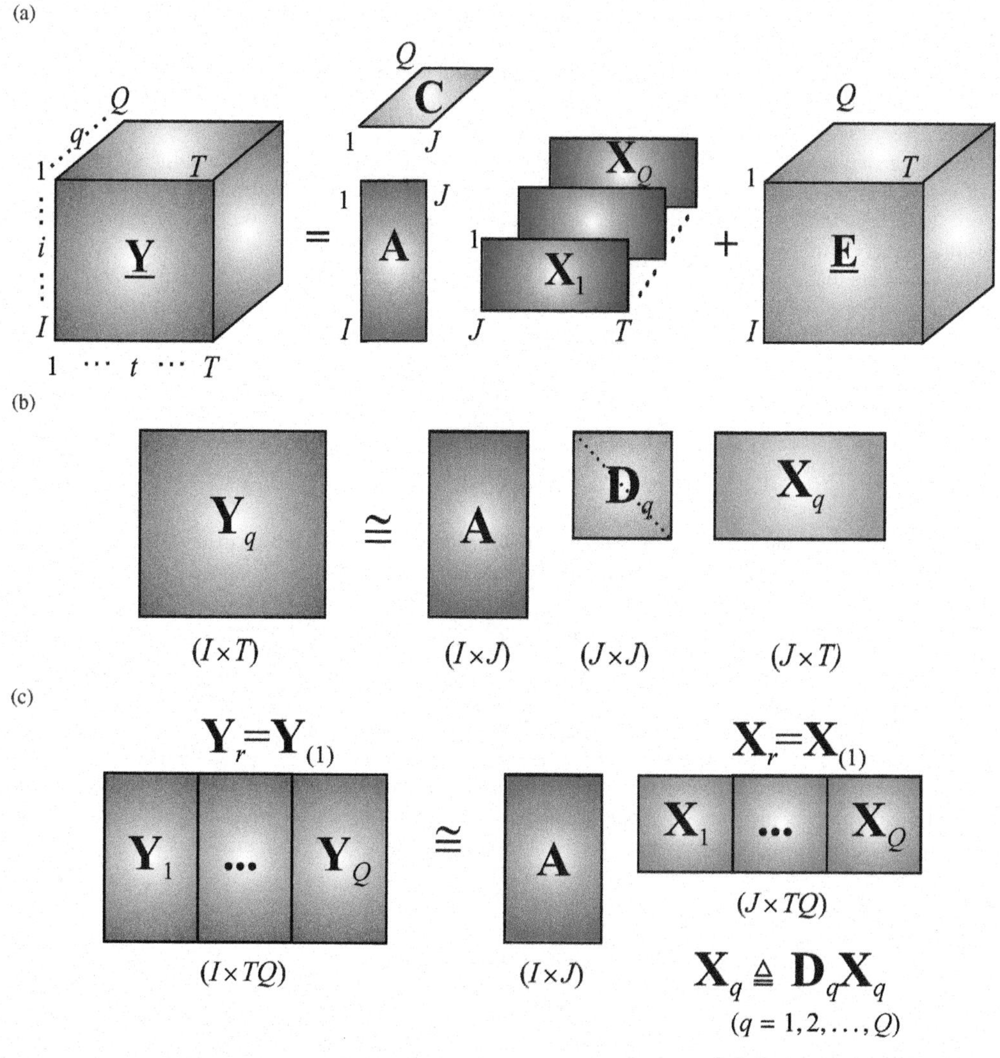

Figure 1.29 (a) NTF1 model that (approximately) decomposes tensor $\underline{\mathbf{Y}} \in \mathbb{R}_+^{I \times T \times Q}$ into a set of nonnegative matrices $\mathbf{A} = [a_{ij}] \in \mathbb{R}_+^{I \times J}$, $\mathbf{C} \in \mathbb{R}_+^{Q \times J}$ and $\{\mathbf{X}_1, \mathbf{X}_2, \ldots, \mathbf{X}_Q\}$, $\mathbf{X}_q = [x_{jtq}] \in \mathbb{R}_+^{J \times T}$, and $\underline{\mathbf{E}} \in \mathbb{R}^{I \times T \times Q}$ is a tensor representing errors. (b) Equivalent representation using joint diagonalization of frontal slices, where $\mathbf{D}_q = \text{diag}(\boldsymbol{c}_q)$ are diagonal matrices. (c) Global matrix representation using row-wise unfolding of the tensor; the sub-matrices are defined as $\mathbf{X}_q \triangleq \mathbf{D}_q \mathbf{X}_q$, $(q = 1, 2, \ldots, Q)$.

For applications where the observed data are incomplete or have different dimensions for each frontal slice (as shown in Figure 1.30) the model can be described as

$$\mathbf{Y}_q = \mathbf{A}\mathbf{D}_q\mathbf{X}_q + \mathbf{E}_q, \qquad (q = 1, 2, \ldots, Q), \tag{1.144}$$

where $\mathbf{Y}_q \in \mathbb{R}_+^{I \times T_q}$ are the frontal slices of the irregular tree-dimensional array, $\mathbf{X}_q = [x_{jtq}] \in \mathbb{R}_+^{J \times T_q}$ are matrices representing sources (or hidden components), and $\mathbf{E}_q = \mathbf{E}_{::q} \in \mathbb{R}^{I \times T}$ is the q-th vertical slice of the multi-way array comprising errors.

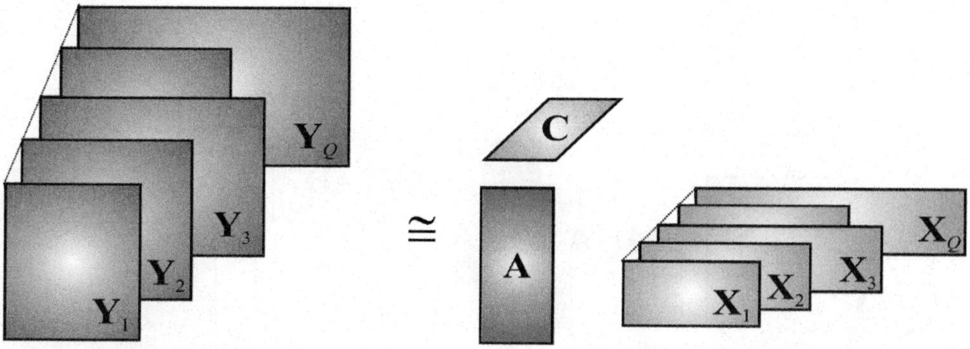

Figure 1.30 Extended NTF1 model for a three-way array. The goal is to estimate the set of nonnegative matrices $\mathbf{A}$, $\mathbf{C}$ and $\{\mathbf{X}_1, \mathbf{X}_2, \ldots, \mathbf{X}_Q\}$.

1.5.4 NTF2 Model

The dual model to the NTF1 is referred to as the 3D NTF2 (by analogy to the PARAFAC2 model [17,78,89,106,127], see Figure 1.31).

A given tensor $\underline{\mathbf{Y}} \in \mathbb{R}_+^{I\times T\times Q}$ is decomposed into a set of matrices $\{\mathbf{A}_1, \mathbf{A}_2, ..., \mathbf{A}_Q\}$, $\mathbf{X} = \mathbf{B}^T$ and $\mathbf{C}$ with nonnegative entries, by the three-way NTF2 model as

$$\boxed{\mathbf{Y}_q = \mathbf{A}_q \mathbf{D}_q \mathbf{X} + \mathbf{E}_q, \qquad (q = 1, 2, \ldots, Q),} \tag{1.145}$$

where $\mathbf{Y}_q = \mathbf{Y}_{::q} = [y_{itq}] \in \mathbb{R}_+^{I\times T}$ are the frontal slices of $\underline{\mathbf{Y}} \in \mathbb{R}_+^{I\times T\times Q}$, Q is the number of frontal slices, $\mathbf{A}_q = [a_{ijq}] \in \mathbb{R}_+^{I\times J}$ are the basis (mixing) matrices, $\mathbf{D}_q \in \mathbb{R}_+^{J\times J}$ is a diagonal matrix that holds the q-th row of $\mathbf{C} \in \mathbb{R}_+^{Q\times J}$ in its main diagonal, $\mathbf{X} = [x_{jt}] \in \mathbb{R}_+^{J\times T}$ is a matrix representing latent sources (or hidden components or common factors), and $\mathbf{E}_q = \mathbf{E}_{::q} \in \mathbb{R}^{I\times T}$ is the q-th frontal slice of a tensor $\underline{\mathbf{E}} \in \mathbb{R}^{I\times T\times Q}$ comprising error or noise depending on the application. The goal is to estimate the set of matrices $\{\mathbf{A}_q\}$, $(q = 1, 2, \ldots, Q)$, $\mathbf{C}$ and $\mathbf{X}$, subject to some nonnegativity constraints and other possible natural constraints such as sparseness and/or smoothness. Since the diagonal matrices $\mathbf{D}_q$ are scaling matrices they can be absorbed into the matrices $\mathbf{A}_q$ by introducing column-normalization, that is, $\mathbf{A}_q := \mathbf{A}_q\mathbf{D}_q$. In BSS applications, therefore, only the matrix $\mathbf{X}$ and the set of scaled matrices $\mathbf{A}_1, \ldots, \mathbf{A}_Q$ need be estimated. This, however, comes at a price, as we may lose the uniqueness of the NTF2 representation ignoring the scaling and permutation ambiguities. The uniqueness can still be preserved by imposing nonnegativity and sparsity constraints.

The NTF2 model is similar to the well-known PARAFAC2 model[18] with nonnegativity constraints and to the Tucker models described in the next section [89,106,127]. In a special case, when all matrices $\mathbf{A}_q$ are identical, the NTF2 model can be simplified into the ordinary PARAFAC model (see Section 1.5.2) with the nonnegativity constraints described by

$$\mathbf{Y}_q = \mathbf{A}\mathbf{D}_q\mathbf{X} + \mathbf{E}_q, \qquad (q = 1, 2, \ldots, Q). \tag{1.146}$$

As shown in Figure 1.32(a) the NTF2 model can be extended to the decomposition of multi-way arrays with different dimensions using the simultaneous factorizations

$$\mathbf{Y}_q = \mathbf{A}_q\mathbf{D}_q\mathbf{X} + \mathbf{E}_q, \qquad (q = 1, 2, \ldots, Q), \tag{1.147}$$

[18] In the PARAFAC2 model we usually assume that $\mathbf{A}_q^T\mathbf{A}_q = \boldsymbol{\Phi} \in \mathbb{R}^{J\times J}$, $\forall q$ (i.e., it is required that the matrix product $\mathbf{A}_q$ with its transpose is invariant for all frontal slices of a core three-way tensor $\underline{\mathbf{A}}$).

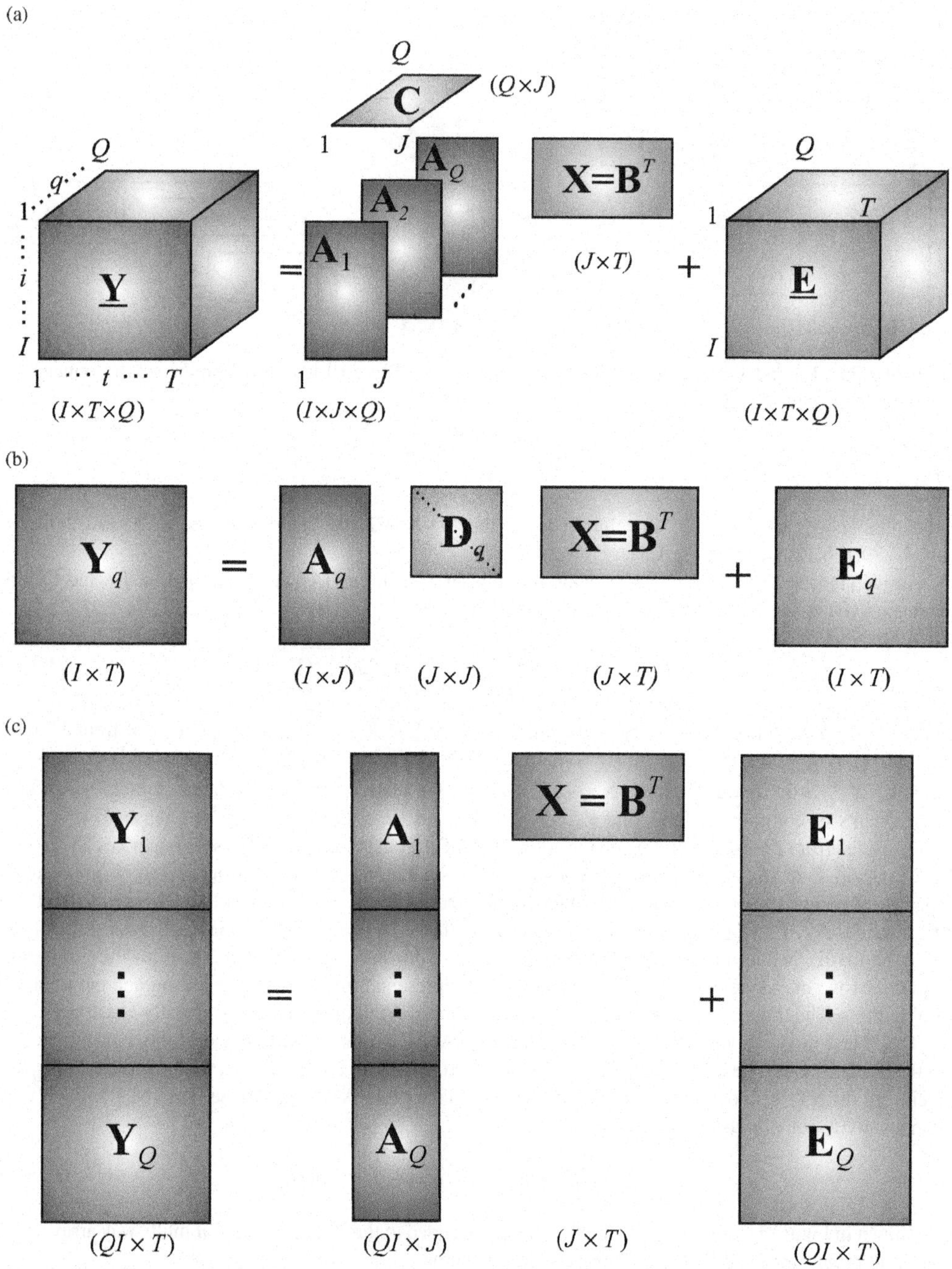

Figure 1.31 (a) NTF2 model in which a third-order tensor is decomposed into a set of nonnegative matrices: $\{\mathbf{A}_1, \ldots, \mathbf{A}_Q\}$, $\mathbf{C}$, and $\mathbf{X}$. (b) Equivalent representation in which the frontal slices of a tensor are factorized by a set of nonnegative matrices using joint diagonalization approach. (c) Global matrix representation using column-wise unfolding with sub-matrices $\mathbf{A}_q \triangleq \mathbf{A}_q \mathbf{D}_q$.

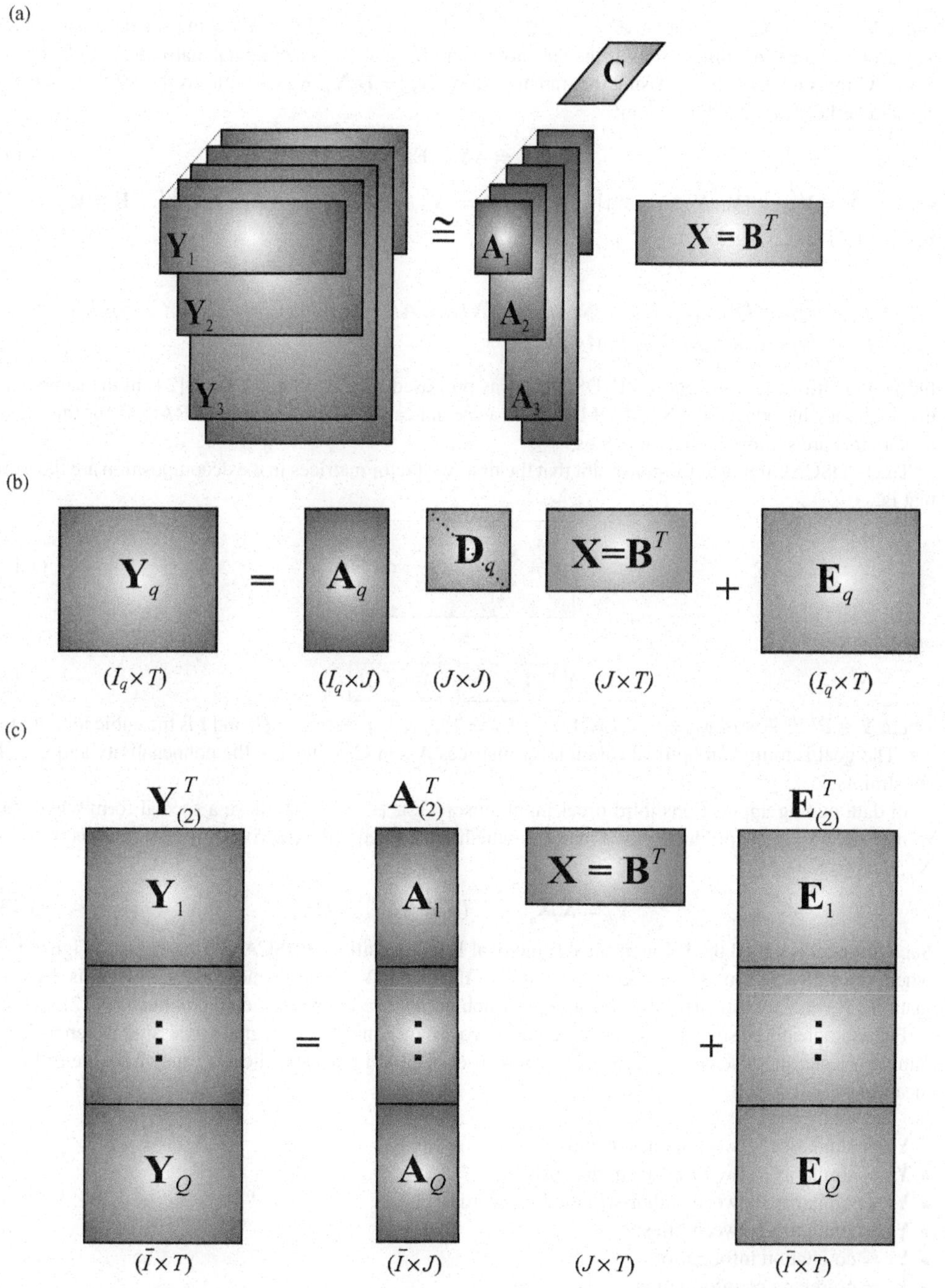

Figure 1.32 (a) Extended NTF2 model. (b) An equivalent representation in which the frontal slices of a three-way array are factorized by a set of nonnegative matrices, (c) Global matrix representation using column-wise unfolding with sub-matrices $\mathbf{A}_q \triangleq \mathbf{A}_q\mathbf{D}_q$.

where $\mathbf{Y} \in \mathbb{R}_+^{I_q \times T}, \mathbf{X} \in \mathbb{R}_+^{J \times T}, \mathbf{C} \in \mathbb{R}_+^{Q \times J}, \mathbf{A}_q \in \mathbb{R}_+^{I_q \times J}\}, \mathbf{E}_q = \mathbf{E}_{::q} \in \mathbb{R}^{I_q \times T}$ is the q-th frontal slice of a three-way array (of the same dimensions as the data array) and $\mathbf{D}_q \in \mathbb{R}_+^{J \times J}$ is a diagonal matrix that holds the q-th row of $\mathbf{C}$ in its main diagonal. Using the transformation $\mathbf{X}_q := \mathbf{D}_q \mathbf{X}_q$, we can convert the NTF2 problem to the standard (2-way) NMF problem:

$$\bar{\mathbf{Y}} = \bar{\mathbf{A}}\mathbf{X} + \bar{\mathbf{E}}, \tag{1.148}$$

where $\bar{\mathbf{Y}} = \mathbf{Y}_{(2)}^T = [\mathbf{Y}_1; \mathbf{Y}_2; \ldots; \mathbf{Y}_Q] \in \mathbb{R}_+^{\bar{I} \times T}$, $\bar{\mathbf{A}} = \mathbf{A}_{(2)}^T = [\mathbf{A}_1; \mathbf{A}_2; \ldots; \mathbf{A}_Q] \in \mathbb{R}_+^{\bar{I} \times J}$, $\bar{\mathbf{E}} = \mathbf{E}_{(2)}^T = [\mathbf{E}_1; \mathbf{E}_2; \ldots; \mathbf{E}_Q] \in \mathbb{R}_+^{\bar{I} \times T}$ and $\bar{I} = \sum_q I_q$.

1.5.5 Individual Differences in Scaling (INDSCAL) and Implicit Slice Canonical Decomposition Model (IMCAND)

Individual Differences in Scaling (INDSCAL) was proposed by Carroll and Chang [24] in the same paper in which they introduced CANDECOMP, and is a special case of the three-way PARAFAC for third-order tensors that are symmetric in two modes.

The INDSCAL imposes the constraint that the first two factor matrices in the decomposition are the same, that is,

$$\underline{\mathbf{Y}} \cong \sum_{j=1}^{J} \boldsymbol{a}_j \circ \boldsymbol{a}_j \circ \boldsymbol{c}_j, \tag{1.149}$$

or equivalently

$$\underline{\mathbf{Y}} \cong \underline{\mathbf{I}} \times_1 \mathbf{A} \times_2 \mathbf{A} \times_3 \mathbf{C}, \tag{1.150}$$

where $\underline{\mathbf{Y}} \in \mathbb{R}^{I \times I \times Q}$ with $y_{itq} = y_{tiq}, i = 1, \ldots, I, t = 1, \ldots, I, q = 1, \ldots, Q$, and $\underline{\mathbf{I}}$ is the cubic identity tensor. The goal is to find an optimal solution for matrices $\mathbf{A}$ and $\mathbf{C}$, subject to the nonnegativity and sparsity constraints.

In data mining applications third-order input tensors $\underline{\mathbf{Y}} \in \mathbb{R}^{I \times I \times Q}$ may have a special form where each frontal slice $\mathbf{Y}_q$ is the product of two matrices which are typically the matrix $\mathbf{X}_q \in \mathbb{R}^{I \times T}$ and its transpose $\mathbf{X}_q^T$, thus yielding

$$\mathbf{Y}_q = \mathbf{X}_q \mathbf{X}_q^T, \qquad (q = 1, 2, \ldots, Q). \tag{1.151}$$

Such a model is called the IMplicit Slice Canonical Decomposition (IMSCAND) Model (see Figure 1.33) where the PARAFAC or Tucker decompositions of the tensor $\underline{\mathbf{Y}}$ are performed implicitly, that is, by using matrices $\mathbf{X}_q$ and not directly the elements y_{itq} (which do not need to be stored on computer) [122].

For example, these slice matrices may represent covariance matrices in signal processing, whereas in text mining (clustering of scientific publications from a set of SIAM journals) slices $\mathbf{Y}_q$ may have the following meanings [122]:

- $\mathbf{Y}_1$ = similarity between names of authors,
- $\mathbf{Y}_2$ = similarity between words in the abstract,
- $\mathbf{Y}_3$ = similarity between author-specified keywords,
- $\mathbf{Y}_4$ = similarity between titles,
- $\mathbf{Y}_5$ = co-citation information,
- $\mathbf{Y}_6$ = co-reference information.

The first four slices are formed from feature-document matrices for the specified similarity. If there exists no similarity between two documents, then the corresponding element in a slice is nonzero. For the fifth slice, the element y_{it5} indicates the number of papers that both documents i and t cite. Whereas the element y_{it6} on the sixth slice is the number of papers cited by both documents i and t.

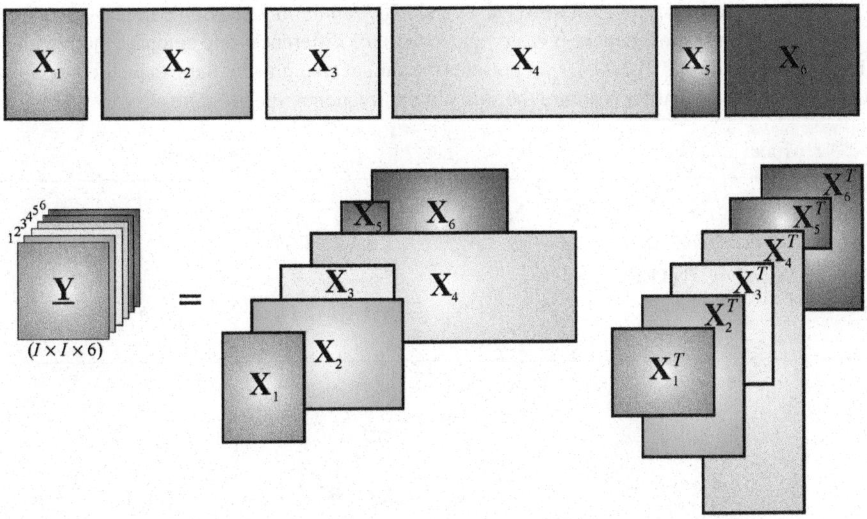

Figure 1.33 Illustration of IMplicit Slice Canonical Decomposition (IMSCAND). The frontal slices $\mathbf{Y}_q$ are not stored directly but rather represented by a set of matrices $\mathbf{X}_q$ as $\mathbf{Y}_q = \mathbf{X}_q\mathbf{X}_q^T$ for $q = 1, 2, \ldots, 6$.

1.5.6 Shifted PARAFAC and Convolutive NTF

Harshman *et al.* [64] introduced the shifted PARAFAC (S-PARAFAC) in order to deal with shifting factors in sequential data such as time series or spectra data. For example, the S-PARAFAC for mode-2 can be described for each entry y_{itq} as

$$\boxed{y_{itq} = \sum_{j=1}^{J} a_{ij}\, b_{(t+s_{qj})j}\, c_{qj} + e_{itq},} \tag{1.152}$$

where the shift parameter s_{qj} gives the shift at column j of the factor $\mathbf{B}$. We can rewrite this model for frontal slices

$$\mathbf{Y}_q = \mathbf{A}\,\mathbf{D}_q\,\mathcal{S}_{s_q}(\mathbf{B})^T + \mathbf{E}_q, \qquad (q = 1, \ldots, Q), \tag{1.153}$$

where the shift operator (or function) $\mathcal{S}_{s_q}(\mathbf{B})$ shifts all elements in each column of the matrix $\mathbf{B}$ by amount s_{qj}. The vector $\mathbf{s}_q$ is a vector of J shift values taken from row q of the (implied) shift matrix $\mathbf{S} \in \mathbb{R}^{Q\times J}$, and the matrix $\mathbf{D}_q$ is a diagonal matrix containing the q-th row of $\mathbf{C}$. One limitation of S-PARAFAC is that it only considers one-dimensional shifts, typically time, but does not handle two-dimensional shifts that might be encountered in neuroimages of brain scans [2,108]

Another extension of PARAFAC is Convolutive PARAFAC (CPARAFAC or CNTF) which is a generalization of CNMF to multiway spectral data. Morup and Schmidt [108] introduced this model with the name Sparse Nonnegative Tensor 2D Deconvolution (SNTF2D). The single convolutive NTF on mode-2 and mode-3 and with rank-J for the nonnegative tensor $\underline{\mathbf{Y}} \in \mathbb{R}_+^{I\times T\times Q}$ returns a factor matrix $\mathbf{A} \in \mathbb{R}_+^{I\times J}$ on the first dimension, a factor matrix $\mathbf{B} \in \mathbb{R}_+^{T\times J}$ on the second dimension, and a set of S factor matrices or tensor $\underline{\mathbf{C}} \in \mathbb{R}_+^{Q\times J\times S}$, and can be expressed as follows

$$\boxed{y_{itq} = \sum_{j=1}^{J}\sum_{s=1}^{S} a_{ij}\, b_{(t-s+1)j}\, c_{qjs} + e_{itq}.} \tag{1.154}$$

Table 1.3 Basic descriptions of PARAFAC (CP) and NTF family models. Some models are expressed in matrix and/or scalar notations to make it easier understand the differences and compare them with standard PARAFAC. For Shifted NTF (S-NTF), s_{qj} represents the shift at column q for the j-th factor. For Convolutive NTF (CNTF), s is used usually to capture the shifts in the frequency spectrogram.

Model	Description
Nonnegative PARAFAC (NTF)	$y_{itq} = \sum_{j=1}^{J} a_{ij}\, b_{tj}\, c_{qj} + e_{itq}$ $\mathbf{Y}_q = \mathbf{A}\mathbf{D}_q\mathbf{B}^T + \mathbf{E}_q = \sum_{j=1}^{J} c_{qj}\boldsymbol{a}_j\, \boldsymbol{b}_j^T + \mathbf{E}_q$
NTF1	$y_{itq} = \sum_{j=1}^{J} a_{ij}\, b_{tjq}\, c_{qj} + e_{itq}$ $\mathbf{Y}_q = \mathbf{A}\,\mathbf{D}_q\,\mathbf{B}_q^T + \mathbf{E}_q = \sum_{j=1}^{J} c_{qj}\,\boldsymbol{a}_j\,(\boldsymbol{b}_j^{(q)})^T + \mathbf{E}_q$
NTF2	$y_{itq} = \sum_{j=1}^{J} a_{ijq}\, b_{tj}\, c_{qj} + e_{itq}$ $\mathbf{Y}_q = \mathbf{A}_q\,\mathbf{D}_q\,\mathbf{B}^T + \mathbf{E}_q = \sum_{j=1}^{J} c_{qj}\,\boldsymbol{a}_j^{(q)}\,\boldsymbol{b}_j^T + \mathbf{E}_q$
Shifted NTF (S-NTF)	$y_{itq} = \sum_{j=1}^{J} a_{ij}\, b_{(t+s_{qj})j}\, c_{qj} + e_{itq}$ $\mathbf{Y}_q = \mathbf{A}\,\mathbf{D}_q\,\mathcal{S}_{s_q}(\mathbf{B})^T + \mathbf{E}_q$
Convolutive NTF (CNTF)	$y_{itq} = \sum_{j=1}^{J}\sum_{s=1}^{S} a_{ij}\, b_{(t-s+1)j}\, c_{qjs} + e_{itq}$ $\mathbf{Y}_q = \mathbf{A}\sum_{s=1}^{S}\mathbf{D}_q^{(s)}\,(\mathbf{T}_{\uparrow(s-1)}\mathbf{B})^T + \mathbf{E}_q$
C2NTF	$y_{itq} = \sum_{j=1}^{J}\sum_{s=1}^{S}\sum_{r=1}^{R} a_{ij}\, b_{(t-s+1)jr}\, c_{(q-r+1)js} + e_{itq}$
INDSCAL	$y_{itq} = \sum_{j=1}^{J} a_{ij}\, a_{tj}\, c_{qj} + e_{itq}$

For $S = 1$, CPARAFAC (CNTF) simplifies to PARAFAC (NTF). Matrix representation of this model via frontal slices $\mathbf{Y}_q$, $(q = 1, \ldots, Q)$ is given by

$$\mathbf{Y}_q \cong \mathbf{A}\sum_{s=0}^{S-1}\mathbf{D}_q^{(s+1)}\,(\mathbf{T}_{\uparrow(s)}\mathbf{B})^T = \mathbf{A}\sum_{s=0}^{S-1}\mathbf{D}_q^{(s+1)}\,\mathbf{B}^T\mathbf{T}_{\underset{\rightarrow}{s}} \tag{1.155}$$

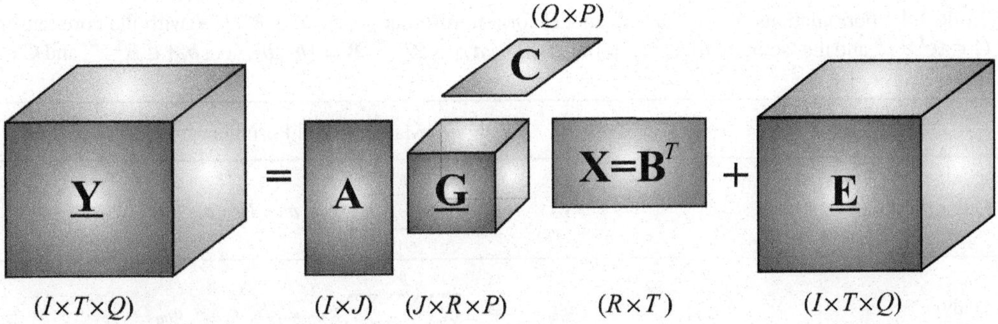

Figure 1.34 Tucker3 model is a weighted sum of the outer product of three vectors (factors) stored as columns of component matrices $\mathbf{A} \in \mathbb{R}^{I\times J}$, $\mathbf{B} = \mathbf{X}^T \in \mathbb{R}^{T\times R}$ and $\mathbf{C} \in \mathbb{R}^{Q\times P}$. The core tensor $\underline{\mathbf{G}} \in \mathbb{R}^{J\times R\times P}$ defines a linking structure between the set of components and J, R, and P denote the number of components. In order to achieve uniqueness for the Tucker models it is necessary to impose additional constraints such as sparsity and nonnegativity.

where $\mathbf{D}_q^{(s)}$ is a diagonal matrix containing the fiber $\boldsymbol{c}_{q:s}$, and the shift operators $\mathbf{T}_{\uparrow(s)}$, $\mathbf{T}_{\underset{\rightarrow}{s}}$ are defined in Section 1.2.12. In the tensor form the CNTF can be described as

$$\underline{\mathbf{Y}} = \sum_{s=1}^{S} \underline{\mathbf{I}} \times_1 \mathbf{A} \times_2 \mathbf{T}_{\uparrow(s-1)}\mathbf{B} \times_3 \mathbf{C}_s + \underline{\mathbf{E}}. \tag{1.156}$$

The CPARAFAC can be extended to the double convolutive model (C2PARAFAC or C2NTF) as

$$y_{itq} \cong \sum_{j=1}^{J}\sum_{s=1}^{S}\sum_{r=1}^{R} a_{ij}\, b_{(t-s+1)jr}\, c_{(q-r+1)js}, \tag{1.157}$$

where the second factor is no longer a matrix but a tensor of size $(T \times J \times R)$. Table 1.3 summarizes various NTF and PARAFAC models.

1.5.7 Nonnegative Tucker Decompositions

The Tucker decomposition, also called the Tucker3 or best rank (J, R, P) approximation, can be formulated as follows[19] (see Figure 1.34) [136,137]:

Given a third-order data tensor $\underline{\mathbf{Y}} \in \mathbb{R}^{I\times T\times Q}$ and three positive indices $\{J, R, P\} << \{I, T, Q\}$, find a core tensor $\mathbf{G} = [g_{jrp}] \in \mathbb{R}^{J\times R\times P}$ and three component matrices called factor or loading matrices or factors: $\mathbf{A} = [\boldsymbol{a}_1, \boldsymbol{a}_2, \ldots, \boldsymbol{a}_J] \in \mathbb{R}^{I\times J}$, $\mathbf{B} = [\boldsymbol{b}_1, \boldsymbol{b}_2, \ldots, \boldsymbol{b}_R] \in \mathbb{R}^{T\times R}$, and $\mathbf{C} = [\boldsymbol{c}_1, \boldsymbol{c}_2, \ldots, \boldsymbol{c}_P] \in \mathbb{R}^{Q\times P}$, which perform the following approximate decomposition:

$$\underline{\mathbf{Y}} = \sum_{j=1}^{J}\sum_{r=1}^{R}\sum_{p=1}^{P} g_{jrp}\, (\boldsymbol{a}_j \circ \boldsymbol{b}_r \circ \boldsymbol{c}_p) + \underline{\mathbf{E}} \tag{1.158}$$

[19]The Tucker3 model with orthogonal factors is also known as three-way PCA (principal component analysis). The model can be naturally extended to N-way Tucker decomposition of arbitrary N-th order tensor.

Table 1.4 Formulations of the Tucker3 model for a third-order tensor $\underline{\mathbf{Y}} \in \mathbb{R}^{I \times T \times Q}$ with the core tensor $\underline{\mathbf{G}} \in \mathbb{R}^{J \times R \times P}$ and the factor matrices: $\mathbf{A} = [\boldsymbol{a}_1, \boldsymbol{a}_2, \ldots, \boldsymbol{a}_J] \in \mathbb{R}^{I \times J}$, $\mathbf{B} = [\boldsymbol{b}_1, \boldsymbol{b}_2, \ldots, \boldsymbol{b}_R] \in \mathbb{R}^{T \times R}$, and $\mathbf{C} = [\boldsymbol{c}_1, \boldsymbol{c}_2, \ldots, \boldsymbol{c}_P] \in \mathbb{R}^{Q \times P}$.

Operator	Mathematical Formula
Outer product	$\underline{\mathbf{Y}} = \sum_{j=1}^{J}\sum_{r=1}^{R}\sum_{p=1}^{P} g_{jrp}\, \boldsymbol{a}_j \circ \boldsymbol{b}_r \circ \boldsymbol{c}_p + \underline{\mathbf{E}}$
Scalar	$y_{itq} = \sum_{j=1}^{J}\sum_{r=1}^{R}\sum_{p=1}^{P} g_{jrp}\, a_{ij}\, b_{tr}\, c_{qp} + e_{itq}$
mode-n multiplications	$\underline{\mathbf{Y}} = \underline{\mathbf{G}} \times_1 \mathbf{A} \times_2 \mathbf{B} \times_3 \mathbf{C} + \underline{\mathbf{E}}$
Slice representation	$\mathbf{Y}_q = \mathbf{A}\mathbf{H}_q\mathbf{B}^T + \mathbf{E}_q, \quad (q = 1, 2, \ldots, Q)$ $\mathbf{H}_q = \sum_{p=1}^{P} c_{qp}\mathbf{G}_p, \quad \mathbf{G}_p \triangleq \mathbf{G}_{::p} \in \mathbb{R}^{J \times R}$
Vector	$\text{vec}(\underline{\mathbf{Y}}) = \text{vec}(\mathbf{Y}_{(1)}) \cong (\mathbf{C} \otimes \mathbf{B} \otimes \mathbf{A})\, \text{vec}(\underline{\mathbf{G}})$
Kronecker product	$\mathbf{Y}_{(1)} \cong \mathbf{A}\, \mathbf{G}_{(1)}\, (\mathbf{C} \otimes \mathbf{B})^T$ $\mathbf{Y}_{(2)} \cong \mathbf{B}\, \mathbf{G}_{(2)}\, (\mathbf{C} \otimes \mathbf{A})^T$ $\mathbf{Y}_{(3)} \cong \mathbf{C}\, \mathbf{G}_{(3)}\, (\mathbf{B} \otimes \mathbf{A})^T$

or equivalently in the element-wise form

$$\boxed{y_{itq} = \sum_{j=1}^{J}\sum_{r=1}^{R}\sum_{p=1}^{P} g_{jrp}\, a_{ij}\, b_{tr}\, c_{qp} + e_{itq},} \tag{1.159}$$

where $\boldsymbol{a}_j \in \mathbb{R}^I$, $\boldsymbol{b}_j \in \mathbb{R}^T$, and $\boldsymbol{c}_j \in \mathbb{R}^Q$, (that is, the vectors within the associated component (factor) matrices $\mathbf{A}$, $\mathbf{B}$ and $\mathbf{C}$,), and g_{jrp} are scaling factors which are the entries of a core tensor $\underline{\mathbf{G}} = [g_{jrp}] \in \mathbb{R}^{J \times R \times P}$.

The original Tucker model makes the assumption of orthogonality of the factor matrices (in analogy to SVD), [17,79,83,84,107,117]. We will, however, ignore these constraints. By imposing nonnegativity constraints the problem of estimating the component matrices and a core tensor is converted into a generalized NMF problem called the Nonnegative Tucker Decomposition (NTD) (see Chapter 7 for details). The first implementations of Tucker decomposition with nonnegativity constraints together with a number of other constraints were given by Kiers, Smilde and Bro in [17,79]. The NTD imposes nonnegativity constraints for all component matrices and a core tensor, while a semi-NTD (in analogy to semi-NMF) imposes nonnegativity constraints to only some components matrices and/or some elements of the core tensor.

There are several equivalent mathematical descriptions for the Tucker model (see Table 1.4). It can be expressed in a compact matrix form using mode-n multiplications

$$\boxed{\underline{\mathbf{Y}} = \underline{\mathbf{G}} \times_1 \mathbf{A} \times_2 \mathbf{B} \times_3 \mathbf{C} + \underline{\mathbf{E}} = [\![\underline{\mathbf{G}}; \mathbf{A}, \mathbf{B}, \mathbf{C}]\!] + \underline{\mathbf{E}},} \tag{1.160}$$

where $\widehat{\underline{\mathbf{Y}}} = [\![\underline{\mathbf{G}}; \mathbf{A}, \mathbf{B}, \mathbf{C}]\!]$ is the shorthand notation for the Tucker3 tensor decomposition.

Using the unfolding approach we can obtain matrix forms expressed compactly by the Kronecker products:

Table 1.5 Tucker models for a third-order tensor $\underline{\mathbf{Y}} \in \mathbb{R}^{I \times T \times Q}$.

Model	Description
Tucker1	$y_{itq} = \sum_{j=1}^{J} g_{jtq}\, a_{ij} + e_{itq}$ $\underline{\mathbf{Y}} = \underline{\mathbf{G}} \times_1 \mathbf{A} + \underline{\mathbf{E}}, \quad (\underline{\mathbf{G}} \in \mathbb{R}^{J \times T \times Q})$ $\mathbf{Y}_q = \mathbf{A}\mathbf{G}_q + \mathbf{E}_q, \quad (q = 1, 2, \ldots, Q)$ $\mathbf{Y}_{(1)} = \mathbf{A}\mathbf{G}_{(1)} + \mathbf{E}_{(1)}$
Tucker2	$y_{itq} = \sum_{j=1}^{J} \sum_{r=1}^{R} g_{jrq}\, a_{ij}\, b_{tr} + e_{itq}$ $\underline{\mathbf{Y}} = \underline{\mathbf{G}} \times_1 \mathbf{A} \times_2 \mathbf{B} + \underline{\mathbf{E}}, \quad (\underline{\mathbf{G}} \in \mathbb{R}^{J \times R \times Q})$ $\underline{\mathbf{Y}} = \sum_{j=1}^{J} \sum_{r=1}^{R} \boldsymbol{a}_j \circ \boldsymbol{b}_r \circ \boldsymbol{g}_{jr} + \underline{\mathbf{E}}$ $\mathbf{Y}_{(3)} = \mathbf{G}_{(3)}\, (\mathbf{B} \otimes \mathbf{A})^T + \mathbf{E}_{(3)}$
Tucker3	$y_{itq} = \sum_{j=1}^{J} \sum_{r=1}^{R} \sum_{p=1}^{P} g_{jrp}\, a_{ij}\, b_{tr}\, c_{qp} + e_{itq}$ $\underline{\mathbf{Y}} = \underline{\mathbf{G}} \times_1 \mathbf{A} \times_2 \mathbf{B} \times_3 \mathbf{C} + \underline{\mathbf{E}}, \quad (\underline{\mathbf{G}} \in \mathbb{R}^{J \times R \times P})$
Shifted Tucker3	$y_{itq} = \sum_{j=1}^{J} \sum_{r=1}^{R} \sum_{p=1}^{P} g_{jrp}\, a_{(i+s_{tj})j}\, b_{tr}\, c_{qp} + e_{itq}$

$$\mathbf{Y}_{(1)} \cong \mathbf{A}\, \mathbf{G}_{(1)}\, (\mathbf{C} \otimes \mathbf{B})^T, \tag{1.161}$$

$$\mathbf{Y}_{(2)} \cong \mathbf{B}\, \mathbf{G}_{(2)}\, (\mathbf{C} \otimes \mathbf{A})^T, \tag{1.162}$$

$$\mathbf{Y}_{(3)} \cong \mathbf{C}\, \mathbf{G}_{(3)}\, (\mathbf{B} \otimes \mathbf{A})^T. \tag{1.163}$$

It is often convenient to represent the three-way Tucker model in its vectorized forms

$$\text{vec}\,(\mathbf{Y}_{(1)}) \cong \text{vec}\,(\mathbf{A}\mathbf{G}_{(1)}(\mathbf{C} \otimes \mathbf{B})^T) = (\mathbf{C} \otimes \mathbf{B}) \otimes \mathbf{A}\ \text{vec}\,(\mathbf{G}_{(1)}), \tag{1.164}$$

$$\text{vec}\,(\mathbf{Y}_{(2)}) \cong \text{vec}\,(\mathbf{B}\mathbf{G}_{(2)}(\mathbf{C} \otimes \mathbf{A})^T) = (\mathbf{C} \otimes \mathbf{A}) \otimes \mathbf{B}\ \text{vec}\,(\mathbf{G}_{(2)}), \tag{1.165}$$

$$\text{vec}\,(\mathbf{Y}_{(3)}) \cong \text{vec}\,(\mathbf{C}\mathbf{G}_{(3)}(\mathbf{B} \otimes \mathbf{A})^T) = (\mathbf{B} \otimes \mathbf{A}) \otimes \mathbf{C}\ \text{vec}\,(\mathbf{G}_{(3)}). \tag{1.166}$$

The Tucker model described above is often called the Tucker3 model because a third-order tensor is decomposed into three factor (loading) matrices (say, $\{\mathbf{A}, \mathbf{B}, \mathbf{C}\}$) and a core tensor $\underline{\mathbf{G}}$. In applications where we have two factor matrices or even only one, the Tucker3 model for a three-way tensor simplifies into the Tucker2 or Tucker1 models (see Table 1.5). The Tucker2 model can be obtained from the Tucker3 model by absorbing one factor by a core tensor (see Figure 1.35(b)), that is,

$$\underline{\mathbf{Y}} \cong \underline{\mathbf{G}} \times_1 \mathbf{A} \times_2 \mathbf{B}. \tag{1.167}$$

$$y_{itq} = \sum_{j=1}^{J}\sum_{r=1}^{R}\sum_{p=1}^{P} g_{jrp}\, a_{ij}\, b_{tr}\, c_{qp}$$

$$y_{itq} = \sum_{j=1}^{J}\sum_{r=1}^{R} g_{jrq}\, a_{ij}\, b_{tr}$$

$$y_{itq} = \sum_{j=1}^{J} g_{jtq}\, a_{ij}$$

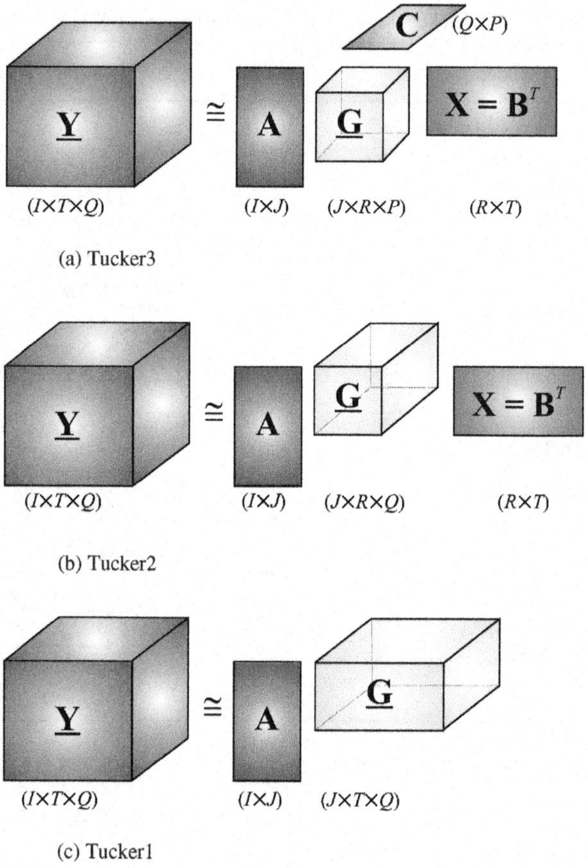

Figure 1.35 Summary of the three related Tucker decompositions.

For the Tucker1 model we have only one factor matrix (while two others are absorbed by a core tensor) which is described as (see also Figure 1.35(c)) (see also Table 1.6)

$$\boxed{\underline{\mathbf{Y}} \cong \underline{\mathbf{G}} \times_1 \mathbf{A}.} \tag{1.168}$$

It is interesting to note that the approximation of a tensor by factor matrices and a core tensor often helps to simplify mathematical operations and reduce the computation cost of some operations in multi-linear (tensor) algebra. For example:

$$\begin{aligned}\underline{\mathbf{Y}} = \underline{\mathbf{X}} \,\bar{\times}_3\, \boldsymbol{a} &\approx (\underline{\mathbf{G}} \times_1 \mathbf{A} \times_2 \mathbf{B} \times_3 \mathbf{C}) \,\bar{\times}_3\, \boldsymbol{a} \\ &= (\underline{\mathbf{G}} \,\bar{\times}_3 \mathbf{C}^T \boldsymbol{a}) \times_1 \mathbf{A} \times_2 \mathbf{B} \\ &= \underline{\mathbf{G}}_{Ca} \times_1 \mathbf{A} \times_2 \mathbf{B},\end{aligned}$$

where $\underline{\mathbf{G}}_{Ca} = \underline{\mathbf{G}} \,\bar{\times}_3\, \mathbf{C}^T \boldsymbol{a}$. Comparison of tensor decomposition models are summarized in Table 1.6.

Table 1.6 Matrix and tensor representations for various factorization models (for most of the models we impose additional nonnegativity constraints).

Model	Matrix Representation	Tensor Representation
NMF	$\mathbf{Y} \cong \mathbf{A}\,\mathbf{X} = \mathbf{A}\,\mathbf{B}^T$	$\mathbf{Y} \cong \mathbf{I} \times_1 \mathbf{A} \times_2 \mathbf{X}^T$
SVD	$\mathbf{Y} \cong \mathbf{U}\,\boldsymbol{\Sigma}\,\mathbf{V}^T = \sum_{r=1}^{R} \sigma_r\, \boldsymbol{u}_r \boldsymbol{v}_r^T$	$\mathbf{Y} \cong \boldsymbol{\Sigma} \times_1 \mathbf{U} \times_2 \mathbf{V} = \sum_{r=1}^{R} \sigma_r\, \boldsymbol{u}_r \circ \boldsymbol{v}_r$
Three-factor NMF	$\mathbf{Y} \cong \mathbf{A}\,\mathbf{S}\,\mathbf{X} = \mathbf{A}\,\mathbf{S}\,\mathbf{B}^T$	$\mathbf{Y} \cong \mathbf{S} \times_1 \mathbf{A} \times_2 \mathbf{X}^T$
NTF (nonnegative PARAFAC)	$\mathbf{Y}_q \cong \mathbf{A}\,\mathbf{D}_q(\boldsymbol{c}_{q:})\,\mathbf{B}^T$ $(q = 1, 2, \ldots, Q)$	$\underline{\mathbf{Y}} \cong \underline{\mathbf{I}} \times_1 \mathbf{A} \times_2 \mathbf{B} \times_3 \mathbf{C} = \sum_{j=1}^{J} \boldsymbol{a}_j \circ \boldsymbol{b}_j \circ \boldsymbol{c}_j$
NTF1	$\mathbf{Y}_q \cong \mathbf{A}\,\mathbf{D}_q(\boldsymbol{c}_{q:})\,\mathbf{B}_q^T = \mathbf{A}\,\mathbf{D}_q(\boldsymbol{c}_{q:})\,\mathbf{X}_q$ $(q = 1, 2, \ldots, Q)$	$\underline{\mathbf{Y}} \cong \underline{\mathbf{X}} \times_1 \mathbf{A} \times_3 \mathbf{C}$
NTF2	$\mathbf{Y}_q \cong \mathbf{A}_q\,\mathbf{D}_q(\boldsymbol{c}_{q:})\,\mathbf{B}^T = \mathbf{A}_q\,\mathbf{D}_q(\boldsymbol{c}_{q:})\,\mathbf{X}$ $(q = 1, 2, \ldots, Q)$	$\underline{\mathbf{Y}} \cong \underline{\mathbf{A}} \times_2 \mathbf{B} \times_3 \mathbf{C}$
Tucker1	$\mathbf{Y}_q \cong \mathbf{A}\,\mathbf{G}_{::q}$ $(q = 1, 2, \ldots, Q)$	$\underline{\mathbf{Y}} = \underline{\mathbf{G}} \times_1 \mathbf{A}$
Tucker2	$\mathbf{Y}_q \cong \mathbf{A}\,\mathbf{G}_{::q}\,\mathbf{B}^T$ $(q = 1, 2, \ldots, Q)$	$\underline{\mathbf{Y}} \cong \underline{\mathbf{G}} \times_1 \mathbf{A} \times_2 \mathbf{B} = \sum_{j=1}^{J}\sum_{r=1}^{R} \boldsymbol{a}_j \circ \boldsymbol{b}_r \circ \boldsymbol{g}_{jr}$
Tucker3	$\mathbf{Y}_q \cong \mathbf{A}\mathbf{H}_q\mathbf{B}^T$, $\mathbf{H}_q = \sum_{p=1}^{P} c_{qp}\,\mathbf{G}_{::p}$	$\underline{\mathbf{Y}} \cong \underline{\mathbf{G}} \times_1 \mathbf{A} \times_2 \mathbf{B} \times_2 \mathbf{C} = \sum_{j=1}^{J}\sum_{r=1}^{R}\sum_{p=1}^{P} g_{jrp}\,(\boldsymbol{a}_j \circ \boldsymbol{b}_r \circ \boldsymbol{c}_p)$

1.5.8 Block Component Decompositions

Block Component Decompositions (BCDs) (also called Block Component Models) introduced by De Lathauwer and Nion for applications in signal processing and wireless communications [46–48,110,111] can be considered as a sum of basic subtensor decompositions (see Figure 1.36). Each basic subtensor in this sum has the same kind of factorization or decomposition, typically, Tucker2 or Tucker3 decomposition, and the corresponding components have a similar structure (regarding dimensions, sparsity profile and nonnegativity constraints) as illustrated in Figures 1.36 (a), (b) and (c).

The model shown in Figure 1.36(a), called the BCD rank-$(J_r, 1)$, decomposes a data tensor $\underline{\mathbf{Y}} \in \mathbb{R}^{I \times T \times Q}$ into a sum of R subtensors $\underline{\mathbf{Y}}^{(r)} \in \mathbb{R}^{I \times T \times Q}$, $(r = 1, 2, \ldots, R)$. Each of the subtensor $\underline{\mathbf{Y}}^{(r)}$ is factorized into three factors $\mathbf{A}_r \in \mathbb{R}^{I \times J_r}$, $\mathbf{B}_r \in \mathbb{R}^{T \times J_r}$ and $\boldsymbol{c}_r \in \mathbb{R}^{Q}$. The mathematical description of this BCD model

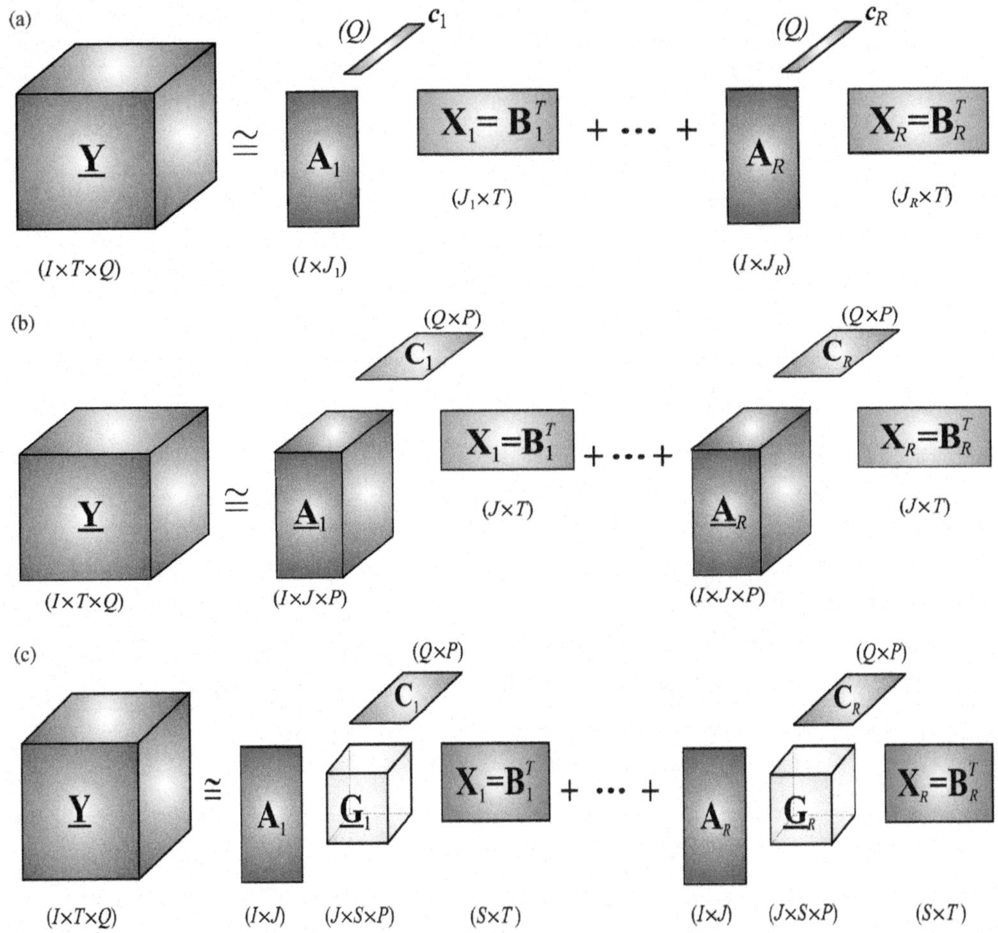

Figure 1.36 Block Component Decompositions (BCD) for a third-order tensor $\underline{\mathbf{Y}} \in \mathbb{R}^{I \times T \times Q}$: (a) BCD with rank-$(J_r, 1)$, (b) BCD with rank -$(J, P)$ and (c) BCD with rank-(J, S, P).

is given by

$$\underline{\mathbf{Y}} = \sum_{r=1}^{R} \underline{\mathbf{Y}}^{(r)} + \underline{\mathbf{E}} = \sum_{r=1}^{R} (\mathbf{A}_r \mathbf{B}_r^T) \circ \boldsymbol{c}_r + \underline{\mathbf{E}}, \tag{1.169}$$

or equivalently

$$\underline{\mathbf{Y}} = \sum_{r=1}^{R} \underline{\mathbf{A}}_r \times_2 \mathbf{B}_r \times_3 \boldsymbol{c}_r + \underline{\mathbf{E}}, \tag{1.170}$$

where tensors $\underline{\mathbf{A}}_r \triangleq \mathbf{A}_r \in \mathbb{R}^{I \times J_r \times 1}$ are three-way tensors with only one frontal slice. With this notation, each subtensor $\underline{\mathbf{Y}}^{(r)}$ is a Tucker2 model with the core tensor $\underline{\mathbf{A}}_{(r)}$, and factors $\mathbf{B}_r$ and $\boldsymbol{c}_r$. Hence, the BCD rank-$(J_r, 1)$ decomposition can be considered as a sum of simplified Tucker-2 models. The objective is to estimate component matrices $\mathbf{A}_r \in \mathbb{R}^{I \times J_r}$, $\mathbf{B}_r \in \mathbb{R}^{T \times J_r}$, $(r = 1, 2, \ldots, R)$ and a factor matrix $\mathbf{C} = [\boldsymbol{c}_1, \boldsymbol{c}_2, \ldots, \boldsymbol{c}_R] \in \mathbb{R}^{Q \times R}$ subject to optional nonnegativity and sparsity constraints.

Using the unfolding approach, the above BCD model can be written in several equivalent forms:

$$\mathbf{Y}_{(1)} \cong \sum_{r=1}^{R} \left[\underline{\mathbf{A}}_r\right]_{(1)} (\boldsymbol{c}_r \otimes \mathbf{B}_r)^T = \sum_{r=1}^{R} \mathbf{A}_r (\boldsymbol{c}_r \otimes \mathbf{B}_r)^T$$

$$= \boxed{\left[\mathbf{A}_1 \ \mathbf{A}_2 \ \cdots \ \mathbf{A}_R\right] \left[\boldsymbol{c}_1 \otimes \mathbf{B}_1 \ \boldsymbol{c}_2 \otimes \mathbf{B}_2 \ \cdots \ \boldsymbol{c}_R \otimes \mathbf{B}_R\right]^T}, \quad (1.171)$$

$$\mathbf{Y}_{(2)} \cong \sum_{r=1}^{R} \mathbf{B}_r \left[\underline{\mathbf{A}}_r\right]_{(2)} (\boldsymbol{c}_r \otimes \mathbf{I}_I)^T = \sum_{r=1}^{R} \mathbf{B}_r \mathbf{A}_r^T (\boldsymbol{c}_r \otimes \mathbf{I}_I)^T$$

$$= \sum_{r=1}^{R} \mathbf{B}_r \left[(\boldsymbol{c}_r \otimes \mathbf{I}_I)\mathbf{A}_r\right]^T = \sum_{r=1}^{R} \mathbf{B}_r (\boldsymbol{c}_r \otimes \mathbf{A}_r)^T$$

$$= \boxed{\left[\mathbf{B}_1 \ \mathbf{B}_2 \ \cdots \ \mathbf{B}_R\right] \left[\boldsymbol{c}_1 \otimes \mathbf{A}_1 \ \boldsymbol{c}_2 \otimes \mathbf{A}_2 \ \cdots \ \boldsymbol{c}_R \otimes \mathbf{A}_R\right]^T}, \quad (1.172)$$

$$\mathbf{Y}_{(3)} \cong \sum_{r=1}^{R} \boldsymbol{c}_r \left[\underline{\mathbf{A}}_r\right]_{(3)} (\mathbf{B}_r \otimes \mathbf{I}_I)^T = \sum_{r=1}^{R} \boldsymbol{c}_r \operatorname{vec}(\mathbf{A}_r)^T (\mathbf{B}_r \otimes \mathbf{I}_I)^T$$

$$= \sum_{r=1}^{R} \boldsymbol{c}_r \left[(\mathbf{B}_r \otimes \mathbf{I}_I) \operatorname{vec}(\mathbf{A}_r)\right]^T = \sum_{r=1}^{R} \boldsymbol{c}_r \operatorname{vec}\left(\mathbf{A}_r \mathbf{B}_r^T\right)^T$$

$$= \boxed{\mathbf{C} \left[\operatorname{vec}\left(\mathbf{A}_1 \mathbf{B}_1^T\right) \ \operatorname{vec}\left(\mathbf{A}_2 \mathbf{B}_2^T\right) \ \cdots \ \operatorname{vec}\left(\mathbf{A}_R \mathbf{B}_R^T\right)\right]^T}. \quad (1.173)$$

A simple and natural extension of the model BCD rank-(J_r, 1) assumes that the tensors $\underline{\mathbf{A}}_r \in \mathbb{R}^{I \times J \times P}$ contains P (instead of one) frontal slices of size ($I \times J$) (see Figure 1.36(b)). This model is referred to as the BCD with rank-(J, P) and is described as follows

$$\boxed{\underline{\mathbf{Y}} = \sum_{r=1}^{R} \left(\underline{\mathbf{A}}_r \times_2 \mathbf{B}_r \times_3 \mathbf{C}_r\right) + \underline{\mathbf{E}}.} \quad (1.174)$$

The objective is to find a set of R tensors $\underline{\mathbf{A}}_r \in \mathbb{R}^{I \times J \times P}$, a tensor $\underline{\mathbf{B}} \in \mathbb{R}^{T \times J \times R}$, and a tensor $\underline{\mathbf{C}} \in \mathbb{R}^{Q \times P \times R}$. By stacking tensors $\underline{\mathbf{A}}_r$ along their third dimension, we form a common tensor $\underline{\mathbf{A}} \in \mathbb{R}^{I \times J \times PR}$. The mode-1 matricization of this BCD model gives an equivalent matrix factorization model [48,111]:

$$\mathbf{Y}_{(1)} \cong \sum_{r=1}^{R} \mathbf{A}_{r(1)} (\mathbf{C}_r \otimes \mathbf{B}_r)^T \quad (1.175)$$

$$= \boxed{\mathbf{A}_{(1)} \left[\mathbf{C}_1 \otimes \mathbf{B}_1 \ \mathbf{C}_2 \otimes \mathbf{B}_2 \ \cdots \ \mathbf{C}_R \otimes \mathbf{B}_R\right]^T}. \quad (1.176)$$

The most general, BCD rank-(J, S, P) model is formulated as a sum of R Tucker3 models of corresponding factors $\mathbf{A}_r \in \mathbb{R}^{I \times J}$, $\mathbf{B}_r \in \mathbb{R}^{T \times S}$, $\mathbf{C}_r \in \mathbb{R}^{Q \times P}$ and core tensor $\underline{\mathbf{G}}_r \in \mathbb{R}^{J \times S \times P}$, and described in a compact form as (see Figure 1.36(c)):

$$\boxed{\underline{\mathbf{Y}} = \sum_{r=1}^{R} \left(\underline{\mathbf{G}}_r \times_1 \mathbf{A}_r \times_2 \mathbf{B}_r \times_3 \mathbf{C}_r\right) + \underline{\mathbf{E}}.} \quad (1.177)$$

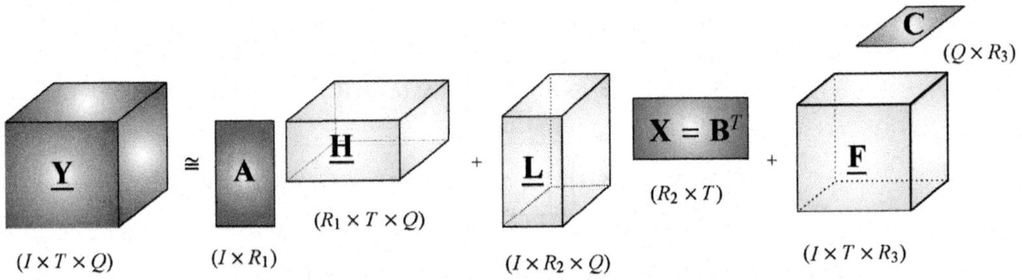

Figure 1.37 Illustration of the Block-Oriented Decomposition (BOD1) for a third-order tensor. Three Tucker1 models express the data tensor along each modes. Typically, core tensors $\underline{\mathbf{H}} \in \mathbb{R}^{R_1 \times T \times Q}$, $\underline{\mathbf{L}} \in \mathbb{R}^{I \times R_2 \times Q}$, $\underline{\mathbf{F}} \in \mathbb{R}^{I \times T \times R_3}$ have much smaller dimensions than a data tensor $\underline{\mathbf{Y}} \in \mathbb{R}^{I \times T \times Q}$, i.e., $R_1 << I$, $R_2 << T$, and $R_3 << Q$.

This model can be also converted in a similar way to a matrix factorization model with set of constrained component matrices.

1.5.9 Block-Oriented Decompositions

A natural extension of the tensor decomposition models discussed in the previous sections will be a decomposition which uses sum of subtensors factorized the data tensor along different modes. Such decompositions will be referred to as Block-Oriented Decompositions (BODs). The key distinction between BOD and BCD models is that subtensors in a BCD model attempt to explain the data tensor in the same modes while BOD models exploit at least two up to all possible separate modes for each subtensors. For example, using Tucker2 model the corresponding BOD2 model can be formulated as follows

$$\underline{\mathbf{Y}} \cong \underline{\mathbf{G}}_1 \times_1 \mathbf{A}_1 \times_2 \mathbf{B}_1 + \underline{\mathbf{G}}_2 \times_1 \mathbf{A}_2 \times_3 \mathbf{C}_1 + \underline{\mathbf{G}}_3 \times_2 \mathbf{B}_2 \times_3 \mathbf{C}_2, \tag{1.178}$$

where core tensors and factor matrices have suitable dimensions.

Analogously, we can define a simpler BOD1 model which is based on the Tucker1 models (see Figure 1.37):

$$\boxed{\underline{\mathbf{Y}} \cong \underline{\mathbf{H}} \times_1 \mathbf{A} + \underline{\mathbf{L}} \times_2 \mathbf{B} + \underline{\mathbf{F}} \times_3 \mathbf{C},} \tag{1.179}$$

where tensors $\underline{\mathbf{H}} \in \mathbb{R}^{R_1 \times T \times Q}$, $\underline{\mathbf{L}} \in \mathbb{R}^{I \times R_2 \times Q}$, $\underline{\mathbf{F}} \in \mathbb{R}^{I \times T \times R_3}$ are core tensors in the Tucker1 models with mode-n, $n = 1, 2, 3$, $\mathbf{A} \in \mathbb{R}^{I \times R_1}$, $\mathbf{B} \in \mathbb{R}^{T \times R_2}$ and $\mathbf{C} \in \mathbb{R}^{Q \times R_3}$ are corresponding factors. The objective is to find three core tensors $\underline{\mathbf{H}}, \underline{\mathbf{L}}, \underline{\mathbf{F}}$ and three corresponding factor matrices $\mathbf{A}, \mathbf{B}$ and $\mathbf{C}$. This model is also called the Slice Oriented Decomposition (SOD) which was recently proposed and investigated by Caiafa and Cichocki [23], and may have various mathematical and graphical representations.

Remark 1.1 *The main motivation to use BOD1 model is to eliminate offset in a data tensor and provide unique and meaningful representation of the extracted components. The BOD1 can be considered as a generalization or extension of the affine NMF model presented in section (1.2.7). As we will show in Chapter 7 the BOD1 model can resolve the problem related with offset degraded by flicker, occlusion or discontinuity.*

Using matricization approach we obtain for BOD1 model several equivalent matrix factorization models:

$$\begin{aligned} \mathbf{Y}_{(1)} &\cong \mathbf{A}\,\mathbf{H}_{(1)} + \mathbf{L}_{(1)} \left(\mathbf{I}_Q \otimes \mathbf{B}\right)^T + \mathbf{F}_{(1)} \left(\mathbf{C} \otimes \mathbf{I}_T\right)^T \\ &= \boxed{\left[\mathbf{A}\ \mathbf{L}_{(1)}\ \mathbf{F}_{(1)}\right] \left[\mathbf{H}_{(1)}\ \mathbf{I}_Q \otimes \mathbf{B}^T\ \mathbf{C}^T \otimes \mathbf{I}_T\right]}, \end{aligned} \tag{1.180}$$

$$\mathbf{Y}_{(2)} \cong \mathbf{H}_{(2)} \left(\mathbf{I}_Q \otimes \mathbf{A}\right)^T + \mathbf{B}\,\mathbf{L}_{(2)} + \mathbf{F}_{(2)}\,(\mathbf{C} \otimes \mathbf{I}_I)^T$$

$$= \boxed{\left[\mathbf{H}_{(2)}\ \mathbf{B}\ \mathbf{F}_{(2)}\right]\left[\mathbf{I}_Q \otimes \mathbf{A}^T\ \mathbf{L}_{(2)}\ \mathbf{C}^T \otimes \mathbf{I}_I\right]}, \tag{1.181}$$

$$\mathbf{Y}_{(3)} \cong \mathbf{H}_{(3)}\,(\mathbf{I}_T \otimes \mathbf{A})^T + \mathbf{L}_{(3)}\,(\mathbf{B} \otimes \mathbf{I}_I)^T + \mathbf{C}\,\mathbf{F}_{(3)}$$

$$= \boxed{\left[\mathbf{H}_{(3)}\ \mathbf{L}_{(3)}\ \mathbf{C}\right]\left[\mathbf{I}_T \otimes \mathbf{A}^T\ \mathbf{B}^T \otimes \mathbf{I}_I\ \mathbf{F}_{(3)}\right]}. \tag{1.182}$$

These matrix representations allow us to compute core tensors and factor matrices via matrix factorization.

Similar, but more sophisticated BOD models can be defined based on a restricted Tucker3 model [129] and also PARATUCK2 or DEDICOM models (see the next section).

1.5.10 PARATUCK2 and DEDICOM Models

The PARATUCK2, developed by Harshman and Lundy [65] is a generalization of the PARAFAC model, that adds some of the flexibility of Tucker2 model while retaining some of PARAFAC's uniqueness properties. The name PARATUCK2 indicates its similarity to both the PARAFAC and the Tucker2 model. The PARATUCK2 model performs decomposition of an arbitrary third-order tensor (see Figure 1.38(a)) $\underline{\mathbf{Y}} \in \mathbb{R}^{I\times T\times Q}$ as follows

$$\boxed{\mathbf{Y}_q = \mathbf{A}\,\mathbf{D}_q^{(A)}\,\mathbf{R}\,\mathbf{D}_q^{(B)}\,\mathbf{B}^T + \mathbf{E}, \qquad (q = 1, 2, \ldots, Q),} \tag{1.183}$$

where $\mathbf{A} \in \mathbb{R}^{I\times J}$, $\mathbf{B} \in \mathbb{R}^{T\times P}$, $\mathbf{R} \in \mathbb{R}^{J\times P}$, $\mathbf{D}_q^{(A)} \in \mathbb{R}^{J\times J}$ and $\mathbf{D}_q^{(B)} \in \mathbb{R}^{P\times P}$ are diagonal matrices representing the q-th frontal slices of the tensors $\underline{\mathbf{D}}^{(A)} \in \mathbb{R}^{J\times J\times Q}$, and $\underline{\mathbf{D}}^{(B)} \in \mathbb{R}^{P\times P\times Q}$, respectively. In fact, tensor $\underline{\mathbf{D}}^{(A)}$ is formed by a matrix $\mathbf{U} \in \mathbb{R}^{J\times Q}$ whose columns are diagonals of the corresponding frontal slices and tensor $\underline{\mathbf{D}}^{(B)}$ is constructed from a matrix $\mathbf{V} \in \mathbb{R}^{P\times Q}$.

$$\mathbf{D}_q^{(A)} = \operatorname{diag}(\boldsymbol{u}_q), \quad \mathbf{D}_q^{(B)} = \operatorname{diag}(\boldsymbol{v}_q). \tag{1.184}$$

The j-th row $\underline{\boldsymbol{u}}_j$ (or $\underline{\boldsymbol{v}}_j$) gives the weights of participation for the corresponding component $\boldsymbol{a}_j$ in the factor $\mathbf{A}$ (or $\boldsymbol{b}_j$ in $\mathbf{B}$) with respect to the third dimension. The terms $\mathbf{D}_q^{(A)}\mathbf{R}\mathbf{D}_q^{(B)\,T}$ correspond to frontal slices of the core tensor $\underline{\mathbf{G}}$ of a Tucker2 model, but due to the restricted structure of the core tensor compared to Tucker2 uniqueness is retained. The core tensor $\underline{\mathbf{G}}$ can be described as

$$\begin{aligned}
\operatorname{vec}\left(\mathbf{G}_q\right) &= \operatorname{vec}\left(\mathbf{D}_q^{(A)}\mathbf{R}\mathbf{D}_q^{(B)}\right) \qquad\qquad = \operatorname{vec}\left(\operatorname{diag}(\boldsymbol{u}_q)\mathbf{R}\operatorname{diag}(\boldsymbol{v}_q)^T\right)\\
&= \left(\operatorname{diag}(\boldsymbol{v}_q) \otimes \operatorname{diag}(\boldsymbol{u}_q)\right)\ \operatorname{vec}(\mathbf{R}) = \operatorname{diag}\left(\boldsymbol{v}_q \otimes \boldsymbol{u}_q\right)\ \operatorname{vec}(\mathbf{R})\\
&= \left(\boldsymbol{v}_q \otimes \boldsymbol{u}_q\right) \circledast\ \operatorname{vec}(\mathbf{R}),
\end{aligned} \tag{1.185}$$

or simply via frontal slices

$$\mathbf{G}_q = \left(\boldsymbol{u}_q\,\boldsymbol{v}_q^T\right) \circledast\ \mathbf{R}, \qquad (q = 1, 2, \ldots, Q). \tag{1.186}$$

This leads to the mode-3 matricization of the core tensor $\underline{\mathbf{G}}$ having following form

$$\begin{aligned}
\mathbf{G}_{(3)} &= [\operatorname{vec}(\mathbf{G}_1), \ldots, \operatorname{vec}(\mathbf{G}_R)]^T\\
&= [(\boldsymbol{v}_1 \otimes \boldsymbol{u}_1) \circledast\ \operatorname{vec}(\mathbf{R}), \ldots, (\boldsymbol{v}_R \otimes \boldsymbol{u}_R) \circledast\ \operatorname{vec}(\mathbf{R})]^T\\
&= [\boldsymbol{v}_1 \otimes \boldsymbol{u}_1, \ldots, \boldsymbol{v}_R \otimes \boldsymbol{u}_R]^T \circledast\ [\operatorname{vec}(\mathbf{R}), \ldots, \operatorname{vec}(\mathbf{R})]^T\\
&= (\mathbf{V} \odot \mathbf{U})^T \circledast\ \mathbf{T}_{(3)}\\
&= \mathbf{Z}_{(3)} \circledast\ \mathbf{T}_{(3)}
\end{aligned} \tag{1.187}$$

or

$$\underline{\mathbf{G}} = \underline{\mathbf{Z}} \circledast\ \underline{\mathbf{T}}, \tag{1.188}$$

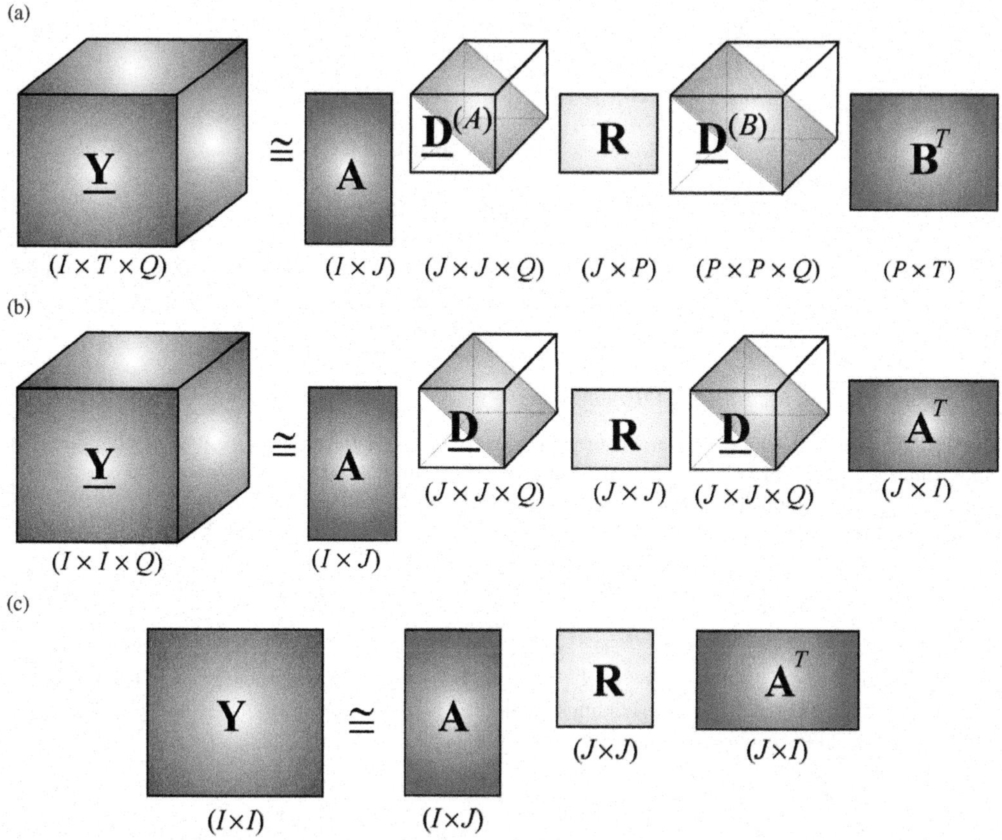

Figure 1.38 (a) PARATUCK2 model performing decomposition of tensor $\underline{\mathbf{Y}} \in \mathbb{R}^{I\times T\times Q}$. (b) DEDICOM model for a symmetric third-order tensor $\underline{\mathbf{Y}} \in \mathbb{R}^{I\times I\times Q}$, (c) DEDICOM model for a symmetric matrix.

where $\underline{\mathbf{Z}} \in \mathbb{R}^{J\times P\times Q}$ is a rank-Q PARAFAC tensor represented by two factors $\mathbf{U}$ and $\mathbf{V}$ (the third factor for the mode-3 is identity matrix $\mathbf{I}_Q$), that is

$$\underline{\mathbf{Z}} = \underline{\mathbf{I}} \times_1 \mathbf{U} \times_2 \mathbf{V}, \tag{1.189}$$

and $\underline{\mathbf{T}} \in \mathbb{R}^{J\times P\times Q}$ is a tensor with identical frontal slices expressed by the matrix $\mathbf{R}$: $\mathbf{T}_q = \mathbf{R}, \;\; \forall q$. Eq. (1.188) indicates that the core tensor $\underline{\mathbf{G}}$ is the Hadamard product of the PARAFAC tensor and a special (constrained) tensor with identity frontal slices. In other words the PARATUCK2 can be considered as the Tucker2 model in which the core tensor has special PARAFAC decomposition as illustrated in Figure 1.38(a). The PARATUCK2 model is well suited for a certain class of multi-way problems that involve interactions between factors.

Figure 1.38(b) illustrates a special form of PARATUCK2 called the three-way DEDICOM (inxDEcomposition into DIrectional COMponents) model [9,10,85]. In this case for a given symmetric third-order data tensor $\underline{\mathbf{Y}} \in \mathbb{R}^{I\times I\times Q}$ with frontal slices $\mathbf{Y}_q = \mathbf{Y}_q^T$ the simplified decomposition is:

$$\boxed{\mathbf{Y}_q = \mathbf{A}\,\mathbf{D}_q\,\mathbf{R}\,\mathbf{D}_q\,\mathbf{A}^T + \mathbf{E}, \qquad (q = 1, 2, \ldots, Q),} \tag{1.190}$$

where $\mathbf{A} \in \mathbb{R}^{I\times J}$ is a matrix of loadings, $\mathbf{D}_q$ is a diagonal matrix representing the q-th frontal slice of the tensor $\underline{\mathbf{D}} \in \mathbb{R}^{J\times J\times Q}$, and $\mathbf{R} \in \mathbb{R}^{J\times J}$ is an asymmetric matrix.

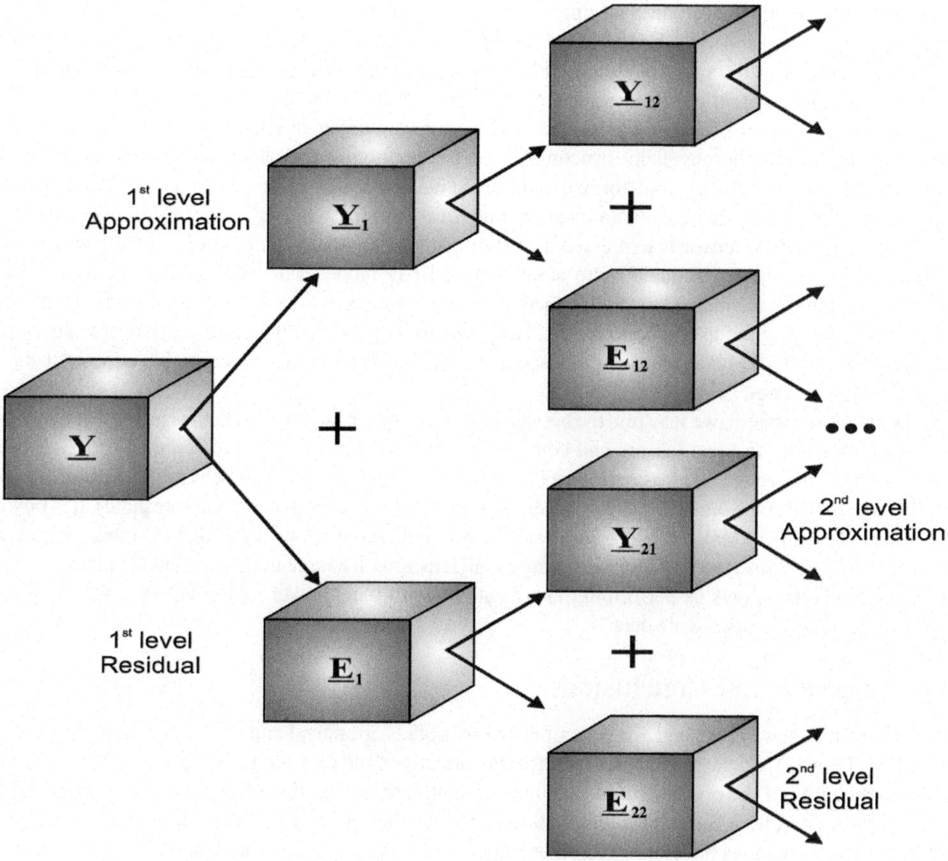

Figure 1.39 Graphical illustration of hierarchical tensor decomposition.

The three-way DEDICOM model can be considered as natural extension of the 2-way DEDICOM model [132]:

$$\boxed{\mathbf{Y} = \mathbf{A}\mathbf{R}\mathbf{A}^T + \mathbf{E},} \tag{1.191}$$

where $\mathbf{Y} \in \mathbb{R}^{I\times I}$ is a given symmetric data matrix, and $\mathbf{E}$ is a matrix representing error not explained by the model. The goal is to estimate the best-fitting matrices: $\mathbf{A} \in \mathbb{R}^{I\times J}$ and $\mathbf{R} \in \mathbb{R}^{J\times J}$. To achieve this goal we usually perform the following optimization problem:

$$\min_{\mathbf{A},\mathbf{R}} \; ||\mathbf{Y} - \mathbf{A}\mathbf{R}\mathbf{A}^T||_F^2. \tag{1.192}$$

The matrix $\mathbf{A} \in \mathbb{R}_+^{I\times J}$ comprises loadings or weights (with $J < I$), and the square matrix $\mathbf{R}$ is a matrix that represents the asymmetric relationships for the latent dimensions of $\mathbf{A}$.

This uniqueness of the three-way DEDICOM gives plausibility to the factors making them a valid description with a high confidence that they can explain more variance than convenient rotated 2-way solutions [85].

1.5.11 Hierarchical Tensor Decomposition

Recently, multi-linear models based on tensor approximation have received much attention as tools for denoising, reduction or compression as they have the potential to produce more compact representations of multi-dimensional data than traditional dimensionality reduction methods. We will exploit the aforementioned characteristics of visual 3D data and develop an analysis and a representation technique based on a hierarchical tensor-based transformation. In this technique, a multi-dimensional dataset is transformed into a hierarchy of signals to reflect multi-scale structures present in the multi-way data (see Figure 1.39). The signal at each level of the hierarchy is further divided into a number of tensors with smaller spatial support to expose spatial inhomogeneity structures. To achieve a highly compact representation these smaller dimensional tensors are further transformed and pruned using a tensor approximation technique [143].

It is interesting to note that the hierarchical scheme is very similar to the BCD model discussed in section (1.5.8). A source (data) tensor is expressed as a sum of multiple tensor decomposition models. However, the hierarchical technique is much more simpler than BCD model. For BCD model, all factors in all subtensors are simultaneously estimated, hence, constraints imposed on these factors such as nonnegative can be assured during the estimation process. However, BCD increases the complexity of the algorithms, especially for very large-scale data set. For a specified data, we can choose an acceptable trade-off between simplicity and accuracy.

In the simplest scenario, we may find the best approximation of the given input data tensor $\underline{\mathbf{Y}}$ by PARAFAC, that is, by a sum of rank-one tensors and compute the residual $\underline{\mathbf{E}} = \underline{\mathbf{Y}} - \hat{\underline{\mathbf{Y}}}$. We can then estimate a rank of tensor which is a rough approximation of the residual, and the process can be repeated until a stopping criterion is satisfied (for example, the residual entries reach values below some small threshold). It is obvious that a sum of two PARAFAC models with rank-J_1 and rank-J_2 gives us a PARAFAC model with rank-$(J_1 + J_2)$. Therefore, instead of factorizing a high rank tensor with a large number of latent components, the hierarchical scheme allows us performing factorizations with much lower ranks. This is a very important feature, especially for large-scale data.

1.6 Discussion and Conclusions

In this chapter we have presented a variety of different models, graphical and mathematical representations for NMF, NTF, NTD and the related matrix/tensor factorizations and decompositions. Our emphasis has been on the formulation of the problems and establishing relationships and links among different models. Each model usually provides a different interpretation of the data and may have different applications. Various equivalent representations have been presented which will serve as a basis for the development of learning algorithms throughout this book.

It has been highlighted that constrained models with nonnegativity and sparsity constraints for real-world data cannot provide a perfect fit to the observed data (i.e., they do not explain as much variance in the input data and may have larger residual errors) as compared to unconstrained factorization and decomposition models. They, however, often produce more meaningful physical interpretations. Although nonnegative factorizations/decompositions already exhibit some degree of sparsity, the combination of both constraints enables a precise control of sparsity.

Appendix 1.A: Uniqueness Conditions for Three-way Tensor Factorizations

The most attractive feature of the PARAFAC model is its uniqueness property. Kruskal [90] has proved that, for fixed error tensor $\mathbf{E}$, the vectors $\boldsymbol{a}_j$, $\boldsymbol{b}_j$, and $\boldsymbol{c}_j$ of component matrices $\mathbf{A}$, $\mathbf{B}$ and $\mathbf{C}$ are unique up to unavoidable scaling and permutation of columns,[20] provided that

$$k_A + k_B + k_C \geq J + 2, \tag{A.1}$$

[20] If a PARAFAC solution is unique up to these indeterminacies, it is called essentially unique. Two PARAFAC solutions that are identical up to the essential uniqueness indeterminacies will be called equivalent.

where k_A, k_B, k_C denote the k-ranks of the component matrices. The k-rank of a matrix is the largest number k such that every subset of k columns of the matrix is linearly independent [131].

Kruskal's uniqueness condition was generalized to N-th order tensors with $N > 3$ by Sidropoulos and Bro [125]. A more accessible proof of the uniqueness condition (A.1) for the PARAFAC model was given by Stegeman and Sidiropoulos [130]. For the case where one of the component matrices **A**, **B** and **C** has full column rank, weaker uniqueness conditions than (A.1) have been derived by Jiang and Sidiropoulos, De Lathauwer, and Stegeman (e.g., see [128,131]). For example, if a component matrix $\mathbf{C} \in \mathbb{R}^{Q \times J}$ is full column rank, and $\mathbf{A} \in \mathbb{R}^{I \times J}$ and $\mathbf{B} \in \mathbb{R}^{T \times J}$ have k-rank at least 2, then Kruskal's condition $k_A + k_B \geq J + 2$ implies uniqueness of a PARAFAC solution [128].

It should be noted that the above conditions are only valid for the unrestricted PARAFAC model. If we impose additional constraints such as nonnegativity, sparsity, orthogonality the conditions for uniqueness can be relaxed[21] and they can be different [21,129]. For example, the NTF, NTF1, NTF2 and NTD models are unique (i.e., without rotational ambiguity) if component matrices and/or core tensor are sufficiently sparse.

Appendix 1.B: Singular Value Decomposition (SVD) and Principal Component Analysis (PCA) with Sparsity and/or Nonnegativity Constraints

SVD and PCA are widely used tools, for example, in medical image analysis for dimension reduction, model building, and data understanding and exploration. They have applications in virtually all areas of science, machine learning, image processing, engineering, genetics, neurocomputing, chemistry, meteorology, computer networks, to name just a few, where large data sets are encountered. If $\mathbf{Y} \in \mathbb{R}^{I \times T}$ is a data matrix encoding T samples of I variables, with I being large, PCA aims at finding a few linear combinations of these variables, called the principal components, which point in orthogonal directions explaining as much of the variance in the data as possible. The purpose of principal component analysis PCA is to derive a relatively small number of uncorrelated linear combinations (principal components) of a set of random zero-mean variables while retaining as much of the information from the original variables as possible. Among the objectives of Principal Components Analysis are the following.

1. Dimensionality reduction.
2. Determination of linear combinations of variables.
3. Feature selection: the choosing of the most useful variables.
4. Visualization of multi-dimensional data.
5. Identification of underlying variables.
6. Identification of groups of objects or of outliers.

The success of PCA/SVD is due to two main optimal properties: Principal components sequentially capture the maximum variability of **Y** thus guaranteeing minimal information loss, and they are mutually uncorrelated. Despite the power and popularity of PCA, one key drawback is its lack of sparseness (i.e., factor loadings are linear combinations of all the input variables), yet sparse representations are generally desirable since they aid human understanding (e.g., with gene expression data), reduce computational costs and promote better generalization in learning algorithms. In other words, the standard principal components (PCs) can sometimes be difficult to interpret, because they are linear combinations of all the original variables. To facilitate better interpretation, sparse and/or nonnegative PCA estimate modified PCs with sparse and/or

[21] Moreover, by imposing the nonnegativity or orthogonality constraints the PARAFAC has an optimal solution, i.e., there is no risk for degenerate solutions [99]. Imposing nonnegativity constraints makes degenerative solutions impossible since no factor can counteract the effect of another factor and usually improves convergence since the search space is greatly reduced.

nonnegative eigenvectors, i.e. loadings with very few nonzero and possibly nonnegative entries. We use the connection of PCA with SVD of the data matrix and extract the PCs through solving a low rank matrix approximation problem. Regularization penalties are usually incorporated to the corresponding minimization problem to enforce sparsity and/or nonnegativity in PC loadings [124,144].

Standard PCA is essentially the same technique as SVD but usually obtained using slightly different assumptions. Usually, in PCA we use normalized data with each variable centered and possibly normalized by the standard deviation.

1.B.1 Standard SVD and PCA

At first let us consider basic properties of standard SVD and PCA. The SVD of a data matrix $\mathbf{Y} \in \mathbb{R}^{I\times T}$ assuming without loss of generality that $T > I$ leads to the following matrix factorization

$$\mathbf{Y} = \mathbf{U}\boldsymbol{\Sigma}\mathbf{V}^T = \sum_{j=1}^{J} \sigma_j \, \boldsymbol{u}_j \, \boldsymbol{v}_j^T, \tag{B.1}$$

where the matrix $\mathbf{U} = [\boldsymbol{u}_1, \boldsymbol{u}_2, \ldots, \boldsymbol{u}_I] \in \mathbb{R}^{I\times I}$ contains the I left singular vectors, $\boldsymbol{\Sigma} \in \mathbb{R}_+^{I\times T}$ with nonnegative elements on the main diagonal representing the singular values σ_j and the matrix $\mathbf{V} = [\boldsymbol{v}_1, \boldsymbol{v}_2, \ldots, \boldsymbol{v}_T] \in \mathbb{R}^{T\times T}$ represents the T right singular vectors called the loading factors. The nonnegative quantities σ_j, sorted as $\sigma_1 \geq \sigma_2 \geq \cdots \geq \sigma_J > \sigma_{J+1} = \sigma_{J+2} = \cdots = \sigma_I = 0$ can be shown to be the square roots of the eigenvalues of the data covariance matrix $(1/T)\mathbf{Y}\mathbf{Y}^T \in \mathbb{R}^{I\times I}$. The term $\boldsymbol{u}_j\boldsymbol{v}_j^T$ is an $I \times T$ rank-one matrix called often the j-th eigenimage of $\mathbf{Y}$. Orthogonality of the SVD expansion ensures that the left and right singular vectors are orthogonal, i.e., $\boldsymbol{u}_i^T\boldsymbol{u}_j = \delta_{ij}$ and $\boldsymbol{v}_i^T\boldsymbol{v}_j = \delta_{ij}$,with δ_{ij} the Kronecker function (or equivalently $\mathbf{U}^T\mathbf{U} = \mathbf{I}$ and $\mathbf{V}^T\mathbf{V} = \mathbf{I}$). In many applications, it is most practical to work with the truncated form of the SVD where only the first $P < J$, (where J is a rank of $\mathbf{Y}$ with $J < I$) singular values are used so that

$$\mathbf{Y} \cong \mathbf{U}_P \, \boldsymbol{\Sigma}_P \, \mathbf{V}_P^T = \sum_{j=1}^{P} \sigma_j \, \boldsymbol{u}_j \, \boldsymbol{v}_j^T, \tag{B.2}$$

where $\mathbf{U}_P = [\boldsymbol{u}_1, \boldsymbol{u}_2, \ldots, \boldsymbol{u}_P] \in \mathbb{R}^{I\times P}$, $\boldsymbol{\Sigma}_P = \operatorname{diag}\{\sigma_1, \sigma_2, \ldots, \sigma_P\}$ and $\mathbf{V} = [\boldsymbol{v}_1, \boldsymbol{v}_2, \ldots, \boldsymbol{v}_P] \in \mathbb{R}^{T\times P}$. This is no longer an exact decomposition of the data matrix $\mathbf{Y}$, but according to the Eckart-Young theorem it is the best rank-P approximation in the least-squares sense and it is still unique (neglecting signs of vectors ambiguity) if the singular values are distinct.

Approximation of the matrix $\mathbf{Y}$ by a rank-one matrix $\sigma\boldsymbol{u}\boldsymbol{v}^T$ of two unknown vectors $\boldsymbol{u} = [u_1, u_2, \ldots, u_I]^T \in \mathbb{R}^I$ and $\boldsymbol{v} = [v_1, v_2, \ldots, v_T]^T \in \mathbb{R}^T$ normalized to unit length with a scaling constant term σ can be presented as follows:

$$\mathbf{Y} = \sigma \, \boldsymbol{u} \, \boldsymbol{v}^T + \mathbf{E}, \tag{B.3}$$

where $\mathbf{E} \in \mathbb{R}^{I\times T}$ is a matrix of the residual errors e_{it}. In order to compute the unknown vectors we minimize the squared Euclidean error as [100]

$$J_1 = ||\mathbf{E}||_F^2 = \sum_{it} e_{it}^2 = \sum_{i=1}^{I}\sum_{t=1}^{T} (y_{it} - \sigma \, u_i \, v_t)^2. \tag{B.4}$$

The necessary conditions for minimization of (B.4) are obtained by equating gradients to zero:

$$\frac{\partial J_1}{\partial u_i} = -2\sigma \sum_{t=1}^{T} (y_{it} - \sigma \, u_i \, v_t) v_t \; = \; 0, \tag{B.5}$$

$$\frac{\partial J_1}{\partial v_t} = -2\sigma \sum_{i=1}^{I} (y_{it} - \sigma \, u_i \, v_t) u_i \; = \; 0, \tag{B.6}$$

These equations can be expressed as follows:

$$\sum_{t=1}^{T} y_{it}\, v_t = \sigma\, u_i \sum_{t=1}^{T} v_t^2, \qquad \sum_{i=1}^{I} y_{it}\, u_i = \sigma\, v_t \sum_{i=1}^{I} u_i^2. \tag{B.7}$$

Taking into account that the vectors are normalized to unit length, that is, $\boldsymbol{u}^T\boldsymbol{u} = \sum_{i=1}^{I} u_i^2 = 1$ and $\boldsymbol{v}^T\boldsymbol{v} = \sum_{t=1}^{T} v_t^2 = 1$, we can write the above equations in a compact matrix form as

$$\mathbf{Y}\,\boldsymbol{v} = \sigma\,\boldsymbol{u}, \qquad \mathbf{Y}^T\,\boldsymbol{u} = \sigma\,\boldsymbol{v} \tag{B.8}$$

or equivalently (by substituting one of Eqs. (B.8) into another)

$$\mathbf{Y}^T\mathbf{Y}\,\boldsymbol{v} = \sigma^2\boldsymbol{v}, \qquad \mathbf{Y}\mathbf{Y}^T\boldsymbol{u} = \sigma^2\boldsymbol{u}, \tag{B.9}$$

which are classical eigenvalue problems which estimate the maximum eigenvalue $\lambda_1 = \sigma_1^2 = \sigma_{max}^2$ with the corresponding eigen vectors $\boldsymbol{u}_1 = \boldsymbol{u}$ and $\boldsymbol{v}_1 = \boldsymbol{v}$. The solutions of these problems give the best first rank-one approximation of Eq. (B.1).

One of the most important results for nonnegative matrices is the following [75]:

Theorem B.1 (Perron-Frobenius) *For a square nonnegative matrix* $\mathbf{Y}$ *there exists a largest modulus eigenvalue of* $\mathbf{Y}$ *which is nonnegative and a corresponding nonnegative eigenvector.*

The eigenvector satisfying the Perron-Frobenius theorem is usually referred to as the Perron vector of a nonnegative matrix. For a rectangular nonnegative matrix, a similar result can be established for the largest singular value and its corresponding singular vector:

Theorem B.2 *The leading singular vectors:* $\boldsymbol{u}_1$, *and* $\boldsymbol{v}_1$ *corresponding to the largest singular value* $\sigma_{max} = \sigma_1$ *of a nonnegative matrix* $\mathbf{Y} = \mathbf{U}\boldsymbol{\Sigma}\mathbf{V}^T = \sum_i \sigma_i \boldsymbol{u}_i \boldsymbol{v}_i^T$ *(with* $\sigma_1 \geq \sigma_2 \geq \cdots \geq \sigma_I$*) are nonnegative.*

Based on this observation, it is straightforward to compute the best rank-one NMF approximation $\sigma_1\boldsymbol{u}_1\boldsymbol{v}_1^T$, this idea can be extended to approximate a higher-order NMF. If we compute the rank-one NMF and subtract it from the original matrix $\hat{\mathbf{Y}}_1 = \mathbf{Y} - \sigma_1\boldsymbol{u}_1\boldsymbol{v}_1^T$, the input data matrix will no longer be nonnegative, however, all negative elements can be forced to be zero or are positive and the procedure can be repeated [12]. In order to estimate the next singular values and the corresponding singular vectors, we may apply a deflation approach, that is,

$$\mathbf{Y}_j = \mathbf{Y}_{j-1} - \sigma_j\boldsymbol{u}_j\boldsymbol{v}_j^T, \qquad (j = 1, 2, \ldots, J), \tag{B.10}$$

where $\mathbf{Y}_0 = \mathbf{Y}$. Solving the same optimization problem (B.4) for the residual matrix $\mathbf{Y}_j$ yields the set of consecutive singular values and corresponding singular vectors. Repeating reduction of the matrix yields the next set of the solution until the deflation matrix $\mathbf{Y}_{J+1}$ becomes the zero.

Using the property of the orthogonality of the eigenvectors, and the equality $\boldsymbol{u}^T\mathbf{Y}\boldsymbol{v} = \boldsymbol{v}^T\mathbf{Y}\boldsymbol{u} = \sigma$ we can estimate the precision of the matrix approximation with the first $P \leq J$ pairs of singular vectors [100]:

$$\begin{aligned} ||\mathbf{Y} - \sum_{j=1}^{P} \sigma_j\boldsymbol{u}_j\boldsymbol{v}_j^T||_F^2 &= \sum_{i=1}^{I}\sum_{t=1}^{T}(y_{it} - \sum_{j=1}^{J}\sigma_j u_{ij} v_{jt})^2 \\ &= ||\mathbf{Y}||_F^2 - \sum_{j=1}^{P}\sigma_j^2, \end{aligned} \tag{B.11}$$

and the residual error reduces exactly to zero with the number of singular values equal to the matrix rank, that is, for $P = J$. Thus, we can write for the rank-J matrix:

$$||\mathbf{Y}||_F^2 = \sum_{j=1}^{J}\sigma_j^2. \tag{B.12}$$

It is interesting to note that (taking into account that $\sigma u_i = \sum_{t=1}^{T} y_{it} v_t / \sum_{t=1}^{T} v_t^2$ (see Eqs. (B.7)) the cost function (B.4) can be expressed as [100]

$$\begin{aligned} J_1 = ||\mathbf{E}||_F^2 &= \sum_{i=1}^{I}\sum_{t=1}^{T}(y_{it} - \sigma\, u_i\, v_t)^2 \\ &= \sum_{i=1}^{I}\sum_{t=1}^{T} y_{it}^2 - 2\sum_{i=1}^{I}(\sigma\, u_i)\sum_{t=1}^{T}(y_{it}v_t) + \sum_{i=1}^{I}(\sigma\, u_i)^2\sum_{t=1}^{T} v_t^2 \\ &= \sum_{i=1}^{I}\sum_{t=1}^{T} y_{it}^2 - \frac{\sum_{i=1}^{I}(\sum_{t=1}^{T} y_{it}v_t)^2}{\sum_{t=1}^{T} v_t^2}. \end{aligned} \tag{B.13}$$

In matrix notation the cost function can be written as

$$\|\mathbf{E}\|_F^2 = \|\mathbf{Y}\|_F^2 - \frac{\boldsymbol{v}^T\mathbf{Y}^T\mathbf{Y}\boldsymbol{v}}{\|\boldsymbol{v}\|_2^2} = \|\mathbf{Y}\|_F^2 - \sigma^2, \tag{B.14}$$

where the second term is called the Rayleigh quotient. The maximum value of the Rayleigh quotient is exactly equal to the maximum eigenvalue $\lambda_1 = \sigma_1^2$.

1.B.2 Sparse PCA

For sparse PCA we may employ many alternative approaches [124]. One of the simplest and most efficient approaches is to apply minimization of the following cost function [124]:

$$J_\rho(\tilde{\boldsymbol{v}}) = \|\mathbf{Y} - \boldsymbol{u}\,\tilde{\boldsymbol{v}}^T\|_F^2 + \rho(\tilde{\boldsymbol{v}}), \tag{B.15}$$

where $\tilde{\boldsymbol{v}} = \sigma\boldsymbol{v}$, $\rho(\tilde{\boldsymbol{v}})$ is the additional penalty term which imposes sparsity. Typically, $\rho(|\tilde{\boldsymbol{v}}|) = 2\lambda||\tilde{\boldsymbol{v}}||_1$ or $\rho(\tilde{\boldsymbol{v}}) = \lambda^2||\tilde{\boldsymbol{v}}||_0$, where λ is the nonnegative coefficient that controls the degree of sparsity.[22] The cost function can be evaluated in scalar form as follows

$$\begin{aligned} J_\rho(\tilde{\boldsymbol{v}}) &= \sum_i\sum_t(y_{it} - u_i\,\tilde{v}_t)^2 + \sum_t \rho(\tilde{v}_t) = \sum_t\left(\sum_i(y_{it} - u_i\,\tilde{v}_t)^2 + \rho(v_t)\right) \\ &= \sum_t\left(\sum_i y_{it}^2 - 2\,[\mathbf{Y}^T\boldsymbol{u}]_t\,\tilde{v}_t + \tilde{v}_t^2 + \rho(\tilde{v}_t)\right), \qquad (t = 1, 2, \ldots, T). \end{aligned} \tag{B.16}$$

It is not difficult to see (see Chapter 4) that the optimal value of $\tilde{v}_t$ depends on the penalty term $\rho(\tilde{v}_t)$, in particular for $\rho(\tilde{v}_t) = 2\lambda||\tilde{\boldsymbol{v}}||_1$ we use soft shrinkage with threshold λ

$$\tilde{v}_t = P_\lambda^{(s)}(x) = \text{sign}(x)[|x| - \lambda]_+ \tag{B.17}$$

and for $\rho(\tilde{v}_t) = \lambda^2||\tilde{\boldsymbol{v}}||_0$ we use hard shrinkage projection

$$\tilde{v}_t = P_\lambda^{(h)}(x) = I(x > \lambda)\,x = \begin{cases} x, & \text{for } |x| \geq \lambda; \\ 0, & \text{otherwise}, \end{cases} \tag{B.18}$$

where $x = [\mathbf{Y}^T\boldsymbol{u}]_t$. This leads to the following iterative algorithm proposed by Shen and Huang [124]

1. Initialize: Apply the regular SVD to data matrix $\mathbf{X}$ and estimate the maximum singular value $\sigma_1 = \sigma_{\max}$ and corresponding singular vectors $\boldsymbol{u} = \boldsymbol{u}_1$ and $\boldsymbol{v} = \boldsymbol{v}_1$, take $\tilde{\boldsymbol{v}}_1 = \sigma_1\boldsymbol{v}_1$. This corresponds to the best rank-one approximation of $\mathbf{Y} = \mathbf{Y}_0$,

[22] We define the degree of sparsity of a PC as the number of zero elements in the corresponding loading vector $\boldsymbol{v}$.

2. Update:

$$\tilde{\boldsymbol{v}}_1 \leftarrow P_\lambda\left(\mathbf{Y}^T \boldsymbol{u}_1\right), \tag{B.19}$$

$$\tilde{\boldsymbol{u}}_1 \leftarrow (\mathbf{Y}\tilde{\boldsymbol{v}}_1)/||\mathbf{Y}\tilde{\boldsymbol{v}}_1||_2, \tag{B.20}$$

3. Repeat Step 2 until convergence,
4. Normalize the vector $\tilde{\boldsymbol{v}}_1$ as $\tilde{\boldsymbol{v}}_1 = \tilde{\boldsymbol{v}}_1/||\tilde{\boldsymbol{v}}_1||_2$.

Note that for $\lambda = 0$ the nonlinear shrinkage function P_λ becomes a linear function and the above procedure simplifies to the well-known standard alternating least squares SVD algorithm. The subsequent pair $\{\boldsymbol{u}_2, \sigma_2 \boldsymbol{v}_2\}$ provides the best rank-one approximation of the corresponding residual matrix $\mathbf{Y}_1 = \mathbf{Y} - \sigma_1 \boldsymbol{u}_1 \boldsymbol{v}_1$. In other words, subsequent sparse loading vectors $\boldsymbol{v}_j$ can be obtained sequentially via a deflation approach and a rank-one approximation of the residual data matrices $\mathbf{Y}_j$.

1.B.3 Nonnegative PCA

In some applications it is necessary to incorporate both nonnegativity and/or sparseness constraints into PCA maintaining the maximal variance property of PCA and relaxing orthogonality constraints. The algorithm described in the previous section can be applied almost directly to nonnegative PCA by applying a suitable nonnegative shrinkage function. However, in such a case, it should be noted that the orthogonality among vectors $\boldsymbol{v}_j$ is completely lost. If we need to control the orthogonality constraint (to some extent) together with nonnegativity constraints, we may alternatively apply the following optimization problem [144]

$$\max_{\mathbf{V} \geq \mathbf{0}} \frac{1}{2} ||\mathbf{V}_J^T \mathbf{Y}||_F^2 - \alpha_o ||\mathbf{I} - \mathbf{V}_J^T \mathbf{V}_J||_F^2 - \alpha_s ||\mathbf{V}_J||_1, \tag{B.21}$$

subject to nonnegativity constraints $\mathbf{V}_J \geq 0$, where $||\mathbf{V}_J||_1 = \mathbf{1}^T \mathbf{V}_J \mathbf{1}$. The nonnegative coefficients α_o and α_s control a level and tradeoff between sparsity and orthogonality.

Alternatively, we can use nonnegative loading parametrization, for example, exponential parametrization [100]

$$v_t = \exp(\gamma_t), \qquad \forall t, \tag{B.22}$$

where γ_t are the estimated parameters. To obtain the loading in the range from zero to one we can use multinomial parametrization

$$v_t = \frac{\exp(\gamma_t)}{\sum_t \exp(\gamma_t)}, \qquad (t = 1, 2, \ldots, T). \tag{B.23}$$

Appendix 1.C: Determining a True Number of Components

Determining the number of components J for the NMF/NTF models, or more generally determining the dimensions of a core tensor, J, R, P, for the Tucker models is very important since the approximately valid model is instrumental in discovering or capturing the underlying structure in the data.

There are several approximative and heuristic techniques for determining the number of components [20,40,42,43,71,109,133]. In an ideal noiseless case when the PARAFAC model is perfectly satisfied, we can apply a specific procedure for calculating the PARAFAC components for $J = 2, 3, \ldots$ until we reach the number of components for which the errors $\mathbf{E} = \mathbf{Y} - \hat{\mathbf{Y}}$ are zero. However, in practice, it is not possible to perfectly satisfy this model. Other proposed methods include: residual analysis, visual appearance of loadings, the number of iterations of the algorithm and core consistency. In this book, we mostly rely on the PCA approach and the core consistency diagnostic developed by Bro and Kiers [20] for finding the number of components and selecting an appropriate (PARAFAC or Tucker) model. The core consistency quantifies the resemblance between the Tucker3 core tensor and the PARAFAC core, which is a super-identity or a superdiagonal core tensor, or in other words, a vector of coefficients. This diagnostic tool suggests whether the PARAFAC model with the specified number of components is a valid model for the data. The core consistency above 90% is often used as an indicator of the trilinear structure of the data, and suggests that the PARAFAC model would be an appropriate model for the data. A core consistency value close to or lower than 50%, on the other hand, would indicate that the PARAFAC-like model is not appropriate. This diagnostic method has been commonly applied in the neuroscience-multi-way literature [57,104], often together with other diagnostic tools, in order to determine the number of components.

An efficient way is to use the PCA/SVD approach, whereby for a three-way data set we first unfold the tensor as matrices $\mathbf{Y}_{(1)}$ and eventually compute covariance matrix $\mathbf{R}_y = (1/T)\mathbf{Y}_{(1)}\mathbf{Y}_{(1)}^T$.

Under the assumption that the power of the signals is larger than the power of the noise, the PCA enables us to divide observed (measured) sensor signals: $\boldsymbol{x}(t) = \boldsymbol{x}_s(t) + \boldsymbol{\nu}(t)$ into two subspaces: the *signal subspace* corresponding to principal components associated with the largest eigenvalues called the principal eigenvalues: $\lambda_1, \lambda_2, ..., \lambda_J$, $(I > J)$ and associated eigenvectors $\mathbf{V}_J = [\boldsymbol{v}_1, \boldsymbol{v}_2, \ldots, \boldsymbol{v}_J]$ called the principal eigenvectors and the *noise subspace* corresponding to the minor components associated with the eigenvalues $\lambda_{J+1}, ..., \lambda_I$. The subspace spanned by the J first eigenvectors $\boldsymbol{v}_i$ can be considered as an approximation of the noiseless signal subspace. One important advantage of this approach is that it enables not only a reduction in the noise level, but also allows us to estimate the number of sources on the basis of distribution of the eigenvalues. However, a problem arising from this approach is how to correctly set or estimate the threshold which divides eigenvalues into the two subspaces, especially when the noise is large (i.e., the SNR is low). The covariance matrix of the observed data can be written as

$$\begin{aligned} \mathbf{R}_y &= E\{\boldsymbol{y}_t \boldsymbol{y}_t^T\} = [\mathbf{V}_S, \mathbf{V}_{\mathcal{N}}] \begin{bmatrix} \boldsymbol{\Lambda}_S & \\ \mathbf{0} & \boldsymbol{\Lambda}_{\mathcal{N}} \end{bmatrix} [\mathbf{V}_S, \mathbf{V}_{\mathcal{N}}]^T \\ &= \mathbf{V}_S \boldsymbol{\Lambda}_S \mathbf{V}_S^T + \mathbf{V}_{\mathcal{N}} \boldsymbol{\Lambda}_{\mathcal{N}} \mathbf{V}_{\mathcal{N}}^T, \end{aligned} \tag{C.1}$$

where $\mathbf{V}_S \boldsymbol{\Lambda}_S \mathbf{V}_S^T$ is a rank-J matrix, $\mathbf{V}_S \in \mathbb{R}^{I \times J}$ contains the eigenvectors associated with J principal (signal+noise subspace) eigenvalues of $\boldsymbol{\Lambda}_S = \text{diag}\{\lambda_1 \geq \lambda_2 \cdots \geq \lambda_J\}$ in a descending order. Similarly, the matrix $\mathbf{V}_{\mathcal{N}} \in \mathbb{R}^{I \times (I-J)}$ contains the $(I - J)$ (noise) eigenvectors that correspond to noise eigenvalues $\boldsymbol{\Lambda}_{\mathcal{N}} = \text{diag}\{\lambda_{J+1}, \ldots, \lambda_I\} = \sigma_e^2 \mathbf{I}_{I-J}$. This means that, theoretically, the $(I - J)$ smallest eigenvalues of $\mathbf{R}_y$ are equal to σ_e^2, so we can determine the dimension of the signal subspace from the multiplicity of the smallest eigenvalues under the assumption that the variance of the noise is relatively low and we have a perfect estimate of the covariance matrix. However, in practice, we estimate the sample covariance matrix from a limited number of samples and the smallest eigenvalues are usually different, so the determination of the dimension of the signal subspace is usually not an easy task.

A crucial problem is to decide how many principal components (PCs) should be retained. A simple ad hoc rule is to plot the eigenvalues in decreasing order and search for an elbow where the signal eigenvalues are on the left side and the noise eigenvalues on the right. Another simple technique is to compute the cumulative percentage of the total variation explained by the PCs and retain the number of PCs that represent, say 95% of the total variation. Such techniques often work well in practice, but their disadvantage is that they need a subjective decision from the user [138]. Many sophisticated methods have been introduced such as a Bayesian model selection method, which is referred to as the Laplace method. It is based on computing the evidence for the data and requires integrating out all the model parameters. Another method is the BIC (Bayesian Information Criterion) method which can be thought of as an approximation of the Laplace criterion.

A simple heuristic method proposed by He and Cichocki [71,109] computes the GAP (smoothness) index defined as

$$GAP(p) = \frac{\text{var}\left[\{\tilde{\lambda}_i\}_{i=p+1}^{I-1}\right]}{\text{var}\left[\{\tilde{\lambda}_i\}_{i=p}^{I-1}\right]}, \qquad (p = 1, 2, \ldots, I-2), \tag{C.2}$$

where $\tilde{\lambda}_i = \lambda_i - \lambda_{i+1}$ and $\lambda_1 \geq \lambda_2 \geq \cdots \geq \lambda_I > 0$ are eigenvalues of the covariance matrix for the noisy data and the sample variance is computed as follows

$$\text{var}\left[\{\tilde{\lambda}_i\}_{i=p}^{I-1}\right] = \frac{1}{I-p}\sum_{i=p}^{I-1}\left(\tilde{\lambda}_i - \frac{1}{I-p}\sum_{i=p}^{I-1}\tilde{\lambda}_i\right)^2. \tag{C.3}$$

The number of components (for each mode) is selected using the following criterion:

$$\hat{J} = \arg \min_{p=1,2,\ldots,I-2} GAP(p). \tag{C.4}$$

Recently, Ulfarsson and Solo [138] proposed a method called SURE (Stein's Unbiased Risk Estimator) which allows the number of PC components to estimate reliably.

The Laplace, BIC and SURE methods are based on the following considerations [138]. The PCA model is given by

$$\boldsymbol{y}_t = \mathbf{A}\boldsymbol{x}_t + \boldsymbol{e}_t + \bar{\boldsymbol{y}} = \boldsymbol{\mu}_t + \boldsymbol{e}_t, \tag{C.5}$$

where $\bar{\boldsymbol{y}} = (1/T)\sum_{t=1}^{T} \boldsymbol{y}_t$ and $\boldsymbol{\mu}_t = \mathbf{A}\boldsymbol{x}_t + \bar{\boldsymbol{y}}$. The maximum-likelihood estimate (MLE) of PCA is given by

$$\widehat{\mathbf{A}} = \mathbf{V}_r(\mathbf{\Lambda}_r - \widehat{\sigma}_r^2 \mathbf{I}_r)^{1/2}\mathbf{Q}, \qquad \widehat{\sigma}_r^2 = \sum_{j=r+1}^{I} \lambda_j, \tag{C.6}$$

where $\mathbf{Q} \in \mathbb{R}^{r \times r}$ is an arbitrary orthogonal rotation matrix, $\mathbf{\Lambda}_r = \text{diag}\{\lambda_1, \lambda_2, \ldots, \lambda_r\}$ is a diagonal matrix with ordered eigenvalues $\lambda_1 > \lambda_2 > \cdots > \lambda_r$ and $\mathbf{V}_r = [\boldsymbol{v}_1, \boldsymbol{v}_2, \ldots, \boldsymbol{v}_r]$ is a matrix of corresponding eigenvectors of the data covariance matrix:

$$\mathbf{R}_y = \frac{1}{T}\sum_{t=1}^{T}(\boldsymbol{y}_t - \bar{\boldsymbol{y}})\,(\boldsymbol{y}_t - \bar{\boldsymbol{y}})^T = \mathbf{V}_r\mathbf{\Lambda}_r\mathbf{V}_r^T. \tag{C.7}$$

Hence, the estimate for $\boldsymbol{\mu}_t = \mathbf{A}\boldsymbol{x}_t + \bar{\boldsymbol{y}}$ is given by

$$\widehat{\boldsymbol{\mu}}_t = \bar{\boldsymbol{y}} + \sum_{j=1}^{r} \boldsymbol{v}_j \frac{\lambda_j - \widehat{\sigma}_r^2}{\lambda_j} \boldsymbol{v}_j^T(\boldsymbol{y}_t - \bar{\boldsymbol{y}}). \tag{C.8}$$

Ideally, we would like to choose $r = J$ that minimizes the risk function $R_r = E(||\boldsymbol{\mu}_t - \widehat{\boldsymbol{\mu}}_t||_2^2)$ which is estimated by the SURE formula

$$\widehat{R}_r = \frac{1}{T}\sum_{t=1}^{T} ||\boldsymbol{e}_t||_2^2 + \frac{2\sigma_e^2}{T}\sum_{t=1}^{T} \text{tr}\left(\frac{\partial \widehat{\boldsymbol{\mu}}_t}{\partial \boldsymbol{y}_t^T}\right) - I\sigma_e^2. \tag{C.9}$$

In practice, the SURE algorithm chooses the number of components to minimize the SURE formula, that is,

$$\widehat{J} = \arg \min_{r=1,2,\ldots,I} \widehat{R}_r, \tag{C.10}$$

where the SURE formula is given by [138]

$$\begin{aligned}\widehat{R}_r =& (I-r)\widehat{\sigma}_r^2 + \widehat{\sigma}_r^4 \sum_{j=1}^{r}\frac{1}{\lambda_j} + 2\sigma_e^2(1-\frac{1}{T})r \\ &-2\sigma_e^2\widehat{\sigma}_r^2(1-\frac{1}{T})\sum_{j=1}^{r}\frac{1}{\lambda_j} + \frac{4(1-1/T)\sigma_e^2\widehat{\sigma}_k^2}{T}\sum_{j=1}^{r}\frac{1}{\lambda_j} + C_r,\end{aligned} \tag{C.11}$$

$$C_r = \frac{4(1-1/T)\sigma_e^2}{T}\sum_{j=1}^{r}\sum_{i=r+1}^{I}\frac{\lambda_j-\widehat{\sigma}_r^2}{\lambda_j-\lambda_i}+\frac{2(1-1/T)\sigma_e^2}{T}r(r+1)$$
$$-\frac{2(1-1/T)\sigma_e^2}{T}(I-1)\sum_{j=1}^{r}\left(1-\frac{\widehat{\sigma}_r^2}{\lambda_j}\right), \tag{C.12}$$
$$\sigma^2 = \frac{\text{median}(\lambda_{r+1},\lambda_{r+2},\ldots,\lambda_I)}{F_{\gamma,1}^{-1}(\frac{1}{2})}, \tag{C.13}$$

and $\gamma = T/I$, $F_{\gamma,1}$ denotes the Marchenko-Pastur (MP) distribution function with parameter "γ".

Our extensive numerical experiments indicate that the Laplace method usually outperforms the BIC method while the SURE method can achieve significantly better performance than the Laplace method for NMF, NTF and Tucker models.

Appendix 1.D: Nonnegative Rank Factorization Using Wedderborn Theorem – Estimation of the Number of Components

Nonnegative Rank Factorization (NRF) is defined as exact bilinear decomposition:

$$\mathbf{Y} = \sum_{j=1}^{J}\boldsymbol{a}_j\,\boldsymbol{b}_j^T, \tag{D.1}$$

where $\mathbf{Y} \in \mathbb{R}^{I\times T}$, $\boldsymbol{a}_j \in \mathbb{R}_+^I$ and $\boldsymbol{b}_j \in \mathbb{R}_+^T$.

In order to perform such decomposition (if it exists), we begin with a simple, but far reaching, result first proved by Wedderburn.

Theorem D.3 *Suppose* $\mathbf{Y} \in \mathbb{R}^{I\times T}$, $\boldsymbol{a} \in \mathbb{R}^I$ *and* $\boldsymbol{b} \in \mathbb{R}^T$. *Then*

$$rank\left(\mathbf{Y}-\sigma^{-1}\mathbf{Y}\,\boldsymbol{b}\,\boldsymbol{a}^T\,\mathbf{Y}\right) = rank\,(\mathbf{Y})-1, \tag{D.2}$$

if and only if $\sigma = \boldsymbol{a}^T\,\mathbf{Y}\,\boldsymbol{b} \neq 0$.

Usually, the Wedderburn theorem is formulated in more general form:

Theorem D.4 *Suppose* $\mathbf{Y}_1 \in \mathbb{R}^{I\times T}$, $\boldsymbol{a} \in \mathbb{R}^I$ *and* $\boldsymbol{b} \in \mathbb{R}^T$. *Then the matrix*

$$\mathbf{Y}_2 = \mathbf{Y}_1 - \sigma^{-1}\boldsymbol{a}\boldsymbol{b}^T \tag{D.3}$$

satisfies the rank subtractivity $rank(\mathbf{Y}_2) = rank(\mathbf{Y}_1) - 1$ *if and only if there are vectors* $\boldsymbol{x} \in \mathbb{R}^T$ *and* $\boldsymbol{y} \in \mathbb{R}^I$ *such that*

$$\boldsymbol{a} = \mathbf{Y}_1\boldsymbol{x}, \quad \boldsymbol{b} = \mathbf{Y}_1^T\,\boldsymbol{y}, \quad \sigma = \boldsymbol{y}^T\,\mathbf{Y}_1\,\boldsymbol{x}\,. \tag{D.4}$$

The Wedderburn rank-one reduction formula (D.4) has led to a general matrix factorization process (e.g., the LDU and QR decompositions, the Lanczos algorithm and the SVD are special cases) [53].

The basic idea for the NRF is that, starting with $\mathbf{Y}_1 = \mathbf{Y}$, then so long as $\mathbf{Y}_j$ are nonnegative, we can repeatedly apply the Wedderburn formula to generate a sequence $\{\mathbf{Y}_j\}$ of matrices by defining

$$\mathbf{Y}_{j+1} = \mathbf{Y}_j - (\boldsymbol{y}_j^T\,\mathbf{Y}_j\,\boldsymbol{x}_j)^{-1}\mathbf{Y}_j\,\boldsymbol{x}_j\,\boldsymbol{y}_j^T\,\mathbf{Y}_j, \qquad (j = 1, 2, \ldots, J) \tag{D.5}$$

for properly chosen nonnegative vectors satisfying $\boldsymbol{y}_j^T\mathbf{Y}_j\boldsymbol{x}_j \neq 0$. We continue such extraction till the residual matrix $\mathbf{Y}_j$ becomes zero matrix or with negative elements. Without loss of generality we assume that $\boldsymbol{y}_j^T\mathbf{Y}_j\,\boldsymbol{x}_j = 1$ and consider the following constrained optimization problem [53]

$$\max_{\mathbf{x}_j\in\mathbb{R}_+^I,\ \mathbf{y}_j\in\mathbb{R}_+^T} \quad \min\left(\mathbf{Y}_j - \mathbf{Y}_j\,\boldsymbol{x}_j\,\boldsymbol{y}_j^T\mathbf{Y}_j\right) \tag{D.6}$$
$$\text{s.t.}\quad \mathbf{Y}_j\,\boldsymbol{x}_j \geq 0, \quad \boldsymbol{y}_j^T\,\mathbf{Y}_j \geq 0, \quad \boldsymbol{y}_j^T\,\mathbf{Y}_j\,\boldsymbol{a}_j = 1.$$

There are some available routines for solving the above optimization problem, especially, the MATLAB routine "fminmax" implements a sequential quadratic programming method. It should be noted that this method does not guarantee finding a global solution, but only a suboptimal local solution. On the basis of this idea Dong, Lin and Chu developed an algorithm for the NRF using the Wedderburn rank reduction formula [53].

- Given a nonnegative data matrix $\mathbf{Y}$ and a small threshold of machine $\varepsilon > 0$ set $j = 1$ and $\mathbf{Y}_1 = \mathbf{Y}$.
- Step 1. If $||\mathbf{Y}_j|| \geq \varepsilon$, go to Step 2. Otherwise, retrieve the following information and stop.
  1. $\text{rank}(\mathbf{Y}) = \text{rank}_+(\mathbf{Y}) = j - 1$.
  2. The NRF of $\mathbf{Y}$ is approximately given by the summation $\mathbf{Y} \cong \sum_{k=1}^{j-1} \mathbf{Y}_k\, \boldsymbol{x}_k\, \boldsymbol{y}_k^T\, \mathbf{Y}_k$ with an error less than ε.
- Step 2. Randomly select a feasible initial value $(\boldsymbol{x}_j^{(0)}, \boldsymbol{y}_j^{(0)})$ satisfying the nonnegativity constraints.
- Step 3. Solve the maximin problem (D.6).
- Step 4. If the objective value at the local maximizer $(\boldsymbol{x}_j, \boldsymbol{y}_j)$ is negative, go to Step 5. Otherwise, do update as follows and go to Step 1.
  1. Define $\mathbf{Y}_{j+1} := \mathbf{Y}_j - \mathbf{Y}_j\, \boldsymbol{x}_j\, \boldsymbol{y}_j^T\, \mathbf{Y}_j$.
  2. Set $j = j + 1$.
- Step 5. Since the algorithm may get stuck at a local minimum try to restart Steps 2 and 3 multiple times. If it is decided within reasonable trials that no initial value can result in nonnegative values, report with caution that the matrix $\mathbf{Y}$ does not have an NRF and stop [53].

References

[1] E. Acar, T.G. Kolda, and D.M. Dunlavy. An optimization approach for fitting canonical tensor decompositions. Technical Report SAND2009-0857, Sandia National Laboratories, Albuquerque, NM and Livermore, CA, February 2009.

[2] E. Acar and B. Yener. Unsupervised multiway data analysis: A literature survey. *IEEE Transactions on Knowledge and Data Engineering*, 21:6–20, 2008.

[3] R. Albright, J. Cox, D. Duling, A. N. Langville, and C. D. Meyer. Algorithms, initializations, and convergence for the nonnegative matrix factorization. Technical report, NCSU Technical Report Math 81706, 2006.

[4] S. Amari and A. Cichocki. Adaptive blind signal processing - neural network approaches. *Proceedings of the IEEE*, 86:1186–1187, 1998.

[5] A.H. Andersen and W.S. Rayens. Structure-seeking multilinear methods for the analysis of fMRI data. *NeuroImage*, 22:728–739, 2004.

[6] C.A. Andersson and R. Bro. The N-way toolbox for MATLAB. *Chemometrics Intell. Lab. Systems*, 52(1):1–4, 2000.

[7] C.J. Appellof and E.R. Davidson. Strategies for analyzing data from video fluoromatric monitoring of liquid chromatographic effluents. *Analytical Chemistry*, 53:2053–2056, 1981.

[8] L. Badea. Extracting gene expression profiles common to Colon and Pancreatic Adenocarcinoma using simultaneous nonnegative matrix factorization. In *Proceedings of Pacific Symposium on Biocomputing PSB-2008*, pages 267–278, World Scientific, 2008.

[9] B.W. Bader, R.A. Harshman, and T.G. Kolda. Pattern analysis of directed graphs using DEDICOM: An application to Enron email. Technical Report SAND2006-7744, Sandia National Laboratories, Albuquerque, NM and Livermore, CA, 2006.

[10] B.W. Bader, R.A. Harshman, and T.G. Kolda. Temporal analysis of semantic graphs using ASALSAN. In *ICDM 2007: Proceedings of the 7th IEEE International Conference on Data Mining*, pages 33–42, October 2007.

[11] M. Berry, M. Browne, A. Langville, P. Pauca, and R. Plemmons. Algorithms and applications for approximate nonnegative matrix factorization. *Computational Statistics and Data Analysis*, 52(1):155–173, 2007.

[12] M. Biggs, A. Ghodsi, and S. Vavasis. Nonnegative matrix factorization via rank-one downdate. In *ICML-2008*, Helsinki, July 2008.

[13] J. Bobin, J.L. Starck, J. Fadili, Y. Moudden, and D.L. Donoho. Morphological component analysis: An adaptive thresholding strategy. *IEEE Transactions on Image Processing*, 16(11):2675–2681, 2007.

[14] C. Boutsidis and E. Gallopoulos. SVD based initialization: A head start for nonnegative matrix factorization. *Pattern Recognition*, 41:1350–1362, 2008.

[15] C. Boutsidis, M.W. Mahoney, and P. Drineas. An improved approximation algorithm for the column subset selection problem. In *Proc. 20-th Annual SODA*, pages 968–977, USA, 2009.

[16] R. Bro. PARAFAC. Tutorial and applications. In *Special Issue 2nd Internet Conf. in Chemometrics (INCINC'96)*, volume 38, pages 149–171. Chemom. Intell. Lab. Syst, 1997.

[17] R. Bro. *Multi-way Analysis in the Food Industry - Models, Algorithms, and Applications*. PhD thesis, University of Amsterdam, Holland, 1998.

[18] R. Bro. Review on multiway analysis in chemistry 2000–2005. *Critical Reviews in Analytical Chemistry*, 36:279–293, 2006.

[19] R. Bro and C.A. Andersson. Improving the speed of multiway algorithms - Part II: Compression. *Chemometrics and Intelligent Laboratory Systems*, 42:105–113, 1998.

[20] R. Bro and H.A.L. Kiers. A new efficient method for determining the number of components in PARAFAC models. *Journal of Chemometrics*, 17(5):274–286, 2003.

[21] A.M. Bruckstein, M. Elad, and M. Zibulevsky. A non-negative and sparse enough solution of an underdetermined linear system of equations is unique. *IEEE Transactions on Information Theory*, 54:4813–4820, 2008.

[22] C. Caiafa and A. Cichocki. CUR decomposition with optimal selection of rows and columns. *(submitted)*, 2009.

[23] C. Caiafa and A. Cichocki. Slice Oriented Decomposition: A new tensor representation for 3-way data. *(submitted to Journal of Signal Processing)*, 2009.

[24] J.D. Carroll and J.J. Chang. Analysis of individual differences in multidimensional scaling via an n-way generalization of Eckart–Young decomposition. *Psychometrika*, 35(3):283–319, 1970.

[25] J.D. Carroll, G. De Soete, and S. Pruzansky. Fitting of the latent class model via iteratively reweighted least squares CANDECOMP with nonnegativity constraints. In R. Coppi and S. Bolasco, editors, *Multiway data analysis.*, pages 463–472. Elsevier, Amsterdam, The Netherlands, 1989.

[26] A. Cichocki and S. Amari. *Adaptive Blind Signal and Image Processing*. John Wiley & Sons Ltd, New York, 2003.

[27] A. Cichocki, S. Amari, R. Zdunek, R. Kompass, G. Hori, and Z. He. Extended SMART algorithms for non-negative matrix factorization. *Springer, LNAI-4029*, 4029:548–562, 2006.

[28] A. Cichocki, S.C. Douglas, and S. Amari. Robust techniques for independent component analysis (ICA) with noisy data. *Neurocomputing*, 23(1–3):113–129, November 1998.

[29] A. Cichocki and P. Georgiev. Blind source separation algorithms with matrix constraints. *IEICE Transactions on Fundamentals of Electronics, Communications and Computer Sciences*, E86-A(1):522–531, January 2003.

[30] A. Cichocki and A.H. Phan. Fast local algorithms for large scale nonnegative matrix and tensor factorizations. *IEICE (invited paper)*, March 2009.

[31] A. Cichocki, A.H. Phan, and C. Caiafa. Flexible HALS algorithms for sparse non-negative matrix/tensor factorization. In *Proc. of 18-th IEEE workshops on Machine Learning for Signal Processing*, Cancun, Mexico, 16–19, October 2008.

[32] A. Cichocki and R. Unbehauen. *Neural Networks for Optimization and Signal Processing*. John Wiley & Sons ltd, New York, 1994.

[33] A. Cichocki and R. Zdunek. Multilayer nonnegative matrix factorization. *Electronics Letters*, 42(16):947–948, 2006.

[34] A. Cichocki and R. Zdunek. NTFLAB for Signal Processing. Technical report, Laboratory for Advanced Brain Signal Processing, BSI, RIKEN, Saitama, Japan, 2006.

[35] A. Cichocki, R. Zdunek, and S. Amari. Csiszar's divergences for non-negative matrix factorization: Family of new algorithms. *Springer, LNCS-3889*, 3889:32–39, 2006.

[36] A. Cichocki, R. Zdunek, and S. Amari. New algorithms for non-negative matrix factorization in applications to blind source separation. In *Proc. IEEE International Conference on Acoustics, Speech, and Signal Processing, ICASSP2006*, volume 5, pages 621–624, Toulouse, France, May 14–19 2006.

[37] A. Cichocki, R. Zdunek, and S.-I. Amari. Hierarchical ALS algorithms for nonnegative matrix and 3D tensor factorization. *Springer, Lecture Notes on Computer Science, LNCS-4666*, pages 169–176, 2007.

[38] A. Cichocki, R. Zdunek, S. Choi, R. Plemmons, and S. Amari. Nonnegative tensor factorization using Alpha and Beta divergencies. In *Proc. IEEE International Conference on Acoustics, Speech, and Signal Processing (ICASSP07)*, volume III, pages 1393–1396, Honolulu, Hawaii, USA, April 15–20 2007.

[39] A. Cichocki, R. Zdunek, S. Choi, R. Plemmons, and S.-I. Amari. Novel multi-layer nonnegative tensor factorization with sparsity constraints. *Springer, LNCS-4432*, 4432:271–280, April 11–14 2007.

[40] J.P.C.L. Da Costa, M. Haardt, and F. Roemer. Robust methods based on the HOSVD for estimating the model order in PARAFAC models. In *Proc. 5-th IEEE Sensor Array and Multich. Sig. Proc. Workshop (SAM 2008)*, pages 510–514, Darmstadt, Germany, July 2008.

[41] S. Cruces, A. Cichocki, and L. De Lathauwer. Thin QR and SVD factorizations for simultaneous blind signal extraction. In *Proc. of the European Signal Processing Conference (EUSIPCO*, pages 217–220. Vienna, Austria, 2004.

[42] E. Cuelemans and H.A.L. Kiers. Selecting among three-mode principal component models of different types and complexities: A numerical convex hull based method. *British Journal of Mathematical and Statistical Psychology*, 59:133–150, 2006.

[43] E. Cuelemans and H.A.L. Kiers. Discriminating between strong and weak structures in three-mode principal component analysis. *British Journal of Mathematical and Statistical Psychology*, page (in print), 2008.

[44] L. De Lathauwer. Parallel factor analysis by means of simultaneous matrix decompositions. In *Proceedings of the First IEEE International Workshop on Computational Advances in Multi-sensor Adaptive Processing (CAMSAP 2005)*, pages 125–128, Puerto Vallarta, Jalisco State, Mexico, 2005.

[45] L. De Lathauwer. A link between the canonical decomposition in multilinear algebra and simultaneous matrix diagonalization. *SIAM Journal on Matrix Analysis and Applications*, 28:642–666, 2006.

[46] L. De Lathauwer. Decompositions of a higher-order tensor in block terms – Part I: Lemmas for partitioned matrices. *SIAM Journal on Matrix Analysis and Applications (SIMAX)*, 30(3):1022–1032, 2008. Special Issue on Tensor Decompositions and Applications.

[47] L. De Lathauwer. Decompositions of a higher-order tensor in block terms – Part II: Definitions and uniqueness. *SIAM Journal of Matrix Analysis and Applications*, 30(3):1033–1066, 2008. Special Issue on Tensor Decompositions and Applications.

[48] L. De Lathauwer and D. Nion. Decompositions of a higher-order tensor in block terms – Part III: Alternating least squares algorithms. *SIAM Journal on Matrix Analysis and Applications (SIMAX)*, 30(3):1067–1083, 2008. Special Issue Tensor Decompositions and Applications.

[49] I. Dhillon and S. Sra. Generalized nonnegative matrix approximations with Bregman divergences. In *Neural Information Proc. Systems*, pages 283–290, Vancouver, Canada, December 2005.

[50] C. Ding, X. He, and H.D. Simon. On the equivalence of nonnegative matrix factorization and spectral clustering. In *Proc. SIAM International Conference on Data Mining (SDM'05)*, pages 606–610, 2005.

[51] C. Ding, T. Li, W. Peng, and M.I. Jordan. Convex and semi-nonnegative matrix factorizations. Technical Report 60428, Lawrence Berkeley National Laboratory, November 2006.

[52] C. Ding, T. Li, W. Peng, and H. Park. Orthogonal nonnegative matrix tri-factorizations for clustering. In *KDD06: Proceedings of the 12th ACM SIGKDD international conference on Knowledge Discovery and Data Mining*, pages 126–135, New York, NY, USA, 2006. ACM Press.

[53] B. Dong, M.M. Lin, and M.T. Chu. Nonnegative rank factorization via rank reduction. *http://www4.ncsu.edu/ mtchu/Research/Papers/Readme.html*, page (submitted), 2008.

[54] D. Donoho and V. Stodden. When does nonnegative matrix factorization give a correct decomposition into parts? In *Neural Information Processing Systems*, volume 16. MIT press, 2003.

[55] P. Drineas, M.W. Mahoney, and S. Muthukrishnan. Relative-error CUR matrix decompositions. *SIAM Journal on Matrix Analysis and Applications*, 30:844–881, 2008.

[56] J. Eggert, H. Wersing, and E. Koerner. Transformation-invariant representation and NMF. USA, 2004. Proceedings of International Joint Conference Neural Networks.

[57] F. Estienne, N. Matthijs, D.L. Massart, P. Ricoux, and D. Leibovici. Multi-way modeling of high-dimensionality electroencephalographic data. *Chemometrics and Intelligent Laboratory Systems*, 58(1):59–72, 2001.

[58] N. Gillis and F. Glineur. Nonnegative matrix factorization and underapproximation. In *9th International Symposium on Iterative Methods in Scientific Computing*, Lille, France, 2008. http://www.core.ucl.ac.be/ ngillis/.

[59] N. Gillis and F. Glineur. Nonnegative factorization and maximum edge biclique problem. In *submitted*, 2009. http://www.uclouvain.be/en-44508.html.

[60] S.A. Goreinov, I.V. Oseledets, D.V. Savostyanov, E.E. Tyrtyshnikov, and N.L. Zamarashkin. How to find a good submatrix. (submitted), 2009.

[61] S.A. Goreinov, E.E. Tyrtyshnikov, and N.L. Zamarashkin. A theory of pseudo-skeleton approximations. *Linear Alegebra and Applications*, 261:1–21, 1997.

[62] R.A. Harshman. Foundations of the PARAFAC procedure: Models and conditions for an explanatory multimodal factor analysis. *UCLA Working Papers in Phonetics*, 16:1–84, 1970.

[63] R.A. Harshman. PARAFAC2: Mathematical and technical notes. *UCLA Working Papers in Phonetics*, 22:30–44, 1972.

[64] R.A. Harshman, S. Hong, and M.E. Lundy. Shifted factor analysis - Part I: Models and properties. *Journal of Chemometrics*, 17(7):363–378, 2003.

[65] R.A. Harshman and M.E. Lundy. *Research Methods for Multimode Data Analysis*. Praeger, New York, USA, 1984.

[66] M. Hasan, F. Pellacini, and K. Bala. Matrix row-column sampling for the many-light problem. In *SIGGRAPH*, page http://www.cs.cornell.edu/ mhasan/, 2007.

[67] M. Hasan, E. Velazquez-Armendariz, F. Pellacini, and K. Bala. Tensor clustering for rendering many-light animations. In *Eurographics Symposium on Rendering*, volume 27, page http://www.cs.cornell.edu/ mhasan/, 2008.

[68] T. Hazan, S. Polak, and A. Shashua. Sparse image coding using a 3D non-negative tensor factorization. In *Proc. Int. Conference on Computer Vision (ICCV)*, pages 50–57, 2005.

[69] Z. He and A. Cichocki. K-EVD clustering and its applications to sparse component analysis. *Lecture Notes in Computer Science*, 3889:438–445, 2006.

[70] Z. He and A. Cichocki. An efficient K-hyperplane clustering algorithm and its application to sparse component analysis. *Lecture Notes in Computer Science*, 4492:1032–1041, 2007.

[71] Z. He and A. Cichocki. Efficient method for estimating the dimension of Tucker3 model. *Journal of Multivariate Analysis*, (submitted), 2009.

[72] Z. He, S. Xie, L. Zhang, and A. Cichocki. A note on Lewicki-Sejnowski gradient for learning overcomplete representations. *Neural Computation*, 20(3):636–643, 2008.

[73] M. Heiler and C. Schnoerr. Controlling sparseness in non-negative tensor factorization. *Springer LNCS*, 3951:56–67, 2006.

[74] F.L. Hitchcock. Multiple invariants and generalized rank of a p-way matrix or tensor. *Journal of Mathematics and Physics*, 7:39–79, 1927.

[75] N.-D. Ho, P. Van Dooren, and V.D. Blondel. Descent methods for nonnegative matrix factorization. *Numerical Linear Algebra in Signals, Systems and Control*, 2008.

[76] P.O. Hoyer. Non-negative matrix factorization with sparseness constraints. *Journal of Machine Learning Research*, 5:1457–1469, 2004.

[77] A. Hyvärinen, J. Karhunen, and E. Oja. *Independent Component Analysis*. John Wiley & Sons Ltd, New York, 2001.

[78] H.A.L. Kiers, J.M.F. Ten Berg, and R. Bro. PARAFAC2–Part I. A direct fitting algorithm for the PARAFAC2 model. *Journal of Chemometrics*, 13(3–4):275–294, 1999.

[79] H.A.L. Kiers and A.K. Smilde. Constrained three-mode factor analysis as a tool for parameter estimation with second-order instrumental data. *Journal of Chemometrics*, 12:125–147, 1998.

[80] H. Kim, L. Eldén, and H. Park. Non-negative tensor factorization based on alternating large-scale non-negativity-constrained least squares. In *Proceedings of IEEE 7th International Conference on Bioinformatics and Bioengineering (BIBE07)*, volume II, pages 1147–1151, 2007.

[81] M. Kim and S. Choi. Monaural music source separation: Nonnegativity, sparseness, and shift-invariance. pages 617–624, USA, Charleston, 2006. Proceedings of the International Conference on Independent Component Analysis and Blind Signal Separation (Springer LNCS 3889).

[82] Y.-D. Kim and S. Choi. A method of initialization for nonnegative matrix factorization. In *Proc. IEEE International Conference on Acoustics, Speech, and Signal Processing (ICASSP07)*, volume II, pages 537–540, Honolulu, Hawaii, USA, April 15–20 2007.

[83] Y.-D. Kim and S. Choi. Nonnegative Tucker Decomposition. In *Proc. of Conf. Computer Vision and Pattern Recognition (CVPR-2007)*, Minneapolis, Minnesota, June 2007.

[84] Y.-D. Kim, A. Cichocki, and S. Choi. Nonnegative Tucker Decomposition with Alpha Divergence. In *Proceedings of the IEEE International Conference on Acoustics, Speech, and Signal Processing, ICASSP2008*, Nevada, USA, 2008.

[85] T.G. Kolda and B.W. Bader. Tensor decompositions and applications. *SIAM Review*, 51(3):(in print), September 2009.

[86] T.G. Kolda and J. Sun. Scalable tensor decompositions for multi-aspect data mining. In *ICDM 2008: Proceedings of the 8th IEEE International Conference on Data Mining*, December 2008.

[87] T.G. Kolda and J. Sun. Thinking sparsely: Scalable tensor decompositions for multi-aspect data mining. Submitted for publication, February 2008.

[88] K. Kreutz-Delgado, J.F. Murray, B.D. Rao, K. Engan, T.-W. Lee, and T.J. Sejnowski. Dictionary learning algorithms for sparse representation. *Neural Computation*, 15(2):349–396, 2003.

[89] P.M. Kroonenberg. *Applied Multiway Data Analysis*. John Wiley & Sons Ltd, New York, 2008.

[90] J.B. Kruskal. Three-way arrays: Rank and uniqueness of trilinear decompositions, with application to arithmetic complexity and statistics. *Linear Algebra Appl.*, 18:95–138, 1977.

[91] J.B. Kruskal. Rank, decomposition, and uniqueness for 3-way and N-way arrays. In R. Coppi and S. Bolasco, editors, *Multiway data analysis.*, pages 8–18. Elsevier, Amsterdam, The Netherlands, 1989.

[92] B. Kulis, M.A. Sustik, and I.S. Dhillon. Learning low-rank kernel matrices. In *Proc. of the Twenty-third International Conference on Machine Learning (ICML06)*, pages 505–512, July 2006.

[93] A. N. Langville, C. D. Meyer, and R. Albright. Initializations for the nonnegative matrix factorization. In *Proc. of the Twelfth ACM SIGKDD International Conference on Knowledge Discovery and Data Mining*, Philadelphia, USA, August 20–23 2006.

[94] D.D. Lee and H.S. Seung. Learning of the parts of objects by non-negative matrix factorization. *Nature*, 401:788–791, 1999.

[95] D.D. Lee and H.S. Seung. *Algorithms for Nonnegative Matrix Factorization*, volume 13. MIT Press, 2001.

[96] Y. Li, S. Amari, A. Cichocki, D. Ho, and S. Xie. Underdetermined blind source separation based on sparse representation. *IEEE Transactions on Signal Processing*, 54:423–437, 2006.

[97] Y. Li, A. Cichocki, and S. Amari. Blind estimation of channel parameters and source components for EEG signals: A sparse factorization approach. *IEEE Transactions on Neural Networks*, 17:419–431, 2006.

[98] Y. Li, Y. Du, and X. Lin. Kernel-based multifactor analysis for image synthesis and recognition. In *Proc. of 10th IEEE Conference on Computer Vision (ICCV05)*, volume 1, pages 114–119, 2005.

[99] L.-H. Lim and P. Comon. Nonnegative approximations of nonnegative tensors. *Journal of Chemometrics*, page (in print), 2009.

[100] S. Lipovetsky. PCA and SVD with nonnegative loadings. *Pattern Recognition*, 99(9):(in print), 2009.

[101] M.W. Mahoney and P. Drineas. CUR matrix decompositions for improved data analysis. *Proc. National Academy of Science*, 106:697–702, 2009.

[102] M.W. Mahoney, M. Maggioni, and P. Drineas. Tensor-CUR decompositions and data applications. *SIAM Journal on Matrix Analysis and Applications*, 30:957–987, 2008.

[103] E. Martínez-Montes, J.M. Sánchez-Bornot, and P.A. Valdés-Sosa. Penalized PARAFAC analysis of spontaneous EEG recordings. *Statistica Sinica*, 18:1449–1464, 2008.

[104] F. Miwakeichi, E. Martnez-Montes, P. Valds-Sosa, N. Nishiyama, H. Mizuhara, and Y. Yamaguchi. Decomposing EEG data into space−time−frequency components using parallel factor analysis. *NeuroImage*, 22(3):1035–1045, 2004.

[105] J. Möck. Topographic components model for event-related potentials and some biophysical considerations. *IEEE Transactions on Biomedical Engineering*, 35:482–484, 1988.

[106] M. Mørup, L. K. Hansen, C. S. Herrmann, J. Parnas, and S. M. Arnfred. Parallel factor analysis as an exploratory tool for wavelet transformed event-related EEG. *NeuroImage*, 29(3):938–947, 2006.

[107] M. Mørup, L.K. Hansen, and S.M. Arnfred. Algorithms for Sparse Nonnegative Tucker Decompositions. *Neural Computation*, 20:2112–2131, 2008.

[108] M. Mørup and M.N. Schmidt. Sparse non-negative tensor factor double deconvolution (SNTF2D) for multi channel time-frequency analysis. Technical report, Technical University of Denmark, DTU, 2006.

[109] J. Niesing. *Simultaneous Component and Factor Analysis Methods for Two or More Groups: A Comparative Study*, volume 2nd ed. Leiden: The Netherlands. DSWO Press, Leiden University, 1997, 1997.

[110] D. Nion and L. De Lathauwer. A Block Component Model based blind DS-CDMA receiver. *IEEE Transactions on Signal Processing*, 56(11):5567–5579, 2008.

[111] D. Nion and L. De Lathauwer. An enhanced line search scheme for complex-valued tensor decompositions. Application in DS-CDMA. *Signal Processing*, 88(3):749–755, 2008.

[112] P. Paatero. Least-squares formulation of robust nonnegative factor analysis. *Chemometrics and Intelligent Laboratory Systems*, 37:23–35, 1997.

[113] P. Paatero. A weighted non-negative least squares algorithm for three-way "PARAFAC" factor analysis. *Chemometrics Intelligent Laboratory Systems*, 38(2):223–242, 1997.

[114] P. Paatero and U. Tapper. Positive matrix factorization: A nonnegative factor model with optimal utilization of error estimates of data values. *Environmetrics*, 5:111–126, 1994.

[115] A. Pascual-Montano, J.M. Carazo, K. Kochi, D. Lehmean, and R. Pacual-Marqui. Nonsmooth nonnegative matrix factorization (nsNMF). *IEEE Transactions Pattern Analysis and Machine Intelligence*, 28(3):403–415, 2006.

[116] A.H. Phan and A. Cichocki. Fast and efficient algorithms for nonnegative Tucker decomposition. In *Proc. of The Fifth International Symposium on Neural Networks, Springer LNCS-5264*, pages 772–782, Beijing, China, 24–28, September 2008.

[117] A.H. Phan and A. Cichocki. Local learning rules for nonnegative Tucker decomposition. *(submitted)*, 2009.

[118] A.H. Phan and A. Cichocki. Multi-way nonnegative tensor factorization using fast hierarchical alternating least squares algorithm (HALS). In *Proc. of The 2008 International Symposium on Nonlinear Theory and its Applications*, Budapest, Hungary, 2008.

[119] R.J. Plemmons and R.E. Cline. The generalized inverse of a nonnegative matrix. *Proceedings of the American Mathematical Society*, 31:46–50, 1972.

[120] R. Rosipal, M. Girolami, L.J. Trejo, and A. Cichocki. Kernel PCA for feature extraction and de-noising in nonlinear regression. *Neural Computing & Applications*, 10:231–243, 2001.

[121] P. Sajda, S. Du, and L. Parra. Recovery of constituent spectra using non-negative matrix factorization. In *Proceedings of SPIE*, volume 5207, pages 321–331, 2003.

[122] T.M. Selee, T.G. Kolda, W.P. Kegelmeyer, and J.D. Griffin. Extracting clusters from large datasets with multiple similarity measures using IMSCAND. In Michael L. Parks and S. Scott Collis, editors, *CSRI Summer Proceedings 2007, Technical Report SAND2007-7977, Sandia National Laboratories, Albuquerque, NM and Livermore, CA*, pages 87–103, December 2007.

[123] A. Shashua, R. Zass, and T. Hazan. Multi-way clustering using super-symmetric non-negative tensor factorization. In *European Conference on Computer Vision (ECCV)*, Graz, Austria, May 2006.

[124] H. Shen and J.-Z Huang. Sparse principal component analysis via regularized low rank matrix approximation. *Journal of Multivariate Analysis*, 99:1015–1034, 2008.

[125] N.D. Sidiropoulos and R. Bro. On the uniqueness of multilinear decomposition of N-way arrays. *Journal of Chemometrics*, 14:229–239, 2000.

[126] P. Smaragdis. Non-negative matrix factor deconvolution; Extraction of multiple sound sources from monophonic inputs. *Lecture Notes in Computer Science*, 3195:494–499, 2004.

[127] A. Smilde, R. Bro, and P. Geladi. *Multi-way Analysis: Applications in the Chemical Sciences*. John Wiley & Sons Ltd, New York, 2004.

[128] A. Stegeman. On uniqueness conditions for Candecomp/Parafac and Indscal with full column rank in one mode. *Linear Algebra and its Applications*, http://www.gmw.rug.nl/ stegeman/:(in print), 2009.

[129] A. Stegeman and A.L.F. de Almeida. Uniqueness conditions for a constrained three-way factor decomposition. *SIAM Journal on Matrix Analysis and Applications*, http://www.gmw.rug.nl/ stegeman/:(in print), 2009.

[130] A. Stegeman and N.D. Sidiropoulos. On Kruskal's uniqueness condition for the Candecomp/Parafac decomposition. *Linear Algebra and its Applications*, 420:540–552, 2007.

[131] A. Stegeman, J.M.F. Ten Berge, and L. De Lathauwer. Sufficient conditions for uniqueness in Candecomp/Parafac and Indscal with random component matrices. *Psychometrika*, 71(2):219–229, 2006.

[132] Y. Takane and H.A.L. Kiers. Latent class DEDICOM. *Journal of Classification*, 14:225–247, 1997.

[133] M.E. Timmerman and H.A.L. Kiers. Three mode principal components analysis: Choosing the numbers of components and sensitivity to local optima. *British Journal of Mathematical and Statistical Psychology*, 53(1):1–16, 2000.

[134] G. Tomasi and R. Bro. PARAFAC and missing values. *Chemometrics Intelligent Laboratory Systems*, 75(2):163–180, 2005.

[135] G. Tomasi and R. Bro. A comparison of algorithms for fitting the PARAFAC model. *Computational Statistics and Data Analysis*, 50(7):1700–1734, April 2006.

[136] L.R. Tucker. The extension of factor analysis to three-dimensional matrices. In H. Gulliksen and N. Frederiksen, editors, *Contributions to Mathematical Psychology*, pages 110–127. Holt, Rinehart and Winston, New York, 1964.

[137] L.R. Tucker. Some mathematical notes on three-mode factor analysis. *Psychometrika*, 31:279–311, 1966.

[138] M.O. Ulfarsson and V. Solo. Dimension estimation in noisy PCA with SURE and random matrix theory. *IEEE Transactions on Signal Processing*, 56(12):5804–5816, December 2008.

[139] H. Wang and N. Ahuja. Compact representation of multidimensional data using tensor rank-one decomposition. In *Proc. of International Conference on Pattern Recognition.*, volume 1, pages 44–47, 2004.

[140] H. Wang and N. Ahuja. A tensor approximation approach to dimensionality reduction. *International Journal of Computer Vision*, 76(3):217–229, 2008.

[141] Z. Wang, A. Maier, N.K. Logothetis, and H. Liang. Single-trial decoding of bistable perception based on sparse nonnegative tensor decomposition. *Journal of Computational Intelligence and Neuroscience*, 30:1–10, 2008.

[142] Y. Washizawa and A. Cichocki. On line K-plane clustering learning algorithm for Sparse Component Analysis. In *Proc. of 2006 IEEE International Conference on Acoustics, Speech and Signal Processing, ICASSP2006*, pages 681–684, Toulouse, France, May 14-19 2006.

[143] Q. Wu, T. Xia, and Y. Yu. Hierarchical tensor approximation of multi-dimensional images. In *14th IEEE International Conference on Image Processing*, volume IV, pages 49–52, San Antonio, 2007.

[144] R. Zass and A. Shashua. Nonnegative sparse PCA. In *Neural Information Processing Systems (NIPS)*, Vancuver, Canada, Dec. 2006.

[145] R. Zdunek and A. Cichocki. Nonnegative matrix factorization with constrained second-order optimization. *Signal Processing*, 87:1904–1916, 2007.

[146] T. Zhang and G.H. Golub. Rank-one approximation to high order tensors. *SIAM Journal of Matrix Analysis and Applications*, 23(2):534–550, 2001.

2

Similarity Measures and Generalized Divergences

In this chapter, we overview and discuss properties of a large family of generalized and flexible divergences or similarity distances between two nonnegative sequences or patterns. They are formulated for probability distributions and for arrays used in applications to development of novel algorithms for Nonnegative Matrix Factorization (NMF) and Nonnegative Tensor Factorization (NTF). Information theory, convex analysis, and information geometry play key roles in the formulation of divergences [2,3,6,9,15,21,23,29,36,38–40,54–58].

Divergences, or their counter part (dis)similarity measures play an important role in the areas of neural computation, pattern recognition, learning, estimation, inference, and optimization. Generally speaking, they measure a quasi-distance or directed difference between two probability distributions $\boldsymbol{p}$ and $\boldsymbol{q}$ which can also be expressed for unconstrained nonnegative arrays and patterns. Divergence measures are commonly used to find a distance or difference between two n-dimensional probability distributions[1] $\boldsymbol{p} = (p_1, p_2, \ldots, p_n)$ and $\boldsymbol{q} = (q_1, q_2, \ldots, q_n)$. They are called nonnegative measures when they are not normalized to $\sum_{i=1}^{n} p_i = 1$, that is, their total masses are not necessarily unity but an arbitrary positive number.

We are mostly interested in distance-type measures which are separable, thus, satisfying the condition

$$D(\boldsymbol{p}\,||\,\boldsymbol{q}) = \sum_{i=1}^{n} d(p_i, q_i) \geq 0, \qquad \text{which equals zero if and only if } \boldsymbol{p} = \boldsymbol{q} \tag{2.1}$$

but are not necessarily symmetric in the sense

$$D(\boldsymbol{p}\,||\,\boldsymbol{q}) = D(\boldsymbol{q}\,||\,\boldsymbol{p}), \tag{2.2}$$

and do not necessarily satisfy the triangular inequality

$$D(\boldsymbol{p}\,||\,\boldsymbol{q}) \leq D(\boldsymbol{p}\,||\,\boldsymbol{z}) + D(\boldsymbol{z}\,||\,\boldsymbol{q}). \tag{2.3}$$

[1] Usually, the vector $\boldsymbol{p}$ corresponds to the observed data and the vector $\boldsymbol{q}$ to estimated or expected data which are subject to constraints imposed on the assumed models. For NMF problem $\boldsymbol{p}$ corresponds to the data matrix $\mathbf{Y}$ and $\boldsymbol{q}$ corresponds to estimated matrix $\hat{\mathbf{Y}} = \mathbf{AX}$. An information divergence is a measure of distance between two probability curves. In this chapter, we discuss only one-dimensional probability curves (represented by nonnegative signals or images). Generalization to two or multidimensional dimensional variables is straightforward; each single subscript is simply replaced by a doubly or triply indexed one.

Nonnegative Matrix and Tensor Factorizations: Applications to Exploratory Multi-way Data Analysis and Blind Source Separation Andrzej Cichocki, Rafal Zdunek, Anh Huy Phan and Shun-ichi Amari

In other words, the distance-type measures under consideration are not necessarily a metric[2] on the space $\mathcal{P}$ of all probability distributions.

The scope of these results is vast since the generalized divergence functions and their variants include quite a large number of useful loss functions including those based on the Relative entropies, generalized Kullback-Leibler or I-divergence, Hellinger distance, Jensen-Shannon divergence, J-divergence, Pearson and Neyman Chi-square divergences, Triangular Discrimination and Arithmetic-Geometric (AG) Taneya divergence. Many of these measures belong to the class of Alpha-divergences and Beta-divergences and have been applied successfully in disciplines such as signal processing, pattern recognition, probability distributions, information theory, finance and economics [37]. In the following chapters we will apply such divergences as cost functions (possibly with additional constraints and regularization terms) to derive novel multiplicative and additive projected gradient and fixed point algorithms. These provide working solutions for the problems where nonnegative latent (hidden) components can be generally statistically dependent, and satisfy some other conditions or additional constraints such as sparsity or smoothness.

Section 2.1 addresses the divergences derived from simple component-wise errors (losses), these include the Euclidean and Minkowski metrics. We show that they are related to robust cost functions in Section 2.2. We then study in Section 2.3 the class of Csiszár f-divergences, which are characterized by the invariance and monotonicity properties. This class includes the Alpha-divergence, in particular the Kullback-Leibler divergence. The Bregman type divergences, derived from convex functions, are studied in Section 2.4. We also discuss divergences between positive-definite matrices. An important class of the Beta-divergences belongs to the class of Bregman divergences and is studied in detail in Section 2.6. They do not satisfy the invariance property except for the special case of the KL-divergence. When we extend divergences to positive measures where the total mass $\sum p_i$ is not restricted to unity, the Alpha-divergences belong to the classes of both Csiszár f-divergences and Bregman divergences. They are studied in detail in Section 2.5. Moreover, in Section 2.7 we discuss briefly Gamma-divergences which have "super robust" properties. Furthermore, in Section 2.8 we derive various divergences from Tsallis and Rényi entropy.

The divergences are closely related to the invariant geometrical properties of the manifold of probability distributions. This is a two-way relation: Divergences, in particular the Alpha-divergences, are naturally induced from geometry, and on the other hand divergences give a geometrical structure to the set of probability distributions or positive measures [4]. A brief introduction to information geometry is given in Appendix A. Information geometry provides mathematical tools to analyze families of probability distributions and positive measures. The structure of a manifold of probability distributions is derived from the invariance principle, and it consists of a Riemannian metric derived from the Fisher information matrix, together with dually coupled $\pm\alpha$ affine connections. An f-divergence always induces such a structure. However, in general, a Bregman divergence is not invariant, and hence gives a Riemannian metric different from the Fisher information matrix and a different pair of dual affine connections. The Alpha-divergences belong to both classes in the space of positive measures, however, the KL-divergence is the only one belonging to the two classes in the case of probability distributions.

2.1 Error-induced Distance and Robust Regression Techniques

For arbitrary sequences described by n-dimensional vectors in the Euclidean space, the Minkowski family of metrics, also known as the ℓ_p-norm, includes most commonly used similarity (error) measures:

$$D_M(\boldsymbol{p}\,||\,\boldsymbol{q}) = \left(\sum_{i=1}^{n} |p_i - q_i|^p\right)^{1/p} = \left(\sum_{i=1}^{n} e_i^p\right)^{1/p}, \tag{2.4}$$

[2]The distance between two pdfs is called a metric if the following conditions hold: $D(\boldsymbol{p}\,||\,\boldsymbol{q}) \geq 0$ with equality iff $\boldsymbol{p} = \boldsymbol{q}$, $D(\boldsymbol{p}\,||\,\boldsymbol{q}) = D(\boldsymbol{q}\,||\,\boldsymbol{p})$ and $D(\boldsymbol{p}\,||\,\boldsymbol{q}) \leq D(\boldsymbol{p}\,||\,\boldsymbol{z}) + D(\boldsymbol{z}\,||\,\boldsymbol{q})$. Distances which are not a metric, are referred to as divergences.

where,

$$e_i = p_i - q_i \tag{2.5}$$

is regarded as the error component when estimating $\boldsymbol{q}$ from observed $\boldsymbol{p}$, where $\boldsymbol{q}$ belongs to a model to which constraints are imposed. The ℓ_2-norm, also known as the Euclidean distance, defined as

$$D_2(\boldsymbol{p}\,||\,\boldsymbol{q}) = \sum_i (p_i - q_i)^2 = \sum_i e_i^2 \tag{2.6}$$

is probably the most frequently used metric for low-dimensional data. Its popularity is due to a number of factors, especially for its intuitive simplicity and the optimality of estimation for the case of Gaussian noise or error. However, if the observations are contaminated not only by Gaussian noise but also by outliers, estimators based on this metric can be strongly biased. The Euclidean distance as well as other Minkowski metrics suffer from the curse of dimensionality when they are applied to high-dimensional data. Namely, in Euclidean spaces sparsity of observations increases exponentially with the number of dimensions, leading to observations becoming equidistant in terms of the Euclidean distance.

The squared Euclidean distance is often parameterized using a symmetric positive definite weighting matrix $\mathbf{W}$ as follows:

$$D_{\mathbf{W}}(\boldsymbol{p}\,||\,\boldsymbol{q}) = ||\boldsymbol{p} - \boldsymbol{q}||^2_{\mathbf{W}} = (\boldsymbol{p} - \boldsymbol{q})^T \mathbf{W} (\boldsymbol{p} - \boldsymbol{q}). \tag{2.7}$$

The weighting can be chosen to emphasize parts of the data to be approximated. This form of the parameterized squared Euclidean distance is equivalent to the Mahalanobis distance for a positive semi-definite weight matrix $\mathbf{W}$. It can be shown that if the parameterized squared Euclidean distance (Mahalanobis distance) is minimized for the Maximum Likelihood (ML) estimate with non-white Gaussian noise, then the matrix $\mathbf{W}$ corresponds to the inverse of the covariance matrix of noise.

Remark 2.1 *In this chapter we consider for simplicity 1D nonnegative patterns. However, all the cost function can be directly extended for nonnegative matrices or tensors. For example, for two matrices* $\mathbf{P} = \mathbf{Y} = [p_{it}]$ *and* $\mathbf{Q} = \mathbf{A}\mathbf{X} = [q_{it}]$*, the squared Euclidean distance is defined as*

$$D_2(\mathbf{P}\,||\,\mathbf{Q}) = ||\mathbf{P} - \mathbf{Q}||^2_F = \sum_{it} |p_{it} - q_{it}|^2. \tag{2.8}$$

In many applications, we use the weighted square distance defined as

$$D_{2W}(\mathbf{P}\,||\,\mathbf{Q}) = ||\mathbf{W}^{1/2} \odot (\mathbf{P} - \mathbf{Q})||^2_F = \sum_{it} w_{it} |p_{it} - q_{it}|^2, \tag{2.9}$$

where $\mathbf{W}$ is a symmetric nonnegative weight matrix and $\odot$ denotes the element-wise (Hadamard) product of two matrices. The weight matrix $\mathbf{W}$ reflects our confidence in the entries of $\mathbf{P}$ and when each p_{it} represents an observation or measurement with variance σ^2_{it}, we may set $w_{it} = 1/\sigma^2_{it}$. In other words, such a distance is obtained when considering the ML estimate with white Gaussian noise, that is, when the covariance matrix of the noise is diagonal and the variance of each associated noise variable n_{it} is $\sigma^2_{it} = 1/w_{it}$. If any entry of $\mathbf{P}$ is missing, one sets the corresponding weight to zero.

It should be noted that any norm (not necessarily the Frobenius norm) can be used as a cost function. For instance, using the ℓ_p-norm we have the following distance measure:

$$D_p(\mathbf{P}\,||\,\mathbf{Q}) = \sum_{it} \left[w_{it}|p_{it} - q_{it}|^p\right]^{1/p}, \qquad 1 < p < \infty \tag{2.10}$$

$$D_\infty(\mathbf{P}\,||\,\mathbf{Q}) = \max_{it}\{w_{it}|p_{it} - q_{it}|\}, \qquad p = \infty. \tag{2.11}$$

The ℓ_p-norm cost functions are strictly convex for $1 < p < \infty$, leading to a unique solution (but with respect to only one set of parameters).

2.2 Robust Estimation

To enhance robustness of estimation techniques, we may formally apply the weighted squared Euclidean distance:

$$D_{2w}(\boldsymbol{p}\,||\,\boldsymbol{q}) = \sum_i w_i(p_i - q_i)^2. \tag{2.12}$$

This leads to the M-estimator introduced by Huber in 1973 and it is the simplest approach both theoretically and computationally for robust statistics. Robust estimators are based on the idea of replacing the squared function of residual errors $e_i = p_i - q_i$ by another function of the errors which is less sensitive to outliers, that is by defining to the following robust cost (loss) function:

$$J_\rho(\boldsymbol{e}) = \sum_{i=1}^{n} \rho[e_i(p_i, q_i)], \tag{2.13}$$

where $\rho[e]$ is a function which a user may choose. From a statistical point of view, this corresponds to the negative log likelihood, when the probability distribution error vector $\boldsymbol{e}$ is expressed as

$$p(\boldsymbol{e}) = c \prod_{i=1}^{n} \exp(-\rho(e_i)). \tag{2.14}$$

The cost (loss) function should satisfy the following properties:

- nonnegative valued, $\rho(e) \geq 0, \;\; \forall e$,
- even symmetric, $\rho(e) = \rho(-e)$,
- nondecreasing for $e \geq 0$,
- has unique minimum at the origin, $\rho(0) = 0$,
- has first two derivatives piecewise continuous.

Figure 2.1 and Figure 2.2 present the plots of some popular cost (loss) functions $\rho(e)$, the corresponding influence (activation) functions $\Psi(e) = d\rho/de$, and weight functions defined by $w(e) = \Psi(e)/e$. Typical robust loss functions and the corresponding influence functions are summarized in Table 2.1.

The influence function is a measure of robustness which intuitively is the change in an estimate caused by insertion of outlying data as a function of the distance of the data from the (uncorrupted) estimate. For example, the influence function of the squared Euclidean estimator is simply proportional to the error i.e. the distance of the point from the estimate. To achieve robustness, the influence function should increase more slowly or tend to zero with increasing distance. Ideally, the influence function, should distinguish inliers from outliers and at least reduce influence of outliers.

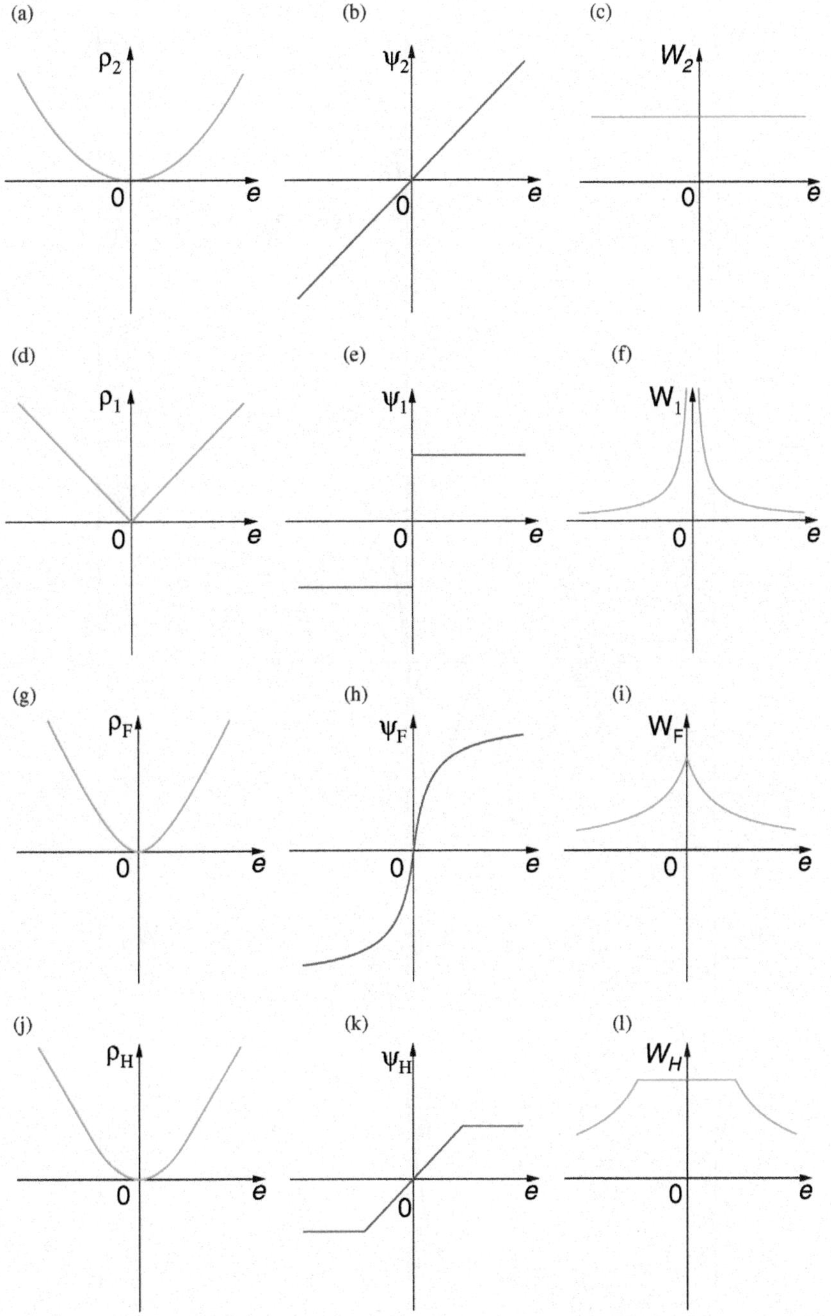

Figure 2.1 Plots of typical M-functions: (a),(b),(c) ℓ_2-norm; (d),(e),(f) ℓ_1-norm; (g),(h),(i) Fair; (j),(k),(l) Huber.

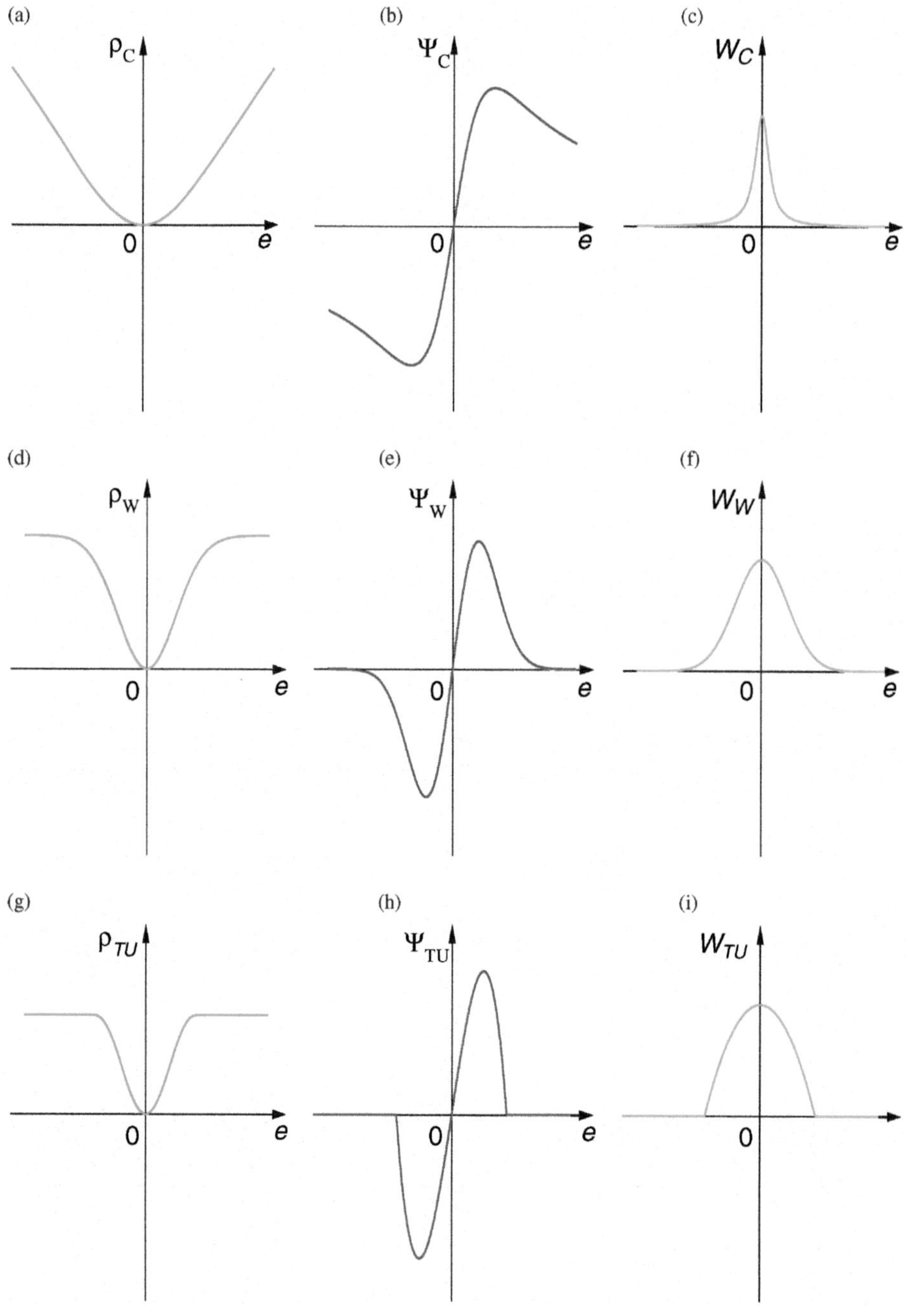

Figure 2.2 Plots of typical M-functions (continued): (a),(b),(c) Cauchy; (d),(e),(f) Welsh; (g),(h),(i) Tukey. ($\rho(e)$ - loss function, $\Psi(e)$ influence function, and $w(e)$-weight function).

Table 2.1 Basic robust loss functions $\rho(e)$ and the corresponding influence functions $\Psi(e) = d\rho(e)/de$.

Name	Loss Function $\rho(e)$	Influence Functions $\Psi(e)$
Logistic	$\rho_L = \frac{1}{\beta}\ln(\cosh(\beta e))$	$\Psi_L = \tanh(\beta e)$
Huber	$\rho_H = \begin{cases} e^2/2, & for\ \lvert e\rvert \le \beta; \\ \beta\lvert e\rvert - \frac{\beta^2}{2}, & otherwise \end{cases}$	$\Psi_H = \begin{cases} e, & for\ \lvert e\rvert \le \beta; \\ \beta sign(e), & otherwise \end{cases}$
ℓ_p-norm	$\rho_{\ell_p} = \frac{1}{p}\lvert e\rvert^p$	$\Psi_{\ell_p} = \lvert e\rvert^{p-1} sign(e)$
Cauchy	$\rho_C = \frac{\sigma^2}{2}\ln\left[1 + \left(\frac{e}{\sigma}\right)^2\right]$	$\Psi_C = \frac{e}{1 + \left(\frac{e}{\sigma}\right)^2}$
Geman, McCulre	$\rho_G = \frac{1}{2}\frac{e^2}{\sigma^2 + e^2}$	$\Psi_G = \frac{\sigma^2 e}{(\sigma^2 + e^2)^2}$
Welsh	$\rho_W = \frac{\sigma^2}{2}\left[1 - \exp\left(-\left(\frac{e}{\sigma}\right)^2\right)\right]$	$\Psi_W = e\exp\left(-\left(\frac{e}{\sigma}\right)^2\right)$
Fair	$\rho_F = \sigma^2\left[\frac{\lvert e\rvert}{\sigma} - ln\left(1 + \frac{\lvert e\rvert}{\sigma}\right)\right]$	$\Psi_F = \frac{e}{1 + \frac{\lvert e\rvert}{\sigma}}$
$\ell_1 - \ell_2$ norm	$\rho_{\ell_2-\ell_1} = 2(\sqrt{1 + e^2/2} - 1)$	$\Psi_{\ell_2-\ell_1} = \frac{e}{\sqrt{1 + e^2/2}}$
Talvar	$\rho_{Ta} = \begin{cases} e^2/2, & for\ \lvert e\rvert \le \beta; \\ \beta^2/2, & otherwise \end{cases}$	$\Psi_{Ta} = \begin{cases} e, & for\ \lvert e\rvert < \beta; \\ 0, & otherwise \end{cases}$
Hampel	$\rho_{Ha} = \begin{cases} \frac{\beta^2}{\pi}\left(1 - \cos\left(\frac{\pi e}{\beta}\right)\right), & for\ \lvert e\rvert \le \beta; \\ \frac{2\beta^2}{\pi}, & otherwise \end{cases}$	$\Psi_{Ha} = \begin{cases} \beta\sin\left(\frac{\pi e}{\beta}\right), \\ 0 \end{cases}$

Example 2.1:

1. For the Welsh function:

$$\rho_W(e) = \frac{\sigma^2}{2}\left(1 - \exp\left(-\left(\frac{e}{\sigma}\right)^2\right)\right) \tag{2.15}$$

the weight function can be defined as

$$w_W(e) = \frac{\Psi_W(e)}{e} = \exp\left(-\left(\frac{e}{\sigma}\right)^2\right), \tag{2.16}$$

and the derivative of the loss function

$$\Psi_W(e) = \frac{d\rho_W(e)}{de} = e\exp\left(-\left(\frac{e}{\sigma}\right)^2\right), \tag{2.17}$$

which called the influence function.[3]

Note that by taking $\rho(e) = e^2/2$, we obtain the ordinary squared Euclidean distance (i.e., linear least-squares problem). Analogously, by choosing $\rho(e) = |e|$, we obtain the total variation metric called also the Least Absolute Deviation (LAD) estimation. In fact, both loss functions belong to a wide family of M-estimators.[4] However, in order to reduce the influence of outliers, other weighting functions can be chosen (see Figure 2.1 and Figure 2.2).

2. A most popular robust loss function is probably the Huber function defined as

$$\rho_H(e) = \begin{cases} \dfrac{e^2}{2}, & \textit{for } |e| \le \beta; \\ \beta|e| - \dfrac{\beta^2}{2}, & \textit{otherwise} \end{cases} \tag{2.18}$$

where the cut-off parameter $\beta > 0$ (typically $\beta \le 2$) tunes efficiency, e.g. $\beta = 1.345$ yields 0.95 asymptotic efficiency[5] with the normally distributed errors (see Figure 2.1 (j)–(l)). In practice, an accurate value of β is not important. The corresponding weight function is defined as

$$w_H(e) = \begin{cases} 1, & \textit{for } |e| \le \beta; \\ \beta/|e|, & \textit{otherwise} \end{cases} \tag{2.19}$$

3. The Tuckey biweight estimator (see Figure 2.2 (g)–(i)) is often used with the loss function:

$$\rho_{Tu}(e) = \begin{cases} \dfrac{1}{2}e^2 - \dfrac{e^4}{4\beta^2}, & \textit{for } |e| \le \beta; \\ \dfrac{\beta^2}{4}, & \textit{otherwise} \end{cases} \tag{2.20}$$

---

[3]It is important to note that the so defined cost function is convex if the loss functions are convex.

[4]The ordinary least-squares error criterion equals the weights for all the errors and may produce biased estimates, if the observed data are contaminated by impulsive noise or large isolated errors.

[5]For an unbiased estimator, asymptotic efficiency is the limit of its efficiency as the sample size tends to infinity. An estimator with asymptotic efficiency 1.0 is said to be an "asymptotically efficient estimator". Roughly speaking, the precision of an asymptotically efficient estimator tends to the theoretical limit as the sample size grows.

where β is a positive tuning parameter which controls asymptotic efficiency (typically, $\beta = 4.685$ for the same 0.95 asymptotic efficiency with a Gaussian distribution), with the weight function:

$$w_{Tu}(e) = \begin{cases} 1 - \dfrac{e^2}{\beta^2}, & \mathit{for}\ \ |e| \le \beta; \\ 0 & \mathit{otherwise} \end{cases} \tag{2.21}$$

4. The logistic function (see Table 2.1) is given by

$$\rho_L[e] = \beta^2 \ln(\cosh(e/\beta)), \tag{2.22}$$

with the weight function

$$w_L(e) = \left(\frac{e}{\beta}\right)^{-1} \tanh\left(\frac{e}{\beta}\right), \tag{2.23}$$

where β is a problem dependent parameter, which is called the cut-off parameter.

5. Andrew's wave function has the form

$$\rho_A(e) = \begin{cases} 1 - \cosh(e/\beta), & \mathit{for}\ \ |e| \le \beta; \\ 2\beta, & \mathit{otherwise} \end{cases} \tag{2.24}$$

with typical value of $\beta = 1.339$ and the weight function

$$w_A(e) = \begin{cases} -\sinh(e/\beta)/(e\beta), & \mathit{for}\ \ |e| \le \beta; \\ 0, & \mathit{otherwise} \end{cases} \tag{2.25}$$

6. Talwar's function is given by

$$\rho_{Ta}(e) = \begin{cases} e^2/2, & \mathit{for}\ \ |e| \le \beta; \\ \beta^2/2, & \mathit{otherwise} \end{cases} \tag{2.26}$$

with the weights

$$w_{Ta}(e) = \begin{cases} 1, & \mathit{for}\ \ |e| \le \beta; \\ 0, & \mathit{otherwise} \end{cases} \tag{2.27}$$

Generally, there are three kinds of loss functions $\Psi(e)$ (see Table 2.1):

1. Monotone nondecreasing $\Psi(e)$, e.g., Huber and Fair functions.
2. Hard redescended $\Psi(e)$ with $\Psi(e) = 0$ for $|e| \ge \beta$, e.g., Talvar, Tuckey, Hampel functions.
3. Soft rediscounted $\Psi(\pm\infty) = 0$, e.g. Cauchy function.

In many applications, such as pattern matching, image analysis, statistical learning, noise is not necessarily Gaussian and we use distances referred to as information divergences which are generally not ℓ_p norms. Several information divergences such as the Pearson Chi-square, Hellinger, Kullback-Leibler divergences have been studied in information theory and statistics. In the next sections, we will discuss two broad classes of parameterized information divergences, Csiszár f-divergences and Bregman divergences which include the Alpha- and Beta-divergences. We discuss their relationships from the point of view of information geometry [6].

2.3 Csiszár Divergences

The f-divergence was proposed by Csiszár in 1963 [20–24] and independently by Ali and Silvey in 1967 [1], and is defined as follows.

Definition 2.1 **Csiszár f-divergence**: *Given two distributions* $\mathbf{p} = (p_1, p_2, \ldots, p_n)$ *and* $\mathbf{q} = (q_1, q_2, \ldots, q_n)$*, the Csiszár f-divergence is defined as*

$$D_f(\mathbf{p} \,||\, \mathbf{q}) = \sum_i q_i f\left(\frac{p_i}{q_i}\right), \tag{2.28}$$

where f is a real-valued convex function over $(0, \infty)$ *and satisfies* $f(1) = 0$. *We define*[6] $0\, f(0/0) = 0$ *and* $0\, f(p/0) = \lim_{q \to 0+} q\, f(\frac{p}{q}) = p\, f'(\infty)$.

It is easy to see that $D_f(\mathbf{p} \,\|\, \mathbf{q}) \geq 0$, with the equality when and only when $\mathbf{p} = \mathbf{q}$. The ratio p_i/q_i is called the "likelihood ratio" and can be seen as the discrete Radon-Nikodym derivative $\mathbf{p}$ with respect to $\mathbf{q}$. The fundamental aspects of these measures are that they are connected with the "ratio test" in the Pearson-Neyman style hypothesis testing, and in many ways are "natural" concerning distributions and statistics. The research of Csiszár, Amari, Liese, and Vajda show that the f-divergences are a unique class of divergences on distributions that arise naturally from a set of axioms, e.g., permutation invariance and nondecreasing local projections [6,20,53].

In the case of continuous random variable x where a probability distribution is represented by a density function $p(x)$, the invariance is shown as follows. Let $y = r(x)$ be a continuous monotone function. Probability density functions $p(x)$ and $q(x)$ are changed to $\bar{p}(y)$ and $\bar{q}(y)$ by this transformation from x to y. The invariance requires

$$D_f(\mathbf{p} \,||\, \mathbf{q}) = D_f(\bar{\mathbf{p}} \,||\, \bar{\mathbf{q}}). \tag{2.29}$$

When x is discrete, it is easy to see that $D_f(\mathbf{p} \,||\, \mathbf{q})$ is invariant under the permutation of indices. We next show its non-monotone characteristic under summarization of indices. Among n indices $1, 2, \ldots, n$, let us summarize some indices, to form new m classes of indices. For example, when $n = 5$, we partition $\{1, 2, 3, 4, 5\}$ into $\{(1, 2), (3, 4), (5)\}$, that is, $m = 3$ in this case. In general let

$$A = \{A_1, A_2, \ldots, A_m\} \tag{2.30}$$

be a partition of $\{1, 2, \ldots, n\}$ such that

$$A_\kappa = \{\kappa_1, \kappa_2, \ldots, \kappa_s\}, \quad |A_\kappa| = s. \tag{2.31}$$

The partition induces a reduced probability distribution over the set A,

$$p_\kappa^A = \sum_{i \in A_\kappa} p_i, \quad \kappa = 1, 2, \ldots, m \leq n. \tag{2.32}$$

For two distributions $\mathbf{p}$ and $\mathbf{q}$, their reduced distributions are denoted by $\mathbf{p}^A$ and $\mathbf{q}^A$, respectively. Information is lost by this summarization. The invariance and monotonicity of the f-divergence is characterized by the following theorem.

[6] We use the standard convention that $0\, f(0/0) = 0$, thus the sums above can be replaced by the sums over $i \in \mathcal{I}$ with $p_i > 0$.

Theorem 2.1 *In general the f-divergence decreases by summarization,*

$$D(\boldsymbol{p}\,||\,\boldsymbol{q}) \geq D(\boldsymbol{p}^A\,||\,\boldsymbol{q}^A) = \sum_{\kappa=1}^{m} p_\kappa^A\, f\left(\frac{q_\kappa^A}{p_\kappa^A}\right). \tag{2.33}$$

The equality holds if and only if

$$\frac{q_i}{p_i} = \lambda_\kappa \quad for\ i \in A_\kappa. \tag{2.34}$$

Conversely, any separable divergence satisfying the above properties is an f-divergence.

For a positive constant c, f and cf give

$$cD_f(\boldsymbol{p}\,||\,\boldsymbol{q}) = D_{cf}(\boldsymbol{p}\,||\,\boldsymbol{q}). \tag{2.35}$$

This is known as the problem of scale, and we may normalize f such that $f''(1) = 1$. Furthermore, for $f \in \mathcal{F}$, and for an arbitrary constant c,

$$\tilde{f}(u) = f(u) - c(u-1) \tag{2.36}$$

gives the same divergence as f, that is,

$$D_f(\boldsymbol{p}\,||\,\boldsymbol{q}) = D_{\tilde{f}}(\boldsymbol{p}\,||\,\boldsymbol{q}). \tag{2.37}$$

For $c = f'(1)$, we have $\tilde{f}'(1) = 0$ and

$$\tilde{f}(u) \geq 0 \tag{2.38}$$

where the equality holds if and only if $u = 1$. We denote by $\mathcal{F}$ the class of convex functions satisfying $f(1) = 1$, $f'(1) = 0$, and $f''(1) = 1$. There is no loss of generality by using convex functions belonging to $\mathcal{F}$, when f is differentiable. Furthermore, if the Csiszár f-divergence is bounded then $\tilde{f}(0) = \lim_{u\to+0} \tilde{f}(u)$ exists.

The Csizár f-divergence is defined originally for two probability distributions. We can extend it for two positive measures $\tilde{\boldsymbol{p}}$ and $\tilde{\boldsymbol{q}}$, where the constraints $\sum \tilde{p}_i = 1$, and $\sum \tilde{q}_i = 1$ are removed. When we use a convex function $f \in \mathcal{F}$, the f-divergence is given by the same form [4]

$$D_f(\tilde{\boldsymbol{p}}\,||\,\tilde{\boldsymbol{q}}) = \sum \tilde{p}_i\, f\left(\frac{\tilde{q}_i}{\tilde{p}_i}\right). \tag{2.39}$$

However, when we use a general f with $f'(1) = c_f \neq 0$, this form is not invariant, and we need to use

$$D_f(\tilde{\boldsymbol{p}}\,||\,\tilde{\boldsymbol{q}}) = c_f \sum (\tilde{q}_i - \tilde{p}_i) + \sum \tilde{p}_i\, f\left(\frac{\tilde{q}_i}{\tilde{p}_i}\right). \tag{2.40}$$

When $\tilde{\boldsymbol{p}}$ and $\tilde{\boldsymbol{q}}$ are probability distributions, the additional term vanishes.

The Csiszár f-divergence corresponds to a generalized entropy of the form

$$H_f(\boldsymbol{p}) = -\sum_i f(p_i), \tag{2.41}$$

for which the Shannon entropy is a special case of $f(p) = p \ln p$.

The class of Csiszár f-divergences includes many popular and important divergences between two probability measures. Notably, the ℓ_1-norm distance is the Csiszár f-divergence with $f(u) = |u-1|$ and referred to as the total variation distance. Analogously, we realize the Hellinger distance with $f(u) = (\sqrt{u}-1)^2$ and the Jensen-Shannon divergence with $f(u) = u \ln u - (u+1)\ln((u+1)/2)$ (see Table 2.3 and Table 2.4

Table 2.2 Basic Csiszár f-divergences $D_f(\boldsymbol{p}\,||\,\boldsymbol{q}) = \sum_i q_i f\left(\frac{p_i}{q_i}\right)$.

Name	Function $f(u)$, $u = \frac{p}{q}$	Divergence $D_f(\boldsymbol{p}\,\|\,\boldsymbol{q})$
Total variation distance	$f(u) = f^*(u) = \|u-1\|$	$\sum_i \|p_i - q_i\|$
Pearson Chi-square distance	$(u-1)^2$	$\sum_i \frac{(p_i - q_i)^2}{q_i}$
Neyman Chi square	$\frac{(u-1)^2}{u}$	$\sum_i \frac{(p_i - q_i)^2}{p_i}$
Rukhin distance	$\frac{(u-1)^2}{a+(1-a)u}$	$\sum_i \frac{(p_i - q_i)^2}{(1-a)p_i + aq_i}$, $a \in [0, 1]$
Triangular Discrimination (TD)	$\frac{(u-1)^2}{u+1}$	$\sum_i \frac{(p_i - q_i)^2}{p_i + q_i}$
Squared Hellinger distance	$(\sqrt{u} - 1)^2$	$\sum_i (\sqrt{p_i} - \sqrt{q_i})^2$
Matsusita distance	$\|u^\alpha - 1\|^{\frac{1}{\alpha}}$ $0 \leq \alpha \leq 1$	$\sum_i \|p_i^\alpha - q_i^\alpha\|^{\frac{1}{\alpha}}$
Piuri and Vinche divergence	$\frac{\|1-u\|^\gamma}{(u+1)^{\gamma-1}}$ $\gamma \geq 1$	$\sum_i \frac{\|p_i - q_i\|^\gamma}{(p_i + q_i)^{\gamma-1}}$
Arimoto distance	$\sqrt{1+u^2} - \frac{1+u}{\sqrt{2}}$	$\sum_i \left(\sqrt{p_i^2 + q_i^2} - \frac{p_i + q_i}{\sqrt{2}}\right)$

for details). The Alpha-divergences, which we discuss later, can be expressed formally as the Csiszár f-divergence, as shown in Table 2.7.

For $f(u) = u(u^{\alpha-1} - 1)/(\alpha^2 - \alpha) + (1-u)/\alpha$, we obtain the Alpha-divergence

$$\boxed{D_A^{(\alpha)}(\boldsymbol{p}\,||\,\boldsymbol{q}) = \frac{1}{\alpha(\alpha-1)} \sum_i \left(p_i^\alpha q_i^{1-\alpha} - \alpha p_i + (\alpha - 1) q_i\right), \quad \alpha \in \mathbb{R}.} \tag{2.42}$$

Analogously, for $f(u) = [(u^\alpha + u^{1-\alpha} - (u+1)]/(\alpha^2 - \alpha)$, we obtain the symmetric Alpha-divergence of type I (2.114) and for $f(u) = [(u^{1-\alpha} + 1)((u+1)/2)^\alpha - (1+u)]/(\alpha^2 - \alpha)$, we obtain the symmetric Alpha-divergence of type II (2.118).

On the other hand, by choosing the following convex function [45] (see Table 2.2)

$$f(u) = \frac{(u-1)^2}{a + u(1-a)}, \qquad a \in [0, 1], \tag{2.43}$$

Table 2.3 Basic symmetric divergences expressed as the Csiszár f-Divergence (see also Figure 2.3).

Divergence $D(\boldsymbol{p}\,\|\,\boldsymbol{q})$	$f(u), \quad u \in (0, \infty)$	$f'(u)$	$f''(u) > 0$
Squared Hellinger distance D_H $\frac{1}{2}\sum_i (\sqrt{p_i} - \sqrt{q_i})^2$	$\frac{1}{2}(\sqrt{u} - 1)$	$\frac{1}{2\sqrt{u}}$	$\frac{1}{4u\sqrt{u}}$
Triangular Discrimination (TD) D_T $\sum_i \frac{(p_i - q_i)^2}{p_i + q_i}$	$\frac{(u-1)^2}{u+1}$	$\frac{(u-1)(u+3)}{(u+1)^2}$	$\frac{8}{(u+1)^3}$
Symmetric Chi-squared divergence D_X $\sum_i \frac{(p_i - q_i)^2 (p_i + q_i)}{p_i q_i}$	$\frac{(u-1)^2(u+1)}{u}$	$\frac{(u-1)(2u^2+u+1)}{u^2}$	$\frac{2(u^3+1)}{u^3}$
J-divergence D_J $\sum_i (p_i - q_i) \ln\left(\frac{p_i}{q_i}\right)$	$(u-1)\ln u$	$1 - u^{-1} + \ln u$	$\frac{u+1}{u^2}$
Jensen-Shannon divergence D_{JS} $\sum_i p_i \ln\left(\frac{2p_i}{p_i + q_i}\right) + q_i \ln\left(\frac{2q_i}{p_i + q_i}\right)$	$\frac{u}{2}\ln u + \frac{u+1}{2}\ln\frac{2}{u+1}$	$\frac{1}{2}\ln\frac{2u}{u+1}$	$\frac{1}{2u(u+1)}$
A-G Mean divergence D_{AG} $\sum_i \left(\frac{p_i + q_i}{2}\right) \ln\left(\frac{p_i + q_i}{2\sqrt{p_i q_i}}\right)$	$\frac{u+1}{2}\ln\left(\frac{u+1}{2\sqrt{u}}\right)$	$\frac{1}{4}\left[1 - u^{-1} + 2\ln\left(\frac{u+1}{2\sqrt{u}}\right)\right]$	$\frac{u^2+1}{4u^2(u+1)}$

Table 2.4 Relative divergences and the corresponding Csiszár functions $f(u)$ (see also Figure 2.3).

Divergence $D(\boldsymbol{p} \,\|\, \boldsymbol{q})$	$f(u), \quad u \in (0, \infty)$	$f'(u)$	$f''(u) > 0$
I-divergence $\sum_i \left(p_i \ln \left(\frac{p_i}{q_i} \right) - p_i + q_i \right)$	$1 - u + u \ln u$	$\ln u$	u^{-1}
Relative J-divergence $\sum_i (p_i - q_i) \ln \left(\frac{p_i + q_i}{2q_i} \right)$	$(u - 1) \ln \left(\frac{u+1}{2} \right)$	$\frac{u-1}{u+1} + \ln \left(\frac{u+1}{2} \right)$	$\frac{u+3}{(u+1)^2}$
Relative Jensen-Shannon divergence $\sum_i p_i \ln \left(\frac{2p_i}{p_i + q_i} \right) - \frac{p_i - q_i}{2}$	$u \ln \left(\frac{2u}{u+1} \right) - \frac{u-1}{2}$	$\frac{1}{2} \frac{u-1}{u+1} + \ln \left(\frac{2u}{u+1} \right)$	$\frac{1}{u(u+1)^2}$
Relative Arithmetic-Geometric divergence $\sum_i \frac{p_i + q_i}{2} \ln \left(\frac{p_i + q_i}{2p_i} \right) + \frac{p_i - q_i}{2}$	$\frac{(u+1)}{2} \ln \left(\frac{u+1}{2u} \right) + \frac{u-1}{2}$	$\frac{1}{2} \left[\ln \left(\frac{u+1}{2u} \right) - \frac{u-1}{u} \right]$	$\frac{1}{2u^2(u+1)}$

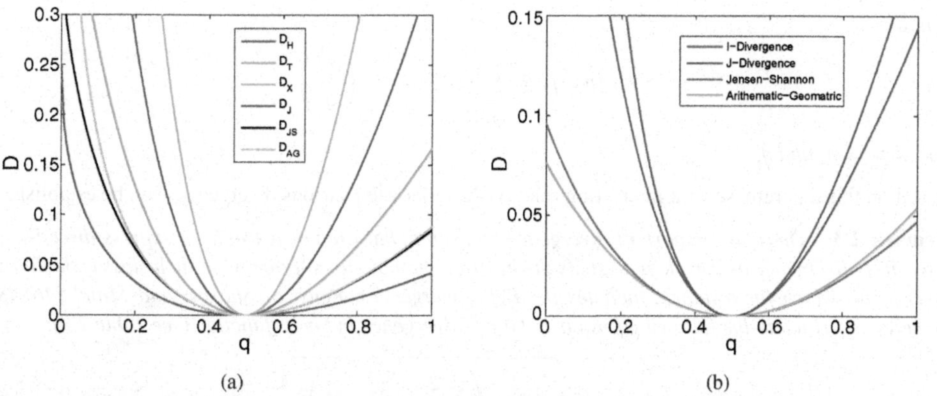

Figure 2.3 2D plots of basic divergences see: (a) Table 2.3 and (b) Table 2.4. All 2D plots of various divergences presented in this chapter were obtained for fixed parameter $p = 0.5$ and we changed only values of parameter q.

we obtain the divergence (called Rukhin distance) which connects smoothly the Pearson and Neyman divergences:

$$D_{Ru}(\mathbf{p} \,||\, \mathbf{q}) = \sum_i \frac{(p_i - q_i)^2}{((1-a)p_i + aq_i)}, \qquad a \in [0, 1]. \tag{2.44}$$

Remark 2.2 *In general, the Csiszár f-divergences are not symmetric, in the sense that $D_f(\mathbf{p} \,||\, \mathbf{q})$ does not necessarily equals $D_f(\mathbf{q}||\mathbf{p})$, unless the function $f(u)$ satisfies:*

$$f(u) = u\, f(1/u) + c(u-1), \tag{2.45}$$

for some fixed constant[7] *c. When the Csiszár f-divergence is not symmetric, we may define the conjugate generated function $f^*(u) = uf(1/u)$ of f such that*

$$D_f(\mathbf{p} \,||\, \mathbf{q}) = D_{f^*}(\mathbf{q} \,||\, \mathbf{p}). \tag{2.46}$$

Similarly, for an arbitrary Csiszár f-divergence, it is possible to construct a symmetric divergence using the convex function $f_s(u) = f(u) + f^*(u)$. For instance, the reversed I-divergence $D_{KL}(\mathbf{q}||\mathbf{p})$ and the corresponding J-divergence are derived in this way. Additionally, we have the following fundamental relationship:

$$0 \le D_f(\mathbf{p} \,||\, \mathbf{q}) \le f(0) + f^*(0), \tag{2.47}$$

where $f^*(0) = \lim_{u\to\infty} f(u)/u$.

If the generated functions $f(u)$ are differentiable, convex and normalized, in the sense $f(1) = 0$, we can then apply the following theorem [27]:

Theorem 2.2 *Let $f : \mathbb{R}_+ \to \mathbb{R}$ be differentiable, convex and normalized, i.e., $f(1) = 0$. Then*

$$0 \le D_f(\mathbf{p} \,||\, \mathbf{q}) \le D_E(\mathbf{p} \,||\, \mathbf{q}), \tag{2.48}$$

[7]For instance, for the triangular discrimination, squared Hellinger distance and total variational distance.

where

$$D_E(\boldsymbol{p}\,||\,\boldsymbol{q}) = \sum_i (p_i - q_i)\, f'\left(\frac{p_i}{q_i}\right) \tag{2.49}$$

for all $p_i > 0$, and $q_i > 0$.

Based on this theorem, several interesting relationships between various divergences can be established.

Remark 2.3 *There are important divergence measures that are not possible to express directly as the Csiszár f-divergence as they do not satisfy the invariant monotone criterion. One such class is the Bergman divergence, which for instance, includes the Beta-divergences, Rényi or Sharma and Mittel [46–48]. A divergence written in the following form is not an f-divergence but is its function (see Table 2.9):*

$$D_{f,h}(\boldsymbol{p}\,||\,\boldsymbol{q}) = h\left(\sum_i q_i\, f(p_i/q_i)\right), \tag{2.50}$$

where $h(x)$ is an increasing differentiable function mapping from $[0, \infty)$ to $[0, \infty)$ with $h(0) = 0$ and $h'(0) > 0$. Such a divergence satisfies the invariance monotone property. In Table 2.5 we present such generalized divergences.

2.4 Bregman Divergence

Another class of divergence functions widely used in optimization and clustering is the so-called Bregman divergence. It plays an essential role in unifying a class of projection and alternating minimization algorithms [7,9,25,26,29,40]. This is due to its dually flat geometrical structure explained in a later section on information geometry, where the generalized Pythagorean theorem and projection theorem play a fundamental role.

Definition 2.2 *When $\boldsymbol{\phi}(p)$ is a strictly convex real-valued function, the generalized ϕ-entropy of a discrete measure $p_i \geq 0$ is defined by*

$$H_\phi(\boldsymbol{p}) = -\sum_i \phi(p_i) \tag{2.51}$$

and the Bregman divergence is given by

$$D_{\boldsymbol{\phi}}(\boldsymbol{p}\,||\,\boldsymbol{q}) = \sum_i \left(\boldsymbol{\phi}(p_i) - \boldsymbol{\phi}(q_i) - \boldsymbol{\phi}'(q_i)(p_i - q_i)\right), \tag{2.52}$$

where $\boldsymbol{\phi}'(q)$ denotes the derivative with respect to q.

In our applications, we will mostly employ separable divergences $D_{\boldsymbol{\phi}}(\boldsymbol{p}||\boldsymbol{q}) = \sum_{i=1}^{n} D_{\boldsymbol{\phi}}(p_i||q_i)$. However, the Bregman divergence can be defined in a more general form as shown below.

Definition 2.3 *If $\boldsymbol{\phi}(\boldsymbol{p}) : \mathbb{R}^n \to \mathbb{R}$ is a strictly convex and C^1 (i.e., first-order differentiable) function, the corresponding Bregman distance (divergence) is defined by*

$$D_{\boldsymbol{\phi}}(\boldsymbol{p}\,||\,\boldsymbol{q}) = \boldsymbol{\phi}(\boldsymbol{p}) - \boldsymbol{\phi}(\boldsymbol{q}) - (\boldsymbol{p} - \boldsymbol{q})^T \nabla\boldsymbol{\phi}(\boldsymbol{q}), \tag{2.53}$$

where $\nabla\boldsymbol{\phi}(\boldsymbol{q})$ represents the gradient of $\boldsymbol{\phi}$ evaluated at $\boldsymbol{q}$. Since the Bregman divergence is generally non symmetric, for our purpose, we consider (2.52) as well as the dual Bregman divergence:

$$D_{\boldsymbol{\phi}}(\boldsymbol{q}||\boldsymbol{p}) = \boldsymbol{\phi}(\boldsymbol{q}) - \boldsymbol{\phi}(\boldsymbol{p}) + \nabla\boldsymbol{\phi}^T(\boldsymbol{p})(\boldsymbol{p} - \boldsymbol{q}). \tag{2.54}$$

Table 2.5 Generalized divergences which do not belong to family of Csiszár *f*-divergences: $D_{h,f}(\boldsymbol{p}||\boldsymbol{q}) = \sum_i h\left(q_i f\left(\frac{p_i}{q_i}\right)\right)$.

Divergence	$h(u)$	$f(u)$
Rényi $\frac{\sum_i \ln\left(p_i^r q_i^{1-r} - r p_i + (r-1)\, q_i + 1\right)}{r(r-1)}$	$\frac{\ln\left(r(r-1)u+1\right)}{r(r-1)}$	$\frac{u^r - r(u-1) - 1}{r(r-1)}, \quad r \neq 0, 1$
Battacharya $-\sum_i \ln\left(1 - \frac{(\sqrt{p_i} - \sqrt{q_i})^2}{2}\right)$	$-\ln(1-u)$	$\frac{1}{2}(u+1) - u^{\frac{1}{2}}$
Sharma-Mittal $\frac{\sum_i \left(\left[p_i^{\alpha} q_i^{1-\alpha} - \alpha p_i + (\alpha-1) q_i + 1\right]^{\frac{\beta-1}{\alpha-1}}\right)}{\beta - 1}$	$\frac{1}{\beta-1}\left[(1+\alpha(\alpha-1)u)^{\frac{\beta-1}{\alpha-1}}\right]$	$\frac{u^{\alpha} - \alpha(u-1) - 1}{\alpha(\alpha-1)}, \quad \alpha, \beta \neq 0, 1$
Tsallis $\frac{1}{\kappa}\sum_i \left(p_i(p_i^{\kappa} - q_i^{\kappa})\right) - \sum_i \left(q_i^{\kappa}(p_i - q_i)\right)$	$q^{\kappa} u$	$\frac{u}{\kappa}(u^{\kappa} - 1) - (u-1), \quad \kappa \neq 0$

We formulate the dual Bregman divergence by using the Legendre transformation, which is a useful tool for analyzing convex problems. It is shown later that this duality is the essence of information geometry [6]. Let $\phi(\mathbf{p})$ be a convex function then,

$$\mathbf{p}^* = \nabla\phi(\mathbf{p}) \tag{2.55}$$

gives a one-to-one correspondence with $\mathbf{p}$. Hence, we may regard $\mathbf{p}^*$ as another dual representation of $\mathbf{p}$. There exists a convex function $\phi^*(\mathbf{p}^*)$ derived from

$$\phi^*(\mathbf{p}^*) = \max_{\mathbf{p}}\{\mathbf{p}^T\mathbf{p}^* - \phi(\mathbf{p})\}. \tag{2.56}$$

The two convex functions satisfy

$$\phi(\mathbf{p}) + \phi^*(\mathbf{p}^*) - \mathbf{p}^T\mathbf{p}^* = 0, \tag{2.57}$$

when $\mathbf{p}$ and $\mathbf{p}^*$ correspond to each other. The reverse transformation is given by

$$\mathbf{p} = \nabla\phi^*(\mathbf{p}^*), \tag{2.58}$$

so that they are dually coupled.

By using the dual representation, the Bregman divergence takes a symmetric form

$$D_\phi(\mathbf{p}\,||\,\mathbf{q}) = \phi(\mathbf{p}) + \phi^*(\mathbf{q}^*) - \mathbf{p}^T\mathbf{q}^*, \tag{2.59}$$

which is the same as (2.53), and is nonnegative. It is zero if and only if $\mathbf{p} = \mathbf{q}$.

Example 2.2 (see also Table 2.6):

1. For $\phi(p) = p\ln p$, the generalized entropy H_ϕ simplifies to the Shannon entropy $H_S(\mathbf{p}) = -\sum_i p_i \ln p_i$ and $D_{\boldsymbol{\phi}}(\mathbf{p}\,||\,\mathbf{q})$ is the I-divergence:

$$D_{KL}(\mathbf{p}\,||\,\mathbf{q}) = \sum_i p_i \ln\left(\frac{p_i}{q_i}\right) - p_i + q_i. \tag{2.60}$$

2. For $\phi(p) = -\ln p$, we obtain the Burg entropy $H_B(\mathbf{p}) = \sum \ln p_i$ and the discrete Itakura-Saito distortion:[8]

$$D_\phi(\mathbf{p}\,||\,\mathbf{q}) = D_{IS}(\mathbf{p}\,||\,\mathbf{q}) = \sum_i \left(\ln\left(\frac{q_i}{p_i}\right) + \frac{p_i}{q_i} - 1\right), \tag{2.61}$$

which often arises in spectral analysis of speech signals.

3. For $\phi(p) = p^2/2$, which represents energy, we have the standard squared Euclidean distance:

$$D_2(\mathbf{p}\,||\,\mathbf{q}) = \frac{1}{2}||\mathbf{p} - \mathbf{q}||_2^2 = \frac{1}{2}\sum_i (p_i - q_i)^2. \tag{2.62}$$

The case $\phi(\mathbf{p}) = \frac{1}{2}\mathbf{p}^T\mathbf{W}\mathbf{p}$, where $\mathbf{W}$ is a symmetric positive definite matrix, yields the weighted squared Euclidean distance

$$D_W(\mathbf{p}\,||\,\mathbf{q}) = \frac{1}{2}(\mathbf{p} - \mathbf{q})^T\mathbf{W}(\mathbf{p} - \mathbf{q}) = ||\mathbf{p} - \mathbf{q}||_W^2. \tag{2.63}$$

[8] We obtain similar cost function for $\phi(p) = (\ln p)^2$, which corresponds to the Gaussian entropy $H_G(\mathbf{p}) = \sum_i (\ln p_i)^2$.

4. Let $\phi(p) = p \ln p + (1 - p)\ln(1 - p)$ be defined in the range $p = [0, 1]$, (thus corresponding to the Bernoulli entropy), then we obtain the Fermi-Dirac distance (Bit entropic loss):

$$D_{FD}(\mathbf{p} \,||\, \mathbf{q}) = \sum_i \left(p_i \ln(p_i/q_i) + (1 - p_i) \ln \frac{1 - p_i}{1 - q_i} \right), \qquad p_i, q_i \in [0, 1]. \tag{2.64}$$

5. For $\phi(p) = p^2 + p \ln p,\ p \geq 0,$ we have the Log-Quad cost function:

$$D_{LQ}(\mathbf{p} \,||\, \mathbf{q}) = \sum_i \left(|p_i - q_i|^2 + p_i \ln(p_i/q_i) + q_i - p_i \right). \tag{2.65}$$

6. Let $\phi(p) = (p^\alpha - p^\beta)$, with $\alpha \geq 1$ and $0 < \beta < 1$, then for $\alpha = 2$ and $\beta = 0.5$ we have

$$D_\phi(\mathbf{p} \,||\, \mathbf{q}) = ||\mathbf{p} - \mathbf{q}||_2^2 + \sum_i \left(\frac{(\sqrt{p_i} - \sqrt{q_i})^2}{2\sqrt{q_i}} \right) \tag{2.66}$$

and for $\alpha = 1,\ \beta = 0.5$ we obtain

$$D_\phi(\mathbf{p} \,||\, \mathbf{q}) = \sum_i \left(\frac{(\sqrt{p_i} - \sqrt{q_i})^2}{2\sqrt{q_i}} \right) \tag{2.67}$$

7. The Beta-divergence (see Section 2.6 for detail): consider a convex function:

$$\phi(p) = \begin{cases} \dfrac{1}{\beta(\beta+1)} \left(p^{\beta+1} - (\beta+1)p + \beta \right), & \beta > -1, \\ p \ln p - p + 1, & \beta = 0, \\ p - \ln p - 1, & \beta = -1, \end{cases} \tag{2.68}$$

which is also related to the Tsallis entropy [42]. Using the definition of the Bregman divergence, for this function, we can construct the asymmetric Beta-divergence (2.123), in a slightly different but equivalent form:

$$D_B^{(\beta)}(\mathbf{p} \,||\, \mathbf{q}) = \begin{cases} \dfrac{1}{\beta(\beta+1)} \displaystyle\sum_i \left(p_i^{\beta+1} - q_i^{\beta+1} - (\beta+1) q_i^\beta (p_i - q_i) \right), & \beta > -1, \\ \displaystyle\sum_i \left(p_i \ln \left(\frac{p_i}{q_i} \right) - p_i + q_i \right), & \beta = 0, \\ \displaystyle\sum_i \left(\ln \left(\frac{q_i}{p_i} \right) + \frac{p_i}{q_i} - 1 \right), & \beta = -1. \end{cases} \tag{2.69}$$

which is an extension of Beta-divergence proposed by Eguchi, Kano, Minami [28,36] and is related to the density power divergence proposed to Basu [8].

The family of Beta-divergences is derived using the Bregman divergence whereas a family of Alpha-divergences is based on the Csiszár f-divergence. Both families of the divergences are complementary and together build up a wide class of useful divergences.

8. The Alpha-divergence (see Section 2.5 for detail): The Alpha-divergence belongs to the class of Csiszár f-divergence. To show that it is also given by the Bregman divergence in the space of positive measures

$\tilde{\mathcal{P}} = \{\tilde{\boldsymbol{p}}\}$, for which the constraint $\sum p_i = 1$ is not imposed, denote

$$\phi(\tilde{\boldsymbol{p}}) = \sum \tilde{p}_i, \tag{2.70}$$

which is not a convex function of $\tilde{\boldsymbol{p}}$. To make it convex, we introduce a new representation of $\tilde{\boldsymbol{p}}$, which we call the α-representation,

$$\tilde{p}_i^{(\alpha)} = \begin{cases} \tilde{p}_i^{\alpha-1}, & \alpha \neq 1, \\ \ln \tilde{p}_i, & \alpha = 1. \end{cases} \tag{2.71}$$

Then ϕ is a convex function with respect to $\tilde{\boldsymbol{p}}^{(\alpha)} = (\tilde{p}_1^{\alpha-1}, \cdots \tilde{p}_n^{\alpha-1})$, which we write as

$$\phi^{(\alpha)}(\tilde{\boldsymbol{p}}^{(\alpha)}) = \phi(\tilde{\boldsymbol{p}}), \tag{2.72}$$

where $\tilde{\boldsymbol{p}}$ is a function of $\boldsymbol{p}^{(\alpha)}$, that is, the inverse transformation of $\boldsymbol{p}^{(\alpha)}$,

$$\tilde{p}_i = \begin{cases} \{\tilde{p}_i^{(\alpha)}\}^{\frac{1}{\alpha-1}}, & \alpha \neq 1 \\ \exp\{\tilde{p}_i^{(\alpha)}\}. & \alpha = 1 \end{cases} \tag{2.73}$$

Indeed,

$$\phi^{(\alpha)} = \begin{cases} \sum \{\tilde{p}_i^{(\alpha)}\}^{\frac{1}{\alpha-1}}, & \alpha \neq 1, \\ \sum \exp\{\tilde{p}_i^{(\alpha)}\}, & \alpha = 1. \end{cases} \tag{2.74}$$

and the divergence derived from the convex function $\phi^{(\alpha)}$ is the Alpha-divergence.

Notice that $\phi^{(\alpha)}$ is convex with respect to $\tilde{\boldsymbol{p}}^{(\alpha)}$, but is not convex in the space $\mathcal{P}$ having the constraint $\sum \tilde{p}_i = 1$, as this constraint

$$\sum \{p_i^{(\alpha)}\}^{\frac{1}{\alpha-1}} = 1, \tag{2.75}$$

is nonlinear with respect to $\tilde{\boldsymbol{p}}^{(\alpha)}$.

The Alpha-divergence is derived in the space of positive measures by using

$$D_A^{(\alpha)}(\tilde{\boldsymbol{p}} \,\|\, \tilde{\boldsymbol{q}}) = \sum \tilde{p}_i \, f_\alpha\left(\frac{\tilde{q}_i}{\tilde{p}_i}\right), \tag{2.76}$$

where

$$f_\alpha(u) = \frac{1}{\alpha(1-\alpha)}(1-u^\alpha) - \frac{1}{1-\alpha}(u-1). \tag{2.77}$$

Thus f_α belongs to $\mathcal{F}$ and can be written in the Bregman form as

$$D_A^{(\alpha)}(\tilde{\boldsymbol{p}} \,||\, \tilde{\boldsymbol{q}}) = \frac{1}{\alpha(1-\alpha)}\left[\sum (1-\alpha)\,\tilde{q}_i + \alpha \tilde{p}_i - \tilde{p}_i^{1-\alpha}\,\tilde{q}_i^{\alpha}\right], \tag{2.78}$$

where the potential function is $\phi(\tilde{\boldsymbol{p}}) = \sum \tilde{p}_i$, and should be interpreted as a convex function of $\tilde{p}^{(\alpha)}$.

The Bregman divergence can be also interpreted as a measure of convexity of $\phi(\boldsymbol{p})$. This is easy to visualize in the one-dimensional case: by drawing a tangent line to the graph of ϕ at the point q, the Bregman divergence $D_\phi(p||q)$ is seen as the vertical distance between this line and the point $\phi(p)$ as illustrated in Figures 2.4 (a)–(c). It is easy to verify that any positive linear combination of the Bregman divergences is the Bregman divergence. Table 2.6 lists some typical convex functions and their corresponding Bregman divergences.

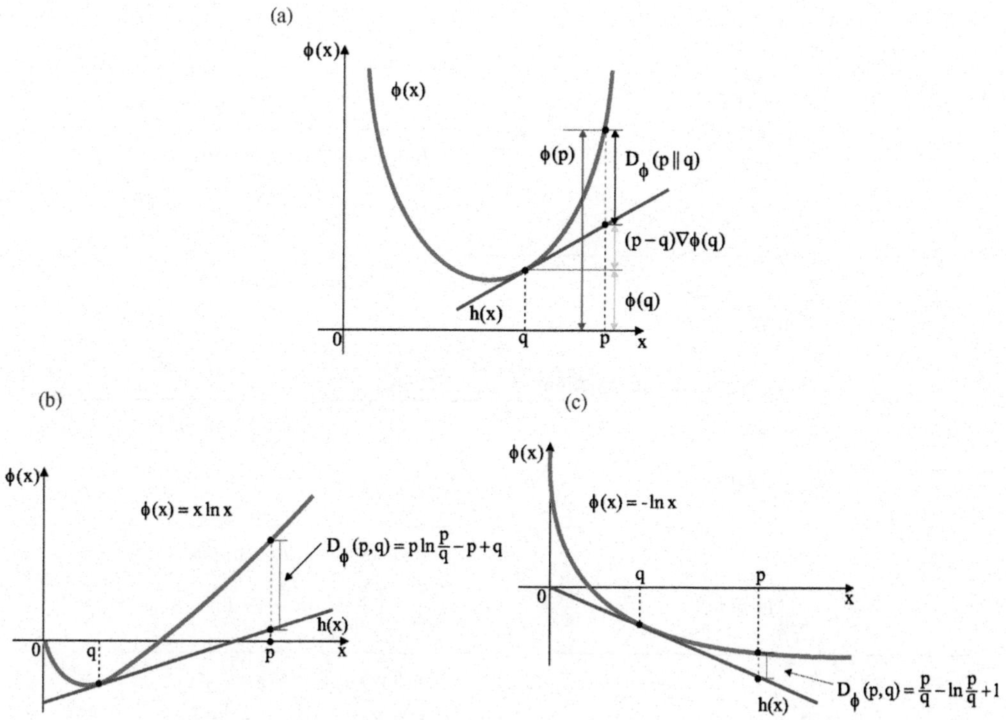

Figure 2.4 Graphical illustration of the Bregman divergence in a general case and three particular Bregman divergences for (a) Squared Euclidean distance, (b) I-divergence, and (c) Itakura-Saito distance.

It is important to mention that the derivative of the Bregman divergence with respect to $\boldsymbol{q}$ can be expressed as

$$\nabla_q D_{\phi}(\boldsymbol{p} \,||\, \boldsymbol{q}) = \nabla_q \left[\boldsymbol{\phi}(\boldsymbol{p}) - \boldsymbol{\phi}(\boldsymbol{q}) - \nabla \boldsymbol{\phi}^T(\boldsymbol{q})\,(\boldsymbol{p}-\boldsymbol{q})\right]$$
$$= -\nabla_q^2 \boldsymbol{\phi}^T(\boldsymbol{q})\,(\boldsymbol{p}-\boldsymbol{q}), \tag{2.79}$$

where $\nabla_q^2(\boldsymbol{\phi}((\boldsymbol{q}))$ is the matrix of the second order derivatives of $\boldsymbol{\phi}$ with respect to $\boldsymbol{q}$. If we assume that $\boldsymbol{\phi}(\boldsymbol{q}) = \sum_i \boldsymbol{\phi}(q_i)$, then $\nabla_q^2(\boldsymbol{\phi}((\boldsymbol{q}))$ is a diagonal matrix with positive diagonal elements.

In summary, the Bregman divergences have the following basic properties:

1. $D_{\phi}(\boldsymbol{p} \,||\, \boldsymbol{q})$ is convex in $\boldsymbol{p}$, but not convex (in general) in the second argument $\boldsymbol{q}$, however, it is convex with respect to the dual variable $\boldsymbol{q}^*$.
2. If $\boldsymbol{\phi}(\boldsymbol{p})$ is strictly convex, then $D_{\phi}(\boldsymbol{p} \,||\, \boldsymbol{q}) \geq 0$ and the equality holds if and only if $\boldsymbol{p} = \boldsymbol{q}$.
3. The Bregman divergence is not a metric (the triangular inequality does not hold in general) and usually not symmetric, that is,

$$D_{\phi}(\boldsymbol{p} \,||\, \boldsymbol{q}) \neq D_{\phi}(\boldsymbol{q} || \boldsymbol{p}). \tag{2.80}$$

4. The gradient can be written as

$$\nabla_p D_{\phi}(\boldsymbol{p} \,||\, \boldsymbol{q}) = \nabla \boldsymbol{\phi}(\boldsymbol{p}) - \nabla \boldsymbol{\phi}(\boldsymbol{q}). \tag{2.81}$$

Table 2.6 Bregman divergences $D_\phi(\boldsymbol{p}\,||\,\boldsymbol{q}) = \sum_{i=1}^{n} \phi(p_i) - \phi(q_i) - \phi'(q_i)\,(p_i - q_i)$ generated from basic convex function $\phi(\boldsymbol{p})$.

Domain	Function $\phi(\boldsymbol{p})$	Divergence $D(\boldsymbol{p}\|\|\boldsymbol{q})$
$\mathbb{R}^n$	$\frac{1}{2}\|\|\boldsymbol{p}\|\|_2^2 = \frac{1}{2}\sum_{i=1}^{n} p_i^2$	*1. Squared Euclidean distance* $\frac{1}{2}\|\|\boldsymbol{p}-\boldsymbol{q}\|\|_2^2 = \frac{1}{2}\sum_{i=1}^{n}(p_i - q_i)^2$
$\mathbb{R}^n$	$\frac{1}{2}\boldsymbol{p}^T\mathbf{W}\boldsymbol{p}$ **W**-symmetric positive definite	*2. Mahalanobis distance* $\frac{1}{2}(\boldsymbol{p}-\boldsymbol{q})^T\mathbf{W}(\boldsymbol{p}-\boldsymbol{q})$
$\mathbb{R}^n_+$	$\sum_{i=1}^{n}(p_i \ln p_i)$	*3. Generalized KL divergence, I-divergence* $\sum_{i=1}^{n}\left(p_i \ln \frac{p_i}{q_i} - p_i + q_i\right)$
$\mathbb{R}^n_{++}$	$-\sum_{i=1}^{n} \ln p_i$	*4. Itakura - Saito distance* $\sum_{i=1}^{n}\left(\frac{p_i}{q_i} - \ln \frac{p_i}{q_i} - 1\right)$
$\mathbb{R}^n_+$	$\sum_{i=1}^{n}\frac{1}{p_i}$	*5. Inverse* $\sum_{i=1}^{n}\left(\frac{p_i}{q_i^2} + \frac{1}{p_i} - \frac{2}{q_i}\right)$
$\mathbb{R}^n$	$\sum_{i=1}^{n}\exp(p_i)$	*6. Exponential* $\sum_{i=1}^{n}(e^{p_i} - (p_i - q_i + 1)e^{q_i})$

5. Linearity: for $a > 0$ we have

$$D_{\phi_1 + a\,\phi_2}(\boldsymbol{p}\,||\,\boldsymbol{q}) = D_{\phi_1}(\boldsymbol{p}||\boldsymbol{q}) + a D_{\phi_2}(\boldsymbol{p}\,||\,\boldsymbol{q}). \tag{2.82}$$

6. The Bregman divergence is invariant up to linear terms, that is,

$$D_{\phi + a\,p + c}(\boldsymbol{p}\,||\,\boldsymbol{q}) = D_{\phi}(\boldsymbol{p}||\boldsymbol{q}). \tag{2.83}$$

7. The three-point property generalizes the "Law of Cosines":

$$D_{\phi}(\boldsymbol{p}\,||\,\boldsymbol{q}) = D_{\phi}(\boldsymbol{p}\,||\,\boldsymbol{z}) + D_{\phi}(\boldsymbol{z}\,||\,\boldsymbol{q}) - (\boldsymbol{p}-\boldsymbol{z})^T(\nabla\phi(\boldsymbol{q}) - \nabla\phi(\boldsymbol{z})). \tag{2.84}$$

8. Generalized Pythagoras Theorem:

$$D_{\phi}(\boldsymbol{p}\,||\,\boldsymbol{q}) \geq D_{\phi}(\boldsymbol{p}\,||\,P_\Omega(\boldsymbol{q})) + D_{\phi}(P_\Omega(\boldsymbol{q})\,||\,\boldsymbol{q}), \tag{2.85}$$

where $P_\Omega(\boldsymbol{q}) = \arg\min_{\omega\in\Omega} D_{\phi}(\omega||\boldsymbol{q})$ is the Bregman projection onto the convex set Ω. When Ω is an affine set then it holds with equality. This is proved to be the generalized Pythagorean relation in terms of information geometry.

2.4.1 Bregman Matrix Divergences

The Bergman divergence can be generalized to give a measure (distance) that compares two real-valued symmetric n by n positive definite matrices which have the following eigenvalue decompositions:

$$\mathbf{P} = \mathbf{V}\boldsymbol{\Lambda}\mathbf{V}^T = \sum_{i=1}^{n} \lambda_i \boldsymbol{v}_i \boldsymbol{v}_i^T, \tag{2.86}$$

$$\mathbf{Q} = \mathbf{U}\tilde{\boldsymbol{\Lambda}}\mathbf{U}^T = \sum_{i=1}^{n} \tilde{\lambda}_i \boldsymbol{u}_i \boldsymbol{u}_i^T, \tag{2.87}$$

In such a case, we can define the Bregman matrix divergences as follows:

$$D_\Phi(\mathbf{P}\,||\,\mathbf{Q}) = \Phi(\mathbf{P}) - \Phi(\mathbf{Q}) - \mathrm{tr}\left((\nabla\Phi(\mathbf{Q}))^T(\mathbf{P}-\mathbf{Q})\right), \tag{2.88}$$

where the operator tr denotes the trace of a matrix [26].

We show a natural way of constructing a convex function of $\mathbf{P}$ since eigenvalue λ_i of $\mathbf{P}$ is a convex function of $\mathbf{P}$. We further note that tr$\mathbf{P}$ depends only on Λ in the decomposition (2.86). Let $f(\lambda)$ be a monotonically increasing function. Then

$$\Phi(\mathbf{P}) = \sum_i f(\lambda_i) = \mathrm{tr} f(\mathbf{P}) \tag{2.89}$$

is a convex function of $\mathbf{P}$, which leads to various types of the matrix Bregman divergence. We give typical examples in the following.

Example 2.3:

1. For $f(\lambda) = \frac{1}{2}\lambda^2$, and $\Phi(\mathbf{P}) = \frac{1}{2}\,\mathrm{tr}(\mathbf{P}^T\mathbf{P})$, we obtain the Squared Frobenius norm:

$$D_F(\mathbf{P}\,||\,\mathbf{Q}) = \frac{1}{2}||\mathbf{P}-\mathbf{Q}||_F^2. \tag{2.90}$$

2. For $f(\lambda) = \lambda\ln(\lambda) - \lambda$, and $\Phi(\mathbf{P}) = \mathrm{tr}(\mathbf{P}\ln\mathbf{P} - \mathbf{P})$, we obtain von Neumann divergence[9]

$$D_{vN}(\mathbf{P}\,||\,\mathbf{Q}) = \mathrm{tr}\left\{\mathbf{P}\ln\mathbf{P} - \mathbf{P}\ln\mathbf{Q} - \mathbf{P} + \mathbf{Q}\right\}, \tag{2.91}$$

   Using the eigenvalue decomposition we can write the von Neumann divergence as follows:

$$D_{vN}(\mathbf{P}\,||\,\mathbf{Q}) = \sum_i \lambda_i \ln\lambda_i - \sum_{ij} (\boldsymbol{v}_i^T\boldsymbol{u}_j)^2 \lambda_i \ln\tilde{\lambda}_j - \sum_i (\lambda_i - \tilde{\lambda}_i). \tag{2.92}$$

3. For $f(\lambda) = -\ln(\lambda)$, and $\Phi(\mathbf{P}) = -\ln(\det\mathbf{P})$, we have the Burg matrix divergence or Itakura-Saito matrix divergence:

$$D_{Burg}(\mathbf{P}\,||\,\mathbf{Q}) = \mathrm{tr}(\mathbf{P}\mathbf{Q}^{-1}) - \ln\det(\mathbf{P}\mathbf{Q}^{-1}) - n \tag{2.93}$$

$$= \sum_{ij} \frac{\lambda_i}{\tilde{\lambda}_i}(\boldsymbol{v}_i^T\boldsymbol{u}_j)^2 - \sum_i \ln\left(\frac{\lambda_i}{\tilde{\lambda}_i}\right) - n. \tag{2.94}$$

[9]This divergence corresponds to the von Neumann entropy defined as $-\,\mathrm{tr}(\mathbf{P}\ln\mathbf{P})$, which is the Shannon entropy of eigenvalues $-\sum_i \lambda_i \ln\lambda_i$.

4. For $f(\lambda) = \frac{1}{\alpha(\alpha-1)}(\lambda^\alpha - \lambda)$, and $\Phi(\mathbf{P}) = \mathrm{tr}(\mathbf{P}^\alpha - \mathbf{P})/(\alpha^2 - \alpha)$, we have the generalized Alpha-divergence,

$$D_A^{(\alpha)}(\mathbf{P}\,\|\,\mathbf{Q}) = \frac{1}{\alpha(\alpha-1)}\,\mathrm{tr}(\mathbf{P}^\alpha\mathbf{Q}^{1-\alpha} - \alpha\mathbf{P} + (\alpha-1)\mathbf{Q}). \tag{2.95}$$

2.5 Alpha-Divergences

The Alpha-divergences can be derived from the Csiszár f-divergence and also from the Bregman divergence [4]. The properties of the Alpha-divergence was proposed by Chernoff [13] and have been extensively investigated and extended by Amari [2,6] and other researchers. For some modifications and/or extensions see also works of Liese & Vajda, Cressie-Read, Zhu-Rohwer, Jun Zhang, Minka and Taneya [19,37,50–52,58].

2.5.1 Asymmetric Alpha-Divergences

The basic Alpha-divergence can be defined as [2]:

$$D_A^{(\alpha)}(\boldsymbol{p}\,||\,\boldsymbol{q}) = \frac{1}{\alpha(\alpha-1)}\sum_i \left(p_i^\alpha q_i^{1-\alpha} - \alpha p_i + (\alpha-1)q_i\right), \qquad \alpha \in \mathbb{R}, \tag{2.96}$$

where the $\boldsymbol{p}$ and $\boldsymbol{q}$ do not need to be normalized.[10]

Remark 2.4 *Various authors use the parameter α in different ways. For example, using the Amari notation, α_A with $\alpha = (1+\alpha_A)/2$, the Alpha-divergence takes the following form [2,3,6,18,19,53–58]:*

$$D_A^{(\alpha_A)}(\boldsymbol{p}\,||\,\boldsymbol{q}) = \frac{4}{1-\alpha_A^2}\sum_i \left(\frac{1-\alpha_A}{2}p_i + \frac{1+\alpha_A}{2}q_i - p_i^{\frac{1-\alpha_A}{2}}\, q_i^{\frac{1+\alpha_A}{2}}\right), \quad \alpha_A \in \mathbb{R} \tag{2.97}$$

When α takes values from 0 to 1, α_A takes values from -1 to 1. The duality exists between α_A and $-\alpha_A$, while the duality is between α and $1-\alpha$.

The Alpha-divergence attains zero if $q_i = p_i$, $\forall i$, and is otherwise positive, thus satisfying the basic property of an error measure. This fundamental property follows from the fact that the Alpha-divergence is a convex function with respect to p_i and q_i.

The basic properties of the Alpha-divergence can be summarized as follows:

- $D_A^{(\alpha)}(\boldsymbol{p}\,||\,\boldsymbol{q}) \geq 0$ and $D_A^{(\alpha)}(\boldsymbol{p}\,||\,\boldsymbol{q}) = 0$ if and only if $\boldsymbol{q} = \boldsymbol{p}$ almost everywhere, i.e., has desirable convexity property over all spaces and the minimum of D_A is relatively easily obtainable for one set of parameters (keeping the other set fixed);

- $D_A^{(\alpha)}(\boldsymbol{p}\,||\,\boldsymbol{q}) = D_A^{(1-\alpha)}(\boldsymbol{q}\,||\,\boldsymbol{p})$;

- $D_A^{(\alpha)}(c\boldsymbol{p}\,||\,c\boldsymbol{q}) = c D_A^{(\alpha)}(\boldsymbol{p}\,||\,\boldsymbol{q})$.

[10] Note that this form of Alpha-divergence differs slightly from the loss function given by Amari in 1985, because it was defined only for probability distributions. It was extended in Amari and Nagaoka in 2000 [2,3,6] for positive measures, by incorporating additional term. This term is needed to allow de-normalized densities (positive measures), in the same way that the extended (generalized) Kullback-Leibler divergences (2.101) or (2.102) differ from the standard forms (without terms $\pm(q_i - p_i)$) [58].

In the special cases for $\alpha = 2, 0.5, -1$, we obtain the well known Pearson Chi-square, Hellinger and inverse Pearson, also called the Neyman Chi-square distances, respectively, given respectively by

$$D_A^{(2)}(\mathbf{p} \,||\, \mathbf{q}) = D_P(\mathbf{p} \,||\, \mathbf{q}) = \frac{1}{2} \sum_i \frac{(p_i - q_i)^2}{q_i}, \tag{2.98}$$

$$D_A^{(1/2)}(\mathbf{p} \,||\, \mathbf{q}) = 2D_H(\mathbf{p} \,||\, \mathbf{q}) = 2 \sum_i (\sqrt{p_i} - \sqrt{q_i})^2, \tag{2.99}$$

$$D_A^{(-1)}(\mathbf{p} \,||\, \mathbf{q}) = D_N(\mathbf{p} \,||\, \mathbf{q}) = \frac{1}{2} \sum_i \frac{(p_i - q_i)^2}{p_i}. \tag{2.100}$$

For the singular values $\alpha = 1$ and $\alpha = 0$ the Alpha-divergences (2.42) have to be defined as limiting cases for respectively $\alpha \to 1$ and $\alpha \to 0$. When these limits are evaluated for $\alpha \to 1$ we obtain the generalized Kullback-Leibler divergence (I-divergence):

$$D_{KL}(\mathbf{p} \,||\, \mathbf{q}) = \lim_{\alpha \to 1} D_A^{(\alpha)}(\mathbf{p} \,||\, \mathbf{q}) = \sum_i \left(p_i \ln \left(\frac{p_i}{q_i} \right) - p_i + q_i \right), \tag{2.101}$$

with the conventions $0/0 = 0$, $0 \ln(0) = 0$ and $p/0 = \infty$ for $p > 0$.

From the inequality $p \ln p \geq p - 1$ it follows that the I-divergence is nonnegative and achieves zero if and only if $\mathbf{p} = \mathbf{q}$.

Similarly, for $\alpha \to 0$, we obtain the dual generalized Kullback-Leibler divergence:

$$D_{KL}(\mathbf{q} || \mathbf{p}) = \lim_{\alpha \to 0} D_A^{(\alpha)}(\mathbf{p} \,||\, \mathbf{q}) = \sum_i \left(q_i \ln \left(\frac{q_i}{p_i} \right) - q_i + p_i \right). \tag{2.102}$$

The Alpha-divergence can be evaluated in a more explicit form as

$$D_A^{(\alpha)} = \begin{cases} \dfrac{1}{\alpha(\alpha-1)} \displaystyle\sum_i \left(q_i \left[\left(\frac{p_i}{q_i} \right)^{\alpha} - 1 \right] - \alpha(p_i - q_i) \right), & \alpha \neq 0, 1 \\[2ex] \displaystyle\sum_i \left(q_i \ln \frac{q_i}{p_i} + p_i - q_i \right), & \alpha = 0 \\[2ex] \displaystyle\sum_i \left(p_i \ln \frac{p_i}{q_i} - p_i + q_i \right), & \alpha = 1. \end{cases} \tag{2.103}$$

In fact, the Alpha-divergence smoothly connects the I-divergence $D_{KL}(\mathbf{p} \,||\, \mathbf{q})$ with the dual I-divergence $D_{KL}(\mathbf{q} \,||\, \mathbf{p})$ and passes through the Hellinger distance [31]. Moreover, it also smoothly connects the Pearson Chi-square and Neyman Chi-square divergences and passes through the I-divergences.

In practice, we can use the regularized form of the Alpha-divergence [8]:

$$D_A^{(\alpha)}(\mathbf{p} \,||\, \mathbf{q}) = \frac{1}{\alpha(\alpha-1)} \sum_{i: p_i \neq 0} \left(p_i^{\alpha} q_i^{1-\alpha} - \alpha p_i + (\alpha - 1) q_i \right) + \frac{1}{\alpha} \sum_{i: p_i = 0} q_i, \tag{2.104}$$

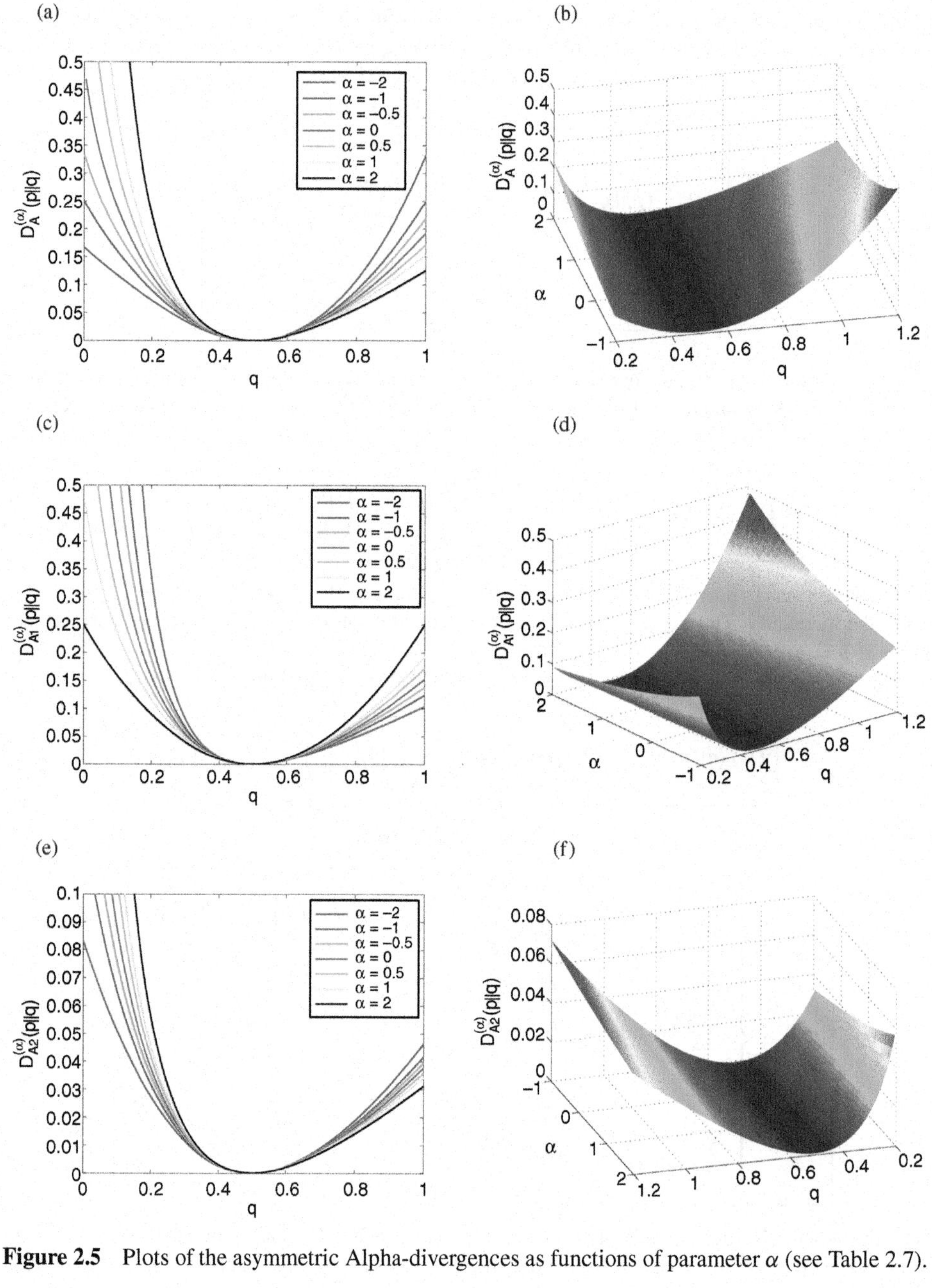

Figure 2.5 Plots of the asymmetric Alpha-divergences as functions of parameter α (see Table 2.7).

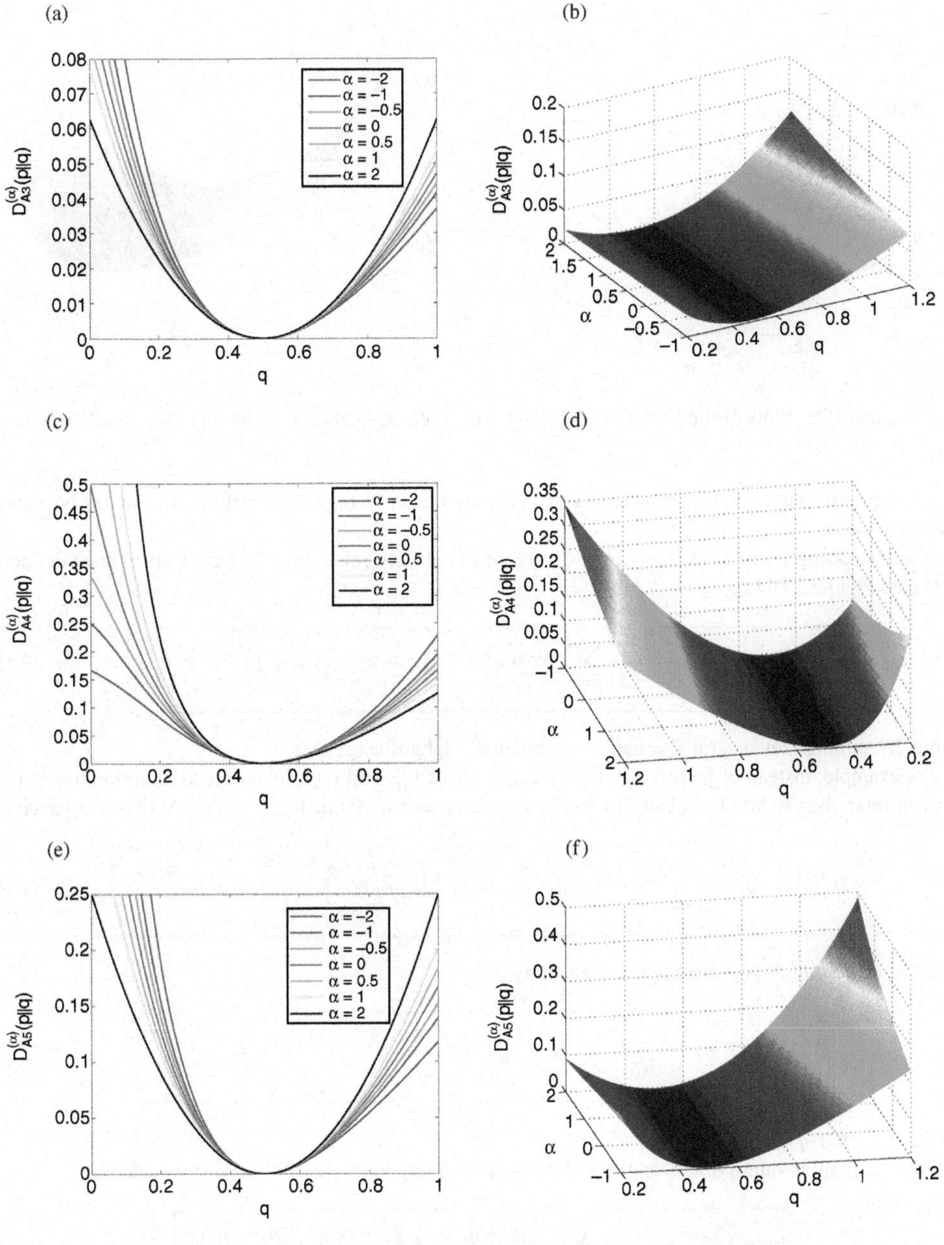

Figure 2.6 Plots of the asymmetric Alpha-divergences (see Table 2.7).

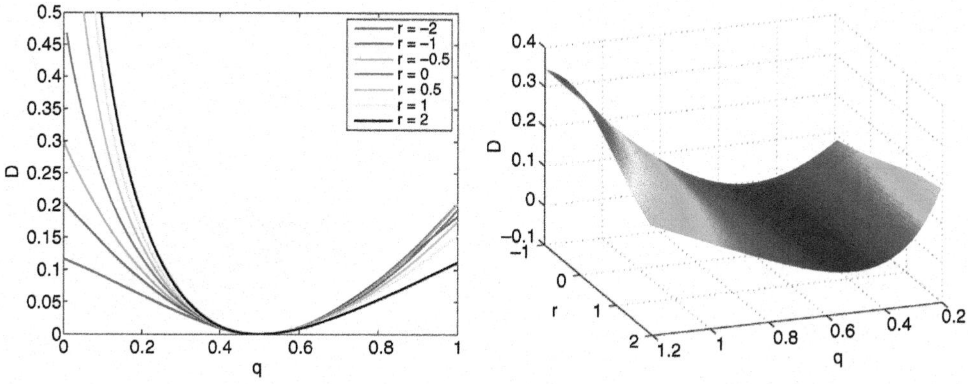

Figure 2.7 Plots of the Rényi divergence for different values of parameters r (see Eq. (2.105)).

where a penalty term on the right hand side often significantly improves performance and robustness of estimators.

It is interesting to note that the Alpha-divergence is closely related to the Rényi divergence defined as [44] (see Figure 2.7)[44]:

$$D_R^{(r)}(\boldsymbol{p}\,||\,\boldsymbol{q}) = \frac{1}{r(r-1)} \sum_i \ln\left(p_i^r q_i^{1-r} - r p_i + (r-1) q_i + 1\right), \quad r \neq 0, 1 \tag{2.105}$$

Table 2.7 summarizes several alternative asymmetric Alpha-divergences.

For example, instead of $\{q_i\}$ we can take the average of $\{q_i\}$ and $\{p_i\}$, under the assumption that if $\boldsymbol{p}$ and $\boldsymbol{q}$ are similar, they should be "close" to their average, so we can define the modified Alpha-divergence as

$$D_{Am1}^{(\alpha)}(\boldsymbol{p}\,||\,\widetilde{\boldsymbol{q}}) = \frac{1}{\alpha(\alpha-1)} \sum_i \left(\left(\frac{p_i + q_i}{2}\right) \left[\left(\frac{2p_i}{p_i + q_i}\right)^{\alpha} - 1 \right] - \alpha\, \frac{p_i - q_i}{2}, \right) \tag{2.106}$$

whereas an adjoint Alpha-divergence is given by

$$D_{Am2}^{(\alpha)}(\widetilde{\boldsymbol{q}}\,||\,\boldsymbol{p}) = \frac{1}{\alpha(\alpha-1)} \sum_i \left(p_i \left[\left(\frac{p_i + q_i}{2p_i}\right)^{\alpha} - 1 \right] + \alpha\, \frac{p_i - q_i}{2} \right), \tag{2.107}$$

where $\widetilde{\boldsymbol{q}} = (\boldsymbol{p} + \boldsymbol{q})/2$.

For the singular values $\alpha = 1$ and $\alpha = 0$, the Alpha-divergences (2.106) can be evaluated as

$$\lim_{\alpha \to 0} D_{Am1}^{(\alpha)}(\widetilde{\boldsymbol{q}}\,||\,\boldsymbol{p}) = \sum_i \left(\frac{p_i + q_i}{2} \ln\left(\frac{p_i + q_i}{2p_i}\right) + \frac{p_i - q_i}{2} \right), \tag{2.108}$$

and

$$\lim_{\alpha \to 1} D_{Am1}^{(\alpha)}(\widetilde{\boldsymbol{q}}\,||\,\boldsymbol{p}) = \sum_i \left(p_i \ln\left(\frac{2p_i}{p_i + q_i}\right) + \frac{q_i - p_i}{2} \right). \tag{2.109}$$

Table 2.7 Asymmetric Alpha-divergences and associated Csiszár functions (see also Figures 2.5 and 2.6).

Divergence $D_A^{(\alpha)}(\boldsymbol{p}\|\|\boldsymbol{q}) = \sum_i q_i f^{(\alpha)}\left(\frac{p_i}{q_i}\right)$	Csiszár function $f^{(\alpha)}(u), \quad u = p/q$
$\begin{cases} \frac{1}{\alpha(\alpha-1)} \sum_i \left(q_i \left[\left(\frac{p_i}{q_i}\right)^{\alpha} - 1 \right] - \alpha(p_i - q_i) \right), \\ \sum_i \left(q_i \ln \frac{q_i}{p_i} + p_i - q_i \right), \\ \sum_i \left(p_i \ln \frac{p_i}{q_i} - p_i + q_i \right). \end{cases}$	$\begin{cases} \frac{u^{\alpha} - 1 - \alpha(u-1)}{\alpha(\alpha-1)}, & \alpha \neq 0, 1 \\ u - 1 - \ln u, & \alpha = 0, \\ 1 - u + u \ln u, & \alpha = 1. \end{cases}$
$\begin{cases} \frac{1}{\alpha(\alpha-1)} \sum_i \left(p_i \left[\left(\frac{q_i}{p_i}\right)^{\alpha} - 1 \right] + \alpha(p_i - q_i) \right), \\ \sum_i \left(p_i \ln \frac{p_i}{q_i} - p_i + q_i \right), \\ \sum_i \left(q_i \ln \frac{q_i}{p_i} + p_i - q_i \right). \end{cases}$	$\begin{cases} \frac{u^{1-\alpha} + (\alpha-1)u - \alpha}{\alpha(\alpha-1)}, & \alpha \neq 0, 1, \\ 1 - u + u \ln u, & \alpha = 0 \\ u - 1 - \ln u, & \alpha = 1. \end{cases}$
$\begin{cases} \frac{1}{\alpha(\alpha-1)} \sum_i \left(q_i \left(\frac{p_i + q_i}{2q_i}\right)^{\alpha} - q_i - \alpha \frac{p_i - q_i}{2} \right), \\ \sum_i \left(q_i \ln \left(\frac{2q_i}{p_i + q_i}\right) + \frac{p_i - q_i}{2} \right), \\ \sum_i \left(\frac{p_i + q_i}{2} \ln \left(\frac{p_i + q_i}{2q_i}\right) - \frac{p_i - q_i}{2} \right). \end{cases}$	$\begin{cases} \frac{\left(\frac{u+1}{2}\right)^{\alpha} - 1 - \alpha\left(\frac{u-1}{2}\right)}{\alpha(\alpha-1)}, & \alpha \neq 0, 1, \\ \frac{u-1}{2} + \ln\left(\frac{2}{u+1}\right), & \alpha = 0, \\ \frac{1-u}{2} + \frac{u+1}{2} \ln\left(\frac{u+1}{2}\right), & \alpha = 1. \end{cases}$
$\begin{cases} \frac{1}{\alpha(\alpha-1)} \sum_i \left(p_i \left(\frac{p_i + q_i}{2p_i}\right)^{\alpha} - p_i + \alpha \frac{p_i - q_i}{2} \right), \\ \sum_i \left(p_i \ln \left(\frac{2p_i}{p_i + q_i}\right) - \frac{p_i - q_i}{2} \right), \\ \sum_i \left(\frac{p_i + q_i}{2} \ln \left(\frac{p_i + q_i}{2p_i}\right) + \frac{p_i - q_i}{2} \right). \end{cases}$	$\begin{cases} \frac{u\left(\frac{u+1}{2u}\right)^{\alpha} - u - \alpha\left(\frac{1-u}{2}\right)}{\alpha(\alpha-1)}, & \alpha \neq 0, 1, \\ \frac{1-u}{2} - u \ln\left(\frac{u+1}{2u}\right), & \alpha = 0, \\ \frac{u-1}{2} + \left(\frac{u+1}{2}\right) \ln\left(\frac{u+1}{2u}\right), & \alpha = 1. \end{cases}$
$\begin{cases} \frac{1}{(\alpha-1)} \sum_i (p_i - q_i) \left[\left(\frac{p_i + q_i}{2q_i}\right)^{\alpha-1} - 1 \right], \\ \sum_i (p_i - q_i) \ln \left(\frac{p_i + q_i}{2q_i}\right). \end{cases}$	$\begin{cases} \frac{(u-1)\left[\left(\frac{u+1}{2}\right)^{\alpha-1} - 1\right]}{(\alpha-1)}, & \alpha \neq 1, \\ (u-1) \ln\left(\frac{u+1}{2}\right), & \alpha = 1. \end{cases}$

(continued)

Table 2.7 (*Continued*)

Divergence $D_A^{(\alpha)}(\boldsymbol{p}\|\boldsymbol{q}) = \sum_i q_i f^{(\alpha)}\left(\frac{p_i}{q_i}\right)$	Csiszár function $f^{(\alpha)}(u), \quad u = p/q$
$\begin{cases} \frac{1}{(\alpha-1)} \sum_i (q_i - p_i) \left[\left(\frac{p_i + q_i}{2p_i} \right)^{\alpha-1} - 1 \right], \\ \sum_i (q_i - p_i) \ln \left(\frac{p_i + q_i}{2p_i} \right). \end{cases}$	$\begin{cases} \frac{(1-u)\left[\left(\frac{u+1}{2u} \right)^{\alpha-1} - 1 \right]}{(\alpha-1)}, & \alpha \neq 1, \\ (1-u) \ln \left(\frac{u+1}{2u} \right), & \alpha = 1. \end{cases}$

It is important to consider the following particular cases:

1. Triangular Discrimination (TD) (Dacunha-Castelle)

$$D_{Am2}^{(-1)}(\widetilde{\boldsymbol{q}} \,||\, \boldsymbol{p}) = \frac{1}{4} D_T(\boldsymbol{p} \,||\, \boldsymbol{q}) = \frac{1}{4} \sum_i \frac{(p_i - q_i)^2}{p_i + q_i}. \tag{2.110}$$

2. Relative Jensen-Shannon divergence (Burbea and Rao, Sgarro, Sibson[10,11,49])

$$\lim_{\alpha \to 0} D_{Am2}^{(\alpha)}(\widetilde{\boldsymbol{q}} \,||\, \boldsymbol{p}) = D_{RJS}(\boldsymbol{p} \,||\, \boldsymbol{q}) = \sum_i \left(p_i \ln \left(\frac{2p_i}{p_i + q_i} \right) - p_i + q_i \right). \tag{2.111}$$

3. Relative Arithmetic-Geometric divergence [50–52]

$$\lim_{\alpha \to 1} D_{Am2}^{(\alpha)}(\widetilde{\boldsymbol{q}} \,||\, \boldsymbol{p}) = \frac{1}{2} D_{RAG}(\boldsymbol{p} \,||\, \boldsymbol{q}) = \sum_i \left((p_i + q_i) \ln \left(\frac{p_i + q_i}{2p_i} \right) + p_i - q_i \right). \tag{2.112}$$

4. Pearson Chi-square divergence

$$D_{Am2}^{(2)}(\widetilde{\boldsymbol{q}} \,||\, \boldsymbol{p}) = \frac{1}{8} D_\chi(\boldsymbol{p} \,||\, \boldsymbol{q}) = \frac{1}{8} \sum_i \frac{(p_i - q_i)^2}{q_i}. \tag{2.113}$$

2.5.2 Symmetric Alpha-Divergences

The standard Alpha-divergence is asymmetric, that is, $D_A^{(\alpha)}(\boldsymbol{p} \,||\, \boldsymbol{q}) \neq D_A^{(\alpha)}(\boldsymbol{q} \,||\, \boldsymbol{p})$. The symmetric Alpha-divergence (Type-1) can be defined as

$$D_{AS1}^{(\alpha)}(\boldsymbol{p} \,||\, \boldsymbol{q}) = D_A^{(\alpha)}(\boldsymbol{p} \,||\, \boldsymbol{q}) + D_A^{(\alpha)}(\boldsymbol{q} \,||\, \boldsymbol{p}) = \sum_i \frac{p_i^\alpha q_i^{1-\alpha} + p_i^{1-\alpha} q_i^\alpha - (p_i + q_i)}{\alpha(\alpha-1)}. \tag{2.114}$$

As special cases, we obtain several well-known symmetric divergences (see also Figure 2.8):

1. Symmetric Chi-Squared divergence [27]

$$D_{AS1}^{(-1)}(\boldsymbol{p}\,||\,\boldsymbol{q}) = D_{AS1}^{(2)}(\boldsymbol{p}\,||\,\boldsymbol{q}) = \frac{1}{2} D_{\chi}(\boldsymbol{p}\,||\,\boldsymbol{q}) = \frac{1}{2}\sum_i \frac{(p_i - q_i)^2 (p_i + q_i)}{p_i q_i}. \tag{2.115}$$

2. J-divergence corresponding to Jeffreys entropy maximization [32,35]

$$\lim_{\alpha \to 0} D_{AS1}^{(\alpha)}(\boldsymbol{p}\,||\,\boldsymbol{q}) = \lim_{\alpha \to 1} D_{AS1}^{(\alpha)}(\boldsymbol{p}\,||\,\boldsymbol{q}) = D_J(\boldsymbol{p}\,||\,\boldsymbol{q}) = \sum_i (p_i - q_i) \ln\left(\frac{p_i}{q_i}\right). \tag{2.116}$$

3. Squared Hellinger distance [31]

$$D_{AS1}^{(1/2)}(\boldsymbol{p}\,||\,\boldsymbol{q}) = 8 D_H(\boldsymbol{p}\,||\,\boldsymbol{q}) = 4 \sum_i (\sqrt{p_i} - \sqrt{q_i})^2. \tag{2.117}$$

An alternative wide class of symmetric divergences can be described by the following symmetric Alpha-divergence (Type-2):

$$\begin{aligned} D_{AS2}^{(\alpha)}(\boldsymbol{p}\,||\,\boldsymbol{q}) &= D_A^{(\alpha)}\left(\frac{\boldsymbol{p}+\boldsymbol{q}}{2}\,||\,\boldsymbol{q}\right) + D_A^{(\alpha)}\left(\frac{\boldsymbol{p}+\boldsymbol{q}}{2}\,||\,\boldsymbol{p}\right) \\ &= \frac{1}{\alpha(\alpha-1)} \sum_i \left((p_i^{1-\alpha} + q_i^{1-\alpha}) \left(\frac{p_i + q_i}{2}\right)^{\alpha} - (p_i + q_i) \right). \end{aligned} \tag{2.118}$$

The above measure admits the following particular cases:

1. Triangular Discrimination

$$D_{AS2}^{(-1)}(\boldsymbol{p}\,||\,\boldsymbol{q}) = \frac{1}{2} D_T(\boldsymbol{p}\,||\,\boldsymbol{q}) = \frac{1}{2} \sum_i \frac{(p_i - q_i)^2}{p_i + q_i}. \tag{2.119}$$

2. Symmetric Jensen-Shannon divergence[11]

$$\lim_{\alpha \to 0} D_{AS2}^{(\alpha)}(\boldsymbol{p}\,||\,\boldsymbol{q}) = D_{JS}(\boldsymbol{p}\,||\,\boldsymbol{q}) = \sum_i \left(p_i \ln\left(\frac{2p_i}{p_i + q_i}\right) + q_i \ln\left(\frac{2q_i}{p_i + q_i}\right)\right). \tag{2.120}$$

3. Arithmetic-Geometric divergence [50–52]

$$\lim_{\alpha \to 1} D_{AS2}^{(\alpha)}(\boldsymbol{p}\,||\,\boldsymbol{q}) = D_{AG}(\boldsymbol{p}\,||\,\boldsymbol{q}) = \sum_i (p_i + q_i) \ln\left(\frac{p_i + q_i}{2\sqrt{p_i q_i}}\right). \tag{2.121}$$

[11] The Jensen-Shannon divergence is a symmetrized and smoothed version of the Kullback-Leibler divergence, i.e., it can be interpreted as the average of the Kullback-Leibler divergences to the average distribution. In other words, the Jensen-Shannon divergence is the entropy of the average from which the average of the Shannon entropies is subtracted: $D_{JS} = H_S((\boldsymbol{p}+\boldsymbol{q})/2) - (H_S(\boldsymbol{p}) + H_S(\boldsymbol{q}))/2$, where $H_S(\boldsymbol{p}) = -\sum_i p_i \ln p_i$.

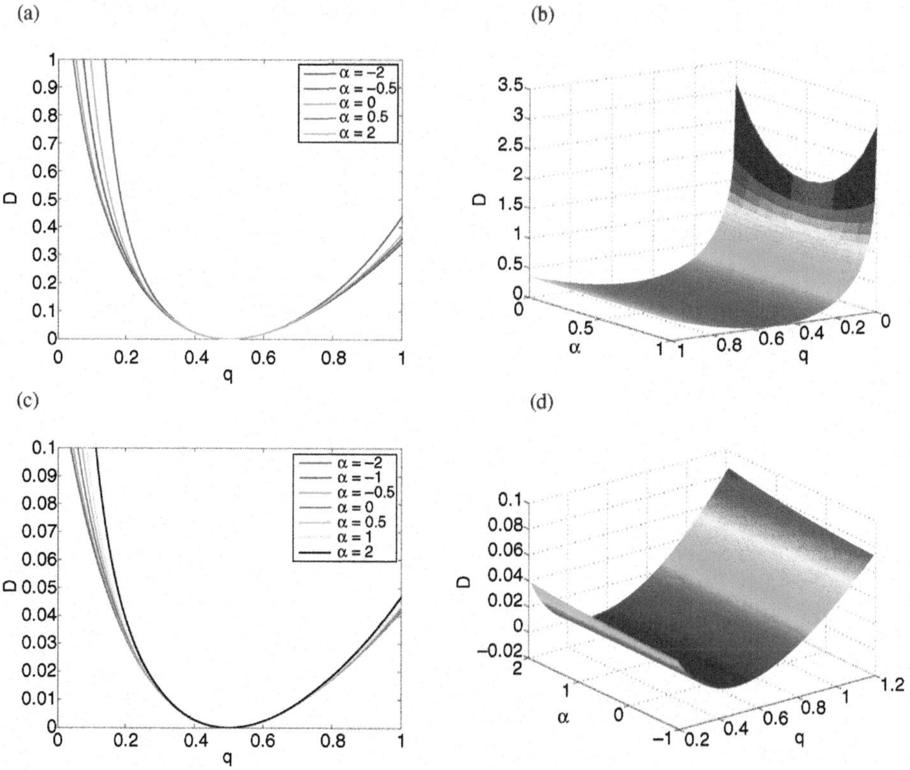

Figure 2.8 2D and 3D plots of Symmetric Alpha-divergences for different values of parameter α (a)–(b) $D_{AS1}^{(\alpha)}(\boldsymbol{p}\,||\,\boldsymbol{q})$ – Eq. (2.114) and (c)–(d) $D_{AS2}^{(\alpha)}(\boldsymbol{p}\,||\,\boldsymbol{q})$ – Eq. (2.118).

4. Symmetric Chi-squared divergence [27]

$$D_{AS2}^{(2)}(\boldsymbol{p}\,||\,\boldsymbol{q}) = \frac{1}{8} D_{\chi}(\boldsymbol{p}\,||\,\boldsymbol{q}) = \frac{1}{8} \sum_i \frac{(p_i - q_i)^2 (p_i + q_i)}{p_i q_i}. \tag{2.122}$$

The above Alpha-divergence is symmetric in its arguments $\boldsymbol{p}$ and $\boldsymbol{q}$, and it is well-defined even if $\boldsymbol{p}$ and $\boldsymbol{q}$ are not absolutely continuous, i.e., D_{AS2} is well-defined even if, for some indexes p_i it vanishes without vanishing q_i or if q_i vanishes without vanishing p_i. It is also lower- and upper-bounded, for example, the Jensen-Shannon divergence is bounded between 0 and 2.

2.6 Beta-Divergences

The Beta-divergence was introduced by Eguchi, Kano, Minami and also investigated by others [8,14–17,34,36,36,38–40]. It has a dually flat structure of information geometry, where the Pythagorean theorem holds. However, it is not invariant under a change of the dominating measure, and not invariance monotone for summarization, except for the special case of $\beta = 0$ which gives the KL-divergence.

First let us define the discrete Beta-divergence between two un-normalized density functions: p_i and q_i by

$$D_B^{(\beta)}(\boldsymbol{p}\,||\,\boldsymbol{q}) = \sum_i \left(p_i \,\frac{p_i^{\beta} - q_i^{\beta}}{\beta} - \frac{p_i^{\beta+1} - q_i^{\beta+1}}{\beta+1} \right), \tag{2.123}$$

where β is a real number ($\beta \neq 0$ and $\beta \neq -1$).

It is interesting to note that, for $\beta = 1$, we obtain the standard squared Euclidean distance, while for the singular cases $\beta = 0$ and $\beta = -1$ the Beta-divergence has to be defined in limiting cases as $\beta \to 0$ and $\beta \to -1$, respectively.

When these limits are evaluated for $\beta \to 0$, we obtain the generalized Kullback-Leibler divergence (called the I-divergence) defined as[12]

$$D_{KL}(\boldsymbol{p}\,||\,\boldsymbol{q}) = \lim_{\beta \to 0} D_B^{(\beta)}(\boldsymbol{p}\,||\,\boldsymbol{q}) = \sum_i \left(p_i \,\ln \frac{p_i}{q_i} - p_i + q_i \right), \tag{2.124}$$

whereas for $\beta \to -1$ the Itakura-Saito distance is obtained as

$$D_{IS}(\boldsymbol{p}\,||\,\boldsymbol{q}) = \lim_{\beta \to -1} D_B^{(\beta)}(\boldsymbol{p}\,||\,\boldsymbol{q}) = \sum_i \left(\ln \left(\frac{q_i}{p_i} \right) + \frac{p_i}{q_i} - 1 \right). \tag{2.125}$$

Remark 2.5 *This distance was first obtained by Itakura and Saito (in 1968) from the maximum likelihood (ML) estimation of short-time speech spectra under autoregressive modeling. It was presented as a measure of the goodness of fit between two spectra and became a standard measure in the speech processing community due to the good perceptual properties of the reconstructed signals. Other important properties of the Itakura-Saito divergence include scale invariance, meaning that low energy components of* **p** *bear the same relative importance as high energy ones. This is relevant to situations where the coefficients of* **p** *have a large dynamic range, such as in short-term audio spectra. The Itakura-Saito divergence also leads to desirable statistical interpretations of the NMF problem and we later explain how under simple Gaussian assumptions NMF can be recast as a maximum likelihood (ML) estimation of matrices* **A** *and* **X**. *Equivalently, we will describe how IS-NMF can be interpreted as ML of* **A** *and* **X** *in multiplicative Gamma noise.*

Hence, the Beta-divergence can be represented in a more explicit form:

$$D_B^{(\beta)}(\boldsymbol{p}\,||\,\boldsymbol{q}) = \begin{cases} \displaystyle\sum_i \left(p_i \,\frac{p_i^{\beta} - q_i^{\beta}}{\beta} - \frac{p_i^{\beta+1} - q_i^{\beta+1}}{\beta+1} \right), & \beta > 0, \\ \displaystyle\sum_i \left(p_i \,\ln \left(\frac{p_i}{q_i} \right) - p_i + q_i \right), & \beta = 0, \\ \displaystyle\sum_i \left(\ln \left(\frac{q_i}{p_i} \right) + \frac{p_i}{q_i} - 1 \right), & \beta = -1. \end{cases} \tag{2.126}$$

Note that the derivative of the Beta-divergence for separable terms $d(p_i||q_i)$ with respect to q_i is also continuous in β, and can be expressed as

$$\nabla_{q_i} d(p_i||q_i) = q_i^{\beta+1}(q_i - p_i) \tag{2.127}$$

[12]It should be noted that $\lim_{\beta \to 0} \frac{p^{\beta} - q^{\beta}}{\beta} = \ln(p/q)$ and $\lim_{\beta \to 0} \frac{p^{\beta} - 1}{\beta} = \ln p$.

Table 2.8 Special cases for Tweedie distributions.

Parameter β	Divergence	Distribution
1	Squared Euclidean distance	Normal
0	KL I-divergence $D_{KL}(\boldsymbol{p}\,\|\,\boldsymbol{q})$	Poisson
$(-1, 0)$		Compound Poisson
-1	Dual KL I-divergence $D_{KL}(\boldsymbol{q}\,\|\,\boldsymbol{p})$	Gamma
-2	Dual KL I-divergence $D_{KL}(\boldsymbol{q}\,\|\,\boldsymbol{p})$	Inverse–Gaussian

This derivative shows that $d(p_i||q_i)$, as a function of q_i, has a single minimum at $q_i = p_i$ and that it increases with $|q_i - p_i|$, justifying its relevance as a measure of fit.

The Beta-divergence smoothly connects the Itakura-Saito distance and the squared Euclidean distance and passes through the KL I-divergence $D_{KL}(\boldsymbol{p}\,||\,\boldsymbol{q})$. Such a parameterized connection is impossible in the family of the Alpha-divergences.

The Beta-divergence is related to the Tweedie distributions [8,33,36]. In probability and statistics, the Tweedie distributions are a family of probability distributions which include continuous distributions such as the normal and gamma, the purely discrete scaled Poisson distribution, and the class of mixed compound Poisson-Gamma distributions which have positive mass at zero, but are otherwise continuous. Tweedie distributions belong to the exponential dispersion model family of distributions, a generalization of the exponential family, which are the response distributions for generalized linear models [33]. Tweedie distributions exist for all real values of β except for $0 < \beta < 1$. Apart from special cases shown in Table 2.8, their probability density function have no closed form. The choice of the parameter β depends on the statistical distribution of data. For example, the optimal choice of the parameter β for the normal distribution is $\beta = 1$, for the gamma distribution it is $\beta = -1$, for the Poisson distribution $\beta = 0$, and for the compound Poisson distribution $\beta \in (-1, 0)$ (see Table 2.8) [14–17,36,38,39].

The asymmetric Beta-divergences can be obtained from the Alpha-divergence (2.42) by applying nonlinear transformations:

$$\boxed{p_i \to p_i^{\beta+1}, \qquad q_i \to q_i^{\beta+1}.} \tag{2.128}$$

Using these substitutions for (2.42) and assuming that $\alpha = (\beta + 1)^{-1}$ we obtain the following divergence

$$D_A^{(\beta)}(\boldsymbol{p}\,||\,\boldsymbol{q}) = (\beta + 1)^2 \sum_i \left[\frac{p_i^{\beta+1}}{\beta(\beta + 1)} - \frac{p_i q_i^{\beta}}{\beta} + \frac{q_i^{1+\beta}}{\beta + 1} \right] \tag{2.129}$$

Observe that, after simple algebraic manipulations and by ignoring the scaling factor $(\beta + 1)^2$, we obtain the Beta-divergence defined by Eq. (2.123).

In fact, there exists the same link between the whole family of Alpha-divergences and the family of Beta-divergences (see Table 2.7). For example, we can derive the symmetric Beta divergences (see also Figure 2.9):

$$\boxed{\begin{aligned} D_{BS1}^{(\beta)}(\boldsymbol{p}\,||\,\boldsymbol{q}) &= D_B^{(\beta)}(\boldsymbol{p}\,||\,\boldsymbol{q}) + D_B^{(\beta)}(\boldsymbol{q}\,||\,\boldsymbol{p}) \\ &= \frac{1}{\beta} \sum_i \left(p_i^{\beta+1} + q_i^{\beta+1} - p_i q_i^{\beta} - p_i^{\beta} q_i \right), \end{aligned}} \tag{2.130}$$

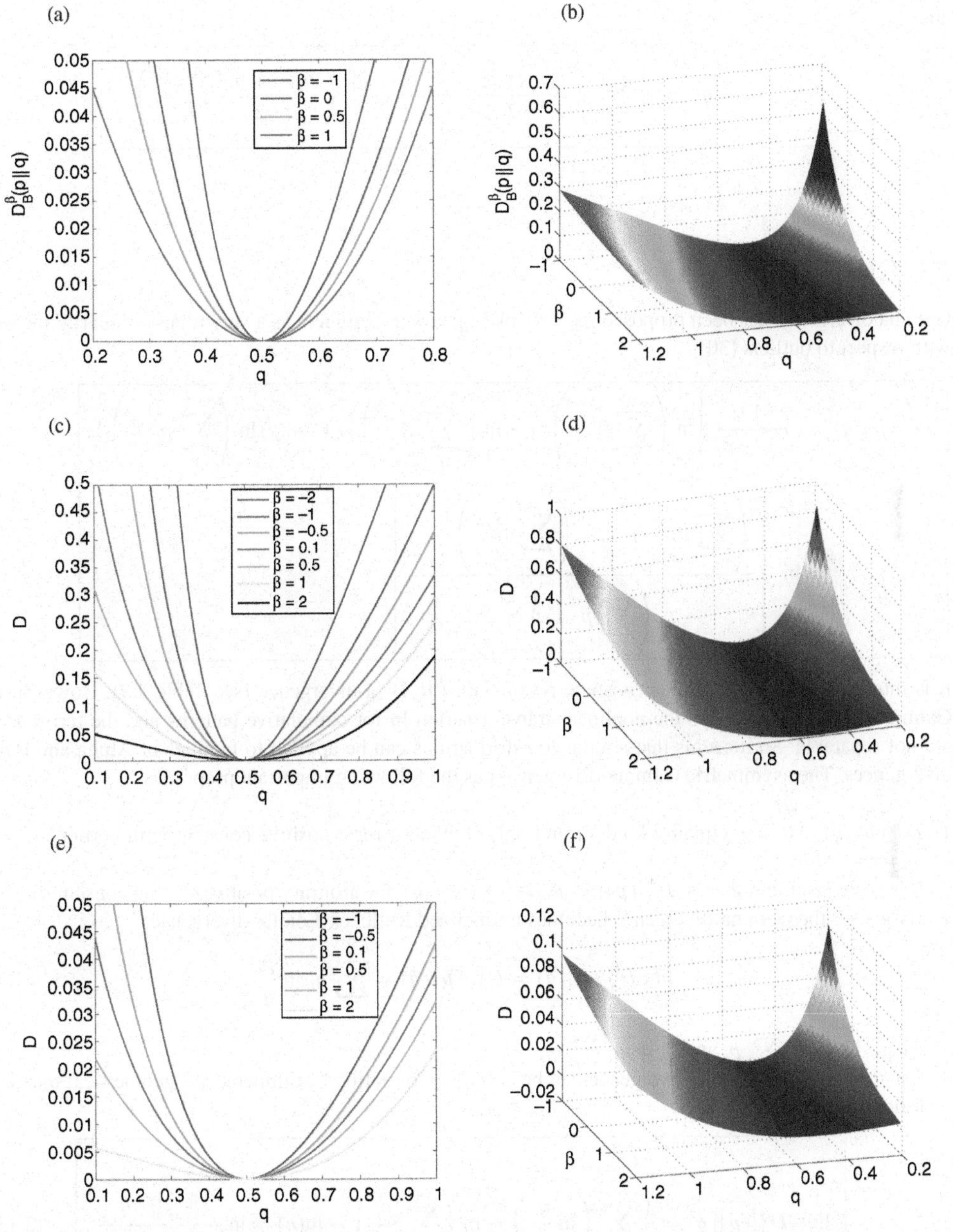

Figure 2.9 2D and 3D plots of Beta-divergences for different values of the parameter β: (a)–(b) for $D_B^{(\beta)}(\boldsymbol{p}\,||\,\boldsymbol{q})$ – Eq. (2.126); (c)–(e) for $D_{BS1}^{(\beta)}(\boldsymbol{p}\,||\,\boldsymbol{q})$ – Eq. (2.130) and (f)–(d) for $D_{BS2}^{(\beta)}(\boldsymbol{p}\,||\,\boldsymbol{q})$ – Eq. (2.131).

and

$$D_{BS2}^{(\beta)}(\boldsymbol{p}\,||\,\boldsymbol{q}) = \frac{1}{\beta}\sum_i \left(\frac{p_i^{\beta+1}+q_i^{\beta+1}}{2} - \frac{p_i^{\beta}+q_i^{\beta}}{2}\left(\frac{p_i^{\beta+1}+q_i^{\beta+1}}{2}\right)^{\frac{1}{\beta+1}}\right). \tag{2.131}$$

2.7 Gamma-Divergences

Gamma divergence has been proposed recently by Fujisawa and Eguchi as a very robust similarity measure with respect to outliers [30]:

$$\begin{aligned} D_A^{(\gamma)}(\boldsymbol{p}\,||\,\boldsymbol{q}) &= \frac{1}{\gamma(1+\gamma)}\left[\ln\left(\sum_i p_i^{1+\gamma}\right) + \gamma \ln\left(\sum_i q_i^{1+\gamma}\right) - (1+\gamma)\ln\left(\sum_i p_i\, q_i^{\gamma}\right)\right] \\ &= \ln\left[\frac{\left(\sum_i p_i^{1+\gamma}\right)^{1/\gamma(1+\gamma)}\left(\sum_i q_i^{1+\gamma}\right)^{1/(1+\gamma)}}{\left(\sum_i p_i\, q_i^{\gamma}\right)^{1/\gamma}}\right], \end{aligned} \tag{2.132}$$

It is interesting to note that there is some resemblance to Beta-divergence (see Table 2.9). However, the Gamma-divergence employs the nonlinear transformation ln for cumulative patterns and the terms p_i, q_i are not separable. So it seems that similar transformations can be applied to families of Alpha and Rényi divergences. The asymmetric Gamma-divergence has the following important properties:

1. $D_A^{(\gamma)}(\boldsymbol{p}\,||\,\boldsymbol{q}) \geq 0$. The equality holds if and only if $\boldsymbol{p} = c\boldsymbol{q}$ for a positive constant c, in particular, $p_i = q_i, \ \forall i$.
2. It is scale invariant, that is, $D_A^{(\gamma)}(\boldsymbol{p}\,||\,\boldsymbol{q}) = D_A^{(\gamma)}(c_1\boldsymbol{p}\,||\,c_2\boldsymbol{q})$, for arbitrary positive scaling constants c_1, c_2.
3. As $\gamma \to 0$ the Gamma-divergence becomes normalized Kullback-Leibler divergence:

$$\lim_{\gamma\to 0} D_A^{(\gamma)}(\boldsymbol{p}\,||\,\boldsymbol{q}) = D_{KL}(\tilde{\boldsymbol{p}}\,||\,\tilde{\boldsymbol{q}}) = \sum_i \tilde{p}_i \ln \frac{\tilde{p}_i}{\tilde{q}_i}, \tag{2.133}$$

   where $\tilde{p}_i = p_i/\sum_i p_i$ and $\tilde{q}_i = q_i/\sum_i q_i$.
4. For $\gamma \to -1$ the Gamma-divergence can be expressed by ratio of arithmetic and geometrical means as follows[13]

$$\lim_{\gamma\to -1} D_A^{(\gamma)}(\boldsymbol{p}\,||\,\boldsymbol{q}) = \frac{1}{n}\sum_{i=1}^{n}\left(\ln\frac{q_i}{p_i}\right) + \ln\left(\sum_{i=1}^{n}\frac{p_i}{q_i}\right) - \ln(n) = \ln\frac{\frac{1}{n}\sum_{i=1}^{n}\frac{p_i}{q_i}}{\left(\prod_{i=1}^{n}\frac{p_i}{q_i}\right)^{1/n}}. \tag{2.134}$$

[13]For $n \to \infty$ the asymmetric Gamma-divergence can be expressed by expectation functions $D_A^{(1)}(\boldsymbol{p}\,||\,\boldsymbol{q}) = \ln(E\{\boldsymbol{u}\}) - E\{\ln(\boldsymbol{u})\}$, where $E\{\cdot\}$ denotes expectation operator and $u_i = \{p_i/q_i\}$.

Table 2.9 The fundamental generalized divergences.

Divergence name	Formula
Alpha-divergence	$D_A^{(\alpha)}(\boldsymbol{p}\|\|\boldsymbol{q}) = \dfrac{\sum_i (p_i^\alpha\, q_i^{1-\alpha} + (\alpha-1)\, q_i - \alpha\, p_i)}{\alpha(\alpha-1)}$
Beta-divergence	$D_B^{(\beta)}(\boldsymbol{p}\|\|\boldsymbol{q}) = \dfrac{\sum_i (p_i^{\beta+1} + \beta\, q_i^{\beta+1} - (\beta+1)\, p_i\, q_i^\beta)}{\beta(\beta+1)}$
Gamma-divergence	$D_A^{(\gamma)}(\boldsymbol{p}\|\|\boldsymbol{q}) = \dfrac{\ln(\sum_i p_i^{\gamma+1}) + \gamma \ln(\sum_i q_i^{\gamma+1}) - (\gamma+1)\ln(\sum_i p_i\, q_i^\gamma)}{\gamma(\gamma+1)}$
Bregman divergence	$D_\phi(\boldsymbol{p}\|\|\boldsymbol{q}) = \sum_i \left(\phi(p_i) - \phi(q_i) - \dfrac{d\phi}{dq_i}(p_i - q_i)\right)$
Csiszár f-divergence	$D_f(\boldsymbol{p}\|\|\boldsymbol{q}) = \sum_i q_i f\left(\dfrac{p_i}{q_i}\right)$
Rényi divergence	$D_R^{(r)}(\boldsymbol{p}\|\|\boldsymbol{q}) = \sum_i \dfrac{\ln(p_i^r q_i^{1-r} - r p_i + (r-1) q_i + 1)}{r(r-1)}$
Rényi-type divergence	$D_\psi(\boldsymbol{p}\|\|\boldsymbol{q}) = \sum_i \ln\left(\psi^{-1}\left(p_i \psi\left(\dfrac{p_i}{q_i}\right)\right)\right)$
Burbea-Rao divergence	$D_{BR}(\boldsymbol{p}\|\|\boldsymbol{q}) = \sum_i \left(\dfrac{h(p_i) + h(q_i)}{2} - h\left(\dfrac{p_i + q_i}{2}\right)\right)$

Similarly to the Alpha and Beta-divergences, we can also define the symmetric Gamma-divergence as

$$D_S^{(\gamma)}(\boldsymbol{p} \,||\, \boldsymbol{q}) = D_A^{(\gamma)}(\boldsymbol{p} \,||\, \boldsymbol{q}) + D_A^{(\gamma)}(\boldsymbol{q} \,||\, \boldsymbol{p}) = \frac{1}{\gamma} \ln \left[\frac{\left(\sum_i p_i^{1+\gamma}\right)\left(\sum_i q_i^{1+\gamma}\right)}{\left(\sum_i p_i^\gamma\, q_i\right)\left(\sum_i p_i\, q_i^\gamma\right)} \right]. \tag{2.135}$$

The symmetric Gamma-divergence has similar properties to the asymmetric Gamma-divergence:

1. $D_S^{(\gamma)}(\boldsymbol{p} \,||\, \boldsymbol{q}) \geq 0$. The equality holds if and only if $\boldsymbol{p} = c\boldsymbol{q}$ for a positive constant c, in particular, $p_i = q_i$, $\forall i$.
2. It is scale invariant, that is,

$$D_S^{(\gamma)}(\boldsymbol{p} \,||\, \boldsymbol{q}) = D_S^{(\gamma)}(c_1 \boldsymbol{p} \,||\, c_2 \boldsymbol{q}), \tag{2.136}$$

for arbitrary positive scaling constants c_1, c_2.

3. For $\gamma \to 0$ it is reduced to special form of the symmetric Kullback-Leibler divergence (also called the J-divergence)

$$\lim_{\gamma\to 0} D_S^{(\gamma)}(\boldsymbol{p}\,||\,\boldsymbol{q}) = \sum_i (\tilde{p}_i - \tilde{q}_i) \ln \frac{\tilde{p}_i}{\tilde{q}_i}, \tag{2.137}$$

where $\tilde{p}_i = p_i / \sum_i p_i$ and $\tilde{q}_i = q_i / \sum_i q_i$.

4. For $\gamma \to -1$ we obtain new divergence[14]

$$\lim_{\gamma\to -1} D_S^{(\gamma)}(\boldsymbol{p}\,||\,\boldsymbol{q}) = \ln \left(\sum_{i=1}^{n} \frac{p_i}{q_i} \sum_{i=1}^{n} \frac{q_i}{p_i} \right) - \ln(n)^2 = \ln \left(\frac{1}{n^2} \left(\sum_{i=1}^{n} \frac{p_i}{q_i} \right) \left(\sum_{i=1}^{n} \frac{q_i}{p_i} \right) \right). \tag{2.138}$$

5. For $\gamma = 1$ the asymmetric Gamma-divergences is symmetric Gamma-divergence, that is,

$$D_A^{(1)}(\boldsymbol{p}\,||\,\boldsymbol{q}) = D_A^{(1)}(\boldsymbol{q}\,||\,\boldsymbol{p}) = \frac{1}{2} D_S^{(1)}(\boldsymbol{p}\,||\,\boldsymbol{q}) = \ln \left[\frac{\left(\sum_i p_i^2 \right)^{1/2} \left(\sum_i q_i^2 \right)^{1/2}}{\sum_i (p_i\, q_i)} \right]. \tag{2.139}$$

2.8 Divergences Derived from Tsallis and Rényi Entropy

Let us define a function $h_r(\tilde{\boldsymbol{p}})$ of probability distribution $\tilde{\boldsymbol{p}}$ by

$$h_r(\tilde{\boldsymbol{p}}) = \sum_i \tilde{p}_i^{\,r}, \qquad 0 < r < 1, \tag{2.140}$$

where $\tilde{p}_i = p_i / \sum_i p_i$. This is a separable convex function of $\tilde{\boldsymbol{p}}$. The Tsallis r-entropy [42] is defined by

$$H_T^{(r)}(\tilde{\boldsymbol{p}}) = \frac{1}{1-r} \{h_r(\tilde{\boldsymbol{p}}) - 1\} \tag{2.141}$$

and the Rényi r-entropy (α-entropy originally) is

$$H_T^{(r)}(\tilde{\boldsymbol{p}}) = \frac{1}{1-r} \ln h_r(\tilde{\boldsymbol{p}}). \tag{2.142}$$

When $r = 1$, we take the limit $r \to 1$, and they are identical to the Shannon entropy.

We show three different approaches to obtaining divergences. The first one is an information theoretic approach. From the entropy we can derive a divergence function. Let us define the r-logarithm function by

$$\ln_r(u) = \begin{cases} \frac{1}{1-r} \left\{ u^{1-r} - 1 \right\}, & r \neq 1, \\ \ln\, u, & r = 1. \end{cases} \tag{2.143}$$

Then, the Tsallis entropy is written as

$$H_T^{(r)}(\tilde{\boldsymbol{p}}) = \sum \tilde{p}_i \, \ln_q \frac{1}{\tilde{p}_i} \tag{2.144}$$

[14] For $n \to \infty$ the symmetric Gamma-divergence can be expressed by expectation functions $D_S^{(1)}(\boldsymbol{p}\,||\,\boldsymbol{q}) = \ln(E\{\boldsymbol{u}\}\; E\{\boldsymbol{u}^{-1}\})$, where $u_i = \{p_i/q_i\}$ and $u_i^{-1} = \{q_i/p_i\}$.

The related divergence, called the r-divergence is defined as

$$D^{(r)}(\tilde{\boldsymbol{p}} \parallel \tilde{\boldsymbol{q}}) = \sum \tilde{p}_i \ \ln_r \frac{\tilde{q}_i}{\tilde{p}_i} = \frac{1}{r-1}\left\{1 - \sum_i \tilde{p}_i^{\,r}\, \tilde{q}_i^{\,1-r}\right\}, \tag{2.145}$$

where $\tilde{q}_i = q_i / \sum_i q_i$. Therefore, this is essentially the same as the Alpha-divergence.

The second one is the Bregman style approach using convex functions. Since H_r^T and H_r^R are convex functions of $\boldsymbol{p}$, the related Bregman divergence is naturally induced. Here $\boldsymbol{p}$ is affine coordinates, and $h_r(\boldsymbol{p})$ is a separable power function of p_i. The Tsallis entropy leads to the Beta-divergence, where $\beta = r - 1$ (see (2.69)). The Rényi divergence is a nonlinear function of $h_r(\boldsymbol{p})$, and is non-separable. This gives the Bregman divergence expressed as

$$D_R^{(r)}(\tilde{\boldsymbol{p}}||\tilde{\boldsymbol{q}}) = \frac{1}{1-r}\left\{\ln h_r(\tilde{\boldsymbol{p}}) - \ln h_r(\tilde{\boldsymbol{q}}) - \frac{r}{h_r(\tilde{\boldsymbol{q}})} \sum_i \tilde{p}_i\, \tilde{q}_i^{\,r-1} + 1.\right\} \tag{2.146}$$

The third approach is due to a statistical approach. A family of probability distributions, called r-exponential functions [41,43], is given in the form

$$\ln_r p(x, \boldsymbol{\theta}) = \sum_i \theta^i\, h_i(x) - \psi(\boldsymbol{\theta}), \tag{2.147}$$

where

$$\theta^i = \frac{1}{1-r}\left(\tilde{p}_i^{\,1-r} - \tilde{p}_n^{\,1-r}\right), \quad i = 1, \cdots, n-1 \tag{2.148}$$

$$\psi_r(\boldsymbol{\theta}) = -\ln_r \tilde{p}_n, \tag{2.149}$$

and $\tilde{p}_n = 1 - \sum_{i=1}^{n-1} \tilde{p}_i$ are functions of $\tilde{p}_1, \tilde{p}_2, \ldots, \tilde{p}_{n-1}$. The function $\psi_r(\boldsymbol{\theta})$ is convex, so that we have the induced Bregman divergence. This is, the set of discrete probability distributions $\{\boldsymbol{p}\}$ is a r-exponential family for any r. Here, $\boldsymbol{\theta}$ is the affine coordinate system. The divergence is non-separable, given by

$$D_B^{(r)}[\tilde{\boldsymbol{p}}||\tilde{\boldsymbol{q}}] = \frac{1}{(1-r)\, h_r(\tilde{\boldsymbol{p}})}\left\{1 - \sum \tilde{p}_i^{\,r}\, \tilde{q}_i^{\,1-r}\right\}. \tag{2.150}$$

and is related to the Alpha-divergences.

2.8.1 Concluding Remarks

In this chapter we have unified and extended several important divergence measures used in information theory, statistical inference and pattern recognition. The parametric families of the generalized divergences presented in this chapter, especially the Alpha- , Beta- and Gamma-divergences enable us to consider smooth connections of various pairs of well-known and frequently used fundamental divergences under one "umbrella". By using such generalized divergences we may find a compromise between the efficiency and robustness and/or compromise between a mean squared error and bias. These families have many desirable properties such as flexibility and they involve only one single parameter (α, β, γ or r). An insight using information geometry elucidates (see Appendix A) the fundamental structures of divergences and geometry.

The extended families of divergences presented in this chapter will be used throughout the monograph to derive various algorithms. The properties of divergence measures discussed in this chapter allowed us to derive a wide family of robust NMF/NTF algorithms presented in next chapters. They have great potential in many applications, especially in blind separation of statistically dependent sparse and/or smooth sources and in feature extraction.

Appendix 2.A: Information Geometry, Canonical Divergence, and Projection

Information geometry studies invariant geometrical structures of the space of probability distributions [6,12]. The space of all the discrete distributions is dually flat, so that the generalized Pythagorean theorem and projection theorem hold. A dually flat space has a canonical divergence, which is the KL I-divergence in the special case. The space of all the positive measures is also dually flat, where the canonical divergence is the Alpha-divergence. The Bregman divergence, in particular the Beta-divergence, gives also a different type of dually flat structure to the space of probability distributions.

2.A.1 Space of Probability Distributions

Consider the space S of probability distributions $\boldsymbol{p} = (p_1, \cdots, p_n)$. Due to the constraint $\sum p_i = 1$, there are $n - 1$ degrees of freedom, and hence S is regarded as a space of $n - 1$ dimensions. We may use any coordinates $\boldsymbol{\xi} = (\xi_1, \cdots, \xi_{n-1})$ to specify the distributions. A simple choice is $\xi_i = p_i$, $i = 1, \cdots, n - 1$, and p_n is given by $p_n = 1 - \sum \xi_i$. A probability distribution $\boldsymbol{p}$ is denoted by $p(x, \boldsymbol{\xi})$, $x = 1, 2, \cdots, n$, that is, $p_i = p(i, \boldsymbol{\xi})$.

We can introduce two geometrical structures in S based on the invariance criterion stated in Section 2.3. Based on the framework by Chentsov [12] (see also Amari and Nagaoka) [6], the first one is a local distance, named the Riemannian metric and the other is the concept of straightness or flatness which defines a geodesic. This is represented by an affine connection [5].

1. Riemannian metric

   Let ds denote the distance between two nearby distributions $p(x, \boldsymbol{\xi})$ and $p(x, \boldsymbol{\xi} + d\boldsymbol{\xi})$. When its square is given by the quadratic form

$$ds^2 = \sum_{i,j} g_{ij}(\boldsymbol{\xi}) d\xi_i d\xi_j, \tag{A.1}$$

   the space is said to be Riemannian. The matrix $g = (g_{ij})$ is called the Riemannian metric. An Euclidean space is a special case of the Riemannian space, where g is equal to the identity matrix at any $\boldsymbol{\xi}$.

   Theorem A.3 *The unique Riemannian metric under the invariance criterion is the Fisher information matrix,*

$$g_{ij}(\boldsymbol{\xi}) = E\left[\frac{\partial}{\partial \xi_i} \ln p(x, \boldsymbol{\xi}) \frac{\partial}{\partial \xi_j} \ln p(x, \boldsymbol{\xi})\right]. \tag{A.2}$$

   and for us the coordinates given by $\xi_i = p_i$, $i = 1, \cdots, n - 1$, *yield*

$$g_{ij} = \frac{1}{p_i}\delta_{ij} + \frac{1}{1 - \sum_{i=1}^{n-1} p_i} \tag{A.3}$$

2. Geodesics and affine connections

   There are many different types of geodesics, i.e., one-parameter family of geodesics. Let the parameter be α. We then have the α-geodesic for each α, connecting two distributions. The $\alpha = 1/2$ connection is the Riemannian (or Levi-Civita) connection, which minimizes the length of the path connecting two points. The α-connection and $(1 - \alpha)$-connection are dually coupled with respect to the Riemannian metric. See Amari and Nagaoka for more details [6].

3. Dually flat space

   The space S of probability distributions $S = \{\boldsymbol{p}\}$ is flat with respect to $\alpha = 0$ and $\alpha = 1$ affine connections. When a space is flat the Riemannian-Christoffel curvature vanishes, we have affine coordinates $\boldsymbol{\theta}$, so that in this coordinate system any geodesic curve $\boldsymbol{\theta}(t)$ is represented by a linear form

$$\theta_i(t) = c_i t + b_i \tag{A.4}$$

   and coordinate axes themselves are geodesics.
   In our case, denote an affine coordinate system of $\alpha = 1$ connection by $\boldsymbol{\theta}$, to have

$$\theta_i = \ln p_i, \quad (i = 1, 2, \cdots, n-1) \tag{A.5}$$

   or

$$p_i = \exp\{\theta_i\}, \quad i = 1, \cdots, n-1, \tag{A.6}$$

   which is referred to as the exponential coordinate system (e-coordinate system). Since the exponential family $\{\boldsymbol{r}(t)\}$ connecting two distributions, $r_i(t) = \exp\{t \ln q_i + (1-t)\ln p_i - c(t)\}$, is an e-geodesic. Similarly, for $\alpha = 0$, the space is flat and we also have an affine coordinate system denoted by $\boldsymbol{\eta}$,

$$\eta_i = p_i, \quad i = 1, \cdots, n-1 \tag{A.7}$$

   which is called the mixture coordinate system (m-coordinate system) since the mixture of two distributions $\boldsymbol{p}$ and $\boldsymbol{q}$,

$$\boldsymbol{r}(t) = t\boldsymbol{q} + (1-t)\boldsymbol{p} \tag{A.8}$$

   is an m-geodesic.
4. Potential functions and canonical divergence

   When a space is α-flat, it is also $(1-\alpha)$-flat and when a space is dually flat, there exist convex functions $\psi(\boldsymbol{\theta})$ and $\varphi(\boldsymbol{\eta})$. In our case, the space is $\alpha = 1$ flat (e-flat) and $\alpha = 0$ flat (m-flat) and the first convex function is

$$\psi(\boldsymbol{\theta}) = \sum_{x=1}^{n} \exp\left\{\sum_{i=1}^{n-1} \theta_i \delta_i(x)\right\}, \tag{A.9}$$

   where $\delta_i(x) = 1$ when $x = i$ and $\delta_i(x) = 0$ otherwise. This function is also known the cumulant generating function, or the free energy in physics.

   The dual convex function is the negative entropy,

$$\phi(\boldsymbol{\eta}) = \sum_{i=1}^{n-1} \eta_i \ln \eta_i + (1 - \sum \eta_i)\ln(1 - \sum \eta_i). \tag{A.10}$$

   The two potential functions are mutually dual, connected by the Legendre transformation,

$$\eta_i = \frac{\partial}{\partial \theta_i}\psi(\boldsymbol{\theta}), \quad \theta_i = \frac{\partial}{\partial \eta_i}\phi(\boldsymbol{\eta}), \tag{A.11}$$

   and

$$\psi(\boldsymbol{\theta}) + \varphi(\boldsymbol{\eta}) - \sum \theta_i \eta_i = 0. \tag{A.12}$$

   The Bregman divergence derived from a dually flat space is

$$D[\boldsymbol{\theta}_p \| \boldsymbol{\eta}_q] = \psi(\boldsymbol{\theta}_p) + \varphi(\boldsymbol{\eta}_q) - \sum \theta_i^p \eta_i^q, \tag{A.13}$$

where $\boldsymbol{\theta}_p$ and $\boldsymbol{\eta}_q$ are the e- and m-coordinates of $\boldsymbol{p}$ and $\boldsymbol{q}$, respectively. In our case, it is nothing but the KL-divergence

$$D[\boldsymbol{\theta}_p \| \boldsymbol{\eta}_q] = \sum_{i=1}^{n} p_i \ln \frac{p_i}{q_i}. \tag{A.14}$$

5. Pythagorean Theorem

   Let P, Q, R be three points in a dually flat space. When the $\boldsymbol{\theta}$-geodesic connecting P and Q is orthogonal to the $\boldsymbol{\eta}$-geodesic connecting Q and R at the intersection Q, we may regard ΔPQR as a generalized orthogonal triangle. The orthogonality is measured by the Riemannian metric. For a curve $\boldsymbol{\theta}(t)$, its tangent vector at t is denoted by $\dot{\boldsymbol{\theta}}(t) = \frac{d}{dt}\theta_i(t)$. The two curves $\boldsymbol{\theta}_1(t)$ and $\boldsymbol{\theta}_2(t)$ intersect orthogonally at $t = 0$, when $\boldsymbol{\theta}_1(0) = \boldsymbol{\theta}_2(0)$ and

$$< \dot{\boldsymbol{\theta}}_1(0), \dot{\boldsymbol{\theta}}_2(0) >= \sum g_{ij}\dot{\theta}_{1i}\dot{\theta}_{2j} = 0. \tag{A.15}$$

   By using the dual $\boldsymbol{\eta}$-coordinates, this can be rewritten in a simpler way as

$$< \dot{\boldsymbol{\theta}}_1(0), \dot{\boldsymbol{\theta}}_2(0) >= \sum_{i=0}^{n-1} \dot{\theta}_{1i}(0)\dot{\eta}_{2i}(0). \tag{A.16}$$

   Theorem A.4 *When ΔPQR forms a generalized orthogonal triangle:*

$$D(Q\|P) + D(R\|Q) = D(R\|P). \tag{A.17}$$

   and when $\boldsymbol{\eta}$-geodesic connecting P and Q is orthogonal to the $\boldsymbol{\theta}$-geodesic connecting Q and R, then

$$D(P\|Q) + D(Q\|R) = D(P\|R). \tag{A.18}$$

   In our case of the space of probability distributions, the canonical divergence is the KL-divergence, and the above theorem is well known. However, this originates from the dually flat structure, and is much more general. It is applicable to many other cases, because the Bregman divergence also gives a dually flat structure.

6. Projection theorem

   In many problems such as estimation, approximation and noise reduction, we need to obtain $\boldsymbol{q}$ from a noisy observed $\boldsymbol{p}$. In such a case, $\boldsymbol{q}$ belongs to a model, which imposes regularity conditions or corresponds to the data generating specific mechanism. In this case, $\boldsymbol{q}$ belongs to M, which is a smooth subset of S. The problem is then to obtain $\boldsymbol{q} \in M$ that minimizes the divergence $D(\boldsymbol{p} \parallel \boldsymbol{q})$ (or $D(\boldsymbol{q} \parallel \boldsymbol{p})$). The optimal point lies on the boundary surface of the area of M, so that we can assume that M is a subspace which may be curved.

   Given $\boldsymbol{p}$, the point $\hat{\boldsymbol{q}} \in M$ is called the m-projection of $\boldsymbol{p}$ to M, when the m-geodesic connecting $\boldsymbol{p}$ and $\hat{\boldsymbol{q}}$ is orthogonal to M. Dually, the point $\hat{\boldsymbol{q}}^*$ is called the e-projection of $\boldsymbol{p}$ to M, when the e-geodesic connecting $\boldsymbol{p}$ and $\hat{\boldsymbol{q}}^*$ is orthogonal to M.

   The optimal point is given by the following projection theorem.

   Theorem A.5 *Given $\boldsymbol{p}$, the point $\boldsymbol{q} \in M$ that minimizes $D(\boldsymbol{p} \parallel \boldsymbol{q})$ is given by the m-projection of $\boldsymbol{p}$ to M, and the point $\boldsymbol{q}^*$ that minimizes $D[\boldsymbol{q}\|\boldsymbol{p}]$ is given by the e-projection of $\boldsymbol{p}$ to M.*

7. Alpha-divergence and invariant geometry

   Given a divergence function $D(\boldsymbol{p}(\boldsymbol{\xi}) \parallel \boldsymbol{q}(\boldsymbol{\xi}'))$ between two distributions, geometrical structures are automatically introduced in M. They are a Riemannian metric and a pair of dual affine connections. The Riemannian metric is given by

$$g_{ij}(\boldsymbol{\xi}) = \sum \frac{1}{2} \frac{\partial^2 D(\boldsymbol{p}\|\boldsymbol{q})}{\partial \xi_i \partial \xi_j} \Big|_{\boldsymbol{p}=\boldsymbol{q}}, \tag{A.19}$$

which is the Fisher information matrix. Similarly, we have a pair of dual affine connections.

The Csiszár f-divergence gives the Fisher information metric and a pair of α and $(1-\alpha)$ connections, where

$$\alpha = 2 + f'''(1). \tag{A.20}$$

Any invariant divergence gives this α-structure. However, it is not dually flat except for $\alpha = 0, 1$ corresponding to the KL-divergence. We can show, for any α, the α-connections are flat in the extended manifold of positive measures. Hence, for the Alpha-divergence ($\alpha \neq 0, 1$), there are no affine coordinate systems. The KL-divergence is the canonical divergence of $\alpha = 0, 1$ with dually flat structure.

8. Bregman divergence and dually flat structure

The Bregman divergence is generated from a convex function ψ. Hence, it gives a Riemannian metric and a pair of dually flat affine connections whose canonical divergence is the Bregman divergence. However, since it is not invariant, the Riemannian metric, in general, is different from the Fisher information matrix and the induced flat affine connection is different from the α-connection. In the following, we study the geometry induced by the Beta-divergence.

2.A.2 Geometry of Space of Positive Measures

Let $\tilde{S}$ be the space of all the positive measures $\tilde{\boldsymbol{p}}$, where $0 < \tilde{p}_i$, and no imposed constraints, such as $\sum \tilde{p}_i = 1$. Some $\tilde{p}_i$ may take zero value, and in such a case $\tilde{\boldsymbol{p}}$ lies on the boundary of $\tilde{S}$. We show that both the Alpha-divergence and Beta-divergence give different dually flat structures, except for the KL-divergence which belongs to both of them.

1. Alpha-divergence geometry

The Alpha-divergence in $\tilde{S}$ is given by

$$D_\alpha(\tilde{\boldsymbol{p}}\|\tilde{\boldsymbol{q}}) = \frac{1}{\alpha(1-\alpha)} \sum \{(1-\alpha)\tilde{p}_i + \alpha\tilde{q}_i - \tilde{p}_i^{1-\alpha}\tilde{q}_i^{\alpha}\} \tag{A.21}$$

when $\alpha \neq 0, 1$. It becomes

$$D_1(\tilde{\boldsymbol{p}}\|\tilde{\boldsymbol{q}}) = D_0(\tilde{\boldsymbol{q}}\|\tilde{\boldsymbol{p}}) = \sum \left(\tilde{q}_i - \tilde{p}_i + \tilde{p}_i \ln \frac{\tilde{p}_i}{\tilde{q}_i} \right) \tag{A.22}$$

in the case of $\alpha = 0, 1$. Let us introduce the following coordinates,

$$\theta_i^\alpha = \tilde{p}_i^{1-\alpha}, \quad \eta_i^\alpha = \tilde{p}_i^\alpha \quad 0 < \alpha < 1 \tag{A.23}$$

and

$$\begin{aligned} \theta_i^1 &= \ln \tilde{p}_i, \; \eta_i^1 = \tilde{p}_i, \\ \theta_i^0 &= \tilde{p}_i \quad\;\; \eta_i^0 = \ln \tilde{p}_i. \end{aligned} \tag{A.24}$$

We can then rewrite the Alpha-divergence as

$$D_\alpha(\tilde{\boldsymbol{p}}\|\tilde{\boldsymbol{q}}) = \psi^{(\alpha)}(\boldsymbol{\theta}_p^\alpha) + \psi^{(1-\alpha)}(\boldsymbol{\theta}_q^{1-\alpha}) - \theta_i^\alpha(p)\eta_i^\alpha(q), \tag{A.25}$$

where

$$\psi^{(\alpha)}(\boldsymbol{\theta}^\alpha) = \frac{1}{\alpha}\sum p_i = \frac{1}{\alpha}\sum (\theta_i^\alpha)^{\frac{1}{1-\alpha}} \tag{A.26}$$

is the potential function and $\psi^{(1-\alpha)}$ is its dual. This shows that $\tilde{S}$ is dually flat for any α, $\boldsymbol{\theta}^\alpha$ and $\boldsymbol{\eta}^\alpha$ are α-affine and $(1-\alpha)$-affine coordinates, and the Alpha-divergence is the canonical divergence.

The Fisher information matrix takes a simple diagonal form

$$g_{ij} = \frac{1}{(1-\alpha)^2}(\theta_i^{\alpha})^{\frac{2\alpha-1}{1-\alpha}}\delta_{ij}, \qquad \alpha \neq 1 \tag{A.27}$$

$$g_{ij}(\boldsymbol{\theta}^1) = \exp(\theta_i^1)\delta_{ij}, \qquad \alpha = 1. \tag{A.28}$$

In particular, when $\alpha = 1/2$, $\boldsymbol{\theta}^{\alpha}$ and $\boldsymbol{\eta}^{\alpha}$ coincide and the space is self-dual and flat. The Fisher information matrix is $4\delta_{ij}$, that is, four times the identity matrix. Hence, it is an Euclidean space. However, we have also the dual α-structure in the same space. The Pythagorean and projection theorems hold for any α and therefore, the α-projection is easily implemented in $\tilde{S}$.

The space S of probability distribution is a subspace of $\tilde{S}$. However, the constraint $\sum \tilde{p}_i = 1$ is not linear in the α-coordinates except for $\alpha = 0$ or $\alpha = 1$. Thus, the space S is curved in $\tilde{S}$ from the point of view of the α-connection, and hence is not α-flat.

2. Geometry of Beta-divergence

We first address the geometry of Eguchi's U-divergence, where U is a convex function by constructing a convex function of $\tilde{\boldsymbol{p}}$ as

$$\psi_U(\tilde{\boldsymbol{p}}) = \sum_{i=1}^{n} U(\tilde{p}_i). \tag{A.29}$$

We can then construct the Bregman divergence, named the U-divergence, whereby in order to rewrite it in a dual form, we introduce two functions

$$u(z) = U'(z), \quad v(z) = u^{-1}(z). \tag{A.30}$$

Apply the Legendre transformation to give

$$\tilde{\boldsymbol{p}}^* = \nabla\psi_U(\boldsymbol{p}), \tag{A.31}$$

with

$$\tilde{p}_i^* = u(\tilde{p}_i) \tag{A.32}$$

and

$$\tilde{p}_i = v(\tilde{p}_i^*). \tag{A.33}$$

The dual convex function is

$$\psi_U^*(\tilde{\boldsymbol{p}}^*) = \sum V(\tilde{p}_i^*), \tag{A.34}$$

where V is the dual of U, and

$$V(p^*) = p^* v(p^*) - U(v(p^*)). \tag{A.35}$$

Now the U-divergence has a simple form

$$D_v[\tilde{\boldsymbol{p}} : \tilde{\boldsymbol{q}}] = \sum U(\tilde{p}_i) + \sum V(\tilde{q}_i^*) - \sum \tilde{p}_i\tilde{q}_i^*. \tag{A.36}$$

and the space $\tilde{S}$ is dually flat, where $\tilde{\boldsymbol{p}}$ is the affine coordinate system ($\boldsymbol{\theta}$-coordinates) and $\tilde{\boldsymbol{p}}^*$ is the dual affine coordinate system. The Riemannian metric (U-metric) which in general is different from the Fisher information matrix is expressed as

$$g_{ij}^U(\tilde{\boldsymbol{p}}) = u'(\tilde{p}_i)\delta_{ij}. \tag{A.37}$$

Since the space is dually flat, both the Pythagorean and projection theorems hold.

For the subspace S of the probability distributions the constraint $\sum \tilde{p}_i = 1$ is linear in the $\boldsymbol{\theta}$-coordinates. Therefore this is a linear subspace, and hence dually flat.

The Beta-divergence is derived from the following U function

$$U_\beta(z) = \begin{cases} \frac{1}{\beta+1}(1+\beta z)^{\frac{\beta+1}{\beta}}, & \beta > 0, \\ \exp(z), & \beta = 0. \end{cases} \tag{A.38}$$

and is expressed as

$$D_\beta(\tilde{\boldsymbol{p}} \| \tilde{\boldsymbol{q}}) = \begin{cases} \frac{1}{\beta+1}(\sum \tilde{q}_i^{\beta+1} - \sum \tilde{p}_i^{\beta+1}) - \frac{1}{\beta}\sum \tilde{p}_i(\tilde{q}_i^{\beta} - \tilde{p}_i^{\beta}), & \beta > 0 \\ KL[\tilde{\boldsymbol{p}} : \tilde{\boldsymbol{q}}], & \beta = 0. \end{cases} \tag{A.39}$$

This leads to a flat structure with $\tilde{\boldsymbol{p}}$ as the affine coordinates, which are given by

$$\tilde{p}_i^* = \begin{cases} (1+\beta\tilde{p}_i)^{\frac{1}{\beta}}, & \beta > 0 \\ \ln \tilde{p}_i, & \beta = 0. \end{cases} \tag{A.40}$$

Appendix 2.B: Probability Density Functions for Various Distributions

Table B.1 Non-Gaussian distributions.

Name	Density	Mean	Variance
Laplacian	$\frac{1}{\sqrt{2\sigma^2}} \exp\left(-\frac{\lvert x-m\rvert}{\sqrt{\frac{\sigma^2}{2}}}\right)$	m	σ^2
Gamma	$\frac{b^a}{\Gamma(a)} x^{a-1} \exp(-bx),\ x > 0;\ a, b > 0$	$\frac{a}{b}$	$\frac{a}{b^2}$
Rayleigh	$2ax \exp(-ax^2), \quad x \geq 0$	$\sqrt{\frac{\pi}{4a}}$	$\frac{1}{a}\left(1-\frac{\pi}{4}\right)$
Uniform	$\frac{1}{b-a}, \quad a \leq x \leq b$	$\frac{a+b}{2}$	$\frac{(b-a)^2}{12}$
Cauchy	$\frac{\frac{a}{\pi}}{(x-m)^2+a^2}$	m	∞
Weibull	$abx^{b-1}\exp(-ax^b),\ x > 0;\ a, b > 0$	$\left(\frac{1}{a}\right)^{\frac{1}{b}} \Gamma\left(1+\frac{1}{b}\right)$	$a^{-\frac{2}{b}}\left[\Gamma\left(1+\frac{2}{b}\right) - \Gamma^2\left(1+\frac{1}{b}\right)\right]$
Arc-Sine	$\frac{1}{\pi\sqrt{x(1-x)}}, \quad 0 < x < 1$	$\frac{1}{2}$	$\frac{1}{8}$

(continued)

Table B.1 (*Continued*)

Name	Density	Mean	Variance
Circular Normal	$\dfrac{\exp\left(a\cos(x-m)\right)}{2\pi I_0(a)}, \quad -\pi < x \leq \pi$	m	$1-\left(\dfrac{I_1(a)}{I_0(a)}\right)^2$

$I_n(a)$ is the modified function of order n

Table B.2 Non-Gaussian distributions (continuation Table A.1).

Name	Density	Mean	Variance
Logistic	$\dfrac{\exp\left(-\dfrac{x-m}{a}\right)}{a\left[1+\exp\left(-\dfrac{(x-m)}{a}\right)\right]^2}, \quad a>0$	m	$\dfrac{a^2\pi^2}{3}$
Gumbel	$\dfrac{\exp\left(-\dfrac{x-m}{a}\right)}{a}\exp\left(-\exp\left(-\dfrac{(x-m)}{a}\right)\right)$ $a>0$	$m+a\gamma$	$\dfrac{a^2\pi^2}{6}$
Pareto	$\dfrac{ab^a}{x^{1-a}}, \quad a>0;\ 0<b\leq x$	$\dfrac{ab}{a-1},$ $a>1$	$\dfrac{ab^2}{(a-2)(a-1)^2},$ $a>2$
Triangular	$\dfrac{2x}{a}, \quad 0\leq x\leq a$ $\dfrac{2(1-x)}{(1-a)}, \quad a\leq x\leq 1$	$\dfrac{1+a}{3}$	$\dfrac{1-a+a^2}{18}$
Exponential	$\lambda\exp(-\lambda x), \quad x\geq 0;$	$\dfrac{1}{\lambda}$	$\dfrac{1}{\lambda^2}$
Lognormal	$\dfrac{1}{\sqrt{2\pi\sigma^2x^2}}\exp\left(-\dfrac{1}{2}\left(\dfrac{\ln x-m}{\sigma}\right)^2\right), \quad x>0$	$e^m+\dfrac{\sigma^2}{2}$	$e^{2m}\left(e^{2\sigma^2}-e^{\sigma^2}\right)$
Maxwell	$\sqrt{\dfrac{2}{\pi}a^{\frac{3}{2}}x^2\exp\left(-\dfrac{ax^2}{2}\right)}, \quad x>0$	$\sqrt{\dfrac{8}{\pi a}}$	$\left(3-\dfrac{8}{\pi}\right)a^{-1}$
Generalized	$\dfrac{1}{2\Gamma(1+\frac{1}{r})A(r)}\exp\left(-\left\lvert\dfrac{x-m}{A(r)}\right\rvert^r\right)$	m	σ^2
Gaussian	$A(r)=\left[\dfrac{\sigma^2\Gamma(\frac{1}{r})}{\Gamma(\frac{3}{r})}\right]^{\frac{1}{2}}$		

Table B.3 Discrete probability distributions.

Name	Probability	Mean	Variance
Discrete Uniform	$\frac{1}{N-M+1}, \quad M \le n \le N$; 0 otherwise	$\frac{M+N}{2}$	$\frac{(N-M+2)(N-M)}{12}$
Poisson	$\frac{\lambda^n e^{-\lambda}}{n!}, \quad n \ge 0$	λ	λ
Bernoulli	$Pr(n=0) = 1-p$; $Pr(n=1) = p$	p	$p(1-p)$
Binomial	$\binom{N}{n} p^n (1-p)^{N-n}, \ n = 0, \ldots, N$	Np	$Np(1-p)$
Geometric	$(1-p)p^n, n \ge 0$	$\frac{p}{1-p}$	$\frac{p}{(1-p)^2}$
Negative Binomial	$\binom{n-1}{N-1} p^N (1-p)^{n-N}, \ n \ge N$	$\frac{N}{p}$	$\frac{N(1-p)}{p^2}$
Hypergeometric	$\frac{\binom{a}{n}\binom{b}{N-n}}{\binom{a+b}{N}}, \quad n = 0, \ldots, N;$ $0 \le n \le a+b; \quad 0 \le N \le a+b$	$\frac{Na}{(a+b)}$	$\frac{Nab(a+b-N)}{(a+b)^2(a+b-1)}$
Logarithmic	$\frac{-p^n}{n \ln(1-p)}$	$\frac{-p}{(1-p)\ln(1-p)}$	$\frac{-p[p+\ln(1-p)]}{(1-p)^2 \ln^2(1-p)}$

References

[1] M.S. Ali and S.D. Silvey. A general class of coefficients of divergence of one distribution from another. *Journal of Royal Statistical Society*, Ser B(28):131–142, 1966.

[2] S. Amari. *Differential-Geometrical Methods in Statistics*. Springer Verlag, 1985.

[3] S. Amari. Dualistic geometry of the manifold of higher-order neurons. *Neural Networks*, 4(4):443–451, 1991.

[4] S. Amari. Information geometry and its applications: Convex function and dually flat manifold. In F. Nielson, editor, *Emerging Trends in Visual Computing*, pages 75–102. Springer Lecture Notes in Computer Science, 2009.

[5] S. Amari, K. Kurata, and H. Nagaoka. Information geometry of Boltzman machines. *IEEE Transactions on Neural Networks*, 3:260–271, 1992.

[6] S. Amari and H. Nagaoka. *Methods of Information Geometry*. Oxford University Press, New York, 2000.

[7] A. Banerjee, I. Dhillon, J. Ghosh, S. Merugu, and D.S. Modha. A generalized maximum entropy approach to Bregman co-clustering and matrix approximation. In *KDD '04: Proceedings of the Tenth ACM SIGKDD International Conference on Knowledge Discovery and Data Mining*, pages 509–514, New York, NY, USA, 2004. ACM Press.

[8] A. Basu, I. R. Harris, N.L. Hjort, and M.C. Jones. Robust and efficient estimation by minimising a density power divergence. *Biometrika*, 85(3):549–559, 1998.

[9] L. Bregman. The relaxation method of finding a common point of convex sets and its application to the solution of problems in convex programming. *Comp. Math. Phys., USSR*, 7:200–217, 1967.

[10] J. Burbea and C.R. Rao. Entropy differential metric, distance and divergence measures in probability spaces: A unified approach. *J. Multi. Analysis*, 12:575–596, 1982.

[11] J. Burbea and C.R. Rao. On the convexity of some divergence measures based on entropy functions. *IEEE Transactions on Information Theory*, IT-28:489–495, 1982.

[12] N.N. Chentsov. *Statistical Decision Rules and Optimal Inference.* AMS (translated from Russian, Nauka, 1972, New York, NY, 1982.

[13] H. Chernoff. A measure of asymptotic efficiency for tests of a hypothesis based on a sum of observations. *Annals of Mathematical Statistics*, 23:493–507, 1952.

[14] A. Cichocki, S. Amari, R. Zdunek, R. Kompass, G. Hori, and Z. He. Extended SMART algorithms for non-negative matrix factorization. *Springer, LNAI-4029*, 4029:548–562, 2006.

[15] A. Cichocki, R. Zdunek, and S. Amari. Csiszar's divergences for non-negative matrix factorization: Family of new algorithms. *Springer, LNCS-3889*, 3889:32–39, 2006.

[16] A. Cichocki, R. Zdunek, S. Choi, R. Plemmons, and S. Amari. Nonnegative tensor factorization using Alpha and Beta divergencies. In *Proc. IEEE International Conference on Acoustics, Speech, and Signal Processing (ICASSP07)*, volume III, pages 1393–1396, Honolulu, Hawaii, USA, April 15–20 2007.

[17] A. Cichocki, R. Zdunek, S. Choi, R. Plemmons, and S.-I. Amari. Novel multi-layer nonnegative tensor factorization with sparsity constraints. *Springer, LNCS-4432*, 4432:271–280, April 11–14 2007.

[18] N. Cressie and T. Read. Multinomial goodness-of-fit tests. *Journal of Royal Statistical Society B*, 46(3):440–464, 1984.

[19] N.A. Cressie and T.C.R. Read. *Goodness-of-Fit Statistics for Discrete Multivariate Data.* Springer, New York, 1988.

[20] I. Csiszár. Eine Informations Theoretische Ungleichung und ihre Anwendung auf den Beweis der Ergodizität von Markoffschen Ketten. *Magyar Tud. Akad. Mat. Kutató Int. Közl*, 8:85–108, 1963.

[21] I. Csiszár. Information measures: A critial survey. In *Transactions of the 7th Prague Conference*, pages 83–86, 1974.

[22] I. Csiszár. A geometric interpretation of darroch and ratcliff's generalized iterative scaling. *The Annals of Statistics*, 17(3):1409–1413, 1989.

[23] I. Csiszár. Axiomatic characterizations of information measures. *Entropy*, 10:261–273, 2008.

[24] I. Csiszár and J. Körner. *Information Theory: Coding Theorems for Discrete Memoryless Systems.* Academic Press, New York, USA, 1981.

[25] I. Dhillon and S. Sra. Generalized nonnegative matrix approximations with Bregman divergences. In *Neural Information Proc. Systems*, pages 283–290, Vancouver, Canada, December 2005.

[26] I.S. Dhillon and J.A. Tropp. Matrix nearness problems with Bregman divergences. *SIAM Journal on Matrix Analysis and Applications*, 29(4):1120–1146, 2007.

[27] S.S. Dragomir. *Inequalities for Csiszár f-Divergence in Information Theory.* Victoria University, Melbourne, Australia, 2000. edited monograph.

[28] S. Eguchi and Y. Kano. Robustifying maximum likelihood estimation. In *Institute of Statistical Mathematics*, Tokyo, 2001.

[29] Y. Fujimoto and N. Murata. A modified EM algorithm for mixture models based on Bregman divergence. *Annals of the Institute of Statistical Mathematics*, 59:57–75, 2007.

[30] H. Fujisawa and S. Eguchi. Robust parameter estimation with a small bias against heavy contamination. *Multivariate Analysis*, 99(9):2053–2081, 2008.

[31] E. Hellinger. Neue Begründung der Theorie Quadratischen Formen von unendlichen vielen Veränderlichen. *Journal Reine Ang. Math.*, 136:210–271, 1909.

[32] H. Jeffreys. An invariant form for the prior probability in estimation problems. *Proc. Roy. Soc. Lon., Ser. A*, 186:453–461, 1946.

[33] B. Jorgensen. *The Theory of Dispersion Models.* Chapman and Hall, London, 1997.

[34] R. Kompass. A generalized divergence measure for nonnegative matrix factorization. *Neural Computation*, 19(3):780–791, 2006.

[35] S. Kullback and R. Leibler. On information and sufficiency. *Annals of Mathematical Statistics*, 22:79–86, 1951.

[36] M. Minami and S. Eguchi. Robust blind source separation by Beta-divergence. *Neural Computation*, 14:1859–1886, 2002.

[37] T.P. Minka. Divergence measures and message passing. *Microsoft Research Technical Report (MSR-TR-2005)*, 2005.

[38] M.N.H. Mollah, S. Eguchi, and M. Minami. Robust prewhitening for ICA by minimizing beta-divergence and its application to FastICA. *Neural Processing Letters*, 25(2):91–110, 2007.
[39] M.N.H. Mollah, M. Minami, and S. Eguchi. Exploring latent structure of mixture ica models by the minimum beta-divergence method. *Neural Computation*, 16:166–190, 2006.
[40] N. Murata, T. Takenouchi, T. Kanamori, and S. Eguchi. Information geometry of U-Boost and Bregman divergence. *Neural Computation*, 16:1437–1481, 2004.
[41] J. Naudats. Generalized exponential families and associated entropy functions. *Entropy*, 10:131–149, 2008.
[42] A. Ohara. Possible generalization of Boltzmann-Gibbs statistics. *Journal Statistics Physics*, 52:479–487, 1988.
[43] A. Ohara. Geometry of distributions associated with Tsallis statistics and properties of relative entropy minimization. *Physics Letters A*, 370:184–193, 2007.
[44] A. Rényi. On measures of entropy and information. In *Proc. 4th Berk. Symp. Math. Statist. and Probl.*, volume 1, pages 547–561, University of California Press, Berkeley, 1961.
[45] A.L. Rukhin. Recursive testing of multiple hypotheses: Consistency and efficiency of the Bayes rule. *Ann. Statist.*, 22(2):616–633, 1994.
[46] B.D. Sharma and D.P. Mittal. New nonadditive measures of inaccuracy. *J. Math. Sci.*, 10:122–133, 1975.
[47] B.D. Sharma and D.P. Mittal. New nonadditive measures of relative information. *J. Comb. Inform. and Syst. Sci.*, 2:122–133, 1977.
[48] B.D. Sharma and I.J. Taneja. Entropy of type (α, β) and other generalized additive measures in information theory. *Metrika*, 22:205–215, 1975.
[49] R. Sibson. Information radius. *Probability Theory and Related Fields*, 14(2):149–160, June 1969.
[50] I.J. Taneja. On measures of information and inaccuarcy. *J. Statist. Phys.*, 14:203–270, 1976.
[51] I.J. Taneja. On generalized entropies with applications. In L.M. Ricciardi, editor, *Lectures in Applied Mathematics and Informatics*, pages 107–169. Manchester University Press, England, 1990.
[52] I.J. Taneja. New developments in generalized information measures. In P.W. Hawkes, editor, *Advances in Imaging and Electron Physics*, volume 91, pages 37–135. 1995.
[53] I. Vajda. *Theory of Statistical Inference and Information*. Kluwer Academic Press, London, 1989.
[54] J. Zhang. Divergence function, duality, and convex analysis. *Neural Computation*, 16(1):159–195, 2004.
[55] J. Zhang. Referential duality and representational duality on statistical manifolds. In *Proceedings of the Second International Symposium on Information Geometry and its Applications)*, pages 58–67, Tokyo, Japan, 2006.
[56] J. Zhang. A note on curvature of a-connections of a statistical manifold. *Annals of the Institute of Statistical Mathematics*, 59:161–170, 2007.
[57] J. Zhang and H. Matsuzoe. Dualistic differential geometry associated with a convex function. In *Springer Series of Advances in Mechanics and Mathematics*, pages 58–67, 2008.
[58] H. Zhu and R. Rohwer. Information geometric measurements of generalization. Technical Report NCRG/4350, Aston University, Birmingham, UK, August 31 1995.

3

Multiplicative Iterative Algorithms for NMF with Sparsity Constraints

In this chapter we introduce a wide family of iterative multiplicative algorithms for nonnegative matrix factorization (NMF) and related problems, subject to additional constraints such as sparsity and/or smoothness. Although a standard multiplicative update rule for NMF achieves a sparse representation[1] of its factor matrices, we can impose control over the sparsity of the matrices by designing a suitable cost function with additional penalty terms. There are several ways to incorporate sparsity constraints. A simple approach is to add suitable regularization or penalty terms to an optimized cost (loss) function. Another alternative approach is to implement at each iteration step a nonlinear projection (shrinkage) or filtering which increases sparseness of the estimated matrices.

We consider a wide class of cost functions or divergences (see Chapter 2), leading to generalized multiplicative algorithms with regularization and/or penalty terms. Such relaxed forms of the multiplicative NMF algorithms usually provide better performance and convergence speed, and allow us to extract desired unique components. The results included in this chapter give a vast scope as the range of cost functions includes a large number of generalized divergences, such as the squared weighted Euclidean distance, relative entropy, Kullback Leibler I-divergence, Alpha- and Beta-divergences, Bregman divergence and Csiszár f-divergence. As special cases we introduce the multiplicative algorithms for the squared Hellinger, Pearson's Chi-squared, and Itakura-Saito distances.

We consider the basic NMF model

$$\mathbf{Y} \cong \mathbf{AX}, \tag{3.1}$$

for which we construct a set of suitable cost functions $D(\mathbf{Y}||\mathbf{AX})$ which measure the distance between data $y_{it} = [\mathbf{Y}]_{it}$ and the set of estimated parameters $q_{it} = [\hat{\mathbf{Y}}]_{it} = [\mathbf{AX}]_{it} = \sum_{j=1}^{J} a_{ij} x_{jt}$. The multiplicative learning algorithms aim at minimizing a specific cost function or a set of cost functions by alternately updating the parameters a_{ij} while keeping x_{jt} fixed, and then updating the parameters x_{jt} while keeping all a_{ij} fixed. In fact, it is often convenient to estimate a set of parameters $\mathbf{A}$ and $\mathbf{X}$ by the sequential minimization of two different cost functions with the same global minima.

[1] We define sparse NMF $\mathbf{Y} \cong \mathbf{AX}$ as approximate nonnegative matrix factorization in which both or at least one factor matrix $\mathbf{A}$ or $\mathbf{X}$ is sparse.

Nonnegative Matrix and Tensor Factorizations: Applications to Exploratory Multi-way Data Analysis and Blind Source Separation Andrzej Cichocki, Rafal Zdunek, Anh Huy Phan and Shun-ichi Amari

For a large-scale NMF problem (with $T \geq I >> J$) we do not need to store and process large data matrices $\mathbf{Y} \in \mathbb{R}^{I\times T}$ and $\hat{\mathbf{Y}} = \mathbf{AX} \in \mathbb{R}_+^{I\times T}$. Instead of the typical alternating minimization of a one global cost function $D(\mathbf{Y}||\mathbf{AX})$ we may perform the following alternating minimization on the subsets:[2]

$$\mathbf{A} = \arg\min_{\mathbf{A}\geq \mathbf{0}} D_1(\mathbf{Y}_c||\mathbf{AX}_c), \quad \text{for fixed} \quad \mathbf{X}, \tag{3.2}$$

$$\mathbf{X} = \arg\min_{\mathbf{X}\geq \mathbf{0}} D_2(\mathbf{Y}_r||\mathbf{A}_r\mathbf{X}), \quad \text{for fixed} \quad \mathbf{A}, \tag{3.3}$$

where $\mathbf{Y}_r \in \mathbb{R}^{R\times T}$ and $\mathbf{Y}_c \in \mathbb{R}^{I\times C}$ comprise respectively the row and column subsets of the matrix $\mathbf{Y}$, whereas $\mathbf{A}_r \in \mathbb{R}_+^{R\times J}$ and $\mathbf{X}_c \in \mathbb{R}_+^{J\times C}$ are the row and column subsets of matrices $\mathbf{A}$ and $\mathbf{X}$. Typically $R << I$ and $C << T$ (see Section 1.3.1 for more detail).

All multiplicative learning rules ensure the nonnegativity of the factor matrices. Obviously, all the successive estimates remain positive if the initial estimate is positive. However, if a component of the solution becomes equal to zero, it remains at zero for all the successive iterations. To circumvent this problem, we usually force the values of the estimates a_{ij} and x_{jt} not to be less than a certain small positive value ε (typically, $\varepsilon = 10^{-9}$), called the threshold constraint, which often determines the noise floor, that is, $x_{jt} = \varepsilon$ if $x_{jt} \leq \varepsilon$, or in vector form $\boldsymbol{x} = \max(\boldsymbol{x}, \varepsilon)$. This means that we need to perform the following optimization problem:

$$\mathbf{A} = \arg\min_{\mathbf{A}\geq \boldsymbol{\varepsilon}} D_1(\mathbf{Y}_c||\mathbf{AX}_c), \quad \text{for fixed} \quad \mathbf{X}, \tag{3.4}$$

$$\mathbf{X} = \arg\min_{\mathbf{X}\geq \boldsymbol{\varepsilon}} D_2(\mathbf{Y}_r||\mathbf{A}_r\mathbf{X}), \quad \text{for fixed} \quad \mathbf{A}, \tag{3.5}$$

which lead in fact to Positive Matrix Factorization (PMF).

3.1 Extended ISRA and EMML Algorithms: Regularization and Sparsity

The most popular algorithms for NMF belong to the class of multiplicative ISRA (Image Space Reconstruction Algorithm) [18,20,26,38] and EMML (Expectation Maximization Maximum Likelihood) [2–4,19,21,22,27,31,32,37,42] update rules (also often referred to as the Lee-Seung algorithms[3] [40,41]). These classes of algorithms have a relative low complexity but are characterized by slow convergence and the risk of converging to spurious local minima. In this chapter, we discuss extensions of this class of multiplicative NMF algorithms by imposing additional constraints and applying more flexible and general cost functions. Moreover, we discuss how to unify and generalize them and how to implement them for large-scale problems.

3.1.1 Multiplicative NMF Algorithms Based on the Squared Euclidean Distance

For $\mathbf{E} = \mathbf{Y} - \mathbf{AX}$ modeled as i.i.d. (independent identically distributed) white Gaussian noise, we can formulate the problem of estimating the matrices $\mathbf{A}$ and $\mathbf{X}$ as that of maximizing the likelihood function:

$$p(\mathbf{Y}|\mathbf{A},\mathbf{X}) = \frac{1}{\sqrt{2\pi}\sigma} \exp\left(-\frac{||\mathbf{Y}-\mathbf{AX}||_F^2}{2\sigma^2}\right), \tag{3.6}$$

subject to $\mathbf{A} \geq \mathbf{0}$ and $\mathbf{X} \geq \mathbf{0}$, element-wise, where σ is the standard deviation of the Gaussian noise.

[2]Generally, we assume that we minimize sequentially one or two different cost functions with the same global minima, depending on statistical distributions of factor matrices. For simplicity, in this chapter we assume in the most cases that D_1 and D_2 are equal. However, in general the "mixture" or combination of updates rules are possible.

[3]However, since these algorithms have a long history, we refer to them as ISRA and EMML algorithms. They have been developed independently in many fields, including emission tomography, image restoration and astronomical imaging.

Maximizing the likelihood is equivalent to minimizing the corresponding negative log-likelihood function, or equivalently, the squared Frobenius norm

$$D_F(\mathbf{Y}||\mathbf{AX}) = \frac{1}{2}||\mathbf{Y} - \mathbf{AX}||_F^2, \tag{3.7}$$

$$\text{subject to} \quad a_{ij} \geq 0, \quad x_{jt} \geq 0, \quad \forall\, i, j, t.$$

Using the gradient descent approach and switching alternatively between the two sets of parameters, we obtain simple multiplicative update formulas (see derivation below):[4]

$$a_{ij} \leftarrow a_{ij} \frac{[\mathbf{Y}\mathbf{X}^T]_{ij}}{[\mathbf{A}\mathbf{X}\mathbf{X}^T]_{ij} + \varepsilon}, \tag{3.8}$$

$$x_{jt} \leftarrow x_{jt} \frac{[\mathbf{A}^T\mathbf{Y}]_{jt}}{[\mathbf{A}^T\mathbf{A}\mathbf{X}]_{jt} + \varepsilon}. \tag{3.9}$$

The above algorithm (3.8)–(3.9), called often Lee-Seung NMF algorithm can be considered as an obvious and natural extension of the well known ISRA algorithm proposed first by Daube-Witherspoon and Muehllehner [18] and investigated by many researchers, especially, De Pierro and Byrne [3,19–21,37,42]. The above update rules can be written in compact matrix form as

$$\mathbf{A} \leftarrow \mathbf{A} \circledast \left[(\mathbf{Y}\mathbf{X}^T) \oslash (\mathbf{A}\mathbf{X}\mathbf{X}^T + \varepsilon)\right], \tag{3.10}$$

$$\mathbf{X} \leftarrow \mathbf{X} \circledast \left[(\mathbf{A}^T\mathbf{Y}) \oslash (\mathbf{A}^T\mathbf{A}\mathbf{X} + \varepsilon)\right], \tag{3.11}$$

where $\circledast$ is the Hadamard (components-wise) product and $\oslash$ is element-wise division between two matrices.

Remark 3.1 *The ISRA NMF algorithm can be extended to weighted squared Euclidean norm (corresponding to colored Gaussian noise) by minimizing the cost functions (see Figure 1.12)*

$$D_{\mathbf{W}}(\mathbf{Y}_c||\mathbf{A}\mathbf{X}_c) = \frac{1}{2}\,\mathrm{tr}(\mathbf{Y}_c - \mathbf{A}\mathbf{X}_c)^T\mathbf{W}_{\mathbf{A}}(\mathbf{Y}_r - \mathbf{A}\mathbf{X}_c) = \frac{1}{2}||\mathbf{W}_2(\mathbf{Y}_c - \mathbf{A}\mathbf{X}_c)||_F^2, \tag{3.12}$$

$$D_{\mathbf{W}}(\mathbf{Y}_r||\mathbf{A}_r\mathbf{X}) = \frac{1}{2}\,\mathrm{tr}(\mathbf{Y}_r - \mathbf{A}_r\mathbf{X})^T\mathbf{W}_{\mathbf{X}}(\mathbf{Y}_r - \mathbf{A}_r\mathbf{X}) = \frac{1}{2}||\mathbf{W}_1(\mathbf{Y}_r - \mathbf{A}_r\mathbf{X})||_F^2, \tag{3.13}$$

where $\mathbf{W}_{\mathbf{A}} = \mathbf{W}_2^T\mathbf{W}_2$ and $\mathbf{W}_{\mathbf{X}} = \mathbf{W}_1^T\mathbf{W}_1$ are symmetric positive-definite weighted matrices, thus giving

$$\mathbf{A} \leftarrow \mathbf{A} \circledast \left[(\mathbf{W}_{\mathbf{A}}\mathbf{Y}_c\mathbf{X}_c^T) \oslash (\mathbf{W}_{\mathbf{A}}\mathbf{A}\mathbf{X}_c\mathbf{X}_c^T)\right]_+, \tag{3.14}$$

$$\mathbf{X} \leftarrow \mathbf{X} \circledast \left[(\mathbf{A}_r^T\mathbf{W}_{\mathbf{X}}\mathbf{Y}_r) \oslash (\mathbf{A}_r^T\mathbf{W}_{\mathbf{X}}\mathbf{A}_r\mathbf{X})\right]_+. \tag{3.15}$$

In practice, the columns of the matrix **A** *should be normalized to the unit ℓ_p-norm (typically, $p = 1$).*

The original ISRA algorithm is relatively slow, and many heuristic approaches have been proposed to speed it up. For example, a relaxation approach rises the multiplicative coefficients to some power $\omega \in (0, 2]$, that is,

$$a_{ij} \leftarrow a_{ij} \left(\frac{[\mathbf{Y}\mathbf{X}^T]_{ij}}{[\mathbf{A}\mathbf{X}\mathbf{X}^T]_{ij}}\right)^{\omega}, \tag{3.16}$$

$$x_{jt} \leftarrow x_{jt} \left(\frac{[\mathbf{A}^T\mathbf{Y}]_{jt}}{[\mathbf{A}^T\mathbf{A}\mathbf{X}]_{jt}}\right)^{\omega}, \tag{3.17}$$

in order to achieve faster convergence.

[4]Small positive constant ε is usually added to denominators to avoid division by zero.

The above learning rules usually provide sparse nonnegative representations of the data, although they do not guarantee the sparsest possible solution (that is, that the solutions contain the largest possible number of zero elements of $\mathbf{X}$ and/or $\mathbf{A}$). Moreover, the solutions are not necessarily unique and the algorithms may converge to local minima or worse to a saddle point. A much better performance (in the sense of convergence) may be achieved by using the multilayer NMF structure as explained in Chapter 1 [10,11,15,17].

To understand the origin of the above update rules, consider the Karush–Kuhn–Tucker (KKT)[5] first-order optimality conditions for NMF [30]:

$$\mathbf{A} \geq \mathbf{0}, \qquad \mathbf{X} \geq \mathbf{0}, \tag{3.18}$$

$$\nabla_{\mathbf{A}} D_F \geq \mathbf{0}, \qquad \nabla_{\mathbf{X}} D_F \geq \mathbf{0}, \tag{3.19}$$

$$\mathbf{A} \circledast \nabla_{\mathbf{A}} D_F = \mathbf{0}, \qquad \mathbf{X} \circledast \nabla_{\mathbf{X}} D_F = \mathbf{0}, \tag{3.20}$$

where the gradient components are expressed as

$$\nabla_{\mathbf{A}} D_F = \mathbf{A}\mathbf{X}\mathbf{X}^T - \mathbf{Y}\mathbf{X}^T, \qquad \nabla_{\mathbf{X}} D_F = \mathbf{A}^T\mathbf{A}\mathbf{X} - \mathbf{A}^T\mathbf{Y}. \tag{3.21}$$

Substituting (3.21) in (3.20) we obtain

$$\mathbf{A} \circledast (\mathbf{A}\mathbf{X}\mathbf{X}^T) = \mathbf{A} \circledast (\mathbf{Y}\mathbf{X}^T), \tag{3.22}$$

$$\mathbf{X} \circledast (\mathbf{A}^T\mathbf{A}\mathbf{X}) = \mathbf{X} \circledast (\mathbf{A}^T\mathbf{Y}). \tag{3.23}$$

Hence, we obtain the multiplicative updated rules (3.10)–(3.11).

It can be proved [30] (see also Appendix C) that for every constant $\varepsilon > 0$ the cost function $D_F = (1/2)||\mathbf{Y} - \mathbf{A}\mathbf{X}||_F^2$ is nonincreasing under update rules

$$\mathbf{A} \leftarrow \max\left\{\varepsilon, \mathbf{A} \circledast \left[(\mathbf{Y}\mathbf{X}^T) \oslash (\mathbf{A}\mathbf{X}\mathbf{X}^T)\right]\right\}, \tag{3.24}$$

$$\mathbf{X} \leftarrow \max\left\{\varepsilon, \mathbf{X} \circledast \left[(\mathbf{A}^T\mathbf{Y}) \oslash (\mathbf{A}^T\mathbf{A}\mathbf{X})\right]\right\}, \tag{3.25}$$

for any $\mathbf{A} \geq \boldsymbol{\varepsilon}$ and $\mathbf{X} \geq \boldsymbol{\varepsilon}$. Moreover, every limit point found by these update rules is a stationary point[6] of the following optimization problem:

$$\min_{\mathbf{A} \geq \boldsymbol{\varepsilon}} D_F(\mathbf{Y}||\mathbf{A}\mathbf{X}) \quad \text{and} \quad \min_{\mathbf{X} \geq \boldsymbol{\varepsilon}} D_F(\mathbf{Y}||\mathbf{A}\mathbf{X}). \tag{3.26}$$

This follows directly from the KKT conditions[7] (3.18)–(3.20).

The multiplicative ISRA update rules assume that a data matrix $\mathbf{Y}$ is nonnegative. However, in some applications $\mathbf{Y}$ is not necessarily nonnegative (for example, due to noise or physical constraints) but still the underlying factor matrices $\mathbf{A}$, $\mathbf{X}$ should be nonnegative. The simplest approach is to replace the negative entries of the data matrix by zeros if their absolute values are relatively small. Alternatively, we can apply the following approach proposed by Gillis and Glineur [30] and referred to as Nonnegative Factorization (NF).

[5] The Karush–Kuhn–Tucker conditions (also known as the Kuhn–Tucker or the KKT conditions) are contained in a system of equations and inequalities which the solution of a nonlinear programming problem must satisfy when the objective function and the constraint functions are differentiable. The KKT conditions are necessary for a solution in nonlinear programming to be optimal, provided some regularity conditions are satisfied. It is a generalization of the method of Lagrange multipliers to inequality constraints which provides a strategy for finding the minimum of a cost function subject to constraints.

[6] A stationary point is an argument of a cost function where the gradient is zero, where the cost function is flat, that is, "stops" increasing or decreasing (hence the name). This point can be a maximum, a minimum, or a point of inflection.

[7] Note, that this follows from the fact that for positive factor matrices, according to (3.20) all gradient components must be zero for the stationary point.

Any real data matrix can be expressed by two nonnegative matrices as $\mathbf{Y} = [\mathbf{Y}]_+ - [\mathbf{Y}]_-$, where nonnegative matrices are defined as

$$[\mathbf{Y}]_+ = \max\{\mathbf{0}, \mathbf{Y}\} \quad \text{and} \quad [\mathbf{Y}]_- = \max\{\mathbf{0}, -\mathbf{Y}\}. \tag{3.27}$$

We try to estimate the nonnegative factor matrices by minimizing alternatively the standard cost functions:

$$\min_{\mathbf{A} \geq \boldsymbol{\varepsilon}} D_F(\mathbf{Y}||\mathbf{AX}), \tag{3.28}$$

$$\min_{\mathbf{X} \geq \boldsymbol{\varepsilon}} D_F(\mathbf{Y}||\mathbf{AX}), \tag{3.29}$$

where $D_F(\mathbf{Y}||\mathbf{AX}) = (1/2)||\mathbf{Y} - \mathbf{AX}||_F^2$. Note that the KKT optimality conditions for this optimization problem are the same as for the regular NMF (see Eqs. (3.22)–(3.23)), which can be expressed as

$$\mathbf{A} \circledast (\mathbf{AXX}^T + [\mathbf{Y}]_-\mathbf{X}^T) = \mathbf{A} \circledast ([\mathbf{Y}]_+\mathbf{X}^T), \tag{3.30}$$

$$\mathbf{X} \circledast (\mathbf{A}^T\mathbf{AX} + \mathbf{A}^T[\mathbf{Y}]_-) = \mathbf{X} \circledast (\mathbf{A}^T[\mathbf{Y}]_+). \tag{3.31}$$

Hence, we obtain the multiplicative updated rules [30]

$$\mathbf{A} \leftarrow \mathbf{A} \circledast \left[([\mathbf{Y}]_+\mathbf{X}^T) \oslash (\mathbf{AXX}^T + [\mathbf{Y}]_-\mathbf{X}^T)\right], \tag{3.32}$$

$$\mathbf{X} \leftarrow \mathbf{X} \circledast \left[(\mathbf{A}^T[\mathbf{Y}]_+) \oslash (\mathbf{A}^T\mathbf{AX} + \mathbf{A}^T[\mathbf{Y}]_-)\right]. \tag{3.33}$$

Remark 3.2 *It can be shown that for every constant $\varepsilon > 0$ and for $\mathbf{Y} = [\mathbf{Y}]_+ - [\mathbf{Y}]_-$ with $\mathbf{Y}_+ \geq \mathbf{0}$ and $\mathbf{Y}_- \geq \mathbf{0}$ the cost function $D_F = (1/2)||\mathbf{Y} - \mathbf{AX}||_F^2$ is nonincreasing under update rules*

$$\mathbf{A} \leftarrow \max\left\{\varepsilon,\ \mathbf{A} \circledast \left[([\mathbf{Y}]_+\mathbf{X}^T) \oslash (\mathbf{AXX}^T + [\mathbf{Y}]_-\mathbf{X}^T)\right]\right\}, \tag{3.34}$$

$$\mathbf{X} \leftarrow \max\left\{\varepsilon,\ \mathbf{X} \circledast \left[(\mathbf{A}^T[\mathbf{Y}]_+) \oslash (\mathbf{A}^T\mathbf{AX} + \mathbf{A}^T[\mathbf{Y}]_-)\right]\right\}, \tag{3.35}$$

for any $\mathbf{A} \geq \boldsymbol{\varepsilon}$ and $\mathbf{X} \geq \boldsymbol{\varepsilon}$. Moreover, every limit point of these update rules is a stationary point of the optimization problems (3.28)–(3.29).

We shall derive now the algorithm described by (3.8)–(3.9) using a different approach for a slightly more general cost function by considering additional penalty terms:

$$D_{Fr}(\mathbf{Y}||\mathbf{AX}) = \frac{1}{2}||\mathbf{Y} - \mathbf{AX}||_F^2 + \alpha_{\mathbf{A}}\, J_{\mathbf{A}}(\mathbf{A}) + \alpha_{\mathbf{X}}\, J_{\mathbf{X}}(\mathbf{X}) \tag{3.36}$$

$$\text{subject to } a_{ij} \geq 0, \quad x_{jt} \geq 0, \quad \forall\, i, j, t,$$

where $\alpha_{\mathbf{A}}$ and $\alpha_{\mathbf{X}}$ are nonnegative regularization parameters and penalty terms $J_{\mathbf{X}}(\mathbf{X})$, $J_{\mathbf{A}}(\mathbf{A})$ are used to enforce a certain application-dependent characteristic of a desired solution. As a special practical case, we have $J_{\mathbf{X}}(\mathbf{X}) = \sum_{jt} \varphi_{\mathbf{X}}(x_{jt})$, where $\varphi(x_{jt})$ are suitably chosen functions which are measures of smoothness or sparsity. To achieve a sparse representation we usually chose $\varphi(x_{jt}) = |x_{jt}|$ or simply $\varphi(x_{jt}) = x_{jt}$ (since, x_{jt} are nonnegative). Similar regularization terms can also be implemented[8] for the matrix $\mathbf{A}$.

Applying the standard gradient descent approach, we have

$$a_{ij} \leftarrow a_{ij} - \eta_{ij} \frac{\partial D_{Fr}(\mathbf{A}, \mathbf{X})}{\partial a_{ij}}, \qquad x_{jt} \leftarrow x_{jt} - \eta_{jt} \frac{\partial D_{Fr}(\mathbf{A}, \mathbf{X})}{\partial x_{jt}}, \tag{3.37}$$

[8]Note that we treat both matrices **A** and **X** in a similar way.

where η_{ij} and η_{jt} are positive learning rates. The gradients above can be expressed in compact matrix forms as

$$\frac{\partial D_{Fr}(\mathbf{A},\mathbf{X})}{\partial a_{ij}} = [-\mathbf{Y}\mathbf{X}^T + \mathbf{A}\mathbf{X}\mathbf{X}^T]_{ij} + \alpha_{\mathbf{A}} \frac{\partial J_{\mathbf{A}}(\mathbf{A})}{\partial a_{ij}}, \tag{3.38}$$

$$\frac{\partial D_{Fr}(\mathbf{A},\mathbf{X})}{\partial x_{jt}} = [-\mathbf{A}^T\mathbf{Y} + \mathbf{A}^T\mathbf{A}\mathbf{X}]_{jt} + \alpha_{\mathbf{X}} \frac{\partial J_{\mathbf{X}}(\mathbf{X})}{\partial x_{jt}}. \tag{3.39}$$

At this point, we can follow the Lee and Seung approach to choose the learning rates [40,41,45,46]

$$\eta_{ij} = \frac{a_{ij}}{[\mathbf{A}\mathbf{X}\mathbf{X}^T]_{ij}}, \qquad \eta_{jt} = \frac{x_{jt}}{[\mathbf{A}^T\mathbf{A}\,\mathbf{X}]_{jt}}, \tag{3.40}$$

leading to multiplicative update rules

$$a_{ij} \leftarrow a_{ij} \frac{\left[[\mathbf{Y}\,\mathbf{X}^T]_{ij} - \alpha_{\mathbf{A}}\, \Psi_{\mathbf{A}}(a_{ij})\right]_+}{[\mathbf{A}\,\mathbf{X}\,\mathbf{X}^T]_{ij}}, \tag{3.41}$$

$$x_{jt} \leftarrow x_{jt} \frac{\left[[\mathbf{A}^T\,\mathbf{Y}]_{jt} - \alpha_{\mathbf{X}}\, \Psi_{\mathbf{X}}(x_{jt})\right]_+}{[\mathbf{A}^T\mathbf{A}\,\mathbf{X}]_{jt}}, \tag{3.42}$$

where the functions $\Psi_{\mathbf{A}}(a_{ij})$ and $\Psi_{\mathbf{X}}(x_{jt})$ are defined as

$$\Psi_{\mathbf{A}}(a_{ij}) = \frac{\partial J_{\mathbf{A}}(\mathbf{A})}{\partial a_{ij}}, \qquad \Psi_{\mathbf{X}}(x_{jt}) = \frac{\partial J_{\mathbf{X}}(\mathbf{X})}{\partial x_{jt}}. \tag{3.43}$$

And the regularization parameters $\alpha_{\mathbf{A}}, \alpha_{\mathbf{X}}$ should be sufficiently small to ensure the nonnegativity in (3.42) in each iteration, or we can avoid this problem by a half-wave rectifier $[x]_+ = \max\{x, \varepsilon\}$ with small ε (typically 10^{-9}).

Remark 3.3 *The above or similar update rules can be derived by using general multiplicative heuristic formulas*

$$a_{ij} \leftarrow a_{ij} \left(\frac{\left[\dfrac{\partial D(\mathbf{Y}||\mathbf{A}\mathbf{X})}{\partial a_{ij}}\right]_-}{\left[\dfrac{\partial D(\mathbf{Y}||\mathbf{A}\mathbf{X})}{\partial a_{ij}}\right]_+} \right)^{\omega}, \qquad x_{jt} \leftarrow x_{jt} \left(\frac{\left[\dfrac{\partial D(\mathbf{Y}||\mathbf{A}\mathbf{X})}{\partial x_{jt}}\right]_-}{\left[\dfrac{\partial D(\mathbf{Y}||\mathbf{A}\mathbf{X})}{\partial x_{jt}}\right]_+} \right)^{\omega}, \tag{3.44}$$

where the nonlinear operator $\left[\dfrac{\partial D(\mathbf{Y}||\mathbf{A}\mathbf{X})}{\partial a_{ij}}\right]_{\mp}$ *denotes the negative or positive parts of the gradient, and* ω *is an over-relaxation positive parameter (typically, in the range of* $[0.5, 2]$*). Substituting the negative and positive terms of the gradients (3.38)–(3.39) into (3.44), we obtain*

$$a_{ij} \leftarrow a_{ij} \left(\frac{[\mathbf{Y}\,\mathbf{X}^T]_{ij}}{[\mathbf{A}\,\mathbf{X}\,\mathbf{X}^T]_{ij} + \alpha_{\mathbf{A}}\, \Psi_{\mathbf{A}}(a_{ij})} \right)^{\omega}, \tag{3.45}$$

$$x_{jt} \leftarrow x_{jt} \left(\frac{[\mathbf{A}^T\,\mathbf{Y}]_{jt}}{[\mathbf{A}^T\mathbf{A}\,\mathbf{X}]_{jt} + \alpha_{\mathbf{X}}\, \Psi_{\mathbf{X}}(x_{jt})} \right)^{\omega} \tag{3.46}$$

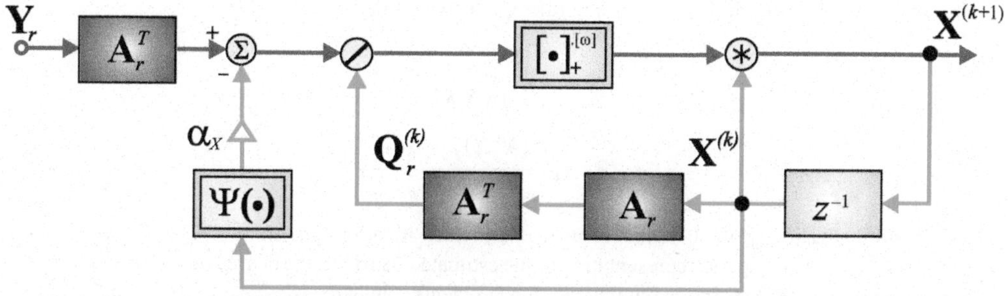

Figure 3.1 Simplified functional block diagram of the extended ISRA algorithm for a large-scale NMF. Learning rule for matrix $\mathbf{X}$: $\mathbf{X}^{(k+1)} = \mathbf{X}^{(k)} \circledast \left[(\mathbf{A}_r^T\mathbf{Y}_r - \alpha_{\mathbf{X}}\, \Psi(\mathbf{X}^{(k)})) \oslash (\mathbf{A}_r^T\mathbf{A}_r\mathbf{X}^{(k)})\right]_+^{.[\omega]}$; and a similar one for matrix $\mathbf{A}$: $\mathbf{A}^{(k+1)} = \mathbf{A}^{(k)} \circledast \left[(\mathbf{Y}_c\mathbf{X}_c^T - \alpha_{\mathbf{A}}\, \Psi(\mathbf{A}^{(k)})) \oslash (\mathbf{A}^{(k)}\mathbf{X}_c\mathbf{X}_c^T)\right]_+^{.[\omega]}$, where the block matrices have the following dimensions: $\mathbf{Y}_c \in \mathbb{R}_+^{I\times C}$, $\mathbf{Y}_r \in \mathbb{R}_+^{R\times T}$, $\mathbf{X}_c \in \mathbb{R}_+^{J\times C}$ and $\mathbf{A}_r \in \mathbb{R}_+^{R\times J}$, with $C << T$ and $R << I$.

or alternatively (see also Figure 3.1)

$$a_{ij} \leftarrow a_{ij} \left[\frac{[\mathbf{Y}\mathbf{X}^T]_{ij} - \alpha_{\mathbf{A}}\, \Psi_{\mathbf{A}}(a_{ij})}{[\mathbf{A}\mathbf{X}\mathbf{X}^T]_{ij}}\right]_+^{\omega}, \tag{3.47}$$

$$x_{jt} \leftarrow x_{jt} \left[\frac{[\mathbf{A}^T\mathbf{Y}]_{jt} - \alpha_{\mathbf{X}}\, \Psi_{\mathbf{X}}(x_{jt})}{[\mathbf{A}^T\mathbf{A}\mathbf{X}]_{jt}}\right]_+^{\omega}. \tag{3.48}$$

The above update rules simplify to the standard Lee-Seung algorithm for $\alpha_{\mathbf{A}} = \alpha_{\mathbf{X}} = 0$ [40,41].

We shall now consider several examples with different penalty/regularization terms. In order to ensure the smoothness or boundedness of the solution, we may apply the standard Tikhonov regularization terms by employing the following cost function:

$$D_{F2}(\mathbf{A},\mathbf{X}) = \frac{1}{2}\left(\|\mathbf{Y}-\mathbf{A}\mathbf{X}\|_F^2 + \alpha_{\mathbf{A}}\, \|\mathbf{A}\|_F^2 + \alpha_{\mathbf{X}}\, \|\mathbf{X}\|_F^2\right), \tag{3.49}$$

$$\text{s. t. } a_{ij} \geq 0, \quad x_{jt} \geq 0, \quad \forall\, i, j, t,$$

where $\alpha_{\mathbf{A}} \geq 0$ and $\alpha_{\mathbf{X}} \geq 0$ are nonnegative regularization parameters. This leads to the following update formulas

$$a_{ij} \leftarrow a_{ij} \frac{\left[[\mathbf{Y}\mathbf{X}^T]_{ij} - \alpha_{\mathbf{A}}\, a_{ij}\right]_+}{[\mathbf{A}\mathbf{X}\mathbf{X}^T]_{ij}}, \tag{3.50}$$

$$x_{jt} \leftarrow x_{jt} \frac{\left[[\mathbf{A}^T\mathbf{Y}]_{jt} - \alpha_{\mathbf{X}}\, x_{jt}\right]_+}{[\mathbf{A}^T\mathbf{A}\mathbf{X}]_{jt}}. \tag{3.51}$$

In order to impose sparsity, instead of using the ℓ_2-norm, we can use the regularization terms based on the ℓ_1-norm, and minimize the cost function:

$$D_{F1}(\mathbf{A},\mathbf{X}) = \frac{1}{2}\|\mathbf{Y}-\mathbf{A}\mathbf{X}\|_F^2 + \alpha_{\mathbf{A}}\, \|\mathbf{A}\|_1 + \alpha_{\mathbf{X}}\, \|\mathbf{X}\|_1, \tag{3.52}$$

$$\text{s. t. } a_{ij} \geq 0, \quad x_{jt} \geq 0, \quad \forall\, i, j, t,$$

where the ℓ_1-norms for matrices are defined as $||\mathbf{A}||_1 = \sum_{ij} |a_{ij}|$ and $||\mathbf{X}||_1 = \sum_{jt} |x_{jt}|$.

This way, the above multiplicative learning rules can be simplified as

$$a_{ij} \leftarrow a_{ij} \frac{\left[[\mathbf{Y}\,\mathbf{X}^T]_{ij} - \alpha_{\mathbf{A}} \right]_+}{[\mathbf{A}\,\mathbf{X}\,\mathbf{X}^T]_{ij}}, \tag{3.53}$$

$$x_{jt} \leftarrow x_{jt} \frac{\left[[\mathbf{A}^T\,\mathbf{Y}]_{jt} - \alpha_{\mathbf{X}} \right]_+}{[\mathbf{A}^T\mathbf{A}\,\mathbf{X}]_{jt}}, \tag{3.54}$$

where the normalization of the columns of matrix $\mathbf{A}$ in each iteration is performed as $a_{ij} \leftarrow a_{ij}/\sum_p a_{pj}$. This algorithm provides a sparse representation of the estimated matrices; the degree of sparseness increases with an increase in the values of the regularization coefficients (typically, $\alpha_{\mathbf{A}} = \alpha_{\mathbf{X}} = 0.01 - 0.5$).

For a large-scale NMF problem with sparsity constraints the above multiplicative algorithm (3.53)–(3.54) can be summarized in the following pseudo-code Algorithm 3.1 (see also Figure 3.2).

Algorithm 3.1: ISRA NMF with Over-relaxation and Sparsity Control

Input: $\mathbf{Y} \in \mathbb{R}_+^{I\times T}$: input data, J: rank of approximation,
ω: over-relaxation, and $\alpha_{\mathbf{A}}$, $\alpha_{\mathbf{X}}$: sparsity levels
Output: $\mathbf{A} \in \mathbb{R}_+^{I\times J}$ and $\mathbf{X} \in \mathbb{R}_+^{J\times T}$ such that cost function (3.36) is minimized.

begin
  initialization for $\mathbf{A}$ and $\mathbf{X}$
  repeat `/* update X and A */`
    select R row indices
    $\mathbf{X} \leftarrow \mathbf{X} \circledast \left[(\mathbf{A}_r^T\,\mathbf{Y}_r - \alpha_{\mathbf{X}}\,\mathbf{1}_{J\times T}) \oslash (\mathbf{A}_r^T\mathbf{A}_r\,\mathbf{X}) \right]_+^{.[\omega]}$ (3.55)
    select C column indices
    $\mathbf{A} \leftarrow \mathbf{A} \circledast \left[(\mathbf{Y}_c\,\mathbf{X}_c^T - \alpha_{\mathbf{A}}\,\mathbf{1}_{I\times J}) \oslash (\mathbf{A}\,\mathbf{X}_c\,\mathbf{X}_c^T) \right]_+^{.[\omega]}$, $\omega \in (0, 2]$ (3.56)
    foreach $\boldsymbol{a}_j$ *of* $\mathbf{A}$ **do** $\boldsymbol{a}_j \leftarrow \boldsymbol{a}_j/\|\boldsymbol{a}_j\|_p$ `/* normalize to` ℓ_p `unit length */`
  until *a stopping criterion is met* `/* convergence condition */`
end

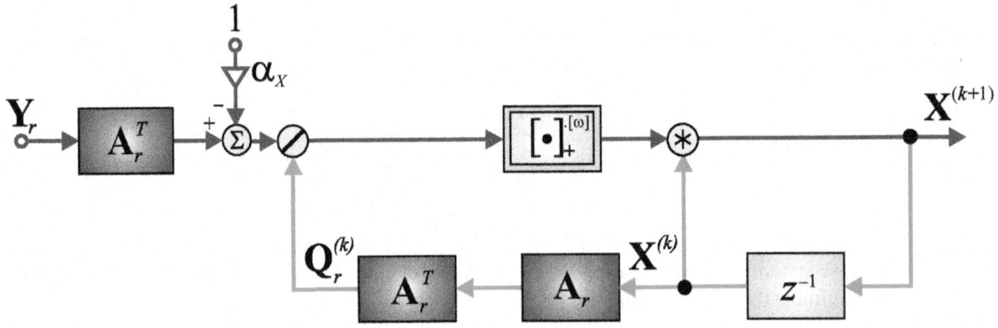

Figure 3.2 Simplified functional block diagram of the ISRA algorithm for a large-scale NMF with over-relaxation parameter ω and sparseness controlled via parameter $\alpha_{\mathbf{X}}$: $\mathbf{X}^{(k+1)} = \mathbf{X}^{(k)} \circledast \left[(\mathbf{A}_r^T\mathbf{Y}_r - \alpha_{\mathbf{X}}\,\mathbf{1}_{J\times T}) \oslash (\mathbf{A}_r^T\mathbf{A}_r\mathbf{X}^{(k)}) \right]_+^{.[\omega]}$. A similar block diagram can be drawn for updating $\mathbf{A}$ as $\mathbf{A}^{(k+1)} = \mathbf{A}^{(k)} \circledast \left[(\mathbf{Y}_c\mathbf{X}_c^T - \alpha_{\mathbf{A}}\,\mathbf{1}_{I\times J}) \oslash (\mathbf{A}^{(k)}\mathbf{X}_c\mathbf{X}_c^T) \right]_+^{.[\omega]}$.

3.1.2 Multiplicative NMF Algorithms Based on Kullback-Leibler I-Divergence

The best known and the most frequently used adaptive multiplicative algorithms for NMF are based on two loss functions: Squared Euclidean distance (discussed in the previous section) and generalized Kullback-Leibler divergence also called I-divergence defined as

$$D_{KL}(\mathbf{Y}||\mathbf{AX}) = \sum_{it} \left(y_{it} \ln \frac{y_{it}}{[\mathbf{AX}]_{it}} - y_{it} + [\mathbf{AX}]_{it} \right). \tag{3.57}$$

Similar to the squared Euclidean cost function, the I-divergence is convex with respect to either **A** or **X**, but it is not convex with respect to **A** and **X** (when the cost function is minimized simultaneously with respect of the both set of paramters), so the minimization of such a cost function can yield many local minima.

We shall now show that the minimization of the I-divergence (3.57) is equivalent to the maximization of the Poisson likelihood:

$$L(\mathbf{X}) = \prod_{it} \left(\frac{q_{it}^{y_{it}}}{y_{it}!} \exp(-q_{it}) \right), \tag{3.58}$$

where $q_{it} = [\mathbf{AX}]_{it}$. Using Stirling's formula: $n! \approx n \ln(n) - n$ for $n \gg 1$ the negative log-likelihood becomes:

$$D_{KL} = -\ln(L(\mathbf{X})) = \sum_{it} \left(q_{it} - y_{it} + y_{it} \ln \frac{y_{it}}{q_{it}} + \ln(\sqrt{2\pi y_{it}}\,) \right). \tag{3.59}$$

Then, the cost function simplifies to

$$D_{KL} = \sum_{it} (q_{it} - y_{it} \ln q_{it}). \tag{3.60}$$

By minimizing the cost function (3.57) subject to nonnegativity constraints, we can derive the following multiplicative learning rule, referred to as the EMML algorithm[9] [3,5,34,38,40,41,49]

$$x_{jt} \leftarrow x_{jt} \frac{\sum_{i=1}^{I} a_{ij}\, (y_{it}/[\mathbf{A\,X}]_{it})}{\sum_{i=1}^{I} a_{ij}}, \tag{3.61}$$

$$a_{ij} \leftarrow a_{ij} \frac{\sum_{t=1}^{T} x_{jt}\, (y_{it}/[\mathbf{A\,X}]_{it})}{\sum_{t=1}^{T} x_{jt}}. \tag{3.62}$$

We will derive the above algorithm in a more general and universal form in the next section.

To accelerate the convergence and enhance control over sparsity, we consider several extensions of the EMML algorithm [12,14]: The sparsity and smoothness can be enforced by adding suitable regularization/penalty terms to the loss functions or generalized divergences, or by suitable nonlinear projections imposed upon the multiplicative updating rules. For example, the following extended EMML algorithm enforces sparse solutions by using nonlinear projection (or shrinkage) functions:

$$x_{jt} \leftarrow \left(x_{jt} \left(\frac{\sum_{i=1}^{I} a_{ij}\, (y_{it}/[\mathbf{A\,X}]_{it})}{\sum_{i=1}^{I} a_{ij}} \right)^{\omega} \right)^{1+\alpha_{sX}}, \tag{3.63}$$

[9]The EMML algorithm is sometimes also called the Richardson-Lucy algorithm (RLA). In fact, the EMML algorithm was developed for a fixed and known **A**. The Lee-Seung algorithm based on I-divergence employs alternative switching between **A** and **X**.

$$a_{ij} \leftarrow \left(a_{ij} \left(\frac{\sum_{t=1}^{T} x_{jt} \, (y_{it}/[\mathbf{A}\,\mathbf{X}]_{it})}{\sum_{t=1}^{T} x_{jt}} \right)^{\omega} \right)^{1+\alpha_{sA}}, \tag{3.64}$$

where the relaxation parameter ω in the range of (0, 2] helps to improve the convergence, and the additional small nonnegative regularization terms $\alpha_{sX} \geq 0$ and $\alpha_{sA} \geq 0$ are introduced in order to enforce sparseness of the solution (if necessary). Typical values of these sparsification coefficients are $\alpha_{sX} = \alpha_{sA} = 0.001 - 0.005$. In an enhanced form to reduce the computational cost, the numerators in (3.63) and (3.64) can be ignored due to normalizing $\boldsymbol{a}_j$ to unit length of ℓ_1-norm:

$$\mathbf{X} \leftarrow \left(\mathbf{X} \circledast (\mathbf{A}\,(\mathbf{Y} \oslash [\mathbf{A}\,\mathbf{X}]))^{.[\omega]}\right)^{.[1+\alpha_{sX}]}, \tag{3.65}$$

$$\mathbf{A} \leftarrow \left(\mathbf{A} \circledast (\mathbf{X}\,(\mathbf{Y} \oslash [\mathbf{A}\,\mathbf{X}]))^{.[\omega]}\right)^{.[1+\alpha_{sA}]}, \tag{3.66}$$

$$a_{ij} \leftarrow a_{ij}/\|\boldsymbol{a}_j\|_1. \tag{3.67}$$

For sparse large-scale NMF with the block-wise technique, the EMML algorithm has a slightly different form as illustrated in Figure 3.3 and as presented in the pseudo-code of the Algorithm 3.2.

Remark 3.4 *The I-divergence and the corresponding learning rules have several interesting properties [33]. It should be noted that the gradient of the I-divergence can be expressed as*

$$\frac{\partial D_{KL}}{\partial x_{jt}} = \sum_{i=1}^{I} a_{ij} \left[1 - \left(\frac{y_{it}}{[\mathbf{AX}]_{it}}\right)\right], \tag{3.70}$$

$$\frac{\partial D_{KL}}{\partial a_{ij}} = \sum_{t=1}^{T} x_{jt} \left[1 - \left(\frac{y_{it}}{[\mathbf{AX}]_{it}}\right)\right]. \tag{3.71}$$

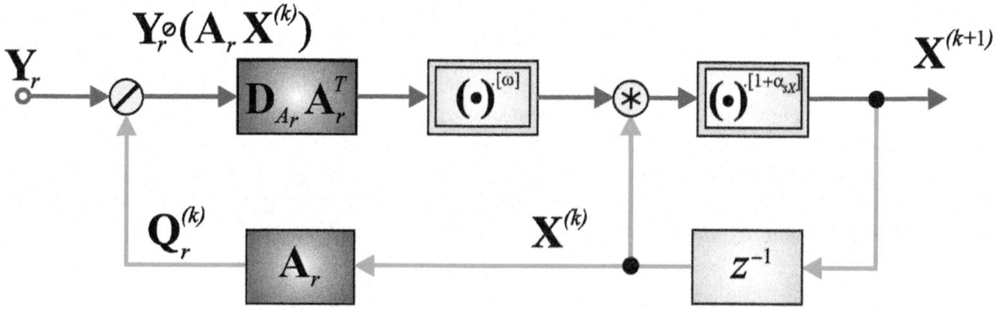

Figure 3.3 Simplified functional block diagram of the multiplicative EMML algorithm for sparse large-scale NMF. Sparsity of $\mathbf{X}$ is controlled via small coefficient α_{sX} and convergence speed via ω an over-relaxation parameter, in the range (0.5, 2]: $\mathbf{X}^{(k+1)} = \left(\mathbf{X}^{(k)} \circledast \left(\mathbf{D}_{A_r}\mathbf{A}_r^T(\mathbf{Y}_r \oslash (\mathbf{A}_r^T\mathbf{X}^{(k)}))\right)^{.[\omega]}\right)^{.[1+\alpha_{sX}]}$, where $\mathbf{D}_{A_r} = \text{diag}(\|\boldsymbol{a}_{r_1}\|_1^{-1}, \|\boldsymbol{a}_{r_2}\|_1^{-1}, \ldots, \|\boldsymbol{a}_{r_J}\|_1^{-1})$, where $\boldsymbol{a}_{r_j}$ are columns of the reduced matrix $\mathbf{A}_r$. A similar block diagram can be drawn for updating matrix $\mathbf{A}$ as $\mathbf{A}^{(k+1)} = \left(\mathbf{A}^{(k)} \circledast \left((\mathbf{Y}_c \oslash (\mathbf{A}^{(k)}\mathbf{X}_c))\mathbf{X}_c^T\right)^{.[\omega]}\right)^{.[1+\alpha_{sA}]}$.

Algorithm 3.2: EMML NMF Algorithm with Over-relaxation and Sparsity Control

Input: $\mathbf{Y} \in \mathbb{R}_+^{I \times T}$: input data, J: rank of approximation
ω: over-relaxation, and α_{sA}, α_{sX}: sparsity levels
Output: $\mathbf{A} \in \mathbb{R}_+^{I \times J}$ and $\mathbf{X} \in \mathbb{R}_+^{J \times T}$ such that cost function (3.57) is minimized.

begin
  initialization for $\mathbf{A}$ and $\mathbf{X}$
  repeat /* update X and A */
    select R row indices

$$\mathbf{X} \longleftarrow \left(\mathbf{X} \circledast \left(\mathbf{D}_{A_r} \mathbf{A}_r^T (\mathbf{Y}_r \oslash (\mathbf{A}_r \mathbf{X}))\right)^{.[\omega]}\right)^{.[1+\alpha_{sX}]}, \tag{3.68}$$

    select C column indices

$$\mathbf{A} \longleftarrow \left(\mathbf{A} \circledast \left((\mathbf{Y}_c \oslash (\mathbf{A}\mathbf{X}_c))\, \mathbf{X}_c^T\right)^{.[\omega]}\right)^{.[1+\alpha_{sA}]}, \tag{3.69}$$

    foreach $\boldsymbol{a}_j$ *of* $\mathbf{A}$ **do** $\boldsymbol{a}_j \leftarrow \boldsymbol{a}_j / \|\boldsymbol{a}_j\|_1$ /* normalize to ℓ_1 unit length */
  until *a stopping criterion is met* /* convergence condition */
end

The KKT optimality conditions for the I-divergence are as follows:

$$a_{ij} \geq 0, \qquad x_{jt} \geq 0, \tag{3.72}$$

$$\frac{\partial D_{KL}}{\partial x_{jt}} \geq 0, \qquad \frac{\partial D_{KL}}{\partial a_{ij}} \geq 0, \tag{3.73}$$

$$x_{jt} \frac{\partial D_{KL}}{\partial x_{jt}} = 0, \qquad a_{ij} \frac{\partial D_{KL}}{\partial a_{ij}} = 0, \qquad \forall i, j, t. \tag{3.74}$$

The above KKT optimality conditions can be generalized to an arbitrary convex cost function $D(\mathbf{Y}||\mathbf{AX})$ *with nonnegativity constraints as follows:*

$$\mathbf{A} \geq \mathbf{0}, \qquad \mathbf{X} \geq \mathbf{0}, \tag{3.75}$$

$$\nabla_{\mathbf{A}} D \geq \mathbf{0}, \qquad \nabla_{\mathbf{X}} D \geq \mathbf{0}, \tag{3.76}$$

$$\mathbf{A} \circledast \nabla_{\mathbf{A}} D = \mathbf{0}, \qquad \mathbf{X} \circledast \nabla_{\mathbf{X}} D = \mathbf{0}, \tag{3.77}$$

These conditions can be written in the equivalent forms:

$$\min\{\mathbf{A}, \nabla_{\mathbf{A}} D(\mathbf{Y}||\mathbf{AX})\} = \mathbf{0} \quad \text{and} \quad \min\{\mathbf{X}, \nabla_{\mathbf{X}} D(\mathbf{Y}||\mathbf{AX})\} = \mathbf{0}, \tag{3.78}$$

and for positive matrices $\mathbf{A} \geq \boldsymbol{\varepsilon}$ *and* $\mathbf{X} \geq \boldsymbol{\varepsilon}$

$$\nabla_{\mathbf{A}} D(\mathbf{Y}||\mathbf{AX}) = \mathbf{0} \quad \text{and} \quad \nabla_{\mathbf{X}} D(\mathbf{Y}||\mathbf{AX}) = \mathbf{0}. \tag{3.79}$$

It can be shown that for every stationary points a_{ij}, x_{jt} *of the cost function (3.57) we have [33]:*

$$\sum_i [\mathbf{AX}]_{it} = \sum_i y_{it}, \quad \text{and} \quad \sum_t [\mathbf{AX}]_{it} = \sum_t y_{it}. \tag{3.80}$$

This property can be formulated based on the following Theorem [33]:

Theorem 3.1 ***(Ho – Van Dooren)*** *Let* $\mathbf{Y} \in \mathbb{R}^{I \times T}$ *be a nonnegative matrix. Then every stationary point* $\mathbf{A}, \mathbf{X}$ *of the cost function (3.57) preserves the rows of* $\mathbf{Y}$ *in the sense that* $\mathbf{1}_{1\times I}\ \mathbf{Y} = \mathbf{1}_{1\times I}\ [\mathbf{AX}]$*, the column sums of* $\mathbf{Y}$*, i.e.,* $\mathbf{Y}\ \mathbf{1}_{T\times 1} = [\mathbf{AX}]\ \mathbf{1}_{T\times 1}$*, and the matrix sum of* $\mathbf{Y}$*, i.e.,* $\mathbf{1}_{1\times I}\ \mathbf{Y}\ \mathbf{1}_{T\times 1} = \mathbf{1}_{1\times I}\ [\mathbf{AX}]\ \mathbf{1}_{T\times 1}$*, where* $\mathbf{1}_{1\times I}$ *is a row vector with all elements equal to* 1.

Finally, the smoothness and sparsity constraints can be imposed on the multiplicative EMML NMF algorithm by incorporating additional terms into the cost function, such as regularization/penalty terms and/or projecting components by suitable nonlinear functions. For example, the regularized Kullback-Leibler I-divergence has the form:

$$D_{KL}(\mathbf{Y}||\mathbf{AX}) = \sum_{it} \left(y_{it} \ln \frac{y_{it}}{[\mathbf{AX}]_{it}} + y_{it} - [\mathbf{AX}]_{it} \right) + \alpha_{\mathbf{X}}\ J(\mathbf{X}) + \alpha_{\mathbf{A}}\ J(\mathbf{A}), \tag{3.81}$$

for which regularized EMML multiplicative update rules are given by:

$$x_{jt} \leftarrow \left(x_{jt} \frac{\sum_{i \in \mathcal{S}_I} a_{ij}\ (y_{it}/[\mathbf{A}_r\ \mathbf{X}]_{it})}{[\sum_{i \in \mathcal{S}_I} a_{ij} + \alpha_{\mathbf{X}}\ \Psi_{\mathbf{X}}(x_{jt})]_+} \right)^{1+\alpha_{sX}}, \tag{3.82}$$

$$a_{ij} \leftarrow \left(a_{ij} \frac{\sum_{t \in \mathcal{S}_T} x_{jt}\ (y_{it}/[\mathbf{A}\ \mathbf{X}_c]_{it})}{[\sum_{t \in \mathcal{S}_T} x_{jt} + \alpha_{\mathbf{A}}\ \Psi_{\mathbf{A}}(a_{ij})]_+} \right)^{1+\alpha_{sA}}, \tag{3.83}$$

where the matrix functions $\Psi_{\mathbf{X}}(\mathbf{X})$ and $\Psi_{\mathbf{A}}(\mathbf{A})$ are defined as follows

$$\Psi_{\mathbf{X}}(\mathbf{X}) = \frac{\partial J(\mathbf{X})}{\partial \mathbf{X}}, \qquad \Psi_{\mathbf{A}}(\mathbf{A}) = \frac{\partial J(\mathbf{A})}{\partial \mathbf{A}}. \tag{3.84}$$

The regularization terms enforce some specific properties of the estimated parameters, in particular:

1. In order to impose **sparsity** on the matrix $\mathbf{A}$, we may use

$$J(\mathbf{A}) = ||\mathbf{A}||_1, \qquad \Psi_{\mathbf{A}}(\mathbf{A}) = \mathbf{1}_{I\times J}. \tag{3.85}$$

2. In order to enforce **orthogonality** of vectors $\boldsymbol{a}_j$ (a very sparse representation of $\mathbf{A}$), we may use

$$J(\mathbf{A}) = \frac{1}{2} \sum_{j \neq p} \boldsymbol{a}_j \boldsymbol{a}_p^T, \qquad \Psi_{\mathbf{A}}(\mathbf{A}) = \mathbf{A}(\mathbf{1}\mathbf{1}^T - \mathbf{I}). \tag{3.86}$$

3. In order to impose **smoothness** of vectors $\boldsymbol{a}_j$, we may use a second order differential operator (see Chapter 4 for more detail)

$$J(\mathbf{A}) = ||\mathbf{LA}||_F^2. \tag{3.87}$$

The multiplicative algorithms presented so far assume that the noise has Gaussian (then we use squared Euclidean distance) or Poisson distribution (then we use I-divergence). However, the error distribution is not confined only to such distributions. In the next sections we adopt exponential family of distributions of noise where Gaussian and Poisson noise are special cases.

3.2 Multiplicative Algorithms Based on Alpha-Divergence

3.2.1 Multiplicative Alpha NMF Algorithm

We shall now consider a more general cost function called the Alpha-divergence [10,12,13,16]),

$$D_A^{(\alpha)}(\mathbf{Y}||\mathbf{AX}) = \frac{1}{\alpha(\alpha-1)} \sum_{it} \left(y_{it}^{\alpha} \, [\mathbf{AX}]_{it}^{1-\alpha} - \alpha \, y_{it} + (\alpha - 1) \, [\mathbf{AX}]_{it} \right). \tag{3.88}$$

Recall that special case of the Alpha-divergence for $\alpha = 2, 0.5, -1$ are respectively the Pearson's Chi squared, squared Hellinger and Neyman's Chi-squared distances, while I-divergences are defined by the limits of (3.88) as $\alpha \to 1$ and $\alpha \to 0$, that is,

- for $\alpha \to 1$, we obtain the generalized Kullback-Leibler divergence (I-divergence)

$$D_{KL}(\mathbf{Y}||\mathbf{AX}) = \lim_{\alpha \to 1} D_A^{(\alpha)}(\mathbf{Y}||\mathbf{AX}) = \sum_{it} \left(y_{it} \, \ln(\frac{y_{it}}{[\mathbf{AX}]_{it}}) - y_{it} + [\mathbf{AX}]_{it} \right). \tag{3.89}$$

- for $\alpha \to 0$, we have the generalized dual KL I-divergence

$$D_{KL}(\mathbf{AX}||\mathbf{Y}) = \lim_{\alpha \to 0} D_A^{(\alpha)}(\mathbf{Y}||\mathbf{AX}) = \sum_{it} \left([\mathbf{AX}]_{it} \, \ln(\frac{[\mathbf{AX}]_{it}}{y_{it}}) + y_{it} - [\mathbf{AX}]_{it} \right). \tag{3.90}$$

The gradient of the Alpha-divergence can be expressed in a compact form as

$$\frac{\partial D_A^{(\alpha)}}{\partial x_{jt}} = \frac{1}{\alpha} \sum_{i=1}^{I} a_{ij} \left[1 - \left(\frac{y_{it}}{[\mathbf{AX}]_{it}} \right)^{\alpha} \right], \qquad \alpha \neq 0, \tag{3.91}$$

$$\frac{\partial D_A^{(\alpha)}}{\partial a_{ij}} = \frac{1}{\alpha} \sum_{t=1}^{T} x_{jt} \left[1 - \left(\frac{y_{it}}{[\mathbf{AX}]_{it}} \right)^{\alpha} \right], \qquad \alpha \neq 0. \tag{3.92}$$

Remark 3.5 *It is not difficult to show that for every stationary point a_{ij}, x_{jt} of the cost function (3.88) the following properties hold:*

$$\mathbf{1}_{1 \times I} \, [\mathbf{AX}] = \sum_{i} (y_{it}^{\alpha} \, [\mathbf{AX}]_{it}^{1-\alpha}), \tag{3.93}$$

$$[\mathbf{AX}] \; \mathbf{1}_{T \times 1} = \sum_{t} (y_{it}^{\alpha} \, [\mathbf{AX}]_{it}^{1-\alpha}). \tag{3.94}$$

In the special case, the I-divergence for $\alpha = 1$ ensures that the row sum and column sum of the data matrix $\mathbf{Y}$ are preserved in approximation of $\hat{\mathbf{Y}} = \mathbf{AX}$. In other words, for a nonnegative matrix $\mathbf{Y}$ we can generally perform the following factorization

$$\mathbf{Y} = \mathbf{ADX}, \tag{3.95}$$

where $\mathbf{D}$ is a scaling diagonal nonnegative matrix with $\sum_i d_{ii} = \sum_{it} y_{it}$.

To derive a multiplicative learning algorithm, instead of applying gradient descent, we use a projected (transformed) gradient approach[10]:

$$\Phi(x_{jt}) \leftarrow \Phi(x_{jt}) - \eta_{jt} \frac{\partial D_A^{(\alpha)}}{\partial \Phi(x_{jt})}, \tag{3.96}$$

$$\Phi(a_{ij}) \leftarrow \Phi(a_{ij}) - \eta_{ij} \frac{\partial D_A^{(\alpha)}}{\partial \Phi(a_{ij})}, \tag{3.97}$$

where $\Phi(x)$ is a suitable chosen function.

Hence, we have

$$x_{jt} \leftarrow \Phi^{-1} \left(\Phi(x_{jt}) - \eta_{jt} \frac{\partial D_A^{(\alpha)}}{\partial \Phi(x_{jt})} \right), \tag{3.98}$$

$$a_{ij} \leftarrow \Phi^{-1} \left(\Phi(a_{ij}) - \eta_{ij} \frac{\partial D_A^{(\alpha)}}{\partial \Phi(a_{ij})} \right). \tag{3.99}$$

In general, such a nonlinear scaling (or transformation) provides a stable solution and the so obtained gradients are much better behaved in the Φ space. We employ $\Phi(x) = x^\alpha$, and choose the learning rates as follows:

$$\eta_{jt} = \alpha^2 \Phi(x_{jt}) / (x_{jt}^{1-\alpha} \sum_{i=1}^{I} a_{ij}), \qquad \eta_{ij} = \alpha^2 \Phi(a_{ij}) / (a_{ij}^{1-\alpha} \sum_{t=1}^{T} x_{jt}). \tag{3.100}$$

This leads directly to the Alpha NMF algorithm,[11] given by

$$x_{jt} \leftarrow x_{jt} \left(\frac{\sum_{i=1}^{I} a_{ij} \, (y_{it}/[\mathbf{AX}]_{it})^\alpha}{\sum_{i=1}^{I} a_{ij}} \right)^{1/\alpha}, \tag{3.101}$$

$$a_{ij} \leftarrow a_{ij} \left(\frac{\sum_{t=1}^{T} (y_{it}/[\mathbf{AX}]_{it})^\alpha \, x_{jt}}{\sum_{t=1}^{T} x_{jt}} \right)^{1/\alpha}, \quad \alpha \neq 0, \tag{3.102}$$

where, at every iteration, the columns of $\mathbf{A}$ are normalized to unit length, typically, $a_{ij} \leftarrow a_{ij} / \sum_p a_{pj}$. For $\alpha = 1$ the algorithm simplifies to the standard EMML algorithm [37,41].

The multiplicative Alpha NMF algorithm for a large-scale problem can be expressed in a compact matrix form given by[12]

$$\mathbf{X} \leftarrow \mathbf{X} \circledast \left(\mathbf{A}_r^T \left((\mathbf{Y}_r + \varepsilon) \oslash (\mathbf{A}_r \mathbf{X} + \varepsilon) \right)^{.[\alpha]} \right)^{.[1/\alpha]}, \tag{3.103}$$

$$\mathbf{A} \leftarrow \mathbf{A} \circledast \left(\left((\mathbf{Y}_c + \varepsilon) \oslash (\mathbf{A} \mathbf{X}_c + \varepsilon) \right)^{.[\alpha]} \mathbf{X}_c^T \right)^{.[1/\alpha]}, \tag{3.104}$$

$$\mathbf{A} \leftarrow \mathbf{A} \, \mathrm{diag}(\|\boldsymbol{a}_1\|_1, \|\boldsymbol{a}_2\|_1, \ldots, \|\boldsymbol{a}_J\|_1)^{-1}. \tag{3.105}$$

To have control over the sparsity of a solution, we can use a nonlinear transformation (or filter) in the form $\mathcal{P}_\Omega(x) = x^{1+\alpha_s}$, which gives the following updates [10,12,13]:

[10] Which can be considered as a generalization of the exponentiated gradient (EG) [35].

[11] For $\alpha = 0$ instead of $\Phi(x) = x^\alpha$ we have used $\Phi(x) = \ln(x)$.

[12] For large matrix $\mathbf{A}$, we recommend to use MATLAB binary singleton expansion function `bsxfun` to normalize $\mathbf{A}$, that is, $\mathbf{A}$ = `bsxfun(@rdivide, A, sum(A, 1))`.

$$x_{jt} \leftarrow \left(x_{jt} \left(\frac{\sum_{i \in \mathcal{S}_I} a_{ij} \, (y_{it}/[\mathbf{AX}]_{it})^{\alpha}}{\sum_{i \in \mathcal{S}_I} a_{ij}} \right)^{1/\alpha} \right)^{1+\alpha_{sX}}, \tag{3.106}$$

$$a_{ij} \leftarrow \left(a_{ij} \left(\frac{\sum_{t \in \mathcal{S}_T} (y_{it}/[\mathbf{AX}]_{it})^{\alpha} \, x_{jt}}{\sum_{t \in \mathcal{S}_T} x_{jt}} \right)^{1/\alpha} \right)^{1+\alpha_{sA}}, \tag{3.107}$$

where additional small nonnegative regularization terms $\alpha_{sA} \geq 0$ and $\alpha_{sX} \geq 0$ are introduced in order to enforce sparseness of the solution (if necessary). Typical values of sparsification coefficients are $\alpha_{sX} = \alpha_{sA} = 0.001 - 0.005$.

A variant of this algorithm that improves the convergence speed can be achieved by over-relaxation:

$$x_{jt} \leftarrow \left(x_{jt} \left(\frac{\sum_{i \in \mathcal{S}_I} a_{ij} \, (y_{it}/[\mathbf{A}\,\mathbf{X}]_{it})^{\alpha}}{\sum_{i \in \mathcal{S}_I} a_{ij}} \right)^{\omega/\alpha} \right)^{1+\alpha_{sX}}, \tag{3.108}$$

$$a_{ij} \leftarrow \left(a_{ij} \left(\sum_{t \in \mathcal{S}_T} x_{jt} \, (y_{it}/[\mathbf{A}\,\mathbf{X}]_{it})^{\alpha} \right)^{\omega/\alpha} \right)^{1+\alpha_{sA}}, \tag{3.109}$$

$$a_{ij} \leftarrow a_{ij} / \sum_{i=1}^{I} a_{ij}, \qquad \alpha \neq 0, \tag{3.110}$$

where the over-relaxation parameter ω accelerates convergence or makes the solution more stable (typically, $\omega \in [0.5, 2]$).

The alternative form of the multiplicative Alpha NMF algorithm for a large-scale problem can be obtained using the block-wise approach (by selecting some rows and columns of data matrix $\mathbf{Y}$) (see also Figure 3.5 and Algorithm 3.3):

$$\mathbf{X} \leftarrow \left(\mathbf{X} \circledast \left(\mathbf{D}_{A_r} \mathbf{A}_r^T \left([\mathbf{Y}_r]_+ \oslash [\mathbf{A}\,\mathbf{X}_r]_+ \right)^{.[\alpha]} \right)^{.[\omega/\alpha]} \right)^{.[1+\alpha_{sX}]}, \tag{3.111}$$

$$\mathbf{A} \leftarrow \left(\mathbf{A} \circledast \left(([\mathbf{Y}_c]_+ \oslash [\mathbf{A}\,\mathbf{X}_c]_+)^{.\alpha} \, \mathbf{X}_c^T \right)^{.[\omega/\alpha]} \right)^{.[1+\alpha_{sA}]}, \tag{3.112}$$

$$\mathbf{A} \leftarrow \mathbf{A} \operatorname{diag}(\|\boldsymbol{a}_1\|_1^{-1}, \|\boldsymbol{a}_2\|_1^{-1}, \ldots, \|\boldsymbol{a}_J\|_1^{-1}), \tag{3.113}$$

where $\mathbf{D}_{A_r} = \operatorname{diag}(\|\boldsymbol{a}_{r_1}\|_1^{-1}, \|\boldsymbol{a}_{r_2}\|_1^{-1}, \ldots, \|\boldsymbol{a}_{r_J}\|_1^{-1})$.

For efficiency, instead of using half-wave rectifying and a simple power function, we can use various nonnegative shrinkage functions where the threshold λ decreases gradually to zero (see Figure 3.4 and Chapter 4 for mathematical description of various shrinkage functions).

(a) Hard shrinkage

$$P(x) = \begin{cases} x, & x \geq \lambda \\ 0, & \text{elsewhere} \end{cases}$$

(b) Soft shrinkage

$$P(x) = \begin{cases} x - \lambda, & x \geq \lambda \\ 0, & \text{elsewhere} \end{cases}$$

(c) Semi-soft shrinkage

$$P(x) = \begin{cases} x, & x \geq \lambda_2 \\ \lambda_2 \dfrac{x - \lambda_1}{\lambda_2 - \lambda_1}, & \lambda_1 \leq x \leq \lambda_2 \\ 0, & \text{elsewhere} \end{cases}$$

(d) Abramovich shrinkage

$$P(x) = \begin{cases} \sqrt{x^2 - \lambda^2}, & x \geq \lambda \\ 0, & \text{elsewhere} \end{cases}$$

(e) Nonnegative garrotte shrinkage

$$P(x) = \begin{cases} x - \lambda^2/x, & x \geq \lambda \\ 0, & \text{elsewhere} \end{cases}$$

(c) N-degree garrotte shrinkage

$$P(x) = \frac{x^{2n+1}}{\lambda^{2n} + x^{2n}}$$

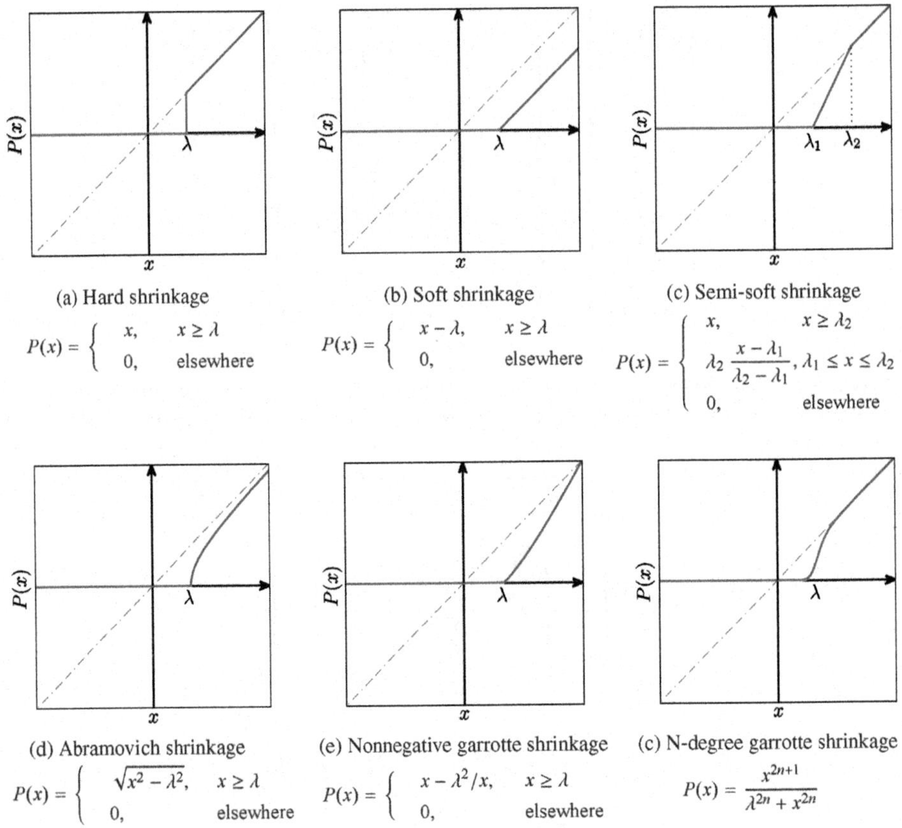

Figure 3.4 Various nonlinear projections (nonnegative shrinkages). Typically, threshold λ decreases gradually to zero along the iteration.

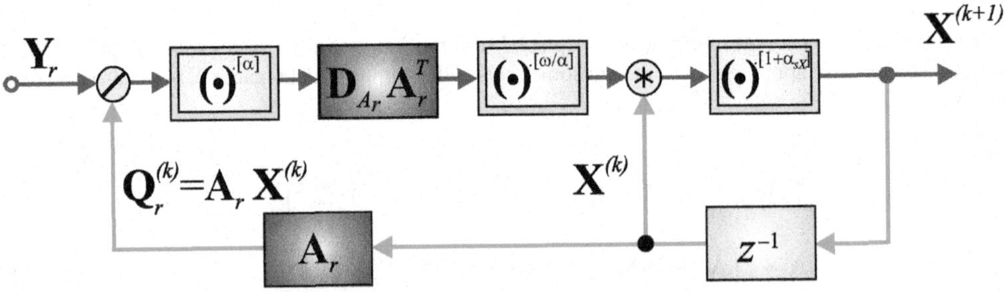

Figure 3.5 Simplified block diagram of the multiplicative Alpha NMF algorithm with sparsity constraint and over-relaxation for a large-scale NMF: $\mathbf{X}^{(k+1)} = \left(\mathbf{X}^{(k)} \circledast \left(\mathbf{D}_{A_r} \mathbf{A}_r^T (\mathbf{Y}_r \oslash (\mathbf{A}_r^T \mathbf{X}^{(k)})) \cdot [\alpha] \right) \cdot [\omega/\alpha] \right)^{\cdot [1+\alpha_s X]}$,

where $\mathbf{D}_{A_r} = \text{diag} \left(\|\boldsymbol{a}_{r_1}\|_1^{-1}, \|\boldsymbol{a}_{r_2}\|_1^{-1}, \dots, \|\boldsymbol{a}_{r_J}\|_1^{-1} \right)$. A similar block diagram can be drawn for updating matrix $\mathbf{A}$ as $\mathbf{A}^{(k+1)} = \left(\mathbf{A}^{(k)} \circledast \left((\mathbf{Y}_c \oslash (\mathbf{A}^{(k)} \mathbf{X}_c)) \cdot [\alpha] \mathbf{X}_c^T \right) \cdot [\omega/\alpha] \right)^{\cdot [1+\alpha_s A]}$.

Algorithm 3.3: Multiplicative Alpha NMF with Over-relaxation and Sparsity Control

Input: $\mathbf{Y} \in \mathbb{R}_+^{I \times T}$: input data, J: rank of approximation, α: order of Alpha divergence
ω: over-relaxation, and α_{sA}, α_{sX}: sparsity levels
Output: $\mathbf{A} \in \mathbb{R}_+^{I \times J}$ and $\mathbf{X} \in \mathbb{R}_+^{J \times T}$ such that cost function (3.88) is minimized.

begin
  initialization for $\mathbf{A}$ and $\mathbf{X}$
  repeat
    select R row indices
    $\mathbf{X} \longleftarrow \left(\mathbf{X} \circledast \left(\mathbf{D}_{A_r} \mathbf{A}_r^T (\mathbf{Y}_r \oslash [\mathbf{A}_r \mathbf{X}]_+)^{.[\alpha]}\right)^{.[\omega/\alpha]}\right)^{.[1+\alpha_{sX}]}$ `/* update X */`
    select C column indices
    $\mathbf{A} \longleftarrow \left(\mathbf{A} \circledast \left((\mathbf{Y}_c \oslash [\mathbf{A}\mathbf{X}_c]_+)^{.[\alpha]} \mathbf{X}_c^T\right)^{.[\omega/\alpha]}\right)^{.[1+\alpha_{sA}]}$ `/* update A */`
    foreach $\boldsymbol{a}_j$ *of* $\mathbf{A}$ **do** $\boldsymbol{a}_j \leftarrow \boldsymbol{a}_j / \|\boldsymbol{a}_j\|_1$ `/* normalize to ℓ1 unit length */`
  until *a stopping criterion is met* `/* convergence condition */`
end

If we apply the exponentiated gradient (EG) approach, we can directly obtain the following alternative multiplicative Alpha NMF algorithm (see the next section):

$$x_{jt} \leftarrow x_{jt} \exp\left\{\eta_{jt} \sum_{i \in S_I} a_{ij} \left[\left(\frac{y_{it}}{[\mathbf{AX}]_{it}}\right)^{\alpha} - 1\right]\right\}, \tag{3.114}$$

$$a_{ij} \leftarrow a_{ij} \exp\left\{\eta_{ij} \sum_{t \in S_T} \left[\left(\frac{y_{it}}{[\mathbf{AX}]_{it}}\right)^{\alpha} - 1\right] x_{jt}\right\}. \tag{3.115}$$

The above multiplicative update rules are valid only for $\alpha \neq 0$. For $\alpha = 0$ the NMF algorithm can take the following special form:

$$x_{jt} \leftarrow x_{jt} \prod_{i \in S_I} \left(\frac{y_{it}}{[\mathbf{AX}]_{it}}\right)^{\eta_{jt} a_{ij}}, \qquad a_{ij} \leftarrow a_{ij} \prod_{t \in S_T} \left(\frac{y_{it}}{[\mathbf{AX}]_{it}}\right)^{\eta_{ij} x_{jt}}, \tag{3.116}$$

where $\eta_{ij} = \omega(\sum_t x_{jt})^{-1}$ and a_{ij} is normalized in each step as $a_{ij} \leftarrow a_{ij}/\sum_p a_{pj}$. This learning rule is referred to as the SMART (Simultaneous Multiplicative Algebraic Reconstruction Technique) algorithm, discussed in the next section [3,5,10].

3.2.2 Generalized Multiplicative Alpha NMF Algorithms

The multiplicative Alpha NMF algorithm can be generalized as follows [10,12,23] (see also Figure 3.6)

$$x_{jt} \leftarrow x_{jt}\, \Psi^{-1}\left(\frac{\sum_{i \in S_I} a_{ij}\, \Psi(y_{it}/q_{it})}{\sum_{i \in S_I} a_{ij}}\right), \qquad a_{ij} \leftarrow a_{ij}\, \Psi^{-1}\left(\frac{\sum_{t \in S_T} x_{jt}\, \Psi(y_{it}/q_{it})}{\sum_{t \in S_T} x_{jt}}\right), \tag{3.117}$$

where $q_{it} = [\mathbf{AX}]_{it}$, and $\Psi(x)$ is a monotonic increasing function such that $\Psi^{-1}(1) = 1$ and $\Psi(1) = 1$.

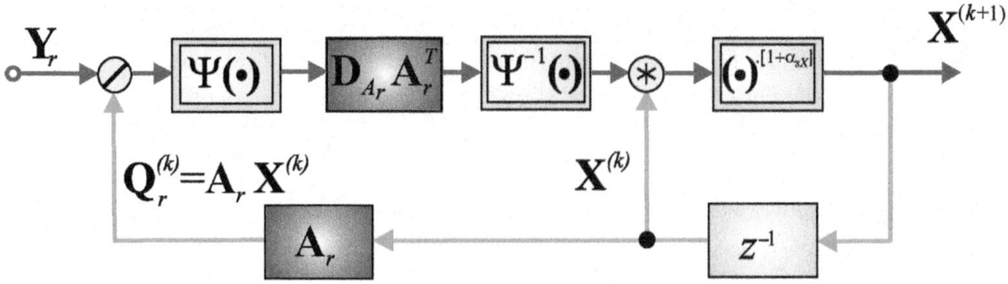

Figure 3.6 Simplified functional block diagram of the generalized multiplicative Alpha NMF algorithm with sparsity control via small parameter α_{sX} for the large-scale NMF: $\mathbf{X}^{(k+1)} = \left(\mathbf{X}^{(k)} \circledast \Psi^{-1}\left(\mathbf{D}_{A_r} \mathbf{A}_r^T \Psi(\mathbf{Y}_r \oslash (\mathbf{A}_r^T \mathbf{X}^{(k)}))\right)\right)^{.[1+\alpha_{sX}]}$, where $\mathbf{D}_{A_r} = \text{diag}(\|\boldsymbol{a}_{r_1}\|_1^{-1}, \|\boldsymbol{a}_{r_2}\|_1^{-1}, \ldots, \|\boldsymbol{a}_{r_J}\|_1^{-1})$. A similar block diagram can be formulated for updating matrix $\mathbf{A}$ as: $\mathbf{A}^{(k+1)} = \left(\mathbf{A}^{(k)} \circledast \Psi^{-1}\left(\Psi(\mathbf{Y}_c \oslash (\mathbf{A}^{(k)}\mathbf{X}_c))\mathbf{X}_c^T\right)\right)^{.[1+\alpha_{sA}]}$.

3.3 Alternating SMART: Simultaneous Multiplicative Algebraic Reconstruction Technique

3.3.1 Alpha SMART Algorithm

The multiplicative Alpha NMF algorithm based on Alpha-divergence is valid for any $\alpha \neq 0$. In order to derive a multiplicative algorithm for $\alpha = 0$, we need to minimize the I-divergence defined as

$$D_{KL}(\mathbf{AX}||\mathbf{Y}) = \sum_{it} \left([\mathbf{AX}]_{it} \ln \left(\frac{[\mathbf{AX}]_{it}}{y_{it}} \right) - [\mathbf{AX}]_{it} + y_{it} \right), \tag{3.118}$$

subject to nonnegativity constraints (compare with Eq. (3.57)).

Remark 3.6 *The alternative minimization of the I-divergence is equivalent to a set of optimization problems maximizing the Shannon entropy:*

$$a) \qquad \min_{\mathbf{X}} \sum_{jt} x_{jt} \ln(x_{jt}) \qquad s.t. \qquad \mathbf{AX} = \mathbf{Y} \qquad \textit{for a fixed matrix } \mathbf{A}, \tag{3.119}$$

$$b) \qquad \min_{\mathbf{A}} \sum_{ij} a_{ij} \ln(a_{ij}) \qquad s.t. \qquad \mathbf{X}^T\mathbf{A}^T = \mathbf{Y}^T \qquad \textit{for a fixed matrix } \mathbf{X}. \tag{3.120}$$

Upon applying the multiplicative exponentiated gradient (EG) descent updates to the cost function (3.118), we have:

$$x_{jt} \leftarrow x_{jt} \exp\left(-\eta_{jt} \frac{\partial D_{KL}}{\partial x_{jt}}\right), \qquad a_{ij} \leftarrow a_{ij} \exp\left(-\eta_{ij} \frac{\partial D_{KL}}{\partial a_{ij}}\right), \tag{3.121}$$

where

$$\frac{\partial D_{KL}}{\partial x_{jt}} = \sum_i \left(a_{ij} \ln[\mathbf{AX}]_{it} - a_{ij} \ln y_{it} \right), \tag{3.122}$$

$$\frac{\partial D_{KL}}{\partial a_{ij}} = \sum_t \left(x_{jt} \ln[\mathbf{AX}]_{it} - x_{jt} \ln y_{it} \right), \tag{3.123}$$

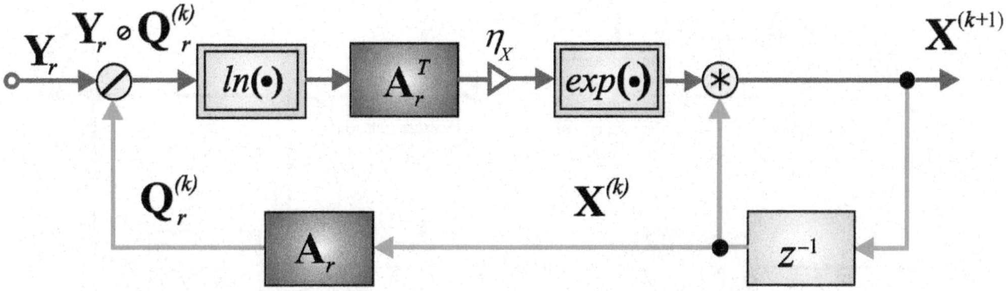

Figure 3.7 Functional block diagram illustrating the SMART NMF algorithm for a large-scale NMF: $\mathbf{X}^{(k+1)} = \mathbf{X}^{(k)} \circledast \exp\left(\boldsymbol{\eta}_\mathbf{X} \circledast \mathbf{A}_r^T \ln(\mathbf{Y}_r \oslash (\mathbf{A}_r^T\mathbf{X}^{(k)}))\right)$. A similar block diagram can be formulated for updating matrix $\mathbf{A}$ as: $\mathbf{A}^{(k+1)} = \mathbf{A}^{(k)} \circledast \exp\left(\ln(\boldsymbol{\eta}_\mathbf{A} \circledast \mathbf{Y}_c \oslash (\mathbf{A}^{(k)}\mathbf{X}_c))\mathbf{X}_c^T\right)$.

which lead to the following simple multiplicative NMF learning rules for large-scale problems:

$$x_{jt} \leftarrow x_{jt} \exp\left(\sum_{i \in \mathcal{S}_I} \eta_{jt} a_{ij} \ln\left(\frac{y_{it}}{[\mathbf{AX}]_{it}}\right)\right) = x_{jt} \prod_{i \in \mathcal{S}_I} \left(\frac{y_{it}}{[\mathbf{AX}]_{it}}\right)^{\eta_{jt} a_{ij}}, \tag{3.124}$$

$$a_{ij} \leftarrow a_{ij} \exp\left(\sum_{t \in \mathcal{S}_T} \eta_{ij} x_{jt} \ln\left(\frac{y_{it}}{[\mathbf{AX}]_{it}}\right)\right) = a_{ij} \prod_{t \in \mathcal{S}_T} \left(\frac{y_{it}}{[\mathbf{AX}]_{it}}\right)^{\eta_{ij} x_{jt}}. \tag{3.125}$$

The nonnegative learning rates η_{jt} and η_{ij} can take different forms. Typically, for simplicity and to guarantee stability of the algorithm, we assume $\eta_{jt} = \eta_\mathbf{X} = \omega/(\sum_i a_{ij})$, $\eta_{ij} = \eta_\mathbf{A} = \omega/(\sum_t x_{jt})$,[13] where $\omega \in (0, 2)$ is an over-relaxation parameter.

For a large-scale NMF problem, the above multiplicative learning rules can be written in a compact matrix form (see Figure 3.7)

$$\mathbf{X} \leftarrow \mathbf{X} \circledast \exp\left(\boldsymbol{\eta}_\mathbf{X} \circledast (\mathbf{A}_r^T \ln(\mathbf{Y}_r \oslash (\mathbf{A}_r\mathbf{X} + \epsilon)))\right), \tag{3.126}$$

$$\mathbf{A} \leftarrow \mathbf{A} \circledast \exp\left(\boldsymbol{\eta}_\mathbf{A} \circledast (\ln(\mathbf{Y}_c \oslash (\mathbf{A}\mathbf{X}_c + \epsilon))\mathbf{X}_c^T)\right), \tag{3.127}$$

$$\mathbf{A} \leftarrow \mathbf{A} \operatorname{diag}(\|\boldsymbol{a}_1\|_1^{-1}, \|\boldsymbol{a}_2\|_1^{-1}, \ldots, \|\boldsymbol{a}_J\|_1^{-1}). \tag{3.128}$$

In practice, a small constant $\varepsilon = 10^{-9}$ is introduced in order to make components positive and/or to avoid possible division by zero, and $\boldsymbol{\eta}_\mathbf{A}$ and $\boldsymbol{\eta}_\mathbf{X}$ are nonnegative scaling matrices representing individual learning rates. The above algorithm may be considered as an alternating version of the SMART [3,5,10], which can also be extended to MART and BI-MART (Block-Iterative Multiplicative Algebraic Reconstruction Technique) [3].

Because the parameters (weights) $\{x_{jt}, a_{ij}\}$ are restricted to positive values, the resulting update rules can be written as:

$$\ln(x_{jt}) \leftarrow \ln(x_{jt}) - \eta_{jt} \frac{\partial D_{KL}}{\partial x_{jt}}, \qquad \ln(a_{ij}) \leftarrow \ln(a_{ij}) - \eta_{ij} \frac{\partial D_{KL}}{\partial a_{ij}}, \tag{3.129}$$

where the projection via the natural logarithm is applied element-wise. Thus, the EG approach has the same general form as the standard gradient descent (GD) but based on the logarithm of the parameters. This offers great potential advantage, as the parameters $\{x_{jt}, a_{ij}\}$ are best adapted in log-space, where their gradients are much better behaved.

[13] Because of ℓ_1 normalization, $\eta_\mathbf{X} = \omega$.

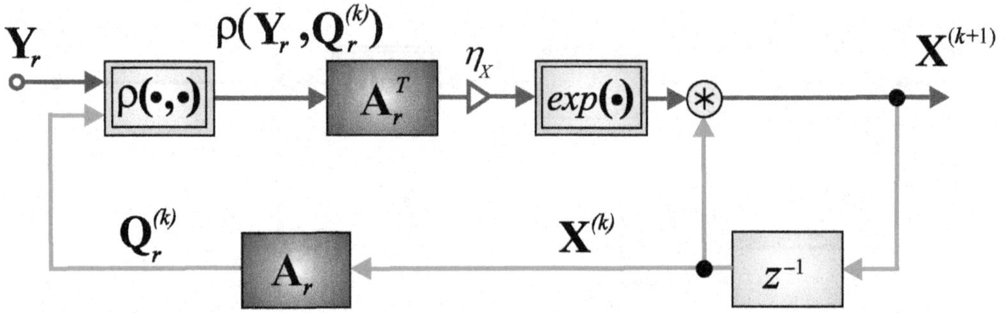

Figure 3.8 Simplified functional block diagram illustrating the generalized SMART algorithm for a large-scale NMF: $\mathbf{X}^{(k+1)} = \mathbf{X}^{(k)} \circledast \exp\left(\eta_{\mathbf{X}} \circledast \left(\mathbf{A}_r^T\, \rho\,(\mathbf{Y}_r,\, \mathbf{A}_r\mathbf{X}^{(k)})\right)\right)$. A similar block diagram can be drawn for updating matrix $\mathbf{A}$ as: $\mathbf{A}^{(k+1)} = \mathbf{A}^{(k)} \circledast \exp\left(\eta_{\mathbf{A}} \circledast \left(\rho\,(\mathbf{Y}_c,\, \mathbf{A}^{(k)}\mathbf{X}_c)\, \mathbf{X}_c^T\right)\right)$.

Charles Byrne showed that the EMML and SMART algorithms are related to likelihood maximization, that is, minimizing the set of cost functions $D_{KL}(\boldsymbol{y}_t||\mathbf{A}\boldsymbol{x}_t)$, $t = 1, 2, \ldots, T$, is equivalent to maximizing the likelihood when $\boldsymbol{y}_t$ are drawn from independent Poisson distributions with means $(\mathbf{A}\boldsymbol{x}_t)$, and the entries of $\boldsymbol{x}_t$ are the parameters to be determined. This situation arises typically in emission tomography [3,5].

3.3.2 Generalized SMART Algorithms

Using various cost functions, we can generalize the SMART algorithm to the following update rules (illustrated in Figure 3.8 and Algorithm 3.4):

$$x_{jt} \leftarrow x_{jt} \exp\left[\sum_{i \in S_I} \eta_{jt}\, a_{ij}\, \rho(y_{it}, q_{it})\right], \tag{3.130}$$

$$a_{ij} \leftarrow a_{ij} \exp\left[\sum_{t \in S_T} \eta_{ij}\, x_{jt}\, \rho(y_{it}, q_{it})\right], \tag{3.131}$$

Algorithm 3.4: Generalized SMART NMF

Input: $\mathbf{Y} \in \mathbb{R}_+^{I \times T}$: input data, J: rank of approximation

$\rho(y_{it}, q_{it}) = -\dfrac{\partial D(\mathbf{Y}||\mathbf{Q})}{\partial q_{it}}$: error function

Output: $\mathbf{A} \in \mathbb{R}_+^{I \times J}$ and $\mathbf{X} \in \mathbb{R}_+^{J \times T}$ such that cost function $D(\mathbf{Y}||\mathbf{AX})$ is minimized.

begin
  initialization for $\mathbf{A}$ and $\mathbf{X}$
  repeat
    $\mathbf{X} \longleftarrow \mathbf{X} \circledast \exp\left(\eta_{\mathbf{X}} \circledast \left(\mathbf{A}^T \rho(\mathbf{Y}, \mathbf{A}\mathbf{X})\right)\right)$ /* update X */
    $\mathbf{A} \longleftarrow \mathbf{A} \circledast \exp\left(\eta_{\mathbf{A}} \circledast \left(\rho(\mathbf{Y}, \mathbf{A}\mathbf{X})\mathbf{X}^T\right)\right)$ /* update A */
    foreach $\boldsymbol{a}_j$ *of* $\mathbf{A}$ **do** $\boldsymbol{a}_j \leftarrow \boldsymbol{a}_j / \|\boldsymbol{a}_j\|_p$ /* normalize to ℓ_p unit length */
  until *a stopping criterion is met* /* convergence condition */
end

where $q_{it} = [\mathbf{AX}]_{it}$, the error functions are

$$\rho(y_{it}, q_{it}) = -\frac{\partial D(\mathbf{Y}||\mathbf{AX})}{\partial q_{it}} \tag{3.132}$$

and can take different forms depending on the chosen or designed loss (cost) function $D(\mathbf{Y}||\mathbf{AX})$ (see Table 3.2). Note that the error function $\rho(y_{it}, q_{it}) = 0$ if and only if $y_{it} = q_{it}$.

As an illustrative example consider the Bose-Einstein divergence:

$$BE_\alpha(\mathbf{Y}||\mathbf{AX}) = \sum_{it} y_{it} \ln\left(\frac{(1+\alpha)y_{it}}{y_{it}+\alpha q_{it}}\right) + \alpha q_{it} \ln\left(\frac{(1+\alpha)q_{it}}{y_{it}+\alpha q_{it}}\right). \tag{3.133}$$

This loss function has many interesting properties:

1. $BE_\alpha(\mathbf{Y}||\mathbf{Q}) = 0$ if $\mathbf{Q} = \mathbf{Y}$ almost everywhere.
2. $BE_\alpha(\mathbf{Y}||\mathbf{Q}) = BE_{1/\alpha}(\mathbf{Q}||\mathbf{Y})$
3. For $\alpha = 1$, BE_α simplifies to the symmetric Jensen-Shannon divergence measure (see Table 3.2).
4. $\lim_{\alpha\to\infty} BE_\alpha(\mathbf{Y}||\mathbf{Q}) = D_{KL}(\mathbf{Y}||\mathbf{Q})$ and for α sufficiently small $BE_\alpha(\mathbf{Y}||\mathbf{Q}) \approx D_{KL}(\mathbf{Q}||\mathbf{Y})$.

The gradient of the Bose-Einstein loss function with respect to q_{it} can be expressed as

$$\frac{\partial BE_\alpha(\mathbf{Y}||\mathbf{AX})}{\partial q_{it}} = -\alpha \ln\left(\frac{y_{it}+\alpha q_{it}}{(1+\alpha)q_{it}}\right), \tag{3.134}$$

whereas the gradients with respect to x_{jt} and a_{ij} are given by

$$\frac{\partial BE_\alpha}{\partial x_{jt}} = \sum_{i=1}^{I} a_{ij}\frac{\partial BE_\alpha}{\partial q_{it}}, \qquad \frac{\partial BE_\alpha}{\partial a_{ij}} = \sum_{t=1}^{T} x_{jt}\frac{\partial BE_\alpha}{\partial q_{it}}. \tag{3.135}$$

Hence, applying the standard (un-normalized) EG approach (3.121) for (3.134) and (3.135), we obtain a SMART algorithm based on the Bose-Einstein divergence given in the form (3.131) with the error function $\rho(y_{it}, q_{it}) = \alpha \ln\left((y_{it} + \alpha q_{it})/((1+\alpha)q_{it})\right)$.

The multiplicative NMF algorithms are listed in Table 3.1 and Table 3.2. It is necessary for practical purposes either to avoid division by zero and $\ln 0$, or to avoid negative and zero values by adding a small positive value ε or by using an operator $[x]_+ = \max\{x, \varepsilon\}$, where ε is a small positive number, typically 10^{-9}.

The EG updates can be further improved in terms of convergence, computational efficiency and numerical stability in several ways. To keep the weight magnitudes bounded, Kivinen and Warmuth [35] proposed a variation of the EG method that applies a normalization step after each weight update. The normalization linearly rescales all weights so that their sum is a constant. Moreover, for practical purposes, we can use the approximation of the exponential function: $e^u \approx \max\{0.5, 1+u\}$. For fast convergence, we may apply individual adaptive learning rates defined as $\eta_{jt} \leftarrow \eta_{jt} c$ if the corresponding gradient component $\partial D(\mathbf{Y}||\mathbf{AX})/\partial x_{jt}$ has the same sign in two consecutive steps and $\eta_{jt} \leftarrow \eta_{jt}/c$ otherwise, where $c > 1$ (typically $c = 1.02 - 1.5$) [48].

3.4 Multiplicative NMF Algorithms Based on Beta-Divergence

3.4.1 Multiplicative Beta NMF Algorithm

We shall now reformulate the NMF problem as that of the alternating minimization of the discrete Beta-divergence between the two un-normalized density variables y_{it} and $q_{it} = [\mathbf{AX}]_{it} = \sum_{j=1}^{J} a_{ij}x_{jt}$. These represent alternative similarity measures between the elements of the matrix $\mathbf{Y}$ and the corresponding elements

Table 3.1 Multiplicative Alpha NMF algorithms (for a large-scale NMF) with over-relaxation and sparsity control for selected Alpha-Divergences. The over-relaxation parameter $\omega \in (0, 2]$, (typically, $\omega \in [0.5, 2]$) is used to accelerate convergence and to stabilize the algorithm, and α_{sX}, α_{sA} are nonnegative parameters used to sparsify the estimated components, typically α_{sX}, $\alpha_{sA} \in [0, 0.01]$).

Cost function	Iterative Learning Algorithm
Basic Alpha-Divergence $D_A^{(\alpha)}(\mathbf{Y}\|\|\mathbf{AX})$, $(\alpha \neq 0)$ $\sum_{it} \frac{y_{it}^{\alpha}[\mathbf{AX}]_{it}^{1-\alpha} - y_{it}}{\alpha(\alpha-1)} + \frac{[\mathbf{AX}]_{it} - y_{it}}{\alpha}$	$x_{jt} \leftarrow \left(x_{jt}\left(\sum_{i\in\mathcal{S}_I} \tilde{a}_{ij}\left(\frac{y_{it}}{[\mathbf{A\,X}]_{it}}\right)^{\alpha}\right)^{\omega/\alpha}\right)^{1+\alpha_{sX}}$ $a_{ij} \leftarrow \left(a_{ij}\left(\sum_{t\in\mathcal{S}_T} x_{jt}\left(\frac{y_{it}}{[\mathbf{A\,X}]_{it}}\right)^{\alpha}\right)^{\omega/\alpha}\right)^{1+\alpha_{sA}}$
Pearson Distance $D_A^{(\alpha)}(\mathbf{Y}\|\|\mathbf{AX})$, $(\alpha = 2)$ $\sum_{it} \frac{(y_{it} - [\mathbf{AX}]_{it})^2}{[\mathbf{AX}]_{it}}$	$x_{jt} \leftarrow \left(x_{jt}\left(\sum_{i\in\mathcal{S}_I} \tilde{a}_{ij}\left(\frac{y_{it}}{[\mathbf{A\,X}]_{it}}\right)^{2}\right)^{\omega/2}\right)^{1+\alpha_{sX}}$ $a_{ij} \leftarrow \left(a_{ij}\left(\sum_{t\in\mathcal{S}_T} x_{jt}\left(\frac{y_{it}}{[\mathbf{A\,X}]_{it}}\right)^{2}\right)^{\omega/2}\right)^{1+\alpha_{sA}}$
(Squared) Hellinger Distance $D_A^{(\alpha)}(\mathbf{Y}\|\|\mathbf{AX})$, $(\alpha = 0.5)$ $\sum_{it} \left(\sqrt{[\mathbf{AX}]_{it}} - \sqrt{y_{it}}\right)^2,$	$x_{jt} \leftarrow \left(x_{jt}\left(\sum_{i\in\mathcal{S}_I} \tilde{a}_{ij}\left(\frac{y_{it}}{[\mathbf{A\,X}]_{it}}\right)^{0.5}\right)^{2\omega}\right)^{1+\alpha_{sX}}$ $a_{ij} \leftarrow \left(a_{ij}\left(\sum_{t\in\mathcal{S}_T} x_{jt}\left(\frac{y_{it}}{[\mathbf{A\,X}]_{it}}\right)^{0.5}\right)^{2\omega}\right)^{1+\alpha_{sA}}$
KL I-Divergence $D_{KL}(\mathbf{Y}\|\|\mathbf{AX})$ $(\alpha \to 1)$ $\sum_{it} (y_{it} \ln \frac{y_{it}}{[\mathbf{AX}]_{it}} - y_{it} + [\mathbf{AX}]_{it})$	$x_{jt} \leftarrow \left(x_{jt}\left(\sum_{i\in\mathcal{S}_I} \tilde{a}_{ij}\frac{y_{it}}{[\mathbf{A\,X}]_{it}}\right)^{\omega}\right)^{1+\alpha_{sX}}$ $a_{ij} \leftarrow \left(a_{ij}\left(\sum_{t\in\mathcal{S}_T} x_{jt}\frac{y_{it}}{[\mathbf{A\,X}]_{it}}\right)^{\omega}\right)^{1+\alpha_{sA}}$
Dual KL I-Divergence $D_{KL}(\mathbf{AX}\|\|\mathbf{Y})$ $(\alpha \to 0)$ $\sum_{it} ([\mathbf{AX}]_{it} \ln \frac{[\mathbf{AX}]_{it}}{y_{it}} + y_{it} - [\mathbf{AX}]_{it})$	$x_{jt} \leftarrow \left(x_{jt}\prod_{i\in\mathcal{S}_I}\left(\frac{y_{it}}{[\mathbf{AX}]_{it}}\right)^{\omega\, a_{ij}}\right)^{1+\alpha_{sX}}$ $a_{ij} \leftarrow \left(a_{ij}\prod_{t\in\mathcal{S}_T}\left(\frac{y_{it}}{[\mathbf{AX}]_{it}}\right)^{\tilde{\eta}_j x_{jt}}\right)^{1+\alpha_{sA}}$ $\tilde{\eta}_j = \omega(\sum_{t\in\mathcal{S}_T} x_{jt})^{-1}$

Table 3.1 (*Continued*)

Triangular Discrimination (TD) $D_{Am2}^{(-1)}(\tilde{\mathbf{Q}}\|\|\mathbf{Y})$,	$x_{jt} \leftarrow \left(x_{jt} \left(\sum_{i \in S_I} \tilde{a}_{ij} \left(\frac{2y_{it}}{y_{it} + [\mathbf{A}\,\mathbf{X}]_{it}} \right)^2 \right)^{\omega} \right)^{1+\alpha_{sX}}$
$\sum_{it} \frac{(y_{it} - [\mathbf{AX}]_{it})^2}{y_{it} + [\mathbf{AX}]_{it}}$	$a_{ij} \leftarrow \left(a_{ij} \left(\sum_{t \in S_T} x_{jt} \left(\frac{2y_{it}}{y_{it} + [\mathbf{A}\,\mathbf{X}]_{it}} \right)^2 \right)^{\omega} \right)^{1+\alpha_{sA}}$

* All the algorithms require ℓ_1 normalization for column vectors $\boldsymbol{a}_j$, that is, $a_{ij} \leftarrow a_{ij}/\sum_{i=1}^{I} a_{ij}$ and $\tilde{a}_{ij} = a_{ij}/\sum_{i \in S_I} a_{ij}$. It should be noted that for large-scale problems the summation can be performed only for the preselected indexes $i \in S_I$ and $t \in S_T$; For example, for randomly selected rows and columns of data matrix $\mathbf{Y}$.

of the estimated matrix $\mathbf{Q} = \mathbf{AX}$ [10,17,36]:

$$D_B^{(\beta)}(\mathbf{Y}||\mathbf{AX}) = \sum_{it} \left(y_{it} \frac{y_{it}^{\beta} - [\mathbf{AX}]_{it}^{\beta}}{\beta} + \frac{[\mathbf{AX}]_{it}^{\beta+1} - y_{it}^{\beta+1}}{\beta+1} \right) + \alpha_{\mathbf{X}} \, \|\mathbf{X}\|_1 + \alpha_{\mathbf{A}} \, \|\mathbf{A}\|_1, \quad (3.136)$$

where the small positive regularization parameters $\alpha_{\mathbf{X}}$ and $\alpha_{\mathbf{A}}$ control the degree of sparseness of the matrices $\mathbf{X}$ and $\mathbf{A}$, respectively, and the ℓ_1-norms $||\mathbf{A}||_1$ and $||\mathbf{X}||_1$ enforce sparse representations of the solutions.

For $\beta = 1$ we obtain the standard squared Euclidean distance expressed by the Frobenius norm (3.36), while for the singular cases $\beta = 0$ and $\beta = -1$, the Beta-divergence is defined by the limits of (3.136) as $\beta \to 0$ and $\beta \to -1$.

- for $\beta \to 0$, we obtain the generalized Kullback-Leibler divergence (I-divergence) defined as (for simplicity assuming that $\alpha_{\mathbf{A}} = \alpha_{\mathbf{X}} = 0$)[14]

$$D_{KL}(\mathbf{Y}||\mathbf{AX}) = \lim_{\beta \to 0} D_B^{(\beta)}(\mathbf{Y}||\mathbf{AX}) = \sum_{it} \left(y_{it} \, \ln\left(\frac{y_{it}}{[\mathbf{AX}]_{it}} \right) - y_{it} + [\mathbf{AX}]_{it} \right). \quad (3.137)$$

- for $\beta \to -1$, we have the Itakura-Saito distance

$$D_{IS}(\mathbf{Y}||\mathbf{AX}) = \lim_{\beta \to 0} D_B^{(\beta)}(\mathbf{Y}||\mathbf{AX}) = \sum_{it} \left[\ln\left(\frac{[\mathbf{AX}]_{it}}{y_{it}} \right) + \frac{y_{it}}{[\mathbf{AX}]_{it}} - 1 \right]. \quad (3.138)$$

In order to derive a multiplicative Beta NMF learning algorithm with sparsity control, we compute the gradient of (3.136) with respect to the elements of matrices $x_{jt} = [\mathbf{X}]_{jt}$ and $a_{ij} = [\mathbf{A}]_{ij}$ as follows

$$\frac{\partial D_B^{(\beta)}}{\partial x_{jt}} = \sum_{i=1}^{I} a_{ij} \left([\mathbf{AX}]_{it}^{\beta} - y_{it} \, [\mathbf{AX}]_{it}^{\beta-1} \right) + \alpha_{\mathbf{X}}, \quad (3.139)$$

$$\frac{\partial D_B^{(\beta)}}{\partial a_{ij}} = \sum_{t=1}^{T} \left([\mathbf{AX}]_{it}^{\beta} - y_{it}[\mathbf{AX}]_{it}^{\beta-1} \right) x_{jt} + \alpha_{\mathbf{A}}. \quad (3.140)$$

Then, using standard gradient descent

$$x_{jt} \leftarrow x_{jt} - \eta_{it} \frac{\partial D_B^{(\beta)}}{\partial x_{jt}}, \qquad a_{ij} \leftarrow a_{ij} - \eta_{ij} \frac{\partial D_B^{(\beta)}}{\partial a_{ij}} \quad (3.141)$$

[14] $\lim_{\beta \to 0} \frac{y^{\beta} - z^{\beta}}{\beta} = \ln(y/z)$.

Table 3.2 Extended Multiplicative SMART NMF Algorithms for Alpha-Divergences for a large-scale NMF (see Chapters 1 and 2).

$$a_{ij} \leftarrow a_{ij} \exp\left(\sum_{t \in S_T} \eta_{ij}\, x_{jt}\, \rho(y_{it}, q_{it})\right), \quad q_{it} = [\mathbf{AX}]_{it} > 0 \qquad x_{jt} \leftarrow x_{jt} \exp\left(\sum_{i \in S_I} \eta_{jt}\, a_{ij}\, \rho(y_{it}, q_{it})\right)$$

$$\|a_j\|_1 = \sum_{i=1}^{I} a_{ij} = 1, \ \ \forall j, \quad a_{ij} \geq 0 \qquad y_{it} > 0, \quad x_{jt} \geq 0$$

Exemplary cost function	Symbol of the SMART Algorithm Corresponding error function $\rho(y_{it}, q_{it})$
1. I-divergence, $D_A^{(\alpha)}(\mathbf{Y}\,\|\,\mathbf{Q})$, $(\alpha \to 0)$ $\sum_{it}\left(q_{it} \ln \frac{q_{it}}{y_{it}} + y_{it} - q_{it}\right)$	S-DKL $\rho(y_{it}, q_{it}) = \ln\left(\frac{y_{it}}{q_{it}}\right)$
2. Relative A-G divergence $D_{Am2}^{(\alpha)}(\tilde{\mathbf{Q}}\,\|\,\mathbf{Y})$, $(\alpha \to 1)$ $\sum_{it}\left((y_{it} + q_{it}) \ln\left(\frac{y_{it} + q_{it}}{2y_{it}}\right) + y_{it} - q_{it}\right)$	S-RAG $\rho(y_{it}, q_{it}) = \ln\left(\frac{2y_{it}}{y_{it} + q_{it}}\right)$
3. Relative Jensen-Shannon divergence $D_{Am2}^{(\alpha)}(\tilde{\mathbf{Q}}\,\|\,\mathbf{Y})$, $(\alpha \to 0)$ $\sum_{it}\left(2y_{it} \ln\left(\frac{2y_{it}}{y_{it} + q_{it}}\right) + q_{it} - y_{it}\right)$	S-RJS $\rho(y_{it}, q_{it}) = \frac{y_{it} - q_{it}}{y_{it} + q_{it}}$
4. Symmetric Jensen-Shannon divergence $D_{AS2}^{(\alpha)}(\tilde{\mathbf{Q}}\,\|\,\mathbf{Y})$, $(\alpha \to 0)$ $\sum_{it} y_{it} \ln\left(\frac{2y_{it}}{y_{it} + q_{it}}\right) + q_{it} \ln\left(\frac{2q_{it}}{y_{it} + q_{it}}\right)$	S-SJS $\rho(y_{it}, q_{it}) = \ln\left(\frac{y_{it} + q_{it}}{2q_{it}}\right)$
5. J-Divergence $D_{AS1}^{(\alpha)}(\mathbf{Y}\,\|\,\mathbf{Q})$ $\sum_{it}\left(\frac{y_{it} - q_{it}}{2} \ln\left(\frac{y_{it}}{q_{it}}\right)\right)$	S-JD $\rho(y_{it}, q_{it}) = \frac{1}{2} \ln\left(\frac{y_{it}}{q_{it}}\right) + \frac{y_{it} - q_{it}}{2q_{it}}$
6. Triangular Discrimination (TD) $D_{AS2}(\mathbf{Y}\,\|\mathbf{Q})$, $(\alpha = -1)$ $\sum_{it}\left\{\frac{(y_{it} - q_{it})^2}{y_{it} + q_{it}}\right\}$	S-TD $\rho(y_{it}, q_{it}) = \left(\frac{2y_{it}}{y_{it} + q_{it}}\right)^2 - 1$
7. Bose-Einstein Divergence $BE(\mathbf{Y}\|\|\mathbf{Q})$ $\sum_{it} y_{it} \ln\left(\frac{(1+\alpha)y_{it}}{y_{it} + \alpha q_{it}}\right) + \alpha q_{it} \ln\left(\frac{(1+\alpha)q_{it}}{y_{it} + \alpha q_{it}}\right)$	S-BE $\rho(y_{it}, q_{it}) = \alpha \ln\left(\frac{y_{it} + \alpha q_{it}}{(1+\alpha)q_{it}}\right)$
8. Basic (Asymmetric) Alpha Divergence $D_A^{(\alpha)}(\mathbf{Y}\,\|\,\mathbf{Q})$, $(\alpha \neq 0)$ $\frac{1}{\alpha(\alpha-1)} \sum_{it}\left(y_{it}^{\alpha} q_{it}^{1-\alpha} - y_{it} + (\alpha - 1)(q_{it} - y_{it})\right)$	S-alpha $\rho(y_{it}, q_{it}) = \frac{1}{\alpha}\left[\left(\frac{y_{it}}{q_{it}}\right)^{\alpha} - 1\right]$

and by choosing suitable learning rates

$$\eta_{jt} = \frac{x_{jt}}{\sum_{i=1}^{I} a_{ij}[\mathbf{AX}]_{it}^{\beta}}, \qquad \eta_{ij} = \frac{a_{ij}}{\sum_{t=1}^{T}[\mathbf{AX}]_{it}^{\beta} x_{jt}}, \tag{3.142}$$

we obtain the multiplicative update rules [12,36]:

$$x_{jt} \leftarrow x_{jt} \frac{[\sum_{i=1}^{I} a_{ij} (y_{it}/[\mathbf{AX}]_{it}^{1-\beta}) - \alpha_{\mathbf{X}}]_{+}}{\sum_{i=1}^{I} a_{ij} [\mathbf{AX}]_{it}^{\beta}}, \tag{3.143}$$

$$a_{ij} \leftarrow a_{ij} \frac{[\sum_{t=1}^{T} (y_{it}/[\mathbf{AX}]_{it}^{1-\beta}) x_{jt} - \alpha_{\mathbf{A}}]_{+}}{\sum_{t=1}^{T}[\mathbf{AX}]_{it}^{\beta} x_{jt}}. \tag{3.144}$$

For a large-scale NMF problem, the above multiplicative learning rules can be generalized in the matrix block-wise form as (we assume here for simplicity that $\alpha_{\mathbf{A}} = \alpha_{\mathbf{X}} = 0$):

$$\mathbf{X} \leftarrow \mathbf{X} \circledast \left(\mathbf{A}_r^T \left(\mathbf{Y}_r \oslash (\mathbf{A}_r\mathbf{X})^{\cdot[1-\beta]}\right)\right) \oslash \left(\mathbf{A}_r^T(\mathbf{A}_r\mathbf{X})^{\cdot[\beta]}\right), \tag{3.145}$$

$$\mathbf{A} \leftarrow \mathbf{A} \circledast \left(\left(\mathbf{Y}_c \oslash (\mathbf{AX}_c)^{\cdot[1-\beta]}\right) \mathbf{X}_c^T\right) \oslash \left((\mathbf{AX}_c)^{\cdot[\beta]} \mathbf{X}_c^T\right), \tag{3.146}$$

$$\mathbf{A} \leftarrow \mathbf{A} \ \mathrm{diag}(\|\mathbf{a}_1\|_1^{-1}, \|\mathbf{a}_2\|_1^{-1}, \dots, \|\mathbf{a}_J\|_1^{-1}) \tag{3.147}$$

and in the enhanced form outlined in Algorithm 3.5.

Algorithm 3.5: Multiplicative Beta NMF with Over-relaxation and Sparsity Control

Input: $\mathbf{Y} \in \mathbb{R}_+^{I \times T}$: input data, J: rank of approximation, β: order of Beta divergence
ω: over-relaxation, and $\alpha_{\mathbf{A}}$, $\alpha_{\mathbf{X}}$: sparsity degrees
Output: $\mathbf{A} \in \mathbb{R}_+^{I \times J}$ and $\mathbf{X} \in \mathbb{R}_+^{J \times T}$ such that cost function (3.136) is minimized.

1 **begin**
2   initialization for $\mathbf{A}$ and $\mathbf{X}$
3   **repeat**   /* update X and A */
4   select R row indices
5         $\mathbf{X} \leftarrow \mathbf{X} \circledast \left(\left[\mathbf{A}_r^T \left(\mathbf{Y}_r \circledast \hat{\mathbf{Y}}_r^{\cdot[\beta-1]}\right) - \alpha_{\mathbf{X}} \mathbf{1}_{J \times T}\right]_+ \oslash (\mathbf{A}_r^T \hat{\mathbf{Y}}_r^{\cdot[\beta]})\right)^{\cdot[\omega]}$ /* $\hat{\mathbf{Y}}_r = \mathbf{A}_r \mathbf{X}$ */
6   select C column indices
7         $\mathbf{A} \leftarrow \mathbf{A} \circledast \left(\left[\left(\mathbf{Y}_c \circledast \hat{\mathbf{Y}}_c^{\cdot[\beta-1]}\right) \mathbf{X}_c^T - \alpha_{\mathbf{A}} \mathbf{1}_{I \times J}\right]_+ \oslash (\hat{\mathbf{Y}}_c^{\cdot[\beta]} \mathbf{X}_c^T)\right)^{\cdot[\omega]}$ /* $\hat{\mathbf{Y}}_c = \mathbf{AX}_c$ */
8   **foreach** $\mathbf{a}_j$ of $\mathbf{A}$ **do** $\mathbf{a}_j \leftarrow \mathbf{a}_j / \|\mathbf{a}_j\|_p$ /* normalize to ℓ_p unit length */
9   **until** *a stopping criterion is met*   /* convergence condition */
10 **end**

3.4.2 Multiplicative Algorithm Based on the Itakura-Saito Distance

A special case of the Beta-divergence for $\beta = -1$ is the Itakura-Saito distance which forms the cost function

$$D_{IS}(\mathbf{Y}||\mathbf{AX}) = \sum_{i=1}^{I}\sum_{t=1}^{T}\left(-\ln(\frac{y_{it}}{q_{it}}) + \frac{y_{it}}{q_{it}}\right), \tag{3.148}$$

and relates to the negative cross Burg entropy or simply cross log entropy. If $\mathbf{X}^*$ satisfies $\mathbf{AX}^* = \mathbf{Y}$, then the Itakura-Saito distance is equal to zero, but not convex since the logarithmic term is concave. However, it is strictly convex on the set $[\mathbf{AX}]_{it} \le 2y_{it}$ since $-\ln(a/t) + a/t$ is convex on $0 < t \le 2a$ for $a > 0$.

The Itakura-Saito distance is optimal for a Gamma distribution, in other words, it corresponds to maximum likelihood estimation using the Gamma likelihood function. To illustrate this, consider the Gamma likelihood of order γ ($\gamma > 0$)

$$L(\mathbf{X}) = \prod_{it} \frac{z_{it}^{-\gamma} y_{it}^{\gamma-1} \exp(-y_{it}/z_{it})}{\Gamma(\gamma)}, \tag{3.149}$$

where $z_{it} = [\mathbf{AX}]_{it}/\gamma$.

The negative log-likelihood is then equal to

$$L_\Gamma(\mathbf{X}) = IT\ln(\Gamma(\gamma)) + \sum_{it}\left(\gamma \ln z_{it} - (\gamma - 1)\ln y_{it} + \frac{y_{it}}{z_{it}}\right). \tag{3.150}$$

Substituting z_{it} and noting that some terms in the above expression do not depend on $\mathbf{A}$ and $\mathbf{X}$, we obtain the Itakura-Saito distance.

The minimization of the above cost function leads to the following algorithm [27]:

$$\mathbf{X} \leftarrow \mathbf{X} \circledast [(\mathbf{A}^T\mathbf{P}) \oslash (\mathbf{A}^T\mathbf{Q} + \varepsilon)]^{.[\omega]}, \tag{3.151}$$

$$\mathbf{A} \leftarrow \mathbf{A} \circledast [(\mathbf{P}\mathbf{X}^T) \oslash (\mathbf{Q}\mathbf{X}^T + \varepsilon)]^{.[\omega]}, \tag{3.152}$$

$$\mathbf{A} \leftarrow \mathbf{A}\ \operatorname{diag}(\|\boldsymbol{a}_1\|_1^{-1}, \|\boldsymbol{a}_2\|_1^{-1}, \ldots, \|\boldsymbol{a}_J\|_1^{-1}), \tag{3.153}$$

where $\omega \in (0.5, 1)$ is a relaxation parameter and

$$\mathbf{P} = \mathbf{Y} \oslash \hat{\mathbf{Y}}^{.[2]}, \qquad \mathbf{Q} = \hat{\mathbf{Y}}^{.[-1]}, \qquad \hat{\mathbf{Y}} = \mathbf{AX} + \varepsilon. \tag{3.154}$$

3.4.3 Generalized Multiplicative Beta Algorithm for NMF

The multiplicative Beta NMF algorithm can be generalized as (see also Figure 3.9) [10,12,23,36]:

$$x_{jt} \leftarrow x_{jt} \frac{\sum_{i \in S_I} a_{ij}\ \Psi(y_{it}, q_{it})}{\sum_{i \in S_I} a_{ij}\ \Psi(q_{it}, q_{it})}, \qquad a_{ij} \leftarrow a_{ij} \frac{\sum_{t \in S_T} x_{jt}\ \Psi(y_{it}, q_{it})}{\sum_{t \in S_T} x_{jt}\ \Psi(q_{it}, q_{it})}, \tag{3.155}$$

where $q_{it} = [\mathbf{AX}]_{it}$, $\Psi(q, q)$ is a nonnegative nondecreasing function, and $\Psi(y, q)$ may take several different forms, for example:

1. $\Psi(y, q) = y, \quad \Psi(q, q) = q$;
2. $\Psi(y, q) = y/q, \quad \Psi(q, q) = 1$;
3. $\Psi(y, q) = y/q^{\beta}, \quad \Psi(q, q) = q^{1-\beta}$;
4. $\Psi(y, q) = y/(c + q), \quad \Psi(q, q) = q/(c + q)$.

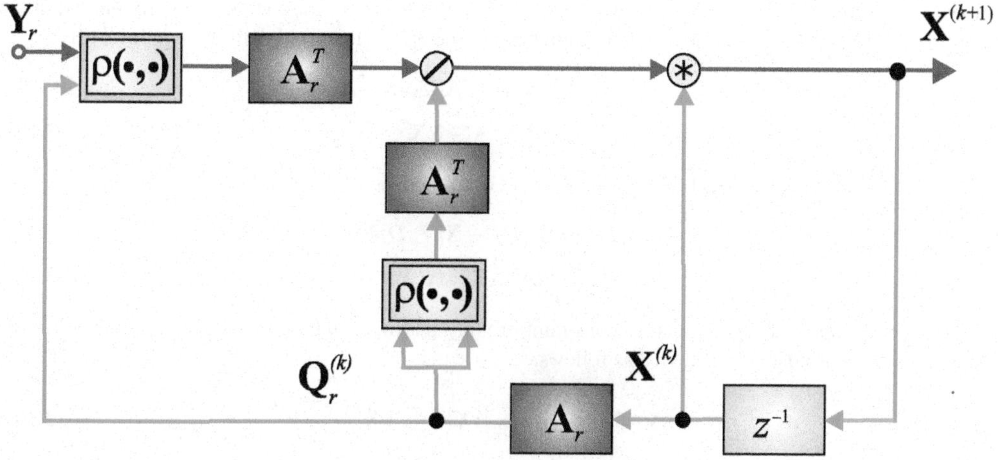

Figure 3.9 Functional block diagram illustrating the generalized multiplicative Beta NMF algorithm for a large-scale NMF: $\mathbf{X}^{(k+1)} = \mathbf{X}^{(k)} \circledast (\mathbf{A}_r^T \, \rho\,(\mathbf{Y}_r\,,\, \mathbf{A}_r^T\mathbf{X}^{(k)})) \oslash (\mathbf{A}_r^T \, \rho\,(\mathbf{A}_r\mathbf{X}^{(k)}\,,\, \mathbf{A}_r^T\mathbf{X}^{(k)}))$. A similar block diagram can be formulated for updating matrix $\mathbf{A}$ as: $\mathbf{A}^{(k+1)} = \mathbf{A}^{(k)} \circledast (\rho\,(\mathbf{Y}_c\,,\, \mathbf{A}^{(k)}\mathbf{X}_c)\, \mathbf{X}_c^T) \oslash (\rho\,(\mathbf{A}^{(k)}\mathbf{X}_c\,,\, \mathbf{A}^{(k)}\mathbf{X}_c)\, \mathbf{X}_c^T)$.

Not all the generalized multiplicative NMF algorithms are expected to work well for any given set of functions and parameters. In practice, in order to ensure stability it is necessary to introduce a suitable scaling and/or a relaxation parameter.[15]

Our main objective here was to unify most existing multiplicative algorithms for the standard NMF problem and to show how to incorporate additional constraints such as sparsity and smoothness.

3.5 Algorithms for Semi-orthogonal NMF and Orthogonal Three-Factor NMF

In many applications we need to impose additional orthogonality constraints which automatically provide very sparse representation of estimated factor matrices [7,24,43]. Let us consider at first the semi-orthogonal NMF model

$$\mathbf{Y} = \mathbf{A}\mathbf{X} + \mathbf{E}, \tag{3.156}$$

where the factor matrices are nonnegative and additionally one of them satisfies an orthogonality constraint, that is, $\mathbf{A}^T\mathbf{A} = \mathbf{I}$ or $\mathbf{X}\mathbf{X}^T = \mathbf{I}$. In order to estimate the factor matrices we can employ constrained minimization of the following cost function

$$D_F(\mathbf{Y}||\mathbf{A}\mathbf{X}) = \frac{1}{2}||\mathbf{Y} - \mathbf{A}\mathbf{X}||_F^2 \quad \text{s.t.} \quad \mathbf{A}^T\mathbf{A} = \mathbf{I} \quad \text{or} \quad \mathbf{X}\mathbf{X}^T = \mathbf{I}. \tag{3.157}$$

[15] It is still an open question and active area of research to decide which generalized NMF algorithms are potentially most useful and practical.

However, in order to impose automatically the orthogonality constraint for one factor matrix instead of the standard gradient descent we use the natural gradient on the Stiefel manifold defined as follows [7]

$$\begin{aligned}\tilde{\nabla}_{\mathbf{A}} D_F &= \nabla_{\mathbf{A}} D_F - \mathbf{A}[\nabla_{\mathbf{A}} D_F]^T \mathbf{A} \\ &= \mathbf{A}\mathbf{X}\mathbf{Y}^T\mathbf{A} - \mathbf{Y}\mathbf{X}^T,\end{aligned} \tag{3.158}$$

if we need to impose a semi-orthogonality constraint on $\mathbf{A}$, and

$$\begin{aligned}\tilde{\nabla}_{\mathbf{X}} D_F &= \nabla_{\mathbf{X}} D_F - \mathbf{X}[\nabla_{\mathbf{X}} D_F]^T \mathbf{X} \\ &= \mathbf{X}\mathbf{Y}^T\mathbf{A}\mathbf{X} - \mathbf{A}^T\mathbf{Y},\end{aligned} \tag{3.159}$$

if we need to impose an orthogonality constraint on only matrix $\mathbf{X}$, where the standard gradients (in the Riemannian manifolds) are expressed as follows:

$$\nabla_{\mathbf{A}} D_F = \frac{\partial D_F}{\partial a_{ij}} = \mathbf{A}\mathbf{X}\mathbf{X}^T - \mathbf{Y}\mathbf{X}^T, \tag{3.160}$$

and

$$\nabla_{\mathbf{X}} D_F = \frac{\partial D_F}{\partial x_{jt}} = \mathbf{A}\mathbf{A}^T\mathbf{X} - \mathbf{A}^T\mathbf{Y}. \tag{3.161}$$

By the heuristic formulas (3.44) we obtain the following set of multiplicative updated rules [7]: Semi-orthogonal NMF: $\mathbf{A}^T\mathbf{A} = \mathbf{I}$

$$\mathbf{A} \leftarrow \mathbf{A} \circledast \left(\left[\mathbf{Y}\mathbf{X}^T\right] \oslash \left[\mathbf{A}\mathbf{X}\mathbf{Y}^T\mathbf{A}\right]\right), \tag{3.162}$$

$$\mathbf{X} \leftarrow \mathbf{X} \circledast \left(\left[\mathbf{A}^T\mathbf{Y}\right] \oslash \left[\mathbf{A}^T\mathbf{A}\mathbf{X})\right]\right). \tag{3.163}$$

Semi-orthogonal NMF: $\mathbf{X}\mathbf{X}^T = \mathbf{I}$

$$\mathbf{A} \leftarrow \mathbf{A} \circledast \left(\left[\mathbf{Y}\mathbf{X}^T\right] \oslash \left[\mathbf{A}\mathbf{X}\mathbf{X}^T)\right]\right), \tag{3.164}$$

$$\mathbf{X} \leftarrow \mathbf{X} \circledast \left(\left[\mathbf{A}^T\mathbf{Y}\right] \oslash \left[\mathbf{X}\mathbf{Y}^T\mathbf{A}\mathbf{X}\right]\right). \tag{3.165}$$

Alternatively, we can use the ALS approach to derive new efficient update rules

Semi-orthogonal NMF: $\mathbf{A}^T\mathbf{A} = \mathbf{I}$

$$\mathbf{A} \leftarrow \left[\mathbf{Y}\mathbf{X}^T(\mathbf{X}\mathbf{Y}^T\mathbf{A})^{-1}\right]_+, \tag{3.166}$$

$$\mathbf{X} \leftarrow \left[(\mathbf{A}^T\mathbf{A})^{-1}\mathbf{A}^T\mathbf{Y})\right]_+. \tag{3.167}$$

Semi-orthogonal NMF: $\mathbf{X}\mathbf{X}^T = \mathbf{I}$

$$\mathbf{A} \leftarrow \left[\mathbf{Y}\mathbf{X}^T(\mathbf{X}\mathbf{X}^T)^{-1}\right]_+, \tag{3.168}$$

$$\mathbf{X} \leftarrow \left[(\mathbf{X}\mathbf{Y}^T\mathbf{A})^{-1}\mathbf{A}^T\mathbf{Y}\right]_+. \tag{3.169}$$

Similar approaches can be applied for Orthogonal Three-factor NMF (called also Tri-NMF [24])

$$\mathbf{Y} = \mathbf{A}\mathbf{S}\mathbf{X} + \mathbf{E}, \tag{3.170}$$

where two nonnegative factor matrices satisfy additional orthogonality constraints, that is, $\mathbf{A}^T\mathbf{A} = \mathbf{I}$ and simultaneously $\mathbf{X}\mathbf{X}^T = \mathbf{I}$. In order to estimate the factor matrices we can employ constrained minimization of the following cost function:

$$\bar{D}_F(\mathbf{Y}||\mathbf{ASX}) = \frac{1}{2}||\mathbf{Y} - \mathbf{ASX}||_F^2 \quad \text{s.t.} \quad \mathbf{A}^T\mathbf{A} = \mathbf{I} \quad \text{and} \quad \mathbf{XX}^T = \mathbf{I}. \tag{3.171}$$

In order to impose automatically the orthogonality constraint for two factor matrix instead of the standard gradient descent we use the natural gradient on the Stiefel manifold defined as follows [9]

$$\tilde{\nabla}_{\mathbf{A}}\bar{D}_F = \nabla_{\mathbf{A}}\bar{D}_F - \mathbf{A}\left[\nabla_{\mathbf{A}}\bar{D}_F\right]^T \mathbf{A} = \mathbf{ASXY}^T\mathbf{A} - \mathbf{YX}^T\mathbf{S}^T, \tag{3.172}$$

$$\tilde{\nabla}_{\mathbf{X}}\bar{D}_F = \nabla_{\mathbf{X}}\bar{D}_F - \mathbf{X}\left[\nabla_{\mathbf{X}}\bar{D}_F\right]^T \mathbf{X} = \mathbf{XY}^T\mathbf{ASX} - \mathbf{S}^T\mathbf{A}^T\mathbf{Y}, \tag{3.173}$$

$$\nabla_{\mathbf{S}}\bar{D}_F = \mathbf{AA}^T\mathbf{SXX}^T - \mathbf{A}^T\mathbf{YX}^T. \tag{3.174}$$

By using the heuristic formulas (3.44) we obtain the following set of multiplicative update rules

$$\mathbf{A} \leftarrow \mathbf{A} \circledast \left(\left[\mathbf{YX}^T\mathbf{S}^T\right] \oslash \left[\mathbf{ASXY}^T\mathbf{A}\right]\right), \tag{3.175}$$

$$\mathbf{X} \leftarrow \mathbf{X} \circledast \left(\left[\mathbf{S}^T\mathbf{A}^T\mathbf{Y}\right] \oslash \left[\mathbf{XY}^T\mathbf{ASX})\right]\right), \tag{3.176}$$

$$\mathbf{S} \leftarrow \mathbf{S} \circledast \left(\left[\mathbf{A}^T\mathbf{YX}^T\right] \oslash \left[\mathbf{A}^T\mathbf{ASXX}^T\right]\right). \tag{3.177}$$

Alternatively, we can use the ALS approach (see Chapter 4 for detail) by equating the gradient to zero to derive efficient update rules

$$\mathbf{A} \leftarrow \left[\mathbf{YX}^T\mathbf{S}^T(\mathbf{SXY}^T\mathbf{A})^{-1}\right]_+, \tag{3.178}$$

$$\mathbf{X} \leftarrow \left[(\mathbf{XY}^T\mathbf{AS})^{-1}\mathbf{S}^T\mathbf{A}^T\mathbf{Y})\right]_+, \tag{3.179}$$

$$\mathbf{S} \leftarrow \left[(\mathbf{A}^T\mathbf{A})^{-1}\mathbf{A}^T\mathbf{YX}^T(\mathbf{XX}^T)^{-1}\right]_+. \tag{3.180}$$

It should be noted that the ALS algorithm usually has better convergence rate. Moreover, it has better flexibility in comparison to multiplicative algorithms since it can be applied to the Three-Factor Semi-NMF model in which it is not necessary for all factor matrices need to be nonnegative. An alternative ALS algorithm for Three-factor Semi-NMF with orthogonal projections will be presented in Chapter 4.

3.6 Multiplicative Algorithms for Affine NMF

The main advantage of NMF is its potential ability to extract latent sparse and meaningful components from observed data. However, if there exits a common component in all or in most observations, for example, DC component,[16] then standard NMF model is not unique and usually it cannot extract the desired components. In such case we should use affine NMF model (see also Chapters 1 and 7). The mathematical model of the affine NMF can be described as follows [39]

$$\boxed{\mathbf{Y} = \mathbf{AX} + \boldsymbol{a}_0\mathbf{1}^T + \mathbf{E},} \tag{3.181}$$

where $\mathbf{1} \in \mathbb{R}^T$ is a vector of all ones and $\boldsymbol{a}_0 \in \mathbb{R}_+^I$ denotes the offset component. The incorporation of the offset term $\mathbf{Y}_0 = \boldsymbol{a}_0\mathbf{1}^T$ together with nonnegativity constraint often ensures the sparseness and uniqueness of factored matrices. In other words, the main role of the offset is to absorb the constant values of a data matrix, thereby making the factorization sparser and therefore improve (relax) conditions for uniqueness of NMF. The learning rule for the factor matrices **A** and **X**, are generally, the same as for the standard NMF,

[16] We call this constant part the offset or base line.

while the learning rule to estimate the base line is given by

$$\boxed{\boldsymbol{a}_0 \leftarrow \boldsymbol{a}_0 \circledast (\mathbf{Y}\,\mathbf{1}) \oslash (\hat{\mathbf{Y}}\,\mathbf{1}),} \tag{3.182}$$

where $\hat{\mathbf{Y}} = \mathbf{A}\mathbf{X} + \boldsymbol{a}_0\mathbf{1}^T$ is the current estimation of $\mathbf{Y}$.

In Algorithm 3.6, we presented the modified ISRA algorithm for aNMF as an example. The optional sparsity and orthogonality constraints can be imposed on the learning rules for $\mathbf{A}$ and $\mathbf{X}$ via the parameters α_{sA}, α_{sX} and α_{oA}, α_{oX}, respectively.[17] The algorithm 3.6 can be easily modified using other optional NMF algorithms discussed in this and next chapters. Steps 5 and 7 can be replaced by suitable alternative learning rules. We illustrated performance of this algorithm for difficult benchmarks in Section (3.9).

Algorithm 3.6: ISRA algorithm for affine NMF

Input: $\mathbf{Y} \in \mathbb{R}_+^{I\times T}$: input data, J: rank of approximation
α_{sA}, α_{sX}: sparsity levels, α_{oA}, α_{oX}: orthogonality levels.
Output: $\mathbf{A} \in \mathbb{R}_+^{I\times J}$, $\mathbf{X} \in \mathbb{R}_+^{J\times T}$ and $\boldsymbol{a}_0 \in \mathbb{R}_+^{I}$ such that cost function (3.36) is minimized.

1 **begin**
2 initialization for $\mathbf{A}$ and $\mathbf{X}$, and $\boldsymbol{a}_0$
3 **repeat** /* update X, A and a_0 */
4 $\hat{\mathbf{Y}} = \mathbf{A}\mathbf{X} + \boldsymbol{a}_0\mathbf{1}^T$
5 $\mathbf{A} \leftarrow \mathbf{A} \circledast (\mathbf{Y}\mathbf{X}^T) \oslash (\hat{\mathbf{Y}}\mathbf{X}^T + \alpha_{sA} + \alpha_{oA}\mathbf{A}(\mathbf{1}_{J\times J} - \mathbf{I}))$
6 **foreach** $\boldsymbol{a}_j$ *of* $\mathbf{A}$ **do** $\boldsymbol{a}_j \leftarrow \boldsymbol{a}_j/\|\boldsymbol{a}_j\|_1$ /* normalize to ℓ_1 unit length */
7 $\mathbf{X} \leftarrow \mathbf{X} \circledast (\mathbf{A}^T\mathbf{Y}) \oslash (\mathbf{A}^T\hat{\mathbf{Y}} + \alpha_{sX} + \alpha_{oX}(\mathbf{1}_{J\times J} - \mathbf{I})\mathbf{X})$
8 $\boldsymbol{a}_0 \leftarrow \boldsymbol{a}_0 \circledast (\mathbf{Y}\,\mathbf{1}) \oslash (\hat{\mathbf{Y}}\,\mathbf{1})$
9 **until** *a stopping criterion is met* /* convergence condition */
10 **end**

3.7 Multiplicative Algorithms for Convolutive NMF

Most of the multiplicative NMF algorithms derived for the standard NMF model can serve as a basic for the development of the convolutive NMF models (CNMF) described in Chapter 4. The simplest form of the CNMF can be described as (see Figure 3.10) [55,56]

$$\mathbf{Y} = \sum_{p=0}^{P-1} \mathbf{A}_p \overset{p\rightarrow}{\mathbf{X}} + \mathbf{E}, \tag{3.183}$$

where $\mathbf{Y} \in \mathbb{R}_+^{I\times T}$ is a given input data matrix, $\mathbf{A}_p \in \mathbb{R}^{I\times J}$ is a set of unknown nonnegative matrices, $\mathbf{X} = \overset{0\rightarrow}{\mathbf{X}} \in \mathbb{R}^{J\times T}$ is the matrix representing coding information of the source (such as position of activation and its amplitude), $\overset{p\rightarrow}{\mathbf{X}}$ is a p column shifted version of $\mathbf{X}$. In other words, $\overset{p\rightarrow}{\mathbf{X}}$ denotes the p positions (columns) shifting operator to the right, with the columns shifted in from outside the matrix set to zero. This shift (sample-delay) is performed by a basic operator $\mathbf{T}$. Analogously, $\overset{\leftarrow p}{\mathbf{Y}}$ means that the columns of $\mathbf{Y}$ are shifted p columns to the left. Note that, $\overset{0\rightarrow}{\mathbf{X}} = \overset{\leftarrow 0}{\mathbf{X}} = \mathbf{X}$.

[17] Detailed derivation is given in 3.85, 3.86

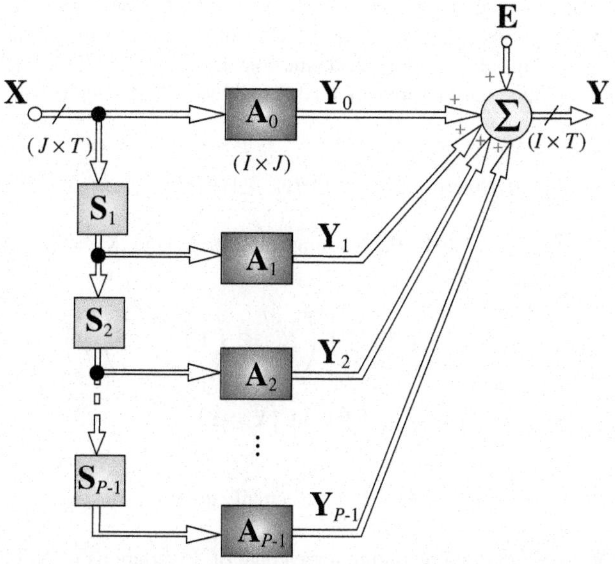

Figure 3.10 Illustration of Convolutive NMF. The goal is to estimate the input sources represented by nonnegative matrix $\mathbf{X} \in \mathbb{R}_+^{J\times T}$ (typically, $T >> I$) and to identify the convoluting system, i.e., to estimate a set of nonnegative matrices $\{\mathbf{A}_0, \mathbf{A}_1, \ldots, \mathbf{A}_{P-1}\}$ ($\mathbf{A}_p \in \mathbb{R}_+^{I\times J}$, $p = 0, 1, \ldots, P-1$) knowing only the input data matrix $\mathbf{Y} \in \mathbb{R}^{I\times T}$. Each operator $\mathbf{S}_p = \mathbf{T}_1$ ($p = 1, 2, \ldots, P-1$) performs a horizontal shift of the columns in $\mathbf{X}$ by one spot.

An example below illustrates the effect of shifting operator on matrix $\mathbf{X}$, the shifting performed by one position right and left

$$\mathbf{X} = \begin{bmatrix} 1 & 2 & 3 \\ 4 & 5 & 6 \end{bmatrix}, \quad \overset{1\rightarrow}{\mathbf{X}} = \begin{bmatrix} 0 & 1 & 2 \\ 0 & 4 & 5 \end{bmatrix}, \quad \overset{\leftarrow 1}{\mathbf{X}} = \begin{bmatrix} 2 & 3 & 0 \\ 5 & 6 & 0 \end{bmatrix}.$$

In most applications, we need to explain the original large-scale data by a small number of basis components and the corresponding activating maps which store the existing positions and the magnitudes of a basis component. For example, for music signals, the note patterns are the basis components, therefore, the activating map is a matrix which expresses the time instant when a note is played (or the presence of its harmonics); for texture images, the texture patterns are stored as the basis components and their positions are saved as the activating map. The CNMF model in the form (3.183) does not always totally reflect these properties. An alternative, equivalent description of the convolutive model, which deals with the underlying physics of the problem more naturally is given by

$$\mathbf{Y} = \sum_{j=1}^{J} \mathbf{A}^{(j)} * \underline{\boldsymbol{x}}_j + \mathbf{E}, \tag{3.184}$$

where symbol "$*$" denotes the 2-D convolution, $\underline{\boldsymbol{x}}_j \in \mathbb{R}^{1\times T}$ is the j-th row of the matrix $\mathbf{X}$, and the nonnegative $I \times P$ matrix $\mathbf{A}^{(j)}$ is the j-th basis component formed by the j-th column vectors of the matrices $\mathbf{A}_p$. In other words, for a set of matrices $\mathbf{A}_p$ considered as a third order tensor $\underline{\mathbf{A}}$ of dimension $I \times J \times P$, the j-th basis component is a lateral slice $\mathbf{A}^{(j)} = \underline{\mathbf{A}}(:, j, :)$, while $\mathbf{A}_p = \underline{\mathbf{A}}(:, :, p+1)$ represents a frontal slice of the same tensor $\underline{\mathbf{A}} \in \mathbb{R}_+^{I\times J\times P}$. Each component $\mathbf{A}^{(j)}$ has its activating map $\underline{\boldsymbol{x}}_j$, whereas each basic term $\mathbf{Y}^{(j)} = \mathbf{A}^{(j)} * \underline{\boldsymbol{x}}_j$

expresses the reconstructed version of each basis $\mathbf{A}^{(j)}$. If $\mathbf{A}^{(j)}$ is, for example, a single note spectral pattern, the time evolution of $\mathbf{Y}^{(j)}$ represents the sound sequence of this note, and summation of all sequences $\mathbf{Y}^{(j)}$ returns the full music song. This also explains the name convolutive NMF (CNMF) [50,54–57].

In the next section we extend the multiplicative Alpha and Beta NMF algorithms to the corresponding CNMF algorithms.

3.7.1 Multiplicative Algorithm for Convolutive NMF Based on Alpha-Divergence

Based on the updates (3.102) and (3.101), the learning rules for $\mathbf{A}_p$ and $\overset{p\rightarrow}{\mathbf{X}}$ can be formulated in a slightly different form as follows:

$$\mathbf{A}_p \leftarrow \mathbf{A}_p \circledast \left(\left((\mathbf{Y} \oslash \hat{\mathbf{Y}})^{.[\alpha]} \right) \left(\overset{p\rightarrow}{\mathbf{X}}^T \oslash \left(\mathbf{1}\,\mathbf{1}^T \overset{p\rightarrow}{\mathbf{X}}^T \right) \right) \right)^{.[1/\alpha]}, \tag{3.185}$$

$$\overset{p\rightarrow}{\mathbf{X}} \leftarrow \overset{p\rightarrow}{\mathbf{X}} \circledast \left(\left(\mathbf{A}_p^T \oslash (\mathbf{A}_p^T \mathbf{1}\,\mathbf{1}^T) \right) (\mathbf{Y} \oslash \hat{\mathbf{Y}})^{.[\alpha]} \right)^{.[1/\alpha]}, \tag{3.186}$$

where $\overset{p\rightarrow}{\mathbf{X}}^T \oslash \left(\mathbf{1}\,\mathbf{1}^T \overset{p\rightarrow}{\mathbf{X}}^T \right)$ and $\mathbf{A}_p^T \oslash \left(\mathbf{A}_p^T \mathbf{1}\,\mathbf{1}^T \right)$ are actually the ℓ_1-norm scaled matrices of $\overset{p\rightarrow}{\mathbf{X}}^T$ and $\mathbf{A}_p^T$.

By transforming $\overset{p\rightarrow}{\mathbf{X}}$ to $\mathbf{X}$, and by computing an average of P variations of $\mathbf{X}$, the learning rule for $\mathbf{X}$ becomes:

$$\mathbf{X} \leftarrow \frac{1}{P} \sum_{p=0}^{P-1} \left[\mathbf{X} \circledast \left(\left(\mathbf{A}_p^T \oslash (\mathbf{A}_p^T \mathbf{1}\,\mathbf{1}^T) \right) \overset{p\leftarrow}{\tilde{\mathbf{Y}}}^{.[\alpha]} \right)^{.[(1/\alpha)]} \right] \tag{3.187}$$

$$= \frac{1}{P} \mathbf{X} \circledast \sum_{p=0}^{P-1} \left(\left(\mathbf{A}_p^T \oslash (\mathbf{A}_p^T \mathbf{1}\,\mathbf{1}^T) \right) \overset{p\leftarrow}{\tilde{\mathbf{Y}}}^{.[\alpha]} \right)^{.[(1/\alpha)]}, \tag{3.188}$$

where $\tilde{\mathbf{Y}} = \mathbf{Y} \oslash \hat{\mathbf{Y}}$.

3.7.2 Multiplicative Algorithm for Convolutive NMF Based on Beta-Divergence

The update is based on the minimization of the Beta-divergence for the CNMF model:

$$D_B^{(\beta)}(\mathbf{Y}||\hat{\mathbf{Y}}) = \sum_{it} \left(y_{it} \frac{y_{it}^{\beta} - \hat{y}_{it}^{\beta}}{\beta} - \frac{y_{it}^{\beta+1} - \hat{y}_{it}^{\beta+1}}{\beta + 1} \right), \tag{3.189}$$

where $\hat{\mathbf{Y}} = [\hat{y}_{it}] = \sum_{p=0}^{P-1} \mathbf{A}_p \overset{p\rightarrow}{\mathbf{X}}$ is the approximation of the observed source $\mathbf{Y} = [y_{it}]$.

For convenience, we shall express the CNMF model (3.183) in the scalar form

$$\hat{y}_{it} = \sum_{p=0}^{P-1} \sum_{j=1}^{J} a_{ij(p+1)} \overset{p\rightarrow}{x}_{jt} = \sum_{p=0}^{P-1} \sum_{j=1}^{J} a_{ij(p+1)}\, x_{j(t-p)}, \tag{3.190}$$

assuming that $x_{j(t-p)} \triangleq 0$ if $t \leq p$. Thus, the gradients of the cost function (3.189) with respect to a_{ijp} and x_{jt} become

$$\frac{\partial D_B^{(\beta)}}{\partial a_{ijp}} = \sum_{t=1}^{T} (\hat{y}_{it}^{\beta} - y_{it}\, \hat{y}_{it}^{\beta-1})\, x_{j(t-p+1)} \tag{3.191}$$

and

$$\frac{\partial D_B^{(\beta)}}{\partial x_{jt}} = \sum_{i=1}^{I} \sum_{h=t}^{t+P-1} a_{ik(h-t+1)} \, (\hat{y}_{it}^{\beta} - y_{it} \, \hat{y}_{it}^{\beta-1})$$

$$= \sum_{i=1}^{I} \sum_{p=0}^{P-1} a_{ik(p+1)} \, (\hat{y}_{i(t+p)}^{\beta} - y_{i(t+p)} \, \hat{y}_{i(t+p)}^{\beta-1}). \tag{3.192}$$

Following the regular multiplicative Beta NMF algorithm, learning rules of the Beta CNMF algorithm for $\mathbf{A}_p$ and $\mathbf{X}$ can be derived in the same manner

$$a_{ijp} \leftarrow a_{ijp} - \eta_{ijp} \frac{\partial D_B^{(\beta)}}{\partial a_{ijp}} \tag{3.193}$$

$$x_{jt} \leftarrow x_{jt} - \eta_{jt} \frac{\partial D_B^{(\beta)}}{\partial x_{jt}}. \tag{3.194}$$

A suitable choice of learning rates, given by

$$\eta_{ijp} = \frac{a_{ijp}}{\sum_{t=1}^{T} \hat{y}_{it}^{\beta} \, x_{j(t-p+1)}}, \qquad \eta_{jt} = \frac{x_{jt}}{\sum_{i=1}^{I} \sum_{p=0}^{P-1} a_{ik(p+1)} \, \hat{y}_{i(t+p)}^{\beta}}, \tag{3.195}$$

leads to simple multiplicative update rules

$$a_{ijp} \leftarrow a_{ijp} \frac{\sum_{t=1}^{T} (y_{it} \, \hat{y}_{it}^{\beta-1}) \, x_{j(t-p+1)}}{\sum_{t=1}^{T} \hat{y}_{it}^{\beta} \, x_{j(t-p+1)}}, \qquad x_{jt} \leftarrow x_{jt} \frac{\sum_{i=1}^{I} \sum_{p=0}^{P-1} a_{ik(p+1)} \, y_{i(t+p)} \, \hat{y}_{i(t+p)}^{(\beta-1)}}{\sum_{i=1}^{I} \sum_{p=0}^{P-1} a_{ik(p+1)} \, \hat{y}_{i(t+p)}^{\beta}}, \tag{3.196}$$

for which the compact matrix form is:

$$\mathbf{A}_p \leftarrow \mathbf{A}_p \circledast \left(\tilde{\mathbf{Y}}_\beta \overset{p\rightarrow}{\mathbf{X}}{}^T \right) \oslash \left(\hat{\mathbf{Y}}^{.[\beta]} \overset{p\rightarrow}{\mathbf{X}}{}^T \right), \tag{3.197}$$

$$\mathbf{X} \leftarrow \mathbf{X} \circledast \left(\sum_{p=0}^{P-1} \mathbf{A}_p^T \overset{p\leftarrow}{\tilde{\mathbf{Y}}_\beta} \right) \oslash \left(\sum_{p=0}^{P-1} \mathbf{A}_p^T \overset{p\leftarrow}{\hat{\mathbf{Y}}}{}^{.[\beta]} \right), \tag{3.198}$$

where $\tilde{\mathbf{Y}}_\beta = \mathbf{Y} \circledast \hat{\mathbf{Y}}^{.[\beta-1]}$.

In practice, we adopt a slightly different procedure for estimation of $\mathbf{X}$

- Estimate each right-shifted matrix $\overset{p\rightarrow}{\mathbf{X}}$ by its gradient descent update rule

$$\overset{p\rightarrow}{\mathbf{X}} \leftarrow \overset{p\rightarrow}{\mathbf{X}} + \eta_{\mathbf{X}} \, \nabla_{\overset{p\rightarrow}{\mathbf{X}}} D_B^{(\beta)}, \tag{3.199}$$

where the gradient of the cost function (3.189) with respect to $\overset{p\rightarrow}{\mathbf{X}}$ is

$$\nabla_{\overset{p\rightarrow}{\mathbf{X}}} D_B^{(\beta)} = \mathbf{A}_p^T (\mathbf{Y} \circledast \hat{\mathbf{Y}}^{.[\beta-1]}) - \mathbf{A}_p^T \hat{\mathbf{Y}}^{.[\beta]} \tag{3.200}$$

and the learning rate

$$\eta_{\mathbf{X}} = \overset{p\rightarrow}{\mathbf{X}} \oslash (\mathbf{A}_p^T \hat{\mathbf{Y}}^{.[\beta]}). \tag{3.201}$$

to give the learning rule for $\overset{p\rightarrow}{\mathbf{X}}$ in the form

$$\overset{p\rightarrow}{\mathbf{X}} \leftarrow \overset{p\rightarrow}{\mathbf{X}} \circledast (\mathbf{A}_p^T \tilde{\mathbf{Y}}_\beta) \oslash (\mathbf{A}_p^T \hat{\mathbf{Y}}^{.[\beta]}). \tag{3.202}$$

- Average P variations of $\mathbf{X}$ from $\overset{p\rightarrow}{\mathbf{X}}$ [51]

$$\mathbf{X} \leftarrow \frac{1}{P}\sum_{p=0}^{P-1} \overset{p\leftarrow}{\overset{p\rightarrow}{\mathbf{X}}} = \frac{1}{P}\sum_{p=0}^{P-1} \mathbf{X} \circledast (\mathbf{A}_p^T \overset{p\leftarrow}{\tilde{\mathbf{Y}}_\beta}) \oslash (\mathbf{A}_p^T \overset{p\leftarrow}{\hat{\mathbf{Y}}}^{.[\beta]}) \tag{3.203}$$

$$= \mathbf{X} \circledast \frac{1}{P}\sum_{p=0}^{P-1} (\mathbf{A}_p^T \overset{p\leftarrow}{\tilde{\mathbf{Y}}_\beta}) \oslash (\mathbf{A}_p^T \overset{p\leftarrow}{\hat{\mathbf{Y}}}^{.[\beta]}). \tag{3.204}$$

Additional requirements such as sparseness, smoothness or uncorrelatedness (orthogonality) can be incorporated into the CNMF model by adding corresponding constraint terms to the cost function:

$$D_{Bc}^{(\beta)}(\mathbf{Y}\|\hat{\mathbf{Y}}) = D_B^{(\beta)} + \lambda_{\mathbf{X}} J_{\mathbf{X}} + \lambda_{\mathbf{A}} J_{\mathbf{A}}, \tag{3.205}$$

where $J_{\mathbf{X}}$ and $J_{\mathbf{A}}$ are suitably designed penalty/regularization terms.

The regularization functions given below are generally different from those used in standard NMF.

- **Sparseness for $\mathbf{A}_p$:** Each factor $\mathbf{A}_p$ can have independent sparsity parameter $\lambda_{\mathbf{A}p}$, however, for simplicity, we can assume that these parameters are equal, thus giving.

$$J_{\mathbf{A}}^{sp} = \sum_p \lambda_{\mathbf{A}p} \|\mathbf{A}_p\|_1, \qquad \Psi_{\mathbf{A}}(\mathbf{A}_p) = \lambda_{\mathbf{A}p} \mathbf{1}_{I\times J}. \tag{3.206}$$

- **Uncorrelatedness (orthogonality) for $\mathbf{A}_p$:** This constraint attempts to reduce correlation between column vectors in a matrix $\mathbf{A}_p$

$$J_{\mathbf{A}}^{o} = \frac{1}{2}\sum_{p\neq q} \|\mathbf{A}_p^T \mathbf{A}_q\|_1, \qquad \Psi_{\mathbf{A}}(\mathbf{A}_p) = \sum_{p\neq q} \mathbf{A}_p. \tag{3.207}$$

For example, the Beta-CNMF algorithm, with a sparsity constraint for $\mathbf{X}$ and uncorrelatedness constraint for $\mathbf{A}_p$, can be formulated as follows (see also Algorithm 3.7):

$$\mathbf{X} \leftarrow \mathbf{X} \circledast \left(\sum_{p=0}^{P-1} \mathbf{A}_p^T \overset{p\leftarrow}{\tilde{\mathbf{Y}}_\beta}\right) \oslash \left(\sum_{p=0}^{P-1} \mathbf{A}_p^T \overset{p\leftarrow}{\hat{\mathbf{Y}}}^{.[\beta]} + \lambda_{\mathbf{X}} \mathbf{1}_{J\times T}\right), \tag{3.208}$$

$$\mathbf{A}_p \leftarrow \mathbf{A}_p \circledast \left(\tilde{\mathbf{Y}}_\beta \overset{p\rightarrow}{\mathbf{X}}^T\right) \oslash \left(\hat{\mathbf{Y}}^{.[\beta]} \overset{p\rightarrow}{\mathbf{X}}^T + \lambda_{\mathbf{A}} \sum_{q\neq p} \mathbf{A}_q\right). \tag{3.209}$$

3.7.3 Efficient Implementation of CNMF Algorithm

By using (3.184), the shifting model can be expressed by the convolution operators; this allows an efficient implementation of CNMF algorithms. The shift operator $\mathbf{T}_p$ can be expressed as a square matrix with ones on the p-th diagonal and zeros elsewhere,[18] as illustrated by a simple example:

$$\mathbf{T}_1 = \begin{bmatrix} 0 & 1 & 0 & 0 \\ 0 & 0 & 1 & 0 \\ 0 & 0 & 0 & 1 \\ 0 & 0 & 0 & 0 \end{bmatrix}, \quad \mathbf{T}_{-1} = \begin{bmatrix} 0 & 0 & 0 & 0 \\ 1 & 0 & 0 & 0 \\ 0 & 1 & 0 & 0 \\ 0 & 0 & 1 & 0 \end{bmatrix} = \mathbf{T}_1^T.$$

Next, we define a flip matrix[19] $\mathbf{F}$ as a square matrix with ones on the anti-diagonal and zeros elsewhere, that is,

$$\mathbf{F} = \begin{bmatrix} 0 & 0 & 0 & 1 \\ 0 & 0 & 1 & 0 \\ 0 & 1 & 0 & 0 \\ 1 & 0 & 0 & 0 \end{bmatrix}. \tag{3.210}$$

Thus, for instance, a right multiplication of a matrix $\mathbf{X}$ with a flip matrix $\mathbf{F}$ flips the columns of $\mathbf{X}$ horizontally

$$\mathbf{X}\,\mathbf{F} = \begin{bmatrix} 1 & 2 & 3 & 4 \\ 5 & 6 & 7 & 8 \\ 9 & 10 & 11 & 12 \end{bmatrix} \begin{bmatrix} 0 & 0 & 0 & 1 \\ 0 & 0 & 1 & 0 \\ 0 & 1 & 0 & 0 \\ 1 & 0 & 0 & 0 \end{bmatrix} = \begin{bmatrix} 4 & 3 & 2 & 1 \\ 8 & 7 & 6 & 5 \\ 12 & 11 & 10 & 9 \end{bmatrix}. \tag{3.211}$$

Analogously, a left multiplication flips rows of $\mathbf{X}$ vertically

$$\mathbf{F}\,\mathbf{X} = \begin{bmatrix} 0 & 0 & 1 \\ 0 & 1 & 0 \\ 1 & 0 & 0 \end{bmatrix} \begin{bmatrix} 1 & 2 & 3 & 4 \\ 5 & 6 & 7 & 8 \\ 9 & 10 & 11 & 12 \end{bmatrix} = \begin{bmatrix} 9 & 10 & 11 & 12 \\ 5 & 6 & 7 & 8 \\ 1 & 2 & 3 & 4 \end{bmatrix}. \tag{3.212}$$

This leads to the following identities:

- $\mathbf{T}_p\,\mathbf{F}$ is a symmetric matrix:

$$\mathbf{T}_p\,\mathbf{F} = (\mathbf{T}_p\,\mathbf{F})^T. \tag{3.213}$$

- Flipping a flip matrix returns an identity matrix:

$$\mathbf{F}\,\mathbf{F} = \mathbf{F}\,\mathbf{F}^T = \mathbf{I}. \tag{3.214}$$

- A shift-left operator can be expressed by linear transformation of a shift-right operator

$$\mathbf{T}_{-p} = \mathbf{T}_{-p}\,\mathbf{F}^T\,\mathbf{F} = \mathbf{T}_p^T\,\mathbf{F}^T\,\mathbf{F} = (\mathbf{F}\,\mathbf{T}_p)^T\,\mathbf{F} = \mathbf{F}\,\mathbf{T}_p\,\mathbf{F}. \tag{3.215}$$

[18] $\mathbf{T}_{-p} = \mathbf{T}_p^T$.

[19] An operator with dynamic size

Using the above properties, especially (3.215), the learning rule for $\mathbf{X}$ (3.198) simplifies as:

$$\sum_{p=0}^{P-1} \mathbf{A}_p^T \overset{p\leftarrow}{\tilde{\mathbf{Y}}_\beta} = \sum_{p=0}^{P-1} \mathbf{A}_p^T \tilde{\mathbf{Y}}_\beta \mathbf{T}_{-p} = \sum_{p=0}^{P-1} \mathbf{A}_p^T \tilde{\mathbf{Y}}_\beta \mathbf{F}\mathbf{T}_p \mathbf{F} \tag{3.216}$$

$$= \left(\sum_{p=0}^{P-1} \mathbf{A}_p^T (\tilde{\mathbf{Y}}_\beta \mathbf{F}) \mathbf{T}_p \right) \mathbf{F}. \tag{3.217}$$

The expression inside the parentheses in (3.217) is the CNMF model of matrices $\mathbf{A}_p^T$ and $\tilde{\mathbf{Y}}_\beta \mathbf{F}$ (a horizontal-flip matrix of $\tilde{\mathbf{Y}}_\beta$). Analogously, the denominator in (3.198) can be expressed by the CNMF model of matrices $\mathbf{A}_p^T$ and $\hat{\mathbf{Y}}^{.[\beta]} \mathbf{F}$. The MATLAB code (Listing 3.1) is listed below for the CNMF model based on the tensor $\underline{\mathbf{A}}$ and matrix $\mathbf{X}$

Listing 3.1 MATLAB function for construction of the CNMF model.

```
function Y = CNMF(A,X)
% Build convolutive source Y from a set of basis features A: A(:,j,:) = A^(j)
% and an activating matrix X
Y = zeros(size(A,1),size(X,2));
for j = 1:size(A,2)
    Y =  Y + conv2(squeeze(A(:,j,:)),X(j,:));
end
```

Hence, the learning rules for $\mathbf{X}$ can be represented simply as in Listing 3.2:

Listing 3.2 An illustrative MATLAB code for the learning rule for $\mathbf{X}$.

```
%Estimate the source Ŷ
Yhat = CNMF(A,X);
Yhatbeta = Yhat.^beta
Ytilde =  Y.* Yhatbeta ./ Yhat;
NumX = CNMF(permute(A,[2 1 3]),fliplr(Ytilde));
DenX = CNMF(permute(A,[2 1 3]),fliplr(Yhatbeta)) + lambdaX  + eps;
X = X .* fliplr(NumX ./ DenX);
```

We illustrate applications of the CNMF and performance of the Beta algorithm in Section 3.10.

3.8 Simulation Examples for Standard NMF

The multiplicative NMF algorithms, especially those based on Alpha and Beta-divergences, have been extensively tested on many difficult benchmarks. Illustrative examples are provided to give insight into the multi-layer techniques, dealing with large-scale data matrices, initialization, and efficient MATLAB implementations.

Example 3.1 *NMF with multi-layer technique*

To show the advantages of multi-layer techniques, simulations were performed on the `X_spectra` dataset [11] which contains five sparse and smooth nonnegative signals shown in Figure 3.11(a), 3.11(c). Ten mixtures shown in Figure 3.11(b) and 3.11(d) were composed by a uniform random mixing matrix $\mathbf{A} \in \mathbb{R}_+^{10\times 5}$. Performance of 17 multiplicative NMF algorithms listed in Table 3.1 and Table 3.2, were analyzed under 100 Monte Carlo trials with 1000 iterations and four layers, and evaluated via PSNR in [dB], SIR in [dB], explained variation (FIT %) and execution time. Both the matrices $\mathbf{A}$ and $\mathbf{X}$ were initialized by uniformly distributed random numbers. However, the IS algorithm ($\beta = -1$) was analyzed for a single layer with 5000 iterations, and we employed the ALS initialization to ensure its convergence.

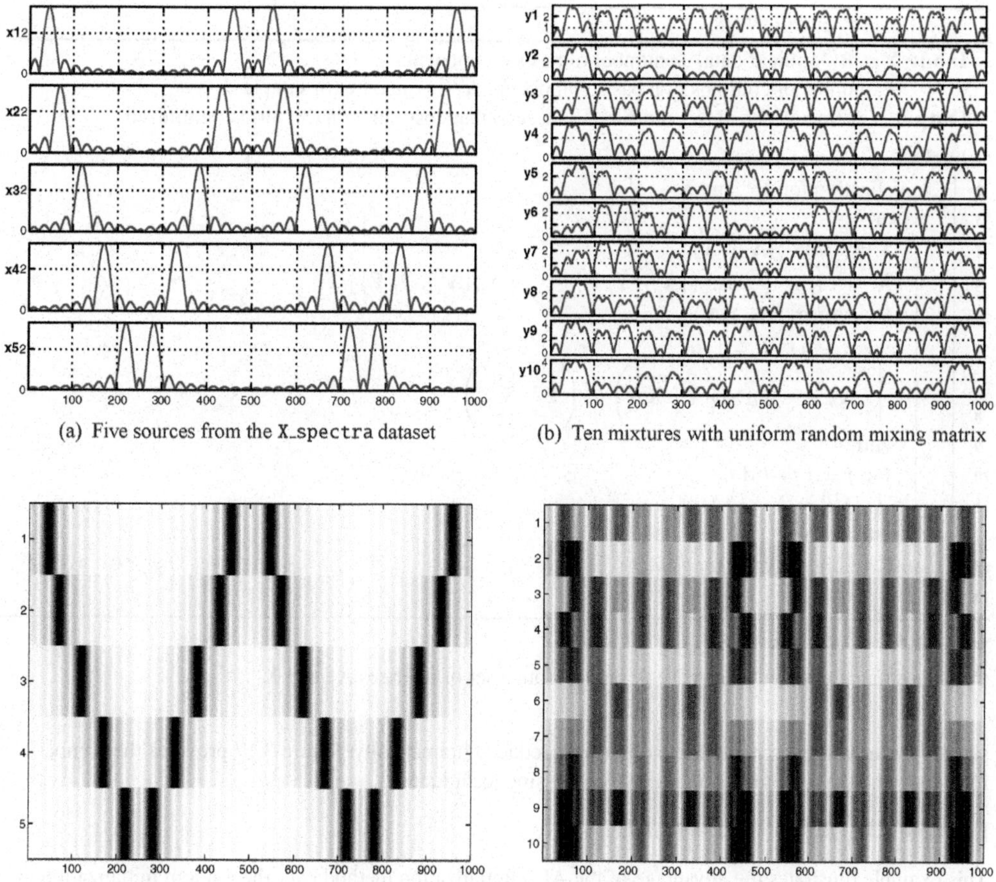

(a) Five sources from the X_spectra dataset

(b) Ten mixtures with uniform random mixing matrix

(c) Five sources from (a) represented as gray-scale images

(d) Images representation of 10 mixtures

Figure 3.11 Illustration for Example 3.1 with five sources (nonnegative components) and their typical mixtures using a randomly generated (uniformly distributed) mixing matrix.

Table 3.3, Table 3.4 and Table 3.5 illustrate the performance results for each layers using PSNR and SIR. All the algorithms achieved better performance by employing the multi-layer technique (see also Figures 3.13 and 3.14). The comparisons of performances of multiplicative NMF algorithms are given in Figures 3.15 and 3.16.

Listing 3.3 illustrates an exemplary usage of the `nmf_multi-layer` (Appendix 3.E.1) for this example.

Example 3.2 *Large-scale data*

The performance of the multiplicative Alpha and Beta NMF algorithms is illustrated for large-scale input data matrix of dimension 1000×1000, $2000 \times 1000, \ldots, 10000 \times 1000$ assuming only five nonnegative components (benchmark: `X_spectra`). The algorithms were compared for the same number of iterations (100) in two processing techniques: full data and block-wise factorization procedures. For the random block-wise procedure, the multi-layer technique with three layers was applied. Table 3.7 shows the differences between execution times of the two processing methods. The performance of the random block-wise procedure is

Algorithm 3.7: Beta CNMF

Input: $\mathbf{Y} \in \mathbb{R}_+^{I \times T}$: input data, J: number of basis components
P: number of columns of a basis component $\mathbf{A}^{(j)}$, or maximum sample-delay
Output: J factors $\mathbf{A}^{(j)} \in \mathbb{R}_+^{I \times P}$ and $\mathbf{X} \in \mathbb{R}_+^{J \times T}$ such that cost function (3.189) is minimized.

begin
  initialization for $\mathbf{A}^{(j)}$ and $\mathbf{X}$
  repeat
    $\hat{\mathbf{Y}} = \mathtt{CNMF}\left(\{\mathbf{A}_p\}, \mathbf{X}\right)$
    $\mathbf{X} \longleftarrow \mathbf{X} \circledast \left(\mathtt{CNMF}\left(\{\mathbf{A}_p^T\}, \tilde{\mathbf{Y}}_\beta \mathbf{F}\right) \oslash \mathtt{CNMF}\left(\{\mathbf{A}_p^T\}, \hat{\mathbf{Y}}_\beta \mathbf{F}\right)\right) \mathbf{F}$ `/* update X */`
    $\hat{\mathbf{Y}} = \mathtt{CNMF}\left(\{\mathbf{A}_p\}, \mathbf{X}\right)$
    for $p = 0$ *to* $P - 1$ **do**
      $\mathbf{A}_p \longleftarrow \mathbf{A}_p \circledast \left(\tilde{\mathbf{Y}}_\beta \overset{p\rightarrow}{\mathbf{X}}{}^T\right) \oslash \left(\hat{\mathbf{Y}}_\beta \overset{p\rightarrow}{\mathbf{X}}{}^T\right)$ `/* update A_p */`
    end
    for $j = 1$ *to* J **do**
      $\mathbf{A}^{(j)} = \mathbf{A}^{(j)} / \|\mathbf{A}^{(j)}\|_2$ `/* normalize to ℓ2-norm unit length */`
    end
  until *a stopping criterion is met* `/* convergence condition */`
end

$\hat{\mathbf{Y}}_\beta = \hat{\mathbf{Y}}^{.[\beta]}, \quad \tilde{\mathbf{Y}}_\beta = \mathbf{Y} \circledast \hat{\mathbf{Y}}_\beta \oslash \hat{\mathbf{Y}}.$
`CNMF` is a function to build the CNMF model from a set of matrices $\mathbf{A}_p$ and $\mathbf{X}$.

slightly better while the computation time was reduced dramatically. Figure 3.17 provides the comparison for PSNR and the run time using the two processing techniques.

Example 3.3 *ALS and random initializations*

This example illustrates the advantage of the ALS initialization method over the random initialization (see Figure 3.12 and Chapter 4 for more detail). Simulations were performed on the `X_sparse` dataset over 5000 iterations and using a single layer. Four NMF algorithms D-KLm ($\alpha = 0$), PS ($\alpha = 2$), EMML ($\beta = 0$) and IS ($\beta = -1$) achieved much better performance when the ALS initialization is used (+13[dB]). However, for the random initialization, the performance can also be improved when combined with the multi-layer

Table 3.3 The performance comparison of multiplicative Alpha and Beta NMF algorithms for the `X_spectra_sparse` dataset with different number of layers in the NMF model. For each layer we performed maximum 1000 iterations.

Algorithm	PSNR [dB]				SIR [dB]			
	Layer 1	Layer 2	Layer 3	Layer 4	Layer 1	Layer 2	Layer 3	Layer 4
$\alpha = 0$	18.39	30.83	32.12	32.04	7.90	20.58	21.98	21.92
$\alpha = 0.5$	18.72	31.45	33.36	33.73	8.25	21.35	23.26	23.65
$\alpha = 1, \beta = 0$	18.64	30.71	32.86	33.65	8.15	20.51	22.71	23.65
$\alpha = 2$	18.38	30.92	33.12	33.31	7.87	20.55	22.76	23.35
$\alpha = 3$	19.19	31.79	33.81	33.74	8.74	21.43	23.99	24.05
$\beta = 0.5$	17.83	27.03	32.59	34.35	7.27	16.62	22.24	24.04
$\beta = 1$	17.23	27.69	32.22	33.23	6.70	17.31	21.97	22.93
TD	22.24	33.33	34.82	35.21	11.81	23.29	25.19	25.72

Table 3.4 MC analysis over 100 trials for four layer NMF. The performance comparison of multiplicative Alpha and Beta NMF algorithms for the `X_spectra_sparse` dataset with four layers and maximum 1000 iterations/layer.

Algorithm	PSNR [dB]			SIR	FIT	Time (seconds)	
	median	worst	best	[dB]	(%)	SC [a]	DC [b]
$\alpha = 0$	32.05	18.35	39.32	21.92	99.91	6.73	3.55
$\alpha = 0.5$	33.73	20.06	38.81	23.65	99.98	8.15	3.31
$\alpha = 1, \beta = 0$	33.65	17.91	40.35	23.65	99.99	**4.18**	**2.51**
$\alpha = 2$	33.31	**21.88**	39.15	23.35	99.99	5.21	2.93
$\alpha = 3$	33.74	15.86	38.06	24.05	99.98	19.01	6.73
$\beta = -1$ [c]	28.81	16.66	32.06	18.74	99.95	15.15	4.37
$\beta = 0.5$	34.35	16.66	38.64	24.04	99.967	6.23	3.09
$\beta = 1$	33.23	15.74	38.97	22.93	99.97	5.77	2.90
TD	**35.21**	21.31	**42.56**	**25.72**	**99.99**	4.66	2.73

[a] Sequential Computing
[b] Distributed Computing

NMF scheme. Example 3.1 illustrates this advantage for setting with the `X_spectra` benchmark and four layers, and 1000 iterations/layer.

Example 3.4 *Monte Carlo (MC) analysis with the MATLAB distributed computing tool*

This example provides an efficient way to perform the MC analysis on multicore or multiprocessor machines based on the advantage of the MATLAB Parallel Computing Toolbox or the MATLAB Distributed Computing Server. Each MC trial is considered as a task of a distributed job. Workers might run several tasks of the same job in succession. The number of workers depends on the number of processor cores on a single chip and/or number of clusters in a system. For instance, a desktop with a quad-core processor may have four up to eight workers. The MATLAB Parallel Computing Toolbox (version 4.1 (R2009a)) enables us to run eight local workers on a multicore desktop through a local scheduler. When integrating with MATLAB Distributed Computing Server, this toolbox allows running applications with any scheduler and any number of workers. We provide the function `mc_nmf` (given in Appendix 3.D.3) which supports MC analysis with any NMF algorithms. For simplicity, this function works with the local scheduler, and returns outputs in two variables:

- `results` is a cell array containing the output of the tested NMF function.
- `rtime` is run time of single trials.

Table 3.5 Performances of SMART NMF algorithms for the `X_spectra_sparse` dataset, different number of layers, and 1000 iterations/layer.

Algorithm	PSNR [dB]				SIR [dB]			
	Layer 1	Layer 2	Layer 3	Layer 4	Layer 1	Layer 2	Layer 3	Layer 4
S-DKL	18.59	31.39	32.87	32.93	8.08	21.42	23.17	23.02
S-RJS	18.67	30.23	32.13	32.55	8.18	19.95	22.02	22.95
S-DRJS	17.88	29.06	31.46	32.20	7.34	18.77	21.29	22.16
S-SJS	18.55	31.54	33.28	33.33	8.06	21.40	23.49	23.61
S-JD	17.94	31.82	32.99	33.17	7.46	21.73	23.15	23.20
S-TD	19.13	30.63	33.19	33.80	8.60	20.77	23.30	24.08
S-BE	17.73	29.09	30.84	31.37	7.24	18.99	20.89	21.42
S-alpha	18.29	32.44	33.33	33.28	7.74	22.60	23.55	23.55

Listing 3.3 MATLAB code for Example 3.1.

```
% This example illustrates how to use the 'nmf_multi_layer' function
% Factorize 10 sources using the ALPHA NMF algorithm under 4 multilayers.
%
%% Generate mixtures
load('X_spectra_sparse_sorted.mat');
[J,T] = size(X);
I = 10;                                                     % number of mixtures
A = rand(I,J);
Y = A*X;
%% Fatorize the source using ALPHA NMF over 4 layers
options = struct('J',J,'alpha',1,'niter',1000,'verbose',0,...
    'algorithm','nmf_alpha_fb','Nlayer',4); % this line for multilayer NMF
[AH,XH,SIR,PSNR,t,Fit]= nmf_multi_layer(Y,options,X,1);
```

An MC example for the Alpha NMF algorithm with 100 trials is given in Listing 3.4.

Table 3.6 shows the average run times per trial of sequential and distributed computing analysis. The `mc_nmf` function can run with the multi-layer technique (see Listing 3.5).

3.9 Examples for Affine NMF

Example 3.5 *The* `swimmer` *benchmark*

We illustrate advantage of employing the affine NMF model over standard NMF by analyzing a `swimmer` dataset [6,8,25,39] shown in Figure 3.18(a). This database consists of a set of 256 binary (black-and-white stick) images ($32 \times 32 pixels$). Each image (picture) representing a view from the top of a "swimmer" consists of a "torso" of 12 pixels in the center and four "limbs" ("arms" and "legs") of six pixels that can be in any of four positions, thus, with limbs in all possible positions, there are a total of 256 pictures of dimension 32×32 pixels. This database was originally proposed by Donoho and Stodden [25] in order to

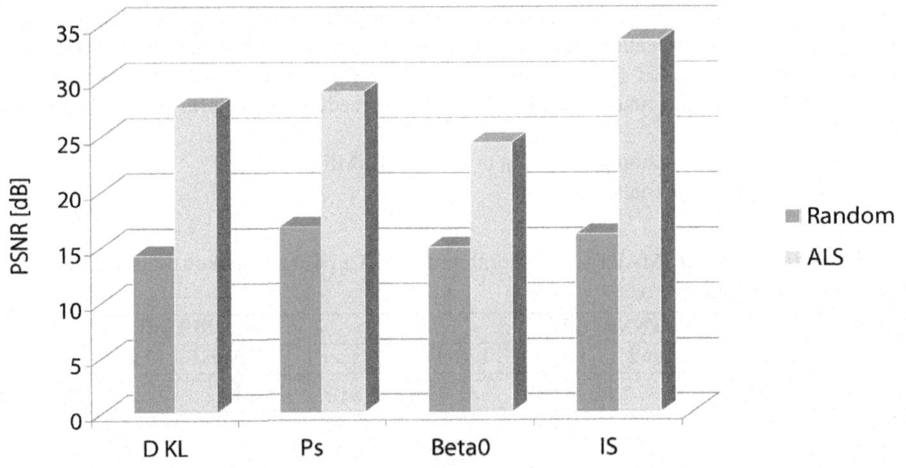

Figure 3.12 Illustration of random and deterministic ALS initializations for the NMF algorithms. Simulations were performed on the `X_sparse` dataset using D-KL ($\alpha = 0$), PS ($\alpha = 2$), Beta NMF ($\beta = 0$, EMML) and IS ($\beta = -1$) algorithms over 5000 iterations and single layer. The nonnegative ALS initialization provided relatively quick convergence. However, random initialization also improved the performance when combined with multilayer NMF.

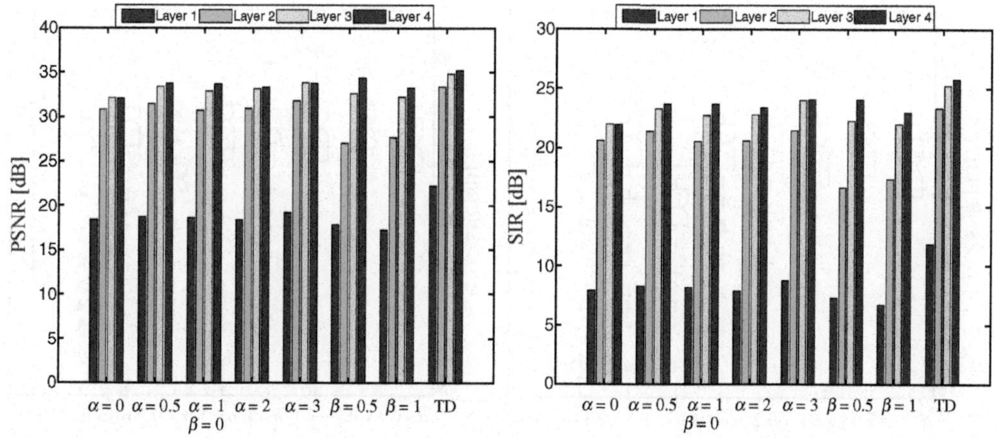

(a) PSNR (b) SIR

Figure 3.13 Performance of multi-layer NMF technique via 100 trial Monte Carlo analysis on the `X_spectra` dataset for the multiplicative Alpha and Beta NMF algorithms. The best and most consistent performance was achieved by the multiplicative NMF algorithm based on the TD (Triangular Discrimination) cost function.

show that standard NMF algorithms do not provide good parts based decomposition due to offset. In fact, the standard NMF algorithm produces parts contaminated with ghosts of the "torso" because it is an invariant part (which can be considered as common factor of offset) [6,8,39]. The images in the dataset should be decomposed into $17(= 4 \times 4 + 1)$ non-overlapping basis images. In other words, the question is: Can the parts be recovered form original dataset?

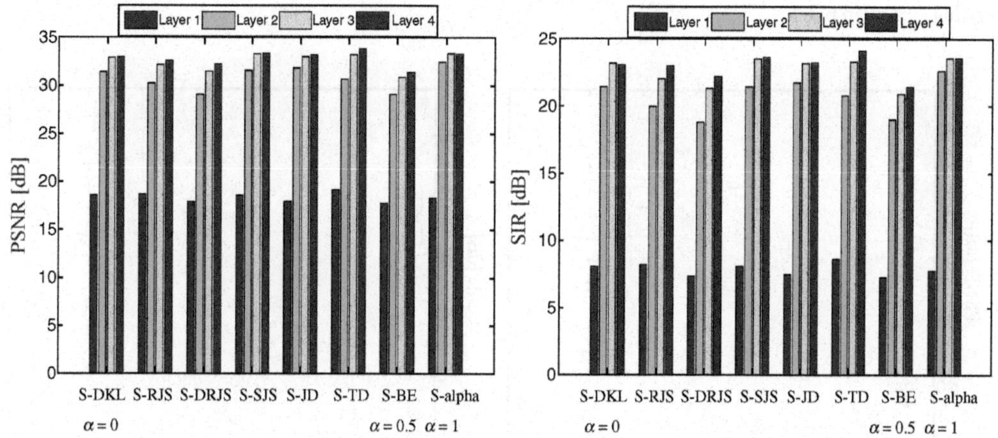

(a) PSNR (b) SIR

Figure 3.14 Performance of multi-layer NMF technique via 100 trial MC analysis for the `X_spectra` dataset for the SMART NMF algorithms. Again, the SMART algorithm based on TD cost function performed slightly better than other algorithms (see Table 3.2).

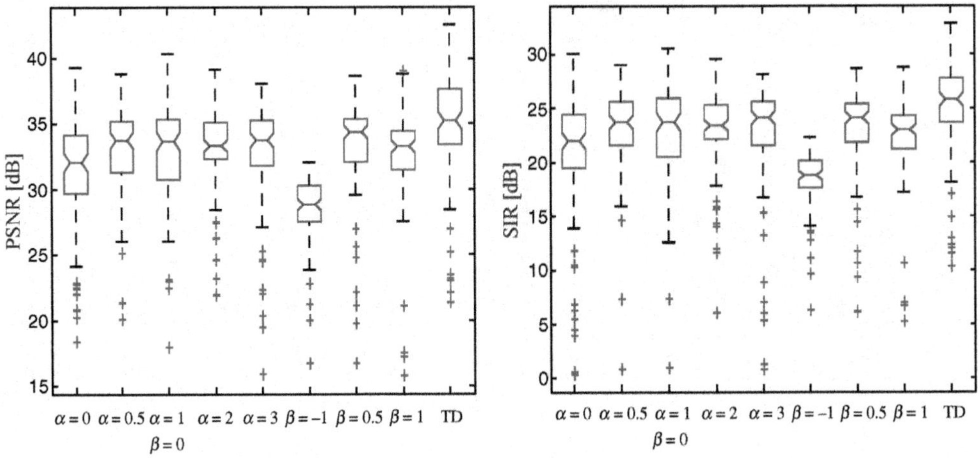

(a) PSNR (b) SIR

Figure 3.15 MC analysis over 100 trials on the `X_spectra` dataset for the multiplicative Alpha and Beta NMF algorithms.

The standard NMF fails to decompose the data into basis components as illustrated in Figure 3.18(b). Note that the "torso" component exists in all extracted components. However by using the modified ISRA algorithm for aNMF, we were able to successfully retrieve all 17 components. Figure 3.18(d) displays 16 "limbs" components in matrix **A** and the last "torso" component as vector $\boldsymbol{a}_0$. All the basis components are rearranged, and displayed in columns. The corresponding coding matrix for all basis components is shown in Figure 3.18(c). At the frames (pictures) that a basis picture is activated, its coding elements are displayed as black, and have value 1. Otherwise, they are zeros, and displayed as white. The offset component has the coding matrix in black.

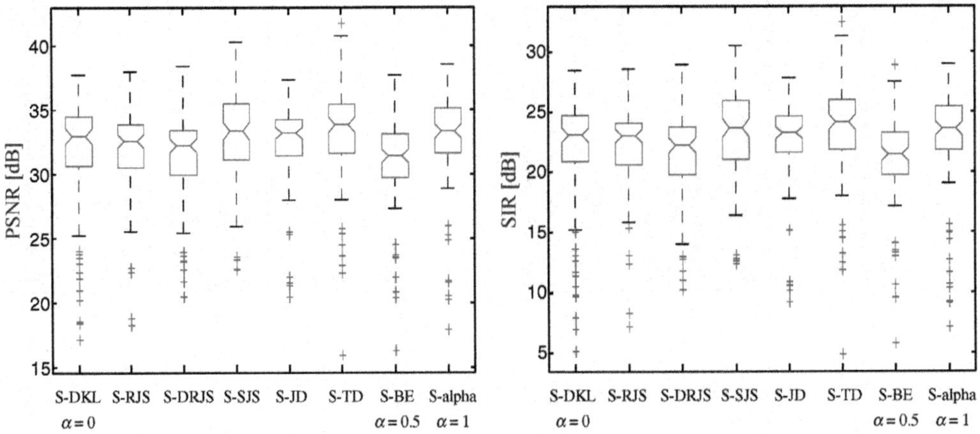

(a) PSNR (b) SIR

Figure 3.16 MC analysis over 100 trials on the `X_spectra` dataset for the SMART NMF algorithms.

Listing 3.4 Monte Carlo Analysis for Alpha NMF algorithm.

```
% This example illustrates Monte Carlo analyis for NMF with the Matlab
% Distributed Computing toolbox
%
% MC analysis with 100 trials for 10 sources using the ALPHA NMF algorithm.
% ALS initialization was employed.
%% Generate mixtures
clear
load('X_spectra_sparse_sorted.mat');
[J,T] = size(X);
I = 10;
A = rand(I,J);
Y = A*X;
%% MC analysis
nmf_algname = 'nmf_alpha';                                      % NMF algorithm
Ntrials = 100;                                                % Number of trials
mc_options = {Y};
nmf_options = struct('J',J,'init',[4 1],'algtype',[1 1],...
    'alpha',1,'tol',1e-6,'niter',1000);
[results,rtime] = mc_nmf(Ntrials,pwd,nmf_algname,nmf_options,mc_options,2);
```

Listing 3.5 Monte Carlo Analysis with the multi-layer technique.

```
%% MC analysis over 100 trials with 4-layer Alpha NMF algorithm
nmf_algname = 'nmf_multi_layer';
Ntrials = 4;                                                  % Number of trials
mc_options = {Y,X,1};
nmf_options = struct('J',J,'init',[1 1],'algtype',[1 1],...
    'alpha',1,'tol',1e-6,'niter',1000,...
    'algorithm','nmf_alpha','Nlayer',4);                % for multilayer NMF
[results,rtime] = mc_nmf(Ntrials,pwd,nmf_algname,nmf_options,mc_options,6);
```

Example 3.6 *Offset data with the* `X_spectra` *dataset*

This example is quite similar to Example 3.1 but with a key difference that one source of the benchmark `X_spectra` took the role of an offset (base line). Fifty randomly generated mixtures were composed from four nonnegative sources, and one base line. In fact, this is a special case of the mixing matrix $\mathbf{A} \in \mathbb{R}_+^{50\times 5}$. All elements of the fifth column vector $\boldsymbol{a}_5$ are represented by the same constant. We fixed $a_{i5} = 10, i = 1, 2, \ldots, I$, whereas the other entries $a_{i,j}$, $(j \neq 5)$ are uniformly distributed random numbers in a range of [0,1]. Figure 3.19(a) shows five typical mixtures randomly picked up from 50 mixtures. Note that the standard

Table 3.6 MC analysis over 100 trials. Performances of SMART NMF algorithms for the `X_spectra_sparse` dataset with 4 layers (maximum 1000 iterations/layer).

Algorithm	PSNR [dB]			SIR	FIT	Time (seconds)	
	median	worst	best	[dB]	(%)	SC [a]	DC [b]
S-DKL	32.93	17.10	37.75	23.02	99.95	8.25	3.50
S-RJS	32.55	18.21	37.98	22.95	99.97	5.37	**2.82**
S-DRJS	32.20	20.41	38.40	22.16	99.97	9.05	3.87
S-SJS	33.33	**22.52**	40.29	23.61	99.99	8.70	3.74
S-JD	33.17	20.40	37.35	23.20	99.97	9.20	3.91
S-TD	**33.80**	15.88	**41.69**	**24.08**	**99.98**	**5.06**	2.91
S-BE	31.37	16.24	37.70	21.42	99.95	8.29	3.64
S-alpha	33.28	17.86	38.52	23.55	99.98	5.38	3.11

[a] Sequential Computing,
[b] Distributed Computing

Table 3.7 Performances of multiplicative Alpha, Beta and TD algorithms for the large dimensional datasets ($1000 \times 1000, 2000 \times 1000, \ldots, 10000 \times 1000$) using both the regular NMF (full matrices) and large-scale NMF (block reduced dimension matrices). Only 20 rows and columns have been randomly selected using uniform distribution.

No. rows	PSNR Full	PSNR Block	Run time Full	Run time Block	PSNR Full	PSNR Block	Run time Full	Run time Block
	Alpha NMF $\alpha = 0.5$				Alpha NMF $\beta = 1$			
1000	34.30	31.16	24.26	5.03	34.42	28.80	9.13	3.61
2000	36.81	30.41	48.72	5.87	34.18	30.82	17.98	4.02
3000	24.18	30.99	72.95	6.26	26.68	28.59	26.92	4.38
4000	28.82	30.43	97.83	7.02	26.34	29.98	35.65	4.74
5000	27.50	30.90	122.94	7.83	24.50	29.69	44.54	5.12
6000	21.32	29.71	148.00	8.58	28.25	28.30	54.17	5.45
7000	25.25	28.77	173.58	9.65	24.70	30.10	62.85	6.14
8000	25.63	30.56	196.55	10.59	26.68	29.61	73.23	6.57
9000	25.05	31.27	220.97	11.30	26.41	29.21	81.45	6.97
10000	28.06	32.02	246.53	12.18	26.22	33.14	89.62	7.40
	Alpha NMF $\alpha = 2$				Alpha NMF $\beta = 0.5$			
1000	33.96	29.67	11.08	3.61	32.74	30.83	12.75	4.40
2000	33.75	31.09	21.91	3.93	34.15	29.55	25.12	4.91
3000	24.21	28.47	34.90	4.23	22.76	29.37	37.64	5.41
4000	24.49	31.38	43.69	4.51	22.1	30.86	50.12	6.00
5000	20.92	30.33	54.83	4.83	27.37	27.30	63.03	6.60
6000	24.83	30.66	66.14	5.09	28.32	30.50	75.49	7.28
7000	27.82	30.68	76.79	5.69	22.92	32.11	88.20	8.21
8000	23.61	29.72	88.03	6.07	25.33	29.51	101.74	9.03
9000	22.99	31.32	99.27	6.38	25.32	29.62	113.99	9.71
10000	21.34	29.10	110.80	6.78	24.72	32.31	127.87	10.71
	Alpha NMF $\beta = 1$				TD NMF			
1000	21.93	31.71	11.64	4.13	26.20	30.09	12.72	3.46
2000	22.05	31.78	23.06	4.61	26.93	31.77	25.27	3.47
3000	25.28	25.90	34.75	5.08	26.10	29.26	33.69	3.87
4000	22.97	30.11	46.30	5.67	27.41	29.16	44.82	4.24
5000	24.61	32.68	58.40	6.24	28.21	29.94	57.01	4.63
6000	26.33	32.27	69.61	6.81	28.58	30.51	67.44	5.11
7000	23.61	29.71	81.37	7.84	28.64	30.80	78.59	5.77
8000	23.22	27.05	93.25	8.59	27.31	30.59	92.57	6.20
9000	25.34	30.05	105.15	9.36	27.01	29.99	100.48	6.63
10000	23.80	30.62	119.03	10.28	28.98	31.91	113.57	7.16

NMF algorithms fail, they cannot reconstruct original sources factorizing this data. The results obtained by using the ISRA algorithm for NMF are displayed in Figure 3.19(b). However, the modified ISRA algorithm for aNMF allows us to find original components almost perfectly as it is illustrated in Figure 3.19(c). The PSNR indices for reconstructed sources are: [25.47, 45.78, 29.95, 50.78, 54.09] dB. MATLAB code is given in Listing 3.7.

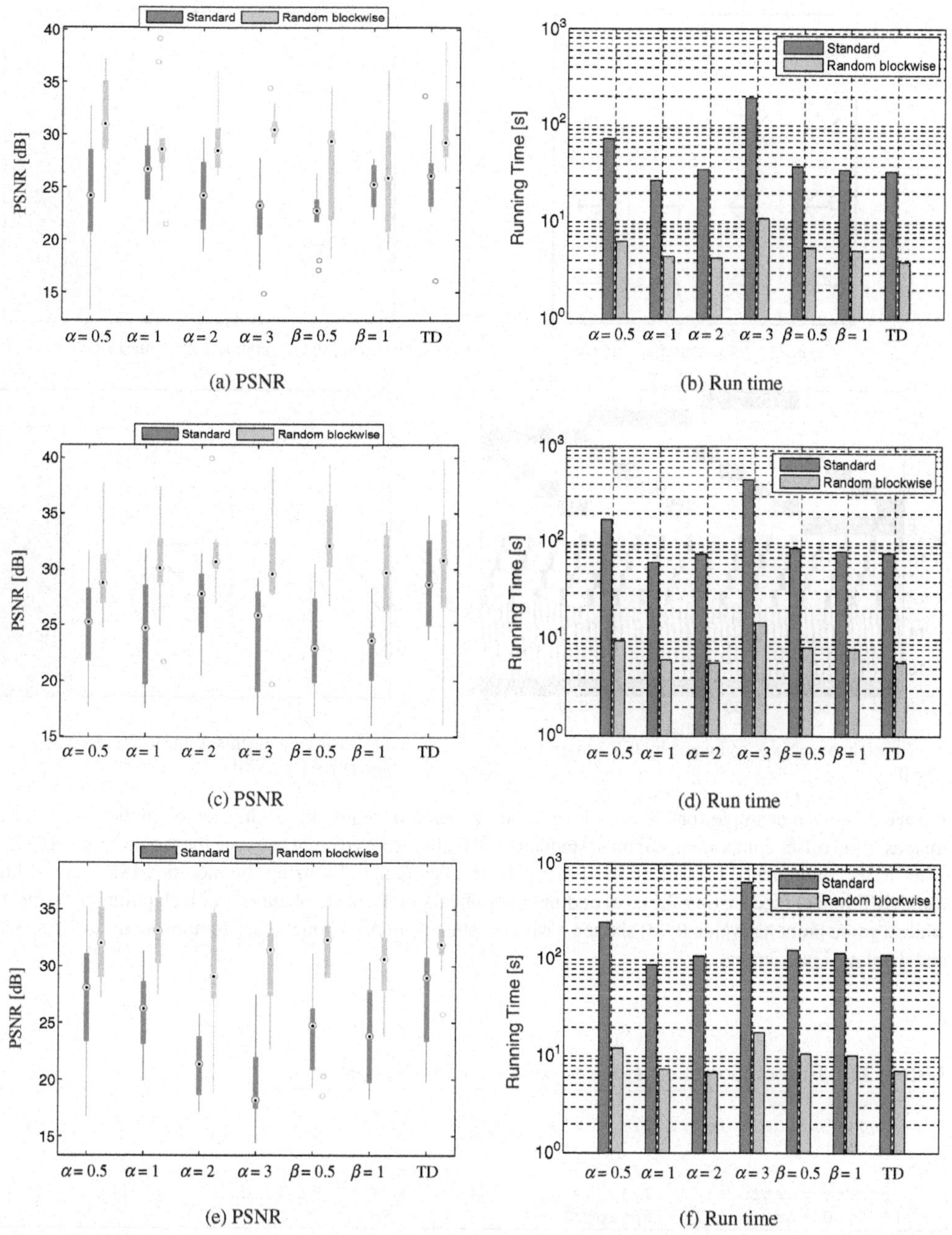

Figure 3.17 MC analysis for a large-scale problem for the multiplicative Alpha, Beta and TD algorithms with full matrix processing procedure and block-wise processing procedure: (top) 3000×1000, (middle) 7000×1000, and (bottom) 10000×1000 dimensional data matrix $\mathbf{Y}$.

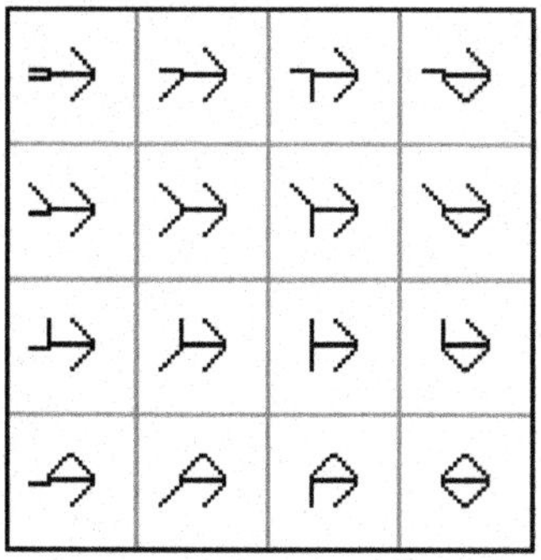

(a) The `swimmer` benchmark

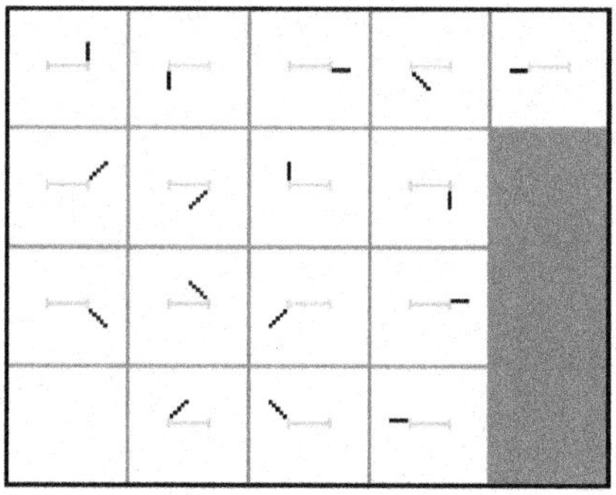

(b) Basis components obtained by using ISRA algorithm

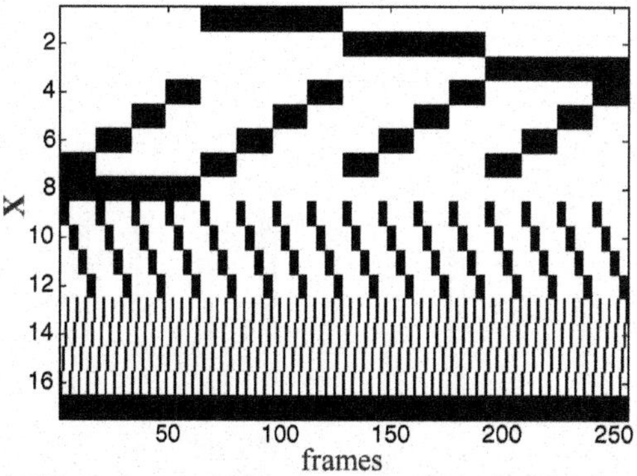

(c) The coding matrix **X** estimated by ISRA algorithm for aNMF

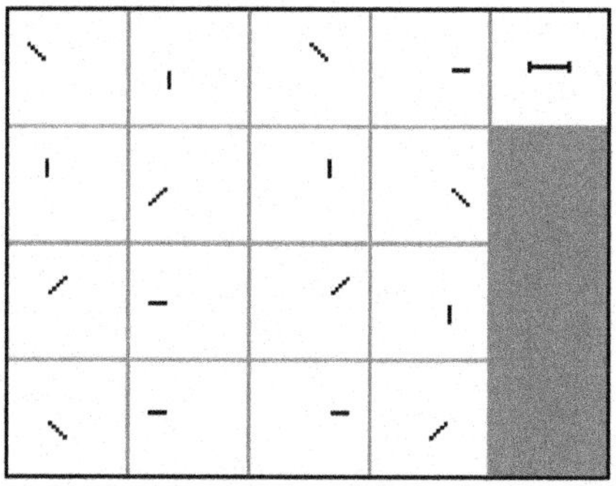

(d) Basis components **A** and $\boldsymbol{a}_0$ obtained by ISRA algorithm for aNMF

Figure 3.18 An example for offset data with the `swimmer` benchmark: (a) the "torso" in the center of all images is the offset component; (b) most standard NMF algorithms cannot separate the "torso" part from the other basis components. Factorization of the `swimmer` benchmark by using the modified ISRA algorithm for aNMF model (see Listing 3.6); (c) coding components of 17 basis pictures. A black point has value 1, while a white point has value 0; (d) sixteen basis components in **A** and one offset component $\boldsymbol{a}_0$ are arranged and displayed in columns.

Listing 3.6 Factorization of the `swimmer` benchmark for Example 3.5.

```
clear
load('swimmer.mat'); Y = max(eps,Y);
J = 17;       %
% Factorize by using the ISRA algorithm for aNMF
options = struct('J',J-1,'maxIter',1000,'lortho',[1 0],'verbose',1);
[A,X,a0]=anmf_ISRA(Y,options);
```

3.10 Music Analysis and Decomposition Using Convolutive NMF

In this section we present application of convolutive NMF to music analysis and decomposition of sound into basic notes [50,55]. The analyzed data for the CNMF model is the spectrogram of the observed sound

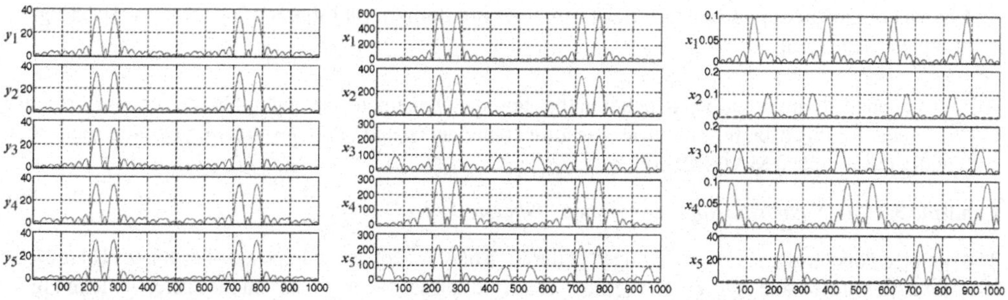

(a) Five samples of data matrix **Y**   (b) ISRA algorithm for NMF   (c) ISRA algorithm for aNMF

Figure 3.19 Blind source separation for 50 nonnegative mixtures of 5 sources. One source took the role of a base line with its fixed mixing values. (a) 5 mixtures; (b) the 5 inaccurate components estimated by the ISRA algorithm with the NMF model. The base line component interferes with all the other ones; (c) results obtained by the modified ISRA algorithm for aNMF.

Listing 3.7 Example 3.6.

```
clear
load('X_spectra_sparse_sorted.mat');
[J,T] = size(X); J = J-1;     % number of basis components
I = 10;                % number of mixtures
A = [rand(I,J) 10*ones(I,1)];
Y = A*X + eps;

% Initialize A, X and a0
Ah = rand(I,J); Xh = rand(J,T); x0 = rand(1,T);
err = inf;
% ISRA algorithm for aNMF
while err > 1e-1
    opts= struct('J',J,'maxIter',1000,'A',Xh','X',Ah','a0',x0','verbose',1);
    [Ah,Xh,x0,err]=anmf_ISRA(Y',opts);
    [Ah,Xh,x0] = deal(Xh',Ah', x0');
end
```

sequence. The purpose of the examples is to explain the sampled sounds based on their spectrograms by as few basis features as possible. The outline of an analysis is as follows:

1. **Compute the spectrogram** of the sound sequence. This spectrogram should be converted to the log-frequency resolution instead of the linear one. The constant Q transform [1] can help to do that. The result of this step is a nonnegative matrix **Y** (the magnitude spectrogram - frequency bin × time frame).
2. **Choose a suitable CNMF model** for the spectrogram involving the number and size of feature components. A basis component $\mathbf{A}^{(j)}$ should cover fully the length of a note: number of rows expresses the number of frequency bins, and number of columns expresses the number of time frames.
3. **Decompose the spectrogram** using a suitable CNMF algorithm with constrainted parameters for sparsity, uncorrelatedness, till we achieve the best FIT ratio possible. The result of this step is that we obtain matrices $\mathbf{A}(j)$ (basis features) and an activating matrix **X** for the 1-D CNMF model or set of matrices $\mathbf{X}^{(j)}$ for the 2-D CNMF model.
4. **Build the individual spectrograms of basis components** $\mathbf{Y}^{(j)} = \mathbf{A}^{(j)} * \underline{\mathbf{x}}_j$ (1-D CNMF) or $\mathbf{Y}^{(j)} = \mathbf{A}^{(j)} * \mathbf{X}^{(j)}$ (2-D CNMF). Convert them into the linear-frequency resolution if necessary.

5. **Convert magnitude spectrograms back into the time-domain real signal** using the phase spectrogram of the original signal.

All steps of analysis are described in detail in the subsequent examples, including: how to choose a suitable CNMF model for this data, how to retrieve feature components, and build the monotonic sound sequences, and finally the complete MATLAB codes.

Example 3.7 *Decomposition of the sequence "London Bridge"*

The sampled popular song "London Bridge" of duration of 4.5 seconds is composed of five notes D4 (62), E4 (64), F4 (65), G4 (67) and A4 (69). In Figure 3.20(a), we can see the five notes on the score, while in Figure 3.20(b), the piano-roll of this segment is displayed. For simplicity and convenience of visualization, the signal was sampled at only 8 kHz and filtered by using a bandpass filter with a bandwidth of 240 − 480 Hz. The spectrogram of this signal was computed by applying short time Fourier transform (STFT) with a 512 point sliding Hanning window and 50% overlap. The obtained 2-D spectrogram had 257 frequency bins in the range 0 Hz to 4000 Hz and 141 time frames. However, the meaningful frequencies are in the range of 240–480 Hz due to the band pass filtering. Figure 3.20(c) shows this spectrogram in the active frequency range; observe that the notes D4, E4, F4, G4 and A4 have their fundamental frequencies at 293.66 Hz, 329.63 Hz, 349.23 Hz, 391.99 Hz and 440.0000 Hz, respectively. Each rising part in the spectrogram (Figure 3.20(c)) corresponds to the note actually played. The note G4 occurs four times in the spectral plane, i.e. exactly four times in the music score (see Figure 3.20(a)) and in the piano roll (Figure 3.20(b)). From the CNMF viewpoint, the spectral part which represents the note G4 can be interpreted as a basis component $\mathbf{A}^{(j)}$, and its piano roll can be represented as an activating map $\underline{\boldsymbol{x}}_j$. Since the music sequence has five notes, we can represent its spectrogram by five matrices $\mathbf{A}^{(j)}$. Every frame window was computed based on 512 samples with 50% overlap (256 samples); hence, in one second interval there are 31.25 (= 8000/256) time frames. Every note lasts for about 1/3 second, or covers 10 ($\approx 31.25/3$) time frames, thus can be expressed by a basis component $\mathbf{A}^{(j)}$ with $P = 10$ columns.

Finally, for such a magnitude spectrogram $\mathbf{Y}$ of 257 frequency bins $\times$ 141 time frames, we can choose a CNMF model with five feature components $\mathbf{A}^{(j)}$ of size 257×10, $j = 1, \ldots, 5$, and $\mathbf{X}$ of size 10×141. MATLAB code of this decomposition is given in Listing 3.8. The results obtained using the Beta CNMF algorithm with the sparsity constraint for $\mathbf{X}$: $\lambda_{\mathbf{X}} = 0.001$ are shown in Figures 3.21(c)–3.21(g) for matrices $\mathbf{A}^{(j)}$, and in Figure 3.21(h) for the coding matrix $\mathbf{X}$. Each component characterizes each individual note; this leads to the ability to recognize notes actually being played in a song by matching the active frequency of a component with a fundamental frequency of a corresponding note. The distribution of the entries in the

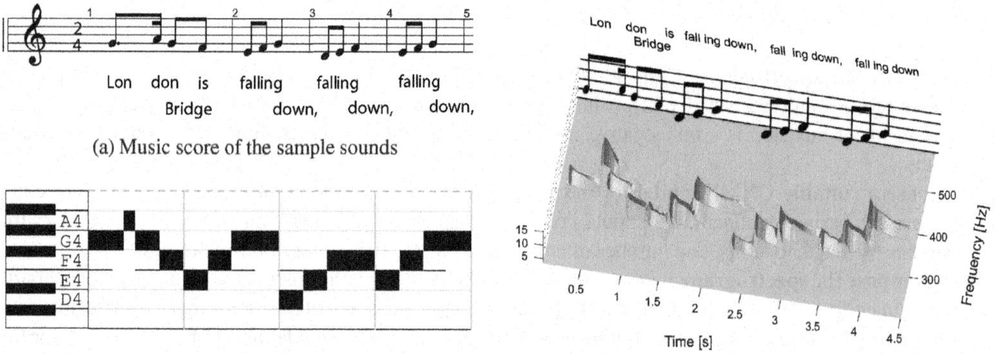

(a) Music score of the sample sounds

(b) Piano roll of the sample sounds (c) 2D spectrum of the analyzed sounds matching with the notes

Figure 3.20 The analysis of the sounds in Example 3.7.

Listing 3.8 Deconvolution of the sample sequence `London_bridge` in Example 3.7.

```
% CNMF example for the sample sequence 'London_bridge.wav', fs = 8kHz.
% The source was composed of 5 notes A4, G4, F4, E4 and D4 in 4.5 seconds.
% The CNMF model was selected with 5 components A^{(j)} having 10 columns.
%% Read sound source (wav file)
filename   = 'Londonbridge_bp_240_480.wav';
[y, fs, nbits] = wavread(filename);y = y(:,1);

%% Set parameters for computing spectrogram
Nfft = 512;   Noverlap = .5  * Nfft;
window = hamming(Nfft, 'periodic');                          % or Hanning window

%% Zeropad source y
leny = length(y); newleny = leny - mod(leny, Nfft - Noverlap) + Nfft;
y(newleny) = 0;

%% Compute spectrogram
[Y, f, t] = spectrogram(y, window, Noverlap, Nfft, fs);
Ym = abs(Y); Ym = Ym/max(Ym(:));                          % magnitude spectrogram
[Nfreqbin,Ntimefr] = size(Ym);

%% Layer 1: decompose the magnitude spectrogram using Beta CNMF
cnmfopt = struct('Nobjects',5,'WinSize',10, 'Niter', 120, 'beta',1,...
          'lambdaX', .001, 'lambdaA',0,'tol',1e-6,'Xrule',2);
cnmfopt.Xinit = rand(cnmfopt.Nobjects,Ntimefr);
cnmfopt.Ainit = rand(Nfreqbin,cnmfopt.Nobjects,cnmfopt.WinSize);

[A,X] = betaCNMF(Ym,cnmfopt);

%% Layer 2: Beta CNMF with update rule 1 for X
cnmfopt.Ainit = A;    cnmfopt.Xinit = X;
cnmfopt.Niter = 500;  cnmfopt.Xrule = 1;

[A,X,objective] = betaCNMF(Ym,cnmfopt);

%% Display the basis components
figure
for k = 1:cnmfopt.Nobjects
    subplot(1,cnmfopt.Nobjects,k)
    ha = imagesc(t(1:cnmfopt.WinSize),f(froi),squeeze(A(froi,k,:)));axis xy;
    colormap(flipud(gray)); xlabel('Time [s]');ylabel('Frequency [Hz]');
    title(sprintf('${\\bf A}^{(%d)}$',k),'interpreter','latex')
end

%% Visualize the coding map X
figure
imagesc(t,1:cnmfopt.Nobjects,X); set(gca,'Ytick',1:cnmfopt.Nobjects);
colormap(flipud(gray)); xlabel('Time [s]'); ylabel('Components')
title('$\bf X$','interpreter','latex')
```

activating matrix **X** reflects the true piano roll as illustrated in Figure 3.20(b). It means that using CNMF we can detect when each individual note was played in a music sequence. In other words, we can code a song by its basis note and therefore, we are able to re-write the score of the song (Figure 3.20(a)). Figure 3.21(b) shows the reconstructed spectrogram from five basis nonnegative components and their coding map is very similar to original music sequence with high performance of reconstruction with PSNR = 51.84 [dB]. This means that the decomposition is very precise in this case. Finally, by using the phase information of the original data, we can convert spectrograms of basis components back into the original waveform. A simple code is given is Listing 3.9.

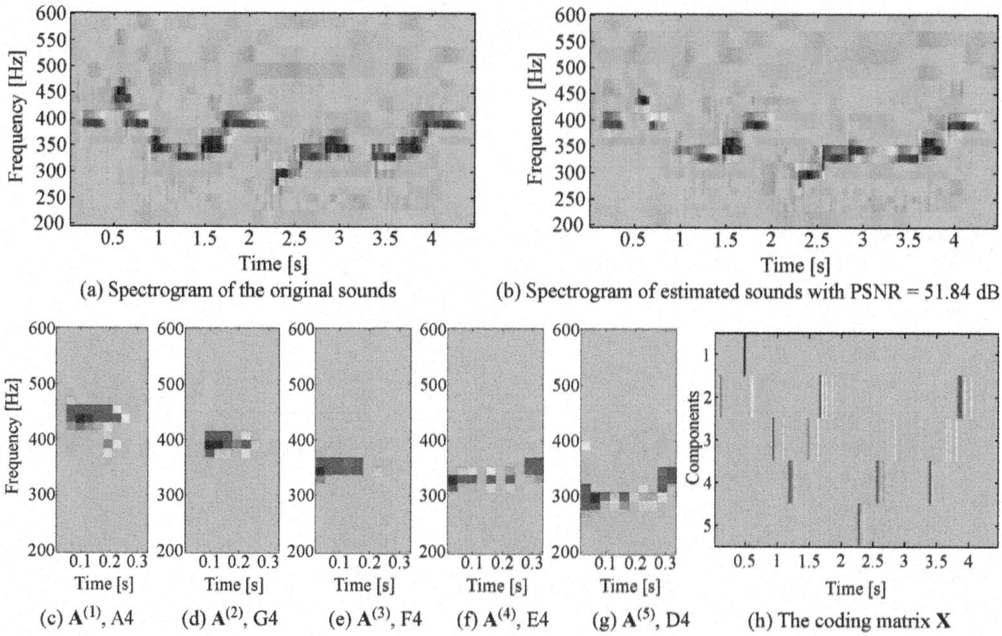

Figure 3.21 Illustration of the basis features for Example 3.7: (a)–(b) spectrogram comparison of the source and the approximation, PSNR = 51.84 dB; (c)–(g) Five basis spectral components $\mathbf{A}^{(j)}$ displayed in the frequency band of 200-600 Hz, characterizing the five notes; (h) Coding matrix $\mathbf{X}$ for five components is very similar to the piano roll of the analyzed song in Figure 3.20(b).

Example 3.8 *Analysis and decomposition for multi harmonic frequencies played on a guitar*

In the second example we perform a similar sound analysis and decomposition for multi harmonic frequencies played on a guitar. The same song as in the previous example was selected but with different basis notes A3, G3, F3, E3 and D3 (sampled at the same frequency rate of 8 kHz). The music score and the piano roll for this example are displayed in Figures 3.22(a) and 3.22(b). The spectrogram of this sound contains the fundamental and harmonic frequencies of the notes (see Figure 3.22(c)). Using the STFT with a 512 point sliding Hanning window and 50% overlap (256 samples), we obtained the spectrogram $\mathbf{Y}$ with 257 linear frequency bins and 152 time frames. The spacing between any two consecutive harmonics is constant for a specified MIDI note, but varies with fundamental frequency. The result is that it is more difficult for the problem of fundamental frequency identification. For the CNMF model, a basis feature component characterizes a spectral component with a fundamental frequency. The relative positions of these frequency components to each other needs to be constant while shifting them along the frequency axis or time axis through the same operator $\mathbf{T}_q$ (see Chapter 3 for more detail). A more suitable spectrogram with log-frequency resolution can be produced by applying the constant Q transform [1]. Or we can use a linear transform $\mathbf{R}$ which maps a linear-frequency spectrogram to log-frequency one [28,47].

The inverse mapping allows us to approximate the original linear spectrogram:

$$\mathbf{Y}_{lg} = \mathbf{R}\,\mathbf{Y} \quad \text{and} \quad \mathbf{Y} = \mathbf{R}^T\,\mathbf{Y}_{lg}. \tag{3.218}$$

The constant Q transform has geometrically spaced center frequencies $f_k = f_0\, 2^{\frac{k}{b}}$, $(k = 0, 1, \ldots)$, where f_0 is the minimum frequency of the log-frequency spectrogram, and b dictates the number of frequency bins

Listing 3.9 Reconstruction of basis sound components of the sample sequence `London_bridge` in Example 3.7.

```
%% Inverse spectrogram and construct the feature components
fname = 'Londonbridge_';
orienergy = norm(y);
phase = angle(Y);                                          % use the original phase
Yhat = 0;
for k = 1:cnmfopt.Nobjects
    Yk = conv2(squeeze(A(:,k,:)), X(k,:))+1e-9;
    Yk = Yk(1:Nfreqbin,1:Ntimefr).* exp(1i*phase);
    Yhat = Yhat + Yk;
    yk = ispecgram(Yk,Nfft,fs,window,Noverlap);
    yk = yk*orienergy/norm(yk);
    wavwrite(yk(1:leny),fs,nbits,sprintf('%s_component_%02d.wav',fname,k));
end
yhat = ispecgram(Yhat,Nfft, fs, window, Noverlap);
yhat = yhat*orienergy/norm(yhat);
wavwrite(yhat(1:leny), fs, nbits, [filename '_full.wav']);

%% Visualize the estimated spectrogram
figure
imagesc(t,f(froi),abs(Yhat(froi,:)));
colormap(flipud(gray)); axis xy;
xlabel('Time [s]');ylabel('Frequency [Hz]');
title('Reconstructed spectrogram $\hat {\bf Y}$','interpreter','latex')
```

per octave. This yields a constant ratio of frequency to resolution Q (otherwise called the "quality factor")

$$Q = \frac{f_k}{f_{k+1} - f_k} = (2^{\frac{1}{b}} - 1)^{-1}. \tag{3.219}$$

The number of frequency bins N or the first dimension of the transformation matrix $\mathbf{R}$ is computed as

$$N = \left\lceil b\,(\log_2(f_N) - \log_2(f_0))\right\rceil, \tag{3.220}$$

where f_N is the maximum frequency. It means that I frequency bins of the linear resolution spectrogram $\mathbf{Y}$ will be composed linearly and return N frequency bins for the log-resolution spectrogram $\mathbf{Y}_{lg}$. For the above spectrogram, a linear transformation with quality factor $Q = 100$ converts 257 linear-frequency bins into $N = 364$ log-frequency bins in the frequency range from $f_0 = 109.4$ Hz (bin 8) to $f_N = f_s/2 = 4000$ Hz (bin 257) with $b = 70$ bins per octave; whereas a transformation with quality factor $Q = 50$ forms a spectrogram

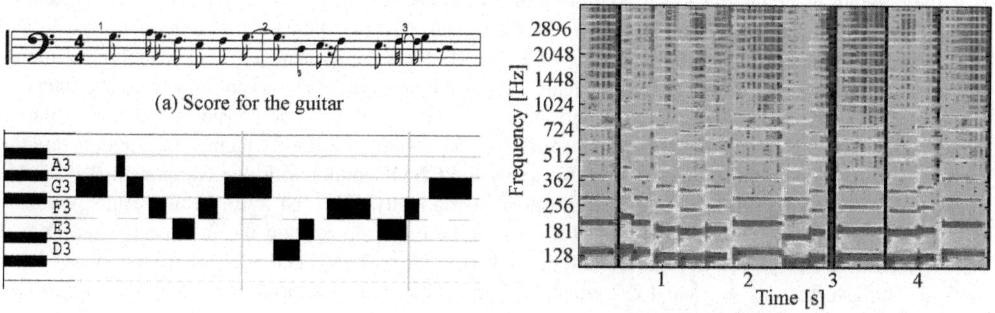

(a) Score for the guitar

(b) Piano roll

(c) Log-frequency spectrogram of the original source

Figure 3.22 The music score, piano roll and log-frequency spectrogram for Example 3.8.

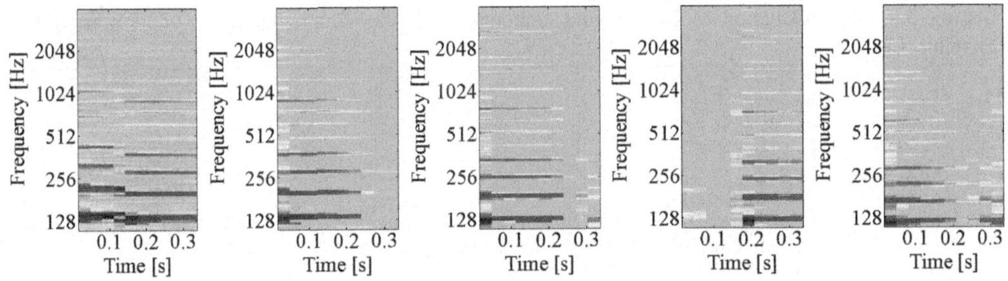

(a) $\mathbf{A}^{(1)}$, A3 (b) $\mathbf{A}^{(2)}$, G3 (c) $\mathbf{A}^{(3)}$, F3 (d) $\mathbf{A}^{(4)}$, E3 (e) $\mathbf{A}^{(5)}$, D3

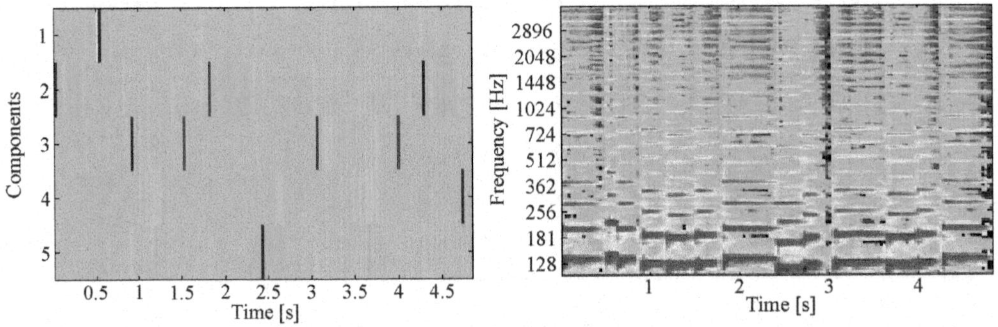

(f) The coding matrix $\mathbf{X}$ (g) Estimated log-frequency spectrogram with PSNR = 42.79 [dB]

Figure 3.23 The decomposition of the spectral data of size 364×152 for Example 3.8: (a)–(e) five basis components $\mathbf{A}^{(j)}$ of size 364×10 in the log-frequency scale correspond to the five notes played in the song A3, G3, F3, E3 and D3; (f) the horizontal coding matrix $\mathbf{X}$ of size of 5×152 for five basis components $\mathbf{A}^{(j)}$, $j = 1, \ldots, 5$ expresses the piano roll in Figure 3.22(b); (g) the spectrogram was reconstructed from its five basis components with high reconstruction performance (PSNR = 42.79[dB]) compared to the original one shown in Figure 3.22(c).

of $N = 187$ log-frequency bins in the similar frequency range. Figure 3.22(c) illustrates the log-frequency spectrogram of the analyzed sound signal. This spectral matrix $\mathbf{Y}$ (364×152) is considered as the input data for the CNMF model for estimating five basis nonnegative components. Using the same parameter values as in the previous example, the 364×10 dimensional matrices $\mathbf{A}^{(j)}$ ($j = 1, \ldots, 5$) are obtained and shown in Figures 3.23(a)–3.23(e). We finally converted them back into the standard linear-frequency scale and saved them as matrices of smaller dimensions of 257×10. Each component $\mathbf{A}^{(j)}$ contains all its harmonic frequencies (see Figures 3.23(a)–3.23(e)) which are shifted in frequency by a multiple of octave intervals higher than the fundamental frequency. Therefore, component $\mathbf{A}^{(j)T}$ can be considered as a convolutive source formed by a basis feature $\mathbf{B}^{(j)}$ which shifts vertically along the log-frequency axis, and covers the fundamental frequency bin. In such a case, we apply the CNMF model with one component $\mathbf{B}^{(j)}$ of size 10×10 and one coding vector $\boldsymbol{x}_v^{(j)}$ to the transpose of the matrix $\mathbf{A}^{(j)}$ 364×10. In the results, we have five small basis components $\mathbf{B}^{(j)}$ of size 10×10 which are able to explain the full spectrogram $\mathbf{Y}_{lg}$ of 364×152, and are displayed in Figures 3.24(a), 3.24(d), 3.24(g), 3.24(j) and 3.24(m). Two horizontal and vertical coding vectors $\boldsymbol{x}_h^{(j)} = \underline{\boldsymbol{x}}_j^T$ and $\boldsymbol{x}_v^{(j)}$, which correspond respectively to the time and frequency axes can be combined to form a common coding matrix:

$$\mathbf{X}^{(j)} = \boldsymbol{x}_v^{(j)} \, \boldsymbol{x}_h^{(j)T}, \quad j = 1, 2, 3, 4, 5. \tag{3.221}$$

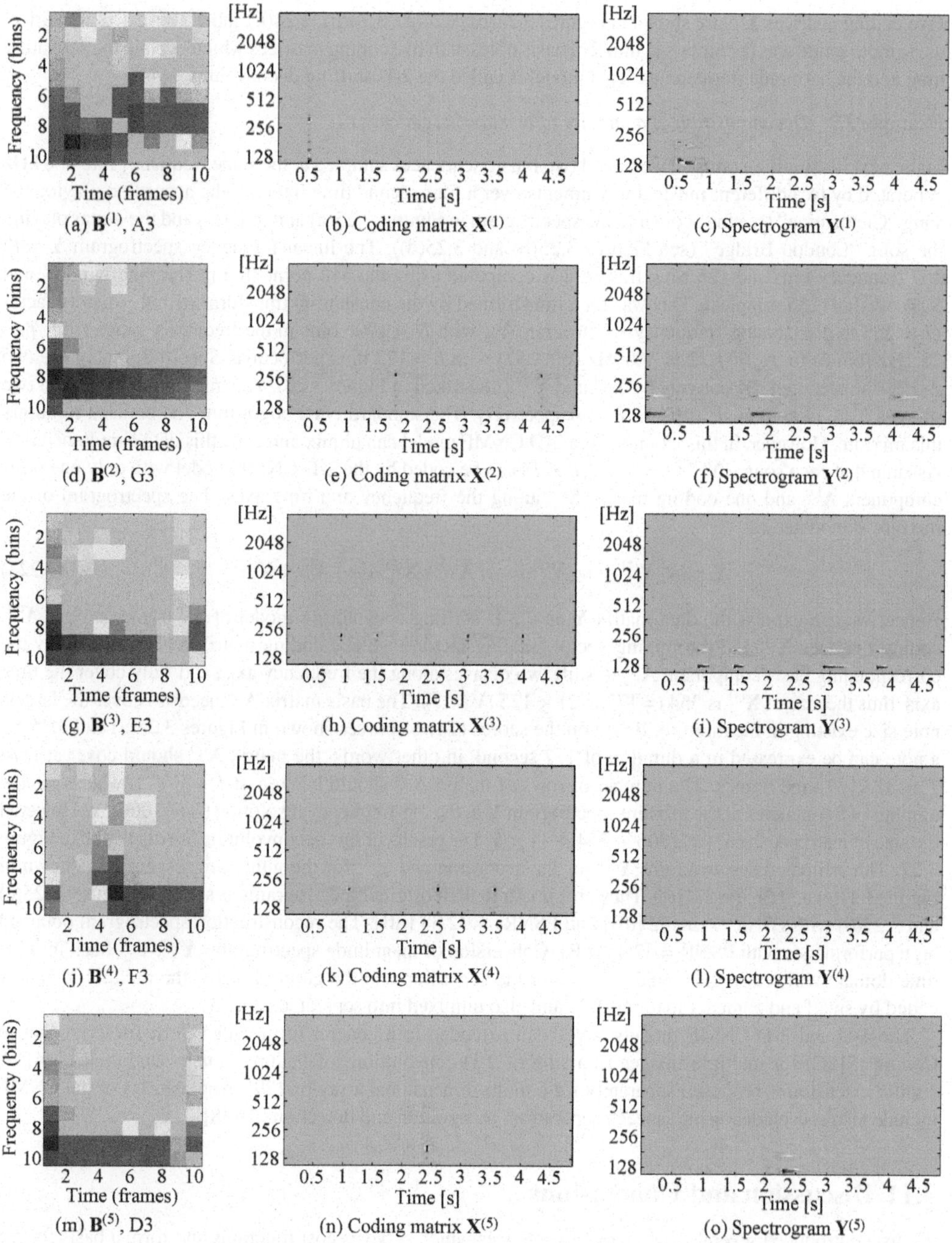

Figure 3.24 The 2-D shifting decomposition for Example 3.8: (left column) the five small basis components $\mathbf{B}^{(j)}$, $j = 1, \ldots, 5$ of size 10×10; (middle column) the 2-D coding matrices $\mathbf{X}^{(j)}$, $j = 1, \ldots, 5$; (right column) the five linear scale spectral components $\mathbf{Y}^{(j)} = \mathbf{B}^{(j)} * \mathbf{X}^{(j)}$, $j = 1, \ldots, 5$.

Five coding matrices $\mathbf{X}^{(j)}$ are shown in Figures 3.24(b), 3.24(e), 3.24(h), 3.24(k) and 3.24(n). To summarize, the sample sound was decomposed into five basis notes with five coding matrices which explain the activating time and the harmonic frequencies. This model is called the 2-D shifting deconvolution [44].

Example 3.9 *Decomposition of a mixture of two music sequences*

In the next illustrative example, we mixed two music sequences sampled at the same frequency rate of 8 kHz generated by two different musical instruments over a four second time interval: the accordion playing the song "Clementine" (see the score and the spectrogram in Figures 3.25(a) and 3.25(c)) and the guitar playing the song "London Bridge" (see Figures 3.25(b) and 3.25(d)). The linear-frequency spectrogram $\mathbf{Y}$ with 257 frequency bins and 125 time frames was computed using the 512 point sliding Hanning window and 50% overlap (256 samples). This was then transformed by the constant-Q transform with the quality factor $Q = 255$ to give the log-frequency spectrogram $\mathbf{Y}_{lg}$ with $N = 1006$ bins in the frequency range from $f_0 =$ 78.1Hz (bin 6) to $f_N = f_s/2 = 4000$Hz (bin 257) with $b = 177$ bins per octave. Spectrograms of the two sources are denoted respectively by $\mathbf{Y}^{(1)}$ and $\mathbf{Y}^{(2)}$ (illustrated in Figure 3.26). The mixing rates between two sources were chosen as $\boldsymbol{w} = [0.3 \quad 0.7]$ respectively. The standard NMF algorithm can be used to unmix this mixture. However, in this example, the 2-D CNMF model can unmix and code this data simultaneously. Assume that each source $\mathbf{Y}^{(j)}$ $(j = 1, 2)$ $(I \times T)$ can be coded by the 2-D CNMF model with only one basis component $\mathbf{A}^{(j)}$ and one coding matrix $\mathbf{X}^{(j)}$ along the frequency and time axes. The spectrogram of the mixture is modeled as:

$$\mathbf{Y} = w_1\, \mathbf{Y}^{(1)} + w_2\, \mathbf{Y}^{(2)} = w_1\, \mathbf{A}^{(1)} * \mathbf{X}^{(1)} + w_2\, \mathbf{A}^{(2)} * \mathbf{X}^{(2)}. \tag{3.222}$$

Hence, we can express the data matrix $\mathbf{Y}$ as a 2-D shifting convoluting model of the basis matrices $\mathbf{A}^{(j)}$, coding matrices $\mathbf{X}^{(j)}$ and the mixing vector (scaling factors) $\boldsymbol{w}$. The coding matrix $\mathbf{X}^{(j)}$ should allow the corresponding basis component $\mathbf{A}^{(j)}$ to shift two octaves along the frequency axis, and fully cover the time axis, thus the size of $\mathbf{X}^{(j)}$ is 354 $(= 177 \times 2) \times 125$ $(V \times T)$. The basis matrix $\mathbf{A}^{(j)}$ needs to cover the largest note size existing in the mixture. Based on the scores of two sources shown in Figures 3.25(a) and 3.25(b), a note can be expressed in a duration of $1/7$ second, in other words, the matrix $\mathbf{A}^{(j)}$ should cover at least $P > 31.25/7$ time frames. The number of rows of matrix $\mathbf{A}^{(j)}$ should be at least $I - V + 1$, where I is the number of frequencies of the mixture spectrogram $\mathbf{Y}$ in the log-frequency resolution ($I = 1006$). We selected the size of matrix $\mathbf{A}^{(j)}$: $653 (= 1006 - 354 + 1) \times 5$. The results of this deconvolution are displayed in Figure 3.27. The estimated spectrograms $\mathbf{Y}^{(1)}$ for the accordion and $\mathbf{Y}^{(2)}$ for the guitar are respectively shown in Figure 3.27(c) and Figure 3.27(f). They are similar to their original spectrograms as shown in Figures 3.25(c) and 3.25(d) with $\text{PSNR}_1 = 26.72$ [dB] and $\text{PSNR}_2 = 32.38$ [dB]. The reconstruction spectrogram obtained high performance with PSNR = 37.42 [dB]. Conversion of magnitude spectrograms $\mathbf{Y}^{(1)}$, $\mathbf{Y}^{(2)}$ back into the time domain will return the sound sources for the accordion and the guitar. Finally, the mixture $\mathbf{Y}$ is now coded by small and compact features $\mathbf{A}^{(j)}$, and also unmixed into separated sources.

The 1-D and 2-D CNMF models have been extended to a general form called Shift Invariant Sparse Coding (SISC) for multiple images as a sum of 2-D convolutions of features images and codes [44]. A similar extension is proposed separately for a multi-dimensional array by [29]. Applications of the CNMF include also speech denoising, speech separation, recognition and detection [51,58].

3.11 Discussion and Conclusions

We have introduced several generalized divergences which serve as cost functions and form a basis for the development of a class of generalized multiplicative algorithms for NMF. These include the Alpha, Beta, Bregman, and Csiszár divergences [10,12,36]. Extensions of Lee-Seung or ISRA and EMML algorithms have been presented, in order to improve performance and reflect the nature of the components, especially sparsity or smoothness. This has allowed us to reconstruct reliably the original signals and to estimate the mixing matrices, even when the original sources are strongly correlated or stochastically dependent.

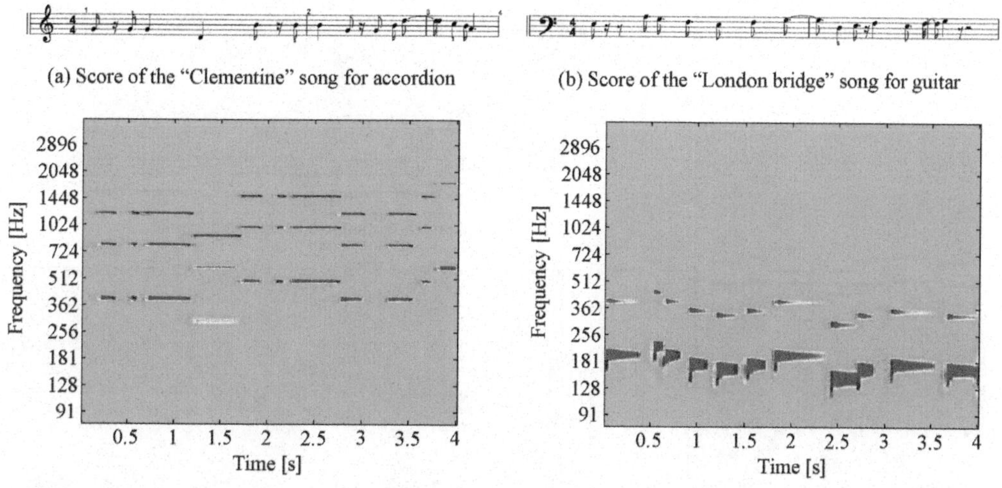

(a) Score of the "Clementine" song for accordion

(b) Score of the "London bridge" song for guitar

(c) Log-frequency spectrogram of the "Clementine"

(d) Log-frequency spectrogram of the "London bridge"

Figure 3.25 Scores and linear scale spectrograms of the two musical sequences in Example 3.9 played by accordion and guitar.

The optimal choice of regularization parameters α and β depends both on the distribution of data and *a priori* knowledge about the hidden (latent) components. The simulation results confirmed that the developed algorithms are efficient and stable for a wide range of parameters. However, if the number of observations is very close to the number of unknown sources the derived algorithms do not guarantee estimation of all the sources and may get stuck in local minima, especially if not regularized. When the number of observations is much larger than the number of sources (nonnegative components) the proposed algorithms give consistent and satisfactory results, especially when the sources or the mixing matrices are sparse. We found that the distributed multi-layer NMF together with the regularization and nonlinear projection terms play a key role in improving the reliability and performance of NMF algorithms.

In summary, the majority of multiplicative algorithms presented in this chapter perform well, especially for very sparse data. The remaining challenges include rigorous proofs of global convergence, and meaningful

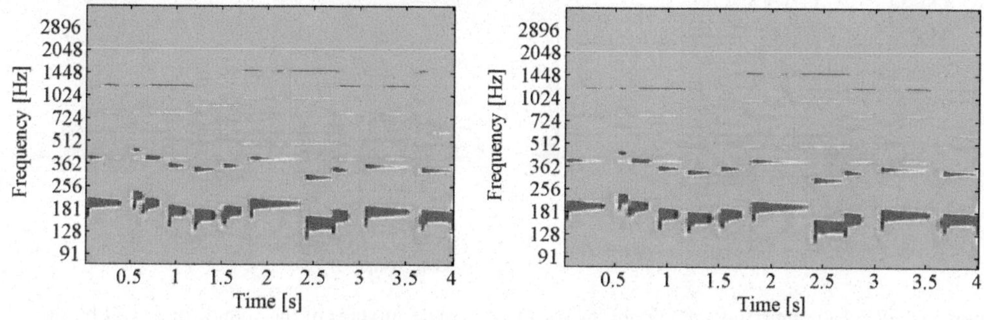

(a) Log-frequency spectrogram of the mixture

(b) Log-frequency spectrogram of the estimation, PSNR = 37.42 [dB]

Figure 3.26 Log-frequency spectrogram of the mixture of sounds in Example 3.9 for the two music sequences: "Clementine" and "London bridge", with the mixing rate $\boldsymbol{w} = [0.3 \quad 0.7]$ during a four seconds period.

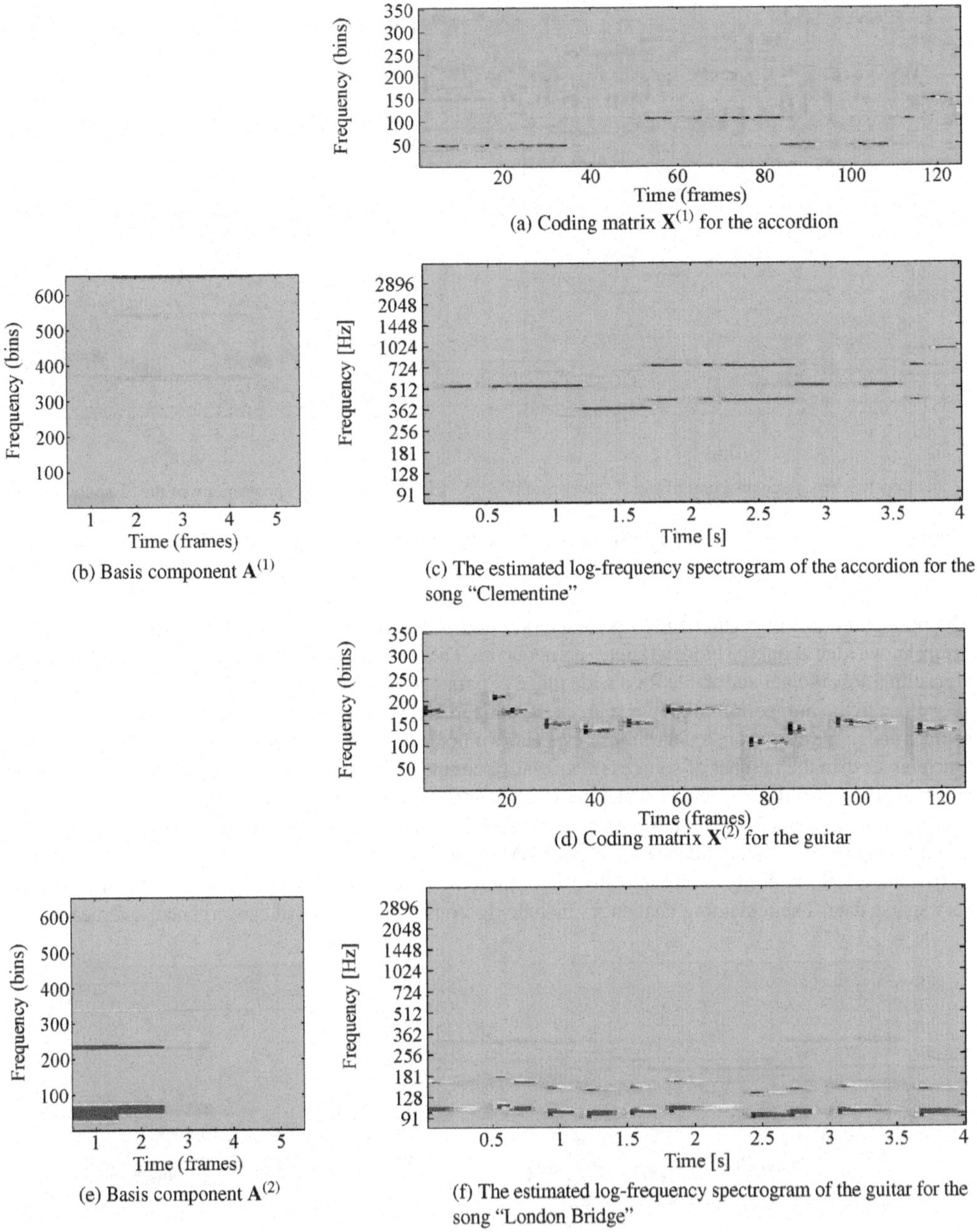

(a) Coding matrix $\mathbf{X}^{(1)}$ for the accordion

(b) Basis component $\mathbf{A}^{(1)}$

(c) The estimated log-frequency spectrogram of the accordion for the song "Clementine"

(d) Coding matrix $\mathbf{X}^{(2)}$ for the guitar

(e) Basis component $\mathbf{A}^{(2)}$

(f) The estimated log-frequency spectrogram of the guitar for the song "London Bridge"

Figure 3.27 The unmixing and coding results for the single mixture of the sounds produced by the two musical instruments in Example 3.9: (a),(b) and (c) the coding matrix $\mathbf{X}^{(1)}$, basis component $\mathbf{A}^{(1)}$ and their reconstruction spectrogram $\mathbf{Y}^{(1)}$ for the accordion in the song "Clementine"; (d),(e) and (f) the same descriptions for the guitar in the song "London Bridge".

physical interpretations of the discovered latent components (or classes of components), especially when the structures of the original sources are completely unknown. Additionally, by extensive computer simulations we have demonstrated that by applying a suitable multi-start initialization and multi-layer approach, we can improve reliability and performance of most multiplicative algorithms. In the following chapters we will show that this approach (the multi-layer system with projected gradient algorithms) has potential to give significantly better performance than the standard multiplicative algorithms, especially if the data are ill-conditioned, badly-scaled, and/or the number of observations is only slightly greater than the number of sources.

Appendix 3.A: Fast Algorithms for Large-scale Data

When the data matrix $\mathbf{Y} \in \mathbb{R}^{I \times T}$ is very large and the number of nonnegative components J is relatively small, that is, $T >> J$ and $I >> J$, we can apply several strategies presented below to accelerate the speed of NMF algorithms and alleviate the problem of storing huge matrices.

In Table 3.7, we demonstrated that for the 9000×1000 dataset stored in $\mathbf{Y}$, the regular multiplicative Beta NMF ($\beta = 1$) algorithm needs 105.15 seconds to meet the stopping criterion, and the multiplicative Alpha NMF (with $\alpha = 3$) algorithm requires 574.57 seconds. In this appendix, we briefly discuss the speed-up strategies.

3.A.1 Random Block-wise Processing Approach – Large-scale NMF

At each iteration step, instead of processing all the entries of data matrix $\mathbf{Y}$ and the estimating matrix $\hat{\mathbf{Y}} = \mathbf{AX}$ we can choose a subset of the columns (or a subset of the rows) of $\mathbf{Y}$ and the same columns and rows for the estimated matrix $\hat{\mathbf{Y}}$. In practice, instead of processing and storing the whole matrix $\mathbf{Y}$ and $\hat{\mathbf{Y}}$ we need to process the matrices $\mathbf{Y}_r$, $\mathbf{Y}_c$, and the corresponding estimating matrices $\hat{\mathbf{Y}}_r = \mathbf{A}_r\mathbf{X}$ and $\hat{\mathbf{Y}}_c = \mathbf{A}\mathbf{X}_c$ of much smaller dimensions. The number of selected columns and rows should be equal or greater than four times the number of basis components J, that is, $C \geq 4J$ and $R \geq 4J$. In such a case, the reconstruction error is usually negligible [53]. The random block-wise (LSNMF) procedure can be summarized as follows:

- Chose a subset of $4J$ indices from a set of I $\{1,2,\dots, I\}$ (or T elements $\{1,2,\dots,T\}$) by the `randperm` MATLAB function.
- Update matrix $\mathbf{X}$ (or $\mathbf{A}$) from $4J$ rows (or columns) of $\mathbf{Y}$ and $\mathbf{A}$ (or $\mathbf{X}$) by a specified NMF algorithm.

The exemplary MATLAB code for the multiplicative Alpha NMF algorithm is given in Listing A.10.

Listing A.10 MATLAB code for block-wise Alpha NMF.

```
% Random block-wise processing for X
Nrow = 4*J;
r = randperm(I);
r = r(1:Nrow);
X = X.*(A(r,:)fl*(Y(r,:)./(A(r,:)*X + eps)).^alphaX).^alphaX;

% Normalize X to unit length
X = bsxfun(@rdivide,X,sum(X,2));

% Random blockwise processing for A
Ncol = 4*J;
c = randperm(T);
c = c(1:Ncol);

A = A.*((Y(:,c)./(A*X(:,c) + eps)).^alphaA *...
    bsxfun(@rdivide,X(:,c),sum(X(:,c),2))fl)^(1/alphaA);
```

3.A.2 Multi-layer Procedure

If we factorize a large data matrix in the first layer into a small number of components and try to achieve the highest FIT ratio, the obtained factors $\mathbf{X}$ (or $\mathbf{A}$) is certainly much smaller dimension than the original data. Processing with the new data (factors) is much faster than dealing with the original data.

- **Dimensionality reduction**. For a given large-scale data matrix $\mathbf{Y}$ and a specified number of basis components J, we perform factorization but with $4J$ basis components and for a very limited number of iterations.
- **Factorization Step**. Apply NMF for the new data to obtain the final basis components.

3.A.3 Parallel Processing

We can exploit the advantage of the system with multicore, multi CPU or GPU by processing data in parallel mode: data or algorithm parallel (see Example 3.4).

- Block-wise processing in many workers simultaneously.
- Average the results.

Appendix 3.B: Performance Evaluation

In order to evaluate the performance and precision of the NMF algorithms we used three different criteria:

1. Signal-to-Interference-Ratio (SIR).
2. Peak Signal-to-Noise-Ratio (PSNR).
3. Fitting or the explained variation (FIT).

3.B.1 Signal-to-Interference-Ratio - SIR

Signal-to-Interference-Ratio (SIR) is often used to evaluate the ratio between the power of the true signal and the power of the estimated signal corrupted by interfrence. A noisy model of scaled signal $\boldsymbol{x}$ can be expressed as follows

$$\hat{\boldsymbol{x}} = a\boldsymbol{x} + e, \tag{B.1}$$

where a is a scaling parameter. The SIR index of $\hat{\boldsymbol{x}}$ and $\boldsymbol{x}$ is given by

$$SIR(\hat{\boldsymbol{x}}, \boldsymbol{x}) = 10\log_{10}\left(\frac{\|a\boldsymbol{x}\|_2^2}{\|\hat{\boldsymbol{x}} - a\boldsymbol{x}\|_2^2}\right). \tag{B.2}$$

For the NMF model, the SIR index can be derived from the estimated matrices $\hat{\mathbf{A}}$ and $\hat{\mathbf{X}}$, since we have

$$\mathbf{Y} = \mathbf{A}\mathbf{X} \approx \hat{\mathbf{A}}\hat{\mathbf{X}} \tag{B.3}$$

$$\hat{\mathbf{X}} = (\hat{\mathbf{A}}^{\dagger}\mathbf{A})\mathbf{X} + \mathbf{E} = \mathbf{G}\mathbf{X} + \mathbf{E} \tag{B.4}$$

Note that for perfect estimation of matrix $\mathbf{A}$ (neglecting arbitrary scaling and permutation of columns), the matrix $\mathbf{G}$ will be a generalized permutation matrix with only one nonzero element in every row and column. Thus, an estimated signal $\hat{\boldsymbol{x}}_j$ can be expressed by an original source $\boldsymbol{x}_k$ and maximal absolute value element g_{jk} of row $\boldsymbol{g}_{j:}$. Therefore, the SIR value of an estimated signal $\hat{\boldsymbol{x}}_j$ can be expressed as (see Listing B.11)

$$SIR(\hat{\boldsymbol{x}}_j, \boldsymbol{x}_k) = 10\log_{10}\left(\frac{\|g_{jk}\boldsymbol{x}_k\|_2^2}{\|\hat{\boldsymbol{x}}_j - g_{jk}\boldsymbol{x}_k\|_2^2}\right). \tag{B.5}$$

Listing B.11 Calculation of the SIR index.

```
function SIR = SIRNMF(AH,XH,A,X)
% Compute SIR value for NMF model via Â X̂
varX = sqrt(sum(X.^2,2));
varXH = sqrt(sum(XH.^2,2));
X = bsxfun(@rdivide,X,varX);
XH = bsxfun(@rdivide,XH,varXH);
A = bsxfun(@times,A,varX');
AH = bsxfun(@times,AH,varXH');
G = (AH'*AH)\(AH'*A) ; %   pinv(AH)*A;
[Gmax,maxind] = max(abs(G),[],2);
SIR = 10*log10(Gmax.^2./sum((XH - bsxfun(@times,X(maxind,:),Gmax)).^2,2));
```

Another approach is to compute the matrix **G** directly as (see Listing B.12)

$$\mathbf{G} = \mathbf{X}\hat{\mathbf{X}}^T(\mathbf{X}\mathbf{X}^T)^{\dagger}. \tag{B.6}$$

For sparse signals, the power is concentrated in a few coefficients, and a large change occurring at few coefficients significantly reduces the SNR index. We illustrate this by a simple example with a sparse signal $\boldsymbol{x}$ with 100 samples from which only samples take the value of unity, as shown in Figure B.1 (in blue). The estimated signal $\hat{\boldsymbol{x}}$ contains an exact source but one zero-value sample has its value changed to unity (in red). In other words, the estimated signal had 99 correct samples and only one incorrect sample, meaning that the interference is the signal with one nonzero value. The SIR value is given by

$$SIR(\hat{\boldsymbol{x}}, \boldsymbol{x}) = 10\log_{10}\left(\frac{\|signal\|_2^2}{\|interference\|_2^2}\right) = 10\log_{10}\left(\frac{1^2+1^2+1^2}{1^2}\right) = 4.77 \text{ dB} \tag{B.7}$$

and does not reflect the performance properly. Using PSNR instead of SIR, the performance index would change to 20[dB] which better reflects the true performance (see the next section).

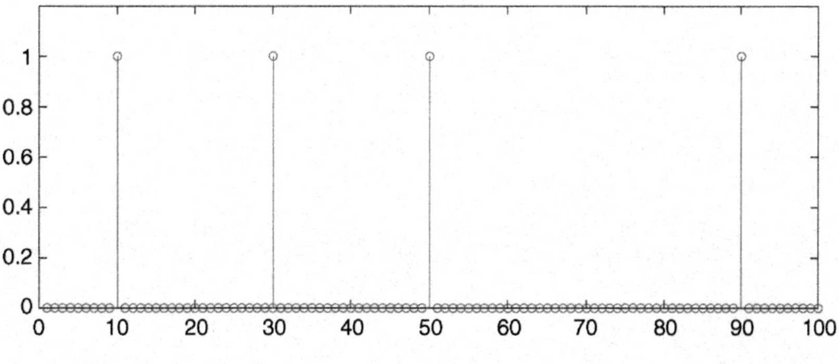

Figure B.1 Illustration of SIR NMF for sparse data of 100 sample length. The estimated signal contains 99 correct samples and only one distorted sample. The very low SIR index of 4.77 dB does not reflect the performance properly.

Listing B.12 Another calculation of the SIR index.

```
function SIR = SIRNMF2(XH,X)
X = bsxfun(@rdivide,X,sqrt(sum(X.^2,2)));
XH = bsxfun(@rdivide,XH,sqrt(sum(XH.^2,2)));
G = X*XH'/ (X*X');    %    X*XH'*pinv(X*X');
[Gmax,maxind] = max(abs(G),[],2);
SIR = 10*log10(Gmax.^2./sum((XH - bsxfun(@times,X(maxind,:),Gmax)).^2,2));
```

3.B.2 Peak Signal-to-Noise-Ratio (PSNR)

In contrast to the SIR index, the $PSNR$ estimates the ratio between a maximum possible value of the normalized signal and its root mean squared error, that is,

$$\begin{aligned} PSNR &= 20\log_{10}\frac{Range\_of\_Signal}{RMSE} \\ &= 10\log_{10}\frac{R^2 \times T}{\|\hat{\boldsymbol{x}}-\boldsymbol{x}\|_2^2}, \end{aligned} \tag{B.8}$$

where R is the maximum possible value of the signal and T is the number of samples. This index is most commonly used as a measure of quality of reconstruction in image compression. For an 8-bit image evaluation, $Range\_of\_Signal$ is 255.

For the NMF model, this index could be used in a slightly modified form as follows. Note that one of the ambiguities of the NMF and BSS models is arbitrary scaling [52], however, the most important information is located in the temporal waveforms or time-frequency patterns of the source signals and not in their amplitudes. Therefore, we can scale signals to either maximum or unit amplitude, for instance.

$$PSNR = 10\log_{10}\frac{T}{\|\hat{\boldsymbol{x}}/R-\boldsymbol{x}/R\|_2^2}. \tag{B.9}$$

Thus, if $\hat{\boldsymbol{x}}$ and $\boldsymbol{x}$ are normalized to have unit intensity (ℓ_∞-norm), PSNR for the NMF has the following form.

$$PSNR = 10\log_{10}\frac{T}{\|\hat{\boldsymbol{x}}-\boldsymbol{x}\|_2^2}. \tag{B.10}$$

The PSNR performance index is evaluated in our MATLAB toolbox and is given in Listing B.13

Listing B.13 Computation of the PSNR index.

```
function PSNR = PSNRNMF(X,XH)
% PSNR for NMF model
X = bsxfun(@minus,X,min(X,[],2));
X = bsxfun(@rdivide,X,max(X,[],2));

XH = bsxfun(@minus,XH,min(XH,[],2));
XH = bsxfun(@rdivide,XH,max(XH,[],2));

[J,T] = size(X);
PSNR = zeros(ns,1);
for j = 1:J
    PSNR(i)= min(sum(bsxfun(@minus,XH,X(j,:)).^2,2));
end
PSNR = -10*log10((PSNR+ eps)/T);
```

In addition to SIR and the PSNR, we can use the explained variation ratio (FIT %) (i.e., how well the approximated signal fits input source):

$$FIT(\%) = \left[1 - \frac{\|\hat{\boldsymbol{x}} - \boldsymbol{x}\|_2^2}{\|\boldsymbol{x} - E(\boldsymbol{x})\|_2^2}\right] \times 100. \tag{B.11}$$

Appendix 3.C: Convergence Analysis of the Multiplicative Alpha NMF Algorithm

We consider the objective function for Alpha NMF, based on the Alpha-divergence between $\mathbf{Y}$ and $\mathbf{Q} = \mathbf{AX}$, given by

$$D_A^{(\alpha)}[\mathbf{Y}||\mathbf{AX}] = \frac{1}{\alpha(\alpha-1)} \sum_{i=1}^{I} \sum_{t=1}^{T} \left(y_{it}^{\alpha} [\mathbf{AX}]_{it}^{1-\alpha} + (\alpha - 1)[\mathbf{AX}]_{it} - \alpha y_{it} \right). \tag{C.1}$$

Definition C.1 *(**Auxiliary function**) A function $G(\mathbf{X}, \widetilde{\mathbf{X}})$ is said to be an auxiliary function for $F(\mathbf{X})$ if the following two conditions are satisfied:*

$$G(\mathbf{X}, \mathbf{X}) = F(\mathbf{X}), \quad \textit{and} \quad G(\mathbf{X}, \widetilde{\mathbf{X}}) \geq F(\mathbf{X}), \quad \textit{for all } \widetilde{\mathbf{X}}. \tag{C.2}$$

Lemma C.1 *The function*

$$G(\mathbf{X}, \widetilde{\mathbf{X}}) = \frac{1}{\alpha(\alpha-1)} \sum_{i,j,t} x_{jt} \zeta_{itj} \left\{ \left(\frac{a_{ij} x_{jt}}{y_{it} \zeta_{itj}} \right)^{(1-\alpha)} + (\alpha - 1) \frac{a_{ij} x_{jt}}{y_{it} \zeta_{itj}} - \alpha \right\}, \tag{C.3}$$

with $\zeta_{itj} = \frac{a_{ij}\widetilde{x}_t}{\sum_j a_{ij}\widetilde{x}_{jt}}$, *is an auxiliary function for*

$$F(\mathbf{X}) = \frac{1}{\alpha(\alpha-1)} \sum_{i,t} \left(y_{it}^{\alpha} [\mathbf{AX}]_{it}^{1-\alpha} + (\alpha - 1)[\mathbf{AX}]_{it} - \alpha y_{it} \right). \tag{C.4}$$

Proof. We need to show that the function $G(\mathbf{X}, \widetilde{\mathbf{X}})$ in (C.3) satisfies two conditions: (i) $G(\mathbf{X}, \mathbf{X}) = F(\mathbf{X})$; (ii) $G(\mathbf{X}, \widetilde{\mathbf{X}}) \geq F(\mathbf{X})$. One can easily see that $G(\mathbf{X}, \mathbf{X}) = F(\mathbf{X})$. Note that $\sum_t \zeta_{itj} = 1$ from the definition and $\zeta_{itj} \geq 0$ for $i = 1, \ldots, I, t = 1, \ldots, T$, and $j = 1, \ldots, J$. In order to prove that the condition (ii) is satisfied, we write $F(\mathbf{X})$ as

$$\begin{aligned} F(\mathbf{X}) &= \frac{1}{\alpha(\alpha-1)} \sum_{i,t} \left(y_{it}^{\alpha} [\mathbf{AX}]_{it}^{1-\alpha} + (\alpha - 1)[\mathbf{AX}]_{it} - \alpha y_{it} \right) \\ &= \sum_{i,t} y_{it} f\left(\frac{\sum_j a_{ij} x_{jt}}{y_{it}} \right), \end{aligned} \tag{C.5}$$

where the Alpha-divergence is written using the convex function for positive α.

$$f(z) = \frac{1}{\alpha(\alpha-1)} \left\{ z^{1-\alpha} + (\alpha - 1)z - \alpha \right\}. \tag{C.6}$$

The Jensen's inequality (due to the convexity of f) leads to

$$f\left(\sum_j a_{ij} x_{jt} \right) \leq \sum_j \zeta_{itj} f\left(\frac{a_{ij} x_{jt}}{\zeta_{itj}} \right). \tag{C.7}$$

Then, it follows from (C.7) that we have

$$F(\mathbf{X}) = \sum_{i,t} y_{it} f\left(\frac{\sum_j a_{ij}x_{jy}}{y_{it}}\right) \le \sum_{i,t,j} y_{it}\zeta_{itj} f\left(\frac{a_{ij}x_{jt}}{y_{it}\zeta_{itj}}\right) = G(\mathbf{X}, \widetilde{\mathbf{X}}), \tag{C.8}$$

which proves the condition (ii). ∎

Lemma C.2 *Reversing the roles of* $\mathbf{X}$ *and* $\mathbf{A}$ *in Lemma C.1, the function*

$$G(\mathbf{A}, \widetilde{\mathbf{A}}) = \frac{1}{\alpha(\alpha-1)} \sum_{i,t,j} y_{it}\xi_{itj} \left\{ \left(\frac{a_{ij}x_{jt}}{y_{it}\xi_{itj}}\right)^{(1-\alpha)} + (\alpha-1)\frac{a_{ij}x_{jt}}{y_{it}\xi_{itj}} - \alpha \right\}, \tag{C.9}$$

with $\xi_{itj} = \frac{\widetilde{a_{ij}x_{jt}}}{\sum_l a_{il}x_{lt}}$, *is an auxiliary function for*

$$F(\mathbf{A}) = \frac{1}{\alpha(\alpha-1)} \sum_{i,j} \left(y_{it}^{\alpha}[\mathbf{AX}]_{it}^{1-\alpha} + (\alpha-1)[\mathbf{AX}]_{it} - \alpha y_{it} \right). \tag{C.10}$$

Proof. This can be easily verified in the same way as for Lemma C.1. ∎

Theorem C.2 *The Alpha-divergence,* $D_A(\mathbf{Y}||\mathbf{AX})$*, is nonincreasing under the following multiplicative update rules*

$$x_{jt} \leftarrow x_{jt} \left[\frac{\sum_j a_{ij}\left(y_{it}/[\mathbf{AX}]_{it}\right)^{\alpha}}{\sum_i a_{ij}} \right]^{\frac{1}{\alpha}}, \tag{C.11}$$

$$a_{ij} \leftarrow a_{ij} \left[\frac{\sum_t x_{jt}\left(y_{it}/[\mathbf{AX}]_{it}\right)^{\alpha}}{\sum_t x_{jt}} \right]^{\frac{1}{\alpha}}. \tag{C.12}$$

Proof. The minimum of (C.3) is determined by setting the gradient to zero:

$$\frac{\partial G(\mathbf{X}, \widetilde{\mathbf{X}})}{\partial x_{jt}} = \frac{1}{\alpha} \sum_i a_{ij} \left\{ 1 - \left(\frac{a_{ij}x_{jt}}{y_{it}\zeta_{itj}}\right)^{-\alpha} \right\} = 0, \tag{C.13}$$

which leads to

$$\left(\frac{x_{jt}}{\widetilde{x}_{jt}}\right)^{\alpha} = \left[\frac{\sum_i a_{ij}\left(\frac{y_{it}}{\sum_i \widetilde{a_{ij}x_{jt}}}\right)^{\alpha}}{\sum_i a_{ij}} \right], \tag{C.14}$$

which suggests an updating rule for x_{jt}:

$$x_{jt} \leftarrow x_{jt} \left[\frac{\sum_i a_{ij}\left(y_{it}/[\mathbf{AX}]_{it}\right)^{\alpha}}{\sum_i a_{ij}} \right]^{\frac{1}{\alpha}}, \tag{C.15}$$

that is, identical to (C.11).

In a similar manner, an updating rule (C.12) is determined by solving $\frac{\partial G(\mathbf{A}, \widetilde{\mathbf{A}})}{\partial \mathbf{A}} = 0$, where $G(\mathbf{A}, \widetilde{\mathbf{A}})$ is of the form (C.9). ∎

Appendix 3.D: MATLAB Implementation of the Multiplicative NMF Algorithms

3.D.1 Alpha Algorithm

```
function [A,X] = nmf_alpha(Y,options)
% Nonnegative Matrix Factorization based on the Amari Alpha-divergence
%
% INPUT
% Y        : source with size of I x T
% options  : structure of optional parameters for algorithm (see defoptions)
%    .tol   : tolerance of stopping criteria (Frobenius distance)       (1e-5)
%    .J:      number of components
%    .algtype vector (1x2) indicates which algorithm will be used      ([1 1])
%        1 :      Alpha      3 :      Hellinger      4 :      KL
%        5 :      Pearson    6 :      ALS  %         7 :      Fixed A or X
%    .init:   vector 1 x 2 defines types of initialization
%             (see nmf_initialize)                                     ([1 1])
%    .nrestart: number of multi initializations                          (1)
%    .niter: number of iterations                                      (300)
%    .omega: over-relaxation parameters                                ([1 1])
%    .lsparse, .lortho:    vector (1x2) indicates degrees of
%             sparsness, orthogonality for each factor A, X
%    .A0, .X0:    initialized matrices for A and X, used in fixed mode
%             (algtype = 7) or for given factors (init = 0)
% OUTPUT
% A and X such that \|Y - A * X\|_alpha is minimized
%
% Copyright 2008 by A. Cichocki and R. Zdunek and A.H. Phan
% Optimized by Anh Huy Phan and Andrzej Cichocki - 04/2008
%% =========================================================================
% Set algorithm parameters from input or by using defaults
defopts = struct('J',size(Y,1),'algtype',[1 1],'init',[1 1],...
    'nrestart',1,'niter',1000,'tol',1e-5,'alpha',1,'omega',[1 1],...
    'lsparse',[0 0],'lortho',[0 0],'alphaS',0,'A0',[],'X0',[],'verbose',0);
if ~exist('options','var'),     options = struct; end
[J,algtype,init,nrestart,niter,tol,alpha,w,lsparse,...
    lcorr,alphaS,A0,X0,verbose] = scanparam(defopts,options);
algname = {'Alpha' 'DualKL','Hellinger','KL','Pearson','ALS','Fix'
          alpha    ,      0,        .5      1      2      0     0};
if (alpha == 0),   algtype(algtype== 1) = 2;  end          % change to D-KL
alpha = [algname{2,algtype}];
Y(Y< 0) = eps;
nr_best = 0;
No_iter = 30;                  % number of iterations to select the best trial
%% Main process
for nr= 0:nrestart
    if (nr == nrestart)&&(nrestart > 0) % initialize
        A = A_best; X = X_best; No_iter = niter;
    else
        [A,X] = nmf_initialize(Y,J,init,{A0 X0'}); X = X';
    end
    cost = costfunction;
    for k  = 1: No_iter
        X = nmfcore(Y,A,X,algtype(2),alpha(2),lsparse(2),lcorr(2),w(2));
        A = nmfcore(Y',X',A',algtype(1),alpha(1),lsparse(1),lcorr(1),w(1))';
        if (algtype(2) ≠ 7)
            A = normalize(A);
```

```
        elseif ismember(algtype(1),[1 3 4 5])                    % fix scaling
            A = bsxfun(@rdivide,A,sum(X,2)fl.^(w(1)*(1+alphaS)/alpha(1)));
        end

        if (nr == nrestart) && ((mod(k,30)==0) || (k == No_iter))
            checkstoppingcondition                        % stopping condition
            if verbose
                fprintf(1,'Best trial %d, step %d, Cost value %d\n',...
                    nr_best+1,k,cost);
            end
            if stop , break; end
        end
    end % k
    if (nr < nrestart)                                     % select best trial
        cost = costfunction;
        if (nr == 0) || (cost < cost_min)
            A_best = A; X_best = X; cost_min= cost; nr_best = nr;
        end
        if verbose
            fprintf(1, 'Trial %d, Cost value = %e\n',    nr+1, cost);
        end
    end
end % nr

    function X = nmfcore(Y,A,X,type_alg,alpha, lsparse,lcorr,w)
        switch type_alg
            case {1, 3 ,4,5}      % Alpha-divergence, Hellinger, KL, Pearson
                Jor = 0;
                if lcorr
                    Jor = lcorr * bsxfun(@minus,X,sum(X,1));
                end
                X = (X.*(A'*(Y./(A*X + eps)).^alpha - lsparse + Jor)...
                    .^(w/alpha)).^(1+alphaS);
            case 2                                          % alpha = 0, D-KL
                X = (X.*exp(w* bsxfun(@rdivide,A,sum(A,1))*...
                    log(Y./(A*X+eps)+ eps))).^(1+alphaS);
            case 6 % ALS
                X = pinv(A)*Y;
        end
        X = max(1E6*eps,X);
    end

    function checkstoppingcondition
        cost_old = cost; cost = costfunction;
        stop = abs(cost_old-cost) ≤ tol*cost;
    end
    function cost = costfunction
        Yhat = A*X+eps;
        if (alpha(1) == alpha(2))&& (alpha(1) ≠0) && (alpha(1) ≠1)
            cost = sum(sum(Y.*((Y./Yhat).^(alpha(1)-1)-1)...
                /(alpha(1)*(alpha(1)-1)) - (Y-Yhat)/alpha(1)));
        else
            cost = sum(sum(Y.*log(Y./Yhat + eps) - Y + Yhat));
        end
    end
    function A = normalize(A)
        A = bsxfun(@rdivide,A,sum(A));
    end
end
```

3.D.2 SMART Algorithm

```
function [A,X]=nmf_SMART(Y,options)
% Nonnegative Matrix Factorization (NMF) with Extended SMART algorithms
%
% INPUT
% Y          :  source with size of I x T
% options :  structure of optional parameters for algorithm (see defoptions)
%    .tol    :  tolerance of stopping criteria (Frobenius distance)     (1e-5)
%    .J:        number of components
%    .algtype vector (1x2) indicates which algorithm will be used    ([1 1])
%    .init:  vector 1 x 2 defines initialization (see nmf_initialize)([1 1])
%    .nrestart: number of multi initializations                            (1)
%    .niter:  number of iterations                                       (300)
%    .niniter:  number of iterations for updating only A or only X        (1)
%    .omega:  over-relaxation parameters                               ([1 1])
%    .A0, .X0:   initialized matrices for A and X, used in fixed mode
%            (algtype = 7) or for given factors (init = 0)
% OUTPUT
% A and X with Y = A * X.
%
% Copyright 2008 by A. Cichocki and R. Zdunek and A.H. Phan
% Optimized by Anh Huy Phan and Andrzej Cichocki - 04/2008
%% =========================================================================
% Set algorithm parameters from input or by using defaults
defoptions = struct('J',size(Y,1),'algtype',[1 1],'init',[1 1],...
    'nrestart',1,'niter',1000,'niniter',1,'tol',1e-5,'alpha',1,...
    'omega',[1 1],'alphaS',0,'A0',[],'X0',[],'verbose',0);
if ~exist('options','var'),       options = struct;  end
[J,algtype,init,nrestart,niter,niniter,tol,alpha,omega,...
    alphaS,A0,X0,verbose] = scanparam(defoptions,options);
if (alpha == 0),   algtype(algtype== 1) = 2;   end           % change to D-KL

Y = max(eps,Y);
nr_best = 0;
No_iter = 30;                    % number of iterations to select the best trial
%% Main process
for nr= 0:nrestart
    if (nr == nrestart)&&(nrestart > 0) % initialize
        A = A_best; X = X_best; No_iter = niter;
    else
        [A,X] = nmf_initialize(Y,J,init,{A0 X0'}); X = X';
    end
    cost = costfunction;
    for k  = 1: No_iter
        for t = 1:niniter                                            % update X
            Q = errorfunction(algtype(2));
            X = (X.*exp(omega(2)*A'*Q)).^(1 + alphaS);
        end
        for t = 1:niniter                                            % update A
            Q = errorfunction(algtype(1));
            A = A.*exp(omega(1)*bsxfun(@rdivide,Q*X',sum(X,2)fl));
            A = normalize(A);
        end
        if (nr == nrestart) && (mod(k,30)==0)         % stopping condition
            checkstoppingcondition
            if verbose
                fprintf(1,'Best trial %d, step %d, Cost value %d\n',...
                    nr_best+1,k,cost);
            end
```

```
            if stop , break; end
        end
    end % k
    if (nr < nrestart)                                    % select best trial
        cost = costfunction;
        if (nr == 0) || (cost < cost_min)
            A_best = A; X_best = X; cost_min= cost; nr_best = nr;
        end
        if verbose
            fprintf(1, 'Trial %d, Cost value = %e\n',   nr+1, cost);
        end
    end
end % nr

    function Q = errorfunction(type_alg)
        Z = A*X + eps;
        switch type_alg
            case 1                              % Asymmetric alpha-divergence
                Q = (1/alpha)*((Y./Z).^alpha - 1);
            case 2                                                  % EG-DKL
                Q = log(Y./Z + eps);
            case 3                                                  % EG-RAG
                Q = log(2*Y./(Z + Y) + eps);
            case 4                                                  % EG-SAG
                Q = log(2*sqrt(Y.*Z)./(Z + Y) + eps) + (Y - Z)./Z/2 ;
            case 5                                                   % EG-DJ
                Q = log(Y./Z/2+1/2) + Y./Z.*(Y - Z)./(Y+Z);             %HP
            case 6                                                  % EG-RJS
                Q = (Y - Z)./(Y + Z) ;
            case 7                                                 % EG-DRJS
                Q = 2*log(.5*(Y + Z)./Z) + (Z - Y)./(Y + Z);
            case 8                                                  % EG-SJS
                Q = log(.5*(Y + Z)./Z ) + (Y-Z)./(Y+Z);
            case 9                                                  % EG-Tri
                Q = (2*Y./(Y + Z)).^2 - 1;
            case 10                                                  % EG-BE
                Q = alpha*log((Y + alpha*Z)./((1 + alpha)*Z) + eps);
        end % switch
    end
    function checkstoppingcondition
        cost_old = cost; cost = costfunction;
        stop = abs(cost_old-cost) ≤ tol*cost;
    end

    function cost = costfunction
        Yhat = A*X+eps;
        if alpha ==1
            cost = sum(sum(Y.*log(Y./Yhat + eps) - Y + Yhat));
        else
            cost = sum(sum(Y.*((Y./Yhat).^(alpha-1)-1)/(alpha*(alpha-1))...
                - (Y - Yhat)/alpha));
        end
    end

    function A = normalize(A)
        A = bsxfun(@rdivide,A,sum(A));
    end
end
```

3.D.3 ISRA Algorithm for NMF

```
function [A,X,a0,err]=anmf_ISRA(Y,options)
% ISRA algorithm for Nonnegative Matrix Factorization (NMF)
% with the offset data
%
% INPUT
% Y        :  source with size of I x T
% options  :  structure of optional parameters for algorithm (see defoptions)
%
% OUTPUT
% A, X and a0 such that \|Y - A * X - a0 * 1^T\|_F is minimized
%
% Copyright 2008 by Anh Huy Phan anh Andrzej Cichocki
%% =========================================================================

[I,T] = size(Y);
% Set algorithm parameters from input or by using defaults
defoptions = struct('J',I,'maxIter',1000,'lspar',[0 0],...
    'lortho',[0 0],'verbose',0,'A',[],'X',[],'a0',[]);

if ~exist('options','var')
    options = struct;
end

[J,maxiter,lspar,lortho,verbose,A,X,a0] = scanparam(defoptions,options);
Y(Y< 0) = eps;

% Initialization
if isempty(A) || isempty(X) || isempty(a0)
    A = rand(I,J); X = rand(J,T); a0 = rand(I,1);
end

errprev = inf;
sY = sum(Y,2);
for k = 1:maxiter
    Yhat = bsxfun(@plus,A*X,a0);

    A = A.* (Y*X') ./ (Yhat*X' + lspar(1) - ...
        lortho(1)*bsxfun(@minus,A,sum(A,2)) +eps);
    A = bsxfun(@rdivide,A,sqrt(sum(A.^2)));

    X = X.* (A' * Y) ./ (A'*Yhat + lspar(2) - ...
        lortho(2)*bsxfun(@minus,X,sum(X,1)) +eps);

    a0 = a0 .* sY./sum(Yhat,2);

    if verbose && ((mod(k,20) ==1 ) || (k == maxiter))
        err = norm(Y(:) - Yhat(:));
        Δ = abs(err - errprev);
        fprintf('Iter %f , error %.5d, del: %.5d\n',k,err,Δ);
        if (err < 1e-5) || Δ < 1e-8
            break
        end
        errprev = err;
    end
end
```

Appendix 3.E: Additional MATLAB Functions

3.E.1 Multi-layer NMF

```
function [AH,XH,SIR,PSNR,rtime,Fit]= nmf_multi_layer(Y,options,X,type)
% Y          :    source with components at matrix rows
% options    :    options for NMF algorithm which is declared as
%                 [A,X] = nmf_algorithmname(Y,options)
%                 Name of NMF function and number of layers need to
%                 be specified by two fields
%                  options.algorithm   = 'nmf_alpha';
%                  options.Nlayer = 3;
%
% X          :    optional variable for computing SIR and PSNR measurements
% type = 1:       X is the expected components
%  otherwise      A
%                 To apply multilayer for X, set : type = 1
%                 Y \S\approx\S Yl =  A(1) A(2) ... A(l)  XH
%                     [AH,XH]= nmf_multi_layer(Y,options,A,0)
%                 Multilayer for A, set : type = 0
%                 and transpose source Y for input
%                     [AH,XH]= nmf_multi_layer(Y',options,A,0)
% Ref.
%   A. Cichocki, R. Zdunek, S. Choi, R. Plemmons, and S.-I. Amari.
%   Novel multi-layer nonnegative tensor factorization with
%   sparsity constraints. Springer LNCS, 4432:271-280, April 11-14 2007.

% Written by Anh Huy Phan and Andrzej Cichocki
% 08-2008
%#########################################################################
if exist('X','var')≠1, X = []; end
if exist('type','var')≠1,  type = 1; end
PSNR = {}; SIR = {}; rtime =  0; AH = 1;
Yn = bsxfun(@rdivide,Y,max(Y,[],2));
for l = 1: options.Nlayer
    fprintf(1,'Layer %d \n',l);
    if l>1
        Yl = XH;
    else
        Yl = Y;
    end
    tic
    [Al,XH] = feval(options.algorithm,Yl,options);
    rtime = rtime + toc;
    if any(isnan(Al(:))), break, end
    AH = AH*Al;

    if ~isempty(X)
        if type==1 %XPN
            Xref = XH;
        else
            Xref = AH';
        end
        Xref = bsxfun(@rdivide,Xref,max(Xref,[],2));
        PSNR{l} = PSNRNMF(X,Xref);
        SIR{l} = SIRNMF2(Xref,X);
        Yhat = AH*XH;
        Yhat = bsxfun(@rdivide,Yhat,max(Yhat,[],2));
        Fit(l) = CalcExpVar(Yhat,Yn);
    end
```

```
    save(sprintf('temp_layer_%d',l));
end % for l

if ~isempty(X)
    fprintf(1,'Finish  Fit: %.2f%%, PSNR: %.2f dB , SIR: %.2f dB\n',...
        Fit(end),mean(PSNR{end}),mean(SIR{end}))
else
    fprintf(1,'Finish  Fit: %.2f%%\n', Fit(end))
end
end
```

3.E.2 MC Analysis with Distributed Computing Tool

```
function [results,rtime] = mc_nmf(Ntrial,nmf_path,nmf_algname,...
    nmf_options,mc_options,Nouts)
% Monte Carlo analysis for NMF algorithm in the distributed enviroment
% with local scheduler. This function runs on MATLAB 7.5 or higher which
% supports the distributed computing toolbox.
% Computer should have multi-core or multi processors.
%
% Ntrial      :    number of trials for MC analysis
% nmf_path    :    path of files and function using during the analysis
% nmf_algname :    file name of NMF algorithm for MC analysis
% nmf_options :    options for NMF function
% mc_options  :    options for MC analysis
% Nouts       :    number of MC outputs

% Copyright 2008 by A.H. Phan and A. Cichocki
% 08-2008
%#########################################################################
jm = findResource('scheduler','type','local');
full_algname = fullfile(nmf_path,nmf_algname);
job = createJob(jm,'FileDependencies',{full_algname},...
    'PathDependencies',{nmf_path});
for k = 1: Ntrial
    createTask(job,  str2func(nmf_algname),Nouts,{mc_options{1},...
        nmf_options,mc_options{2:end}});
end
submit(job)
waitForState(job,'finished')
results = getAllOutputArguments(job);
ftime   = get(job,'FinishTime');
sttime  = get(job,'StartTime');
sidx = strfind(sttime,' ');
sttime([1:4 sidx(4):sidx(5)-1]) = [];
ftime([1:4 sidx(4):sidx(5)-1]) = [];
% Convert to a date number
fmt = 'mmm dd HH:MM:SS yyyy';
sttime = datenum(sttime,fmt);
ftime = datenum(ftime,fmt);
rtime = etime(datevec(datenum(ftime)), datevec(datenum(sttime)));
destroy(job)
```

References

[1] J. Brown. Calculation of a constant Q spectral transform. *Journal of the Acoustical Society of America*, 89(1):425–434, 1991.

[2] J. Browne and A. De Pierro. A row-action alternative to the EM algorithm for maximizing likelihoods in emission tomography. *IEEE Transactions on Medical Imaging*, 15:687–699, 1996.
[3] C.L. Byrne. Accelerating the EMML algorithm and related iterative algorithms by rescaled block-iterative (RBI) methods. *IEEE Transactions on Image Processing*, IP-7:100–109, 1998.
[4] C.L. Byrne. Choosing parameters in block-iterative or ordered subset reconstruction algorithms. *IEEE Transactions on Image Processing*, 14(3):321–327, 2005.
[5] C.L. Byrne. *Signal Processing: A Mathematical Approach.* A.K. Peters, Publ., Wellesley, MA, 2005.
[6] C. Caiafa and A. Cichocki. Slice Oriented Decomposition: A new tensor representation for 3-way data. *(submitted to Journal of Signal Processing)*, 2009.
[7] S. Choi. Algorithms for orthogonal nonnegative matrix factorization. Hong-Kong, 2007. Proceedings of the International Joint Conference on Neural Networks (IJCNN-2008).
[8] M. Chu and M.M. Lin. Low dimensional polytype approximation and its applications to nonnegative matrix factorization. *SIAM Journal on Scientific Computing*, 30:1131–1151, 2008.
[9] A. Cichocki and S. Amari. *Adaptive Blind Signal and Image Processing*. John Wiley & Sons Ltd, New York, 2003.
[10] A. Cichocki, S. Amari, R. Zdunek, R. Kompass, G. Hori, and Z. He. Extended SMART algorithms for non-negative matrix factorization. *Springer, LNAI-4029*, 4029:548–562, 2006.
[11] A. Cichocki and R. Zdunek. NMFLAB for Signal and Image Processing. Technical report, Laboratory for Advanced Brain Signal Processing, BSI, RIKEN, Saitama, Japan, 2006.
[12] A. Cichocki, R. Zdunek, and S. Amari. Csiszar's divergences for non-negative matrix factorization: Family of new algorithms. *Springer, LNCS-3889*, 3889:32–39, 2006.
[13] A. Cichocki, R. Zdunek, and S. Amari. New algorithms for non-negative matrix factorization in applications to blind source separation. In *Proc. IEEE International Conference on Acoustics, Speech, and Signal Processing, ICASSP2006*, volume 5, pages 621–624, Toulouse, France, May 14–19 2006.
[14] A. Cichocki, R. Zdunek, and S.-I. Amari. Hierarchical ALS algorithms for nonnegative matrix and 3D tensor factorization. *Springer, Lecture Notes on Computer Science, LNCS-4666*, pages 169–176, 2007.
[15] A. Cichocki, R. Zdunek, and S.-I. Amari. Nonnegative matrix and tensor factorization. *IEEE Signal Processing Magazine*, 25(1):142–145, 2008.
[16] A. Cichocki, R. Zdunek, S. Choi, R. Plemmons, and S. Amari. Nonnegative tensor factorization using Alpha and Beta divergencies. In *Proc. IEEE International Conference on Acoustics, Speech, and Signal Processing (ICASSP07)*, volume III, pages 1393–1396, Honolulu, Hawaii, USA, April 15–20 2007.
[17] A. Cichocki, R. Zdunek, S. Choi, R. Plemmons, and S.-I. Amari. Novel multi-layer nonnegative tensor factorization with sparsity constraints. *Springer, LNCS-4432*, 4432:271–280, April 11–14 2007.
[18] M.E. Daube-Witherspoon and G. Muehllehner. An iterative image space reconstruction algorthm suitable for volume ECT. *IEEE Transactions on Medical Imaging*, 5:61–66, 1986.
[19] A. R. De Pierro. A modified expectation maximization algorithm for penalized likelihood estimation in emission tomography. *IEEE Transactions on Medial Imaging*, 14(1):132–137, March 1995.
[20] A.R. De Pierro. On the relation between the ISRA and the EM algorithm for positron emission tomography. *IEEE Transactions on Medial Imaging*, 12(2):328–333, June 1993.
[21] A.R. De Pierro and M.E. Beleza Yamagishi. Fast iterative methods applied to tomography models with general Gibbs priors. In *Proc. SPIE Conf. Mathematical Modeling, Bayesian Estimation and Inverse Problems*, volume 3816, pages 134–138, 1999.
[22] A.P. Dempster, N.M. Laird, and D.B. Rubin. Maximum likelihood from incomplete data via the EM algorithm. *Journal of the Royal Statistical Society*, 39(1):1–38, 1977.
[23] I. Dhillon and S. Sra. Generalized nonnegative matrix approximations with Bregman divergences. In *Neural Information Proc. Systems*, pages 283–290, Vancouver, Canada, December 2005.
[24] C. Ding, T. Li, W. Peng, and H. Park. Orthogonal nonnegative matrix tri-factorizations for clustering. In *KDD06: Proceedings of the 12th ACM SIGKDD international conference on Knowledge Discovery and Data Mining*, pages 126–135, New York, NY, USA, 2006. ACM Press.
[25] D. Donoho and V. Stodden. When does nonnegative matrix factorization give a correct decomposition into parts? In *Neural Information Processing Systems*, volume 16. MIT press, 2003.
[26] P.P.B. Eggermont and V.N. LaRiccia. Maximum smoothed likelihood density estimation for inverse problems. *Ann. Statist.*, 23(1):199–220, 1995.
[27] P.P.B. Eggermont and V.N. LaRiccia. On EM-like algorithms for minimum distance estimation. Technical report, Mathematical Sciences, University of Delaware, 1998.

[28] D. Ellis. Spectrograms: Constant-Q (log-frequency) and conventional (linear), 05 2004.

[29] D. FitzGerald, M. Cranitch, and E. Coyle. Extended nonnegative tensor factorisation models for musical sound source separation. *Computational Intelligence and Neuroscience*, 2008.

[30] N. Gillis and F. Glineur. Nonnegative factorization and maximum edge biclique problem. In *submitted*, 2009. http://www.uclouvain.be/en-44508.html.

[31] P.J. Green. Bayesian reconstruction from emission tomography data using a modified EM algorithm. *IEEE Transactions on Medical Imaging*, 9:84–93, 1990.

[32] T. Hebert and R. Leahy. A generalized EM algorithm for 3-D Bayesian reconstruction from Poisson data using Gibbs priors. *IEEE Transactions on Medical Imaging*, 8:194–202, 1989.

[33] N.-D. Ho and P. Van Dooren. Nonnegative matrix factorization with fixed row and column sums. *Linear Algebra and its Applications*, 429(5-6):1020–1025, 2007.

[34] L. Kaufman. Maximum likelihood, least squares, and penalized least squares for PET. *IEEE Transactions on Medical Imaging*, 12(2):200–214, 1993.

[35] J. Kivinen and M.K. Warmuth. Exponentiated gradient versus gradient descent for linear predictors. *Information and Computation*, 132, 1997.

[36] R. Kompass. A generalized divergence measure for nonnegative matrix factorization. *Neural Computation*, 19(3): 780–791, 2006.

[37] H. Lantéri, M. Roche, and C. Aime. Penalized maximum likelihood image restoration with positivity constraints: multiplicative algorithms. *Inverse Problems*, 18:1397–1419, 2002.

[38] H Lantéri, R. Soummmer, and C. Aime. Comparison between ISRA and RLA algorithms: Use of a Wiener filter based stopping criterion. *Astronomy and Astrophysics Supplemantary Series*, 140:235–246, 1999.

[39] H. Laurberg and L.K. Hansen. On affine non-negative matrix factorization. In *Proceedings of the IEEE International Conference on Acoustics, Speech, and Signal Processing. ICASSP 2007*, pages 653–656, 2007.

[40] D.D. Lee and H.S. Seung. Learning of the parts of objects by non-negative matrix factorization. *Nature*, 401:788–791, 1999.

[41] D.D. Lee and H.S. Seung. *Algorithms for Nonnegative Matrix Factorization*, volume 13. MIT Press, 2001.

[42] R.M. Lewitt and G. Muehllehner. Accelerated iterative reconstruction for positron emission tomography based on the EM algorithm for maximum-likelihood estimation. *IEEE Transactions on Medical Imaging*, MI-5:16–22, 1986.

[43] H. Li, T. Adali, W. Wang, D. Emge, and A. Cichocki. Non-negative matrix factorization with orthogonality constraints and its application to Raman spectroscopy. *The Journal of VLSI Signal Processing*, 48(1-2):83–97, 2007.

[44] M. Mørup, M.N. Schmidt, and L.K. Hansen. Shift invariant sparse coding of image and music data. Technical report, 2008.

[45] P. Sajda, S. Du, T.R. Brown, R. Stoyanova, D.C. Shungu, X. Mao, and L.C. Parra. Nonnegative matrix factorization for rapid recovery of constituent spectra in magnetic resonance chemical shift imaging of the brain. *IEEE Transactions on Medical Imaging*, 23(12):1453–1465, 2004.

[46] P. Sajda, S. Du, and L. Parra. Recovery of constituent spectra using non-negative matrix factorization. In *Proceedings of SPIE*, volume 5207, pages 321–331, 2003.

[47] M.N. Schmidt. Single-channel source separation using non-negative matrix factorization, 2008.

[48] N.N. Schraudolph. Gradient-based manipulation of nonparametric entropy estimates. *IEEE Transactions on Neural Networks*, 15(4):828–837, July 2004.

[49] L. Shepp and Y. Vardi. Maximum likelihood reconstruction for emission tomography. *IEEE Transactions on Medical Imaging*, MI-1:113–122, 1982.

[50] P. Smaragdis. Non-negative matrix factor deconvolution; Extraction of multiple sound sources from monophonic inputs. *Lecture Notes in Computer Science*, 3195:494–499, 2004.

[51] P. Smaragdis. Convolutive speech bases and their application to supervised speech separation. *IEEE Transactions on Audio, Speech and Language Processing*, 15(1):1–12, 2007.

[52] L. Tong, R.-W. Liu, V.-C. Soon, and Y.-F. Huang. Indeterminacy and identifiability of blind identification. *IEEE Transactions on Circuits and Systems*, 38(5):499–509, 1991.

[53] Y. Tsaig and D.L. Donoho. Extensions of compressed sensing. *Signal Processing*, 86(3):549–571, 2006.

[54] W. Wang. Squared Euclidean distance based convolutive non-negative matrix factorization with multiplicative learning rules for audio pattern separation. In *Proc. 7th IEEE International Symposium on Signal Processing and Information Technology (ISSPIT 2007)*, Cairo, Egypt, December 15-18 2007.

[55] W. Wang, A. Cichocki, and J. Chambers. Note onset detection via nonnegative factorization of magnitude spectrum. *IEEE Transactions on Signal Processing*, page (in print), 2009.

[56] W. Wang, Y. Luo, J. Chambers, and S. Sanei. Non-negative matrix factorization for note onset detection of audio signals. In *Proc. IEEE International Workshop on Machine Learning for Signal Processing (MLSP 2006)*, pages 447–452, Maynooth, Ireland, September 6–8 2006.
[57] W. Wang, Y. Luo, J. Chambers, and S. Sanei. Note onset detection via nonnegative factorization of magnitude spectrum. *EURASIP Journal on Advances in Signal Processing*, ISSN 1687 6172, 2008.
[58] K. Wilson, B. Raj, P. Smaragdis, and A. Divakaran. Speech denoising using non-negative matrix factorization with priors. In *In proceedings IEEE International Conference on Audio and Speech Signal Processing*, Las Vegas, Nevada, USA, April 2008.

4

Alternating Least Squares and Related Algorithms for NMF and SCA Problems

In this chapter we derive and overview Alternating Least Squares algorithms referred to as ALS algorithms for Nonnegative Matrix Factorization (NMF) and Sparse Component Analysis (SCA). This is important as many existing NMF/SCA techniques are prohibitively slow and inefficient, especially for very large-scale problems. For such problems a promising approach is to apply the ALS algorithms [2,47]. Unfortunately, the standard ALS algorithm and its simple modifications suffer from unstable convergence properties, they often return suboptimal solutions, are quite sensitive with respect to noise, and can be relatively slow for nearly collinear data [2,43,47,64].

As explained in Chapter 1 solutions obtained by NMF algorithms may not be unique, and to this end it is often necessary to impose additional constraints (which arise naturally from the data considered) such as sparsity or smoothness. Therefore, special emphasis in this chapter is put on various regularization and penalty terms together with local learning rules in which we update sequentially one-by-one vectors of factor matrices. By incorporating the regularization and penalty terms into the weighted Frobenius norm, we show that it is possible to achieve sparse, orthogonal, or smooth representations thus helping to obtain a desired global solution.

The main objective of this chapter is to develop efficient and robust regularized ALS (RALS) algorithms. For this purpose, we use several approaches from constrained optimization and regularization theory, and introduce in addition several heuristic algorithms. The algorithms are characterized by improved efficiency and very good convergence properties, especially for large-scale problems. The RALS and HALS algorithms were implemented in our NMFLAB/NTFLAB MATLAB toolboxes, and compared with standard NMF algorithms [19]. Moreover, we have applied the ALS approach for semi-NMF, symmetric NMF, and NMF with orthogonality constraints.

4.1 Standard ALS Algorithm

Consider the standard NMF model, given by:[1]

$$\mathbf{Y} = \mathbf{A}\mathbf{X} + \mathbf{E} = \mathbf{A}\mathbf{B}^T + \mathbf{E}, \qquad \mathbf{A} \geq \mathbf{0} \text{ and } \mathbf{X} \geq \mathbf{0}. \tag{4.1}$$

[1] We use a simplified notation: $\mathbf{A} \geq \mathbf{0}$ which is used to denote component-wise relations, that is, $a_{ij} \geq 0$.

Nonnegative Matrix and Tensor Factorizations: Applications to Exploratory Multi-way Data Analysis and Blind Source Separation Andrzej Cichocki, Rafal Zdunek, Anh Huy Phan and Shun-ichi Amari

The problem of estimating the nonnegative elements in **A** and **X** can be formulated as the minimization of the standard squared Euclidean distance (Frobenius norm):

$$D_F(\mathbf{Y}||\mathbf{AX}) = \frac{1}{2}||\mathbf{Y} - \mathbf{AX}||_F^2 = \frac{1}{2}\,\mathrm{tr}(\mathbf{Y} - \mathbf{AX})^T(\mathbf{Y} - \mathbf{AX}), \tag{4.2}$$

$$\text{subject to } a_{ij} \geq 0, \quad x_{jt} \geq 0, \quad \forall\, i, j, t.$$

In such a case the basic approach is to perform the alternating minimization or alternating projection: the above cost function can be alternately minimized with respect to the two sets of parameters $\{x_{jt}\}$ and $\{a_{ij}\}$, each time optimizing one set of arguments while keeping the other one fixed [13,49]. This corresponds to the following set of minimization problems:

$$\mathbf{A}^{(k+1)} = \arg\min_{\mathbf{A}} ||\mathbf{Y} - \mathbf{AX}^{(k)}||_F^2, \qquad \text{s.}t.\ \mathbf{A} \geq \mathbf{0}, \tag{4.3}$$

$$\mathbf{X}^{(k+1)} = \arg\min_{\mathbf{X}} ||\mathbf{Y}^T - \mathbf{X}^T[\mathbf{A}^{(k+1)}]^T||_F^2, \qquad \text{s.}t.\ \mathbf{X} \geq \mathbf{0}. \tag{4.4}$$

Instead of applying the gradient descent technique, we rather estimate directly the stationary points and thereby exploit the fixed point approach. According to the Karush-Kuhn-Tucker (KKT) optimality conditions, $\mathbf{A}^*$ and $\mathbf{X}^*$ are stationary points of the cost function (4.2) if and only if

$$\mathbf{A}^* \geq \mathbf{0}, \qquad \mathbf{X}^* \geq \mathbf{0}, \tag{4.5}$$

$$\nabla_{\mathbf{A}} D_F(\mathbf{Y}||\mathbf{A}^*\mathbf{X}^*) = \mathbf{A}^*\mathbf{X}^*\mathbf{X}^{*T} - \mathbf{YX}^{*T} \geq \mathbf{0}, \qquad \mathbf{A} \circledast \nabla_{\mathbf{A}} D_F(\mathbf{Y}||\mathbf{A}^*\mathbf{X}^*) = \mathbf{0}, \tag{4.6}$$

$$\nabla_{\mathbf{X}} D_F(\mathbf{Y}||\mathbf{A}^*\mathbf{X}^*) = \mathbf{A}^{*T}\mathbf{A}^*\mathbf{X}^* - \mathbf{A}^{*T}\mathbf{Y} \geq \mathbf{0}, \qquad \mathbf{X} \circledast \nabla_{\mathbf{X}} D_F(\mathbf{Y}||\mathbf{A}^*\mathbf{X}^*) = \mathbf{0}. \tag{4.7}$$

Assuming that the factor matrices **A** and **X** are positive (with zero entries replaced by e.g., $\varepsilon = 10^{-9}$), the stationary points can be found by equating the gradient components to zero:

$$\nabla_{\mathbf{A}} D_F(\mathbf{Y}||\mathbf{AX}) = \frac{\partial D_F(\mathbf{Y}||\mathbf{AX})}{\partial \mathbf{A}} = [-\mathbf{YX}^T + \mathbf{AXX}^T] = \mathbf{0}, \tag{4.8}$$

$$\nabla_{\mathbf{X}} D_F(\mathbf{Y}||\mathbf{AX}) = \frac{\partial D_F(\mathbf{Y}||\mathbf{AX})}{\partial \mathbf{X}} = [-\mathbf{A}^T\mathbf{Y} + \mathbf{A}^T\mathbf{AX}] = \mathbf{0} \tag{4.9}$$

or equivalently in a scalar form

$$\frac{\partial D_F(\mathbf{Y}||\mathbf{AX})}{\partial a_{ij}} = [-\mathbf{YX}^T + \mathbf{AXX}^T]_{ij} = 0, \qquad \forall ij, \tag{4.10}$$

$$\frac{\partial D_F(\mathbf{Y}||\mathbf{AX})}{\partial x_{jt}} = [-\mathbf{A}^T\mathbf{Y} + \mathbf{A}^T\mathbf{AX}]_{jt} = 0, \qquad \forall jt. \tag{4.11}$$

Assuming that the estimated components are nonnegative we obtain the simple nonnegative ALS update rules:

$$\mathbf{A} \leftarrow \left[\mathbf{YX}^T(\mathbf{XX}^T)^{-1}\right]_+ = \left[\mathbf{YX}^\dagger\right]_+, \tag{4.12}$$

$$\mathbf{X} \leftarrow \left[(\mathbf{A}^T\mathbf{A})^{-1}\mathbf{A}^T\mathbf{Y}\right]_+ = \left[\mathbf{A}^\dagger\mathbf{Y}\right]_+, \tag{4.13}$$

where $\mathbf{A}^{\dagger}$ denotes the Moore-Penrose pseudo inverse, and $[x]_+ = \max\{\varepsilon, x\}$ is a half-wave rectifying nonlinear projection to enforce nonnegativity or strictly speaking positive constraints.

In the special case, for a symmetric NMF model, given by

$$\mathbf{Y} = \mathbf{A}\mathbf{A}^T, \tag{4.14}$$

where $\mathbf{Y} \in \mathbb{R}^{I\times I}$ is a symmetric nonnegative matrix, and $\mathbf{A} \in \mathbb{R}^{I\times J}$ with $I \geq J$, we obtain a simplified algorithm

$$\boxed{\mathbf{A} \leftarrow \left[\mathbf{Y}\mathbf{A}(\mathbf{A}^T\mathbf{A})^{-1}\right]_+ = \left[\mathbf{Y}[\mathbf{A}^T]^{\dagger}\right]_+ ,} \tag{4.15}$$

subject to additional normalization of the columns of matrix $\mathbf{A}$.

It is interesting to note that the modified ALS algorithm (4.12) – (4.13) can also be derived from Newton's method based on the second-order gradient descent approach, that is, based on the Hessian.[2] Applying the gradient descent approach, we have

$$\mathrm{vec}\,(\mathbf{X}) \leftarrow \left[\mathrm{vec}\,(\mathbf{X}) - \boldsymbol{\eta}_{\mathbf{X}}\mathrm{vec}\,(\nabla_{\mathbf{X}} D_F(\mathbf{Y}||\mathbf{A}\mathbf{X}))\right]_+ , \tag{4.16}$$

where $\boldsymbol{\eta}_{\mathbf{X}}$ is no longer a positive scalar, but a symmetric positive-definite matrix comprising the learning rates, expressed by inverse of a Hessian matrix of the cost function:

$$\boldsymbol{\eta}_{\mathbf{X}} = \eta_0(\nabla^2_{\mathbf{X}} D_F(\mathbf{Y}||\mathbf{A}\mathbf{X}))^{-1}, \tag{4.17}$$

where $\eta_0 \leq 1$ is a scaling factor (typically $\eta_0 = 1$). The gradient and Hessian of the cost function (4.2) with respect to $\mathbf{X}$ are given by

$$\nabla_{\mathbf{X}} D_F(\mathbf{Y}||\mathbf{A}\mathbf{X}) = \mathbf{A}^T\mathbf{A}\mathbf{X} - \mathbf{A}\mathbf{Y}, \tag{4.18}$$

$$\nabla^2_{\mathbf{X}} D_F(\mathbf{Y}||\mathbf{A}\mathbf{X}) = \mathbf{I}_T \otimes \mathbf{A}^T\mathbf{A}. \tag{4.19}$$

Hence, we obtain the learning rule for $\mathbf{X}$:

$$\mathrm{vec}\,(\mathbf{X}) \leftarrow \left[\mathrm{vec}\,(\mathbf{X}) - \eta_0 \left(\mathbf{I}_T \otimes \mathbf{A}^T\mathbf{A}\right)^{-1} \mathrm{vec}\left(\mathbf{A}^T\mathbf{A}\mathbf{X} - \mathbf{A}\mathbf{Y}\right)\right]_+ \tag{4.20}$$

$$= \left[\mathrm{vec}\,(\mathbf{X}) - \eta_0\, \mathrm{vec}\left(\left(\mathbf{A}^T\mathbf{A}\right)^{-1}\left(\mathbf{A}^T\mathbf{A}\mathbf{X} - \mathbf{A}\mathbf{Y}\right)\right)\right]_+ \tag{4.21}$$

or equivalently in the matrix form as

$$\mathbf{X} \leftarrow \left[(1-\eta_0)\mathbf{X} + \eta_0(\mathbf{A}^T\mathbf{A})^{-1}\mathbf{A}^T\mathbf{Y}\right]_+ . \tag{4.22}$$

In a similar way, assuming that

$$\boldsymbol{\eta}_{\mathbf{A}} = \eta_0(\nabla^2_{\mathbf{A}} D_F(\mathbf{Y}||\mathbf{A}\mathbf{X}))^{-1} = \eta_0(\mathbf{X}\mathbf{X}^T \otimes \mathbf{I}_I)^{-1}, \tag{4.23}$$

we obtain:

$$\mathrm{vec}\,(\mathbf{A}) \leftarrow \left[\mathrm{vec}\,(\mathbf{A}) - \eta_0 \left(\left(\mathbf{X}\mathbf{X}^T\right)^{-1} \otimes \mathbf{I}_I\right) \mathrm{vec}\left(\mathbf{A}\mathbf{X}\mathbf{X}^T - \mathbf{Y}\mathbf{X}^T\right)\right]_+ , \tag{4.24}$$

$$= \left[\mathrm{vec}\,(\mathbf{A}) - \eta_0\, \mathrm{vec}\left(\left(\mathbf{A}\mathbf{X}\mathbf{X}^T - \mathbf{Y}\mathbf{X}^T\right)\left(\mathbf{X}\mathbf{X}^T\right)^{-1}\right)\right]_+ , \tag{4.25}$$

[2] See details in Chapter 6.

or in the matrix form as

$$\mathbf{A} \leftarrow \left[(1-\eta_0)\mathbf{A} + \eta_0 \mathbf{Y}\mathbf{X}^T(\mathbf{X}\mathbf{X}^T)^{-1}\right]_+ . \tag{4.26}$$

For $\eta_0 = 1$ the above updating rules simplify to the standard ALS illustrating that the ALS algorithm is in fact the Newton method with a relatively good convergence rate since it exploits information not only about gradient but also Hessian.

The main problem with the standard ALS algorithm is that it often cannot escape local minima. In order to alleviate this problem we introduce additional regularization and/or penalty terms and introduce novel cost functions to derive local hierarchical ALS (HALS) algorithms.

4.1.1 Multiple Linear Regression – Vectorized Version of ALS Update Formulas

The minimization problems (4.12)– (4.13) can also be formulated using multiple linear regression by vectorizing matrices, leading to the minimization of the following two cost functions:

$$\min_{\boldsymbol{x}\geq 0} ||\boldsymbol{y} - \bar{\mathbf{A}}\boldsymbol{x}||_2^2 \tag{4.27}$$

$$\min_{\boldsymbol{a}\geq 0} ||\bar{\boldsymbol{y}} - \overline{\mathbf{X}}\boldsymbol{a}||_2^2, \tag{4.28}$$

where

$$\boldsymbol{y} = \text{vec}(\mathbf{Y}) \in \mathbb{R}^{IT}, \qquad \bar{\boldsymbol{y}} = \text{vec}(\mathbf{Y}^T) \in \mathbb{R}^{IT},$$

$$\boldsymbol{x} = \text{vec}(\mathbf{X}) \in \mathbb{R}^{JT}, \qquad \bar{\boldsymbol{a}} = \text{vec}(\mathbf{A}^T) \in \mathbb{R}^{IJ},$$

$$\bar{\mathbf{A}} = \text{diag}\{\mathbf{A}, \mathbf{A}, \ldots, \mathbf{A}\} \in \mathbb{R}^{IT\times JT}, \qquad \overline{\mathbf{X}} = \text{diag}\{\mathbf{X}^T, \mathbf{X}^T, \ldots, \mathbf{X}^T\} \in \mathbb{R}^{IJ\times JT}.$$

The solution to the above optimization problem can be expressed as:

$$\boldsymbol{x} \leftarrow \left[(\bar{\mathbf{A}}^T\bar{\mathbf{A}})^{-1}\bar{\mathbf{A}}^T\boldsymbol{y}\right]_+ , \tag{4.29}$$

$$\bar{\boldsymbol{a}} \leftarrow \left[(\overline{\mathbf{X}}^T\overline{\mathbf{X}})^{-1}\overline{\mathbf{X}}^T\bar{\boldsymbol{y}}\right]_+ . \tag{4.30}$$

Such representations are not computationally optimal, since for large-scale problems the block-diagonal matrices $\bar{\mathbf{A}}$ and $\overline{\mathbf{X}}$ are very large dimensions, which makes the inversion of these matrices in each iteration step very time consuming.

4.1.2 Weighted ALS

The ALS algorithm can be generalized to include additional information, for example, about an error variance and a covariance by incorporating a weighting matrix into the cost function. Thus, as a cost function, instead of the standard squared Euclidean distance we can use a weighted squared distance that corresponds to the Mahalanobis distance:

$$D_{\mathbf{W}}(\mathbf{Y}||\mathbf{A}\mathbf{X}) = \frac{1}{2}||\mathbf{Y} - \mathbf{A}\mathbf{X}||_{\mathbf{W}}^2 = \frac{1}{2}\,\text{tr}\left\{(\mathbf{Y} - \mathbf{A}\mathbf{X})^T\mathbf{W}(\mathbf{Y} - \mathbf{A}\mathbf{X})\right\}, \tag{4.31}$$

where $\mathbf{W}$ is a symmetric nonnegative weighting matrix.

This form of the parameterized squared Euclidean distance is equivalent to the Mahalanobis distance with an arbitrary positive semi-definite weight matrix $\mathbf{W}$ replacing the inverse of the covariance matrix.

Let $\mathbf{W}_1$ be a square symmetric matrix which is a square root of the weighted matrix $\mathbf{W}$, that is a matrix[3] for which $\mathbf{W} = \mathbf{W}_1^T \mathbf{W}_1$ and $\mathbf{W}_1 = \mathbf{W}^{1/2}$. Upon the multiplication of each side of the equation $\mathbf{Y} = \mathbf{AX} + \mathbf{E}$ by $\mathbf{W}_1$, we have a new scaled system of equations:

$$\mathbf{W}_1 \mathbf{Y} = \mathbf{W}_1 \mathbf{AX} + \mathbf{W}_1 \mathbf{E}. \tag{4.32}$$

Applying the linear transformations $\mathbf{Y} \leftarrow \mathbf{W}_1 \mathbf{Y}$ and $\mathbf{A} \leftarrow \mathbf{W}_1 \mathbf{A}$, we obtain directly the weighted alternating least squares (WALS) update rules:

$$\mathbf{A} \leftarrow \left[\mathbf{YX}^T(\mathbf{XX}^T)^{-1}\right]_+, \tag{4.33}$$

$$\mathbf{X} \leftarrow \left[(\mathbf{A}^T\mathbf{WA})^{-1}\mathbf{A}^T\mathbf{WY}\right]_+. \tag{4.34}$$

Another form of the weighted Frobenius norm cost function is given by

$$D_{\tilde{W}}(\mathbf{Y}||\mathbf{AX}) = \frac{1}{2}||\tilde{\mathbf{W}} \circledast (\mathbf{Y} - \mathbf{AX})||_F^2 = \frac{1}{2}\sum_{it} \tilde{w}_{it}^2 \, (y_{it} - \sum_j a_{ij}x_{jt})^2, \tag{4.35}$$

where $\tilde{\mathbf{W}} \in \mathbb{R}^{I \times T}$ is a weighting matrix with nonnegative elements which reflects our confidence in the entries of $\mathbf{Y}$. Each weighting coefficient might be chosen as a standard deviation of its corresponding measurement $\tilde{w}_{it} = 1/\sigma_{it}$ or set to be zero value for the missing ones. The following weighted ALS update rules minimizes the above cost function with nonnegativity constraints given by

$$\mathbf{A} \leftarrow \left[((\tilde{\mathbf{W}} \circledast \mathbf{Y})\mathbf{X}^T)(\mathbf{XX}^T)^{-1}\right]_+, \tag{4.36}$$

$$\mathbf{X} \leftarrow \left[(\mathbf{A}^T\mathbf{A})^{-1}(\mathbf{A}^T(\tilde{\mathbf{W}} \circledast \mathbf{Y}))\right]_+. \tag{4.37}$$

4.2 Methods for Improving Performance and Convergence Speed of ALS Algorithms

There are some drawbacks of the standard ALS algorithm. It can stuck in a local minimum and for some data it could be slow when the factor matrices are ill-conditioned or when collinearity occurs in the columns of these matrices.[4] Moreover, the complexity of the ALS algorithm can be high for very large-scale problem. We can reduce the complexity of the ALS algorithm, or improve its performance and increase convergence speed by using several simple techniques especially for large-scale data or ill-conditioned data.

4.2.1 ALS Algorithm for Very Large-scale NMF

Instead of alternately minimizing the cost function:

$$D_F(\mathbf{Y} \,||\, \mathbf{AX}) = \frac{1}{2}\,||\mathbf{Y} - \mathbf{AX}||_F^2, \tag{4.38}$$

[3] $\mathbf{W}^{1/2} = \mathbf{V}\Lambda^{1/2}\mathbf{V}^T$ in MATLAB notation means $\mathbf{W}^{1/2} = (sqrtm(\mathbf{W}))$ and $||\mathbf{W}^{1/2}(\mathbf{Y} - \mathbf{AX})||_F^2 = \mathrm{tr}\left((\mathbf{Y} - \mathbf{AX})^T\mathbf{W}(\mathbf{Y} - \mathbf{AX})\right)$.

[4] Such behavior of the ALS algorithm for ill-conditioned data with a long plateau, that is, many iterations with convergence speed almost null, after which convergence resumes, is called swamp.

consider sequential minimization of the following two cost functions (see also Section 1.3.1):

$$D_F(\mathbf{Y}_r \,||\, \mathbf{A}_r\mathbf{X}) = \frac{1}{2} \|\mathbf{Y}_r - \mathbf{A}_r\mathbf{X}\|_F^2, \qquad \text{for fixed} \quad \mathbf{A}_r, \tag{4.39}$$

$$D_F(\mathbf{Y}_c \,||\, \mathbf{A}\mathbf{X}_c) = \frac{1}{2} \|\mathbf{Y}_c - \mathbf{A}\mathbf{X}_c\|_F^2, \qquad \text{for fixed} \quad \mathbf{X}_c, \tag{4.40}$$

where $\mathbf{Y}_r \in \mathbb{R}_+^{R\times T}$ and $\mathbf{Y}_c \in \mathbb{R}_+^{I\times C}$ are the matrices constructed from the selected rows and columns of the matrix $\mathbf{Y}$, respectively. Analogously, the reduced matrices: $\mathbf{A}_r \in \mathbb{R}^{R\times J}$ and $\mathbf{X}_c \in \mathbb{R}^{J\times C}$ are constructed by using the same indices for the columns and rows as those used for the construction of the data sub-matrices $\mathbf{Y}_c$ and $\mathbf{Y}_r$.

Upon imposing nonnegativity constraints, this leads to the following simple nonnegative ALS update formulas (see Figure 1.12):

$$\mathbf{A} \leftarrow \left[\mathbf{Y}_c\mathbf{X}_c^{\dagger}\right]_+ = \left[\mathbf{Y}_c\mathbf{X}_c^T(\mathbf{X}_c\mathbf{X}_c^T)^{-1}\right]_+, \tag{4.41}$$

$$\mathbf{X} \leftarrow \left[\mathbf{A}_r^{\dagger}\mathbf{Y}_r\right]_+ = \left[(\mathbf{A}_r^T\mathbf{A}_r)^{-1}\mathbf{A}_r^T\mathbf{Y}_r\right]_+ . \tag{4.42}$$

A similar approach can be applied for other cost functions. The details will be given later in this chapter.

4.2.2 ALS Algorithm with Line-Search

The line search procedure consists of the linear interpolation of the unknown factor matrices from their previous estimates [54,57,58]:

$$\mathbf{A}^{(new)} = \mathbf{A}^{(k-2)} + \eta_{\mathbf{A}}^{(k)} \left(\mathbf{A}^{(k-1)} - \mathbf{A}^{(k-2)}\right), \tag{4.43}$$

$$\mathbf{X}^{(new)} = \mathbf{X}^{(k-2)} + \eta_{\mathbf{X}}^{(k)} \left(\mathbf{X}^{(k-1)} - \mathbf{X}^{(k-2)}\right), \tag{4.44}$$

where $\mathbf{A}^{(k-1)}$ and $\mathbf{X}^{(k-1)}$ are the estimates of $\mathbf{A}$ and $\mathbf{X}$, respectively, obtained from the $(k-1)$-th ALS iteration; and $\eta_{\mathbf{A}}^{(k)}$, $\eta_{\mathbf{X}}^{(k)}$ are the step size (or relaxation factors) in the search directions. The known matrices defined as $\mathbf{S}_{\mathbf{A}}^{(k)} = \mathbf{A}^{(k-1)} - \mathbf{A}^{(k-2)}$ and $\mathbf{S}_{\mathbf{X}}^{(k)} = \mathbf{X}^{(k-1)} - \mathbf{X}^{(k-2)}$ represent the search directions in the k-th iteration. The interpolated matrices are used before as input of the ALS updates. In other words, this line search step is performed before applying each ALS iteration formula and the interpolated matrices $\mathbf{A}^{(new)}$ and $\mathbf{X}^{(new)}$ are then used to start the k-th iteration of the ALS algorithm. The main task of line search is to find an optimal or close to optimal step size η in the search directions in order to maximally speed up convergence. In the simplest scenario, the step size η has a fixed value prescribed (typically, $\eta = [1.1 - 1.4]$). In [54] $\eta = k^{1/3}$ is heuristically and experimentally used and the line search step is accepted only if the interpolated value of the cost function in the next iteration is less than its current value. For enhanced line search (ELS) technique we calculate the optimal step size by rooting a polynomial [54,58]. As a result, the convergence speed can be considerably improved by increasing complexity of the ALS algorithm.

4.2.3 Acceleration of ALS Algorithm via Simple Regularization

Alternatively, in order to accelerate the ALS algorithm instead of use the standard cost function we may try to minimize regularized cost functions [52]

$$\mathbf{A}^{(k+1)} = \arg\min_{\mathbf{A}} \left(||\mathbf{Y} - \mathbf{A}\mathbf{X}||_F^2 + \alpha^{(k)} \,||\mathbf{A} - \mathbf{A}^{(k)}||_F^2\right), \tag{4.45}$$

$$\mathbf{X}^{(k+1)} = \arg\min_{\mathbf{X}} \left(||\mathbf{Y} - \mathbf{A}\mathbf{X}||_F^2 + \alpha^{(k)} \,||\mathbf{X} - \mathbf{X}^{(k)}||_F^2\right), \tag{4.46}$$

where $\mathbf{A}^{(k)}$ and $\mathbf{X}^{(k)}$ are the estimates of $\mathbf{A}$ and $\mathbf{X}$, respectively, obtained from the (k)-th iteration and $\alpha^{(k)} > 0$ is regularization parameters which must be suitably chosen. The regularization terms $\alpha^{(k)} \; ||\mathbf{A} - \mathbf{A}^{(k)}||_F^2$ and $\alpha^{(k)} \; ||\mathbf{X} - \mathbf{X}^{(k)}||_F^2$ are Tikhonov-type regularization which stabilize the ALS algorithm for ill conditioning problems and simultaneously force some continuity of the solutions in the sense that factor (component) matrices $\mathbf{A}^{(k)}$ and $\mathbf{X}^{(k)}$ have the similar scaling and permutation of the previous iterations [52].

The choice of the regularization parameter α depends on level of noise $\mathbf{E} = \mathbf{Y} - \hat{\mathbf{Y}}$ which is unknown but often can be estimated or assumed *a priori*. If the level of noise can be estimated, we usually take $\alpha^* \cong \sigma_E^2$, where σ_E^2 is a variance of noise. In general, iterated Tikhonov regularization can be performed in two ways: Either $\alpha^{(k)} = \alpha^*$ is fixed and the iteration number is the main steering parameter or the maximum iteration number K_{max} is set *a priori* and $\alpha^{(k)}$ is tuned. Usually, we assume that the α decaying geometrically $\alpha^{(k)} = c\alpha^{(k-1)}$ with $0 < c < 1$ and we stop when $||\mathbf{Y} - \hat{\mathbf{Y}}|| \cong \sigma_E^2$ [52].

4.3 ALS Algorithm with Flexible and Generalized Regularization Terms

In practice, it is often convenient to use the squared Euclidean distance (squared Frobenius norm) with additional generalized regularization terms:

$$D_{Fr}(\mathbf{Y}||\mathbf{AX}) = \frac{1}{2}||\mathbf{Y} - \mathbf{AX}||_F^2 + \alpha_{\mathbf{A}} J_{\mathbf{A}}(\mathbf{A}) + \alpha_{\mathbf{X}} J_{\mathbf{X}}(\mathbf{X}) \tag{4.47}$$

$$\text{subject to } \; a_{ij} \geq 0, \quad x_{jt} \geq 0, \quad \forall \, i, \, j, \, t,$$

where $\alpha_{\mathbf{A}}$ and $\alpha_{\mathbf{X}}$ are nonnegative regularization parameters and the terms $J_{\mathbf{X}}(\mathbf{X})$ and $J_{\mathbf{A}}(\mathbf{A})$ are chosen to enforce certain desired properties of the solution (depending on the application considered).[5] As a special practical case, we have $J_{\mathbf{X}}(\mathbf{X}) = \sum_{jt} \varphi_{\mathbf{X}}(x_{jt})$, where $\varphi_{\mathbf{X}}(x_{jt})$ are suitably chosen functions which measure smoothness, sparsity or continuity. In order to achieve sparse representations, we usually choose $\varphi(x_{jt}) = |x_{jt}|$ or simply $\varphi(x_{jt}) = x_{jt}$ (due to nonnegativity constraints) or alternatively $\varphi(x_{jt}) = x_{jt} \ln(x_{jt})$ with constraints $x_{jt} \geq \varepsilon$. Similar regularization terms can also be implemented for matrix $\mathbf{A}$.[6]

By setting the gradient components to zero

$$\frac{\partial D_{Fr}(\mathbf{Y}||\mathbf{AX})}{\partial a_{ij}} = [-\mathbf{YX}^T + \mathbf{AXX}^T]_{ij} + \alpha_{\mathbf{A}} \frac{\partial J_{\mathbf{A}}(\mathbf{A})}{\partial a_{ij}} = 0, \tag{4.48}$$

$$\frac{\partial D_{Fr}(\mathbf{Y}||\mathbf{A}, \mathbf{X})}{\partial x_{jt}} = [-\mathbf{A}^T\mathbf{Y} + \mathbf{A}^T\mathbf{AX}]_{jt} + \alpha_{\mathbf{X}} \frac{\partial J_{\mathbf{X}}(\mathbf{X})}{\partial x_{jt}} = 0 \tag{4.49}$$

and additionally by applying the half-wave rectifying nonlinear projection (in order to enforce nonnegativity), we obtain the generalized form of regularized ALS (RALS) updating rules:

$$\mathbf{A} \leftarrow \left[(\mathbf{YX}^T - \alpha_{\mathbf{A}}\boldsymbol{\Psi}_{\mathbf{A}})(\mathbf{XX}^T)^{-1}\right]_+ , \tag{4.50}$$

$$\mathbf{X} \leftarrow \left[(\mathbf{A}^T\mathbf{A})^{-1}(\mathbf{A}^T\mathbf{Y} - \alpha_{\mathbf{X}}\boldsymbol{\Psi}_{\mathbf{X}})\right]_+ , \tag{4.51}$$

with the additional regularization terms defined as

$$\boldsymbol{\Psi}_{\mathbf{A}} = \frac{\partial J_{\mathbf{A}}(\mathbf{A})}{\partial \mathbf{A}} \in \mathbb{R}^{I \times J}, \qquad \boldsymbol{\Psi}_{\mathbf{X}} = \frac{\partial J_{\mathbf{X}}(\mathbf{X})}{\partial \mathbf{X}} \in \mathbb{R}^{J \times T}. \tag{4.52}$$

We next consider several special cases.

---

[5]Within the regularization framework, the minimization of the above cost function can be considered as a way to alleviate the ill-conditioned or near singular nature of matrix $\mathbf{A}$.

[6]Note that we treat both matrices $\mathbf{A}$ and $\mathbf{X}$ in a symmetric way.

4.3.1 ALS with Tikhonov Type Regularization Terms

For ill-conditioned problems in order to impose smoothness and/or boundness on the solution, we may apply the standard Tikhonov (ℓ_2-norm) regularization, or more generally, we may employ the following cost function:

$$D_{F2}(\mathbf{Y}||\mathbf{A},\mathbf{X}) = \frac{1}{2}\left(\|\mathbf{Y}-\mathbf{A}\mathbf{X}\|_F^2 + \alpha_{\mathbf{A}r}\,\|\mathbf{A}\|_F^2 + \alpha_{\mathbf{X}r}\,\|\mathbf{X}\|_F^2\right), \tag{4.53}$$

$$\text{subject to } a_{ij}\geq 0, \quad x_{jt}\geq 0, \quad \forall\, i, j, t,$$

where $\alpha_{\mathbf{A}r}\geq 0$ and $\alpha_{\mathbf{X}r}\geq 0$ are the regularization parameters. The minimization of this cost function leads to the following update rules:

$$\mathbf{A} \leftarrow \left[(\mathbf{Y}\mathbf{X}^T)(\mathbf{X}\mathbf{X}^T + \alpha_{\mathbf{A}r}\mathbf{I})^{-1}\right]_+, \tag{4.54}$$

$$\mathbf{X} \leftarrow \left[(\mathbf{A}^T\mathbf{A} + \alpha_{\mathbf{X}r}\mathbf{I})^{-1}(\mathbf{A}^T\mathbf{Y})\right]_+. \tag{4.55}$$

The Tikhonov-type regularization terms may take more general forms:

$$D_{FT}^{(\alpha)}(\mathbf{Y}||\mathbf{A}\mathbf{X}) = \frac{1}{2}||(\mathbf{Y}-\mathbf{A}\mathbf{X})||_F^2 + \frac{\alpha_{\mathbf{A}r}}{2}\,||\mathbf{A}\mathbf{L}_\mathbf{A}||_F^2 + \frac{\alpha_{\mathbf{X}r}}{2}\,||\mathbf{L}_\mathbf{X}\mathbf{X}||_F^2, \tag{4.56}$$

where the regularization matrices $\mathbf{L}_\mathbf{A}$ and $\mathbf{L}_\mathbf{X}$ are used to enforce certain application-dependent characteristics of the solution. These matrices are typically unit diagonal matrices or discrete approximations to some derivative operators.

Typical examples of the matrix $\mathbf{L}$ are: the $1st$ derivative approximation $\mathbf{L}_1 \in \mathbb{R}^{(J-1)\times J}$ and the $2nd$ derivative approximation $\mathbf{L}_2 \in \mathbb{R}^{(J-2)\times J}$, given by

$$\mathbf{L}_1 = \begin{bmatrix} 1 & -1 & & & 0\\ & 1 & -1 & & \\ \vdots & & \ddots & \ddots & \vdots \\ 0 & & & 1 & -1\end{bmatrix}, \qquad \mathbf{L}_2 = \begin{bmatrix} -1 & 2 & -1 & & & 0\\ & -1 & 2 & -1 & & \\ \vdots & & \ddots & \ddots & \ddots & \\ 0 & & & -1 & 2 & -1\end{bmatrix}. \tag{4.57}$$

Another option is to use the following setting: $\alpha_{\mathbf{X}r}\mathbf{L}_\mathbf{X}^T\mathbf{L}_\mathbf{X} = \mathbf{A}_\mathbf{Y}^T(\mathbf{I}-\mathbf{A}_\mathbf{Y}\mathbf{A}_\mathbf{Y}^T)\mathbf{A}_\mathbf{Y}$, where $\mathbf{A}_\mathbf{Y}$ contains the first J principal eigenvectors of the data covariance matrix $\mathbf{R}_\mathbf{Y} = \mathbf{Y}^T\mathbf{Y} = \mathbf{U}\boldsymbol{\Sigma}\mathbf{U}^\mathbf{T}$ associated with the largest J singular values [13] (see also Appendix in Chapter 1). It is worth noting that both matrices $\mathbf{L}_\mathbf{X}^T\mathbf{L}_\mathbf{X} \in \mathbb{R}^{J\times J}$ and $\mathbf{L}_\mathbf{A}\mathbf{L}_\mathbf{A}^T \in \mathbb{R}^{J\times J}$ are general symmetric and positive-definite matrices.

The gradients of the cost function (4.56) with respect to the unknown matrices $\mathbf{A}$ and $\mathbf{X}$ are expressed by

$$\frac{\partial D_{FT}(\mathbf{Y}||\mathbf{A}\mathbf{X})}{\partial \mathbf{A}} = (\mathbf{A}\mathbf{X}-\mathbf{Y})\mathbf{X}^T + \alpha_{\mathbf{A}r}\,\mathbf{A}\,\mathbf{L}_\mathbf{A}\mathbf{L}_\mathbf{A}^T, \tag{4.58}$$

$$\frac{\partial D_{FT}(\mathbf{Y}||\mathbf{A}\mathbf{X})}{\partial \mathbf{X}} = \mathbf{A}^T(\mathbf{A}\mathbf{X}-\mathbf{Y}) + \alpha_{\mathbf{X}r}\,\mathbf{L}_\mathbf{X}^T\mathbf{L}_\mathbf{X}\,\mathbf{X}, \tag{4.59}$$

and by setting them to zero, we obtain the following regularized ALS algorithm

$$\mathbf{A} \leftarrow \left[(\mathbf{Y}\,\mathbf{X}^T)(\mathbf{X}\mathbf{X}^T + \alpha_{\mathbf{A}r}\,\mathbf{L}_\mathbf{A}\mathbf{L}_\mathbf{A}^T)^{-1}\right]_+, \tag{4.60}$$

$$\mathbf{X} \leftarrow \left[(\mathbf{A}^T\mathbf{A} + \alpha_{\mathbf{X}r}\,\mathbf{L}_\mathbf{X}^T\mathbf{L}_\mathbf{X})^{-1}\mathbf{A}^T\mathbf{Y}\right]_+. \tag{4.61}$$

4.3.2 ALS Algorithms with Sparsity Control and Decorrelation

To enforce sparsity, we can use ℓ_1-norm regularization terms and minimize the following cost function:

$$D_{F1}(\mathbf{Y}||\mathbf{A},\mathbf{X}) = \frac{1}{2}||\mathbf{Y}-\mathbf{A}\mathbf{X}||_F^2 + \alpha_{\mathbf{A}s}||\mathbf{A}||_1 + \alpha_{\mathbf{X}s}||\mathbf{X}||_1, \tag{4.62}$$

$$\text{subject to } a_{ij} \geq 0, \quad x_{jt} \geq 0, \quad \forall\, i, j, t,$$

where ℓ_1-norm of matrices are defined as $||\mathbf{A}||_1 = \sum_{ij} |a_{ij}| = \sum_{ij} a_{ij}$ and $||\mathbf{X}||_1 = \sum_{jt} |x_{jt}| = \sum_{jt} x_{jt}$ and $\alpha_{\mathbf{A}s}$, $\alpha_{\mathbf{X}s}$ are nonnegative parameters which control sparsity levels. In this special case, by using the ℓ_1-norm regularization terms for both matrices $\mathbf{X}$ and $\mathbf{A}$, the above ALS update rules can be simplified as follows:

$$\mathbf{A} \leftarrow \left[(\mathbf{Y}\mathbf{X}^T - \alpha_{\mathbf{A}s}\mathbf{1}_{I\times J})(\mathbf{X}\mathbf{X}^T)^{-1}\right]_+, \tag{4.63}$$

$$\mathbf{X} \leftarrow \left[(\mathbf{A}^T\mathbf{A})^{-1}(\mathbf{A}^T\mathbf{Y} - \alpha_{\mathbf{X}s}\mathbf{1}_{J\times T})\right]_+, \tag{4.64}$$

with the normalization of the columns of matrix $\mathbf{A}$ in each iteration given by $a_{ij} \leftarrow a_{ij}/\sum_i a_{ij}$, where $\mathbf{1}_{I\times J}$ denotes a $I \times J$ matrix with all entries one.[7]

The above algorithm provides a sparse representation of the estimated matrices. The sparsity level is pronounced with increase in the values of regularization coefficients (typically, $\alpha_{\mathbf{A}} = \alpha_{\mathbf{X}} = 0.01 - 0.5$). The above update rules provide sparse nonnegative representations of data, although they do not necessarily guarantee the sparsest possible solution. A much better performance can be achieved by applying the multilayer NMF structure, as explained in Chapter 1.

In practice, it is often necessary to enforce uncorrelatedness among all the nonnegative components by employing the following cost function:

$$\begin{aligned} D_{FD}(\mathbf{Y}||\mathbf{A}\mathbf{X}) = \;& \frac{1}{2}\left[||\mathbf{Y}-\mathbf{A}\mathbf{X}||_F^2 + \alpha_{\mathbf{A}r}\,\mathrm{tr}\{\mathbf{A}\mathbf{1}_{J\times J}\mathbf{A}^T\} + \alpha_{\mathbf{X}r}\,\mathrm{tr}\{\mathbf{X}^T\mathbf{1}_{J\times J}\mathbf{X}\}\right] + \\ & + \alpha_{\mathbf{A}s}||\mathbf{A}||_1 + \alpha_{\mathbf{X}s}||\mathbf{X}||_1, . \end{aligned} \tag{4.65}$$

where the regularization terms ensure that individual nonnegative components (the rows of matrix $\mathbf{X}$ and the columns of matrix $\mathbf{A}$) are as uncorrelated as possible. To understand the uncorrelated penalty term, we consider the following derivation

$$\begin{aligned} J_{cr} &= \frac{1}{2}\alpha_{\mathbf{A}r}\,\mathrm{tr}\{\mathbf{A}\mathbf{1}_{J\times J}\mathbf{A}^T\} = \frac{1}{2}\alpha_{\mathbf{A}r}\,\mathrm{tr}\{\mathbf{A}\,\mathbf{1}\,\mathbf{1}^T\,\mathbf{A}^T\} = \frac{1}{2}\alpha_{\mathbf{A}r}||\mathbf{A}\,\mathbf{1}||_2^2 \\ &= \frac{1}{2}\alpha_{\mathbf{A}r}\sum_i ||\underline{\boldsymbol{a}}_i||_1^2 = \frac{1}{2}\alpha_{\mathbf{A}r}\sum_i\left(\sum_j a_{ij}^2 + 2\sum_{k\neq l} a_{ik}a_{il}\right) \\ &= \frac{1}{2}\alpha_{\mathbf{A}r}\left(\sum_j\sum_i a_{ij}^2 + 2\sum_{k\neq l}\sum_i a_{ik}a_{il}\right) = \frac{1}{2}\alpha_{\mathbf{A}r}\left(\sum_j ||\boldsymbol{a}_j||_2^2 + 2\sum_{k\neq l}\boldsymbol{a}_k^T\,\boldsymbol{a}_l\right) \\ &= \frac{1}{2}\alpha_{\mathbf{A}r}J + \alpha_{\mathbf{A}r}\sum_{k\neq l}\boldsymbol{a}_k^T\,\boldsymbol{a}_l. \end{aligned} \tag{4.66}$$

The column vectors $\boldsymbol{a}_j$ are usually normalized in terms of the ℓ_p-norm (typically, $p = 2$): $||\boldsymbol{a}_j||_2^2 = 1$, therefore, the first term in (4.66) is considered as a constant, whereas the second term expresses the correlation

[7] In general, the columns of a matrix should be normalized in sense of the ℓ_p-norm for any $p \in \{1, \infty\}$.

of these vectors. This helps to explain that the penalty term J_{cr} can impose the uncorrelatedness constraints on the cost function (4.65).

For the cost function (4.65) gradient components can be expressed as

$$\nabla_{\mathbf{A}} D_{FD} = -\mathbf{Y}\mathbf{X}^T + \mathbf{A}\mathbf{X}\mathbf{X}^T + \alpha_{\mathbf{A}r}\mathbf{A}\mathbf{1}_{J\times J} + \alpha_{\mathbf{A}s}\mathbf{1}_{I\times J}, \tag{4.67}$$

$$\nabla_{\mathbf{X}} D_{FD} = \mathbf{A}^T(\mathbf{A}\mathbf{X} - \mathbf{Y}) + \alpha_{\mathbf{X}r}\mathbf{1}_{J\times J}\mathbf{X} + \alpha_{\mathbf{X}s}\mathbf{1}_{J\times T}, \tag{4.68}$$

which lead to the following regularized ALS (RALS) update rules:

$$\mathbf{A} \leftarrow \left[(\mathbf{Y}\mathbf{X}^T - \alpha_{\mathbf{A}s}\mathbf{1}_{I\times J})(\mathbf{X}\mathbf{X}^T + \alpha_{\mathbf{A}r}\mathbf{1}_{J\times J})^{-1}\right]_+, \tag{4.69}$$

$$\mathbf{X} \leftarrow \left[(\mathbf{A}^T\mathbf{A} + \alpha_{\mathbf{X}r}\mathbf{1}_{J\times J})^{-1}(\mathbf{A}^T\mathbf{Y} - \alpha_{\mathbf{X}s}\mathbf{1}_{J\times T})\right]_+. \tag{4.70}$$

In order to obtain a good performance, the regularization parameters $\alpha_{\mathbf{A}r} \geq 0$ and $\alpha_{\mathbf{X}r} \geq 0$ should not be fixed but rather change during iteration process, depending on how far the current solution is from the desired solution. We have found by numerical experiments that a good performance for small-scale problems can be achieved by choosing $\alpha_{\mathbf{A}r}(k) = \alpha_{\mathbf{X}r}(k) = \alpha_0 \exp(-k/\tau)$ with typical values of $\alpha_0 = 20$ and $\tau = 50$ [18,65].

4.4 Combined Generalized Regularized ALS Algorithms

In practice, we may combine several regularization terms to satisfy specific constraints together with relaxing nonnegativity constraints by restricting them to only a preselected set of parameters (**A** and/or **X**), if necessary.

Consider the minimization of a more general and flexible cost function in the form of a regularized weighted least-squares function with additional smoothing and sparsity penalties:

$$D_F^{(\alpha)}(\mathbf{Y}||\mathbf{A}\mathbf{X}) = \frac{1}{2}||\mathbf{W}^{1/2}(\mathbf{Y} - \mathbf{A}\mathbf{X})||_F^2 + \alpha_{\mathbf{A}s}\,||\mathbf{A}||_1 + \alpha_{\mathbf{X}s}\,||\mathbf{X}||_1 + \\ + \frac{\alpha_{\mathbf{A}r}}{2}\,||\mathbf{W}^{1/2}\mathbf{A}\mathbf{L}_{\mathbf{A}}||_F^2 + \frac{\alpha_{\mathbf{X}r}}{2}\,||\mathbf{L}_{\mathbf{X}}\mathbf{X}||_F^2, \tag{4.71}$$

where $\mathbf{W} \in \mathbb{R}^{I\times I}$ is a predefined symmetric positive-definite weighting matrix, $\alpha_{\mathbf{A}s} \geq 0$ and $\alpha_{\mathbf{X}s} \geq 0$ are parameters controlling the degree of sparsity of the matrices, and $\alpha_{\mathbf{A}r} \geq 0$, $\alpha_{\mathbf{X}r} \geq 0$ are the regularization coefficients controlling the smoothness and continuity of the solution. The penalty terms $||\mathbf{A}||_1 = \sum_{ij} |a_{ij}|$ and $||\mathbf{X}||_1 = \sum_{jt} |x_{jt}|$ enforce sparsity in **A** and **X**, respectively.

The gradients of the cost function (4.71) with respect to the unknown matrices **A** and **X** are given by

$$\nabla_{\mathbf{A}} D_F^{(\alpha)}(\mathbf{Y}||\mathbf{A}\mathbf{X}) = \mathbf{W}(\mathbf{A}\mathbf{X} - \mathbf{Y})\mathbf{X}^T + \alpha_{\mathbf{A}s}\,\mathbf{S}_{\mathbf{A}} + \alpha_{\mathbf{A}r}\,\mathbf{W}\mathbf{A}\,\mathbf{L}_{\mathbf{A}}\mathbf{L}_{\mathbf{A}}^T, \tag{4.72}$$

$$\nabla_{\mathbf{X}} D_F^{(\alpha)}(\mathbf{Y}||\mathbf{A}\mathbf{X}) = \mathbf{A}^T\mathbf{W}(\mathbf{A}\mathbf{X} - \mathbf{Y}) + \alpha_{\mathbf{X}s}\,\mathbf{S}_{\mathbf{X}} + \alpha_{\mathbf{X}r}\,\mathbf{L}_{\mathbf{X}}^T\mathbf{L}_{\mathbf{X}}\,\mathbf{X}, \tag{4.73}$$

where $\mathbf{S}_{\mathbf{A}} = \text{sign}(\mathbf{A})$ and $\mathbf{S}_{\mathbf{X}} = \text{sign}(\mathbf{X})$.[8] For a particular case of the NMF problem, the matrices $\mathbf{S}_{\mathbf{A}}$, $\mathbf{S}_{\mathbf{X}}$ are the matrices $\mathbf{1}_{\mathbf{A}} = \mathbf{1}_{I\times J}$ and $\mathbf{1}_{\mathbf{X}} = \mathbf{1}_{J\times T}$ of the same dimension, for which all the entries have unit values.

By equating the gradients (4.72)-(4.73) to zero, we obtain the following regularized ALS algorithm

$$\mathbf{A} \leftarrow (\mathbf{Y}\mathbf{X}^T - \alpha_{\mathbf{A}s}\mathbf{W}^{-1}\mathbf{S}_{\mathbf{A}})(\mathbf{X}\mathbf{X}^T + \alpha_{\mathbf{A}r}\,\mathbf{L}_{\mathbf{A}}\mathbf{L}_{\mathbf{A}}^T)^{-1}, \tag{4.74}$$

$$\mathbf{X} \leftarrow (\mathbf{A}^T\mathbf{W}\mathbf{A} + \alpha_{\mathbf{X}r}\,\mathbf{L}_{\mathbf{X}}^T\mathbf{L}_{\mathbf{X}})^{-1}(\mathbf{A}^T\mathbf{W}\mathbf{Y} - \alpha_{\mathbf{X}s}\,\mathbf{S}_{\mathbf{X}}). \tag{4.75}$$

In order to achieve a high performance, the regularization parameters $\alpha_{\mathbf{A}r} \geq 0$ and $\alpha_{\mathbf{X}r} \geq 0$ are usually not fixed but are made adaptive, for example, they exponentially decrease during the convergence process.

[8] The symbol sign(**X**) denotes a component-wise sign operation (or its robust approximation).

4.5 Wang-Hancewicz Modified ALS Algorithm

An alternative approach is to keep the regularization parameters fixed and try to compensate (reduce) their influence by additional terms as the algorithm converges to the desired solution. To achieve this consider the following approach [43,64]: From (4.75), we have:

$$(\mathbf{A}^T\mathbf{W}\mathbf{A} + \alpha_{\mathbf{X}_r}\,\mathbf{L}_{\mathbf{X}}^T\mathbf{L}_{\mathbf{X}})\mathbf{X}_{new} = \mathbf{A}^T\mathbf{W}\mathbf{Y} - \alpha_{\mathbf{X}_s}\,\mathbf{S}_{\mathbf{X}}. \tag{4.76}$$

In order to compensate for the regularization term $\alpha_{\mathbf{X}_r}\,\mathbf{L}_{\mathbf{X}}^T\mathbf{L}_{\mathbf{X}}\mathbf{X}_{new}$, we can add a similar term $\alpha_{\mathbf{X}_r}\,\mathbf{L}_{\mathbf{X}}^T\mathbf{L}_{\mathbf{X}}\mathbf{X}_{old}$ to the right-hand side, which gradually compensates the regularization term when $\mathbf{X} \rightarrow \mathbf{X}^*$, that is

$$(\mathbf{A}^T\mathbf{W}\mathbf{A} + \alpha_{\mathbf{X}_r}\,\mathbf{L}_{\mathbf{X}}^T\mathbf{L}_{\mathbf{X}})\mathbf{X}_{new} = \mathbf{A}^T\mathbf{W}\mathbf{Y} - \alpha_{\mathbf{X}_s}\,\mathbf{S}_{\mathbf{X}} + \alpha_{\mathbf{X}_r}\,\mathbf{L}_{\mathbf{X}}^T\mathbf{L}_{\mathbf{X}}\mathbf{X}_{old}.$$

The value of the bias (or influence of the regularization term) is a function of the difference between $\mathbf{X}_{old}$ and $\mathbf{X}_{new}$. As the algorithm converges to the desired solution $\mathbf{X}^*$, this difference is gradually decreasing along with the effect of regularization and bias (which becomes smaller and smaller).

After some simple mathematical manipulations, the regularized modified ALS (RMALS) algorithm takes the following general (without nonnegativity constraints) form:

$$\mathbf{A} \leftarrow (\mathbf{Y}\mathbf{X}^T - \alpha_{\mathbf{A}_s}\mathbf{W}^{-1}\mathbf{S}_{\mathbf{A}} + \alpha_{\mathbf{A}_r}\,\mathbf{A}\mathbf{L}_{\mathbf{A}}\mathbf{L}_{\mathbf{A}}^T)(\mathbf{X}\mathbf{X}^T + \alpha_{\mathbf{A}_r}\,\mathbf{L}_{\mathbf{A}}\mathbf{L}_{\mathbf{A}}^T)^{-1}, \tag{4.77}$$

$$\mathbf{X} \leftarrow (\mathbf{A}^T\mathbf{W}\mathbf{A} + \alpha_{\mathbf{X}_r}\mathbf{L}_{\mathbf{X}}^T\mathbf{L}_{\mathbf{X}})^{-1}(\mathbf{A}^T\mathbf{W}\mathbf{Y} - \alpha_{\mathbf{X}_s}\mathbf{S}_{\mathbf{X}} + \alpha_{\mathbf{X}_r}\mathbf{L}_{\mathbf{X}}^T\mathbf{L}_{\mathbf{X}}\mathbf{X}). \tag{4.78}$$

For $\mathbf{W} = \mathbf{I}_I \in \mathbb{R}^{I\times I}$ and for all the regularization coefficients set to zero ($\alpha_{\mathbf{A}_s} = \alpha_{\mathbf{A}_r} = \alpha_{\mathbf{X}_s} = \alpha_{\mathbf{X}_r} = 0$), this algorithm simplifies into the standard ALS. On the other hand, if all the regularization parameters are set to zero and $\mathbf{W} = \mathbf{R}_E^{-1} = (\mathbf{E}\mathbf{E}^T/I)^{-1}$, where the error matrix $\mathbf{E} = \mathbf{Y} - \mathbf{A}\mathbf{X}$ is evaluated for each iteration step, we obtain the extended BLUE (Best Linear Unbiased Estimator) ALS algorithm. Finally, in the special case when $\mathbf{W} = \mathbf{I}_I$ and matrices $\mathbf{L}_{\mathbf{A}}\mathbf{L}_{\mathbf{A}}^T$ and $\mathbf{L}_{\mathbf{X}}^T\mathbf{L}_{\mathbf{X}}$ are diagonal, this algorithm is similar to the MALS (modified ALS) proposed by Hancewicz and Wang [43] and Wang *et al.* [64].

Finally, for the standard NMF, the RMALS update rules take the following forms:

$$\mathbf{A} \leftarrow \left[(\mathbf{Y}\mathbf{X}^T - \alpha_{\mathbf{A}_s}\mathbf{W}^{-1}\mathbf{1}_{I\times J} + \alpha_{\mathbf{A}_r}\,\mathbf{A}\mathbf{L}_{\mathbf{A}}\mathbf{L}_{\mathbf{A}}^T)(\mathbf{X}\mathbf{X}^T + \alpha_{\mathbf{A}_r}\,\mathbf{L}_{\mathbf{A}}\mathbf{L}_{\mathbf{A}}^T)^{-1}\right]_+, \tag{4.79}$$

$$\mathbf{X} \leftarrow \left[(\mathbf{A}^T\mathbf{W}\mathbf{A} + \alpha_{\mathbf{X}_r}\mathbf{L}_{\mathbf{X}}^T\mathbf{L}_{\mathbf{X}})^{-1}(\mathbf{A}^T\mathbf{W}\mathbf{Y} - \alpha_{\mathbf{X}_s}\mathbf{1}_{J\times T} + \alpha_{\mathbf{X}_r}\mathbf{L}_{\mathbf{X}}^T\mathbf{L}_{\mathbf{X}}\mathbf{X})\right]_+. \tag{4.80}$$

4.6 Implementation of Regularized ALS Algorithms for NMF

Based on the algorithms introduced in this chapter, Algorithm 4.1 gives the pseudo-code for the RALS algorithm which is also implemented in the NMFLAB and NTFLAB toolboxes [20].

Further improvement of the RALS algorithm can be achieved by applying the multi-stage or multi-layer system with multi-start initialization [14,20], implemented as follows: In the first step, we perform the basic decomposition (factorization) $\mathbf{Y} = \mathbf{A}_1\mathbf{X}_1$ using the RALS algorithm. In the second stage, the results obtained from the first stage are used to perform a similar decomposition: $\mathbf{X}_1 = \mathbf{A}_2\mathbf{X}_2$ using the same or different set of parameters, and so on. We continue our factorization taking into account only the last obtained components. The process can be arbitrarily repeated many times until some stopping criteria are met. In each step, we usually obtain gradual improvements of the performance. Thus, our model has the form: $\mathbf{Y} = \mathbf{A}_1\mathbf{A}_2\cdots\mathbf{A}_L\mathbf{X}_L$, with the basis nonnegative matrix defined as $\mathbf{A} = \mathbf{A}_1\mathbf{A}_2\cdots\mathbf{A}_L$.

Remark 4.1 *An open theoretical issue is to prove mathematically or explain more rigorously why the multi-layer distributed NMF/NTF system results in considerable improvement in performance and reduces the risk of getting stuck at local minima. An intuitive explanation is as follows: the multi-layer system provides a*

Algorithm 4.1: RALS

Input: $\mathbf{Y} \in \mathbb{R}_+^{I \times T}$: input data, J: rank of approximation,
$\mathbf{W}$: weighting matrix, $\mathbf{L_A}$, $\mathbf{L_X}$: regularization matrices for $\mathbf{A}$ and $\mathbf{X}$
$\alpha_{\mathbf{A}_s}$, $\alpha_{\mathbf{X}_s}$: sparsity levels, $\alpha_{\mathbf{A}_r}$, $\alpha_{\mathbf{X}_r}$: smoothness, continuity levels
Output: $\mathbf{A} \in \mathbb{R}_+^{I \times J}$ and $\mathbf{X} \in \mathbb{R}_+^{J \times T}$ such that the cost function (4.71) is minimized.

begin
  initialization for $\mathbf{A}$
  $\mathbf{C_A} = \alpha_{\mathbf{A}_r} \mathbf{L_A} \mathbf{L_A}^T$
  $\mathbf{C_X} = \alpha_{\mathbf{X}_r} \mathbf{L_X}^T \mathbf{L_X}$
  repeat
    $\mathbf{X} \leftarrow \left[(\mathbf{A}^T\mathbf{W}\mathbf{A} + \mathbf{C_X})^{-1}(\mathbf{A}^T\mathbf{W}\mathbf{Y} - \alpha_{\mathbf{X}_s}\mathbf{1}_{J\times T} + \mathbf{C_X}\mathbf{X})\right]_+$ /* update X */
    $\mathbf{A} \leftarrow \left[(\mathbf{Y}\mathbf{X}^T - \alpha_{\mathbf{A}_s}\mathbf{W}^{-1}\mathbf{1}_{I\times J} + \mathbf{A}\,\mathbf{C_A})(\mathbf{X}\mathbf{X}^T + \mathbf{C_A})^{-1}\right]_+$ /* update A */
    foreach $\boldsymbol{a}_j$ *of* $\mathbf{A}$ **do** $\boldsymbol{a}_j \leftarrow \boldsymbol{a}_j / \|\boldsymbol{a}_j\|_p$ /* normalize to ℓ_p unit length */
  until *a stopping criterion is met* /* convergence condition */
end

sparse distributed representation of basis matrix **A**, *which in general, can be a dense matrix. So, even if a true basis matrix* **A** *is not sparse, it can still be represented by a product of sparse factors, thus in each layer we force (or encourage) a sparse representation. On the other hand, we found by extensive experimentation that if the bases matrix is very sparse, most NTF/NMF algorithms exhibit improved performance (see next section). However, not all real-world data provide sufficiently sparse representations, so the main idea is to model arbitrary data by a distributed sparse multi-layer system. It is also interesting to note that such multi-layer systems are biologically motivated and plausible.*

4.7 HALS Algorithm and its Extensions

4.7.1 *Projected Gradient Local Hierarchical Alternating Least Squares (HALS) Algorithm*

This section focuses on derivation and analysis of a simple local ALS method referred to as the HALS (Hierarchical Alternating Least Squares).[9] We highlight the suitability of this method for large-scale NMF problems, and sparse nonnegative coding or representation [16,17,22]. To derive the algorithm we use a set of local cost functions such as the squared Euclidean distance, Alpha- and Beta-divergences and perform simultaneous or sequential (one-by-one) minimization of these local cost functions, for instance, using a projected gradient or some nonlinear transformations. The family of HALS algorithms is shown to work well not only for the over-determined case, but also (under some weak conditions) for an underdetermined (over-complete) case (i.e., for a system which has less number of sensors than the sources). The extensive experimental results confirm the validity and high performance of the family of HALS algorithms, especially for the multi-layer approach (see Chapter 1) [16,17,22].

[9] We call the developed ALS algorithm hierarchical, since we minimize sequentially a set of simple cost functions which are linked to each other hierarchically via residual matrices $\mathbf{Y}^{(j)}$ which approximate rank-one bilinear decomposition. Moreover, the HALS algorithm is usually used for multi-layer model in order to improve performance.

Denote $\mathbf{A} = [\boldsymbol{a}_1, \boldsymbol{a}_1, \ldots, \boldsymbol{a}_J]$ and $\mathbf{B} = \mathbf{X}^T = [\boldsymbol{b}_1, \boldsymbol{b}_2, \ldots, \boldsymbol{b}_J]$, to express the squared Euclidean cost function as

$$\begin{aligned} J(\boldsymbol{a}_1, \ldots, \boldsymbol{a}_J, \boldsymbol{b}_1, \ldots, \boldsymbol{b}_J) &= \frac{1}{2}||\mathbf{Y} - \mathbf{A}\mathbf{B}^T||_F^2 \\ &= \frac{1}{2}||\mathbf{Y} - \sum_{j=1}^{J} \boldsymbol{a}_j \boldsymbol{b}_j^T||_F^2. \end{aligned} \tag{4.81}$$

A basic idea is to define the residues as:

$$\begin{aligned} \mathbf{Y}^{(j)} &= \mathbf{Y} - \sum_{p \neq j} \boldsymbol{a}_p \boldsymbol{b}_p^T = \mathbf{Y} - \mathbf{A}\mathbf{B}^T + \boldsymbol{a}_j \boldsymbol{b}_j^T \\ &= \mathbf{E} + \boldsymbol{a}_j \boldsymbol{b}_j^T \qquad (j = 1, 2, \ldots, J), \end{aligned} \tag{4.82}$$

and minimize alternatively the set of cost functions (with respect to a set of parameters $\{\boldsymbol{a}_j\}$ and $\{\boldsymbol{b}_j\}$):

$$D_A^{(j)}(\boldsymbol{a}_j) = \frac{1}{2}||\mathbf{Y}^{(j)} - \boldsymbol{a}_j \boldsymbol{b}_j^T||_F^2, \qquad \text{for a fixed } \boldsymbol{b}_j, \tag{4.83}$$

$$D_B^{(j)}(\boldsymbol{b}_j) = \frac{1}{2}||\mathbf{Y}^{(j)} - \boldsymbol{a}_j \boldsymbol{b}_j^T||_F^2, \qquad \text{for a fixed } \boldsymbol{a}_j, \qquad (j = 1, 2, \ldots, J), \tag{4.84}$$

subject to $\boldsymbol{a}_j \geq \mathbf{0}$ and $\boldsymbol{b}_j \geq \mathbf{0}$.

In other words, we perform alternatively minimization of the set of cost functions

$$D_F^{(j)}(\mathbf{Y}^{(j)}||\boldsymbol{a}_j \boldsymbol{b}_j^T) = \frac{1}{2}||\mathbf{Y}^{(j)} - \boldsymbol{a}_j \boldsymbol{b}_j^T||_F^2, \tag{4.85}$$

for $j = 1, 2, \ldots, J$ subject to $\boldsymbol{a}_j \geq \mathbf{0}$ and $\boldsymbol{b}_j \geq \mathbf{0}$.

The KKT optimal conditions for the set of cost functions (4.85) can be expressed as

$$\boldsymbol{a}_j \geq \mathbf{0}, \qquad \boldsymbol{b}_j \geq \mathbf{0}, \tag{4.86}$$

$$\nabla_{\boldsymbol{a}_j} D_F^{(j)}(\mathbf{Y}^{(j)}||\boldsymbol{a}_j \boldsymbol{b}_j^T) \geq \mathbf{0}, \qquad \nabla_{\boldsymbol{b}_j} D_F^{(j)}(\mathbf{Y}^{(j)}||\boldsymbol{a}_j \boldsymbol{b}_j^T) \geq \mathbf{0}, \tag{4.87}$$

$$\boldsymbol{a}_j \circledast \nabla_{\boldsymbol{a}_j} D_F^{(j)}(\mathbf{Y}^{(j)}||\boldsymbol{a}_j \boldsymbol{b}_j^T) = \mathbf{0}, \qquad \boldsymbol{b}_j \circledast \nabla_{\boldsymbol{b}_j} D_F^{(j)}(\mathbf{Y}^{(j)}||\boldsymbol{a}_j \boldsymbol{b}_j^T) = \mathbf{0}. \tag{4.88}$$

In order to estimate the stationary points, we simply compute the gradients of the local cost functions (4.85) with respect to the unknown vectors $\boldsymbol{a}_j$ and $\boldsymbol{b}_j$ (assuming that other vectors are fixed) as follows

$$\nabla_{\boldsymbol{a}_j} D_F^{(j)}(\mathbf{Y}^{(j)}||\boldsymbol{a}_j \boldsymbol{b}_j^T) = \frac{\partial D_F^{(j)}(\mathbf{Y}^{(j)}||\boldsymbol{a}_j \boldsymbol{b}_j^T)}{\partial \boldsymbol{a}_j} = \boldsymbol{a}_j \boldsymbol{b}_j^T \boldsymbol{b}_j - \mathbf{Y}^{(j)} \boldsymbol{b}_j, \tag{4.89}$$

$$\nabla_{\boldsymbol{b}_j} D_F^{(j)}(\mathbf{Y}^{(j)}||\boldsymbol{a}_j \boldsymbol{b}_j^T) = \frac{\partial D_F^{(j)}(\mathbf{Y}^{(j)}||\boldsymbol{a}_j \boldsymbol{b}_j^T)}{\partial \boldsymbol{b}_j} = \boldsymbol{a}_j^T \boldsymbol{a}_j \boldsymbol{b}_j - \mathbf{Y}^{(j)T} \boldsymbol{a}_j. \tag{4.90}$$

Assuming that the entries of vectors $\boldsymbol{a}_j$ and $\boldsymbol{b}_j \; \forall j$ have positive entries the stationary points can be estimated via the following simple updates and illustrated in Figure 4.1:

$$\boldsymbol{b}_j \leftarrow \frac{1}{\boldsymbol{a}_j^T \boldsymbol{a}_j} \left[\mathbf{Y}^{(j)T} \boldsymbol{a}_j\right]_+ = \frac{1}{\boldsymbol{a}_j^T \boldsymbol{a}_j} \max\{\varepsilon, \mathbf{Y}^{(j)T} \boldsymbol{a}_j\}, \tag{4.91}$$

$$\boldsymbol{a}_j \leftarrow \frac{1}{\boldsymbol{b}_j^T \boldsymbol{b}_j} \left[\mathbf{Y}^{(j)} \boldsymbol{b}_j\right]_+ = \frac{1}{\boldsymbol{b}_j^T \boldsymbol{b}_j} \max\{\varepsilon, \mathbf{Y}^{(j)} \boldsymbol{b}_j\}, \quad (j = 1, 2, \ldots, J). \tag{4.92}$$

We refer to these update rules as the HALS algorithm which was first introduced for the NMF by Cichocki *et al.* in [22]. The same or similar update rules for the NMF have been proposed, extended or rediscovered independently in [3,39,40,44,45]. However, our practical implementations of the HALS algorithm are quite different and allow various extensions to sparse and smooth NMF, and also for the N-order NTF (see Chapter 7). It has been recently proved by Gillis and Glineur [39,40] that for every constant $\varepsilon > 0$ the limit points of the HALS algorithm initialized with positive matrices and applied to the optimization problem (4.83)–(4.84) are stationary points (see also [44]) which can also be deduced from the KKT conditions (4.88).

The nonlinear projections can be imposed individually for each source $\underline{\boldsymbol{x}}_j = \boldsymbol{b}_j^T$ and/or vector $\boldsymbol{a}_j$, so the algorithm can be directly extended to a semi-NMF, in which some parameters are relaxed to be unconstrained (by removing the half-wave rectifying $[\cdot]_+$ operator, if necessary). In practice it is necessary to normalize the column vectors $\boldsymbol{a}_j$ and/or $\boldsymbol{b}_j$ to unit length vectors (in ℓ_p-norm sense ($p = 1, 2, \ldots, \infty$)) at each iteration step. In the special case of ℓ_2-norm, the above algorithm can be further simplified by ignoring the denominator in (4.92) and imposing a vector normalization after each iterative step, to give a simplified scalar form of the HALS algorithm:

$$b_{tj} \leftarrow \left[\sum_{i=1}^{I} a_{ij}\, y_{it}^{(j)}\right]_+, \qquad a_{ij} \leftarrow \left[\sum_{t=1}^{T} b_{tj}\, y_{it}^{(j)}\right]_+, \tag{4.93}$$

with $a_{ij} \leftarrow a_{ij}/||\boldsymbol{a}_j||_2$, where $y_{it}^{(j)} = [\mathbf{Y}^{(j)}]_{it} = y_{it} - \sum_{p \neq j} a_{ip} b_{tp}$.

4.7.2 Extensions and Implementations of the HALS Algorithm

The above simple algorithm can be further extended or improved with respect to the convergence rate and performance by imposing additional constraints such as sparsity and smoothness.

Firstly, observe that the residual matrix $\mathbf{Y}^{(j)}$ can be rewritten as

$$\begin{aligned} \mathbf{Y}^{(j)} &= \mathbf{Y} - \sum_{p \neq j} \boldsymbol{a}_p \boldsymbol{b}_p^T = \mathbf{Y} - \mathbf{A}\mathbf{B}^T + \boldsymbol{a}_j \boldsymbol{b}_j^T, \\ &= \mathbf{Y} - \mathbf{A}\mathbf{B}^T + \boldsymbol{a}_{j-1} \boldsymbol{b}_{j-1}^T - \boldsymbol{a}_{j-1} \boldsymbol{b}_{j-1}^T + \boldsymbol{a}_j \boldsymbol{b}_j^T. \end{aligned} \tag{4.94}$$

It then follows that instead of computing explicitly the residual matrix $\mathbf{Y}^{(j)}$ at each iteration step, we can just perform a smart update [55]. An efficient implementation of the HALS algorithm (4.92) is given in the detailed pseudo-code in Algorithm 4.2.

Different cost functions can be used for the estimation of the rows of the matrix $\mathbf{X} = \mathbf{B}^T$ and the columns of matrix $\mathbf{A}$ (possibly with various additional regularization terms [21,24,55]). Furthermore, the columns of $\mathbf{A}$ can be estimated simultaneously, and the rows in $\mathbf{X}$ sequentially. In other words, by minimizing the set of cost functions in (4.85) with respect to $\boldsymbol{b}_j$, and simultaneously the cost function (4.2) with normalization of the columns $\boldsymbol{a}_j$ to unit ℓ_2-norm, we obtain a very efficient NMF learning algorithm in which the individual

Algorithm 4.2: HALS

Input: $\mathbf{Y} \in \mathbb{R}_+^{I \times T}$: input data, J: rank of approximation
Output: $\mathbf{A} \in \mathbb{R}_+^{I \times J}$ and $\mathbf{X} = \mathbf{B}^T \in \mathbb{R}_+^{J \times T}$ such that the cost function (4.85) is minimized.

begin
  ALS or random nonnegative initialization for $\mathbf{A}$ and $\mathbf{X} = \mathbf{B}^T$
  foreach $\boldsymbol{a}_j$ *of* $\mathbf{A}$ **do** $\boldsymbol{a}_j \leftarrow \boldsymbol{a}_j / \|\boldsymbol{a}_j\|_2$ /* normalize to ℓ_2 unit length */
  $\mathbf{E} = \mathbf{Y} - \mathbf{A}\mathbf{B}^T$ /* residue */
  repeat
    for $j = 1$ *to* J **do**
      $\mathbf{Y}^{(j)} \leftarrow \mathbf{E} + \boldsymbol{a}_j \boldsymbol{b}_j^T$
      $\boldsymbol{b}_j \leftarrow \left[\mathbf{Y}^{(j)T} \boldsymbol{a}_j\right]_+$ /* update $\boldsymbol{b}_j$ */
      $\boldsymbol{a}_j \leftarrow \left[\mathbf{Y}^{(j)} \boldsymbol{b}_j\right]_+$ /* update $\boldsymbol{a}_j$ */
      $\boldsymbol{a}_j \leftarrow \boldsymbol{a}_j / \|\boldsymbol{a}_j\|_2$
      $\mathbf{E} \leftarrow \mathbf{Y}^{(j)} - \boldsymbol{a}_j \boldsymbol{b}_j^T$ /* update residue */
    end
  until *a stopping criterion is met* /* convergence condition */
end

vectors of $\mathbf{B} = [\boldsymbol{b}_1, \boldsymbol{b}_2, \ldots, \boldsymbol{b}_J]$ are locally updated (column-by-column) and the matrix $\mathbf{A}$ is globally updated using the global nonnegative ALS (all columns $\boldsymbol{a}_j$ simultaneously) (see also [21]):

$$\boldsymbol{b}_j \leftarrow \left[\mathbf{Y}^{(j)\,T} \boldsymbol{a}_j\right]_+, \qquad \mathbf{A} \leftarrow \left[\mathbf{Y}\mathbf{B}(\mathbf{B}^T\mathbf{B})^{-1}\right]_+ = \left[\mathbf{Y}\mathbf{X}^T(\mathbf{X}\mathbf{X}^T)^{-1}\right]_+, \tag{4.95}$$

with the normalization (scaling) of the columns in $\mathbf{A}$ to unit length in the sense of the ℓ_2-norm after each iteration.

4.7.3 Fast HALS NMF Algorithm for Large-scale Problems

Alternatively, an even more efficient approach is to perform a factor-by-factor procedure instead of updating column-by-column vectors [55]. This way, from (4.92), we obtain the following update rule for $\boldsymbol{b}_j = \underline{\boldsymbol{x}}_j^T$

$$\begin{aligned} \boldsymbol{b}_j &\leftarrow \mathbf{Y}^{(j)T} \boldsymbol{a}_j / (\boldsymbol{a}_j^T \boldsymbol{a}_j) = \left(\mathbf{Y} - \mathbf{A}\mathbf{B}^T + \boldsymbol{a}_j \boldsymbol{b}_j^T\right)^T \boldsymbol{a}_j / (\boldsymbol{a}_j^T \boldsymbol{a}_j) \\ &= (\mathbf{Y}^T \boldsymbol{a}_j - \mathbf{B}\mathbf{A}^T \boldsymbol{a}_j + \boldsymbol{b}_j \boldsymbol{a}_j^T \boldsymbol{a}_j) / (\boldsymbol{a}_j^T \boldsymbol{a}_j) \\ &= \left(\left[\mathbf{Y}^T\mathbf{A}\right]_j - \mathbf{B}\left[\mathbf{A}^T\mathbf{A}\right]_j + \boldsymbol{b}_j \boldsymbol{a}_j^T \boldsymbol{a}_j\right) / (\boldsymbol{a}_j^T \boldsymbol{a}_j), \end{aligned} \tag{4.96}$$

with the nonlinear projection $\boldsymbol{b}_j \leftarrow \left[\boldsymbol{b}_j\right]_+$ at each iteration step to impose the nonnegativity constraints. Since $\|\boldsymbol{a}\|_2^2 = 1$, the learning rule for $\boldsymbol{b}_j$ has a simplified form as

$$\boldsymbol{b}_j \leftarrow \left[\boldsymbol{b}_j + \left[\mathbf{Y}^T\mathbf{A}\right]_j - \mathbf{B}\left[\mathbf{A}^T\mathbf{A}\right]_j\right]_+, \tag{4.97}$$

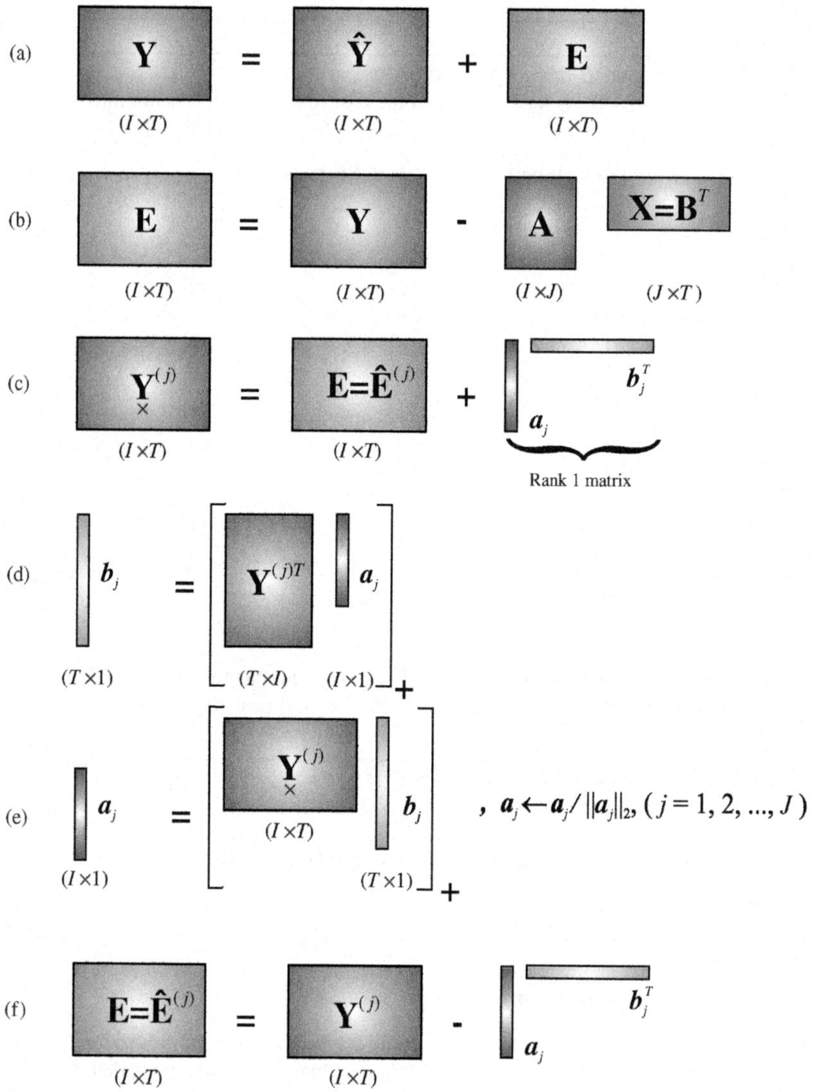

Figure 4.1 Illustration of the basic (local) HALS algorithm and its comparison with the standard (global) ALS algorithm. In the standard ALS algorithm we minimize the mean squared error of the cost function $\|\mathbf{E}\|_F^2 = \|\mathbf{Y} - \hat{\mathbf{Y}}\|_F^2$, where $\hat{\mathbf{Y}} = \mathbf{A}\mathbf{B}^T$ and the target (desired) data $\mathbf{Y}$ is known and fixed (Figures (a)–(b)). In the HALS algorithm the target residual matrices $\mathbf{Y}^{(j)}$, $(j = 1, 2, \ldots, J)$ (Figures (c),(f)) are not fixed but they are estimated during an iterative process via the HALS updates (Figures (d)–(e)) and they converge to rank-one matrices.

and analogously, for vector $\boldsymbol{a}_j$:

$$\begin{aligned}
\boldsymbol{a}_j \leftarrow \mathbf{Y}^{(j)}\, \boldsymbol{b}_j &= \left(\mathbf{Y} - \mathbf{A}\mathbf{B}^T + \boldsymbol{a}_j \boldsymbol{b}_j^T\right) \boldsymbol{b}_j \\
&= \mathbf{Y}\boldsymbol{b}_j - \mathbf{A}\mathbf{B}^T \boldsymbol{b}_j + \boldsymbol{a}_j \boldsymbol{b}_j^T \boldsymbol{b}_j \\
&= [\mathbf{Y}\mathbf{B}]_j - \mathbf{A}\left[\mathbf{B}^T\mathbf{B}\right]_j + \boldsymbol{a}_j \boldsymbol{b}_j^T \boldsymbol{b}_j \\
&= \boldsymbol{a}_j \boldsymbol{b}_j^T \boldsymbol{b}_j + [\mathbf{Y}\mathbf{B}]_j - \mathbf{A}\left[\mathbf{B}^T\mathbf{B}\right]_j .
\end{aligned} \tag{4.98}$$

Hence, by imposing the nonnegativity constraints, we finally have

$$\boldsymbol{a}_j \leftarrow \left[\boldsymbol{a}_j \boldsymbol{b}_j^T \boldsymbol{b}_j + [\mathbf{Y}\mathbf{B}]_j - \mathbf{A}\left[\mathbf{B}^T\mathbf{B}\right]_j\right]_+ , \tag{4.99}$$

$$\boldsymbol{a}_j \leftarrow \boldsymbol{a}_j / \|\boldsymbol{a}_j\|_2 . \tag{4.100}$$

Based on these expressions, the improved and modified HALS NMF algorithm is given in the pseudo-code Algorithm 4.3.

Algorithm 4.3: FAST HALS for Large Scale NMF

Input: $\mathbf{Y} \in \mathbb{R}_+^{I\times T}$: input data, J: rank of approximation
Output: $\mathbf{A} \in \mathbb{R}_+^{I\times J}$ and $\mathbf{X} = \mathbf{B}^T \in \mathbb{R}_+^{J\times T}$ such that the cost function (4.85) is minimized.

begin
  ALS or random nonnegative initialization for $\mathbf{A}$ and $\mathbf{X} = \mathbf{B}^T$
  foreach $\boldsymbol{a}_j$ *of* $\mathbf{A}$ **do** $\boldsymbol{a}_j \leftarrow \boldsymbol{a}_j/\|\boldsymbol{a}_j\|_2$ /* normalize to ℓ_2 unit length */
  repeat
    $\mathbf{W} = \mathbf{Y}^T\mathbf{A}$; $\mathbf{V} = \mathbf{A}^T\mathbf{A}$
    for $j = 1$ *to* J **do**
      $\boldsymbol{b}_j \leftarrow \left[\boldsymbol{b}_j + \boldsymbol{w}_j - \mathbf{B}\, \boldsymbol{v}_j\right]_+$ /* update $\boldsymbol{b}_j$ */
    end
    $\mathbf{P} = \mathbf{Y}\mathbf{B}$; $\mathbf{Q} = \mathbf{B}^T\mathbf{B}$
    for $j = 1$ *to* J **do**
      $\boldsymbol{a}_j \leftarrow \left[\boldsymbol{a}_j\, q_{jj} + \boldsymbol{p}_j - \mathbf{A}\, \boldsymbol{q}_j\right]_+$ /* update $\boldsymbol{a}_j$ */
      $\boldsymbol{a}_j \leftarrow \boldsymbol{a}_j/\|\boldsymbol{a}_j\|_2$
    end
  until *a stopping criterion is met*   /* convergence condition */
end

The NMF problem is often highly redundant for $I >> J$, thus, for large-scale problems in order to estimate the vectors $\boldsymbol{a}_j$ and $\boldsymbol{b}_j = \underline{\boldsymbol{x}}_j^T \; \forall j$, we can use only some selected vectors and/or rows of the data input matrix $\mathbf{Y}$. For large-scale data and a block-wise update strategy (see Chapter 1), the fast HALS learning rule for $\boldsymbol{b}_j$ (4.96) can be rewritten as follows

$$\begin{aligned}
\boldsymbol{b}_j &\leftarrow \left[\boldsymbol{b}_j + \left[\mathbf{Y}_r^T\mathbf{A}_r\right]_j / \|\tilde{\boldsymbol{a}}_j\|_2^2 - \mathbf{B}\left[\mathbf{A}_r^T\mathbf{A}_r\right]_j / \|\tilde{\boldsymbol{a}}_j\|_2^2\right]_+ \\
&= \left[\boldsymbol{b}_j + \left[\mathbf{Y}_r^T\mathbf{A}_r\mathbf{D}_{A_r}\right]_j - \mathbf{B}\left[\mathbf{A}_r^T\mathbf{A}_r\mathbf{D}_{A_r}\right]_j\right]_+ ,
\end{aligned} \tag{4.101}$$

where $\mathbf{D}_{A_r} = \text{diag}(\|\tilde{\boldsymbol{a}}_1\|_2^{-2}, \|\tilde{\boldsymbol{a}}_2\|_2^{-2}, \ldots, \|\tilde{\boldsymbol{a}}_J\|_2^{-2})$ is a diagonal matrix, and $\tilde{\boldsymbol{a}}_j$ is the j-th column vector of the reduced matrix $\mathbf{A}_r \in \mathbb{R}_+^{R\times J}$.

The update rule for $\boldsymbol{a}_j$ takes a similar form

$$\boldsymbol{a}_j \leftarrow \left[\boldsymbol{a}_j + \left[\mathbf{Y}_c \mathbf{B}_c \mathbf{D}_{B_c}\right]_j - \mathbf{A}\left[\mathbf{B}_c^T \mathbf{B}_c \mathbf{D}_{B_c}\right]_j\right]_+, \tag{4.102}$$

where $\mathbf{D}_{B_c} = \text{diag}(\|\tilde{\boldsymbol{b}}_1\|_2^{-2}, \|\tilde{\boldsymbol{b}}_2\|_2^{-2}, \ldots, \|\tilde{\boldsymbol{b}}_J\|_2^{-2})$ and $\tilde{\boldsymbol{b}}_j$ is the j-th column vector of the reduced matrix $\mathbf{B}_c = \mathbf{X}_c^T \in \mathbb{R}_+^{C\times J}$.

In order to estimate all the vectors $\boldsymbol{a}_j$ and $\underline{\boldsymbol{x}}_j$ we only need to take into account the selected rows and columns of the residual matrices $\mathbf{Y}^{(j)}$ and the input data matrix $\mathbf{Y}$. To estimate precisely all $\boldsymbol{a}_j, \underline{\boldsymbol{x}}_j, \ \forall j$ we need to select at least J rows and columns of $\mathbf{Y}$. Moreover, the computations are performed only for the nonzero elements, thus allowing the computation time to be dramatically reduced for sparse and very large-scale problems. The rows and columns of the data matrix $\mathbf{Y}$ can be selected using different criteria. For example, we can choose only those rows and columns which provide the highest normalized squared Euclidean norms. Alternatively, instead of removing completely some rows and/or columns of $\mathbf{Y}$, we can merge (collapse) them into some clusters by adding them together or computing their averages. In this case we can select the rows and columns uniformly. Recently, extensive research is performed how to choose the optimal number of rows and vectors of data matrix [7,10,32,42,50,51].

4.7.4 HALS NMF Algorithm with Sparsity, Smoothness and Uncorrelatedness Constraints

In order to impose sparseness and smoothness constraints for vectors $\boldsymbol{b}_j$ (source signals), we can minimize the following set of cost functions [15,44]:

$$D_F^{(j)}(\mathbf{Y}^{(j)}\|\boldsymbol{a}_j\boldsymbol{b}_j^T) = \frac{1}{2}\|\mathbf{Y}^{(j)} - \boldsymbol{a}_j\boldsymbol{b}_j^T\|_F^2 + \alpha_{sp}\ \|\boldsymbol{b}_j\|_1 + \alpha_{sm}\ \|\varphi(\mathbf{L}\,\boldsymbol{b}_j)\|_1, \tag{4.103}$$

for $j = 1, 2, \ldots, J$ subject to $\boldsymbol{a}_j \geq \mathbf{0}$ and $\boldsymbol{b}_j \geq \mathbf{0}$, where $\alpha_{sp} > 0$, $\alpha_{sm} > 0$ are regularization parameters controlling the levels of sparsity and smoothness, respectively, $\mathbf{L}$ is a suitably designed matrix (the Laplace operator) which measures the smoothness[10] (by estimating the differences between neighboring samples of $\boldsymbol{b}_j$) and $\varphi : \mathbb{R} \to \mathbb{R}$ is an edge-preserving function applied component-wise. Although this edge-preserving nonlinear function may take various forms [53]:

$$\varphi(t) = |t|^\alpha/\alpha,\ 1 \leq \alpha \leq 2, \tag{4.104}$$

$$\varphi(t) = \sqrt{\alpha + t^2}, \tag{4.105}$$

$$\varphi(t) = 1 + |t|/\alpha - \log(1 + |t|/\alpha), \tag{4.106}$$

we restrict ourself to simple cases, where $\varphi(t) = |t|^\alpha/\alpha$ for $\alpha = 1$ or 2 and $\mathbf{L}$ is the difference operator of the first or second order (4.57). For example, the first order difference operator $\mathbf{L}$ with T points can take the

[10] In the special case for $\mathbf{L} = \mathbf{I}_T$ and $\varphi(t) = |t|$, the smoothness regularization term becomes a sparsity term.

form:

$$\mathbf{L} = \begin{bmatrix} 1 & -1 & & & \\ & 1 & -1 & & \\ & & \ddots & \ddots & \\ & & & 1 & -1 \end{bmatrix}, \tag{4.107}$$

and the cost function (4.103) becomes similar to the total-variation (TV) regularization (which is often used in signal and image recovery) but with additional sparsity constraints:

$$D_F^{(j)}(\mathbf{Y}^{(j)}\|\boldsymbol{a}_j \boldsymbol{b}_j^T) = \frac{1}{2}\left\|\mathbf{Y}^{(j)} - \boldsymbol{a}_j\boldsymbol{b}_j^T\right\|_F^2 + \alpha_{sp}\,\|\boldsymbol{b}_j\|_1 + \alpha_{sm}\sum_{t=1}^{T-1}|b_{tj} - b_{(t+1)j}|. \tag{4.108}$$

Another important case assumes that $\varphi(t) = \frac{1}{2}|t|^2$, and $\mathbf{L}$ is the second order difference operator with K points. In such a case, we obtain the Tikhonov type regularization:

$$D_F^{(j)}(\mathbf{Y}^{(j)}\|\boldsymbol{a}_j \boldsymbol{b}_j^T) = \frac{1}{2}\|\mathbf{Y}^{(j)} - \boldsymbol{a}_j\boldsymbol{b}_j^T\|_F^2 + \alpha_{sp}\,\|\boldsymbol{b}_j\|_1 + \frac{1}{2}\alpha_{sm}\,\|\mathbf{L}\boldsymbol{b}_j\|_2^2. \tag{4.109}$$

It is easy to find that in such a case the update rule for $\boldsymbol{a}_j$ is the same as in (4.92), whereas the update rule for $\boldsymbol{b}_j$ is given by (see also (4.61))

$$\boldsymbol{b}_j = \left[(\mathbf{I} + \alpha_{sm}\,\mathbf{L}^T\mathbf{L})^{-1}([\mathbf{Y}^{(j)}]^T\boldsymbol{a}_j - \alpha_{sp}\,\mathbf{1}_{T\times 1})\right]_+. \tag{4.110}$$

This learning rule is robust to noise, however, it involves a rather high computational cost due to the calculation of an inverse of a large matrix in each iteration. To circumvent this problem and to considerably reduce the complexity of the algorithm we present a second-order smoothing operator $\mathbf{L}$ in the following form:

$$\begin{aligned}\mathbf{L} &= \begin{bmatrix} -2 & 2 & & & \\ 1 & -2 & 1 & & \\ & \ddots & \ddots & \ddots & \\ & & 1 & -2 & 1 \\ & & & 2 & -2 \end{bmatrix} = \begin{bmatrix} -2 & & & & \\ & -2 & & & \\ & & \ddots & & \\ & & & -2 & \\ & & & & -2 \end{bmatrix} + \begin{bmatrix} 0 & 2 & & & \\ 1 & 0 & 1 & & \\ & \ddots & \ddots & \ddots & \\ & & 1 & 0 & 1 \\ & & & 2 & 0 \end{bmatrix} \\ &= -2\mathbf{I} + 2\mathbf{S}.\end{aligned} \tag{4.111}$$

However, instead of computing directly $\mathbf{L}\boldsymbol{b}_j = -2\mathbf{I}\boldsymbol{b}_j + 2\mathbf{S}\boldsymbol{b}_j$, in the second term we can approximate $\boldsymbol{b}_j$ by its estimation $\hat{\boldsymbol{b}}_j$ obtained from the previous update. Hence, the smoothing regularization term with $\varphi(t) = |t|^2/8$ takes a simplified and computationally more efficient form:

$$J_{sm} = \|\varphi(-2\boldsymbol{b}_j + 2\mathbf{S}\hat{\boldsymbol{b}}_j)\|_1 = \frac{1}{2}\|\boldsymbol{b}_j - \mathbf{S}\hat{\boldsymbol{b}}_j\|_2^2. \tag{4.112}$$

Finally, the learning rule for the regularized (smoothed) HALS algorithm takes the following form:

$$\boldsymbol{b}_j \leftarrow \frac{\left[\mathbf{Y}^{(j)T}\,\boldsymbol{a}_j - \alpha_{sp}\,\mathbf{1}_{T\times 1} + \alpha_{sm}\,\mathbf{S}\,\hat{\boldsymbol{b}}_j\right]_+}{(\boldsymbol{a}_j^T\,\boldsymbol{a}_j + \alpha_{sm})}. \tag{4.113}$$

Alternatively, for a relatively small dimension of matrix **A**, an efficient solution is based on a combination of a local learning rule for the vectors of **B** and a global one for **A**, based on the nonnegative ALS algorithm:

$$\boldsymbol{b}_j \leftarrow \frac{1}{1+\alpha_{sm}} \left[\mathbf{Y}^{(j)T}\boldsymbol{a}_j - \alpha_{sp}\,\mathbf{1}_{T\times 1} + \alpha_{sm}\,\mathbf{S}\hat{\boldsymbol{b}}_j\right]_+, \tag{4.114}$$

$$\mathbf{A} \leftarrow \left[\mathbf{Y}\mathbf{B}(\mathbf{B}^T\mathbf{B})^{-1}\right]_+, \tag{4.115}$$

where the columns of **A** are normalized to ℓ_2-norm unit length.

To impose the uncorrelatedness constraints on the vectors $\boldsymbol{b}_j$ the following additional penalty term is introduced in the cost function (4.103):

$$J_{cr} = \alpha_{cr} \sum_{k \neq l} \boldsymbol{b}_k^T \boldsymbol{b}_l. \tag{4.116}$$

In fact, this regularization term has relation to the one in (4.65) and (4.66). Computing the gradient of this penalty

$$\frac{\partial J_{cr}}{\partial \boldsymbol{b}_j} = \alpha_{cr} \sum_{l \neq j} \boldsymbol{b}_l = \alpha_{cr}\,\mathbf{B}\,\mathbf{1}_{J\times 1} - \alpha_{cr}\boldsymbol{b}_j, \tag{4.117}$$

and taking into account the derivation of the learning rule for $\boldsymbol{b}_j$ in (4.89), (4.114), we obtain the new learning rule with uncorrelatedness constraints given by

$$\boldsymbol{b}_j \leftarrow \frac{\left[\mathbf{Y}^{(j)T}\boldsymbol{a}_j - \alpha_{sp}\,\mathbf{1}_{T\times 1} + \alpha_{sm}\,\mathbf{S}\hat{\boldsymbol{b}}_j - \alpha_{cr}\,\mathbf{B}\,\mathbf{1}_{J\times 1}\right]_+}{\boldsymbol{a}_j^T\boldsymbol{a}_j + \alpha_{sm} + \alpha_{cr}}. \tag{4.118}$$

Through normalizing vector $\boldsymbol{b}_j$ to unit length, we obtain the simplified update rule given by

$$\boxed{\boldsymbol{b}_j \leftarrow \left[\mathbf{Y}^{(j)T}\boldsymbol{a}_j - \alpha_{sp}\,\mathbf{1}_{T\times 1} + \alpha_{sm}\,\mathbf{S}\hat{\boldsymbol{b}}_j - \alpha_{cr}\,\mathbf{B}\,\mathbf{1}_{J\times 1}\right]_+.} \tag{4.119}$$

The additional constraints can be imposed on the vectors $\boldsymbol{a}_j$, and we can easily adopt above approach to derive suitable update rules.

4.7.5 HALS Algorithm for Sparse Component Analysis and Flexible Component Analysis

In some applications source signals $\underline{\boldsymbol{x}}_j = \boldsymbol{b}_j^T = [x_{j1}, x_{j2}, \ldots, x_{jT}]$ are sparse but not necessarily nonnegative. We can deal with this problem by applying a shrinkage function with an adaptive threshold instead of simple half-wave-rectifying. Furthermore, a shrinkage function with a decreasing threshold will also improve robustness with respect to noise [25].

In sparse component analysis (SCA) our objective is to estimate the sources $\underline{\boldsymbol{x}}_j$ which are sparse and usually with a prescribed or specified sparsification profile and also with additional constraints such as local smoothness. To make the estimated sources sufficiently sparse, we need to adaptively sparsify the data by applying a suitable nonlinear projection or filtering function. A simple nonlinear projection which enforces some sparsity on the normalized data is the following weakly nonlinear element-wise projection:

$$P_{\Omega_j}(x_{jt}) = \text{sign}(x_{jt})\,|x_{jt}|^{1+\alpha_j} \quad \forall t, \tag{4.120}$$

where α_j is a small parameter which controls sparsity. Such a nonlinear projection can be considered as a simple (trivial) shrinking operation.

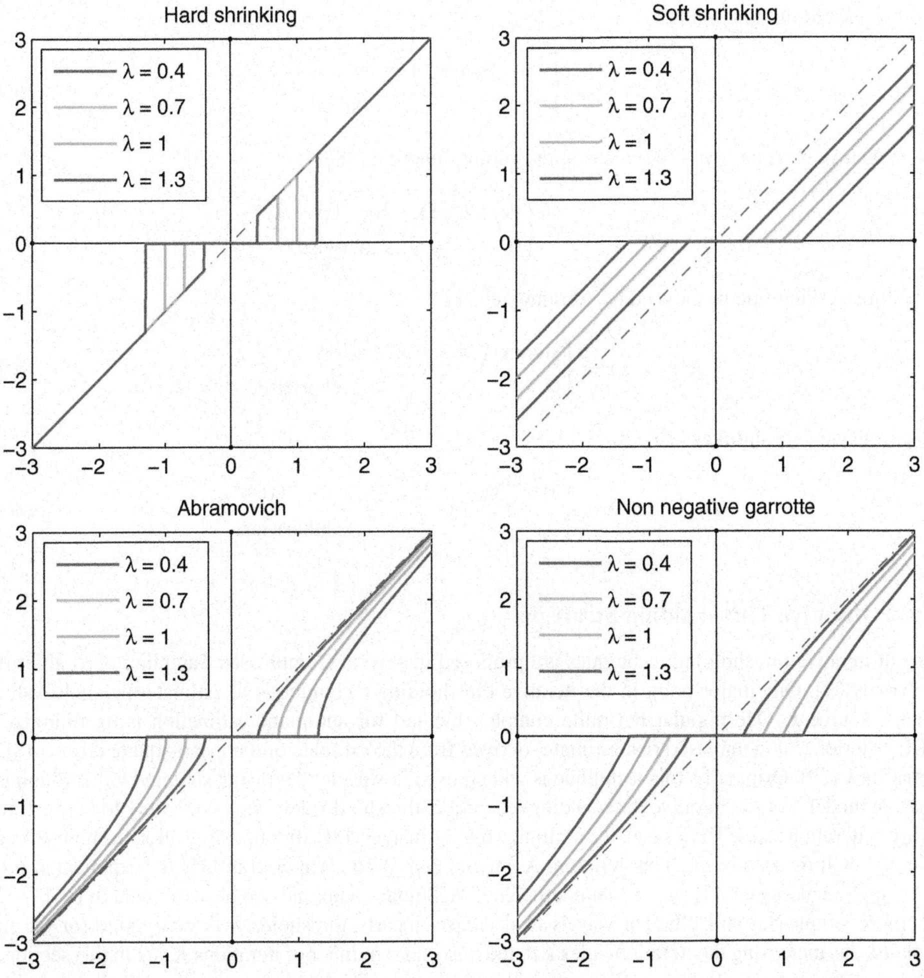

Figure 4.2 Some examples of shrinkage rules for different values of λ. It should be noted that for noisy data, shrinkage parameter λ does not decrease to zero but to the noise threshold which is proportional to an estimate of the noise standard deviation (such as the universal threshold $\sqrt{2\log(I)}\,\sigma_n$, ($\sigma_n$ is the standard deviation of the noise) [25]).

4.7.5.1 Simple Shrinkage Rules

In practice, we use different adaptive local soft or hard shrinkages in order to sparsify the individual sources. The most popular shrinkage rules are as follows (see also Figure 4.2):

- Hard thresholding function [25]

$$P_\lambda^h(x,\lambda) = \begin{cases} x, & \text{for } \ |x| \geq \lambda; \\ 0, & \text{otherwise.} \end{cases} \tag{4.121}$$

- Soft thresholding function [25]

$$P_\lambda^s(x,\lambda) = \begin{cases} (x - \text{sign}(x)\lambda), & \text{for } |x| \geq \lambda; \\ 0, & \text{otherwise.} \end{cases} \tag{4.122}$$

- Nonnegative garrotte $(nn - garrotte)$ thresholding function [38]

$$P_\lambda^{nng}(x,\lambda) = \begin{cases} (x - \lambda^2/x), & \text{for } |x| \geq \lambda; \\ 0, & \text{otherwise.} \end{cases} \tag{4.123}$$

- The thresholding rule introduced by Abramovich [1]

$$P_\lambda^{Abr}(x,\lambda) = \begin{cases} \text{sign}(x)\sqrt{(x^2 - \lambda^2)}, & \text{for } |x| \geq \lambda; \\ 0, & \text{otherwise.} \end{cases} \tag{4.124}$$

- n-degree garrotte shrinkage rule [8]

$$P_\lambda^{n-gar}(x,\lambda) = \begin{cases} x^{(2n+1)}/(\lambda^{(2n)} + x^{(2n)}), & \text{for } |x| \geq \lambda; \\ 0, & \text{otherwise.} \end{cases} \tag{4.125}$$

4.7.5.2 Iterative Thresholding Strategies

Efficient iterative thresholding techniques are proposed in several Compressed Sampling (CS) algorithms [6,12,30,31,37]. One major point in the iterative thresholding technique is to estimate threshold values λ for each source $\boldsymbol{x}$. One popular but quite complex method for automatic estimation is to minimize the SURE criterion (Stein unbiased risk estimate) derived from the estimate of the mean-square error (MSE) of estimators $\hat{\boldsymbol{x}}$ [29]. Originally, this technique is widely used in wavelet denoising strategy with assuming for Gaussian model. For the iterative thresholding technique, threshold values vary according to new estimates $\boldsymbol{x}$ at every iteration steps. To this end, we minimize the estimated MSE after updating all $\boldsymbol{x}$, the new estimated value of λ will be used in the next iteration. A generalized SURE proposed in [36] for exponential family can be applied more suitably to wavelet coefficients. A similar technique can be also found in [41].

A more simple but still efficient way is to decrease linearly thresholds λ to zero value (or the noise thresholds) as increasing the iteration index k to the maximum number of iterations K_{max}. In this section, we list briefly some simple adaptive nonlinear thresholding strategies which can be applied for all the shrinkage rules mentioned above:

- Linear Decreasing Strategy [4]

$$\lambda^{(k)} = \lambda^{(0)} - k\frac{\lambda^{(0)} - \lambda_{\min}}{K_{\max}}. \tag{4.126}$$

- Adaptive Linear Decreasing Strategy [5]

$$\lambda^{(k)} = \alpha^{(k)} \, \text{mad}(\boldsymbol{x}), \tag{4.127}$$

$$\alpha^{(k+1)} = \alpha^{(k)} - \Delta_\alpha . \tag{4.128}$$

- Adaptive Nonlinear Decreasing Strategy [56]

$$\lambda^{(k)} = \alpha^{(k)} \text{mad}(\boldsymbol{x}), \tag{4.129}$$

$$\alpha^{(k)} = \psi(k/K_{\max}), \tag{4.130}$$

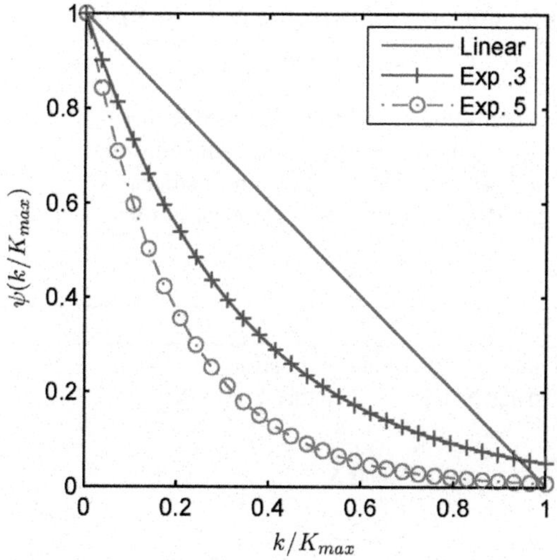

Figure 4.3 Illustration of simple heuristic linear and exponential decreasing strategies for threshold λ.

where $\lambda^{(0)} = \lambda_{\max}$ is the first threshold, $K_{\max}$ is the number of iterations, k is the k-th iteration index, Δ_α is a decreasing step to enforce $\lambda^{(k)}$ towards λ_{min} which is often 0; $\mathrm{mad}(\boldsymbol{x}) = \mathrm{median}(\|\boldsymbol{x} - \mathrm{median}(\boldsymbol{x})\|)$, which is the Median Absolute Deviation function.[11]

For an exact representation with sparse components $\lambda_{\min}$ must be set to zero. When additive noise is contained in the data, $\lambda_{\min}$ should be set to several times the noise standard deviation (typically, $\lambda_{\min} = 3\sigma_n$). Although the simplicity of the linear strategy is an advantage this strategy is not optimal in the sense of convergence speed and it is quite difficult to estimate optimally the maximum number of iterations, which depends on the data (typically, a few hundreds of iterations are needed). A small number of iterations lead to a poor performance whilst a large number of iterations are computationally expensive. Therefore, the threshold strategy for decreasing λ is a key factor in achieving a good performance.

The estimated values $\hat{\boldsymbol{x}}$ approach to its true sources $\boldsymbol{x}$ when the iteration index k increases to the maximum number of iteration K_{max}. For few first iterations, the threshold λ_t needs to be set by a large enough value to extract significant coefficients, and to form a rough shape for $\boldsymbol{x}$. However, because sources $\boldsymbol{x}$ are sparse, distribution of their coefficients will decrease fast at large intensities, and more slowly for small ones. It means that threshold λ should also decrease fast in some first iterations, then vary slowly and tend to zero in successive iterations. This step fine-tunes the rough estimate $\boldsymbol{x}$ to achieve close to perfect reconstruction.

For this reason, λ is considered to adapt to the nonlinear decreasing strategy in which the function $\psi(\cdot)$ in (4.130) will take a key role. The function $\psi(\cdot)$ is a monotonically nonincreasing function which satisfies the following conditions: $\psi(0) = 1$, $\psi(1) \approx 0$. In a special case, for $\psi(x) = 1 - x$, we obtain the linear strategy. The function $\psi(x)$ has a form : decrease fast as $x \approx 0$ or in some first iterations $k = 1, 2, \ldots$, but should be rather flat as $x \to 1$ or $k \to K_{max}$. In practice, this function can be easily designed, for example, by the shape-preserving piece-wise cubic Hermite interpolation with only four initial points:

x	0.0	0.1	0.2	1
y	1.0	0.8	0.4	0

[11]Or the Mean Absolute Deviation function $\mathrm{mad}(\boldsymbol{x}) = \mathrm{mean}(\|\boldsymbol{x} - \mathrm{mean}(\boldsymbol{x})\|)$.

Alternatively, we can apply the exponential function (denoted briefly EXP(η)) defined as

$$\psi(x) = \exp(-\eta\, x), \tag{4.131}$$

where $\eta \geq 0$ is a parameter controlling the decreasing speed. For $\eta \to 0$, λ_t follows the linear decreasing strategy (see Figure 4.3). In the simulation section, we will illustrate some relations between η and the number of iterations K_{max}, and between linear and nonlinear strategies.

The pseudo-code for the HALS algorithm for Sparse Component Analysis (SCA) is given in Algorithm 4.4.

Algorithm 4.4: HALS for SCA

Input: $\mathbf{Y} \in \mathbb{R}^{I\times T}$: input data, J: rank of approximation
Output: $\mathbf{A} \in \mathbb{R}^{I\times J}$ and sparse $\mathbf{X} \in \mathbb{R}^{J\times T}$ such that $\mathbf{Y} = \mathbf{A}\mathbf{X}$.

```
begin
    ALS or random initialization for A and X
    foreach a_j of A do   a_j ← a_j/||a_j||_2          /* normalize to ℓ2 unit length */
    E = Y − AX                                          /* residue */
    Initialize the threshold λ = [λ_1, λ_2, ..., λ_J]
    repeat
        for j = 1 to J do
            Y^(j) ← E + a_j x_j
            x_j ← P_{λ_j} [a_j^T Y^(j)]                 /* update x_j */
            a_j ← [Y^(j) x_j^T]                         /* update a_j */
            a_j ← a_j/||a_j||_2
            E ← Y^(j) − a_j x_j                         /* update residue */
        end
        Decrease thresholds λ following a given strategy
    until a stopping criterion is met                   /* convergence condition */
end
```

The above HALS approach for SCA with an iterative shrinkage projection can be extended to so called Flexible Component Analysis (FCA) when we exploit various intrinsic and morphological properties of the estimated source signals, such as smoothness, continuity, sparseness and nonnegativity [5,17,26,34].

The FCA is more general than NMF or SCA, since it is not only limited to a nonnegative and/or sparse representation via shrinking and linear transformations but allows us to impose general and flexible (soft and hard) constraints, nonlinear projections, transformations, and filtering.[12]

We can outline the HALS FCA method as follows:

1. Set the initial values of the matrix **A** and the matrices **X**, and normalize the vectors $\boldsymbol{a}_j$ to unit ℓ_2-norm length.
2. Calculate the new estimate $\underline{\boldsymbol{x}}_j$ of the matrices **X** using the iterative formula:

$$\underline{\boldsymbol{x}}_j \leftarrow \boldsymbol{a}_j^T \mathbf{Y}^{(j)} - \alpha_{\mathbf{X}} \Psi_{\mathbf{X}}(\underline{\boldsymbol{x}}_j), \qquad (j = 1, 2, \ldots, J), \tag{4.132}$$

[12]In the FCA, we use two kinds of constraints: soft (or weak) constraints via penalty and regularization terms in the local cost functions, and hard (strong) constraints via iteratively adaptive postprocessing using nonlinear projections or filtering. Furthermore, in contrast to many alternative algorithms which process the columns of **X**, we process their rows which directly represent the source signals.

where $\alpha_{\mathbf{X}} \geq 0$ is a regularization coefficient, $\Psi_{\mathbf{X}}(\underline{\boldsymbol{x}}_j) = \partial J_{\mathbf{X}}(\underline{\boldsymbol{x}}_j)/\partial \underline{\boldsymbol{x}}_j$ is a regularization function, and $J_{\mathbf{X}}(\underline{\boldsymbol{x}}_j)$ is a suitably designed regularization (penalty) term (see, for example Chapter 3).

3. If necessary, enforce the nonlinear projection or filtering by imposing natural constraints on individual sources (the rows of $\mathbf{X}$), such as nonnegativity, boundness, smoothness, and/or sparsity:

$$\underline{\boldsymbol{x}}_j \leftarrow \mathcal{P}_{\Omega}(\underline{\boldsymbol{x}}_j), \qquad (j = 1, 2, \ldots, J), \tag{4.133}$$

where $\mathcal{P}_{\Omega}(\underline{\boldsymbol{x}}_j)$ is a suitable nonlinear projection or filtering.

4. Calculate the new estimate of $\mathbf{A}$, normalize each column of $\mathbf{A}$ to unit length, and if necessary impose additional constraints on the columns of $\mathbf{A}$:

$$\boldsymbol{a}_j \leftarrow \mathbf{Y}^{(j)} \underline{\boldsymbol{x}}_j^T - \alpha_{\mathbf{A}} \Psi_{\mathbf{A}}(\boldsymbol{a}_j), \tag{4.134}$$

$$\boldsymbol{a}_j \leftarrow \tilde{\mathcal{P}}_{\Omega}(\boldsymbol{a}_j), \qquad \boldsymbol{a}_j \leftarrow \boldsymbol{a}_j / ||\boldsymbol{a}_j||_2, \qquad (j = 1, 2, \ldots, J), \tag{4.135}$$

5. Repeat steps (2) and (4) until the chosen convergence criterion is satisfied.

4.7.6 Simplified HALS Algorithm for Distributed and Multi-task Compressed Sensing

Another promising application of the modified HALS approach is the solving of under-determined systems of linear equations and related problems. In fact, many problems in science involve finding solutions to under-determined and ill-conditioned linear systems of equations. Prior knowledge is required to solve such a problem and one of the most common forms of structure exploited is sparsity. Another central problem in signal processing is sampling. Recently, it has been shown that it is possible to sample well below the Nyquist limit whenever the signal has additional sparsity structure. This theory is known as compressed sensing (CS) or compressive sampling and a wealth of theoretical insight has been gained for signals that permit a sparse representation [6,11,26,37,63]. If signal $\boldsymbol{x}_t \in \mathbb{R}^J$ is sufficiently sparse, perfect reconstruction from $\boldsymbol{y}_t = \mathbf{A}\boldsymbol{x}_t \in \mathbb{R}^I$, $(J >> I)$ with sampling rate much lower than conventional Shannon-Nyquist bound. Crucially, we assume $\boldsymbol{x}_t$ to be sparse, i.e., we assume that only a small number of elements in $\boldsymbol{x}_t$ are nonzero or, more generally, that most of the energy in $\boldsymbol{x}_t$ is concentrated in a few coefficients. It has recently been shown that, if a signal has such a sparse representation, then it is possible to take less samples (or measurements) from the signal than would be suggested by the Nyquist limit. Furthermore, one is able to reconstruct the original signal using convex optimization techniques [11,37,63].[13]

In this section, we consider very briefly the special form of the CS: Distributed multi-task CS model [33]:

$$\mathbf{Y} = \mathbf{A}\boldsymbol{\Psi}\mathbf{S}, \tag{4.136}$$

where the matrix $\mathbf{S} = [\boldsymbol{s}_1, \boldsymbol{s}_2, \ldots, \boldsymbol{s}_T] \in \mathbb{R}^{J \times T}$ represents T unknown source signals with J samples, $\boldsymbol{\Psi}$ denotes an orthogonal transform which is able to sparsify each signal $\boldsymbol{s}_t$ on its domain, for example: DCT, wavelets or curvelets. The compressed signals $\mathbf{Y} = [\boldsymbol{y}_1, \boldsymbol{y}_2, \ldots, \boldsymbol{y}_T] \in \mathbb{R}^{I \times T}$ are obtained by applying a known projection matrix $\mathbf{A} = [\boldsymbol{a}_1, \boldsymbol{a}_2, \ldots, \boldsymbol{a}_J] \in \mathbb{R}^{I \times J}$ (with $I << J$) on the sparse signals $\mathbf{X} = \boldsymbol{\Psi}\mathbf{S}$. The projection matrix $\mathbf{A}$ can be generated by randomly sampling the columns (*uniform spherical ensemble*), with different columns independently identically distributed (i.i.d.) on the unit sphere S^{I-1} in Euclidean I-space [28,63]. In an optimized strategy, the projection matrix could be selected adaptively to the dictionary $\boldsymbol{\Psi}$ [35].

Actually, compressed sensing of a signal $\boldsymbol{x}_t \in \mathbb{R}^J$ (J samples) to a compressed one $\boldsymbol{y}_t \in \mathbb{R}^I$ (I measurements) by a projection $\mathbf{A} \in \mathbb{R}^{I \times J}$ does not reduce the number of samples stored. To reconstruct $\boldsymbol{x}_t$, the matrix $\mathbf{A}$ also needs to be stored in addition to vectors $\boldsymbol{y}_t$. Therefore, the true number of samples stored is usually much greater than the number of samples of the source: $I + IJ >> J$. Moreover, compression and

[13]The field has existed for at least four decades, but recently the field has exploded, in part due to several important theoretical results.

reconstruction on a very long signal, say, $J > 10^9$ may consume much resources (memory) due to a large projection matrix $\mathbf{A}$. To solve this problem, we can split the source $\boldsymbol{x}_t$ to a set of short signals, then apply the DCS model.

The primary objective is to reconstruct the sparse sources represented by the columns of $\mathbf{X}$ from $\mathbf{Y}$ by exploiting sparsity properties in a suitable transform domain. Signals $\mathbf{S}$ are reconstructed using the inverse transform $\boldsymbol{\Psi}$: $\mathbf{S} = \boldsymbol{\Psi}^T\mathbf{X}$. We assume that in our distributed compressed sensing (DCS) the original source signals have similar characteristics or properties regarding sparsity profile which are expressed by the same redundant dictionary $\boldsymbol{\Psi}$. In other words, we assume that projections are the same for all $\boldsymbol{s}_t$.

This problem is nontrivial since the system of linear equations (4.136) is undetermined and generally provides an infinite number of solutions. Moreover, in the CS the matrix $\mathbf{A}$ is large and cannot be stored explicitly, and it is rather costly and impractical to access the whole or significant portions of $\mathbf{A}$ or $\mathbf{A}^T\mathbf{A}$. However, a matrix-vector product involving $\mathbf{A}$ and $\boldsymbol{\Psi}$ can be computed quite efficiently [6,11,37]. For example, if the columns of $\boldsymbol{\Psi}$ represent the wavelets bases, then the multiplications such as $\boldsymbol{\Psi}\boldsymbol{v}$ and/or $\boldsymbol{\Psi}^T\boldsymbol{v}$ can be performed by the fast wavelet transform. Similarly, if $\boldsymbol{\Psi}$ represents a convolution, then the multiplications $\boldsymbol{\Psi}\boldsymbol{v}$ and/or $\boldsymbol{\Psi}^T\boldsymbol{v}$ can be performed by using the fast Fourier transform (FFT).

For real world data such as images and EEG data we have a large number of samples (J). Computations on large matrices requires significant computational resources (memory and CPU time). In an illustrative example (**Example 4.7**), we compressed a 512×512 image (262144 samples) to only 2713 samples. For large-scale problems, it is impossible to process directly the whole matrices $\mathbf{A}$ and $\mathbf{X}$. This means that we cannot recover the sources based on their compressed versions with global learning rules. To this end, we may form a set of local learning rules which allows us to perform the estimation sequentially (row-by-row). We present here a simple approach and the corresponding learning algorithm which involves a sequential use of multiplications of the vectors of $\mathbf{A}$.

The iterative thresholding algorithm for linear inverse problems has been first proposed by Daubechies [26] and has been extended to source separation [4,5] and for compressed sensing [6,37].

We derive an efficient algorithm which employs an iterative thresholding strategy, thus allowing almost the perfect reconstruction of sparse sources in many difficult cases. For convenience, the above distributed multi-task CS model (4.136) can be represented in the following form:[14]

$$\tilde{\mathbf{Y}} = \mathbf{A}^T\mathbf{Y} = \mathbf{A}^T\mathbf{A}\mathbf{X} = \mathbf{G}\mathbf{X}, \tag{4.137}$$

where $\tilde{\mathbf{Y}} = [\tilde{\mathbf{y}}_1, \tilde{\mathbf{y}}_2, \ldots, \tilde{\mathbf{y}}_T] \in \mathbb{R}^{J\times T}$, $\mathbf{G} = [\boldsymbol{g}_1, \boldsymbol{g}_2, \ldots, \boldsymbol{g}_J] = \mathbf{A}^T\mathbf{A} \in \mathbb{R}^{J\times J}$. Without loss of generality, we assume that the columns $\boldsymbol{a}_j$ of the matrix $\mathbf{A}$ have the unit lengths, that is, all the diagonal elements g_{jj} equal one: $g_{jj} = \boldsymbol{a}_j^T\boldsymbol{a}_j = 1$. Otherwise, we may normalize all the columns of the matrix $\mathbf{G}$ as

$$\boldsymbol{g}_j = \boldsymbol{a}_j^T\mathbf{A}/\boldsymbol{a}_j^T\boldsymbol{a}_j \qquad (j = 1, 2, \ldots, J). \tag{4.138}$$

Consequently, the rows $\underline{\boldsymbol{x}}_j$ of the final result need to be re-scaled by the same corresponding factors as $\underline{\boldsymbol{x}}_j \leftarrow \underline{\boldsymbol{x}}_j/\boldsymbol{a}_j^T\boldsymbol{a}_j$. Hence, our new objective becomes to estimate the matrix $\mathbf{X}$ from the known matrix $\tilde{\mathbf{Y}}$ and the known matrix $\mathbf{G}$. To solve the problem we define the following set of regularized cost functions [17,56]:

$$D^{(j)}\left(\underline{\boldsymbol{y}}^{(j)} \,||\, \underline{\boldsymbol{x}}_j\right) = \frac{1}{2}\|\underline{\boldsymbol{y}}^{(j)} - \underline{\boldsymbol{x}}_j\|_2^2 + \sum_t \lambda_t J_x(\boldsymbol{x}_t) \qquad (j = 1, 2, \ldots, J), \tag{4.139}$$

where $\underline{\boldsymbol{y}}^{(j)} = \underline{\tilde{\boldsymbol{y}}}_j - \sum_{p\neq j} g_{jp}\underline{\boldsymbol{x}}_p$, $(j = 1, 2, \ldots, J)$, $\boldsymbol{\lambda} = [\lambda_1, \lambda_2, \ldots, \lambda_T] \geq \mathbf{0}$ is a row vector of regularization coefficients, and $J_x(\boldsymbol{x}_t)$ are optional regularization or penalty terms which enforce additional constraints such as sparsity or smoothness on the sources $\boldsymbol{x}_t$.

[14] However, it should be noted that matrix $\mathbf{G}$ can be very large and sparse for $J >> I$, so sometimes it is better to use the model $\mathbf{Y} = \mathbf{A}\mathbf{X}$ directly (see the previous sections of this chapter).

Algorithm 4.5: HALS-CS

Input: $\mathbf{Y} \in \mathbb{R}^{I \times T}$: compressed data, $\mathbf{A} \in \mathbb{R}^{I \times J}$: projection matrix
Output: T sparse signals $\mathbf{X} \in \mathbb{R}^{J \times T}$ such that $\mathbf{Y} = \mathbf{A}\mathbf{X}$.

begin
  $\mathbf{X} = \mathbf{0}$ /* initialize X */
  $\tilde{\mathbf{Y}} = \mathbf{A}^T \mathbf{Y}$
  $\mathbf{G} = \mathbf{A}^T \mathbf{A}$, $\boldsymbol{d} = [d_1, d_2, \ldots, d_J]^T = diag(\mathbf{G})$
  foreach $\boldsymbol{g}_j$ *of* $\mathbf{G}$ **do** $\boldsymbol{g}_j \leftarrow \boldsymbol{g}_j / d_j$ /* normalize */
  $\mathbf{E} = [\underline{\boldsymbol{e}}_1^T, \underline{\boldsymbol{e}}_2^T, \ldots, \underline{\boldsymbol{e}}_J^T]^T = \tilde{\mathbf{Y}}$ /* initialize the residue */
  $\boldsymbol{\lambda} = \boldsymbol{\lambda}_{\max} = [\|\boldsymbol{y}_1\|_\infty, \|\boldsymbol{y}_2\|_\infty, \ldots, \|\boldsymbol{y}_T\|_\infty]$ /* initialize the threshold */
  repeat
    for $j = 1$ *to* J **do**
      $\underline{\boldsymbol{x}}_j \leftarrow \underline{\boldsymbol{x}}_j + \underline{\boldsymbol{e}}_j$ /* update $\underline{\boldsymbol{x}}_j$ */
      $\underline{\boldsymbol{x}}_j \leftarrow P_{\boldsymbol{\lambda}}(\underline{\boldsymbol{x}}_j)$ /* a shrinkage or an iterative formula */
      $\mathbf{E} \leftarrow \mathbf{E} - \boldsymbol{g}_j \, \Delta \underline{\boldsymbol{x}}_j$ /* update residue */
    end
    Decrease thresholds $\boldsymbol{\lambda}$ following a given strategy
  until *a stopping criterion is met* /* convergence condition */
end

The gradients of (4.139) are expressed as follows

$$\frac{\partial D^{(j)}}{\partial x_{jt}} = x_{jt} - y_t^{(j)} + \lambda_t \, \Psi(x_{jt}), \tag{4.140}$$

where $\Psi(x_{jt}) = \partial J_x(x_{jt}) / \partial x_{jt}$.

By setting the gradient to zero, we obtain the learning rule:

$$x_{jt} \leftarrow y_t^{(j)} - \lambda_t \Psi(x_{jt}), \tag{4.141}$$

or in the row vector form

$$\underline{\boldsymbol{x}}_j \leftarrow \underline{\boldsymbol{y}}^{(j)} - \Psi(\underline{\boldsymbol{x}}_j) \ \mathrm{diag}\,\{\boldsymbol{\lambda}\}. \tag{4.142}$$

In general, such a nonlinear equation does not have a closed-form (analytical) solution, and iterative updates are necessary to find an optimal solution. However, in some special cases closed-form solutions are possible, for example, for some shrinkage functions.

Let us consider several special simple cases:

1. For the minimum energy term: $J_x(\boldsymbol{x}_t) = \frac{1}{2}||\boldsymbol{x}_t||_2^2$, which is the Tikhonov type regularization, we have an explicit solution

$$\boxed{\underline{\boldsymbol{x}}_j \leftarrow \underline{\boldsymbol{y}}^{(j)} \oslash (\mathbf{1}^T + \boldsymbol{\lambda}),} \tag{4.143}$$

which can be interpreted as a trivial shrinkage operation since the samples $\underline{\boldsymbol{x}}_j$ are attenuated.

2. For the smoothing penalty term: $J_x(\boldsymbol{x}_t) = \frac{1}{2}||\mathbf{L}\,\boldsymbol{x}_t||_2^2$, where $\mathbf{L}$ is a matrix approximating the Laplace operator, which leads to

$$\Psi(x_{jt}) = \partial J_x(x_{jt}) / \partial x_{jt} = \boldsymbol{l}_j^T \, \mathbf{L}\, \boldsymbol{x}_t, \tag{4.144}$$

we obtain the filtering form:

$$\boxed{\underline{\boldsymbol{x}}_j \leftarrow \underline{\boldsymbol{y}}^{(j)} - \boldsymbol{l}_j^T \, \mathbf{L} \, \mathbf{X} \; \text{diag}\{\boldsymbol{\lambda}\}.} \tag{4.145}$$

3. In general, we do not have a closed-form solution, for the sparsity term $J_x(\boldsymbol{x}_t) = ||\mathbf{L}_{sp}\boldsymbol{x}_t||_1$ and $\Psi(x_{jt}) = \boldsymbol{l}_{sp_j}^T \; \text{sign}(\mathbf{L}_{sp}\boldsymbol{x}_t)$, we may resort to the iterative updating rule:

$$\boxed{\underline{\boldsymbol{x}}_j \leftarrow (1-\eta)\,\underline{\boldsymbol{x}}_j + \eta\,\underline{\boldsymbol{y}}^{(j)} - \eta\,\boldsymbol{l}_{sp_j}^T \; \text{sign}\,(\mathbf{L}_{sp}\,\mathbf{X})\;\text{diag}\{\boldsymbol{\lambda}\},} \tag{4.146}$$

where $\eta > 0$ is a learning rate.
However, in the special case when $\mathbf{L}_{sp} = \mathbf{I}$ and $\eta = 1$ (4.146) reduces into a simple nonlinear equation

$$\underline{\boldsymbol{x}}_j \leftarrow \underline{\boldsymbol{y}}^{(j)} - \text{sign}(\underline{\boldsymbol{x}}_j) \circledast \boldsymbol{\lambda}, \tag{4.147}$$

which leads to the soft shrinkage $\underline{\boldsymbol{x}}_j \leftarrow P^S_{\boldsymbol{\lambda}}(\underline{\boldsymbol{y}}^{(j)}) = \text{sign}(\underline{\boldsymbol{y}}^{(j)}) \circledast [|\underline{\boldsymbol{y}}^{(j)}| - \boldsymbol{\lambda}]_+$ with gradually decreasing thresholds $\boldsymbol{\lambda} = [\lambda_1, \lambda_2, \ldots, \lambda_T]$. Analogously, for $J_x(\boldsymbol{x}_t) = ||\boldsymbol{x}_t||_0$, where $||\boldsymbol{x}||_0$ means ℓ_0-norm (that is the number of nonzero elements in the vector $\boldsymbol{x}$), we obtain the hard shrinkage.

4. Applying the total variation (TV) penalty term, we can achieve smoothness while allowing for sharp changes in the estimated sources

$$J_x(\boldsymbol{x}_t) \;=\; ||\nabla \boldsymbol{x}_t||_1 = \sum_{j=1}^{J-1} |x_{jt} - x_{(j+1)t}| = ||\mathbf{L}_1\,\boldsymbol{x}_t||_1, \tag{4.148}$$

where $\mathbf{L}_1$ is the first order difference operator (4.57).
This case does not allow for explicit estimates, and we use the iterative update rule which is similar to (4.146):

$$\boxed{\underline{\boldsymbol{x}}_j \leftarrow (1-\eta)\,\underline{\boldsymbol{x}}_j + \eta\,\underline{\boldsymbol{y}}^{(j)} - \eta\,\boldsymbol{l}_{1j}^T \; \text{sign}(\mathbf{L}_1\,\mathbf{X})\;\text{diag}\{\boldsymbol{\lambda}\}.} \tag{4.149}$$

In order to develop an efficient and flexible iterative algorithm for CS, we can express the residual vector $\underline{\boldsymbol{y}}^{(j)}$ as follows:

$$\begin{aligned} \underline{\boldsymbol{y}}^{(j)} &= \tilde{\underline{\boldsymbol{y}}}_j - \sum_{p \neq j} g_{jp}\,\underline{\boldsymbol{x}}_p \;=\; \tilde{\underline{\boldsymbol{y}}}_j - \sum_{p=1}^{J} g_{jp}\,\underline{\boldsymbol{x}}_p + g_{jj}\,\underline{\boldsymbol{x}}_j \\ &= \underline{\boldsymbol{x}}_j + \tilde{\underline{\boldsymbol{y}}}_j - \underline{\boldsymbol{g}}_j\,\mathbf{X} \;=\; \underline{\boldsymbol{x}}_j + \underline{\boldsymbol{e}}_j, \end{aligned} \tag{4.150}$$

where $\underline{\boldsymbol{e}}_j$ is the j-th row of the error matrix $\mathbf{E} = \mathbf{Y} - \mathbf{G}\mathbf{X}$. Assuming that at the k-th iteration only the j-th row $\underline{\boldsymbol{x}}_j$ is updated, the error matrix can be updated as :

$$\begin{aligned} \mathbf{E}^{(k+1)} &= \tilde{\mathbf{Y}} - \mathbf{G}\mathbf{X}^{(k+1)} = \tilde{\mathbf{Y}} - \sum_{p=1}^{J} \boldsymbol{g}_p\,\underline{\boldsymbol{x}}_p^{(k+1)} \\ &= \tilde{\mathbf{Y}} - \sum_{p=1}^{J} \boldsymbol{g}_p\,\underline{\boldsymbol{x}}_p^{(k)} + \boldsymbol{g}_j\,\underline{\boldsymbol{x}}_j^{(k)} - \boldsymbol{g}_j\,\underline{\boldsymbol{x}}_j^{(k+1)} \\ &= \tilde{\mathbf{Y}} - \mathbf{G}\mathbf{X}^{(k)} - \boldsymbol{g}_j\left(\underline{\boldsymbol{x}}_j^{(k+1)} - \underline{\boldsymbol{x}}_j^{(k)}\right) \\ &= \mathbf{E}^{(k)} - \boldsymbol{g}_j\left(\underline{\boldsymbol{x}}_j^{(k+1)} - \underline{\boldsymbol{x}}_j^{(k)}\right). \end{aligned} \tag{4.151}$$

In the special case, imposing sparse constraints with ℓ_0-norm or ℓ_1-norm leads to a simplified algorithm, referred to as the HALS-CS [56], with the hard or the soft shrinkage rule outlined as follows based on the iterative thresholding strategy

$$\underline{\boldsymbol{x}}_j \leftarrow \underline{\boldsymbol{x}}_j + \underline{\boldsymbol{e}}_j, \tag{4.152}$$

$$\underline{\boldsymbol{x}}_j \leftarrow P_{\boldsymbol{\lambda}}(\underline{\boldsymbol{x}}_j), \tag{4.153}$$

$$\mathbf{E} \leftarrow \mathbf{E} - \boldsymbol{g}_j\, \Delta\underline{\boldsymbol{x}}_j \qquad (j = 1, 2, \ldots, J), \tag{4.154}$$

where $\Delta^{(k+1)}\underline{\boldsymbol{x}}_j = \underline{\boldsymbol{x}}_j^{(k+1)} - \underline{\boldsymbol{x}}_j^{(k)}$ and $P_\lambda(\cdot)$ represents the hard (4.121) or soft (4.122) shrinkage function applied component-wise. The pseudocode of this algorithm is given in Algorithm 4.5.

The hard and soft shrinkages are two well-known threshold functions in wavelet denoising, linear inverse problem and many other problems. In the next section we will derive some alternative shrinkage rules (see Figure 4.2) which can be applied efficiently for HALS-CS algorithm.

Replacing $\boldsymbol{x}_t$ by $z_t = \boldsymbol{x}_t^{.[2]}$, and $\underline{\boldsymbol{y}}_t^{(j)}$ by $\left(\underline{\boldsymbol{y}}_t^{(j)}\right)^{.[2]}$ (component-wise) in objective functions (4.139) leads to

$$\underline{\boldsymbol{x}}_j^{.[2]} = \underline{\boldsymbol{z}}_j = \left(\underline{\boldsymbol{y}}_t^{(j)}\right)^{.[2]} - \text{sign}\left(\underline{\boldsymbol{z}}_j\right) \circledast \boldsymbol{\lambda}, \tag{4.155}$$

then dividing two sides by $\underline{\boldsymbol{x}}_j$

$$\underline{\boldsymbol{x}}_j = \left(\left(\underline{\boldsymbol{y}}_t^{(j)}\right)^{.[2]} - \boldsymbol{\lambda}\right) \oslash \underline{\boldsymbol{x}}_j \tag{4.156}$$

will form a new learning rule for $\underline{\boldsymbol{x}}_j$ with the nn-garrotte shrinkage function (4.123) with threshold $\sqrt{\boldsymbol{\lambda}}$, where $\circledast$ and $\oslash$ denote component-wise multiplication and division. Taking square root[15] for (4.155), we obtain

$$\underline{\boldsymbol{x}}_j = \text{sign}\left(\underline{\boldsymbol{x}}_j\right) \sqrt{\left(\underline{\boldsymbol{y}}_t^{(j)}\right)^{.[2]} - \boldsymbol{\lambda}} \tag{4.157}$$

leading to the new learning rule for $\underline{\boldsymbol{x}}_j$ with the Abramovich shrinkage function given in (4.124).

In practice, $P_\lambda(\cdot)$ can be a suitable shrinkage rule with adaptive threshold, such as: nonnegative garotte (4.123), Abramovich (4.124) or $n-$degree garrotte (4.125) rules with threshholds $\boldsymbol{\lambda} = [\lambda_1, \lambda_2, \ldots, \lambda_T]$. Roughly speaking, the nonlinear shrinkage transformation P_λ is employed in order to achieve a unique sparse solution and to make the algorithm robust to noise [4,5,17,26]. For large-scale data when the sources $\mathbf{X}$ have a large number of samples ($J = 10^6 - 10^8$), $\boldsymbol{g}_j$ should be replaced by $\boldsymbol{g}_j = \mathbf{A}^T \boldsymbol{a}_j$.

4.7.7 Generalized HALS-CS Algorithm

This section will present a generalized version for the HALS-CS algorithm based on the Beta-divergence. Instead of minimizing the objective function (4.139), we minimize a flexible and general objective function

[15] $f(x) = \sqrt{x}$ denotes a component-wise function.

using the Beta-divergence defined as follows [16]:

$$D_\beta^{(j)}(\underline{y}^{(j)} \,||\, \underline{x}_j) = \begin{cases} \sum_t \left(y_t^{(j)} \frac{[y_t^{(j)}]^\beta - [x_{jt}]^\beta}{\beta} - \frac{[y_t^{(j)}]^{\beta+1} - [x_{jt}]^{\beta+1}}{\beta+1} \right), & \beta > 0, \quad (4.158a) \\ \sum_t \left(y_t^{(j)} \ln \left(\frac{y_t^{(j)}}{x_{jt}} \right) - y_t^{(j)} + x_{jt} \right), & \beta = 0, \quad (4.158b) \\ \sum_t \left(\ln \left(\frac{x_{jt}}{y_t^{(j)}} \right) + \frac{y_t^{(j)}}{x_{jt}} - 1 \right), & \beta = -1, \quad (4.158c) \end{cases}$$

where β is the degree parameter of the Beta-divergence.

We recall here that the Beta-divergence in special cases includes the standard squared Euclidean distance (for $\beta = 1$), the Itakura-Saito distance ($\beta = -1$), and the generalized Kullback-Leibler I-divergence ($\beta = 0$). Again, the choice of the parameter β depends on the statistics of the data and the Beta-divergence corresponds to the Tweedie models [21,23,24,46,60]. For example, the optimal choice of the parameter β for a normal distribution is $\beta = 1$, for the gamma distribution is $\beta = -1$, for the Poisson distribution is $\beta = 0$, and for a compound Poisson $\beta \in (-1, 0)$.

Imposing additional constraints on the objective function (4.158) via regularized functions $J_x(\boldsymbol{x}_t)$ as in (4.139), and zeroing the gradient of this modified function with respect to x_{jt} leads to the equation

$$\frac{\partial D_\beta^{(j)}}{\partial x_{jt}} = x_{jt}^\beta - [y_t^{(j)}]^\beta + \lambda_t \, \Psi(x_{jt}) = 0, \tag{4.159}$$

where $\Psi(x_{jt}) = \partial J_x(x_{jt})/\partial x_{jt}$.

Using the same penalty function for sparsity constraints for deriving the HALS-CS algorithm,

$$J_x(\boldsymbol{x}_t) = ||\boldsymbol{x}_t||_1, \qquad \Psi(\boldsymbol{x}_t) = \text{sign}(\boldsymbol{x}_t), \tag{4.160}$$

we obtain a generalized iterative update rule:

$$\boxed{x_{jt} \leftarrow \text{sign}(y_t^{(j)}) \left[\left(y_t^{(j)}\right)^\beta - \lambda_t \right]^{1/\beta} \mathbf{1}(|y_{jt}| \geq \lambda_t^{1/\beta}).} \tag{4.161}$$

For $\beta = 2$, the above learning rule simplifies to the HALS-CS with the Abramovich shrinkage function. Actually, we have just formed a new shrinkage rule called as the beta-Abramovich shrinkage function $P_\beta^{Abra}(x, \lambda)$ defined as

$$\boxed{P_\beta^{Abra}(x, \lambda) = \text{sign}(x)(x^\beta - \lambda^\beta)^{1/\beta} \, \mathbf{1}(|x| \geq \lambda)}. \tag{4.162}$$

The learning rule (4.161) can be rewritten in the compact vector form

$$\boxed{\underline{\boldsymbol{x}}_j \leftarrow P_\beta^{Abra}(\underline{\boldsymbol{y}}^{(j)}, \boldsymbol{\lambda}^{1/\beta}).} \tag{4.163}$$

Analogously, dividing the equation (4.159) by $x_{jt}^{\beta-1}$ gives us a generalized function of the nonnegative garrotte shrinkage rule

$$\boxed{P_\beta^{nng}(x, \lambda) = (x - \lambda^\beta \, \text{sign}(x)/x^\beta) \, \mathbf{1}(|x| > \lambda)} \tag{4.164}$$

where $\tilde{\beta} = \beta - 1$, and the new learning rule is given in the form as

$$\underline{\boldsymbol{x}}_j \leftarrow P_{\tilde{\beta}}^{nng}(\underline{\boldsymbol{y}}^{(j)}, \boldsymbol{\lambda}^{1/\tilde{\beta}}). \tag{4.165}$$

For $\beta = 2$, we obtain the HALS-CS algorithm with the $nn - garrotte$ shrinkage function. For $\beta = 1$, the HALS-CS algorithm is with the $soft$ shrinkage function.

Updating row-by-row the estimate $\underline{\boldsymbol{x}}_j$ as in learning rule (4.154) has advantage for a system with limited resources. However, for a very long source (a lot of samples), it takes time to update all the rows because of updating residue $\mathbf{E}$ in each iteration step. One approach to circumvent this limitation and to decrease running time is that we update $\boldsymbol{x}_t$ with block-wise technique and in parallel mode. In other words, instead of updating one row $\underline{\boldsymbol{x}}_j$, we update simultaneously a block of B rows $[\underline{\boldsymbol{x}}_j^T, \underline{\boldsymbol{x}}_{j+1}^T, \ldots, \underline{\boldsymbol{x}}_{j+B-1}^T]^T$, then update residue, and process the consecutive blocks.

$$\mathbf{E} \leftarrow \mathbf{E} - \mathbf{G}_{j:j+B-1}\, \Delta\underline{\mathbf{X}}_{j:j+B-1}. \tag{4.166}$$

4.7.8 Generalized HALS Algorithms Using Alpha-Divergence

The Alpha-divergence for HALS algorithms can be defined as follows [16]:

$$D_\alpha^{(j)}\left(([\mathbf{Y}^{(j)}]_+)||\boldsymbol{a}_j\, \boldsymbol{b}_j^T\right) = \begin{cases} \displaystyle\sum_{it}\left(\frac{q_{it}^{(j)}}{\alpha(\alpha+1)}\left[\left(\frac{q_{it}^{(j)}}{y_{it}^{(j)}}\right)^\alpha - 1\right] - \frac{q_{it}^{(j)} - y_{it}^{(j)}}{\alpha+1}\right), & \alpha \neq -1, 0, \quad (4.167a)\\[2ex] \displaystyle\sum_{it}\left((q_{it}^{(j)})\, \ln\left(\frac{q_{it}^{(j)}}{y_{it}^{(j)}}\right) - q_{it}^{(j)} + y_{it}^{(j)}\right), & \alpha = 0, \quad (4.167b)\\[2ex] \displaystyle\sum_{it}\left(y_{it}^{(j)}\ln\left(\frac{y_{it}^{(j)}}{q_{it}^{(j)}}\right) + q_{it}^{(j)} - y_{it}^{(j)}\right), & \alpha = -1, \quad (4.167c) \end{cases}$$

where $y_{it}^{(j)} = \left[[\mathbf{Y}]_{it} - \sum_{p\neq j} a_{ip}b_{tp}\right]_+$ and $q_{it}^{(j)} \equiv \hat{y}_{it}^{(j)} = a_{ij}b_{tj}$ for $j = 1, 2, \ldots, J$.

Recall here that the choice of parameter $\alpha \in \mathbb{R}$ depends on statistical distributions of the noise and data. In special cases of the Alpha-divergence for $\alpha = \{1, -0.5, -2\}$, we obtain respectively the Pearson's chi squared, Hellinger's, and Neyman's chi-square distances while for the cases $\alpha = 0$ and $\alpha = -1$, the divergence has to be defined by the limits of (4.167 (a)) as $\alpha \to 0$ and $\alpha \to -1$, respectively. When these limits are evaluated for $\alpha \to 0$ we obtain the generalized Kullback-Leibler I-divergence defined by Eq. (4.167(b)), whereas for $\alpha \to -1$ we have the dual generalized Kullback-Leibler I-divergence given in Eq. (4.167(c)) [21,23,24].

The gradient of the Alpha-divergence (4.167) for $\alpha \neq -1$ with respect to a_{ij} and b_{tj} can be expressed in a compact form as:

$$\frac{\partial D_\alpha^{(j)}}{\partial b_{tj}} = \frac{1}{\alpha}\sum_i a_{ij}\left[\left(\frac{q_{it}^{(j)}}{y_{it}^{(j)}}\right)^\alpha - 1\right], \qquad \frac{\partial D_\alpha^{(j)}}{\partial a_{ij}} = \frac{1}{\alpha}\sum_t b_{tj}\left[\left(\frac{q_{it}^{(j)}}{y_{it}^{(j)}}\right)^\alpha - 1\right]. \tag{4.168}$$

By equating the gradients to zeros we obtain [16]

$$b_{tj} \leftarrow \left(\frac{\sum_i a_{ij}\left(y_{it}^{(j)}\right)^\alpha}{\sum_{i=1}^I a_{ij}^{\alpha+1}}\right)^{1/\alpha}, \qquad a_{ij} \leftarrow \left(\frac{\sum_{t=1}^T b_{tj}\left(y_{it}^{(j)}\right)^\alpha}{\sum_{t=1}^T b_{tj}^{\alpha+1}}\right)^{1/\alpha}, \qquad \forall i, j, t. \tag{4.169}$$

The above local update rules referred to as the Alpha HALS algorithm can be written in a compact matrix-vector form:

$$\boldsymbol{b}_j \leftarrow \left(\frac{\left[\mathbf{Y}^{(j)\,T}\right]_+^{.[\alpha]} \boldsymbol{a}_j}{\boldsymbol{a}_j^T \, \boldsymbol{a}_j^{.[\alpha]}} \right)^{.[1/\alpha]}, \qquad \boldsymbol{a}_j \leftarrow \left(\frac{\left[\mathbf{Y}^{(j)}\right]_+^{.[\alpha]} \boldsymbol{b}_j}{\boldsymbol{b}_j^T \, \boldsymbol{b}_j^{.[\alpha]}} \right)^{.[1/\alpha]}, \qquad (j = 1, \ldots, J), \tag{4.170}$$

where the "rise to the power" operations $\boldsymbol{x}^{.[\alpha]}$ are performed component-wise and

$$\mathbf{Y}^{(j)} = \mathbf{Y} - \sum_{p \neq j} \boldsymbol{a}_p \boldsymbol{b}_p^T. \tag{4.171}$$

These update rules can be reformulated in a more general and flexible form:

$$\boldsymbol{b}_j \leftarrow \Psi^{-1} \left(\frac{\Psi\left(\left[\mathbf{Y}^{(j)T}\right]_+\right) \boldsymbol{a}_j}{\boldsymbol{a}_j^T \, \Psi(\boldsymbol{a}_j)} \right), \qquad \boldsymbol{a}_j \leftarrow \Psi^{-1} \left(\frac{\Psi\left(\left[\mathbf{Y}^{(j)}\right]_+\right) \boldsymbol{b}_j}{\boldsymbol{b}_j^T \, \Psi(\boldsymbol{b}_j)} \right), \qquad (j = 1, \ldots, J), \tag{4.172}$$

where $\Psi(\boldsymbol{x})$ is a suitable chosen function, for example, $\Psi(\boldsymbol{b}) = \boldsymbol{b}^{.[\alpha]}$, applied component-wise.

In the special case of $\alpha = 1$, we obtain the HALS algorithm presented in the previous section (4.92). In a similar way, we can derive the HALS algorithm for $\alpha = 0$ (see Chapter 3 for more detail).

4.7.9 Generalized HALS Algorithms Using Beta-Divergence

The Beta-divergence can be considered as a flexible and complementary cost function to the Alpha-divergence and it is defined as follows [16]:

$$D_\beta^{(j)}([\mathbf{Y}^{(j)}]_+ || \boldsymbol{a}_j \boldsymbol{b}_j^T) = \begin{cases} \displaystyle\sum_{it} \left(y_{it}^{(j)} \frac{[y_{it}^{(j)}]^\beta - [q_{it}^{(j)}]^\beta}{\beta} - \frac{[y_{it}^{(j)}]^{\beta+1} - [q_{it}^{(j)}]^{\beta+1}}{\beta+1} \right), & \beta > 0, \quad (4.173a) \\ \displaystyle\sum_{it} \left(y_{it}^{(j)} \ln\left(\frac{y_{it}^{(j)}}{q_{it}^{(j)}}\right) - y_{it}^{(j)} + q_{it}^{(j)} \right), & \beta = 0, \quad (4.173b) \\ \displaystyle\sum_{it} \left(\ln\left(\frac{q_{it}^{(j)}}{y_{it}^{(j)}}\right) + \frac{y_{it}^{(j)}}{q_{it}^{(j)}} - 1 \right), & \beta = -1, \quad (4.173c) \end{cases}$$

where $y_{it}^{(j)} = \left[y_{it} - \sum_{p \neq j} a_{ip} b_{tp}\right]_+$ and $q_{it}^{(j)} \equiv \hat{y}_{it}^{(j)} = a_{ij} b_{tj}$ for $j = 1, 2, \ldots, J$.

Again, the choice of the parameter β depends on the statistics of the data and the Beta-divergence corresponds to the Tweedie models [21,23,24,46,60]. For example, the optimal choice of the parameter β for a normal distribution is $\beta = 1$, for the gamma distribution is $\beta = -1$, for the Poisson distribution is $\beta = 0$, and for a compound Poisson $\beta \in (-1, 0)$.

In order to derive the multiplicative learning algorithm, we compute the gradient of (4.173) with respect to the elements $x_{jt} = b_{tj},\ a_{ij}$:

$$\frac{\partial D_\beta^{(j)}}{\partial b_{tj}} = \sum_i \left([q_{it}^{(j)}]^\beta - y_{it}^{(j)} \, [q_{it}^{(j)}]^{\beta-1} \right) a_{ij}, \tag{4.174}$$

$$\frac{\partial D_\beta^{(j)}}{\partial a_{ij}} = \sum_t \left([q_{it}^{(j)}]^\beta - y_{it}^{(j)} \, [q_{it}^{(j)}]^{\beta-1} \right) b_{tj}. \tag{4.175}$$

By equating the gradient components to zero, we obtain a set of simple HALS updating rules referred to as the Beta HALS algorithm:

$$b_{tj} \leftarrow \frac{1}{\sum_{i=1}^{I} a_{ij}^{\beta+1}} \sum_{i=1}^{I} a_{ij}^{\beta}\, y_{it}^{(j)}, \tag{4.176}$$

$$a_{ij} \leftarrow \frac{1}{\sum_{t=1}^{T} b_{tj}^{\beta+1}} \sum_{t=1}^{T} b_{tj}^{\beta}\, y_{it}^{(j)}, \qquad \forall i, j, t. \tag{4.177}$$

The above local learning rules can be written in a compact matrix-vector form:

$$\boldsymbol{b}_j \leftarrow \frac{([\mathbf{Y}^{(j)\,T}]_+)\, \boldsymbol{a}_j^{.[\beta]}}{\boldsymbol{a}_j^T\, \boldsymbol{a}_j^{.[\beta]}}, \qquad \boldsymbol{a}_j \leftarrow \frac{([\mathbf{Y}^{(j)}]_+)\, \boldsymbol{b}_j^{.[\beta]}}{(\boldsymbol{b}_j^T)^{.[\beta]}\, \boldsymbol{b}_j} \qquad (j = 1, 2, \ldots, J), \tag{4.178}$$

where the "rise to a power" operations are performed element-wise and $\mathbf{Y}^{(j)} = \mathbf{Y} - \sum_{p \neq j} \boldsymbol{a}_p \boldsymbol{b}_p^T$.

The above Beta HALS update rules can be represented in more general and flexible forms as

$$\boldsymbol{b}_j \leftarrow \frac{([\mathbf{Y}^{(j)\,T}]_+)\, \Psi(\boldsymbol{a}_j)}{\Psi(\boldsymbol{a}_j^T)\, \boldsymbol{a}_j}, \qquad \boldsymbol{a}_j \leftarrow \frac{([\mathbf{Y}^{(j)}]_+)\, \Psi(\boldsymbol{b}_j)}{\Psi(\boldsymbol{b}_j^T)\, \boldsymbol{b}_j}, \tag{4.179}$$

where $\Psi(\boldsymbol{b})$ is a suitably chosen convex function (e.g., $\Psi(\boldsymbol{b}) = \boldsymbol{b}^{.[\beta]}$) and the nonlinear operations are performed element-wise.

In the special cases the Beta HALS algorithm simplifies as follows:

- For $\beta = 1$

$$b_{tj} \leftarrow \frac{1}{||\boldsymbol{a}_j||_2^2} \sum_{i=1}^{I} a_{ij}\, y_{it}^{(j)}, \qquad a_{ij} \leftarrow \frac{1}{||\boldsymbol{b}_j||_2^2} \sum_{t=1}^{T} b_{tj}\, y_{it}^{(j)}, \qquad \forall i, j, t; \tag{4.180}$$

- For $\beta = 0$

$$b_{tj} \leftarrow \frac{1}{||\boldsymbol{a}_j||_1} \sum_{i=1}^{I} y_{it}^{(j)}, \qquad a_{ij} \leftarrow \frac{1}{||\boldsymbol{b}_j||_1} \sum_{t=1}^{T} y_{it}^{(j)}, \qquad \forall i, j, t; \tag{4.181}$$

- For $\beta = -1$

$$b_{tj} \leftarrow \sum_{i=1}^{I} \frac{y_{it}^{(j)}}{a_{ij}}, \qquad a_{ij} \leftarrow \sum_{t=1}^{T} \frac{y_{it}^{(j)}}{b_{tj}}, \qquad \forall i, j, t. \tag{4.182}$$

For extreme values of β the local beta NMF algorithm will pick up maximum or minimum values of $y_{it}^{(j)}$ from the corresponding rows or columns of the matrix $[\mathbf{Y}^{(j)}]$.

The derived algorithms can be further generalized as follows:

$$b_{tj} \leftarrow b_{tj} \Psi^{-1} \left(\frac{\sum_{i=1}^{I} a_{ij}\, \Psi(y_{it}^{(j)})}{\sum_{i=1}^{I} a_{ij} \Psi(a_{ij} b_{tj})} \right), \tag{4.183}$$

$$a_{ij} \leftarrow a_{ij} \Psi^{-1} \left(\frac{\sum_{t=1}^{T} b_{tj}\, \Psi(y_{it}^{(j)})}{\sum_{t=1}^{T} b_{tj} \Psi(a_{ij} b_{tj})} \right), \tag{4.184}$$

where $\Psi(x)$ is an increasing monotonically function such that $\Psi^{-1}(1) = 1$ or

$$b_{tj} \leftarrow b_{tj} \frac{\sum_{i=1}^{I} a_{ij}\, \Phi(y_{it}^{(j)}, q_{it}^{(j)})}{\sum_{i=1}^{I} a_{ij} \Phi(q_{it}^{(j)}, q_{it}^{(j)})}, \tag{4.185}$$

$$a_{ij} \leftarrow a_{ij} \frac{\sum_{t=1}^{T} b_{tj}\, \Phi(y_{it}^{(j)}, q_{it}^{(j)})}{\sum_{t=1}^{T} b_{tj} \Phi(q_{it}^{(j)}, q_{it}^{(j)})}, \tag{4.186}$$

and $\Phi(q, q)$ is a nonnegative nondecreasing function, and $\Phi(y, q)$ may take different forms, for example, $\Phi(y, q) = y$, $\Phi(q, q) = q$; $\Phi(y, q) = y/q$, $\Phi(q, q) = 1$; $\Phi(y, q) = y/q^{\beta}$, $\Phi(q, q) = q^{1-\beta}$, $\Phi(y, q) = y/(c + q)$, and $\Phi(q, q) = q/(c + q)$. It is not expected that all the generalized HALS algorithms presented here will perform well for any set of functions. In some cases, to ensure stability, it is necessary to use suitable scaling and/or relaxation parameters. However, the simple learning rules above have interesting biological links: the nonnegative synaptic weights a_{ij} and x_{jt} are updated locally on the basis of (a normalized) weighted mean (row-wise or columns-wise) values of nonnegative signals $[y_{it}^{(j)}]_+$, which can be interpreted as outputs of neurons with some synaptic inhibitions [48,61]. Taking into account that in each iterative step we normalize the vectors $\boldsymbol{a}_j$ and $\boldsymbol{b}_j$ to unit length, the local algorithms perform a weighting average. This can also be considered as Bayesian model averaging or a generalized mean of nonnegative signals.

In order to avoid local minima we have also developed simple heuristics for local Alpha and Beta HALS NMF algorithms combined with multi-start initializations using the standard ALS as follows:

1. Perform factorization of a matrix for any value of the parameters α or β; preferably, set the value of the parameters to unity for simplicity and a high speed.
2. If the algorithm converges, but a desirable fit value (fit max) is not achieved, restart the factorization, keeping the previously estimated factors as the initial matrices for the ALS initialization. This helps with the problem of local minima.
3. If the algorithm does not converge, the value of α or β should be changed incrementally to help to jump over local minima.
4. Repeat the procedure until a desired fit value is achieved, or the difference in the values of the factor matrices in successive iterations satisfies a given criterion.

4.8 Simulation Results

We now provide simulation studies for several benchmarks and typical applications to illustrate validity and performance of algorithms described in this chapter.

4.8.1 Underdetermined Blind Source Separation Examples

Example 4.1 *Underdetermined blind source separation for nonoverlapped nonnegative sources*

In this section we illustrate performance of the HALS algorithm for the underdetermined NMF problem for only two mixtures of a large number of sparse sources. Simulations were performed for 50, 100 and 1000 sparse nonoverlapped nonnegative sources with 1000 samples and with only two mixtures. The full rank mixing matrix $\mathbf{A} \in \mathbb{R}_+^{2\times J}$ was randomly drawn from an uniform distribution, where J is the number of unknown sources. We have chosen a hybrid algorithm combining the Fast HALS learning rule with orthogonality constraints for the estimation sources $\mathbf{X}$ and the regular ALS learning rule for estimating the mixing matrix $\mathbf{A}$. In order to achieve a good performance it was necessary to employ a multilayer NMF with at least 25 layers (see Listing 4.1).

Listing 4.1 Example 4.1 for underdetermined NMF problem.

```
% Example 4.1 for underdetermined NMF problem with only 2 mixtures
clear
I = 2;                                              %number of mixtures
J = 50;                              %number of sources (50, 100, 1000)
T = 1000;                                            %number of samples
X = nngorth(J,T,0.5);                 %generate non-overlapped sources
A = rand(I,J);                                %generate mixing matrix A
Y = A*X;                                               % build mixtures
%% Find the original sources by using the multilayer NMF
options = struct('tol',1e-6,'J',J,'algtype',[6 5],'init',[4 4],...
    'nrestart',1,'niter',1000,'ellnorm',1,'beta',1,'lortho',[0 0],...
    'algorithm','nmf_ghals','Nlayer',25);

[AH,XH,SIR,PSNR,rtime,Fit]= nmf_multi_layer(Y,options,X,1);

%% Compute and visualize the correlation matrix G
[G,mapind] = nmf_G(XH,X);
figure; hinton(G)
```

We have also evaluated the performance of the estimation of nonoverlapped (orthogonal) sources via a correlation matrix defined as $\mathbf{G} = \hat{\mathbf{X}}\mathbf{X}^T$. For perfect estimation, this matrix should be a diagonal matrix for which the performance is visualized as a deviation from the diagonal matrix (see Figures 4.4(a) and 4.4(b)). The majority of the sources were retrieved correctly and with high PSNR indices (see Table 4.1). In addition to PSNR, we computed the correlation index, CR, defined as in Listing 4.2.

Listing 4.2 Computation of the correlation matrix **G**.

```
% correlation matrix  G is computed in advanced with the normalized sources
CR = -10*log10(sum(G-diag(diag(G)),2));
```

4.8.2 NMF with Sparseness, Orthogonality and Smoothness Constraints

Example 4.2 *The 16 segment display dataset*

In this example, our goal was to decompose a 16 segment display dataset containing 40 grayscale alphanumeric images of dimension 60×43 (see Figure 4.5(a)). Each symbol (letter or digit) was formed by activating the selected segments from 16 segments. Since two segments in the upper part and two segments in the lower part are activated simultaneously, the number of separable components of this dataset is only 14. Including a dot and border part, this dataset has altogether 15 distinct basis components as shown in Figure 4.5(d).

Although this dataset is very sparse with nonoverlapping bases, most standard multiplicative NMF algorithms cannot successfully retrieve all the basis components. Figure 4.6(a) displays 15 components estimated by the multiplicative Beta algorithm (with $\beta = 1$). To decompose successfully all segments, we needed to

Table 4.1 HALS NMF algorithm for an underdetermined NMF problem with only two mixtures and 50 and 100 sparse nonoverlapped sources. Performance indexes are evaluated over 100 Monte Carlo trials and a median of all the trials.

	PSNR [dB]	CR [dB]	Fit (%)	Running time (seconds)
2×50	134.89	53.17	99.98	216.79
2×100	105.33	40.49	99.95	915.18

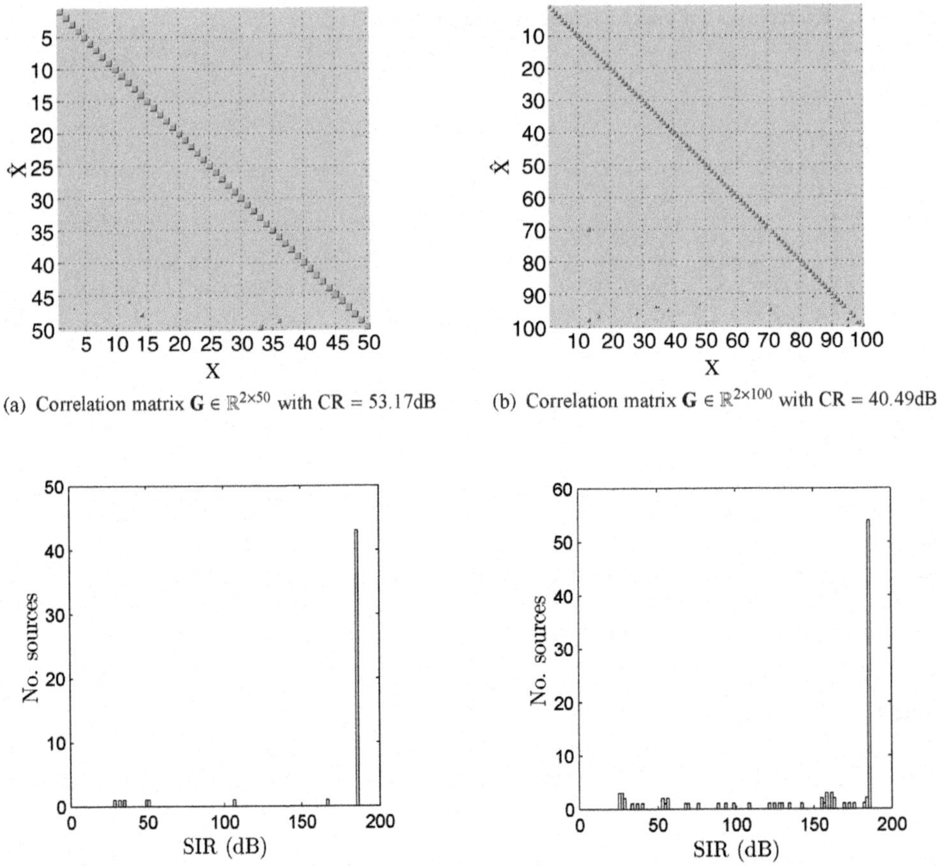

(a) Correlation matrix $\mathbf{G} \in \mathbb{R}^{2\times 50}$ with CR = 53.17dB

(b) Correlation matrix $\mathbf{G} \in \mathbb{R}^{2\times 100}$ with CR = 40.49dB

(c) SIR histogram for 50 estimated sources

(d) SIR histogram for 100 sources

Figure 4.4 Visualization of the performance for Example 4.1 for 50 and 100 sources. (a)–(b) Hinton graphs of weight matrices **G**, (c)–(d) SIR histograms. Almost all sources were retrieved perfectly.

impose an additional orthogonality constraint for **A**, combined with a sparseness constraint for **X**. Figure 4.6(b) shows the results for the Beta HALS algorithm ($\beta = 1$) with the orthogonality and sparseness constraints, using multilayer NMF with three layers with 500 iterations for each layer. Note that this benchmark includes real-world images, and that original basis components are unknown. Therefore, we have evaluated qualitatively the performance using the correlation ratio (CR) index.

Example 4.3 *NMF with smooth signals*

We used five noisy mixtures of three smooth sources (benchmark signals `X_5smooth` [19]) for testing the HALS algorithm with the smoothness constraint. The mixed signals were corrupted by additive Gaussian noise with SNR = 10 dB (Figure 4.7(a)). Figure 4.7 (c) illustrates the efficiency of the HALS NMF algorithm with smoothness constraints using the update rules (4.114), (4.115) with the second-order Laplace operator **L**. The estimated components by using the HALS algorithm with three layers [23] are depicted in Figure 4.7(b), whereas the results obtained by the same algorithm but with smooth constraints ($SIR\_\mathbf{A} = 29.22$ dB and $SIR\_\mathbf{X} = 15.53$ dB) are shown in Figure 4.7(c).

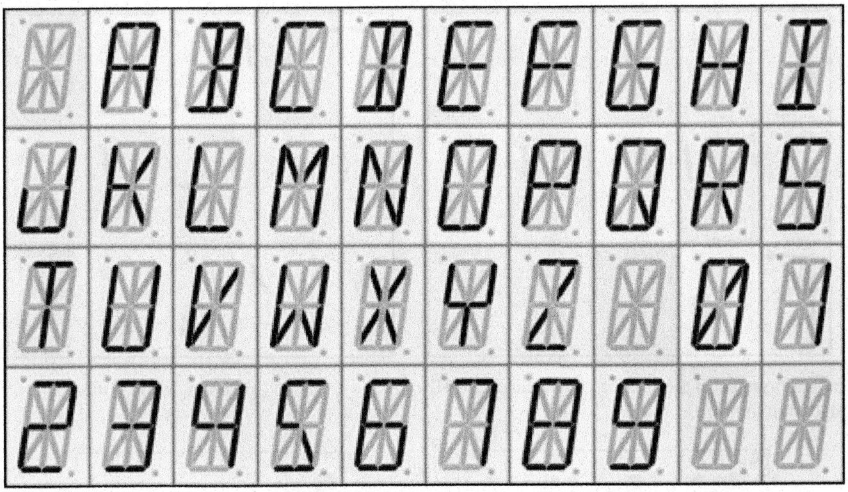

(a) Set of 40 images generated by 16 segment display

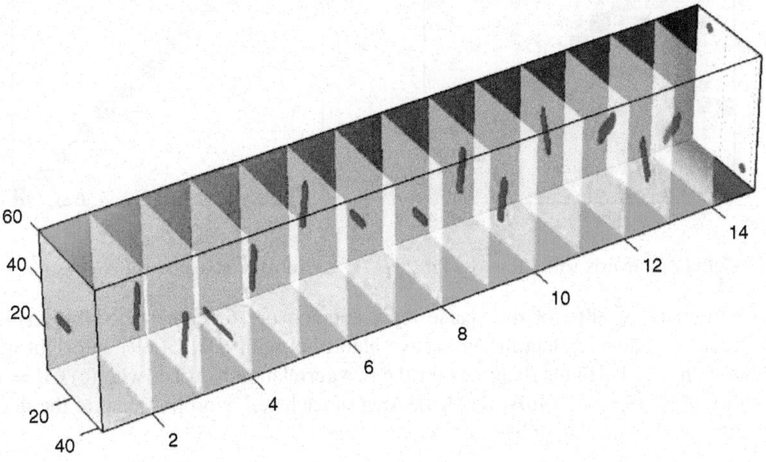

(b) Fifteen basis pictures of the 16 segment displays

Figure 4.5 The Dataset for Example 4.2. Our objective is to perform sparse representation of all the symbols.

4.8.3 Simulations for Large-scale NMF

Example 4.4 *Large-scale NMF*

We performed experiments for three large-scale dataset: data matrices of dimension 1000×1000, 5000×1000 and 10000×1000. The objective was to estimate 10 sources with 1000 sampling points as displayed in Figure 4.8(a). The mixing matrix was randomly generated using a uniform distribution. Figure 4.8(d) presents the comparison of the PSNR obtained by the Beta HALS algorithm with $\beta = 0.5, 1, 1.5, 2$. The algorithms were run over 500 iterations with three layers NMF and by employing the nonnegative ALS initialization.

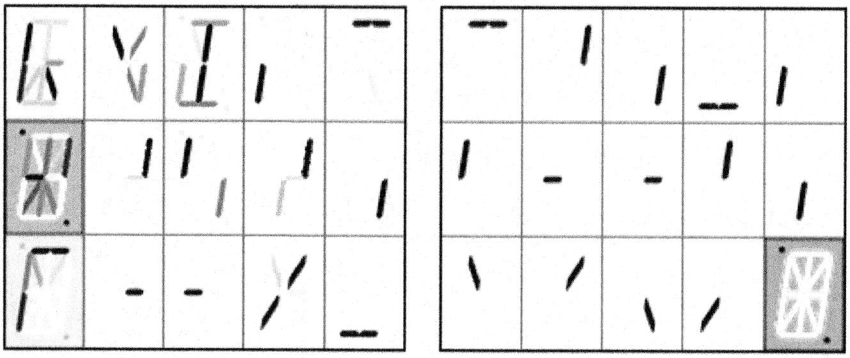

(a) Performance of the standard multiplicative NMF algorithm for $\beta = 1$

(b) HALS NMF, $\beta = 1$

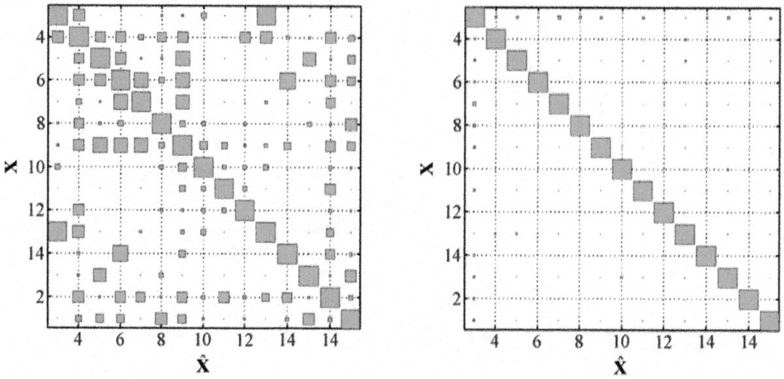

(c) Correlation matrix with CR = -1.14[dB]

(d) Correlation matrix with CR = 14.31[dB]

Figure 4.6 Performance results for the sparse representation of 16 segment displays using two NMF algorithm: Left column standard Beta multiplicative algorithm with $\beta = 1$ (left) and right column using HALS algorithm with $\beta = 1$. Hinton diagrams for the two correlation matrices with (c) CR $= -1.14$ [dB] (global algorithm), (d) CR $= 14.31$ [dB] (HALS). Area of each cell is proportional to the intensity of the corresponding entry in the correlation matrix.

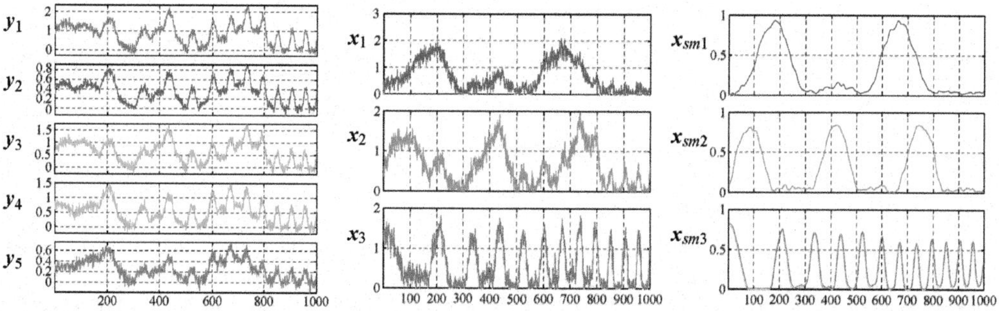

(a) Noisy mixtures with SNR= 10 dB and Gaussian noise

(b) Estimated components using HALS algorithm

(c) Estimated components using the HALS with smoothness constraints

Figure 4.7 Illustration of Example 4.3. Estimation of smooth sources using the standard HALS algorithm (b) and the HALS with additional smoothness constraints (c).

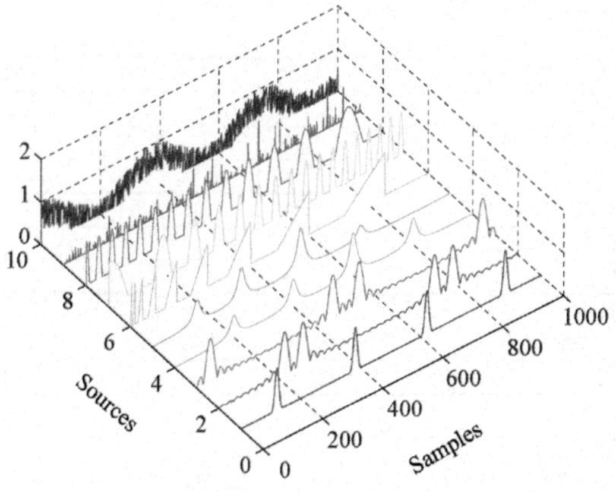

(a) Original 10 sources

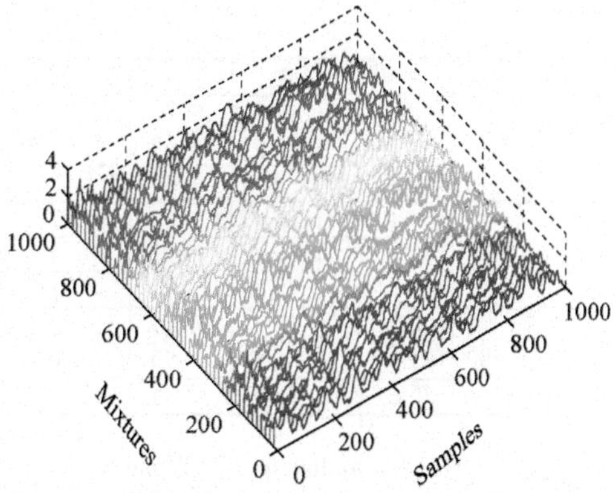

(b) Typical 1000 mixtures

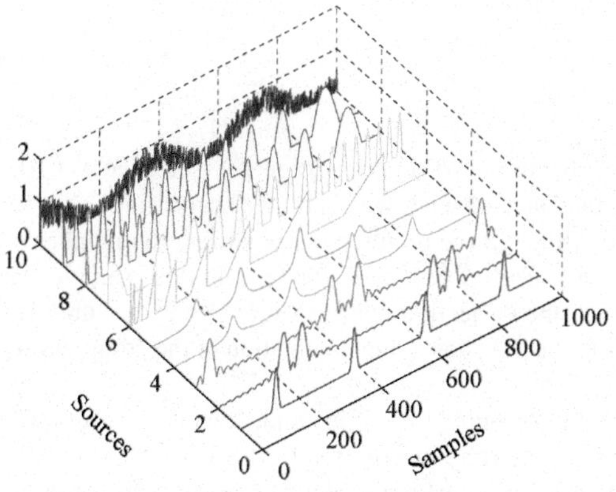

(c) Ten estimated components by using Fast-HALS

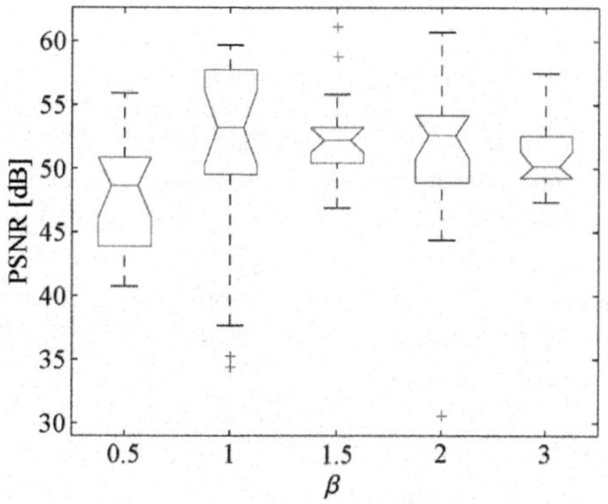

(d) PSNR using Beta HALS for various values of β

Figure 4.8 Illustration for (a) benchmark used in large-scale experiments with 10 nonnegative sources; (b) Typical 1000 mixtures; (c) Ten estimated components by using FAST HALS NMF from the observations matrix $\mathbf{Y}$ of dimension 1000×1000. (d) Performance expressed via the PSNR using the Beta HALS NMF algorithm for $\beta = 0.5$, 1, 1.5, 2 and 3.

Table 4.2 presents the results of three experiments using the Beta HALS, the ALS, and combination of the ALS for $\mathbf{A}$ and the HALS for $\mathbf{X}$ with and without the block-wise approach (see also Figure 4.9). Observe that when the number of components is relatively small, the ALS learning rule is suitable for estimating the mixing matrix $\mathbf{A}$, because the pseudo-inverse of the small matrix $\mathbf{X}\mathbf{X}^T$ can be obtained with a low computational complexity. The multiplicative NMF algorithm was also verified under the same condition in this example; however, to achieve an acceptable performance (PSNR $\approx$ 30 dB), it required 5000 iterations, see in Table 4.2. The results in Table 4.2 show that the block-wise technique reduces the running time considerably while preserving very good PSNR.

4.8.4 Illustrative Examples for Compressed Sensing

We shall now show the benefits of HALS technique on several difficult examples for which the most known and efficient Compressed Sensing (CS) algorithms fail whereas the HALS-CS algorithm reconstructs the desired sources with a high precision.

Table 4.2 HALS NMF algorithms applied to a large-scale problem (β =1 and α = 1.5).

Data size	Index	$\beta_{0.5}$	blk-$\beta_{0.5}$	β_1	blk-β_1	β_2	blk-β_2	ALS	blk-ALS	A+H	blk-A+H
1000×1000	PSNR [dB]	40.31	34.79	44.75	41.99	40.99	45.47	62.39	60.01	45.3	49.75
	FIT (%)	99.91	99.67	99.65	99.92	97.07	99.60	99.54	98.08	98.94	99.40
	Time (sec)	7.31	4.5	5.72	4.56	6.75	4.69	5.38	4.35	5.87	4.78
5000×1000	PSNR [dB]	35.69	34.19	44.77	43.05	40.81	46.57	66.84	66.09	48.47	49.69
	FIT (%)	99.83	99.81	99.54	99.83	96.23	99.56	100.00	99.54	99.13	99.42
	Time (sec)	29.56	18.5	24.51	17.56	28.13	18.75	23.63	17.3	24.51	17.75
10000×1000	PSNR [dB]	38.32	36.77	45.17	41.91	41.7	44.72	70.24	65.06	47.71	52.55
	FIT (%)	99.89	99.90	99.70	99.88	98.00	99.36	100.00	99.42	99.08	99.62
	Time (sec)	59.94	38.5	49.71	34.87	57.5	37.94	48.31	35.62	47.75	35.32

β – The Beta HALS algorithm for both **A** and **X**
blk-β – The block-wise Beta HALS algorithm for both **A** and **X**
bll-ALS – The block-wise ALS algorithm for both **A** and **X**
A-H – The ALS algorithm for **A** and The HALS algorithm for **X**
bll-A+H – The block-wise ALS for **A** and the block-wise HALS algorithm for **X**

Example 4.5 *The* `Blocks` *signal*

An original signal `Blocks` (2048 samples) was taken from the Wavelab package [9] (see Figure 4.10). Upon transforming this signal into a sparse domain by the Haar wavelet, we obtain the sparse signal **X** with total 2048 entries and only $m = 77$ nonzero elements. Although the requirement regarding the number of measurements I is not less than $2m$, the error could become negligible for $I > 4m$ [63]. In this example, we use only $I = 256$ compressed samples ($\approx 3m$) for which the projection matrix $\mathbf{A} \in \mathbb{R}^{256\times 2048}$ is a uniform spherical ensemble (depicted in Figure 4.10) [28,63]. Listing 4.3 shows implementation of the compression of the `Blocks` signal.

Most known CS algorithms (such as BP [9] and GPSR [37]) failed for this example, that is, they returned nonflat and very noisy reconstructed signals (see Figure 4.11). Here, we use the HALS-CS algorithm with the Abramovich shrinkage rule and the exponential decreasing threshold strategy ($\beta = 4$) in 30 iterations giving an excellent performance with SNR = 87.02 dB (see Figure 4.10 and also Figure 4.11). The MATLAB code for the HALS-CS function is provided in Appendix 4.9.

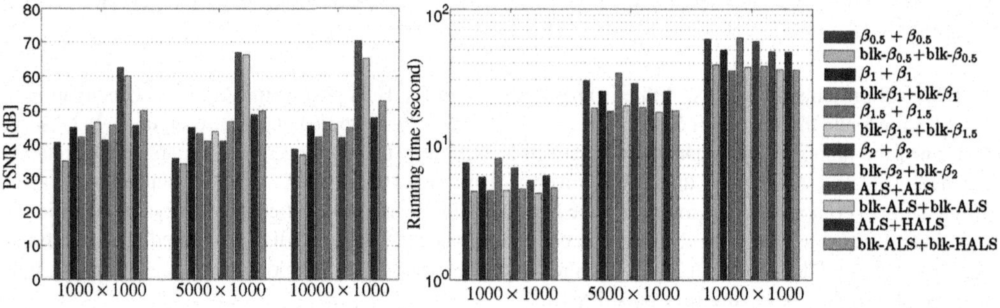

(a) PSNR for NMF algorithms (b) Running time for NMF algorithms (c) Figure legends

Figure 4.9 Comparison of HALS NMF algorithms for large-scale problem with data matrices of dimension: 1000×1000, 5000×1000 and 10000×1000. Hybrid algorithm of block-wise ALS learning rule for **A** and HALS learning rule for **X** achieves a good performance with relatively short running time.

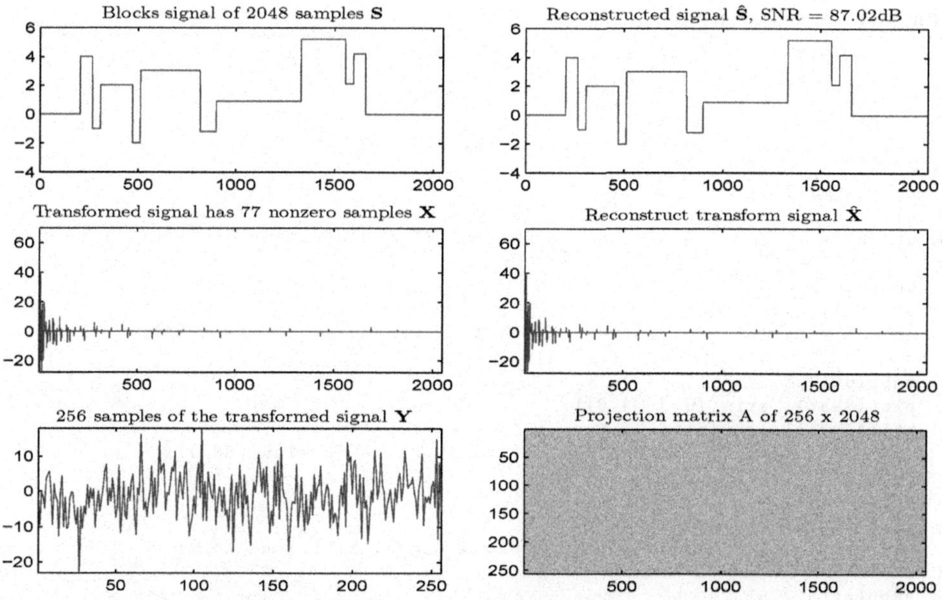

Figure 4.10 Illustration of CS reconstruction of `Blocks` signal [9] with $I = 256$ compressed samples ($\approx 3m, m = 77$). Most Compressed Sensing algorithms (such as BP [9], GPSR BB [37]) return noisy results, however, the HALS-CS algorithm reconstructed the data successfully.

A comparison and visualization of the original and reconstructed signals is performed with Listing 4.4 giving results similar to those in Figure 4.10.

The performances illustrated as in Figure 4.11 correspond to the following strategies:

- HALS-CS algorithm with the linear strategy over 100 iterations with the hard shrinkage rule.
- HALS-CS algorithm with the nonlinear strategy over 30 iterations also with the hard shrinkage rule.

Listing 4.3 Example 4.5 with the `Blocks` signal of 2048 samples.

```
% Compressed sensing signal has I = 256 samples
% This example uses the SparseLab and Wavelab toolboxes
J = 2^11;                                          % number of samples
I = 256;                                      % number of measurements
S = MakeBlocks(J)';                           % Generate Blocks signal

% Transform signal to sparse domain using Haar wavelet (Wavelab toolbox)
qmf = MakeONFilter('Haar',1);
X = FWT_PO(S, 0, qmf);
A = MatrixEnsemble(I,J);                  % Generate Random Dictionary
Y = A * X;                              % Generate compressed signal Y

%% Reconstruct the compressed signal Y by HALS-CS algorithm with
% exponential decreasing strategy and Abramovich shrinkage rule
options = struct('A',A,'J',J,'Niter',30,'nbrestart',0,'Lmax',8,'Lmin',0,...
    'psitype','exp','betarate',4,'shrinkage',4); %Abramovich's rule
Xhat = CS_HALS(Y,options) ;

% Transform the reconstructed signal to the signal domain
Shat = IWT_PO(Xhat,0,qmf);
SNR = 20*log10(norm(S)/norm(Shat-S));
```

Listing 4.4 Visualization code for Example 4.5.

```
%% Visualizate the result for Example 4.5
figure
subplot(3,2,1)
plot(S); axis([0 J -4 6]);
htl(1) = title('Blocks signal of 2048 samples $\bf S$');

subplot(3,2,3)
plot(X); axis tight
htl(2) = title('Transformed signal has 77 nonzero samples $\sbX$');

subplot(3,2,5)
plot(Y); axis tight
htl(3) = title('256 samples of the transformed signal $\sbY$');

subplot(3,2,2)
plot(Shat); axis([0 J -4 6]);
htl(4)=title(sprintf(...
    'Reconstructed signal $\\bf\\hat S$,SNR = %0.2fdB',SNR));

subplot(3,2,4)
plot(Xhat); axis tight
htl(5) = title('Reconstruct transform signal $\bf \hat X$');

subplot(3,2,6)
imagesc(A); axis tight
htl(6) = title('Projection matrix A of 256 x 2048');
set(htl,'interpreter','latex');
```

- HALS-CS algorithm with the adaptive nonlinear strategy over 30 iterations with hard shrinkage rule, nonnegative garrotte (NNG) shrinkage rule, and nonnegative Abramovich (NNA) shrinkage rule, respectively.
- BP algorithm [28].
- GPSR BB algorithm [37].

In Figure 4.12, we illustrate the relation of the SNR performance as a function of the parameter η and the number of iterations K_{max}. The parameter η was varied in the range of [0, 20], whereas the maximum number of iterations K_{max} was in the range of [10, 200]. For each pair of parameters (η, K_{max}), we reconstruct the source and compute the SNR index. As see in Figure 4.12(a), in order to achieve a desired performance with SNR = 80 [dB], with the Abramovich rule and the linear decreasing strategy the HALS-CS ran 80 iterations and took 11 seconds. However, with the pair of parameter (η, K_{max}) = (2.5, 25), we can achieve a similar result but only in 3.5 seconds. Contours for the same SNR levels in Figures 4.12(a), 4.12(b), 4.12(c) and 4.12(d) indicate that the linear decreasing strategy can help to achieve a desired performance, but we can always obtain the same (even better) result by a high parameter η and smaller number of iterations K_{max} in the nonlinear decreasing strategy. One most important result is that we can reconstruct exactly and perfectly the original `Blocks` signal with SNR > 250 [dB], for example with $\eta = 20$ and $K_{max} = 100$.

To reduce the running time of the HALS-CS, we apply the block-wise update with $B = 50$ and 100 rows. The results are given in Figures 4.12(c) and 4.12(d). We also compare running time for the normal row-by-row update and the block-wise $B = 50$ rows as in Figure 4.12(f). Running time with $B = 50$ mostly reduced 7 times.

This benchmark was also tested with the HALS-CS and the beta-Abramovich shrinkage rule (4.163) ($\beta = 4$). The performance given in Figure 4.12(e) is better than the one that we obtained with the standard Abramovich shrinkage in Figures 4.12(a), 4.12(b).

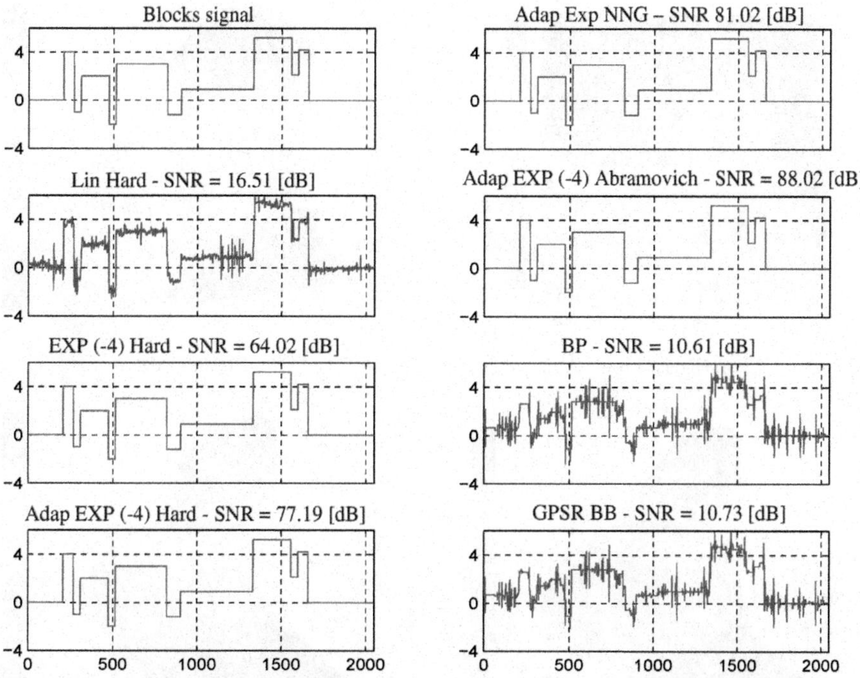

Figure 4.11 Illustration of CS reconstruction of `Blocks` signal [9] with $I = 256$ compressed samples ($\approx 3m$). Most CS algorithms (such as BP [28], GPSR BB [37]) return noisy results, however, the HALS-CS algorithm reconstructed the data successfully. From top to bottom, and left to right, original signal and the signals reconstructed using the algorithms in Table 4.3.

For a large-scale example, we compressed the `Blocks` signal with $J = 8192$ samples. in the Haar wavelet domain, this signal has $m = 92$ nonzero elements. The compressed signal has $I = 3m = 276$ measurements. With $(\eta, K_{max}) = (5{,}1800)$, the HALS-CS using the Abramovich rule achieved SNR $= 195.67$ [dB].

Example 4.6 *The* `Bumps` *signal*

The signal `Bumps` ($J = 1024$ samples) (Wavelab package [9]) was compressed with $I = 512$ measurements (Figure 4.13). Note that most coefficients of this signal are not actually precisely zero value in the time domain and also in the Daubechies "symmlet8" wavelet domain (see Figures 4.13(top-left) and 4.13(top-right)). For example we have 583 coefficients with values below $\epsilon = 10^{-5}$. For such data the BP algorithm returned a "noisy" signal (see Figure 4.13(left)). Even though the reconstructed "noisy" signals were processed with (T-I) [25], the final result was still much poorer than obtained using the HALS-CS algorithm. By selecting suitable values for (η, K_{max}), we obtained precise reconstruction, for example: $(\eta, K_{max}) = (7, 60)$ returned SNR $= 40.28$ [dB], or $(\eta, K_{max}) = (8{,}110)$ returned SNR $= 60$ [dB], or $(\eta, K_{max}) = (14, 400)$ returned SNR $= 90.07$ [dB]. More results are plotted in Figures 4.14 for $\eta \in [0, 15]$, and $K_{max} \in [10, 500]$. Once again, a linear decreasing strategy allows an acceptable result to be obtained, but a nonlinear one helps us to find an exact reconstruction. The performance was still perfect when applying the block-wise update for 50 rows and 100 rows (see in Figures 4.14(c) and 4.14(d)).

Example 4.7 *Multi-scale compressed sampling for the* `Mondrian` *image*

We applied the HALS-CS algorithm to images and compared its performance to those of the BP algorithm, the StOMP algorithm with the CFDR and CFAR thresholdings [28], and the BCS algorithm [59]. The

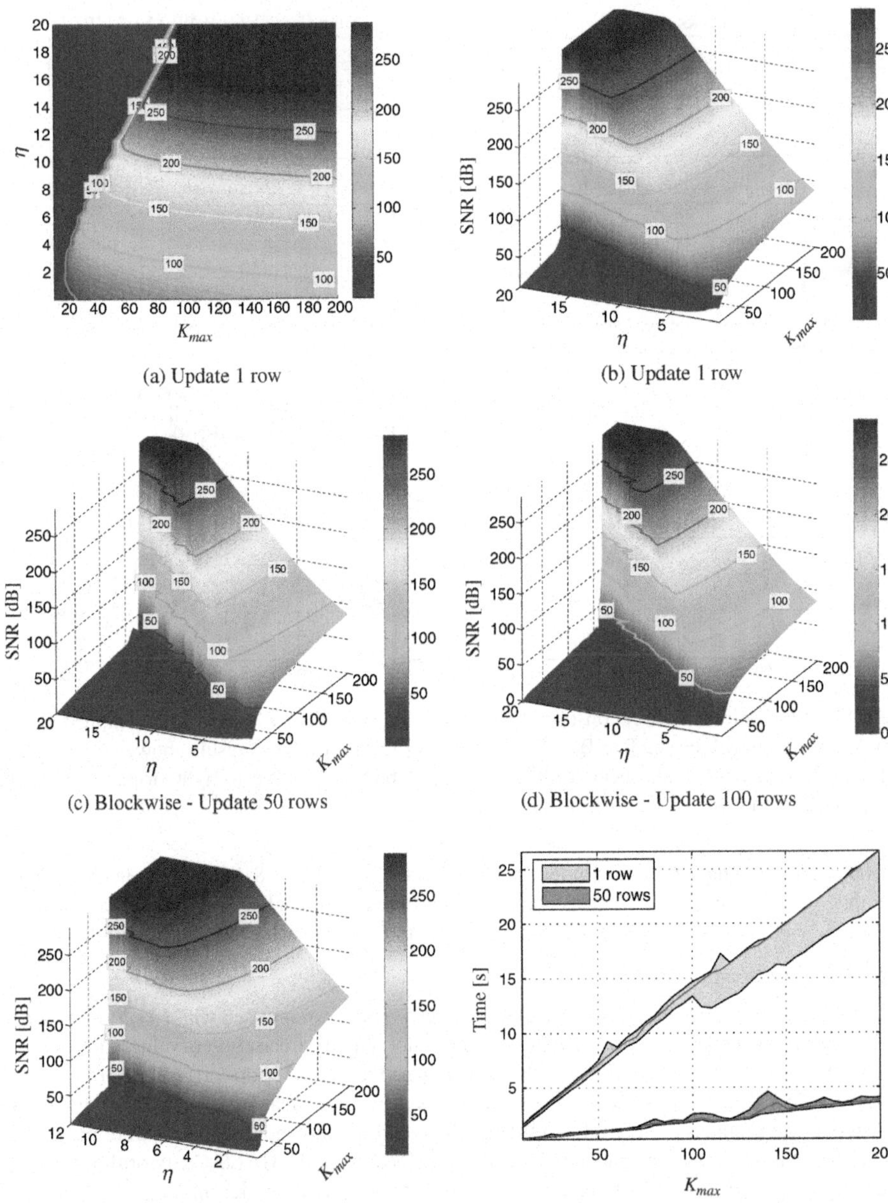

(a) Update 1 row
(b) Update 1 row
(c) Blockwise - Update 50 rows
(d) Blockwise - Update 100 rows
(e) beta-Abramovich shrinkage rule, update 1 row
(f) MC analysis for running time

Figure 4.12 Illustration of the SNR performance as function of $\eta \in [0, 20]$, and $K_{max} \in [10, 200]$ for `Blocks` signal in Example 4.5: (a)–(d) Block-wise technique is applied with $B = 1$ (a)–(b), 50 (c) and 100 (d) rows. Linear decreasing strategy can help us to obtain a desired reconstruction, however, nonlinear adaptive decreasing strategy allows not only to achieve an excellent result, but also to reduce running time; (e) the HALS-CS with the new beta-Abramovich shrinkage rule ($\beta = 4$); (f) Monter-Carlo analysis for running time of block-wise technique for one row, and for simultaneously 50 rows when increasing the number of iterations K_{max}.

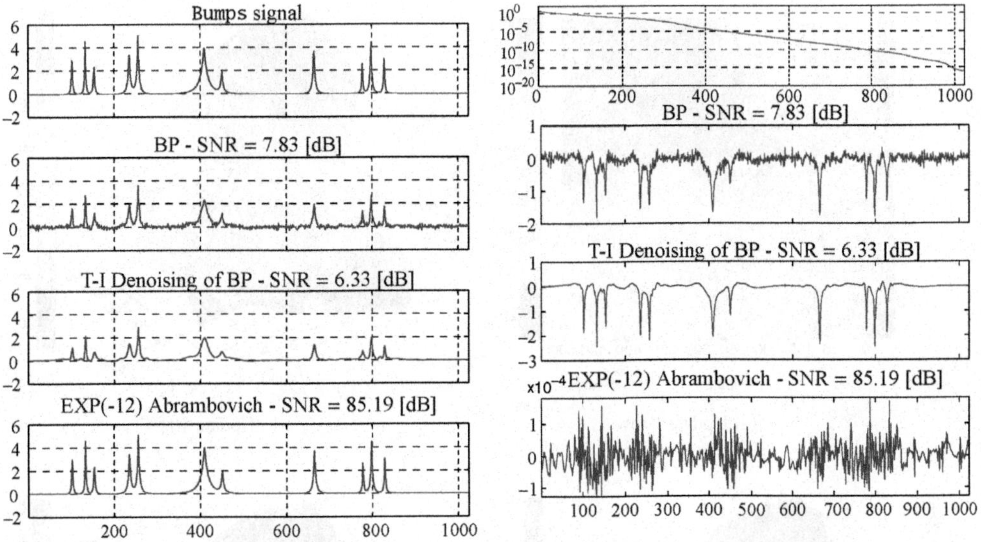

Figure 4.13 Illustration of CS reconstruction of `Bumps` signal [9] with $I = 512$ compressed samples: (top right) distribution of coefficient magnitudes on a log-scale, (from top left) original signal, reconstructed signals by BP algorithm, T-I denoising of BP and HALS-CS with EXP(-12) NN Abramovich rule respectively and the corresponding residue signals (right bottom).

experiment was performed on the `Mondrian` image (Figure 4.16(a)) with a multi-scale CS scheme [28,63] and is similar to the 2nd example analyzed in [59] based on the "symmlet8" wavelet with the coarsest scale at 4 and the finest scale at 6. The result of linear reconstruction with $I = 4096$ samples is displayed in Figure 4.16(b) with the reconstruction error[16] $\epsilon_{\text{LIN}} = \|\boldsymbol{x}_{\text{LIN}} - \boldsymbol{x}\|_F / \|\boldsymbol{x}\|_2 = 0.13$, and the corresponding PSNR value $\text{PSNR}_{\text{LIN}} = 20 \log (255 \times 512 / \|\boldsymbol{x}_{\text{LIN}} - \boldsymbol{x}\|_2) = 21.55$ [dB]. Note that the best performance was achieved using the same samples (wavelet coefficients).

In this example, the multi-scale CS scheme was applied scale-by-scale with $I_4^d = 614$ ($\approx 0.8 \times J_4^d \approx 0.8 \times 4^4 \times 3$) detail coefficients at level 4 and $I_5^d = 1843$ ($\approx 0.6 \times J_5^d \approx 0.6 \times 4^5 \times 3$) detail coefficients at level 5. Including $I^a = 256$ coarse-scale samples, we obtain thereby a total of $I_{CS} = 256 + 614 + 1843 = 2713$ samples, compared to the 262.144 coefficients in total. For image reconstruction, the missing detail coefficients were padded by zeros up to the image size.

The detailed flow of operations for the Multi-scale CS scheme is given below.

1. Decompose the Mondrian image into the "symmlet8" domain at the coarsest scale 4.
2. Hold $J^a = 256$ approximation coefficients for the compressed image.
3. Construct a vector of detail coefficients from the detail coefficient matrices at scale 4. This vector contains $J_4^d = 3 \times 4^4 = 768$ elements and is considered the sparse signal to be compressed.
4. Compress the above signal with sample reducing rate of 0.8 by a projection matrix $\mathbf{A}_4$ of 614×768 ($\approx 0.8 \times 768 \times 768$).
5. Perform steps 3 and 4 for $J_5^d = 3072 (= 3 \times 4^5)$ detail coefficients at scale 5 with a reducing rate of 0.6 and projection matrix $\mathbf{A}_5$ of $1843 \times 3072 (\approx 0.6 \times 3072 \times 3072)$.

[16] SNR ([dB]) $= -20 \log(\epsilon)$.

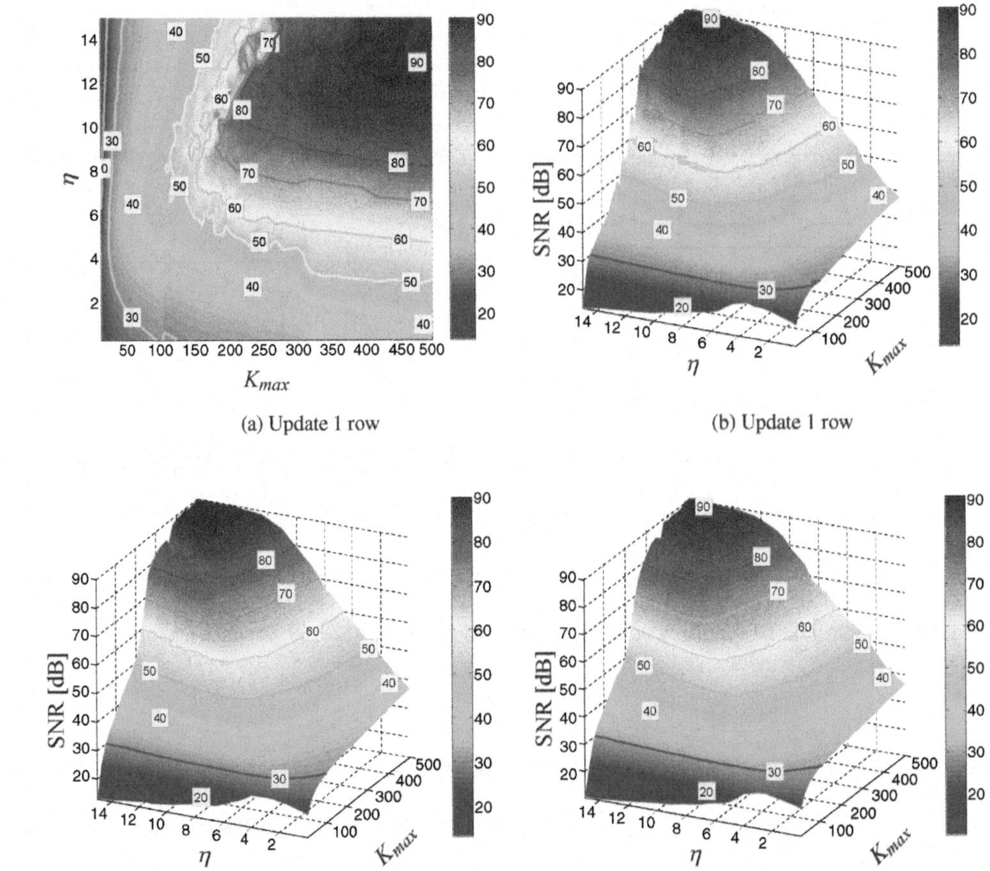

(a) Update 1 row

(b) Update 1 row

(c) Blockwise - 50 rows

(d) Blockwise - 100 rows

Figure 4.14 Illustration of the SNR performance as function of $\eta \in [0, 15]$, and $K_{max} \in [10, 500]$ for `Bumps` signal in Example 4.6. Block-wise technique was applied with $B = 1, 50$ and 100 rows.

6. Store the total compressed sensing image which has $I_{CS} = J^a + 0.8 \times J_4^d + 0.6 \times J_5^d = 256 + 614 + 1843 = 2713$ elements.

The MATLAB code illustrating how to perform this example is given in Listing 4.5. The resulting images by the HALS-CS, BP, StOMP with CFDR, StOMP with CFAR and BCS are displayed in Figures 4.16(c)-(g), respectively. The HALS-CS algorithm returned the smallest error $\epsilon_{\text{HALS}} = 0.14$ and the largest PSNR value $\text{PSNR}_{\text{HALS}} = 21.17$ [dB] in 9.77 seconds, while the errors of all other algorithms (BP, CFDR, CFAR and BCS) were 0.14, 0.18, 0.15 and 0.15, and their PSNR values were 21.02 [dB], 18.81 [dB], 20.48 [dB] and 20.53 [dB], respectively. The HALS-CS algorithm provided the best performance.

The performance evaluation of the reconstructed signals in our experiments is summarized in Table 4.3.

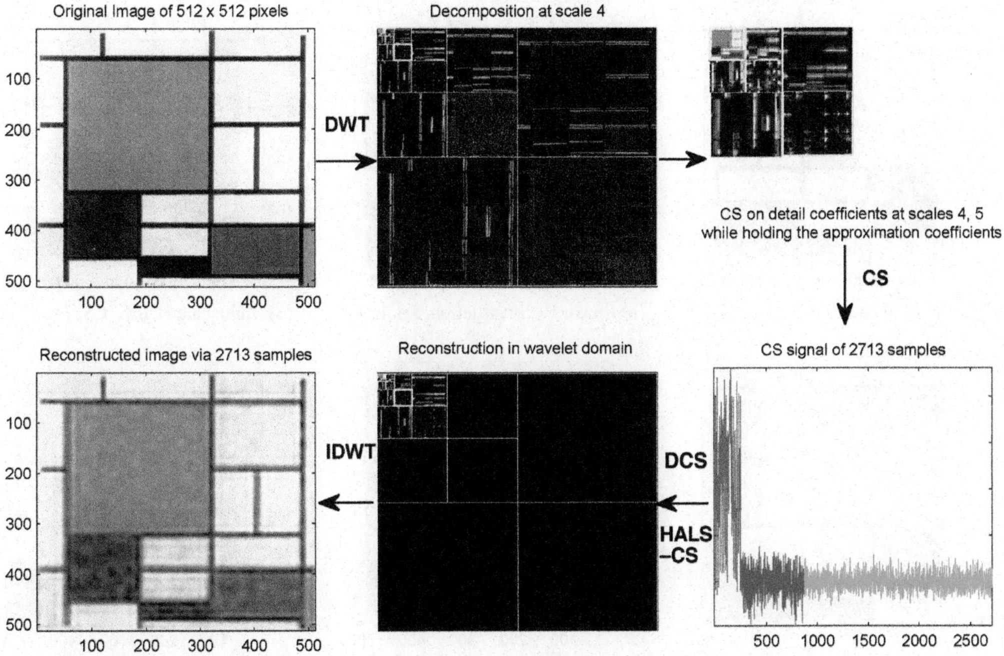

Figure 4.15 Multi-scale CS of the Mondrian image with the HALS-CS algorithm on the detail coefficients at scales 4 and 5.

4.9 Discussion and Conclusions

In this chapter we have presented the extended and modified class of Alternating Least Squares (ALS) algorithms for Nonnegative Matrix Factorization (NMF) and semi-NMF with improved robustness and convergence properties by means of additional constraints. It has been shown that the ALS algorithms can be directly extended to semi-NMF models (in which some parameters are relaxed to be unconstrained), three-NMF models, and orthogonal NMF. We have also introduced a generalized and flexible cost function (controlled by sparsity penalty and flexible multiple regularization terms) that allows us to derive a family of robust and efficient alternating least squares algorithms for NMF, together with the method which allows us to automatically self-regulate or self-compensate the regularization terms. This is an important modification of the standard ALS algorithm. The RALS algorithms with sparsity constraints have some potentials for many applications such as the Sparse Component Analysis and Factor Analysis. They are able to overcome the problems associated with the standard ALS, as the solution which is obtained employing the RALS tends not to be trapped in local minima and usually converges to the desired global solution.

Furthermore, we have presented the local cost functions whose simultaneous or sequential (one-by-one) minimization leads to simple local ALS algorithms which under some sparsity constraints work both for the under-determined (a system which has less sensors than sources) and over-determined model. The proposed algorithms have been implemented in MATLAB in the toolboxes NMFLAB/NTFLAB [20], and compared with the regular multiplicative NMF algorithms in aspects of performance, speed, convergence properties and ability to deal with large-scale problems.

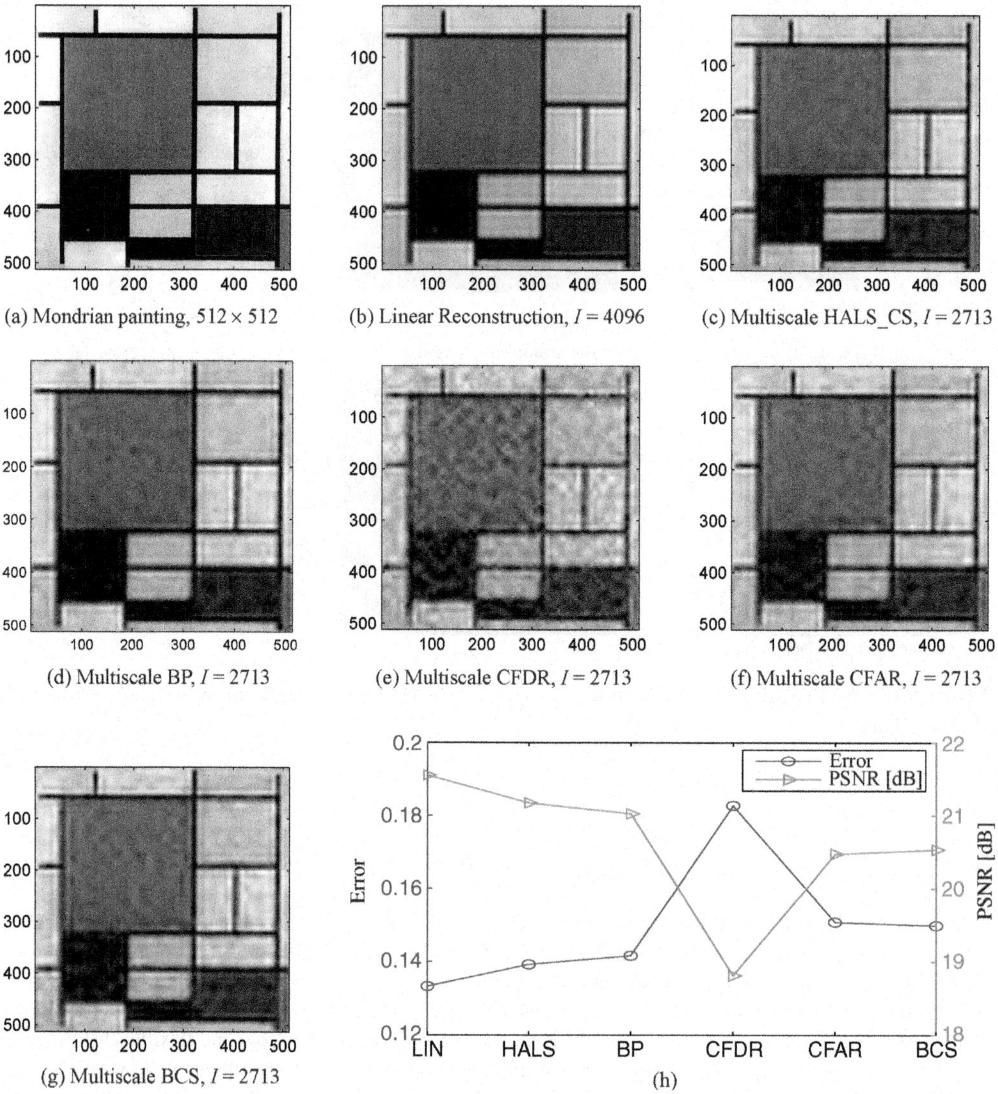

Figure 4.16 Reconstruction of the `Mondrian` image with a multi-scale CS. (a) `Mondrian` image, (b) Linear reconstruction with $I = 4096$ samples, $\epsilon_{\mathrm{LIN}} = 0.1333$ and $\mathrm{PSNR}_{\mathrm{LIN}} = 21.55$ dB; (c)–(g) the results of HALS-CS, BP, CFDR, CFAR and BCS algorithms from $I = 2713$ compressed samples with running time t [s] = [9.77, 66.63, 7.74, 22.22, 11.64]; (h) reconstruction errors $\epsilon = [0.14\ 0.14\ 0.18, 0.15\ 0.15]$, and PSNR [dB] = [21.17, 21.02 , 18.81 , 20.48, 20.53], respectively.

We have identified potential applications in three areas of multidimensional data analysis (especially, EEG and fMRI) and signal/image processing: (i) multi-way blind source separation, (ii) model reduction and selection, and (iii) sparse image coding. The algorithms can be further extended by taking into account prior knowledge about specific components and by imposing some additional natural constraints such as orthogonality, continuity, closure, unimodality and selectivity [27,62].

Listing 4.5 Multi-scale compressed sampling for the Mondrian image.

```
% See example 2 in [59], and example of Sparselab.
%--------------------------------------------------------------------------
%% Compress Mondrian image by MCS scheme
S = double(imread('Mondrian.tif'));
% Sample reduction rate at scale 4 and 5
rate = [0.8, 0.6];
% Step 1. Transform image to wavelet domain
j0 = 4;   %coarsest scale
qmf = MakeONFilter('Symmlet',8);
X = FTWT2_PO(S, j0, qmf);

% Step 2. Hold the approximation coefficients
j1 = 6;
CS.coef = cell(1,j1-j0+1); CS.coef{1} = reshape(X(1:2^j0,1:2^j0),[],1);
CS.size = [2^j0 2^j0];     CS.samples = numel(CS.coef{1});

% Compress the detail coefficients at the scales 4 and 5
A = cell(1,2); t_HALS = 0; % running time
for jj = j0:j1-1
    % Step 3. Construct the vector of detail coefficients
    xV = X((2^jj+1):2^(jj+1),1:2^jj);          % vertical detail coefficients
    xHD = X(1:2^(jj+1),(2^jj+1):2^(jj+1)); % horizontal and diagonal ones
    x = [xV(:); xHD(:)];
    Jdetail = 4^(jj+1) - 4^jj; Idetail = floor(rate(jj-j0+1)*Jdetail);

    % Step 4. Compress Jdetail samples by CS operator
    randn('state',7256157);
    A{jj-j0+1} = MatrixEnsemble(Idetail, Jdetail, 'USE');
    y = A{jj-j0+1} * x;

    CS.coef{jj-j0+2} = y; CS.size(jj-j0+2,1) = size(xV,1)+ size(xHD,1);
    CS.size(jj-j0+2,2) = size(xV,2); CS.samples(jj-j0+2) = Idetail;
end
CS.coef = cell2mat(CS.coef');
%  Reconstruct CS signal using HALS CS algorithm
X_HALS = zeros(size(X));
X_HALS(1:2^j0,1:2^j0)=reshape(CS.coef(1:prod(CS.size(1,:))),CS.size(1,:));
ind = cumsum(CS.samples);

for jj = 2:size(CS.size,1)
    y = CS.coef(ind(jj-1)+1:ind(jj));
    % Reconstruct the CS coefficient y with HALS
    options = struct('Index_norm',1,'J',size(A{jj-1},2),'type_alg_X',4,...
        'max_restart',0,'Niter',60,'Lmax',12,'Lmin',0,'shrinkage',4,...
        'phitype','exp','betarate',5,'A',A{jj-1},'Nrst',1);
    tic; xhat = CS_HALS(y,options); t_HALS = t_HALS + toc;
    X_HALS((2^(jj+2)+1):2^(jj+2+1),1:2^(jj+2)) = ...
        reshape(xhat(1:prod(CS.size(jj,:))/3), [], CS.size(jj,2));
    X_HALS(1:2^(jj+3),2^(jj+2)+1:2^(jj+3)) = ...
        reshape(xhat(prod(CS.size(jj,:))/3+1:end), [], CS.size(jj,2));
end

%% Invert Symmlet transform and compute error
S_HALS = ITWT2_PO(X_HALS, j0, qmf);
E_HALS = norm(S - S_HALS,'fro') / norm(S,'fro');

figure; imagesc(S_HALS);colormap gray;axis image
title(sprintf(['Multiscale CS with HALS\\_CS, %d samples,'...
    '$\\epsilon$ = %.4f'],sum(CS.samples),E_HALS),'interpreter','latex');
```

Table 4.3 Performance comparison of the various CS algorithms.

Dataset	Algorithm	No. Iter.	Strategy η	Shrinking	SNR[(a)] [dB]
Blocks signal 2048 samples / 256 measurements	HALS-CS	100	$0^{(b)}$	Hard	16.51
		30	$3^{(b)}$	Hard	64.02
		30	3	Hard	77.19
		30	5	nn-garrote	81.02
		30	5	Abramovich	87.02
		> 80	> 14	Abramovich	> 250
	BP[(c)]	25			10.61
	GPSR BB	5			10.73
8192 samples / 276 measurements	HALS-CS	2000	0		138.96
		1800	5	Abramovich	195.67
Bumps signal 1024 samples / 512 measurements	HALS-CS	30	5	Abramovich	36.33
		> 200	>8	Abramovich	> 70
	BP	15			7.83
	T-I	1			6.34
	GPSR BB	5			21.49
Mondrian image 512 × 512 pixels / 2713 measurements	HALS-CS	40	8	Abramovich	21.17
	LIN	1			21.55
	BP	100			21.02
	CFDR	30			18.81
	CFAR	30			20.47
	BCS[(d)]	328 + 450			20.53

[(a)] For image compression, we use the PSNR index instead of SNR.
[(b)] non-adaptive decreasing rule
[(c)] For BP, 'No. Iter' is the number of iteration that this algorithm converged.
[(d)] number of iterations for each layers in the MCS scheme.

Appendix 4.A: MATLAB Source Code for ALS Algorithm

```
function [A,X]=nmf_als(Y,opts)
%
% Non-negative Matrix Factorization (NMF) with standard
% Alternating Least-Squares (ALS) algorithm
%
% INPUT
% Y       :   source with size of I x T
% opts    :   structure of optional parameters for algorithm (see defoptions)
%    .tol:    tolerance of stopping criteria (Frobenius distance)      (1e-5)
%    .nrestart: number of multi initializations                          (1)
%    .niter:   number of iterations                                     (50)
%    .init:   vector 1 x 2 for initialization types (see nmf_initialize)
%
% Written by Anh Huy Phan, R. Zdunek and Andrzej Cichocki
%% =========================================================================
Y(Y< 0) = eps; I = size(Y,1);
% Set algorithm parameters from input or by using defaults
```

```
defoptions = struct('tol',1e-5,'J',I,'init',[1 1],'nrestart',1,...
     'niter',50,'verbose',0);
if ~exist('opts','var'),  opts = struct; end
opts = scanparam(defoptions,opts);
No_iter = 30;                    % number of iterations to select the best trial

for nr= 0:opts.nrestart
    if (nr == opts.nrestart)&&(opts.nrestart > 0) % initialize
        A = A_best; X = X_best; No_iter = opts.niter;
    else
        [A,X] = nmf_initialize(Y,opts.J,opts.init); X = X';
    end
    cost = sum(sum((Y-A*X).^2));

    for k  = 1: No_iter
        X = max(eps,pinv(A'*A)*A'*Y);
        A = max(eps,Y*X'*pinv(X*X'));
        A = bsxfun(@rdivide,A,sum(A));

        if (nr == opts.nrestart) && ((mod(k,30)==0) || (k == No_iter))
            [cost_old,cost] = deal(cost,sum(sum((Y-A*X).^2)));
            stop = abs(cost_old-cost) ≤ opts.tol*cost;
            if opts.verbose
                fprintf(1,'Best trial %d, step %d, Cost value %d\n',...
                    nr_best+1,k,cost);
            end
            if stop , break; end
        end
    end % k
    if (nr < opts.nrestart)                              % select best trial
        cost = sum(sum((Y-A*X).^2));
        if (nr == 0) || (cost < cost_min)
            A_best = A; X_best = X; cost_min= cost; nr_best = nr;
        end
        if opts.verbose
            fprintf(1, 'Trial %d, Cost value = %e\n',    nr+1, cost);
        end
    end
end % nr
```

Appendix 4.B: MATLAB Source Code for Regularized ALS Algorithms

```
function [A,X]=nmf_fpals(Y,opts)
%
% Non-negative Matrix Factorization (NMF) with FP ALS algorithms
%
% INPUT
% Y       :   source with size of I x T
% opts    :   structure of optional parameters for algorithm (see defoptions)
%    .J:      number of components
%    .algtype vector (1x2) indicates algorithm type                 ([1 1])
%             1 - ALS with additional constraints,
%             2 - BLUE ALS, 3 - RALS, 4 - SALS, 5 - MALS
%    .init:   vector 1 x 2 defines initialization types
%             (see nmf_initialize)                                  ([1 1])
%    .nrestart: number of multi initializations                        (1)
%    .niter:   number of iterations                                  (300)
```

```
%    .tol:    tolerance of stopping criteria (Frobenius distance)      (1e-5)
%    .reg_oper:    define regularized matrix C in RALS, MALS, SALS    ([1 1])
%    .reg_param:   value of regularized parameters                    ([1 1])
%    .weight:      matrix W                                            ([1 1])
%    .lsparse:     level of sparseness constraints                     ([0 0])
%    .Atrue and .Xtrue:  used for a fixed matrix A or X
% OUTPUT
% A and X with Y = A * X.
%
% Copyright 2008 by A. Cichocki and R. Zdunek and A.H. Phan
% Optimized by Anh Huy Phan and Andrzej Cichocki - 04/2008
%% Setting parameters
Y(Y< 0) = eps;   [I,T]=size(Y);
% Set algorithm parameters from input or by using defaults
defoptions = struct('J',I,'algtype',[1 4],'init',[1 1],'nrestart',1,...
    'niter',300,'tol',1e-5,'reg_oper',[1 1],'reg_param',[1 1],...
    'weight',[1 1],'lsparse',[0 0],'A0',[],'X0',[],'verbose',0);
if ~exist('opts','var'), opts = struct; end
[J,algtype,init,nrestart,niter,tol,reg_oper,reg_param,weight,...
    lsparse,A0,X0,verbose] = scanparam(defoptions,opts);
No_iter = 30;              % number of iterations to select the best trial
traceYY = trace(Y*Y')/T;
%% algtype = 'ALS + sparse' 'BLUE' RALS' 'SALS' 'MALS'
Δ = [0 0];
for k = 1:2
    switch algtype(k)
        case 2 % BLUE
            weight(k) = 2;
        case 4 %SALS
            weight(k) = 1;
            Δ(k) = 1;
        case 5 %MALS
            weight(k) = 1;
            reg_oper(k) = 6;
            Δ(k) = 1;
    end
end
%%
for nr= 0:nrestart
    pause(.0001)
    if (nr == nrestart)&&(nrestart > 0) % initialize
        A = A_best; X = X_best; No_iter = niter;
    else
        [A,X] = nmf_initialize(Y,J,init,{A0 X0'}); X = X';
    end
    cost = costfunction;

    for k  = 1: No_iter
        CA = fCX(reg_oper(1),Y',X',A);
        alphaA_reg = falphaX_reg(reg_param(1),0);
        W = fW(weight(1));

        if algtype(1)≠ 0
            den = bsxfun(@minus,Y*X',lsparse(1)*sum(W,2))+ alphaA_reg*A*CA;
            num = X*X' + Δ(1) * alphaA_reg*CA;
            A = max(1E6*eps,den/num);
            if algtype(2)≠ 0
                A = bsxfun(@rdivide,A,sum(A));   %normalize
            end
        end % for A
```

```
        CX = fCX(reg_oper(2),Y,A,X');
        alphaX_reg = falphaX_reg(reg_param(2));
        W = fW(weight(2));

        if algtype(2)≠ 0
            den = A'*W*Y - lsparse(2) + alphaX_reg*CX*X;
            num = A'*W*A + Δ(2) * alphaX_reg*CX;
            X = max(1E6*eps,num\den);
        end % for X

        if (nr == nrestart) && (mod(k,30)==0)          % stopping condition
            [cost_old, cost] = deal(cost,costfunction);
            stop = abs(cost_old-cost) ≤ tol*cost;
            if verbose
                fprintf(1,'Best trial %d, step %d, Cost value %d\n',...
                    nr_best+1,k,cost);
            end
            if stop , break; end
        end
    end % k
    if (nr < nrestart)                                  % select best trial
        cost = costfunction;
        if (nr == 0) || (cost < cost_min)
            A_best = A; X_best = X; cost_min= cost; nr_best = nr;
        end
        if opts.verbose
            fprintf(1, 'Trial %d, Cost value = %e\n',   nr+1, cost);
        end
    end
end % nr
    function CX = fCX(reg_oper_X ,Y,A,X)
        switch reg_oper_X
            case 1 % Ones
                CX = ones(J);
            case 2 % Identity
                CX = eye(J);
            case 3 % Diff L1
                CX = get_l(J,1);
                CX = CX'*CX;
            case 4 % Diff L2
                CX = get_l(J,2);
                CX = CX'*CX;
            case 5 % PCA-Cov
                Cr = cov((Y - A*X)');
                [V,D] = eig(Cr);
                Er = V*D(:,end:-1:1);
                CF = Er'*Er/T;
                CX = CF(1:J,1:J);
            case 6
                CX = diag(min(sum(abs(Y'*A))./(sum(X)+eps),sum(A'*A)));
        end
    end

    function alphaX_reg = falphaX_reg(reg_param_X ,fX)
        switch reg_param_X
            case 1 % Zero
                alphaX_reg = 0;
            case 2 % Exp
                alphaX_reg = alphaX0*exp(-k/tauX);
```

```
            case 3 % Const
                alphaX_reg = alphaX_reg_const;
            case 4 % Alpha_reg res(k)
                e_res = 0.5*norm(Y - A*X,'fro');
                alphaX_reg = alphaX_reg_const*e_res;
            case 5 % EM-PCA
                alphaX_reg = (traceYY - trace(A*X*Y')/T);
                if fX
                    alphaX_reg = alphaX_reg /I;
                end
        end
    end

    function W = fW(weight)
        if weight == 2% BLUE
            W = pinv(cov((Y - A*X)'));
        else
            W = eye(I);
        end
    end

    function cost = costfunction
        Yhat = A*X+eps;
        cost = norm(Y(:) - Yhat(:),2);
    end
end
```

Appendix 4.C: MATLAB Source Code for Mixed ALS-HALS Algorithms

```
function [A,X]=nmf_ghals(Y,opts)
%
% Non-negative Matrix Factorization (NMF) with Hierarchical
% Alternating Least-Squares (ALS) algorithms
%
% INPUT
% Y      :   source with size of I x T
% opts   :   structure of optional parameters for algorithm (see defoptions)
%   .tol:    tolerance of stopping criteria (Frobenius distance)      (1e-5)
%   .J:      number of components
%   .algtype vector (1x2) indicates which algorithm will be used   ([6 6])
%   .init:   vector 1 x 2 defines initialization types             ([1 1])
%            (see nmf_initialize)
%   .nrestart: number of multi initializations                          (1)
%   .niter:  number of iterations                                     (300)
%   .niniter:  number of iterations for updating only A or only X       (1)
%   .ellnorm:  ell_p norm
%   .Anorm and Xnorm:   normalize A, X
%   .lsparse, .lcorr, .lsmth:     vector (1x2) indicates degrees of
%            sparsness, orthogonality, smoothness for each factor A, X
% OUTPUT
% A and X with Y = A * X.
%
% Copyright 2008 by A. Cichocki and A.H. Phan and R. Zdunek
% Optimized by Anh Huy Phan and Andrzej Cichocki - 01/2008
%% =====================================================================
Y(Y< 0) = eps;  [I,T]=size(Y);
% Set algorithm parameters from input or by using defaults
```

```
defopts = struct('tol',1e-5,'J',I,'algtype',[1 4],'init',[1 1],...
    'nrestart',1,'niter',300,'niniter',1,'ellnorm',2,'Xnorm',0,...
    'Anorm',1,'alpha',1,'beta',1,'lsparse',[0 0],'lcorr',[0 0],...
    'lsmooth',[0 0],'tau',50,'alpha0',20,'A0',[],'X0',[],'verbose',0);
if ~exist('opts','var'),     opts = struct;  end
[tol,J,algtype,init,nrestart,niter,niniter,ellnorm,Xnorm,Anorm,...
    alpha,beta,lspar,lcorr,lsmth,tau,alpha0,A0,X0,verbose] ...
    = scanparam(defopts,opts);
No_iter = 30;                % number of iterations to select the best trial
%%
for nr= 0:nrestart
    pause(.0001)
    if (nr == nrestart)&&(nrestart > 0) % initialize
        A = A_best; X = X_best; No_iter = niter;
    else
        [A,X] = nmf_initialize(Y,J,init,{A0 X0'}); X = X';
    end
    cost = costfunction;
    % Compute the error
    Res = Y - A* X;  Ainew  = zeros(I,1); Xinew  = zeros(1,T);

    for k  = 1: No_iter
        if (algtype(1) == 2) && (algtype(2) == 2)  % alpha-HALS-1
            for j = 1:J
                Res = Res + [-Ainew A(:,j)] * [Xinew; X(j,:)];
                Resalpha = phi(Res,alpha);

                Xinew = max(eps,phi((A(:,j)'*Resalpha - lspar(2))...
                    ./sum(A(:,j).^(alpha+1)),1/alpha));
                Ainew = max(eps,phi(Resalpha*X(j,:)fl-lspar(1),1/alpha));
                if ellnorm ≠ 0
                    Ainew = Ainew/norm(Ainew,ellnorm);
                end
                A(:,j) = Ainew; X(j,:) = Xinew;
            end
        elseif (algtype(1) == 4) && (algtype(2) == 4) % beta-HALS-1
            for j = 1:J
                Res = Res + [ -Ainew A(:,j)] * [Xinew ;X(j,:)];
                Xinew = max(eps,((A(:,j).^beta)'*Res-lspar(2))...
                    /sum(A(:,j).^(beta+1)));
                Ainew = max(eps,(Res*(X(j,:).^beta)'-lspar(1)));
                if ellnorm ≠ 0
                    Ainew = Ainew/norm(Ainew,ellnorm);
                end
                A(:,j) = Ainew; X(j,:) = Xinew;
            end
        else
            % Update for A
            for t = 1:niniter % inner iterations
               A=nmfcore(Y',X',A',algtype(1),lspar(1),lcorr(1),lsmth(1))';
            end % t for A
            if (algtype(2) ≠ 0) && Anorm
                A = normalize(A);
            end

            % Update for X
            for t = 1:niniter % inner iterations
                X = nmfcore(Y,A,X,algtype(2),lspar(2),lcorr(2),lsmth(2));
            end
```

```
            if (algtype(1) ≠ 0) && Xnorm
                X = normalize(X')';
            end
        end

        if (nr == nrestart) && (mod(k,30)==0)           % stopping condition
            checkstoppingcondition
            if verbose
                fprintf(1,'Best trial %d, step %d, Cost value %d\n',...
                    nr_best+1,k,cost);
            end
            if stop , break; end
        end
    end % k
    if (nr < nrestart)                                    % select best trial
        cost = costfunction;
        if (nr == 0) || (cost < cost_min)
            A_best = A; X_best = X; cost_min= cost; nr_best = nr;
        end
        if verbose
            fprintf(1, 'Trial %d, Cost value = %e\n',nr+1, cost);
        end
    end
end % nr

    function X = nmfcore(Y,A,X,type_alg, lspar,lcorr,lsmth)
        Xsm = lsmth * [X(:,2) (X(:,1:end-2)+X(:,3:end))/2  X(:,end-1)];
        switch type_alg
            case 1 % Fast HALS
                AA = A'*A;
                normA = diag(1./(diag(AA)+lsmth+lcorr));
                AA = normA*AA;
                AAX = normA*(A'*Y) - AA * X + Xsm;
                for j = 1:J
                    Xn= max(eps, X(j,:) + AAX(j,:) -lspar -lcorr*sum(X));
                    AAX = AAX - AA(:,j) * (Xn - X(j,:)) ;
                    X(j,:) = Xn;
                end

            case {2,3} % Alpha-HALS
                Res = Y-A*X;
                Ainew = zeros(size(Y,1),1);
                Xinew = zeros(1,size(Y,2));
                for j = 1:J
                    Res = Y-A*X + A(:,j)*X(j,:);
                    Resalpha = real(Res.^alpha);
                    Xinew = max(eps,phi((A(:,j)'*Resalpha - lspar ...
                        -lcorr*sum(X)+ Xsm(j,:))...
                        ./(sum(A(:,j).^(alpha+1))+lsmth),1/alpha));
                    X(j,:) = Xinew;
                    Ainew = A(:,j);
                end

            case {4,5} % Fast Beta-HALS
                normA = diag(1./(sum(A.^(beta+1))+lsmth + lcorr));
                Abeta = real(A.^beta);
                AY = normA*Abeta'*Y;
                AA = normA *Abeta'*A;
                for j = 1:J
                    X(j,:) = max(eps, X(j,:) + AY(j,:) - AA(j,:) * X - ...
```

```
                        lspar − lcorr*sum(X) + Xsm(j,:));
                end

            case 6 % ALS
                X = max(eps,pinv(A'*A)*A'*Y);

            case 7 % Regularized ALS
                alpha_reg = alpha0*exp(-k/tau);
                if cond(A) > 1E6 , lambdaX = 1E-6; else lambdaX = 0; end
                X = max(eps,pinv(A'*A + alpha_reg + lambdaX*I)*A'*Y);
        end
    end

    function y = phi(x,alpha)
        y = real(x.^alpha);
    end

    function A = normalize(A)
        A = bsxfun(@rdivide,A,sum(A.^ellnorm).^(1/ellnorm));
    end

    function checkstoppingcondition
        cost_old = cost; cost = costfunction;
        stop = abs(cost_old-cost) ≤ tol*cost;
    end

    function cost = costfunction
        Yhat = A*X+eps;
        cost = sum(sum(Y.*log(Y./Yhat + eps) − Y + Yhat));
    end
end
```

Appendix 4.D: MATLAB Source Code for HALS CS Algorithm

```
function X = CS_HALS(Y,options)
%
% Hierarchical ALS algorithm for Compressed sensing
% using linear and nonlinear thresholding techniques
% with Hard, Soft, Non negative garrotte, Abramovich,
% N-degree garrotte shrinkage rules
%
% INPUT
% Y          :   T compressed signals of I samples as columns (I × T)
% options    :   structure of optional parameters
%    .A      :   projection matrix
%    .Xinit  :   initialization for sources X, zeros matrix is default
%    .J      :   number of samples of original source X
%    .Niter  :   number of iteration                              (1000)
%    .Nrst   :   number of restart                                   (1)
%    .Lmax   :   maximum threshold of lambda                        (20)
%    .Lmin   :   minimum threshold of lambda                         (0)
%    .psitype:   decreasing funtion type for threshold λ             (2)
%                1. shape-preserving piecewise cubic Hermite interpolation
%                (2). exponential function
%    .Hshape:    initial points (at least two)(x,y) y = ψ(x) forming
%                the decreaseing line used for shape-preserving
```

```
%                  Hermite interpolation.
%                  Ex. 2 points (0.2,0.8) and (0.4, 0.2)
%                  Hshape = [0.2 .8; .4  .2]
%    .betarate: decreasing speed for exponential decreasing strategy     (4)
%    .shrinkage: shrinkage rule                                           (1)
%                  (1). HARD                       2. SOFT
%                  3. Non negative garrotte      4 Abramovich's rule
%                  5. N-degree garrotte
% OUTPUTS:
% X          :     reconstructed signals (columns)
%
% Copyright 2008 by A.H. Phan and A. Cichocki
% 04/2008
% #######################################################################
%%
[I,T]=size(Y);
defoptions = struct('A',[],'J',I,'Niter',300,...
    'Xinit',[],'psitype','exp','betarate',4,'Hshape',[.2 .8; .8 .2],...
    'Lmax',20,'Lmin',0,'shrinkage',1,'nbrestart',1);
if ~exist('options','var')
    options = struct;
end
[A,J,Niter,X,psitype,betarate,Hshape,Lmax,Lmin,shrinkrule,Nrst] = ...
    scanparam(defoptions,options);
if isempty(X)
    X = zeros(J,T);
end
Hshape = [0 1;Hshape;1 0];
tol = 1e-5;
alpha = .6;                                                 % reduction rate
%% Normalization of initial guess
normA = sqrt(sum(A.^2,1));
A = bsxfun(@rdivide,A,normA);
G = A'*A;
E = A'* (Y-A*X); %-AA * X + AY;
%%
normY = sqrt(sum(Y.^2,1));
actind = true(1,T);
itt = 0;
while (itt ≤ Nrst) && any(actind) %(fit > tol)
    itt = itt + 1;

    switch psitype
        case 'Hermite'             % piecewise cubic Hermite interpolation
            psif = Hshape(:,2) * alpha^(itt-1)* (Lmax-Lmin) + Lmin;
            psif = pchip(Hshape(:,1)*(Niter-1),psif,0:1:Niter-1);

        case 'exp'                                    % exponential function
            psif = exp(-betarate*(0:Niter-1)/(Niter-1));
            psif = alpha^(itt-1)*Lmax*...
                (psif-psif(end))/(psif(1)-psif(end))+Lmin;
    end

    k = 0;
    while (k < Niter) && any(actind)%(fit > tol)
        k = k + 1;
        lambda =  max(eps,psif(k)*mad(X(:,actind),0,1));
        for j = 1:J
            Xn= X(j,actind) + E(j,actind);
            switch shrinkrule
```

```
                    case 1                                        % Hard rule
                        Xn(abs(Xn)<lambda) = 0;

                    case 2                                        % Soft rule
                        Xn = (Xn - sign(Xn).* lambda) .* (abs(Xn)>lambda);

                    case 3                          % Nonnegative garrotte rule
                        Xn = (Xn - lambda.^2./Xn) .* (abs(Xn)>lambda);

                    case 4                                   % Abramovich rule
                        Xn =sign(Xn).*sqrt(Xn.^2-lambda.^2).*(abs(Xn)>lambda);

                    case 5                             % n-degree garrotte rule
                        Xn = Xn.^3./(lambda.^2+Xn.^2);
                end
                E(:,actind)= E(:,actind) - G(:,j) * (Xn - X(j,actind));
                X(j,actind) = Xn;
            end
            fit = sqrt(sum((A*X-Y).^2,1))./normY;
            actind = fit> tol;
        end % while (k)
    end % while (restarts)
    fprintf(1,'Number of iterations %f %f ',k,itt);
    X = bsxfun(@rdivide,X,normA');
```

Appendix 4.E: Additional MATLAB Functions

```
function varargout = scanparam(defoptions,options)
% Copyright 2008 by A.H. Phan and A. Cichocki
allfields = fieldnames(defoptions);
opts = defoptions;
for k = 1:numel(allfields)
    if isfield(options,allfields{k})
        if numel(options.(allfields{k}))<numel(defoptions.(allfields{k}))
            opts.(allfields{k}) = repmat(options.(allfields{k}),...
                1,numel(defoptions.(allfields{k})));
        else
            opts.(allfields{k}) = options.(allfields{k});
        end
    end
end
if nargout > 1
    varargout = struct2cell(opts);
else
    varargout = opts;
end
```

```
function varargout = nmf_initialize(Y,J,inittype,Fact)
% Initialization for NMF
% Copyright 2008 by A.H. Phan and A. Cichocki and R. Zdunek
if numel(inittype) ==1
    inittype = inittype([1 1]);
end
factorsize = size(Y);
Nfact = 2;
```

```
if exist('Fact','var')≠1
    Fact = cell(1,Nfact);
end
if ~any(inittype ==5)
    for k = 1:Nfact
        switch inittype(k)
            case {1,4,'random'} % rand
                Fact{k}= rand(factorsize(k),J);
            case {2,'sin'} % sin
                [J_sx , I_sx ] = meshgrid(1:J,1:factorsize(k));
                Fact{k} = abs(.1*sin(2*pi*.1*I_sx)+1*cos(2*pi*.1*J_sx)+...
                    cos(2*pi*.471*I_sx) + sin(2*pi*.471*J_sx));
            case {3,'constant'} % constant 1
                Fact{k}=ones(factorsize(k),J);
        end
    end
    if inittype(1) == 4          % ALS init
        Fact{1} = max(eps,(Y*Fact{2})*pinv(Fact{2}'*Fact{2}));
    end
    if inittype(2) == 4
        Fact{2} = max(eps,(Y'*Fact{1})*pinv(Fact{1}'*Fact{1}));
    end
else
    [Fact{1},Fact{2}] = lsv_NMF(Y,J);
end
varargout = Fact;
```

```
function y = nngorth(J,T,p)
% Generate J nonnegative orthogonal signals of T samples in [0,1]
%
% INPUTS:
% J    :  Number of signals
% T    :  Number of samples
% p    :  Probability of active samples
%
% OUTPUTS:
% y    :  J x T matrix
% Copyright 2008 by A.H. Phan and A. Cichocki
setJ = 1:J;
y =zeros(J,T); % output vector
while 1
    actsample = find(rand(1,T)≤p);
    N = numel(actsample);
    srcidx = ceil(J*rand(1,N));
    if isempty(setdiff(setJ,unique(srcidx))), break; end
end
y(sub2ind([ J,T],srcidx,actsample)) = rand(1,N);
```

```
function [G,mapind] = nmf_G(X,XH)
% This function builds the correlation matrix G of two matrices X and XH,
% for evaluating the performance of uncorrelated sources.
% Output:
% G       :  correlation matrix with high entries in the top left positions
% mapind:    matrix with 2 columns for rearranging rows of X and XH
%            G = X(mapind(:,1),:) * XH(mapind(:,2),:)fl
```

```
% Copyright 2008 by A.H. Phan and A. Cichocki  % 04/2008
XHn = bsxfun(@rdivide,XH,sqrt(sum(XH.^2,2)));
Xbn = bsxfun(@rdivide,X,sqrt(sum(X.^2,2)));
G = Xbn*XHn';
N = size(G,1);

[Gs,inds] = sort(G(:),'descend');
[rs,cs] = ind2sub(size(G),inds);
[reorder_r,i] = unique(rs(1:N),'first'); reorder_r = rs(sort(i));
unc_ind = setdiff(1:N,reorder_r);
if ~isempty(unc_ind)
    k_ind = zeros(1,numel(unc_ind));
    for k = 1: numel(unc_ind)
        k_ind(k) = N+find(rs(N+1:end) == unc_ind(k),1);
    end
    reorder_r(end+1:end+numel(k_ind)) = rs(sort(k_ind));
end

G = G(reorder_r,:); Gn = abs(G);
reorder_c = (1:N)';
for k = 1:N
    [val,ind] = max(abs(Gn(k,k:end)));
    if ind ≠ 1
        Gn(:,[k k+ind-1]) = Gn(:,[k+ind-1 k]);
        reorder_c([k k+ind-1]) = reorder_c([k+ind-1 k]);
    end
end
G = Gn;
mapind = [reorder_r,reorder_c];
```

```
function hinton(w)
% Plot Hinton diagram for a weight matrix.
% Examples
%     W = rand(4,5);
%     hinton(W)
% See also the Matlab function HINTONW
% Written by A. H. Phan and A. Cichocki
% 12/2008
szw = size(w);
w = w(:)';  sgnw = sign(w);
max_w = max(abs(w));
blobsize = sqrt(abs(w)/max_w)*0.45;
blobsize(abs(w)<max_w/100)= 0;
[cx,cy] = ind2sub(szw,1:numel(w));
px = bsxfun(@plus,[-1 ;-1;1; 1 ] * blobsize,cx);
py = bsxfun(@plus,[-1 ;1 ;1;-1 ] * blobsize,cy);
hold on;
patch(py,px,sgnw)
plot(py([4 1 2],sgnw>0),px([4 1 2],sgnw>0),'w')
plot(py(2:4,sgnw<0),px(2:4,sgnw<0),'w')
xticks = get(gca,'xtick'); set(gca,'xtick',xticks(xticks == floor(xticks)))
yticks = get(gca,'ytick'); set(gca,'ytick',yticks(yticks == floor(yticks)))
set(gca,'ydir','reverse','linewidth',1.5,'color',[1 1 1]*.8);
set(gcf,'color',[1 1 1]);
xlim([0 szw(2)]+0.5);ylim([0 szw(1)]+0.5);
colormap([.8 0 0;0 .8 0]); caxis([-max_w max_w])
grid on; box on; axis image
```

References

[1] F. Abramovich, T. Sapatinas, and B. Silverman. Wavelet thresholding via a Bayesian approach. *Journal of the Royal Statistical Society: Series B (Statistical Methodology)*, 60(4):725–749, 1998.

[2] R. Albright, J. Cox, D. Duling, A. N. Langville, and C. D. Meyer. Algorithms, initializations, and convergence for the nonnegative matrix factorization. Technical report, NCSU Technical Report Math 81706, 2006.

[3] M. Biggs, A. Ghodsi, and S. Vavasis. Nonnegative matrix factorization via rank-one downdate. In *ICML-2008*, Helsinki, July 2008.

[4] J. Bobin, Y. Moudden, J. Fadili, and J.L. Starck. Morphological diversity and sparsity in blind source separation. In *Proc. ICA-2007*, volume 4666 of *Lecture Notes in Computer Science*, pages 349–356. Springer, 2007.

[5] J. Bobin, J.L. Starck, J. Fadili, Y. Moudden, and D.L. Donoho. Morphological component analysis: An adaptive thresholding strategy. *IEEE Transactions on Image Processing*, 16(11):2675–2681, 2007.

[6] J. Bobin, J.L. Starck, and R. Ottensamer. Compressed sensing in astronomy. *Submitted to IEEE Journal on Selected Topics in Signal Processing*, February 2008.

[7] C. Boutsidis, M.W. Mahoney, and P. Drineas. An improved approximation algorithm for the column subset selection problem. In *Proc. 20-th Annual SODA*, pages 968–977, USA, 2009.

[8] L. Breiman. Heuristics of instability and stabilization in model selection. *Annals of Statistics*, 24(6):2350–2383, 1996.

[9] J. Buckheit and D. Donoho. Wavelab and reproducible research. Stanford University, Stanford, CA 94305, USA, 1995.

[10] C. Caiafa and A. Cichocki. CUR decomposition with optimal selection of rows and columns. *(submitted)*, 2009.

[11] E. J. Candés, J. Romberg, and T. Tao. Robust uncertainty principles: Exact signal reconstruction from highly incomplete frequency information. *IEEE Transactions on Information Theory*, 52(2):489–509, February 2006.

[12] R. Chartrand and W. Yin. Iteratively reweighted algorithms for compressive sensing. In *in 33rd International Conference on Acoustics, Speech, and Signal Processing (ICASSP)*, 2008.

[13] A. Cichocki and S. Amari. *Adaptive Blind Signal and Image Processing*. John Wiley & Sons Ltd, New York, 2003.

[14] A. Cichocki, S. Amari, R. Zdunek, R. Kompass, G. Hori, and Z. He. Extended SMART algorithms for non-negative matrix factorization. *Springer, LNAI-4029*, 4029:548–562, 2006.

[15] A. Cichocki and A.H. Phan. Fast local algorithms for large scale nonnegative matrix and tensor factorizations. *IEICE (invited paper)*, March 2009.

[16] A. Cichocki, A.H. Phan, and C. Caiafa. Flexible HALS algorithms for sparse non-negative matrix/tensor factorization. In *Proc. of 18-th IEEE workshops on Machine Learning for Signal Processing*, Cancun, Mexico, 16–19, October 2008.

[17] A. Cichocki, A.H. Phan, R. Zdunek, and L.-Q. Zhang. Flexible component analysis for sparse, smooth, nonnegative coding or representation. In *Lecture Notes in Computer Science, LNCS-4984*, volume 4984, pages 811–820. Springer, 2008.

[18] A. Cichocki and R. Zdunek. Multilayer nonnegative matrix factorization. *Electronics Letters*, 42(16):947–948, 2006.

[19] A. Cichocki and R. Zdunek. NMFLAB for Signal and Image Processing. Technical report, Laboratory for Advanced Brain Signal Processing, BSI, RIKEN, Saitama, Japan, 2006.

[20] A. Cichocki and R. Zdunek. NTFLAB for Signal Processing. Technical report, Laboratory for Advanced Brain Signal Processing, BSI, RIKEN, Saitama, Japan, 2006.

[21] A. Cichocki and R. Zdunek. Regularized alternating least squares algorithms for non-negative matrix/tensor factorizations. *Springer, LNCS-4493*, 4493:793–802, June 3–7 2007.

[22] A. Cichocki, R. Zdunek, and S.-I. Amari. Hierarchical ALS algorithms for nonnegative matrix and 3D tensor factorization. *Springer, Lecture Notes on Computer Science, LNCS-4666*, pages 169–176, 2007.

[23] A. Cichocki, R. Zdunek, S. Choi, R. Plemmons, and S. Amari. Nonnegative tensor factorization using Alpha and Beta divergencies. In *Proc. IEEE International Conference on Acoustics, Speech, and Signal Processing (ICASSP07)*, volume III, pages 1393–1396, Honolulu, Hawaii, USA, April 15–20 2007.

[24] A. Cichocki, R. Zdunek, S. Choi, R. Plemmons, and S.-I. Amari. Novel multi-layer nonnegative tensor factorization with sparsity constraints. *Springer, LNCS-4432*, 4432:271–280, April 11–14 2007.

[25] R.R. Coifman and D.L. Donoho. Translation-invariant de-noising. In *Wavelets and Statistics*, pages 125–150. Springer-Verlag, 1995.

[26] I. Daubechies, M. Defrise, and C. De Mol. An iterative thresholding algorithm for linear inverse problems with a sparsity constraint. *Communications on Pure and Applied Mathematics*, 57(11):1413–1457, 2004.

[27] A. de Juan and R. Tauler. Chemometrics applied to unravel multicomponent processes and mixtures: Revisiting latest trends in multivariate resolution. *Analitica Chemica Acta*, 500:195–210, 2003.
[28] D.L. Donoho, I. Drori, V. Stodden, and Y. Tsaig. Sparselab. *http://sparselab.stanford.edu/*, December 2005.
[29] D.L. Donoho and I.M. Johnstone. Adapting to unknown smoothness via wavelet shrinkage. *J. Am. Stat. Assoc*, 90(432):1200–1224, 1995.
[30] D.L. Donoho and J. Tanner. Thresholds for the recovery of sparse solutions via l1 minimization. In *Proceedings of the Conference on Information Sciences and Systems.*
[31] D.L. Donoho and Y. Tsaig. Sparse solution of underdetermined linear equations. by stagewise orthogonal matching pursuit. *IEEE Transactions on Information Theory*, March 2006. submitted.
[32] P. Drineas, M.W. Mahoney, and S. Muthukrishnan. Relative-error CUR matrix decompositions. *SIAM Journal on Matrix Analysis and Applications*, 30:844–881, 2008.
[33] M.F. Duarte, S. Sarvotham, D. Baron, M.B. Wakin, and R.G. Baraniuk. Distributed compressed sensing of jointly sparse signals. In *Proc. 39th Asilomar Conf. Signals, Sys., Comput.*, pages 1537– 1541, 2005.
[34] M. Elad. Why simple shrinkage is still relevant for redundant representations? *IEEE Transactions On Information Theory*, 52:5559–5569, 2006.
[35] M. Elad. Optimized projections for compressed sensing. *IEEE Transactions on Signal Processing*, 55(12):5695 – 5702, Dec 2007.
[36] Y.C. Eldar. Generalized SURE for exponential families: Applications to regularization. *IEEE Transactions on Signal Processing*, 2008. Forthcoming Articles.
[37] M.A.T. Figueiredo, R.D. Nowak, and S.J. Wright. Gradient projection for sparse reconstruction: Application to compressed sensing and other inverse problems. *IEEE Journal of Selected Topics in Signal Processing*, 1(4):586–597, Dec 2007.
[38] H.-Y. Gao. Wavelet shrinkage denoising using the non-negative Garrote. *Journal of Computational and Graphical Statistics*, 7(4):469–488, 1998.
[39] N. Gillis and F. Glineur. Nonnegative matrix factorization and underapproximation. In *9th International Symposium on Iterative Methods in Scientific Computing*, Lille, France, 2008. http://www.core.ucl.ac.be/ ngillis/.
[40] N. Gillis and F. Glineur. Nonnegative factorization and maximum edge biclique problem. In *submitted*, 2009. http://www.uclouvain.be/en-44508.html.
[41] R. Giryes, M. Elad, and Y.C. Eldar. Automatic parameter setting for iterative shrinkage methods. *Electrical and Electronics Engineers in Israel, 2008. IEEEI 2008. IEEE 25th Convention of*, pages 820–824, 2008.
[42] S.A. Goreinov, I.V. Oseledets, D.V. Savostyanov, E.E. Tyrtyshnikov, and N.L. Zamarashkin. How to find a good submatrix. (submitted), 2009.
[43] T.M. Hancewicz and J.-H. Wang. Discriminant image resolution: a novel multivariate image analysis method utilizing a spatial classification constraint in addition to bilinear nonnegativity. *Chemometrics and Intelligent Laboratory Systems*, 77:18–31, 2005.
[44] N.-D. Ho. *Nonnegative Matrix Factorization - Algorithms and Applications.* These/dissertation, Universite Catholique de Louvain, Belgium, FSA/INMA - Departement d'ingenierie mathematique, 2008.
[45] N.-D. Ho, P. Van Dooren, and V.D. Blondel. Descent methods for nonnegative matrix factorization. *Numerical Linear Algebra in Signals, Systems and Control*, 2008.
[46] B. Jorgensen. *The Theory of Dispersion Models.* Chapman and Hall, London, 1997.
[47] A. N. Langville, C. D. Meyer, and R. Albright. Initializations for the nonnegative matrix factorization. In *Proc. of the Twelfth ACM SIGKDD International Conference on Knowledge Discovery and Data Mining*, Philadelphia, USA, August 20–23 2006.
[48] D.D. Lee and H.S. Seung. Learning of the parts of objects by non-negative matrix factorization. *Nature*, 401:788–791, 1999.
[49] D.D. Lee and H.S. Seung. *Algorithms for Nonnegative Matrix Factorization*, volume 13. MIT Press, 2001.
[50] M.W. Mahoney and P. Drineas. CUR matrix decompositions for improved data analysis. *Proc. National Academy of Science*, 106:697–702, 2009.
[51] M.W. Mahoney, M. Maggioni, and P. Drineas. Tensor-CUR decompositions for tensor-based data. In *Proc. of the 12th ACM SIGKDD International Conference on Knowledge Discovery and Data Mining*, pages 327–336, Philadelphia, USA, August 20–23 2006.
[52] C. Navasca, L. De Lathauwer, and S. Kinderman. Swamp reducing technique for tensor decomposition. In *Proc. of the European Signal Processing*, EUSIPCO, Lausanne, Switzerland, 2008.

[53] M. Nikolova. Minimizers of cost-functions involving nonsmooth data-fidelity terms. application to the processing of outliers. *SIAM Journal on Numerical Analysis*, 40(3):965–994, 2002.

[54] D. Nion and L. De Lathauwer. An enhanced line search scheme for complex-valued tensor decompositions. Application in DS-CDMA. *Signal Processing*, 88(3):749–755, 2008.

[55] A.H. Phan and A. Cichocki. Multi-way nonnegative tensor factorization using fast hierarchical alternating least squares algorithm (HALS). In *Proc. of The 2008 International Symposium on Nonlinear Theory and its Applications*, Budapest, Hungary, 2008.

[56] A.H. Phan, A. Cichocki, and K.S. Nguyen. Simple and efficient algorithm for distributed compressed sensing. In *Machine Learning for Signal Processing*, Cancun, 2008.

[57] M. Rajih and P. Comon. Enhanced line search: A novel method to accelerate PARAFAC. Technical report, Laboratoire I3S Sophia Antipolis, France, 2005.

[58] M. Rajih and P. Comon. Enhanced line search: A novel method to accelerate PARAFAC. *SIAM Journal on Matrix Analysis and Applications*, 30(3):1128–1147, 2008.

[59] J. Shihao, Y. Xue, and L. Carin. Bayesian compressive sensing. *IEEE Transactions on Signal Processing*, 56(6), 2008.

[60] G. K. Smyth. Fitting Tweedie's compound Poisson model to insurance claims data: dispersion modelling. *ASTIN Bulletin*, 32:2002.

[61] M.W. Spratling and M.H. Johnson. A feedback model of perceptual learning and categorisation. *Visual Cognition*, 13:129–165, 2006.

[62] R. Tauler and A. de Juan. *Multivariate Curve Resolution–Practical Guide to Chemometrics, Chapter 11*. Taylor and Francis Group, 2006.

[63] Y. Tsaig and D.L. Donoho. Extensions of compressed sensing. *Signal Processing*, 86(3):549–571, 2006.

[64] J.H. Wang, P.K. Hopke, T.M. Hancewicz, and S.-L. Zhang. Application of modified alternating least squares regression to spectroscopic image analysis. *Analytica Chimica Acta*, 476:93–109, 2003.

[65] R. Zdunek and A. Cichocki. Nonnegative matrix factorization with constrained second-order optimization. *Signal Processing*, 87:1904–1916, 2007.

5

Projected Gradient Algorithms

In contrast to the multiplicative NMF algorithms discussed in Chapter 3, this class of Projected Gradient (PG) algorithms has additive updates. The algorithms discussed here provide approximate solutions to Non-negative Least Squares (NLS) problems, and are based on the alternating minimization technique:

$$\min_{\mathbf{x}_t \geq 0} D_F(\mathbf{y}_t||\mathbf{A}\mathbf{x}_t) = \frac{1}{2}||\mathbf{y}_t - \mathbf{A}\mathbf{x}_t||_2^2, \qquad (t = 1, 2, \ldots, T), \tag{5.1}$$

$$\min_{\underline{\mathbf{a}}_i \geq 0} D_F(\underline{\mathbf{y}}_i||\mathbf{X}^T\underline{\mathbf{a}}_i) = \frac{1}{2}||\underline{\mathbf{y}}_i - \mathbf{X}^T\underline{\mathbf{a}}_i||_2^2, \qquad (i = 1, 2, \ldots, I). \tag{5.2}$$

This can also be written in equivalent matrix forms

$$\min_{x_{jt} \geq 0} D_F(\mathbf{Y}||\mathbf{A}\mathbf{X}) = \frac{1}{2}||\mathbf{Y} - \mathbf{A}\mathbf{X}||_F^2, \tag{5.3}$$

$$\min_{a_{ij} \geq 0} D_F(\mathbf{Y}^T||\mathbf{X}^T\mathbf{A}^T) = \frac{1}{2}||\mathbf{Y}^T - \mathbf{X}^T\mathbf{A}^T||_F^2, \tag{5.4}$$

where $\mathbf{A} = [\mathbf{a}_1, \ldots, \mathbf{a}_J] \in \mathbb{R}_+^{I \times J}$, $\mathbf{A}^T = [\underline{\mathbf{a}}_1, \ldots, \underline{\mathbf{a}}_I] \in \mathbb{R}_+^{J \times I}$, $\mathbf{X} = [\mathbf{x}_1, \ldots, \mathbf{x}_T] \in \mathbb{R}_+^{J \times T}$, $\mathbf{X}^T = [\underline{\mathbf{x}}_1, \ldots, \underline{\mathbf{x}}_J] \in \mathbb{R}_+^{T \times J}$, $\mathbf{Y} = [\mathbf{y}_1, \ldots, \mathbf{y}_T] \in \mathbb{R}^{I \times T}$, $\mathbf{Y}^T = [\underline{\mathbf{y}}_1, \ldots, \underline{\mathbf{y}}_I] \in \mathbb{R}^{T \times I}$, and usually $I \geq J$. The matrix $\mathbf{A}$ is assumed to be full-rank, thus providing the existence of a unique solution $\mathbf{X}^* \in \mathbb{R}^{J \times T}$. Since the NNLS problem (5.1) is strictly convex with respect to one set of variables $\{\mathbf{X}\}$, a unique solution exists for any matrix $\mathbf{X}$, and the solution $\mathbf{x}_t^*$ satisfies the Karush-Kuhn-Tucker (KKT) conditions:

$$\mathbf{x}_t^* \geq \mathbf{0}, \quad \mathbf{g}_\mathbf{X}(\mathbf{x}_t^*) \geq \mathbf{0}, \quad \mathbf{g}_\mathbf{X}(\mathbf{x}_t^*)^T\mathbf{x}_t^* = 0 \tag{5.5}$$

or in an equivalent compact matrix form as:

$$\mathbf{X}^* \geq \mathbf{0}, \quad \mathbf{G}_\mathbf{X}(\mathbf{X}^*) \geq \mathbf{0}, \quad \mathrm{tr}\{\mathbf{G}_\mathbf{X}(\mathbf{X}^*)^T\mathbf{X}^*\} = 0, \tag{5.6}$$

where the symbols $\mathbf{g}_\mathbf{X}$ and $\mathbf{G}_\mathbf{X}$ denote the corresponding gradient vector and gradient matrix:

$$\mathbf{g}_\mathbf{X}(\mathbf{x}_t) = \nabla_{\mathbf{x}_t} D_F(\mathbf{y}_t||\mathbf{A}\mathbf{x}_t) = \mathbf{A}^T(\mathbf{A}\mathbf{x}_t - \mathbf{y}_t), \tag{5.7}$$

$$\mathbf{G}_\mathbf{X}(\mathbf{X}) = \nabla_\mathbf{X} D_F(\mathbf{Y}||\mathbf{A}\mathbf{X}) = \mathbf{A}^T(\mathbf{A}\mathbf{X} - \mathbf{Y}). \tag{5.8}$$

Similarly, the KKT conditions for the solution $\underline{\mathbf{a}}^*$ to (5.2), and the solution $\mathbf{A}^*$ to (5.4) are as follows:

$$\underline{\mathbf{a}}_i^* \geq \mathbf{0}, \quad \mathbf{g}_\mathbf{A}(\underline{\mathbf{a}}_i^*) \geq \mathbf{0}, \quad \mathbf{g}_\mathbf{A}(\underline{\mathbf{a}}_i^*)^T\underline{\mathbf{a}}_i^* = 0, \tag{5.9}$$

Nonnegative Matrix and Tensor Factorizations: Applications to Exploratory Multi-way Data Analysis and Blind Source Separation Andrzej Cichocki, Rafal Zdunek, Anh Huy Phan and Shun-ichi Amari

and the corresponding conditions in (5.6) are

$$\mathbf{A}^* \geq \mathbf{0}, \quad \mathbf{G_A}(\mathbf{A}^*) \geq \mathbf{0}, \quad \mathrm{tr}\{\mathbf{A}^*\mathbf{G_A}(\mathbf{A}^*)^T\} = 0, \tag{5.10}$$

where $\boldsymbol{g}_\mathbf{A}$ and $\mathbf{G_A}$ are the gradient vector and gradient matrix of the objective function:

$$\boldsymbol{g}_\mathbf{A}(\underline{\boldsymbol{a}}_i) = \nabla_{\underline{\boldsymbol{a}}_i} D_F(\underline{\boldsymbol{y}}_i||\mathbf{X}^T\underline{\boldsymbol{a}}_i) = \mathbf{X}(\mathbf{X}^T\underline{\boldsymbol{a}}_i - \underline{\boldsymbol{y}}_i), \tag{5.11}$$

$$\mathbf{G_A}(\mathbf{A}) = \nabla_\mathbf{A} D_F(\mathbf{Y}^T||\mathbf{X}^T\mathbf{A}^T) = (\mathbf{AX} - \mathbf{Y})\mathbf{X}^T. \tag{5.12}$$

There are many approaches to solve the minimization problems (5.1) and (5.2), or equivalently (5.3) and (5.4). In this chapter, we shall discuss several projected gradient methods which take a general form of iterative updates:

$$\mathbf{X}^{(k+1)} = \left[\mathbf{X}^{(k)} - \eta_\mathbf{X}^{(k)}\,\mathbf{P}_\mathbf{X}^{(k)}\right]_+, \tag{5.13}$$

$$\mathbf{A}^{(k+1)} = \left[\mathbf{A}^{(k)} - \mathbf{P}_\mathbf{A}^{(k)}\,\eta_\mathbf{A}^{(k)}\right]_+, \tag{5.14}$$

where $[\mathbf{X}]_+ = \mathcal{P}_\Omega[\mathbf{X}]$ denotes a projection of entries of $\mathbf{X}$ onto a convex "feasible" set $\Omega = \left\{x_{jt} \in \mathbb{R} : x_{jt} \geq 0\right\}$ – namely, the nonnegative orthant $\mathbb{R}_+$ (the subspace of nonnegative real numbers), $\mathbf{P}_\mathbf{X}^{(k)}$ and $\mathbf{P}_\mathbf{A}^{(k)}$ are descent directions for $\mathbf{X}$ and $\mathbf{A}$ in the k-th inner iterative step, and $\eta_\mathbf{X}^{(k)}$ and $\eta_\mathbf{A}^{(k)}$ are the learning rate scalars or the diagonal matrices of positive learning rates.

The projection $[\mathbf{X}]_+$ can be performed in many ways.[1] One straightforward way is to replace all the negative entries in $\mathbf{X}$ by zero, or for practical purposes, by a small positive number ε in order to avoid numerical instabilities, thus giving (component-wise)

$$[\mathbf{X}]_+ = \max\{\varepsilon, \mathbf{X}\}. \tag{5.15}$$

Alternatively, it may be more efficient to preserve the nonnegativity of the solutions by an optimal choice of the learning rates $\eta_\mathbf{X}^{(k)}$ and $\eta_\mathbf{A}^{(k)}$, or by solving least-squares problems subject to the constraints (5.6) and (5.10). We here present exemplary PG methods which are proven to be very efficient for NMF problems, and are all part of our MATLAB toolbox: NMFLAB/NTFLAB for Signal and Image Processing [8,9,30].

5.1 Oblique Projected Landweber (OPL) Method

The Landweber method [3] performs gradient descent minimization based on the following iterative scheme:

$$\mathbf{X}^{(k+1)} = \mathbf{X}^{(k)} - \eta_\mathbf{X}\,\mathbf{G}_\mathbf{X}^{(k)}, \tag{5.16}$$

where the descent direction $\mathbf{P}_\mathbf{X}^{(k)}$ is replaced with the gradient $\mathbf{G_X}$ given in (5.8), and the range of the learning rate $\eta_\mathbf{X} \in (0, \eta_{max})$. This update ensures the asymptotic convergence to the minimum-norm least squares solution, with the convergence radius defined by

$$\eta_{max} = \frac{2}{\lambda_{max}(\mathbf{A}^T\mathbf{A})}, \tag{5.17}$$

where $\lambda_{max}(\mathbf{A}^T\mathbf{A})$ is the maximum eigenvalue of $\mathbf{A}^T\mathbf{A}$. Since $\mathbf{A}$ is nonnegative, for its eigenvalues we have $\lambda_{max}(\mathbf{A}^T\mathbf{A}) \leq \max(\mathbf{A}^T\mathbf{A}\mathbf{1}_J)$, where $\mathbf{1}_J = [1, \ldots, 1]^T \in \mathbb{R}^J$, and the modified Landweber method can be

[1] Although in this chapter we use only the simple nonlinear projection "half-wave rectifying", which replaces negative values by small positive constant ε, the PG algorithms discussed here can be easily adopted to the factors which are upper and/or lower bounded at any specific level, for example, $l_j \leq x_{jt} \leq u_j, \;\; \forall j,t$. This can be achieved by applying a suitable projection function $\mathcal{P}_\Omega[x_{jt}]$ which transforms the updated factors to the feasible region.

expressed as:

$$\mathbf{X}^{(k+1)} = \left[\mathbf{X}^{(k)} - \boldsymbol{\eta}_{\mathbf{X}} \, \mathbf{G}_{\mathbf{X}}^{(k)}\right]_+, \quad \text{where} \quad \boldsymbol{\eta}_{\mathbf{X}} = \text{diag}\{\eta_1, \eta_2, \dots, \eta_J\}, \quad \eta_j < \frac{2}{(\mathbf{A}^T\mathbf{A}\mathbf{1}_J)_j}. \quad (5.18)$$

One variant of the method is the Oblique Projected Landweber (OPL) method [18], which can be regarded as a particular case of the PG iterative formula (5.13)–(5.14), where at each iterative step the solution obtained by (5.16) is projected onto the feasible set. Based on these, the method can be implemented for the standard NMF problem as shown in Algorithm 5.1.

Algorithm 5.1: OPL-NMF

Input: $\mathbf{Y} \in \mathbb{R}_+^{I \times T}$: input data, J: rank of approximation
Output: $\mathbf{A} \in \mathbb{R}_+^{I \times J}$ and $\mathbf{X} \in \mathbb{R}_+^{J \times T}$ such that the cost functions (5.3) and (5.4) are minimized.

1 **begin**
2   initialization for $\mathbf{A}, \mathbf{X}$
3   **repeat**
4       $\mathbf{X} \leftarrow \text{OPL}(\mathbf{Y}, \mathbf{A}, \mathbf{X})$ `/* Update X */`
5       $\mathbf{A} \leftarrow \text{OPL}(\mathbf{Y}^T, \mathbf{X}^T, \mathbf{A}^T)^T$ `/* Update A */`
6   **until** *a stopping criterion is met*   `/* convergence condition */`
7 **end**

8 **function** $\mathbf{X} = \text{OPL}(\mathbf{Y}, \mathbf{A}, \mathbf{X})$
9 **begin**
10    $\mathbf{R} = \mathbf{A}^T\mathbf{A}$,
11    $\mathbf{Z} = \mathbf{A}^T\mathbf{Y}$,
12    $\boldsymbol{\eta}_{\mathbf{X}} = \text{diag}\left(\mathbf{1}_J \oslash (\mathbf{R}\mathbf{1}_J)\right)$
13   **repeat**
14       $\mathbf{G}_{\mathbf{X}} = \mathbf{R}\mathbf{X} - \mathbf{Z}$ `/* Gradient with respect to X */`
15       $\mathbf{X} \leftarrow \left[\mathbf{X} - \boldsymbol{\eta}_X \, \mathbf{G}_{\mathbf{X}}\right]_+$ `/* Update X */`
16   **until** *a stopping criterion is met*   `/* convergence condition */`
17 **end**

The MATLAB implementation of the OPL algorithm is given in Listing 5.1.

Listing 5.1   OPL-NMF algorithm.

```
1   function [X] = nmf_opl(A,Y,X,no_iter)
2   %
3   % INPUTS:
4   % A - fixed matrix of dimension [I by J]
5   % Y - data matrix of dimension [I by T]
6   % X - initial solution matrix of dimension [J by T]
7   % no_iter - maximum number of iterations
8   %
9   % OUTPUTS:
10  % X - estimated matrix of dimension [J by T]
11  %
12  % ###################################################################
13  R = A'*A; Z = A'*Y; eta = 1./sum(R,2),
```

```
for k=1:no_iter
      G = R*X - Z;
      X = max(eps, X - bsxfun(@times,G,eta));
end % for k
```

5.2 Lin's Projected Gradient (LPG) Algorithm with Armijo Rule

A typical representative of PG algorithms in applications to NMF is Chih-Jen Lin's algorithm [21], which is given by the iterative formula (5.13)–(5.14) with $\mathbf{P}_{\mathbf{X}}^{(k)}$ and $\mathbf{P}_{\mathbf{A}}^{(k)}$ expressed by the gradients (5.8) and (5.12), respectively, and the projection rule (5.15). In contrast to the OPL algorithm given in Section 5.1 the learning rates $\eta_{\mathbf{X}}^{(k)}$ and $\eta_{\mathbf{A}}^{(k)}$ in Lin's PG algorithm in the inner iterations are not fixed diagonal matrices, but are scalars computed by inexact estimation techniques. Lin considered two options for estimating the learning rules: the Armijo rule along the projective arc of the algorithm proposed by Bertsekas [4,5], and the modified Armijo rule.

In the first case, for every inner iterative step of the algorithm, the value of the learning rate $\eta_{\mathbf{X}}^{(k)}$ is given by

$$\eta_{\mathbf{X}}^{(k)} = \beta^{m_k}, \tag{5.19}$$

where m_k is the first nonnegative integer m for which

$$D_F(\mathbf{Y}||\mathbf{A}\mathbf{X}^{(k+1)}) - D_F(\mathbf{Y}||\mathbf{A}\mathbf{X}^{(k)}) \leq \sigma \operatorname{tr}\left\{\nabla_{\mathbf{X}} D_F(\mathbf{Y}||\mathbf{A}\mathbf{X}^{(k)})^T(\mathbf{X}^{(k+1)} - \mathbf{X}^{(k)})\right\}, \tag{5.20}$$

with $\beta \in (0, 1)$ and $\sigma \in (0, 1)$. The learning rate $\eta_{\mathbf{A}}^{(k)}$ for computing $\mathbf{A}$ is calculated in a similar way.

In the modified Armijo rule, the initial value $\eta_{\mathbf{X}}^{(0)}$ does not necessarily start from β, and the value of the learning rate can decrease (as in the previous case) or even increase, which may greatly increase convergence speed. It was observed by Lin and Moré [22] that $\eta_{\mathbf{X}}^{(k)}$ and $\eta_{\mathbf{X}}^{(k-1)}$ may be very similar, and they proposed to start the estimation from $\eta_{\mathbf{X}}^{(k-1)}$ and to decrease (or increase) the learning rate depending on whether the condition (5.20) is met (or not).

Lin [21] also proposed a very robust stopping criterion which not only examines a statistical distance between the consecutive updates, but also how far the current update is from the stationary point. The iterative process stops when the following conditions are met:

$$||\nabla_{\mathbf{X}}^P D_F(\mathbf{Y}||\mathbf{A}^{(s)}\mathbf{X}^{(s+1)})||_F \leq \max(10^{-3}, \bar{\epsilon}_{\mathbf{X}})||\nabla_{\mathbf{X}} D_F(\mathbf{Y}||\mathbf{A}^{(1)}\mathbf{X}^{(1)})||_F, \tag{5.21}$$

$$||\nabla_{\mathbf{A}}^P D_F(\mathbf{Y}||\mathbf{A}^{(s+1)}\mathbf{X}^{(s+1)})||_F \leq \max(10^{-3}, \bar{\epsilon}_{\mathbf{A}})||\nabla_{\mathbf{A}} D_F(\mathbf{Y}||\mathbf{A}^{(1)}\mathbf{X}^{(1)})||_F, \tag{5.22}$$

where $\bar{\epsilon}_{\mathbf{A}}, \bar{\epsilon}_{\mathbf{X}}$ are tolerance rates, and

$$[\nabla_{\mathbf{X}}^P D_F]_{jt} = \begin{cases} [\nabla_{\mathbf{X}} D_F]_{jt} & \text{if } x_{jt} > 0, \\ \min\left\{0, [\nabla_{\mathbf{X}} D_F(\mathbf{X})]_{jt}\right\} & \text{if } x_{jt} = 0. \end{cases} \tag{5.23}$$

In the case that the iterative process stops just after the first iteration, the tolerance rates are decreased as follows:

$$\bar{\epsilon}_{\mathbf{X}} \leftarrow \frac{\bar{\epsilon}_{\mathbf{X}}}{10}, \quad \bar{\epsilon}_{\mathbf{A}} \leftarrow \frac{\bar{\epsilon}_{\mathbf{A}}}{10}. \tag{5.24}$$

Lin's PG algorithm can be easily extended to Algorithm 5.2 that works for other divergences $D_F(\mathbf{Y}||\mathbf{A}\mathbf{X})$, such as the Kullback-Leibler, dual Kullback-Leibler or Alpha-divergences [10], in order to deal with the noisy disturbances which are not Gaussian. The MATLAB code for the modified Lin's PG (LPG-NMF) algorithm is listed in Appendix 5.B.

Algorithm 5.2: LPG-NMF

Input: $\mathbf{Y} \in \mathbb{R}_+^{I \times T}$: input data, J: rank of approximation
Output: $\mathbf{A} \in \mathbb{R}_+^{I \times J}$ and $\mathbf{X} \in \mathbb{R}_+^{J \times T}$ such that the cost functions (5.3) and (5.4) are minimized.

begin
  initialization for **A**, **X**
  repeat
    $\mathbf{X} \leftarrow \texttt{LPG}(\mathbf{Y}, \mathbf{A}, \mathbf{X}, \beta)$ /* Update **X** */
    $\mathbf{A} \leftarrow \texttt{LPG}(\mathbf{Y}^T, \mathbf{X}^T, \mathbf{A}^T, \beta)^T$ /* Update **A** */
  until *a stopping criterion is met* /* convergence condition */
end

function $\mathbf{X} = \texttt{LPG}(\mathbf{Y}, \mathbf{A}, \mathbf{X}, \beta)$
begin
  $\eta_{\mathbf{X}} \leftarrow 1$
  $\tilde{\mathbf{X}} \leftarrow [\mathbf{X} - \eta_{\mathbf{X}} \, \mathbf{P}_{\mathbf{X}}]_+$
  repeat
    if $D_F(\mathbf{Y}||\mathbf{A}\tilde{\mathbf{X}}) - D_F(\mathbf{Y}||\mathbf{A}\mathbf{X}) \leq \sigma \operatorname{tr}\left\{\nabla_X D_F(\mathbf{Y}||\mathbf{A}\mathbf{X})^T(\tilde{\mathbf{X}} - \mathbf{X})\right\}$, **then**
      repeat
        $\eta_{\mathbf{X}} \leftarrow \dfrac{\eta_{\mathbf{X}}}{\beta}$
        $\tilde{\mathbf{X}} \leftarrow [\mathbf{X} - \eta_{\mathbf{X}} \, \mathbf{P}_{\mathbf{X}}]_+$ /* Update **X** */
      until $D_F(\mathbf{Y}||\mathbf{A}\tilde{\mathbf{X}}) - D_F(\mathbf{Y}||\mathbf{A}\mathbf{X}) > \sigma \operatorname{tr}\left\{\nabla_X D_F(\mathbf{Y}||\mathbf{A}\mathbf{X})^T(\tilde{\mathbf{X}} - \mathbf{X})\right\}$ *or* $\tilde{\mathbf{X}} = \mathbf{X}$
      /* convergence condition */
    else
      repeat
        $\eta_{\mathbf{X}} \leftarrow \eta_{\mathbf{X}} \beta$
        $\tilde{\mathbf{X}} \leftarrow [\mathbf{X} - \eta_{\mathbf{X}} \, \mathbf{P}_{\mathbf{X}}]_+$ /* Update **X** */
      until $D_F(\mathbf{Y}||\mathbf{A}\tilde{\mathbf{X}}) - D_F(\mathbf{Y}||\mathbf{A}\mathbf{X}) \leq \sigma \operatorname{tr}\left\{\nabla_X D_F(\mathbf{Y}||\mathbf{A}\mathbf{X})^T(\tilde{\mathbf{X}} - \mathbf{X})\right\}$ *or* $\tilde{\mathbf{X}} = \mathbf{X}$
      /* convergence condition */
    end
  until *a stopping criterion is met* /* convergence condition */
end

5.3 Barzilai-Borwein Gradient Projection for Sparse Reconstruction (GPSR-BB)

The Barzilai-Borwein gradient projection for sparse reconstruction (GPSR) method [1,11] is motivated by the quasi-Newton approach, whereby the inverse of the Hessian is replaced with an identity matrix, that is, $\mathbf{H}_k = \eta_k \mathbf{I}$. The scalar η_k is selected so that the inverse Hessian in the k-th inner iterative step is locally a good approximation to the inverse of the true Hessian for subsequent iterations. Thus

$$\mathbf{X}^{(k+1)} - \mathbf{X}^{(k)} \approx \eta^{(k)} \left(\nabla_{\mathbf{X}} D_F(\mathbf{Y}||\mathbf{A}^{(k)}\mathbf{X}^{(k+1)}) - \nabla_{\mathbf{X}} D_F(\mathbf{Y}||\mathbf{A}^{(k)}\mathbf{X}^{(k)})\right). \tag{5.25}$$

The computation of **A** is performed in a similar way. Although this method does not ensure that the objective function decreases at every iteration, its convergence has been proven analytically [1]. The Barzilai-Borwein gradient projection algorithm for updating **X** is summarized in Algorithm 5.3.

Algorithm 5.3: GPSR-BB

Input: $\mathbf{Y} \in \mathbb{R}_+^{I\times T}$: input data, J: rank of approximation, η_{min}, η_{max}: limit parameters for Hessian approximation,
Output: $\mathbf{A} \in \mathbb{R}_+^{I\times J}$ and $\mathbf{X} \in \mathbb{R}_+^{J\times T}$ such that the cost functions (5.3) and (5.4) are minimized.

```
begin
    initialization for A, X
    repeat
        X ← GPSR(Y, A, X)                                   /* Update X */
        A ← GPSR(Y^T, X^T, A^T)^T                           /* Update A */
    until a stopping criterion is met                       /* convergence condition */
end

function X = GPSR(Y, A, X)
begin
    η = max{η_min, ½η_max}1_T
    repeat
        Δ = [X − ∇_X D_F(Y||AX) diag{η}]_+ − X
        λ = arg min_{λ∈[0,1]^T} D_F(Y||A(X + Δ diag{λ}))     /* λ = [λ_t] ∈ R^T */
        X ← [X + Δ diag{λ}]_+
        η = min{η_max, max{η_min, diag{Δ^T Δ} ⊘ diag{Δ^T A^T A Δ}}}   /* element-wise */
    until a stopping criterion is met                       /* convergence condition */
end
```

Since $D_F(\mathbf{Y}||\mathbf{AX})$ is a quadratic function, the line search parameter $\boldsymbol{\lambda}^{(k)} = [\lambda_t^{(k)}]$ admits a closed-form solution:

$$\lambda_t^{(k)} = \max\left\{0, \min\left\{1, -\frac{\left((\boldsymbol{\Delta}^{(k)})^T \nabla_{\mathbf{X}} D_F(\mathbf{Y}||\mathbf{AX})\right)_{tt}}{\left((\boldsymbol{\Delta}^{(k)})^T \mathbf{A}^T \mathbf{A} \boldsymbol{\Delta}^{(k)}\right)_{tt}}\right\}\right\}, \tag{5.26}$$

where

$$\boldsymbol{\Delta}^{(k)} = \left[\mathbf{X} - \nabla_{\mathbf{X}} D_F(\mathbf{Y}||\mathbf{AX})\, \boldsymbol{\eta}_{\mathbf{X}}\right]_+ - \mathbf{X}. \tag{5.27}$$

The MATLAB implementation of the GPSR-BB algorithm with the line search determined by (5.26) is given in Listing 5.2.

Listing 5.2 GPSR-BB algorithm.

```
function [X] = nmf_gpsr_bb(A,Y,X,no_iter)
%
% INPUTS:
% A - fixed matrix of dimension [I by J]
% Y - data matrix of dimension [I by T]
% X - initial solution matrix of dimension [J by T]
% no_iter - maximum number of iterations
%
% OUTPUTS:
% X - estimated matrix of dimension [J by T]
%
% #######################################################################
eta_min = 1E-8; eta_max = 1;
eta = max(eta_min,.5*eta_max)*ones(1,size(Y,2));
```

```
15    B = A'*A; Yt = A'*Y;
16
17    for k=1:no_iter
18        G = B*X - Yt;
19        Δ = max(eps, X - bsxfun(@times,G,eta)) - X;
20        ΔB = B*Δ;
21        lambda = max(0, min(1, -sum(Δ.*G,1)./(sum(Δ.*ΔB,1) + eps)));
22        X = max(eps,X + bsxfun(@times,Δ,lambda));
23        eta = min(eta_max,max(eta_min,(sum(Δ.^2,1)+eps)./(sum(Δ.*ΔB,1)+ eps)));
24    end % for k
```

The update for $\mathbf{A}$ can be performed by applying the above algorithm to the transposed system: $\mathbf{X}^T\mathbf{A}^T = \mathbf{Y}^T$, with $\mathbf{X}$ computed in the previous alternating step.

5.4 Projected Sequential Subspace Optimization (PSESOP)

The Projected SEquential Subspace OPtimization (PSESOP) method [12,25] performs a projected minimization of a smooth objective function over a subspace spanned by several directions. These include the current gradient and the gradient from previous iterations, together with the Nemirovski directions. Nemirovski [26]

Algorithm 5.4: PSESOP-NMF

Input: $\mathbf{Y} = [\mathbf{y}_1, \ldots, \mathbf{y}_T] \in \mathbb{R}_+^{I \times T}$: input data, $\quad J$: rank of approximation
Output: $\mathbf{A} \in \mathbb{R}_+^{I \times J}$ and $\mathbf{X} = [\mathbf{x}_1, \ldots, \mathbf{x}_T] \in \mathbb{R}_+^{J \times T}$ such that the cost functions (5.3) and (5.4) are
        minimized.

1 **begin**
2   initialization for $\mathbf{A}, \mathbf{X}$
3   **repeat**
4      $\mathbf{X} \leftarrow$ PSESOP$(\mathbf{Y}, \mathbf{A}, \mathbf{X})$ `/* Update X */`
5      $\mathbf{A} \leftarrow$ PSESOP$(\mathbf{Y}^T, \mathbf{X}^T, \mathbf{A}^T)^T$ `/* Update A */`
6   **until** *a stopping criterion is met*   `/* convergence condition */`
7 **end**

8 **function** $\mathbf{X} =$ PSESOP$(\mathbf{Y}, \mathbf{A}, \mathbf{X})$
9 **foreach** $t = 1, \ldots, T$ **do**
10   **begin**
11      $\mathbf{H} = \mathbf{A}^T\mathbf{A}, \mathbf{z}_t = \mathbf{A}^T \mathbf{y}_t$
12      $\mathbf{x}_t = \mathbf{x}_t^{(0)}, \tilde{\mathbf{p}}_t = \mathbf{0}, w_t = 0,$ `/* initialize, J = card(x) */`
13   **repeat**
14        $\bar{\mathbf{p}}_t = \mathbf{x}_t - \mathbf{x}_t^{(0)}$
15        $\mathbf{g}_t = \mathbf{H}\mathbf{x}_t - \mathbf{z}_t$
16        $w_t \leftarrow \frac{1}{2} + \sqrt{\frac{1}{4} + w_t^2}$
17        $\tilde{\mathbf{p}}_t \leftarrow \tilde{\mathbf{p}}_t + w_t \mathbf{g}_t$
18        $\mathbf{P}_{\mathbf{X}} = \left[\bar{\mathbf{p}}_t, \tilde{\mathbf{p}}_t, \mathbf{g}_t\right]$
19        $\boldsymbol{\eta}_* = \arg\min_{\boldsymbol{\eta}} D_F(\mathbf{y}_t || \mathbf{A}(\mathbf{x}_t + \mathbf{P}_{\mathbf{X}}\boldsymbol{\eta}))$ `/* see Eq. (5.30) */`
20        $\mathbf{x}_t \leftarrow [\mathbf{x}_t + \mathbf{P}_{\mathbf{X}}\boldsymbol{\eta}_*]_+$ `/* update x_t */`
21   **until** *a stopping criterion is met*   `/* convergence condition */`
22   **end**
23 **end**

---

showed that convex smooth unconstrained optimization is optimal if the optimization in the k-th iterative step is performed over a subspace which includes the current gradient $\mathbf{g}^{(k)}$, the directions $\bar{\boldsymbol{p}}^{(k)} = \boldsymbol{x}^{(k)} - \boldsymbol{x}^{(0)}$, and the linear combination of the previous gradients $\tilde{\boldsymbol{p}}_1^{(k)} = \sum_{n=0}^{k-1} w_n \mathbf{g}^{(n)}$ with the coefficients

$$w_0 = 0, \tag{5.28}$$

$$w_n = \frac{1}{2} + \sqrt{\frac{1}{4} + w_{n-1}^2}, \qquad n = 1, \ldots, k-1. \tag{5.29}$$

These directions should be orthogonal to the current gradient.[2] Finally, the PSESOP-NMF is expressed by Algorithm 5.4.

A closed-form of the line search vector $\boldsymbol{\eta}_*^{(k)}$ for the objective function $D_F(\boldsymbol{y}_t||\mathbf{A}\boldsymbol{x}_t)$ is given by:

$$\boldsymbol{\eta}_*^{(k)} = -((\mathbf{P}_{\mathbf{X}}^{(k)})^T \mathbf{A}^T \mathbf{A} \mathbf{P}_{\mathbf{X}}^{(k)} + \lambda \mathbf{I})^{-1} (\mathbf{P}_{\mathbf{X}}^{(k)})^T \nabla_{\boldsymbol{x}_t} D_F(\boldsymbol{y}_t||\mathbf{A}\boldsymbol{x}_t), \qquad (t = 1, 2, \ldots, T), \tag{5.30}$$

where $\mathbf{P}_{\mathbf{X}}^{(k)} = [\bar{\boldsymbol{p}}^{(k)}, \tilde{\boldsymbol{p}}^{(k)}, \mathbf{g}^{(k)}]$. The regularization parameter λ is set to a very small value to avoid rank-deficiency in case $\mathbf{P}_{\mathbf{X}}^{(k)}$ has zero-value or dependent columns.

The row vectors of **A** are updated in a similar way. The MATLAB code of the PSESOP algorithm for computing one column vector of **X** is given in Listing 5.3.

Listing 5.3 PSESOP algorithm

```
function x = nmf_psesop(A,y,x0,no_iter)

% Sequential Coordinate-wise Algorithm
%
% [X]=nmf_psesop(A,y,x0,no_iter) finds
% such x that solves the equation Ax = y.
%
% INPUTS:
% A - fixed matrix of dimension [I by J]
% y - data matrix of dimension [J by 1]
% x0 - initial guess vector of dimension [J by 1]
% no_iter - maximum number of iterations
%
% OUTPUTS:
% x - estimated matrix of dimension [J by 1]
%
% ##########################################################################
H = A'*A; z = A'*y;
x = x0; w = 0;
p_tilde = zeros(numel(x),1);
for k=1:no_iter
    g = H*x - z;
    p_bar = x - x0;
    w = .5 + sqrt(.25 + w^2);
    p_tilde = p_tilde + w * g;
    P = [p_bar, p_tilde, g];
    P = bsxfun(@rdivide,P,sqrt(sum(P.^2,1)+ 1E6*eps));
    eta = (P'*H*P + 1E-4*eye(size(P,2)))\(P'*g);
    x = max(1e-10,x - P*eta);
end % for k
```

[2] Narkiss and Zibulevsky [25] proposed to include another direction: $\boldsymbol{p}^{(k)} = \boldsymbol{x}^{(k)} - \boldsymbol{x}^{(k-1)}$, which is motivated by a natural extension of the Conjugate Gradient (CG) method to a nonquadratic case. However, in our extension this direction does not have a strong impact on the NMF components, and was neglected in our PSESOP-NMF algorithm.

5.5 Interior Point Gradient (IPG) Algorithm

The Interior Point Gradient (IPG) algorithm was proposed by Merritt and Zhang in [23] for solving the NNLS problems, and in this section it is applied to NMF. This algorithm is based on the scaled gradient descent method, where the descent direction $\mathbf{P_X}$ for $\mathbf{X}$ is determined using a negative scaled gradient, that is,

$$\mathbf{P_X} = -\mathbf{D} \circledast \nabla_{\mathbf{X}} D_F(\mathbf{Y}||\mathbf{AX}), \tag{5.31}$$

with the scaling vector

$$\mathbf{D} = \mathbf{X} \oslash (\mathbf{A}^T\mathbf{AX}), \tag{5.32}$$

where the symbols $\circledast$ and $\oslash$ denote component-wise multiplication and division, respectively. The IPG algorithm can be written in the compact forms

$$\mathbf{A} \leftarrow \mathbf{A} - \eta_{\mathbf{A}}\ [\mathbf{A} \oslash (\mathbf{AXX}^T)] \circledast [(\mathbf{Y} - \mathbf{AX})\mathbf{X}^T], \tag{5.33}$$

$$\mathbf{X} \leftarrow \mathbf{X} - \eta_{\mathbf{X}}\ [\mathbf{X} \oslash (\mathbf{A}^T\mathbf{AX})] \circledast [\mathbf{A}^T(\mathbf{Y} - \mathbf{AX})], \tag{5.34}$$

where $\eta_{\mathbf{A}} > 0$ and $\eta_{\mathbf{X}} > 0$ are suitably chosen learning rates. Finally, we have the IPG-NMF given in Algorithm 5.5.

Algorithm 5.5: IPG-NMF

Input: $\mathbf{Y} \in \mathbb{R}_+^{I\times T}$: input data, J: rank of approximation
Output: $\mathbf{A} \in \mathbb{R}_+^{I\times J}$ and $\mathbf{X} \in \mathbb{R}_+^{J\times T}$ such that the cost functions (5.3) and (5.4) are minimized.

begin
  initialization for $\mathbf{A}, \mathbf{X}$
  repeat
    $\mathbf{X} \leftarrow \mathtt{IPG}(\mathbf{Y}, \mathbf{A}, \mathbf{X})$ `/* Update X */`
    $\mathbf{A} \leftarrow \mathtt{IPG}(\mathbf{Y}^T, \mathbf{X}^T, \mathbf{A}^T)^T$ `/* Update A */`
  until *a stopping criterion is met* `/* convergence condition */`
end

function $\mathbf{X} = \mathtt{IPG}(\mathbf{Y}, \mathbf{A}, \mathbf{X})$
begin
  repeat
    $\nabla_{\mathbf{X}} D_F(\mathbf{Y}||\mathbf{AX}) \leftarrow \mathbf{A}^T(\mathbf{AX} - \mathbf{Y})$
    $\mathbf{D} \leftarrow \mathbf{X} \oslash (\mathbf{A}^T\mathbf{AX})$
    $\mathbf{P_X} \leftarrow -\mathbf{D} \circledast \nabla_{\mathbf{X}} D_F(\mathbf{Y}||\mathbf{AX})$
    $\eta_{\mathbf{X}}^* = -\dfrac{\mathrm{vec}(\mathbf{P_X})^T \mathrm{vec}(\nabla_{\mathbf{X}} D_F(\mathbf{Y}||\mathbf{AX}))}{\mathrm{vec}(\mathbf{AP_X})^T \mathrm{vec}(\mathbf{AP_X})}$
    $\hat{\eta}_{\mathbf{X}} = \max\{\eta_{\mathbf{X}} : \mathbf{X} + \eta_{\mathbf{X}}\mathbf{P_X} \geq \mathbf{0}\}$
    $\eta_{\mathbf{X}} = \min(\hat{\tau}\hat{\eta}_{\mathbf{X}}, \eta_{\mathbf{X}}^*)$ `/*` $\hat{\tau} \in [0, 1)$ `*/`
    $\mathbf{X} \leftarrow \mathbf{X} + \eta_{\mathbf{X}}\mathbf{P_X}$
  until *a stopping criterion is met* `/* convergence condition */`
end

In interior-point gradient methods, the learning rates are adjusted in each iteration to keep the matrices positive. They are chosen so as to be close to $\eta_{\mathbf{X}}^*$ and $\eta_{\mathbf{A}}^*$, which are the exact minimizers of $D_F(\mathbf{Y}||\mathbf{A}(\mathbf{X} + \eta_{\mathbf{X}}\mathbf{P_X}))$ and $D_F(\mathbf{Y}||(\mathbf{A} + \eta_{\mathbf{A}}\mathbf{P_A})\mathbf{X})$, and also to keep some distance from the boundary of the nonnegative orthant. The matrices $\mathbf{P_A}$ and $\mathbf{P_X}$ determine descent directions, $\hat{\eta}_\mathbf{A}$ and $\hat{\eta}_\mathbf{X}$ are the longest step-sizes that assure the nonnegative solutions, and $\eta_\mathbf{A}$ and $\eta_\mathbf{X}$ are current step-sizes.

The MATLAB code for the IPG algorithm is given in Listing 5.4.

Listing 5.4 Interior-Point Gradient algorithm.

```
function X = ipg(A,Y,X,IterMax)
%
% Interior-Point Method
% initialization
B = A'*A; BX = B*X; Z = A' * Y;
%
for k = 1:IterMax
%    step computation
     G = BX - Z;
     P = (X./max(BX,eps)).*G; BP = B*P;

%    steplengths
     eta1 = sum(P.*G,1)./max(sum(P.*BP,1), eps);
     eta2 = -1./min(-eps,min(-P./max(X,eps),[ ],1) );
     eta = min([.99*eta2; eta1],[ ],1);

%    updates
     X = X - bsxfun(@times,P,eta);
     BX = BX - bsxfun(@times,BP,eta);
end
```

5.6 Interior Point Newton (IPN) Algorithm

The Interior Point Newton (IPN) algorithm [2] solves the NNLS problem (5.1) by searching for the solution that satisfies the KKT conditions (5.5). This can be equivalently expressed by the nonlinear equation:

$$\mathbf{D}(\boldsymbol{x}_t)\boldsymbol{g}(\boldsymbol{x}_t) = \mathbf{0}, \tag{5.35}$$

where $\mathbf{D}(\boldsymbol{x}_t) = \text{diag}\{d_1(\boldsymbol{x}_t), \ldots, d_J(\boldsymbol{x}_t)\}$, $\boldsymbol{x}_t \geq \mathbf{0}$, and

$$d_j(\boldsymbol{x}_t) = \begin{cases} x_{jt}, & \text{if} \quad g_j(\boldsymbol{x}_t) \geq 0, \\ 1, & \text{otherwise.} \end{cases} \tag{5.36}$$

Applying the Newton method to (5.35) for the k-th iterative step we have:

$$(\mathbf{D}_k(\boldsymbol{x}_t)\mathbf{A}^T\mathbf{A} + \mathbf{E}_k(\boldsymbol{x}_t))\boldsymbol{p}_k = -\mathbf{D}_k(\boldsymbol{x}_t)\boldsymbol{g}_k(\boldsymbol{x}_t), \tag{5.37}$$

where

$$\mathbf{E}_k(\boldsymbol{x}_t) = \text{diag}\{e_1(\boldsymbol{x}_t), \ldots, e_J(\boldsymbol{x}_t)\}. \tag{5.38}$$

Following [2], the entries of the matrix $\mathbf{E}_k(\boldsymbol{x}_t)$ are defined by

$$e_j(\boldsymbol{x}_t) = \begin{cases} g_j(\boldsymbol{x}_t), & \text{if} \qquad 0 \leq g_j(\boldsymbol{x}_t) < x_{jt}^{\gamma}, \text{ or } (g_j(\boldsymbol{x}_t))^{\gamma} > x_{jt}, \\ 0, & \text{otherwise} \end{cases} \tag{5.39}$$

for $1 < \gamma \leq 2$.

If the solution is degenerate, i.e. $t = 1, \ldots, T,\ \exists j : x_{jt}^* = 0,$ and $g_{jt} = 0$, the matrix $\mathbf{D}_k(\boldsymbol{x}_t)\mathbf{A}^T\mathbf{A} + \mathbf{E}_k(\boldsymbol{x}_t)$ may be singular. To circumvent this problem, the system of equations can be re-scaled to the following form:

$$\mathbf{W}_k(\boldsymbol{x}_t)\mathbf{D}_k(\boldsymbol{x}_t)\mathbf{M}_k(\boldsymbol{x}_t)\boldsymbol{p}_k = -\mathbf{W}_k(\boldsymbol{x}_t)\mathbf{D}_k(\boldsymbol{x}_t)\boldsymbol{g}_k(\boldsymbol{x}_t), \tag{5.40}$$

with

$$\mathbf{M}_k(\boldsymbol{x}_t) = \mathbf{A}^T\mathbf{A} + \mathbf{D}_k(\boldsymbol{x}_t)^{-1}\mathbf{E}_k(\boldsymbol{x}_t), \tag{5.41}$$

$$\mathbf{W}_k(\boldsymbol{x}_t) = \mathrm{diag}\{w_1(\boldsymbol{x}_t), \ldots, w_J(\boldsymbol{x}_t)\}, \quad w_j(\boldsymbol{x}_t) = (d_j(\boldsymbol{x}_t) + e_j(\boldsymbol{x}_t))^{-1}, \tag{5.42}$$

for $\boldsymbol{x}_t > 0$. In [2], the system (5.40) is solved by the inexact Newton method, which leads to the following updates:

$$\mathbf{W}_k(\boldsymbol{x}_t)\mathbf{D}_k(\boldsymbol{x}_t)\mathbf{M}_k(\boldsymbol{x}_t)\boldsymbol{p}_k = -\mathbf{W}_k(\boldsymbol{x}_t)\mathbf{D}_k(\boldsymbol{x}_t)\boldsymbol{g}_k(\boldsymbol{x}_t) + \boldsymbol{r}_k(\boldsymbol{x}_t), \tag{5.43}$$

$$\hat{\boldsymbol{p}}_k = \max\{\sigma, 1 - ||[\boldsymbol{x}_t^{(k)} + \boldsymbol{p}_k]_+ - \boldsymbol{x}_t^{(k)}||_2\}\left([\boldsymbol{x}_t^{(k)} + \boldsymbol{p}_k]_+ - \boldsymbol{x}_t^{(k)}\right), \tag{5.44}$$

$$\boldsymbol{x}_t^{(k+1)} = \boldsymbol{x}_t^{(k)} + \hat{\boldsymbol{p}}_k, \tag{5.45}$$

where $\sigma \in (0, 1)$, $\boldsymbol{r}_k(\boldsymbol{x}_t) = \mathbf{A}^T(\mathbf{A}\boldsymbol{x}_t - \boldsymbol{y}_t)$.

The transformation of the normal matrix $\mathbf{A}^T\mathbf{A}$ by the matrix $\mathbf{W}_k(\boldsymbol{x}_t)\mathbf{D}_k(\boldsymbol{x}_t)$ in (5.40) causes the system matrix $\mathbf{W}_k(\boldsymbol{x}_t)\mathbf{D}_k(\boldsymbol{x}_t)\mathbf{M}_k(\boldsymbol{x}_t)$ to be no longer symmetric and positive-definite. There are many methods for handling such systems of linear equations, including QMR [15], BiCG [13,20], BiCGSTAB [29], or GMRES-like methods [28]. They are, however, more complicated and computationally demanding than, e.g. the basic CG algorithm [17]. To apply the CG algorithm, the system matrix in (5.40) must be converted to a positive-definite symmetric matrix, for instance by the normal equations. The transformed system has the form:

$$\mathbf{Z}_k(\boldsymbol{x}_t)\tilde{\boldsymbol{p}}_k = -\mathbf{S}_k(\boldsymbol{x}_t)\boldsymbol{g}_k(\boldsymbol{x}_t) + \tilde{\boldsymbol{r}}_k(\boldsymbol{x}_t), \tag{5.46}$$

$$\mathbf{S}_k(\boldsymbol{x}_t) = \sqrt{\mathbf{W}_k(\boldsymbol{x}_t)\mathbf{D}_k(\boldsymbol{x}_t)}, \tag{5.47}$$

$$\mathbf{Z}_k(\boldsymbol{x}_t) = \mathbf{S}_k(\boldsymbol{x}_t)\mathbf{M}_k(\boldsymbol{x}_t)\mathbf{S}_k(\boldsymbol{x}_t) = \mathbf{S}_k(\boldsymbol{x}_t)\mathbf{A}^T\mathbf{A}\mathbf{S}_k(\boldsymbol{x}_t) + \mathbf{W}_k(\boldsymbol{x}_t)\mathbf{E}_k(\boldsymbol{x}_t), \tag{5.48}$$

with $\tilde{\boldsymbol{p}}_k = \mathbf{S}_k^{-1}(\boldsymbol{x}_t)\boldsymbol{p}_k$ and $\tilde{\boldsymbol{r}}_k = \mathbf{S}_k^{-1}(\boldsymbol{x}_t)\boldsymbol{r}_k(\boldsymbol{x}_t)$. The methods such as CGLS [16] or LSQR [27] are naturally suited for these tasks.

Since our cost function is quadratic, its minimization in a single step is performed by combining the projected Newton step with the constrained scaled Cauchy step, given by:

$$\boldsymbol{p}_k^{(C)} = -\tau_k\mathbf{D}_k(\boldsymbol{x}_t)\boldsymbol{g}_k(\boldsymbol{x}_t), \quad \tau_k > 0. \tag{5.49}$$

Assuming $\boldsymbol{x}_t^{(k)} + \boldsymbol{p}_k^{(C)} > \mathbf{0}$, τ_k is chosen to be either an unconstrained minimizer of the quadratic function $\psi_k(-\tau_k\mathbf{D}_k(\boldsymbol{x}_t)\boldsymbol{g}_k(\boldsymbol{x}_t))$ or a scalar proportional to the distance from the boundary along $-\mathbf{D}_k(\boldsymbol{x}_t)\boldsymbol{g}_k(\boldsymbol{x}_t)$, where

$$\begin{aligned}\psi_k(\boldsymbol{p}) &= \frac{1}{2}\boldsymbol{p}^T\mathbf{M}_k(\boldsymbol{x}_t)\boldsymbol{p} + \boldsymbol{p}^T\boldsymbol{g}_k(\boldsymbol{x}_t)\\ &= \frac{1}{2}\boldsymbol{p}^T(\mathbf{A}^T\mathbf{A} + \mathbf{D}_k^{-1}(\boldsymbol{x}_t)\mathbf{E}_k(\boldsymbol{x}_t))\boldsymbol{p} + \boldsymbol{p}^T\mathbf{A}^T(\mathbf{A}\boldsymbol{x}_t^{(k)} - \boldsymbol{y}_t).\end{aligned} \tag{5.50}$$

Thus

$$\tau_k = \begin{cases} \tau_1 = \arg\min_\tau \psi_k(-\tau_k\mathbf{D}_k(\boldsymbol{x}_t)\boldsymbol{g}_k(\boldsymbol{x}_t)) \;\text{ if }\; \boldsymbol{x}_t^{(k)} - \tau_1\mathbf{D}_k(\boldsymbol{x}_t)\boldsymbol{g}_k(\boldsymbol{x}_t) > \mathbf{0}, \\ \tau_2 = \theta\min\limits_j\left\{\dfrac{x_{jt}^{(k)}}{\left(\mathbf{D}_k(\boldsymbol{x}_t)\boldsymbol{g}_k(\boldsymbol{x}_t)\right)_j} : (\mathbf{D}_k(\boldsymbol{x}_t)\boldsymbol{g}_k(\boldsymbol{x}_t))_j > \mathbf{0}\right\} \;\text{ otherwise}\end{cases} \tag{5.51}$$

where $\psi_k(-\tau_k \mathbf{D}_k(\boldsymbol{x}_t)\boldsymbol{g}_k(\boldsymbol{x}_t)) = \dfrac{\left(\boldsymbol{g}_k(\boldsymbol{x}_t)\right)^T \mathbf{D}_k(\boldsymbol{x}_t)\boldsymbol{g}_k(\boldsymbol{x}_t)}{\left(\mathbf{D}_k(\boldsymbol{x}_t)\boldsymbol{g}_k(\boldsymbol{x}_t)\right)^T \mathbf{M}_k(\boldsymbol{x}_t)\mathbf{D}_k(\boldsymbol{x}_t)\boldsymbol{g}_k(\boldsymbol{x}_t)}$ with $\theta \in (0, 1)$. For $\psi_k(\boldsymbol{p}_k^{(C)}) < 0$, global convergence is achieved if $\text{red}(\boldsymbol{x}_t^{(k+1)} - \boldsymbol{x}_t^{(k)}) \geq \beta$, $\beta \in (0, 1)$, with

$$\text{red}(\boldsymbol{p}) = \frac{\psi_k(\boldsymbol{p})}{\psi_k(\boldsymbol{p}_k^{(C)})}. \tag{5.52}$$

The usage of the constrained scaled Cauchy step leads to the following updates:

$$\boldsymbol{s}_t^{(k)} = t(\boldsymbol{p}_k^{(C)} - \hat{\boldsymbol{p}}_k) + \hat{\boldsymbol{p}}_k, \tag{5.53}$$

$$\boldsymbol{x}_t^{(k+1)} = \boldsymbol{x}_t^{(k)} + \boldsymbol{s}_t^{(k)}, \tag{5.54}$$

where $\hat{\boldsymbol{p}}_k$ and $\boldsymbol{p}_k^{(C)}$ are given in (5.44) and (5.49), and $t \in [0, 1)$ is the smaller square root of the quadratic equation:

$$\pi(t) = \psi_k\left(t(\boldsymbol{p}_k^{(C)} - \hat{\boldsymbol{p}}_k) + \hat{\boldsymbol{p}}_k\right) - \beta\psi_k(\boldsymbol{p}_k^{(C)}) = 0. \tag{5.55}$$

The MATLAB code for the IPN algorithm which computes a single column vector of $\mathbf{X}$ is given in Listing 5.5

Listing 5.5 Interior-Point Newton (IPN) algorithm.

```
function x = nmf_ipn(A,y,x,no_iter)

%% Interior-Point Newton (IPN) algorithm solves the equation Ax = y
%% for nonnegative data.
%% INPUTS:
%% A - fixed matrix of dimension [I by J]
%% y - data vector of dimension [I by 1]
%% x - initial guess of dimension [J by 1]
%% no_iter - maximum number of iterations
%%
%% OUTPUTS:
%% x - estimated solution vector of dimension [J by 1]
%% #########################################################################
H = A'*A; z = A'*y; J = size(x,1);
% Parameters
s = 1.8; theta = 0.5; rho = .1; beta = 1;

for k=1:no_iter
    g = H*x - z;
    d = ones(J,1);  d(g ≥ 0) = x(g ≥ 0);
    ek = zeros(J,1); ek(g ≥0 & g < x.^s) = g(g ≥0 & g < x.^s);

    M = H + diag(ek./d);
    dg = d.*g;

    tau1 = (g'*dg)/(dg'*M*dg); tau = tau1(ones(J,1));
    tau(x - tau1*dg ≤ 0) = theta*min(x(dg > 0)./dg(dg > 0));

    sk2 = d./(d + ek); sk = sqrt(sk2);
    pc = - tau.*dg;

    Z = bsxfun(@times,bsxfun(@times,M,sk),sk');
    [p,flag]= lsqr(Z,-g.*(1+sk2),1E-8);

```

```
    phx = max(0, x + p) - x; ph = max(rho, 1 - norm(phx))*phx;

    Phi_pc = .5*pc'*M*pc + pc'*g;
    Phi_ph = .5*ph'*M*ph + ph'*g;
    red_p = Phi_ph/Phi_pc;
    dp = pc - ph;

    t = 0;
    if red_p < beta
        ax = .5*dp'*M*dp;
        bx = dp'*(M*ph + g);
        cx = Phi_ph - beta*Phi_pc;
        Deltas = sqrt(bx^2 - 4*ax*cx);

        t1 = .5*(-bx + Deltas)/ax;
        t2 = .5*(-bx - Deltas)/ax;

        t = min(t1, t2);
        if (t ≤ 0)
            t =(t1 > 0 & t1 < 1) | (t2 > 0 & t2 < 1);
        end
    end
    x = x + t*dp + ph;
end % for k
```

5.7 Regularized Minimal Residual Norm Steepest Descent Algorithm (RMRNSD)

The Minimal Residual Norm Steepest Descent (MRNSD) algorithm, proposed by Nagy and Strakos [24] for image restoration problems is based on the EMLS algorithm developed by Kaufman [19]. The original MRNSD solves the following problem (assuming that the basis matrix $\mathbf{A}$ is known)

$$D_F(\boldsymbol{y}_t||\mathbf{A}\boldsymbol{x}_t) = \frac{1}{2}||\boldsymbol{y}_t - \mathbf{A}\boldsymbol{x}_t||_2^2,$$

$$\text{subject to} \quad \boldsymbol{x}_t \geq \mathbf{0}, \quad t = 1, \ldots, T. \tag{5.56}$$

The nonnegativity is preserved by the nonlinear transformation $\boldsymbol{x}_t = \exp\{\boldsymbol{z}_t\}$, so that a stationary point of $D_F(\boldsymbol{y}_t||\mathbf{A}\exp\{\boldsymbol{z}_t\})$

$$\begin{aligned} \nabla_{\boldsymbol{z}_t} D_F(\boldsymbol{y}_t||\mathbf{A}\boldsymbol{x}_t) &= \operatorname{diag}(\boldsymbol{x}_t)\nabla_{\boldsymbol{x}_t} D_F(\boldsymbol{y}_t||\mathbf{A}\boldsymbol{x}_t) \\ &= \operatorname{diag}(\boldsymbol{x}_t)\mathbf{A}^T(\mathbf{A}\boldsymbol{x}_t - \boldsymbol{y}_t) = \mathbf{0} \end{aligned} \tag{5.57}$$

satisfies the KKT conditions. The solution is updated as:

$$\boldsymbol{p}_t \leftarrow \operatorname{diag}(\boldsymbol{x}_t)\mathbf{A}^T(\mathbf{A}\boldsymbol{x}_t - \boldsymbol{y}_t), \tag{5.58}$$

$$\boldsymbol{x}_t \leftarrow \boldsymbol{x}_t + \eta_{\mathbf{X}}\boldsymbol{p}_t, \tag{5.59}$$

where $\eta_{\mathbf{X}} > 0$ is a learning rate scalar, adjusted in each iteration to maintain the nonnegativity and ensure a decrease in the cost function. Similarly to the IPG algorithm described in Section 5.5, the learning rate is selected as the smallest value to provide the steepest descent along the direction $\boldsymbol{p}_t$, that is,

$$\min_{\eta_{\mathbf{X}}} \left\{ D_F(\boldsymbol{y}_t||\mathbf{A}(\boldsymbol{x}_t + \eta_{\mathbf{X}}\boldsymbol{p}_t)) \right\},$$

and the other is selected as

$$\theta \min_{p_j>0} \left\{ -\frac{x_j}{p_j} \right\} = \theta \min_{g_j>0} \{g_j\} \tag{5.60}$$

for $\theta \approx 0.99$ which keeps the positive updates close to the nonnegative orthant.

The MRNSD algorithm can also be regarded as the linear residual norm steepest descent algorithm applied to a "Jacobi-like preconditioned" (right-side preconditioned) system:

$$\hat{\mathbf{A}}\hat{\boldsymbol{x}}_t = \boldsymbol{y}_t, \quad \hat{\mathbf{A}} = \mathbf{A}\sqrt{\mathbf{D}}, \quad \boldsymbol{x}_t = \sqrt{\mathbf{D}}\hat{\boldsymbol{x}}_t, \tag{5.61}$$

where $\mathbf{D} = \operatorname{diag}\{\boldsymbol{x}_t\}$.

We shall now apply the MRNSD approach to the regularized cost function:

$$D_F^{(\alpha_X)}(\mathbf{Y}||\mathbf{AX}) = \frac{1}{2}\|\mathbf{Y} - \mathbf{AX}\|_F^2 + \alpha_X J_X(\mathbf{X}), \tag{5.62}$$

where both the matrices $\mathbf{A}$ and $\mathbf{X}$ are unknown, and the regularization term $J_X(\mathbf{X}) = \sum_{j=1}^{J}\sum_{t=1}^{T} x_{jt}$ is introduced in order to increase the degree of sparsity of the solution matrix $\mathbf{X}$.

The RMRNSD algorithm for NMF is summarized in details in Algorithm 5.6.

Algorithm 5.6: MRNSD-NMF

Input: $\mathbf{Y} \in \mathbb{R}_+^{I\times T}$: input data,
J: rank of approximation, α_A, α_X: regularized parameters
Output: $\mathbf{A} \in \mathbb{R}_+^{I\times J}$ and $\mathbf{X} \in \mathbb{R}_+^{J\times T}$ such that the cost functions (5.3) and (5.4) are minimized.

begin
  initialization for $\mathbf{A}$, $\mathbf{X}$
  repeat
    $\mathbf{X} \leftarrow \text{MRNSD}(\mathbf{Y}, \mathbf{A}, \mathbf{X}, \alpha_X)$ /* Update X */
    $\mathbf{A} \leftarrow \text{MRNSD}(\mathbf{Y}^T, \mathbf{X}^T, \mathbf{A}^T, \alpha_A)^T$ /* Update A */
  until *a stopping criterion is met* /* convergence condition */
end

function $\mathbf{X} = \text{MRNSD}(\mathbf{Y}, \mathbf{A}, \mathbf{X}, \alpha)$
begin
  $\mathbf{G} = \nabla_X D_F(\mathbf{Y}||\mathbf{AX}) = \mathbf{A}^T(\mathbf{AX} - \mathbf{Y}) + \alpha$
  repeat
    $\gamma = \mathbf{1}_J^T[\mathbf{G} \circledast \mathbf{X} \circledast \mathbf{G}]\mathbf{1}_T$
    $\mathbf{P}_X = -\mathbf{X} \circledast \mathbf{G}, \ \mathbf{U} = \mathbf{AP}_X$
    $\eta_X = \min\left\{\gamma/\|\mathbf{U}\|_F^2, \min_{g_{jt}>0}(\mathbf{G})\right\}$
    $\mathbf{X} \leftarrow \mathbf{X} + \eta_X \mathbf{P}_X$
    $\mathbf{G} \leftarrow \mathbf{G} + \eta_X \mathbf{A}^T\mathbf{U}$
  until *a stopping criterion is met* /* convergence condition */
end

5.8 Sequential Coordinate-Wise Algorithm (SCWA)

Another way to express the NNLS problem (5.1) is as a constrained Quadratic Problem (QP) [14]:

$$\min_{\boldsymbol{x}_t \geq \mathbf{0}} \Psi(\boldsymbol{x}_t), \quad (t = 1, \dots, T), \tag{5.63}$$

where

$$\Psi(\boldsymbol{x}_t) = \frac{1}{2} \boldsymbol{x}_t^T \mathbf{H} \boldsymbol{x}_t + \boldsymbol{c}_t^T \boldsymbol{x}_t, \tag{5.64}$$

with $\mathbf{H} = \mathbf{A}^T \mathbf{A}$ and $\boldsymbol{c}_t = -\mathbf{A}^T \boldsymbol{y}_t$.

The Sequential Coordinate-Wise Algorithm (SCWA) proposed by V. Franc. *et al.* [14] solves the QP problem given by (5.63) updating only a single variable x_{jt} at each iterative step. The sequential updates can be performed easily after rewriting the function $\Psi(\boldsymbol{x}_t)$ as:

$$\begin{aligned}
\Psi(\boldsymbol{x}_t) &= \frac{1}{2} \sum_{p \in \mathcal{I}} \sum_{r \in \mathcal{I}} x_{pt} x_{rt} (\mathbf{A}^T \mathbf{A})_{pr} + \sum_{p \in \mathcal{I}} x_{pt} (\mathbf{A}^T \boldsymbol{y}_t)_{pt} \\
&= \frac{1}{2} x_{jt}^2 (\mathbf{A}^T \mathbf{A})_{jj} + x_{jt} (\mathbf{A}^T \boldsymbol{y}_t)_{jt} + x_{jt} \sum_{p \in \mathcal{I} \setminus \{j\}} x_{pt} (\mathbf{A}^T \mathbf{A})_{pj} + \sum_{p \in \mathcal{I} \setminus \{j\}} x_{pt} (\mathbf{A}^T \boldsymbol{y}_t)_{pt} \\
&+ \frac{1}{2} \sum_{p \in \mathcal{I} \setminus \{j\}} \sum_{r \in \mathcal{I} \setminus \{j\}} x_{pt} x_{rt} (\mathbf{A}^T \mathbf{A})_{pr} = \frac{1}{2} x_{jt}^2 h_{jj} + x_{jt} \beta_{jt} + \gamma_{jt},
\end{aligned} \tag{5.65}$$

where $\mathcal{I} = \{1, \dots, J\}$, $\mathcal{I} \setminus \{j\}$ denotes all the entries of $\mathcal{I}$ except for the j-th entry, and

$$h_{jj} = (\mathbf{A}^T \mathbf{A})_{jj}, \tag{5.66}$$

$$\beta_{jt} = (\mathbf{A}^T \boldsymbol{y}_t)_{jt} + \sum_{p \in \mathcal{I} \setminus \{j\}} x_{pt} (\mathbf{A}^T \mathbf{A})_{pj} = [\mathbf{A}^T \mathbf{A} \boldsymbol{x}_t + \mathbf{A}^T \boldsymbol{y}_t]_{jt} - (\mathbf{A}^T \mathbf{A})_{jj} x_{jt}, \tag{5.67}$$

$$\gamma_{jt} = \sum_{p \in \mathcal{I} \setminus \{j\}} x_{pt} (\mathbf{A}^T \boldsymbol{y}_t)_{pt} + \frac{1}{2} \sum_{p \in \mathcal{I} \setminus \{j\}} \sum_{r \in \mathcal{I} \setminus \{j\}} x_{pt} x_{rt} (\mathbf{A}^T \mathbf{A})_{pr}. \tag{5.68}$$

The optimization of $\Psi(\boldsymbol{x}_t)$ with respect to the selected variable x_{jt} gives the following analytical solution:

$$\begin{aligned}
x_{jt}^* &= \arg\min_{x_{jt}} \Psi\left([x_{1t}, \dots, x_{jt}, \dots, x_{Jt}]^T\right) \\
&= \arg\min_{x_{jt}} \frac{1}{2} x_{jt}^2 h_{jj} + x_{jt} \beta_{jt} + \gamma_{jt} \\
&= \max_{x_{jt}} \left(0, -\frac{\beta_{jt}}{h_{jj}}\right) \\
&= \max_{x_{jt}} \left(0, x_{jt} - \frac{[\mathbf{A}^T \mathbf{A} \boldsymbol{x}_t]_{jt} + [\mathbf{A}^T \boldsymbol{y}_t]_{jt}}{(\mathbf{A}^T \mathbf{A})_{jj}}\right).
\end{aligned} \tag{5.69}$$

When updating only a single variable x_{jt} in one iterative step, we have

$$x_{pt}^{(k+1)} = x_{pt}^{(k)}, \quad \forall p \in \mathcal{I} \setminus \{j\} \quad \text{and} \quad x_{jt}^{(k+1)} \neq x_{jt}^{(k)}. \tag{5.70}$$

It should be noted that every optimal solution to the QP (5.63) satisfies the KKT conditions given in (5.5) and the stationarity condition of the following Lagrange function:

$$\mathcal{L}(\boldsymbol{x}_t, \boldsymbol{\lambda}_t) = \frac{1}{2}\boldsymbol{x}_t^T \mathbf{H} \boldsymbol{x}_t + \boldsymbol{c}_t^T \boldsymbol{x}_t - \boldsymbol{\lambda}_t^T \boldsymbol{x}_t, \tag{5.71}$$

where $\boldsymbol{\lambda}_t \in \mathbb{R}^J$ is a vector of Lagrange multipliers (or dual variables) corresponding to the vector $\boldsymbol{x}_t$, thus $\nabla_{\boldsymbol{x}_t} \mathcal{L}(\boldsymbol{x}_t, \boldsymbol{\lambda}_t) = \mathbf{H}\boldsymbol{x}_t + \boldsymbol{c}_t - \boldsymbol{\lambda}_t = \mathbf{0}$. In the SCWA, the Lagrange multipliers are updated for each iteration according to:

$$\boldsymbol{\lambda}_t^{(k+1)} = \boldsymbol{\lambda}_t^{(k)} + \left(x_{jt}^{(k+1)} - x_{jt}^{(k)}\right) \boldsymbol{h}_j, \tag{5.72}$$

where $\boldsymbol{h}_j$ is the j-th column of $\mathbf{H}$, and $\boldsymbol{\lambda}_t^{(0)} = \boldsymbol{c}_t$.

We finally arrive at the SCWA updates:

$$x_{jt}^{(k+1)} = \left[x_{jt}^{(k)} - \frac{\lambda_j^{(k)}}{(\mathbf{A}^T\mathbf{A})_{jj}} \right]_+ \quad \text{and} \quad x_{pt}^{(k+1)} = x_{pt}^{(k)}, \quad \forall p \in \mathcal{I}\backslash\{j\}, \tag{5.73}$$

$$\boldsymbol{\lambda}_t^{(k+1)} = \boldsymbol{\lambda}_t^{(k)} + \left(x_{jt}^{(k+1)} - x_{jt}^{(k)}\right) \boldsymbol{h}_j, \tag{5.74}$$

The SCWA algorithm is very similar (and in fact equivalent) to standard HALS algorithm presented in Chapter 4. This algorithm can be extended to handle systems with Multiple Measurement Vectors (MMV), which is the case in NMF. The MATLAB code of the batch-mode algorithm for NMF is listed in Listing 5.6.

Listing 5.6 Sequential Coordinate-Wise (SCW) Algorithm.

```
function X = nmf_scwa(A,Y,no_iter)
%
% Sequential Coordinate-Wise Algorithm
% [X]=nmf_scwa(A,Y,no_iter) finds
% such a nonnegative matrix X that solves the equation AX = Y.
%
% INPUTS:
% A - fixed matrix of dimension [I by J]
% Y - data matrix of dimension [I by T]
% no_iter - maximum number of iterations
%
% OUTPUTS:
% X - estimated solution matrix of dimension [J by T]
%
% #########################################################################
H = A'*A; h_diag = diag(H);
X = zeros(size(A,2),size(Y,2)); lambda=-A'*Y;

for k=1:no_iter
    for j = 1:size(A,2)
        xj = X(j,:);
        X(j,:) = max(eps,X(j,:) - lambda(j,:)./h_diag(j));
        lambda = lambda + H(:,j)*(X(j,:) - xj);
    end
end % for k
```

5.9 Simulations

All the algorithms were implemented in NMFLAB [8], and evaluated using the numerical tests related to standard BSS problems. We used three types of synthetic nonnegative signals to test the PG algorithms:

- **Benchmark A**: Four partially dependent nonnegative signals (with only $T = 1000$ samples) illustrated in Figure 5.1(a). The signals were mixed using a random, uniformly distributed, nonnegative matrix $\mathbf{A} \in \mathbb{R}^{8\times4}$ with the condition number cond$\{\mathbf{A}\} = 4.11$, given by:

$$\mathbf{A} = \begin{bmatrix} 0.0631 & 0.7666 & 0.0174 & 0.6596 \\ 0.2642 & 0.6661 & 0.8194 & 0.2141 \\ 0.9995 & 0.1309 & 0.6211 & 0.6021 \\ 0.2120 & 0.0954 & 0.5602 & 0.6049 \\ 0.4984 & 0.0149 & 0.2440 & 0.6595 \\ 0.2905 & 0.2882 & 0.8220 & 0.1834 \\ 0.6728 & 0.8167 & 0.2632 & 0.6365 \\ 0.9580 & 0.9855 & 0.7536 & 0.1703 \end{bmatrix}. \tag{5.75}$$

  The mixing signals are shown in Figure 5.1(b).
- **Benchmark B**: Five spectral signals (`X_spectra` dataset) that are presented in Figure 3.11(a) in Chapter 3. The source signals were mixed with a random, uniformly distributed, nonnegative matrix. We analyzed two cases of the number of observations: $I = 10$ and $I = 1000$.
- **Benchmark C**: Ten strongly dependent signals that are illustrated in Figure 4.8 in Chapter 4. Similarly as above, a random, uniformly distributed, nonnegative matrix was used to mix the signals.

For the benchmark A and the mixing matrix in (5.75), the number of variables in $\mathbf{X}$ is much higher than that in $\mathbf{A}$, that is, $I \times J = 32$ and $J \times T = 4000$, hence we tested the projected gradient algorithms only on

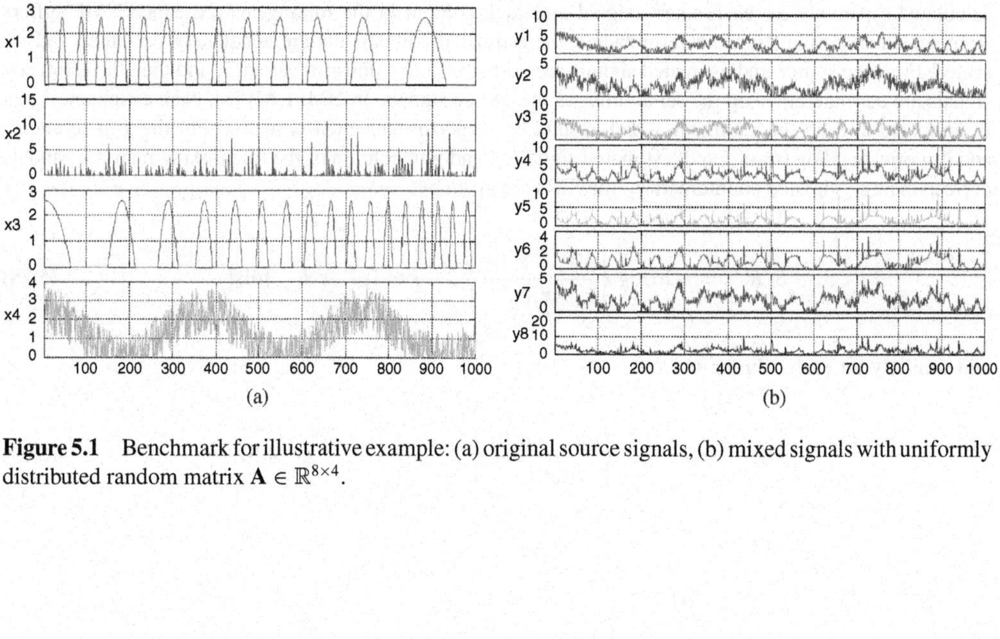

Figure 5.1 Benchmark for illustrative example: (a) original source signals, (b) mixed signals with uniformly distributed random matrix $\mathbf{A} \in \mathbb{R}^{8\times4}$.

the updates of **A**. The variables in **X** were updated using the standard projected ALS algorithm, analyzed in Chapter 4.

In general, the ALS algorithm solves the least-squares problem

$$\mathbf{X}^* = \arg\min_{\mathbf{X}} \left\{ \frac{1}{2} ||\mathbf{Y} - \mathbf{A}\mathbf{X}||_F^2 \right\} \tag{5.76}$$

based on the Moore-Penrose pseudo-inverse of the system matrix, which in our case is the matrix **A**. Since in NMF usually $I \geq J$, we can formulate normal equations as $\mathbf{A}^T\mathbf{A}\mathbf{X} = \mathbf{A}^T\mathbf{Y}$, and the least-squares solution of the minimum ℓ_2-norm is $\mathbf{X}_{LS} = (\mathbf{A}^T\mathbf{A})^{-1}\mathbf{A}^T\mathbf{Y} = \mathbf{A}^{\dagger}\mathbf{Y}$, where $\mathbf{A}^{\dagger}$ is the Moore-Penrose pseudo-inverse of **A**. The projected ALS algorithm is obtained by a simple "half-wave-rectified" projection, that is

$$\mathbf{X} = \left[\mathbf{A}^{\dagger}\mathbf{Y}\right]_+ . \tag{5.77}$$

The alternating minimization is globally nonconvex although the cost function is convex with respect to each particular set of variables. Thus, such NMF algorithms are likely to converge to local minima, and hence, initialization is a prerequisite to successful performance. In the simulations, we used the multi-start initialization described in Chapter 1, with the following parameters: $N = 10$ (number of restarts), $K_i = 30$ (number of initial alternating steps), and $K_f = 1000$ (number of final alternating steps). Each initial sample of **A** and **X** was randomly generated from a uniform distribution.

Basically, the algorithms were tested for two different values of inner iterations: $k = 1$ and $k = 5$. However, we observed that for large-scale problems the SCWA needs more than five inner iterations to give satisfactory results. The inner iterations refer to the number of iterative steps that are performed to update only **A** (before proceeding to update **X**). We also employed the multi-layer technique [6,7,9,30], described in Chapter 1, where L stands for the number of layers.

There are many possible stopping criteria for the alternating steps. We may stop the iterations if $s \geq K_f$ or if the condition $||\mathbf{A}^{(s)} - \mathbf{A}^{(s-1)}||_F < \epsilon$ holds, where s stands for the number of alternating steps, and ϵ is the threshold, e.g. $\epsilon = 10^{-5}$. The condition (5.20) can be also used as a stopping criterion, especially as the gradient is computed in each iteration of the PG algorithms. In the experiments, we used the first stopping criterion with the fixed number of alternation steps, where $K_f = 1000$.

The algorithms were evaluated based on the Signal-to-Interference Ratio (SIR) performance measures, calculated separately for each source signal and each column in the mixing matrix. Since NMF suffers from scale and permutation indeterminacies, the estimated components are adequately permuted and rescaled. First, the source and estimated signals are normalized to a uniform variance, and then the estimated signals are permuted to keep the same order as the source signals. In NMFLAB [8], each estimated signal is compared to each source signal, which results in the performance (SIR) matrix. For the j-th source $\boldsymbol{x}_j$ and its corresponding (re-ordered) estimated signal $\hat{\boldsymbol{x}}_j$, and j-th column of the true mixing matrix $\boldsymbol{a}_j$ and its corresponding estimated counterpart $\hat{\boldsymbol{a}}_j$, the SIRs are given by:

$$SIR_j^{(X)} = -20 \log \left\{ \frac{||\hat{\boldsymbol{x}}_j - \boldsymbol{x}_j||_2}{||\boldsymbol{x}_j||_2} \right\}, \quad j = 1, \dots, J, \quad [\text{dB}] \tag{5.78}$$

and similarly for each column in **A** by:

$$SIR_j^{(A)} = -20 \log \left\{ \frac{||\hat{\boldsymbol{a}}_j - \boldsymbol{a}_j||_2}{||\boldsymbol{a}_j||_2} \right\}, \quad j = 1, \dots, J, \quad [\text{dB}]. \tag{5.79}$$

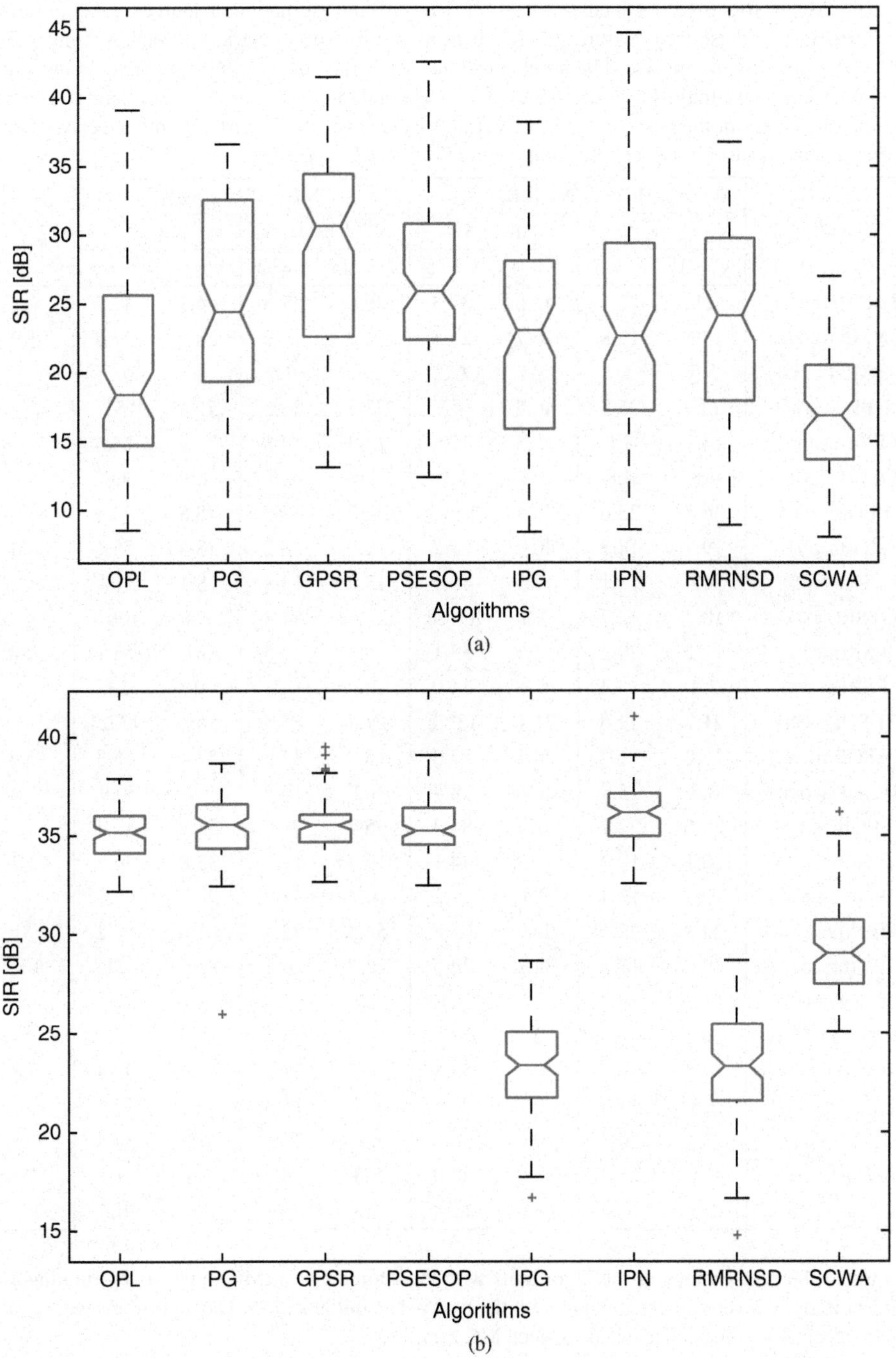

Figure 5.2 The statistics of the $mean - SIR_X$ values obtained from 100 MC trials on the `X_spectra` dataset (benchmark B) for the selected PG algorithms: (a) $I = 10,\ J = 5,\ L = 5,\ k = 5$; (b) $I = 1000,\ J = 5,\ L = 1,\ k = 5$.

Table 5.1 Mean-SIRs [dB] obtained with 100 MC trials for the estimation of sources and columns of the mixing matrix from noise-free mixtures of signals in Figure 5.1 (benchmark A). Sources **X** are estimated with the projected pseudo-inverse. The number of inner iterations for updating **A** is denoted by k, and the number of layers (in the multi-layer technique) by L. The notation *"best"* (or *"worst"*) stands for the mean-SIR value calculated for the best (or worst) trial from MC analysis. In the last column, the elapsed running time [in seconds] is given for every algorithm with $k = 1$ and $L = 1$.

Algorithm	Mean-SIR_A [dB]				Mean-SIR_X [dB]				Time [s]
	$L = 1$		$L = 3$		$L = 1$		$L = 3$		
	$k = 1$	$k = 5$	$k = 1$	$k = 5$	$k = 1$	$k = 5$	$k = 1$	$k = 5$	
M-NMF (best)	21	22.1	42.6	37.3	26.6	27.3	44.7	40.7	
M-NMF (mean)	13.1	13.8	26.7	23.1	14.7	15.2	28.9	27.6	1.9
M-NMF (worst)	5.5	5.7	5.3	6.3	5.8	6.5	5	5.5	
OPL(best)	22.9	25.3	46.5	42	23.9	23.5	55.8	51	
OPL(mean)	14.7	14	25.5	27.2	15.3	14.8	23.9	25.4	1.9
OPL(worst)	4.8	4.8	4.8	5.0	4.6	4.6	4.6	4.8	
LPG(best)	36.3	23.6	78.6	103.7	34.2	33.3	78.5	92.8	
LPG(mean)	19.7	18.3	40.9	61.2	18.5	18.2	38.4	55.4	8.8
LPG(worst)	14.4	13.1	17.5	40.1	13.9	13.8	18.1	34.4	
GPSR(best)	18.2	22.7	7.3	113.8	22.8	54.3	9.4	108.1	
GPSR(mean)	11.2	20.2	7	53.1	11	20.5	5.1	53.1	2.4
GPSR(worst)	7.4	17.3	6.8	24.9	4.6	14.7	2	23	
PSESOP(best)	21.2	22.6	71.1	132.2	23.4	55.5	56.5	137.2	
PSESOP(mean)	15.2	20	29.4	57.3	15.9	34.5	27.4	65.3	5.4
PSESOP(worst)	8.3	15.8	6.9	28.7	8.2	16.6	7.2	30.9	
IPG(best)	20.6	22.2	52.1	84.3	35.7	28.6	54.2	81.4	
IPG(mean)	20.1	18.2	35.3	44.1	19.7	19.1	33.8	36.7	2.7
IPG(worst)	10.5	13.4	9.4	21.2	10.2	13.5	8.9	15.5	
IPN(best)	20.8	22.6	59.9	65.8	53.5	52.4	68.6	67.2	
IPN(mean)	19.4	17.3	38.2	22.5	22.8	19.1	36.6	21	14.2
IPN(worst)	11.7	15.2	7.5	7.1	5.7	2	1.5	2	
RMRNSD(best)	24.7	21.6	22.2	57.9	30.2	43.5	25.5	62.4	
RMRNSD(mean)	14.3	19.2	8.3	33.8	17	21.5	8.4	33.4	3.8
RMRNSD(worst)	5.5	15.9	3.6	8.4	4.7	13.8	1	3.9	
SCWA(best)	12.1	20.4	10.6	24.5	6.3	25.6	11.9	34.4	
SCWA(mean)	11.2	16.3	9.3	20.9	5.3	18.6	9.4	21.7	2.5
SCWA(worst)	7.3	11.4	6.9	12.8	3.8	10	3.3	10.8	

We tested the performance of the algorithms with the Monte Carlo (MC) analysis by running 100 independent trials for every algorithm, and with the multi-start initialization. The algorithms were evaluated using the mean-SIR values, calculated for each MC sample as:

$$Mean - SIR_X = \frac{1}{J}\sum_{j=1}^{J} SIR_j^{(X)}, \tag{5.80}$$

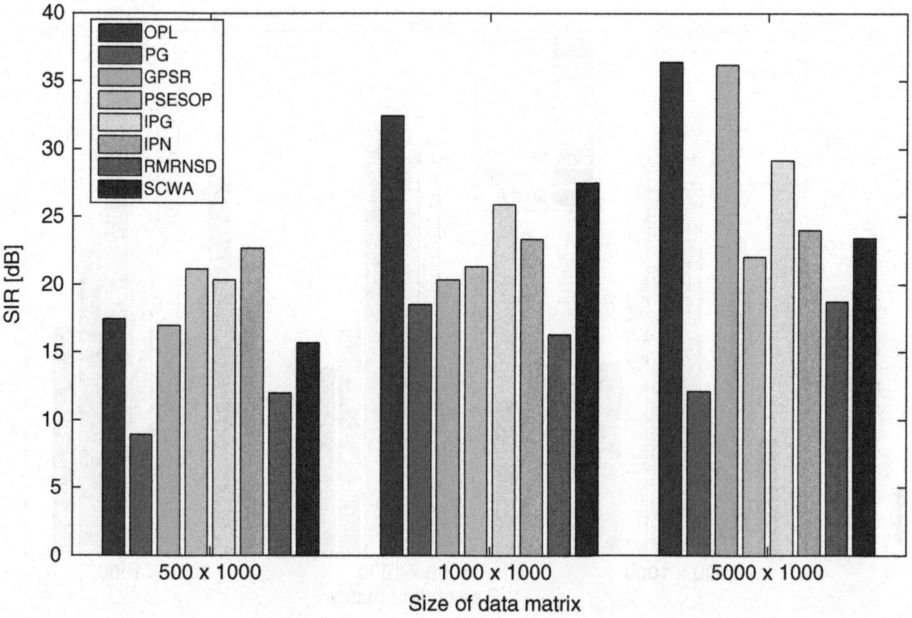

Figure 5.3 Mean-SIRs [dB] obtained by the selected PG algorithms applied for large-scale BSS problems where the benchmark signals (benchmark C: $\mathbf{X} \in \mathbb{R}^{10\times 1000}$) illustrated in Figure 4.8 (Chapter 4) were mixed with the uniformly distributed random mixing matrices of sizes: 500 × 1000, 1000 × 1000, and 5000 × 1000. The number of inner iterations was set to $k = 10$ for the SCWA, and to $k = 5$ for other PG algorithms.

$$Mean - SIR_A = \frac{1}{J}\sum_{j=1}^{J} SIR_j^{(A)}. \tag{5.81}$$

Figure 5.2 illustrates the boxplot for 100 samples of $Mean - SIR_X$ values for the estimation of the signals from the X_spectra dataset (benchmark B). Figure 5.2(a) refers to the case when only 10 mixed signals were observed but Figure 5.2(b) presents the case of 1000 observations.

The mean-SIRs for the worst (with the lowest mean-SIR values), mean, and best (with the highest mean-SIR values) trials are given in Table 5.1 for the estimation of the signals in 5.1 (benchmark A). The number k denotes the number of inner iterations for updating $\mathbf{A}$, and L denotes the number of layers in the multi-layer technique [6,7,9,30]. The case $L = 1$ indicates that the multi-layer technique is not used. The elapsed running time [in seconds] was measured in MATLAB, and it reveals a degree of complexity of the algorithm.

For a comprehensive comparison, Table 5.1 also contains the results obtained for the standard multiplicative NMF algorithm (denoted as M-NMF) that minimizes the squared Euclidean distance.

The results from the tests performed on the benchmark C ($\mathbf{X} \in \mathbb{R}^{10\times 1000}$) are illustrated in Figures 5.3–5.6. The mean-SIR values obtained by the selected PG algorithms applied to the datasets of a different size are plotted in Figure 5.3. The corresponding elapsed runtime, measured in MATLAB 2008a on CPU 9650, Intel Quad Core, 3GHz, 8GB RAM, is presented in Figure 5.4.

In the large-scale experiments, we generate the dataset of size 10000 × 1000 from the benchmark C. Since the number of observations and samples is large with respect to the number of sources, the block-wise data selection technique described in Section 1.3.1 was used to considerably speed up the computations. For updating the source matrix $\mathbf{X}$ we randomly select 50 rows ($R = 50$) from 10000 observations, and thus the reduced-size system (1.45) is solved with the normal projected ALS algorithm. For updating the

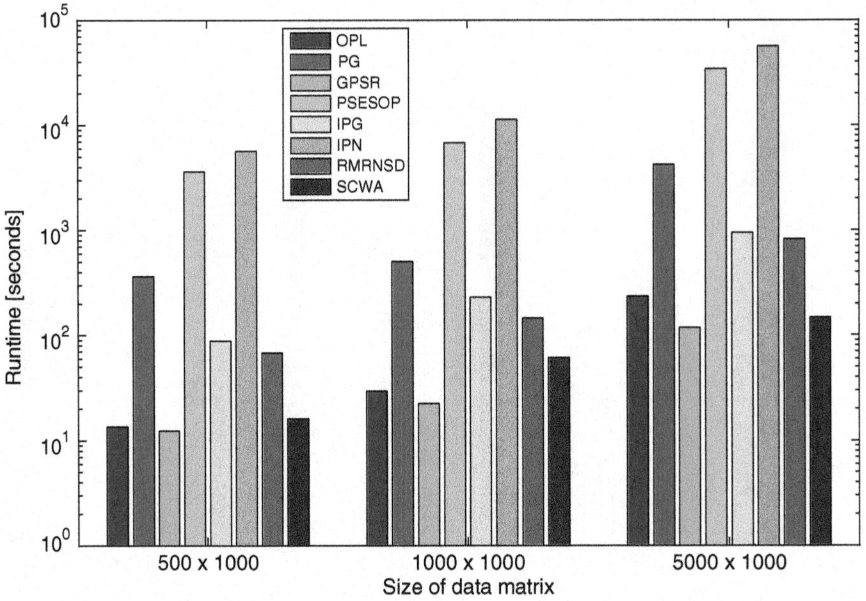

Figure 5.4 Runtime of the selected PG algorithms applied for large-scale BSS problems where the benchmark signals (benchmark C: $\mathbf{X} \in \mathbb{R}^{10\times1000}$) illustrated in Figure 4.8 (Chapter 4) were mixed with the uniformly distributed random mixing matrices of sizes: 500×1000, 1000×1000, and 5000×1000. The computations were performed with MATLAB 2008a on a PC with CPU 9650, Intel Quad Core, 3GHz, 8GB RAM. The number of inner iterations was set to $k = 10$ for the SCWA, and to $k = 5$ for other PG algorithms.

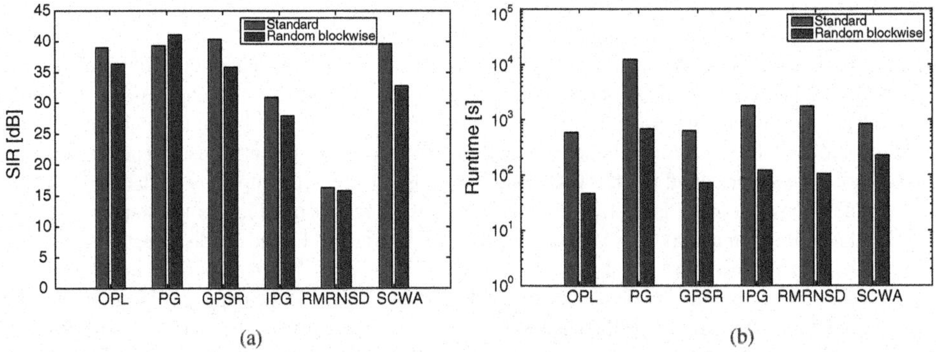

Figure 5.5 Performance of the selected PG algorithms applied for solving a highly redundant large-scale BSS problem where $\mathbf{A} \in \mathbb{R}^{10000\times10}$ and $\mathbf{X} \in \mathbb{R}^{10\times1000}$ (benchmark C): (a) mean-SIR [dB] for $\mathbf{X}$; (b) Runtime [seconds]. In the `Random block-wise` technique we set $C = 100$ (number of columns), and $R = 50$ (number of rows). The number of inner iterations was set to $k = 15$ for the SCWA, and to $k = 5$ for other PG algorithms. The matrix $\mathbf{X}$ is updated with the normal ALS algorithm.

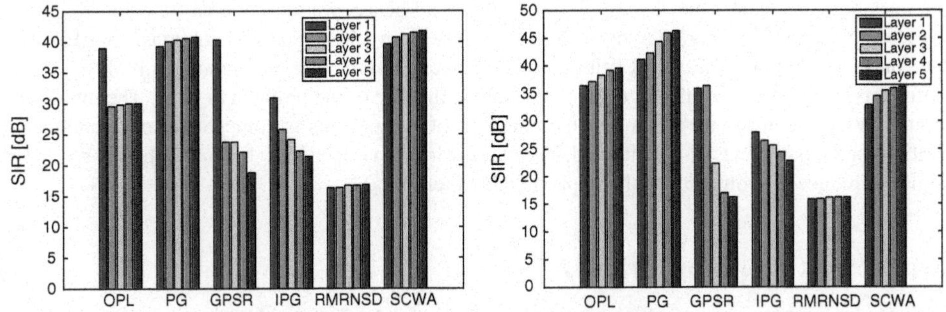

(a) Full data update

(b) Block-wise update $C = 100$ columns, $R = 50$ rows

Figure 5.6 Performance of the selected PG algorithms combined with the multi-layer technique for solving a highly redundant large-scale BSS problem where $\mathbf{A} \in \mathbb{R}^{10000\times 10}$ and $\mathbf{X} \in \mathbb{R}^{10\times 1000}$ (benchmark C): (a) Normal data selection; (b) Random block-wise data selection with $C = 100$ (number of columns) and $R = 50$ (number of rows). The number of inner iterations was set to $k = 15$ for the SCWA, and to $k = 5$ for other PG algorithms. The matrix $\mathbf{X}$ is updated with the normal ALS algorithm.

mixing matrix $\mathbf{A}$, 100 columns from the data matrix $\mathbf{Y}$ are selected randomly ($C = 100$), and consequently, the reduced-size system (1.46) is solved with the selected PG algorithm. The comparison of the mean-SIR values between the normal and random block-wise data selection is shown in Figure 5.5(a). The corresponding comparison of the runtime is given in Figure 5.5(b).

Figure 5.6 presents the mean-SIR values obtained with the selected PG algorithms in consecutive layers of the multi-layer technique.

5.10 Discussions

The MC results in Figure 5.2 and in Table 5.1 show that the algorithms are sensitive to initialization, thus indicating sensitivity to local minima. Moreover, the PG algorithms give the best performance for small-scale data, especially when the number of observations is not very large.

According to Table 5.1, the PSESOP algorithm provides the best estimation (the sample which has the highest mean-SIR value), however, it is not optimal for the worst case scenario. The trial with the highest value of SIR for the worst case was obtained by the LPG algorithm. Considering the complexity (the elapsed running time), the OPL, GPSR, SCWA, and IPG techniques are fastest, and the LPG and IPN algorithms are the slowest.

A somewhat different conclusion can be drawn from Figure 5.2, where the IPN algorithm gives the best performance. Thus we cannot definitely point at the particular PG algorithm that exhibits a strictly dominant performance for small-scale data. Nevertheless, we observed (see Table 5.1) that the multi-layer technique improves the performance of all the tested algorithms for small-scale data, especially when the number of inner iterations is greater than one.

When the number of observations is much larger than the number of sources, the multi-layer technique is not necessary, and in such a case, the choice of the algorithm is important. The results shown in Figure 5.2(b), Figure 5.3, and Figures 5.5–5.6 consistently demonstrate that the RMRNSD algorithm gives the lowest mean-SIR values. The best performances both in terms of the mean-SIR values and runtime are obtained by the OPL and GPSR algorithms. The SCWA can also give satisfactory results but only if the number of inner iterations is relatively large (usually more than 10), which considerably slows down the computations.

Figure 5.5 shows that the random block-wise data selection technique is very useful for all the PG NMF algorithms with multiple right-hand side processing. The PSESOP and IPN algorithms process the successive column vectors of the data matrix by using a FOR loop, and hence, we did not analyze these algorithms in such tests. As shown in Figure 5.5, by selecting 50 rows from 10000, and 100 columns from 1000, we obtained only slightly lower mean-SIR values, but the reduction in the runtime is about tenfold.

In summary, the selection of the algorithm depends on the size of the problem in hand. The multi-layer technique is very useful for the cases in which number of components is comparable to the number of observations or the problem is ill-conditioned. The block-wise data selection technique for highly redundant problems can reduce the computational complexity about tenfold.

Appendix 5.A: Stopping Criteria

In this appendix we present some efficient stopping criteria for NMF algorithms based on the difference between two recent estimated factor matrices using the following alternative criteria:

- Frobenius norm of cost function

$$D_F^{(k)} = \left\|\mathbf{Y} - \hat{\mathbf{Y}}^{(k)}\right\|_F^2 \leq \varepsilon, \qquad (\hat{\mathbf{Y}}^{(k)} = \mathbf{A}^{(k)}\mathbf{X}^{(k)}). \tag{A.1}$$

- Kullback-Leibler distance of the cost function

$$D_{KL}^{(k)} = \sum\sum \mathbf{Y} \log\left(\frac{\mathbf{Y}}{\hat{\mathbf{Y}}^{(k)}}\right) - \mathbf{Y} + \hat{\mathbf{Y}}^{(k)}. \tag{A.2}$$

- Frobenius norm of the estimated matrices

$$D_{\hat{\mathbf{Y}}^{(k)}}^{(k)} = \left\|\hat{\mathbf{Y}}^{(k)} - \hat{\mathbf{Y}}^{(k-1)}\right\|_F^2 \leq \varepsilon. \tag{A.3}$$

- Ratio of the distances

$$\frac{|D^{(k)} - D^{(k-1)}|}{D_{\hat{\mathbf{Y}}^{(k)}}^{(k)}} \leq \varepsilon. \tag{A.4}$$

- Projected Gradient stopping criteria
  The Projected Gradient stopping criteria in [21] is described as

$$\left\|\nabla^P D_F\left(\mathbf{Y}||\mathbf{A}^{(k)}\,\mathbf{X}^{(k)}\right)\right\|_F^2 \leq \varepsilon \left\|\nabla^P D_F\left(\mathbf{Y}||\mathbf{A}^{(1)}, \mathbf{X}^{(1)}\right)\right\|_F^2. \tag{A.5}$$

where ε is the tolerance rate, the projected gradient matrix $((I+T) \times J)$ is defined as

$$\nabla D_F(\mathbf{Y}||\mathbf{A}\,\mathbf{X}) = \begin{bmatrix} \nabla_{\mathbf{A}} D_F(\mathbf{Y}||\mathbf{A}\mathbf{X}) \\ \nabla_{\mathbf{X}} D_F(\mathbf{Y}||\mathbf{A}\mathbf{X})^T \end{bmatrix} = \begin{bmatrix} \mathbf{G}_{\mathbf{A}} \\ \mathbf{G}_{\mathbf{X}^T} \end{bmatrix} = \begin{bmatrix} \mathbf{A}\mathbf{X}\mathbf{X}^T - \mathbf{Y}\mathbf{X}^T \\ \mathbf{X}^T\mathbf{A}^T\mathbf{A} - \mathbf{Y}^T\mathbf{A} \end{bmatrix}, \tag{A.6}$$

and

$$\nabla_{a_{ij}}^P D_F = \left[\mathbf{G}_{\mathbf{A}}^P\right]_{jt} = \begin{cases} [\mathbf{G}_{\mathbf{A}}]_{jt} & \text{if } a_{ij} > 0\,, \quad \text{(A.7a)} \\ \min\left([\mathbf{G}_{\mathbf{A}}]_{ij}\,, 0\right) & \text{if } a_{ij} = 0. \quad \text{(A.7b)} \end{cases}$$

- ℓ_2- Equalized Projected Gradient stopping criteria

We assume that $\mathbf{A}$ and $\mathbf{X}$ are two estimates which satisfy the PG stopping criteria (A.5). Due to scaling ambiguity in NMF, for any nonzero vector $\boldsymbol{d} \in \mathbb{R}^J$: $d_j \neq 0$, the two matrices $\mathbf{A}_d = \mathbf{A}\,\mathrm{diag}\{\boldsymbol{d}\}$ and $\mathbf{X}_d = \mathrm{diag}\{\boldsymbol{d}\}^{-1}\mathbf{X}$ are also considered to be solutions of NMF. By choosing a suitable scaling vector $\boldsymbol{d}$, we obtain a new lower value $\left\|\nabla^P D_F\right\|_F$ with $\mathbf{A}_d$ and $\mathbf{X}_d$ so that the PG stopping criteria (A.5) can be met earlier. The gradients of the cost function with respect to $\mathbf{A}_d$ and $\mathbf{X}_d$ are given by

$$\mathbf{G}_{\mathbf{A}_d} = \mathbf{A}_d\mathbf{X}_d\mathbf{X}_d^T - \mathbf{Y}\mathbf{X}_d^T = \left[\mathbf{A}\mathbf{X}\mathbf{X}^T - \mathbf{Y}\mathbf{X}^T\right]\mathrm{diag}\{\boldsymbol{d}\}^{-1} = \mathbf{G}_{\mathbf{A}}\,\mathrm{diag}\{\boldsymbol{d}\}^{-1}, \tag{A.8}$$

and

$$\mathbf{G}_{\mathbf{X}_d^T} = \mathbf{X}_d^T\mathbf{A}_d^T\mathbf{A}_d - \mathbf{Y}^T\mathbf{A}_d = \left[\mathbf{X}^T\mathbf{A}^T\mathbf{A} - \mathbf{Y}^T\mathbf{A}\right] = \mathbf{G}_{\mathbf{X}^T}\,\mathrm{diag}\{\boldsymbol{d}\}. \tag{A.9}$$

On the basis of the definition of the projected gradient, we obtain the following conditions

$$[\mathbf{G}^P_{\mathbf{A}_d}]_j = \frac{[\mathbf{G}^P_{\mathbf{A}}]_j}{d_j}, \qquad [\mathbf{G}^P_{\mathbf{X}_d^T}]_j = d_j[\mathbf{G}^P_{\mathbf{X}^T}]_j, \tag{A.10}$$

and

$$[\mathbf{G}^P_{\mathbf{A}_d}]_j\,[\mathbf{G}^P_{\mathbf{X}_d^T}]_j = [\mathbf{G}^P_{\mathbf{A}}]_j\,[\mathbf{G}^P_{\mathbf{X}^T}]_j, \quad \forall j. \tag{A.11}$$

By applying Cauchy's inequality on every columns of the projected gradient matrices $\mathbf{G}_{\mathbf{A}}$ and $\mathbf{G}_{\mathbf{X}^T}$, we obtain

$$\begin{aligned}\left\|\nabla^P D_F\right\|_F^2 &= \left\|\begin{matrix}\mathbf{G}^P_{\mathbf{A}}\\ \mathbf{G}^P_{\mathbf{X}^T}\end{matrix}\right\|_F^2 = \sum_{j=1}^{J}\left\|[\mathbf{G}^P_{\mathbf{A}}]_j\right\|_2^2 + \left\|[\mathbf{G}^P_{\mathbf{X}^T}]_j\right\|_2^2 \\ &\geq \sum_{j=1}^{J} 2\left\|[\mathbf{G}^P_{\mathbf{A}}]_j\right\|_2\left\|[\mathbf{G}^P_{\mathbf{X}^T}]_j\right\|_2 .\end{aligned} \tag{A.12}$$

The equality (A.12) holds if and only if

$$\left\|[\mathbf{G}^P_{\mathbf{A}}]_j\right\|_2 = \left\|[\mathbf{G}^P_{\mathbf{X}^T}]_j\right\|_2, \quad \forall j. \tag{A.13}$$

By choosing the scaling coefficients d_j which satisfies the condition

$$\frac{\left\|[\mathbf{G}^P_{\mathbf{A}}]_j\right\|_2}{d_j} = d_j\left\|[\mathbf{G}^P_{\mathbf{A}}]_j\right\|_2, \quad \forall j, \tag{A.14}$$

or

$$d_j = \sqrt{\frac{\left\|[\mathbf{G}^P_{\mathbf{X}}]_j\right\|_2}{\left\|[\mathbf{G}^P_{\mathbf{A}^T}]_j\right\|_2}}, \quad \forall j, \tag{A.15}$$

the projected gradients $\mathbf{G}^P_{\mathbf{A}}$ and $\mathbf{G}^P_{\mathbf{X}^T}$ of the two normalized matrices $\mathbf{A}_d$ and $\mathbf{X}_d$ have columns equalized in the sense of ℓ_2-norm.

Finally, the new stopping criteria referred to as the ℓ_2-Equalized Projected Gradient is summarized as follows:

1. Update matrices $\mathbf{A}$ and $\mathbf{X}$ by any specified NMF algorithm.
2. Scale (normalize) matrices $\mathbf{A}$ and $\mathbf{X}$ by the weighting vector $\boldsymbol{d} = [d_j]_{j=1,\ldots,J}$ (A.15).
3. Repeat Steps 1, 2 until the stopping condition (A.5) is met.

Appendix 5.B: MATLAB Source Code for Lin's PG Algorithm

```
function [X,grad_x,k]=nmf_proj_grad(Y,A,X,tol,no_iter,eta_dec,loss_fun,alpha)
%
% Projected Gradient Algorithm based on the Armijo rule
%
% [X, grad_x, k]=nmf_proj_grad(Y,A,X,tol,no_iter,eta_dec,loss_fun,alpha)
% finds such nonneagative X that solves the equation AX = Y.
%
% INPUTS:
% Y - data matrix of dimension [I by T]
% A - fixed matrix of dimension [I by J]
% X - initial guess
% tol - tolerance
% no_iter - maximum number of iterations
% eta_dec - multiplier of learning rate in descent gradient direction
% loss_fun - index for the loss (cost) function (1--4)
% alpha - parameter "alpha" in the alpha-divergence
%
% OUTPUTS:
% X - estimated matrix of dimension [J by T]
% grad_x - gradient of X
% k - iterative step
%
% ##########################################################################

if alpha == 0, loss_fun = 2; end,
sigma = 0.01;  eta = 1; kinc = 1; lower_bound = 1E-6;
Yx = 0;
for k=1:no_iter
    F = costfunction;
    grad_x = gradX;

    tol_grad = norm(grad_x(grad_x < lower_bound | X > lower_bound));
    if tol_grad < tol,
        break
    end

    % search step size
    while (beta > 1E-15) && (beta < 1E8)
        Xn = max(X - eta*grad_x,lower_bound);
        Δ = Xn - X;

        Fn = costfunction;
        cond_alpha = ((Fn - F) ≤ sigma*grad_x(:)'*Δ(:));
        if cond_alpha || (Xn == X)
            % Increase
            if ~kinc && (s > 1),
                X = Xn;
                break;
            end
            eta = eta/eta_dec;
            kinc = 1; Xp = Xn;
        else
            % Decrease
            eta = eta*eta_dec;
            if kinc && (s > 1),
                X = Xp;
                break;
```

```
            end
            kinc = 0;
        end
    end % while
end % for k
    function F = costfunction
        Yx = A*X;
        switch loss_fun
            case 1 % Frobenius
                F = .5*norm(Y-Yx,'fro')^2;
            case 2 % KL
                F = sum(sum(Y.*log(Y./(Yx + eps)) + Yx - Y));
            case 3 % Dual KL
                F = sum(sum(Yx.*log(Yx./Y) + Y - Yx));
            case 4 % Alpha divergence
                F = sum(sum(Y.*((Y./Yx).^(alpha-1)-1)/(alpha^2-alpha) ...
                    + (Yx - Y)/alpha));
        end  % switch
    end
    function grad_x = gradX
        switch loss_fun
            case 1 % Frobenius
                grad_x = -A'*(Y- Yx);
            case 2 % KL
                grad_x = A'*(1 - Y./(Yx + eps));
            case 3 % Dual KL
                grad_x = A'*log(Yx./Y);
            case 4 % Alpha divergence
                grad_x = (1/alpha)*A'*(1 - (Y./Yx).^alpha);
        end  % switch
    end
end
```

References

[1] J. Barzilai and J.M. Borwein. Two-point step size gradient methods. *IMA Journal of Numerical Analysis*, 8(1):141–148, 1988.

[2] S. Bellavia, M. Macconi, and B. Morini. An interior point Newton-like method for non-negative least-squares problems with degenerate solution. *Numererical Linear Algebra and Applications*, 13:825–846, 2006.

[3] M. Bertero and P. Boccacci. *Introduction to Inverse Problems in Imaging*. Institute of Physics Publishing, Bristol, UK, January 1998.

[4] D. P. Bertsekas. *Nonlinear Programming*. Athena Scientific, Belmont, MA, 1999.

[5] D.P. Bertsekas. On the Goldstein-Levitin-Polyak gradient projection method. *IEEE Transations on Automatic Control*, 21(2):174–184, 1976.

[6] A. Cichocki and R. Zdunek. Multilayer nonnegative matrix factorization. *Electronics Letters*, 42(16):947–948, 2006.

[7] A. Cichocki and R. Zdunek. Multilayer nonnegative matrix factorization using projected gradient approaches. In *The 13th International Conference on Neural Information Processing (ICONIP06)*, Hong Kong, October 3–6 2006. http://iconip2006.cse.cuhk.edu.hk/.

[8] A. Cichocki and R. Zdunek. NMFLAB for Signal and Image Processing. Technical report, Laboratory for Advanced Brain Signal Processing, BSI, RIKEN, Saitama, Japan, 2006.

[9] A. Cichocki and R. Zdunek. Multilayer nonnegative matrix factorization using projected gradient approaches. *International Journal of Neural Systems*, 17(6):431–446, 2007.

[10] A. Cichocki, R. Zdunek, and S. Amari. Csiszar's divergences for non-negative matrix factorization: Family of new algorithms. *Springer, LNCS-3889*, 3889:32–39, 2006.

[11] Y.-H. Dai and R. Fletcher. Projected Barzilai-Borwein methods for large-scale box-constrained quadratic programming. *Numer. Math.*, 100(1):21–47, 2005.

[12] M. Elad, B. Matalon, and M. Zibulevsky. Coordinate and subspace optimization methods for linear least squares with non-quadratic regularization. *Journal on Applied and Computational Harmonic Analysis.*, 23:346–367, 2007.
[13] R. Fletcher. Conjugate gradient methods for indefinite systems. In *Proc. of the Dundee Biennial Conference on Numerical Analysis*, pages 73–89, Castello de la Plana, Spain, 1975. Springer-Verlag.
[14] V. Franc, V. Hlaváč, and M. Navara. Sequential coordinate-wise algorithm for the non-negative least squares problem. In André Gagalowicz and Wilfried Philips, editors, *CAIP 2005: Computer Analysis of Images and Patterns*, volume 1 of *LNCS*, pages 407–414, Berlin, Germany, September 2005. Springer-Verlag.
[15] R.W. Freund and N.M. Nachtigal. QMR: A quasi-minimal residual method for non-Hermitian linear systems. *Numerische Mathematik*, 60(1):315–339, 1991.
[16] P. C. Hansen. *Rank-Deficient and Discrete Ill-Posed Problems.* SIAM, Philadelphia, 1998.
[17] M.R. Hestenes and E. Stiefel. Method of conjugate gradients for solving linear systems. *J. Res. Nat. Bur. Standards*, 49:409–436, 1952.
[18] B. Johansson, T. Elfving, V. Kozlov, T. Censor, P-E. Forsséna, and G. Granlund. The application of an oblique-projected landweber method to a model of supervised learning. *Mathematical and Computer Modelling*, 43(7–8):892–909, April 2006.
[19] L. Kaufman. Maximum likelihood, least squares, and penalized least squares for PET. *IEEE Transactions on Medical Imaging*, 12(2):200–214, 1993.
[20] C. Lanczos. Solution of systems of linear equations by minimized iterations. *Journal of Research of the National Bureau of Standards*, 49(1):33–53, 1952.
[21] Ch-J. Lin. Projected gradient methods for non-negative matrix factorization. *Neural Computation*, 19(10):2756–2779, October 2007.
[22] Ch.J. Lin and J.J. Moré. Newton's method for large bound-constrained optimization problems. *SIAM Journal on Optimization*, 9(4):1100–1127, 1999.
[23] M. Merritt and Y. Zhang. An interior-point gradient method for large-scale totally nonnegative least squares problems. *J. Optimization Theory and Applications*, 126(1):191–202, 2005.
[24] J.G. Nagy and Z. Strakos. Enforcing nonnegativity in image reconstruction algorithms. volume 4121 of *Mathematical Modeling, Estimation, and Imaging*, pages 182–190, 2000.
[25] G. Narkiss and M. Zibulevsky. Sequential subspace optimization method for large-scale unconstrained problems. *Optimization Online*, page 26, October 2005.
[26] A. Nemirovski. Orth-method for smooth convex optimization (in russian). *Izvestia AN SSSR, Ser. Tekhnicheskaya Kibernetika (the journal is translated into English as Engineering Cybernetics. Soviet J. Computer and Systems Science)*, 2, 1982.
[27] C.C. Paige and M.A. Saunders. LSQR: An algorithm for sparse linear equations and sparse least squares. *ACM Transactions on Math. Software*, 8:43–71, 1982.
[28] Y. Saad and M.H. Schultz. GMRES: a generalized minimal residual algorithm for solving nonsymmetric linear systems. *SIAM J. Sci. Stat. Comput.*, 7(3):856–869, 1986.
[29] H.A. van der Vorst. Bi-CGSTAB: A fast and smoothly converging variant of Bi-CG for the solution of nonsymmetric linear systems. *SIAM J. Sci. Stat. Comput.*, 13(2):631–644, 1992.
[30] R. Zdunek and A. Cichocki. Fast nonnegative matrix factorization algorithms using projected gradient approaches for large-scale problems. *Journal of Computational Intelligence and Neuroscience*, 2008(939567), 2008.

6

Quasi-Newton Algorithms for Nonnegative Matrix Factorization

So far we have discussed the NMF algorithms which perform optimization by searching for stationary points of a cost function based on first-order approximations, that is, using the gradient. In consequence, the additive learning algorithms have the following general form:

$$\mathbf{A} \leftarrow \mathcal{P}_{\Omega}\left[\mathbf{A} - \boldsymbol{\eta}_{\mathbf{A}}\, \nabla_{\mathbf{A}} D(\mathbf{Y}||\mathbf{A}\mathbf{X})\right], \tag{6.1}$$

$$\mathbf{X} \leftarrow \mathcal{P}_{\Omega}\left[\mathbf{X} - \boldsymbol{\eta}_{\mathbf{X}} \nabla_{\mathbf{X}} D(\mathbf{Y}||\mathbf{A}\mathbf{X})\right], \tag{6.2}$$

where $\mathcal{P}_{\Omega}[\xi]$ denotes the projection of ξ onto the set Ω of feasible solutions,[1] and the learning rates $\boldsymbol{\eta}_{\mathbf{A}}$ and $\boldsymbol{\eta}_{\mathbf{X}}$ are either fixed or iteratively updated scalars or diagonal matrices.

Let us consider a Taylor series expansion

$$\begin{aligned} D(\mathbf{Y}||(\mathbf{A}+\Delta\mathbf{A})\mathbf{X}) &= D(\mathbf{Y}||\mathbf{A}\mathbf{X}) + \operatorname{vec}\left(\nabla_{\mathbf{A}} D(\mathbf{Y}||\mathbf{A}\mathbf{X})\right)^T \operatorname{vec}(\Delta\mathbf{A}) \\ &+ \frac{1}{2}\operatorname{vec}(\Delta\mathbf{A})^T\, \mathbf{H}_{\mathbf{A}} \operatorname{vec}(\Delta\mathbf{A}) + \mathcal{O}\left((\Delta\mathbf{A})^3\right), \end{aligned} \tag{6.3}$$

$$\begin{aligned} D(\mathbf{Y}||\mathbf{A}(\mathbf{X}+\Delta\mathbf{X})) &= D(\mathbf{Y}||\mathbf{A}\mathbf{X}) + \operatorname{vec}\left(\nabla_{\mathbf{X}} D(\mathbf{Y}||\mathbf{A}\mathbf{X})\right)^T \operatorname{vec}(\Delta\mathbf{X}) \\ &+ \frac{1}{2}\operatorname{vec}(\Delta\mathbf{X})^T\, \mathbf{H}_{\mathbf{X}} \operatorname{vec}(\Delta\mathbf{X}) + \mathcal{O}\left((\Delta\mathbf{X})^3\right), \end{aligned} \tag{6.4}$$

where $\mathbf{H}_{\mathbf{A}} = \nabla^2_{\mathbf{A}} D(\mathbf{Y}||\mathbf{A}\mathbf{X}) \in \mathbb{R}^{IJ\times IJ}$ and $\mathbf{H}_{\mathbf{X}} = \nabla^2_{\mathbf{X}} D(\mathbf{Y}||\mathbf{A}\mathbf{X}) \in \mathbb{R}^{JT\times JT}$ are Hessians with respect to $\mathbf{A}$ and $\mathbf{X}$.

In this chapter, we introduce learning algorithms for the NMF problem using second-order approximations, i.e. the third-order term in the above Taylor series expansion. In consequence, the learning rates $\boldsymbol{\eta}_{\mathbf{A}}$ and $\boldsymbol{\eta}_{\mathbf{X}}$

[1] Typically, the set Ω in NMF is the nonnegative orthant of the space of real numbers, that is, $\mathcal{P}_{\Omega}[\xi] = [\xi]_+$, however, other sets can also be used. For example, the updated factors can be bounded by a box rule, that is, $\Omega = \{\xi : l_{min} \le \xi \le u_{max}\}$.

Nonnegative Matrix and Tensor Factorizations: Applications to Exploratory Multi-way Data Analysis and Blind Source Separation Andrzej Cichocki, Rafal Zdunek, Anh Huy Phan and Shun-ichi Amari

in (6.1)–(6.2) become the inverses of the Hessian, thus yielding the following projected Newton updating rules:

$$\text{vec}(\mathbf{A}) \leftarrow \mathcal{P}_\Omega \left[\text{vec}(\mathbf{A}) - \mathbf{H}_\mathbf{A}^{-1}\, \text{vec}(\mathbf{G}_\mathbf{A})\right], \tag{6.5}$$

$$\text{vec}(\mathbf{X}) \leftarrow \mathcal{P}_\Omega \left[\text{vec}(\mathbf{X}) - \mathbf{H}_\mathbf{X}^{-1}\, \text{vec}(\mathbf{G}_\mathbf{X})\right], \tag{6.6}$$

where $\mathbf{G_A} = \nabla_\mathbf{A} D(\mathbf{Y}||\mathbf{AX})$ and $\mathbf{G_X} = \nabla_\mathbf{X} D(\mathbf{Y}||\mathbf{AX})$. The symbol vec($\mathbf{G}$) denotes the vectorized version of the matrix $\mathbf{G} \in \mathbb{R}^{J\times T}$, that is, $\text{vec}(\mathbf{G}) = [g_{11}, g_{21}, \ldots, g_{J1}, g_{12}, \ldots, g_{JT}]^T \in \mathbb{R}^{JT}$.

Using the information about the curvature of the cost function, which is intimately related to second-derivatives, the convergence can be considerably accelerated. This, however, also introduces many related practical problems that must be addressed prior to applying learning algorithms. For example, the Hessian $\mathbf{H_A}$ and $\mathbf{H_X}$ must be positive-definite to ensure the convergence of approximations of (6.5)–(6.6) to a local minimum of $D(\mathbf{Y}||\mathbf{AX})$. Unfortunately, this is not guaranteed using the NMF alternating minimization rule, and we need to resort to some suitable Hessian approximation techniques (referred to as Quasi-Newton methods). In addition, the Hessian values may be very large, and of severely ill-conditioned nature (in particular for large-scale problems), which gives rise to many difficult problems related to its inversion.

This chapter provides a comprehensive study on the solutions to the above-mentioned problems. We also give some heuristics on the selection of a cost function and related regularization terms which restrict the area of feasible solutions, and help to converge to the global minimum of the cost function.

The layout of the chapter is as follows: first, we discuss the simplest approach to the projected quasi-Newton optimization using the Levenberg-Marquardt regularization of the Hessian. For generality, as a cost function, we consider the Alpha- and Beta-divergences [11,22,29] that unify many well-known cost functions (see the details in Chapter 2), we then discuss the reduced quasi-Newton optimization that involves the Gradient Projection Conjugate Gradient (GPCG) algorithm [1,2,24,34], followed by the FNMA method proposed by Kim, Sra and Dhillon [21]. Further, as a special case of the quasi-Newton method, we present one quadratic programming method [35]. The simulations that conclude the chapter are performed on the same benchmark data (mixed signals) as those in other chapters.

6.1 Projected Quasi-Newton Optimization

In Chapter 2, we have demonstrated that the Bregman, Alpha- or Beta-divergences [11,15] are particularly useful for dealing with non-Gaussian noisy disturbances. In this section, we discuss the projected quasi-Newton method in the context of application to alternating minimization of these functions. First, the basic computations of the gradient and Hessian matrices are presented, and then the efficient method for computing the inverse to the Hessian is discussed.

6.1.1 Projected Quasi-Newton for Frobenius Norm

First, we will present the projected quasi-Newton method that minimizes the standard Euclidean distance in NMF. This case deserves special attention since normally distributed white noise is a common assumption in practice. For the squared Euclidean distance:

$$D_F(\mathbf{Y}||\mathbf{AX}) = \frac{1}{2}||\mathbf{Y} - \mathbf{AX}||_F^2, \tag{6.7}$$

the gradients and Hessians have the following forms:

$$\mathbf{G_X} = \mathbf{A}^T(\mathbf{AX} - \mathbf{Y}) \in \mathbb{R}^{J\times T}, \qquad \mathbf{G_A} = (\mathbf{AX} - \mathbf{Y})\mathbf{X}^T \in \mathbb{R}^{I\times J}, \tag{6.8}$$

$$\mathbf{H_X} = \mathbf{I}_T \otimes \mathbf{A}^T\mathbf{A} \in \mathbb{R}^{JT\times JT}, \qquad \mathbf{H_A} = \mathbf{XX}^T \otimes \mathbf{I}_I \in \mathbb{R}^{IJ\times IJ}, \tag{6.9}$$

where $\mathbf{I}_T \in \mathbb{R}^{T\times T}$ and $\mathbf{I}_I \in \mathbb{R}^{I\times I}$ are identity matrices, and the symbol $\otimes$ stands for the Kronecker product.

The update rule (6.6) for $\mathbf{X}$ can be reformulated as follows

$$\begin{aligned} \mathrm{vec}(\mathbf{X}) &\leftarrow \mathcal{P}_\Omega\left[\mathrm{vec}(\mathbf{X}) - \eta_0\, (\mathbf{H_X})^{-1}\, \mathrm{vec}(\mathbf{G_X})\right] \\ &= \mathcal{P}_\Omega\left[\mathrm{vec}(\mathbf{X}) - \eta_0\, \left(\mathbf{I}_T \otimes \mathbf{A}^T\mathbf{A}\right)^{-1}\, \mathrm{vec}(\mathbf{G_X})\right] \\ &= \mathcal{P}_\Omega\left[\mathrm{vec}(\mathbf{X}) - \eta_0\, \left(\mathbf{I}_T \otimes \left(\mathbf{A}^T\mathbf{A}\right)^{-1}\right)\, \mathrm{vec}(\mathbf{G_X})\right], \end{aligned} \tag{6.10}$$

thus it is re-written in the matrix form as

$$\begin{aligned} \mathbf{X} &\leftarrow \mathcal{P}_\Omega\left[\mathbf{X} - \eta_0\, \left(\mathbf{A}^T\mathbf{A}\right)^{-1}\, \mathbf{G_X}\right] \\ &= \mathcal{P}_\Omega\left[\mathbf{X} - \eta_0\, \left(\mathbf{A}^T\mathbf{A}\right)^{-1}\, \mathbf{A}^T(\mathbf{AX} - \mathbf{Y})\right] \end{aligned} \tag{6.11}$$

or

$$\boxed{\mathbf{X} \leftarrow \mathcal{P}_\Omega\left[(1 - \eta_0)\mathbf{X} + \eta_0\, \mathbf{A}^\dagger\mathbf{Y}\right]} \tag{6.12}$$

where $\mathbf{A}^\dagger = \left(\mathbf{A}^T\mathbf{A}\right)^{-1}\mathbf{A}^T$ is the Moore-Penrose inverse of the matrix $\mathbf{A}$, and the parameter η_0 controls the relaxation and it is normally set to $\eta_0 \simeq 1$.

Similarly for $\mathbf{A}$, the learning rule is derived as

$$\begin{aligned} \mathrm{vec}(\mathbf{A}) &\leftarrow \mathcal{P}_\Omega\left[\mathrm{vec}(\mathbf{A}) - \eta_0\, (\mathbf{H_A})^{-1}\, \mathrm{vec}(\mathbf{G_A})\right] \\ &= \mathcal{P}_\Omega\left[\mathrm{vec}(\mathbf{X}) - \eta_0\, \left(\mathbf{XX}^T \otimes \mathbf{I}_I\right)^{-1}\, \mathrm{vec}(\mathbf{G_A})\right] \\ &= \mathcal{P}_\Omega\left[\mathrm{vec}(\mathbf{X}) - \eta_0\, \left(\left(\mathbf{XX}^T\right)^{-1} \otimes \mathbf{I}_I\right)\, \mathrm{vec}(\mathbf{G_A})\right], \end{aligned} \tag{6.13}$$

and simplifies to

$$\begin{aligned} \mathbf{A} &\leftarrow \mathcal{P}_\Omega\left[\mathbf{A} - \eta_0\, \mathbf{G_A}\left(\mathbf{XX}^T\right)^{-1}\right] \\ &= \mathcal{P}_\Omega\left[\mathbf{A} - \eta_0\, (\mathbf{AX} - \mathbf{Y})\mathbf{X}^T\left(\mathbf{XX}^T\right)^{-1}\right], \end{aligned} \tag{6.14}$$

or

$$\boxed{\mathbf{A} \leftarrow \mathcal{P}_\Omega\left[(1 - \eta_0)\mathbf{A} + \eta_0\, \mathbf{YX}^\dagger\right].} \tag{6.15}$$

The final learning rules (6.15) and (6.12) avoid computing the inverse of the whole Hessians which are usually very large matrices. In a particular case that $\eta_0 = 1$, these learning rules simplify to the ALS algorithm (4.12),

(4.13) presented in Chapter 4

$$\mathbf{A} = \mathcal{P}_{\Omega}\left[\mathbf{Y}\mathbf{X}^{\dagger}\right], \tag{6.16}$$

$$\mathbf{X} = \mathcal{P}_{\Omega}\left[\mathbf{A}^{\dagger}\mathbf{Y}\right]. \tag{6.17}$$

The QNE algorithm (Quasi-Newton method based on Euclidean distance) is given in Algorithm 6.1. Practical implementation will be discussed in Section 6.1.4.

Algorithm 6.1: QNE-NMF

Input: $\mathbf{Y} \in \mathbb{R}_+^{I \times T}$: input data, J: rank of approximation
Output: $\mathbf{A} \in \mathbb{R}_+^{I \times J}$ and $\mathbf{X} \in \mathbb{R}_+^{J \times T}$ such that the cost function (6.7) is minimized.

begin
  initialization for $\mathbf{A}, \mathbf{X}$
  repeat
    $\mathbf{X} \leftarrow \text{QNE}(\mathbf{Y}, \mathbf{A}, \mathbf{X})$ /* Update X */
    $\mathbf{A} \leftarrow \text{QNE}(\mathbf{Y}^T, \mathbf{X}^T, \mathbf{A}^T)^T$ /* Update A */
  until *a stopping criterion is met* /* convergence condition */
end

function $\mathbf{X} = \text{QNE}(\mathbf{Y}, \mathbf{A}, \mathbf{X})$
begin
  $\mathbf{R} = \mathbf{A}^{\dagger}\mathbf{Y}$
  repeat
    $\mathbf{X} \leftarrow [(1-\eta_0)\mathbf{X} + \eta_0 \mathbf{R}]_+$ /* Update X */
  until *a stopping criterion is met* /* convergence condition */
end

6.1.2 Projected Quasi-Newton for Alpha-Divergence

The idea for second-order NMF algorithms for the minimization of Alpha-divergence was originally presented in [33,34] and is based on the projected quasi-Newton method, whereby the adaptively-regularized Hessian within the Levenberg-Marquardt approach is inverted using the QR decomposition.

The Alpha-divergence [8,11] can be expressed as follows:

$$D_A^{(\alpha)}(\mathbf{Y}||\mathbf{AX}) = \sum_{it} y_{it}\frac{(y_{it}/q_{it})^{\alpha-1}-1}{\alpha(\alpha-1)} + \frac{q_{it}-y_{it}}{\alpha}, \quad q_{it} = [\mathbf{AX}]_{it}, \quad y_{it} = [\mathbf{Y}]_{it}. \tag{6.18}$$

The generalized Kullback-Leibler (KL) divergence (I-divergence) can be obtained for $\alpha \to 1$, however, for $\alpha \to 0$ we have the dual KL divergence. The Pearson's, Hellinger's, and Neyman's chi-square distances are obtained for $\alpha = 2, 0.5, -1$, respectively.

For $\alpha \neq 0$, the gradient $\mathbf{G}_{\mathbf{X}}^{(\alpha)} \in \mathbb{R}^{J \times T}$ with respect to $\mathbf{X}$ can be expressed as

$$\mathbf{G}_{\mathbf{X}}^{(\alpha)} = \nabla_{\mathbf{X}} D_A^{(\alpha)} = \frac{1}{\alpha}\mathbf{A}^T\left(\mathbf{1}_{I\times T} - (\mathbf{Y} \oslash \mathbf{AX})^{.[\alpha]}\right), \tag{6.19}$$

where the symbol $\oslash$ and $[\cdot]^{[.\alpha]}$ denote element-wise division and element-wise raised to power α, respectively, thus yielding the Hessian in the form:

$$\begin{aligned}
\mathbf{H}_{\mathbf{X}}^{(\alpha)} &= \nabla_{\mathbf{X}} \mathbf{G}_{\mathbf{X}} = \nabla_{\mathbf{U}} \mathbf{G}_{\mathbf{X}} \, \nabla_{\mathbf{Q}} \mathbf{U} \, \nabla_{\mathbf{X}} \mathbf{Q} \\
&= \frac{-1}{\alpha} \left(\mathbf{I}_T \otimes \mathbf{A}^T\right) (-\alpha) \operatorname{diag}\left\{\operatorname{vec}\left(\mathbf{Y}^{.[\alpha]} \oslash \mathbf{Q}^{.[\alpha+1]}\right)\right\} (\mathbf{I}_T \otimes \mathbf{A}) \\
&= \left(\mathbf{I}_T \otimes \mathbf{A}^T\right) \operatorname{diag}\left\{\operatorname{vec}\left(\mathbf{Y}^{.[\alpha]} \oslash \mathbf{Q}^{.[\alpha+1]}\right)\right\} (\mathbf{I}_T \otimes \mathbf{A}), \qquad (6.20)
\end{aligned}$$

where $\mathbf{U} = (\mathbf{Y} \oslash \mathbf{AX})^{.[\alpha]}$.

Similarly for $\mathbf{A}$, we have

$$\mathbf{G}_{\mathbf{A}}^{(\alpha)} = \nabla_{\mathbf{A}} D_A^{(\alpha)} = \frac{1}{\alpha} \left(\mathbf{1}_{I \times T} - (\mathbf{Y} \oslash \mathbf{AX})^{.[\alpha]}\right) \mathbf{X}^T \in \mathbb{R}^{I \times J}, \qquad (6.21)$$

and the Hessian $\mathbf{H}_{\mathbf{A}}^{(\alpha)} \in \mathbb{R}^{IJ \times IJ}$ becomes:

$$\begin{aligned}
\mathbf{H}_{\mathbf{A}}^{(\alpha)} &= \nabla_{\mathbf{A}} \mathbf{G}_{\mathbf{A}}^{(\alpha)} = \nabla_{\mathbf{U}} \mathbf{G}_{\mathbf{A}}^{(\alpha)} \, \nabla_{\mathbf{Q}} \mathbf{U} \, \nabla_{\mathbf{A}} \mathbf{Q} \\
&= \frac{-1}{\alpha} (\mathbf{X} \otimes \mathbf{I}_I) (-\alpha) \operatorname{diag}\left\{\operatorname{vec}\left(\mathbf{Y}^{.[\alpha]} \oslash \mathbf{Q}^{.[\alpha+1]}\right)\right\} \left(\mathbf{X}^T \otimes \mathbf{I}_I\right) \\
&= (\mathbf{X} \otimes \mathbf{I}_I) \operatorname{diag}\left\{\operatorname{vec}\left(\mathbf{Y}^{.[\alpha]} \oslash \mathbf{Q}^{.[\alpha+1]}\right)\right\} \left(\mathbf{X}^T \otimes \mathbf{I}_I\right). \qquad (6.22)
\end{aligned}$$

Although there are some methods for reducing computational cost of computing the inverse of Hessian matrices, such as the Q-less QR factorization, we will derive fast learning rules which avoid computing the inverse of the large Hessian matrices in the projected quasi-Newton method applied to the Alpha-divergence.

Note that the Hessian for $\mathbf{X}$ is a partitioned matrix that has the block diagonal form

$$\mathbf{H}_{\mathbf{X}}^{(\alpha)} = \begin{bmatrix} \mathbf{H}_{\mathbf{X}}^{(1)} & & & 0 \\ & \mathbf{H}_{\mathbf{X}}^{(2)} & & \\ & & \ddots & \\ 0 & & & \mathbf{H}_{\mathbf{X}}^{(T)} \end{bmatrix}, \qquad (6.23)$$

where the submatrices are defined as

$$\mathbf{H}_{\mathbf{X}}^{(t)} = \mathbf{A}^T \operatorname{diag}\{v_t\} \mathbf{A} \quad \in \mathbb{R}^{J \times J}, \qquad (t = 1, 2, \ldots, T), \qquad (6.24)$$

$$\mathbf{V} = [v_1, \ldots, v_T] = \mathbf{Y}^{.[\alpha]} \oslash \mathbf{Q}^{.[\alpha+1]} \quad \in \mathbb{R}^{I \times T}. \qquad (6.25)$$

Thus inverse of these large matrices can be expressed in terms of their submatrices as follows

$$\left(\mathbf{H}_{\mathbf{X}}^{(\alpha)}\right)^{-1} = \begin{bmatrix} \mathbf{H}_{\mathbf{X}}^{(1)} & & & 0 \\ & \mathbf{H}_{\mathbf{X}}^{(2)} & & \\ & & \ddots & \\ 0 & & & \mathbf{H}_{\mathbf{X}}^{(T)} \end{bmatrix}^{-1} = \begin{bmatrix} \left(\mathbf{H}_{\mathbf{X}}^{(1)}\right)^{-1} & & & 0 \\ & \left(\mathbf{H}_{\mathbf{X}}^{(2)}\right)^{-1} & & \\ & & \ddots & \\ 0 & & & \left(\mathbf{H}_{\mathbf{X}}^{(T)}\right)^{-1} \end{bmatrix}. \qquad (6.26)$$

By considering vec $(\mathbf{G_X})$ as a partitioned matrix formed by its column vectors $\boldsymbol{g}_t$, and based on the multiplicative property of partitioned matrices $\mathbf{A}(I \times J)$, $\mathbf{B}(M \times N)$, $\mathbf{C}(J \times K)$ and $\mathbf{D}(N \times K)$

$$\begin{bmatrix} \mathbf{A} & \\ & \mathbf{B} \end{bmatrix} \begin{bmatrix} \mathbf{C} \\ \mathbf{D} \end{bmatrix} = \begin{bmatrix} \mathbf{AC} \\ \mathbf{BD} \end{bmatrix}, \tag{6.27}$$

we draw the learning rule for each column vector $\boldsymbol{x}_t$ as follows

$$\boxed{\boldsymbol{x}_t \leftarrow [\boldsymbol{x}_t - (\mathbf{H}_{\mathbf{X}}^{(t)})^{-1} \boldsymbol{g}_t]_+. \qquad (t = 1, 2, \ldots, T).} \tag{6.28}$$

Indeed, $\mathbf{H}_{\mathbf{X}}^{(t)}$ (6.24) is the positive definite matrix of second-order partial derivatives of the Alpha-divergence with respect to column vector $\boldsymbol{x}_t$. Thus, each linear system

$$\mathbf{H}_{\mathbf{X}}^{(t)} \tilde{\boldsymbol{g}}_t = \boldsymbol{g}_t \tag{6.29}$$

can be solved by the Conjugate Gradients (CG) method or Preconditioned CG (PCG) method. In MATLAB, this can be done with the `pcg` function

$$\tilde{\boldsymbol{g}}_t = \texttt{pcg}(\mathbf{H}_{\mathbf{X}}^{(t)}, \boldsymbol{g}_t).$$

This type of learning rule is suitable for large scale systems $I > 1000$, $T > 1000$, or for a computer with limited resources. However, for a small data, global learning rule as in (6.6) is always faster.

We note that the Hessian for $\mathbf{A}$ (6.22) does not have the block diagonal form, therefore, a local learning rule for $\mathbf{A}$ is slightly different. Actually, the Hessian $\mathbf{H_A}$ is a permutated version of the block diagonal matrix. To prove this property, we first consider the permutation matrix $\mathbf{P}_{I,T}$ that transforms vec $(\mathbf{A})$ into vec $\left(\mathbf{A}^T\right)$:

$$\text{vec}\left(\mathbf{A}^T\right) = \mathbf{P}_{I,J}\text{vec}\,(\mathbf{A}). \tag{6.30}$$

This matrix operator $\mathbf{P}_{I,J}$ has the size of $(IJ \times IJ)$, depends only on the dimensions I and J, and it is expressed as

$$\mathbf{P}_{I,J} = \sum_{i=1}^{I} \sum_{j=1}^{J} \left(\mathbf{E}_{ij} \otimes \mathbf{E}_{ij}^T\right), \tag{6.31}$$

where each $\mathbf{E}_{ij}$ $(I \times J)$ has entry 1 in the position (i, j), and all other entries are zero. Moreover, $\mathbf{P}_{I,J}$ is an orthogonal matrix

$$\mathbf{P}_{I,T} = \mathbf{P}_{T,I}^T = \mathbf{P}_{T,I}^{-1} \tag{6.32}$$

and also related to the Kronecker product

$$\mathbf{A} \otimes \mathbf{B} = \mathbf{P}_{I,M}\,(\mathbf{B} \otimes \mathbf{A})\,\mathbf{P}_{N,J}, \tag{6.33}$$

where $\mathbf{B} \in \mathbb{R}^{M \times N}$.

Let $\boldsymbol{\Psi} = \text{diag}\,\{\text{vec}\,(\mathbf{Y})\}$ be the diagonal matrix $(IT \times IT)$. It is clear from the definition (6.30) that the premultiplication of $\boldsymbol{\Psi}$ by $\mathbf{P}_{I,T}$, that is $\mathbf{P}_{I,T}\,\boldsymbol{\Psi}$, commutes the $k_{i,t}(= i + (t-1)I)$-th row to the $l_{t,i}(= t + (i-1)T)$-th row, $\forall i, t$, whereas the postmultiplication of $\boldsymbol{\Psi}$ by $\mathbf{P}_{T,I}$ commutes the $k_{i,t}$-th column to the $l_{t,i}$-th column, $\forall i, t$. The result of both row and column permutations: $\mathbf{P}_{I,T}\,\boldsymbol{\Psi}\,\mathbf{P}_{T,I}$, is that the diagonal entries $\psi_{k_{i,t}k_{i,t}} = [\mathbf{Y}]_{it}$ and $\psi_{l_{t,i}l_{t,i}} = [\mathbf{Y}^T]_{ti}$ exchange their positions

$$\mathbf{P}_{I,T}\ \text{diag}\,\{\text{vec}\,(\mathbf{Y})\}\ \mathbf{P}_{T,I} = \text{diag}\left\{\text{vec}\left(\mathbf{Y}^T\right)\right\}. \tag{6.34}$$

This relationship is illustrated in Figure 6.1.

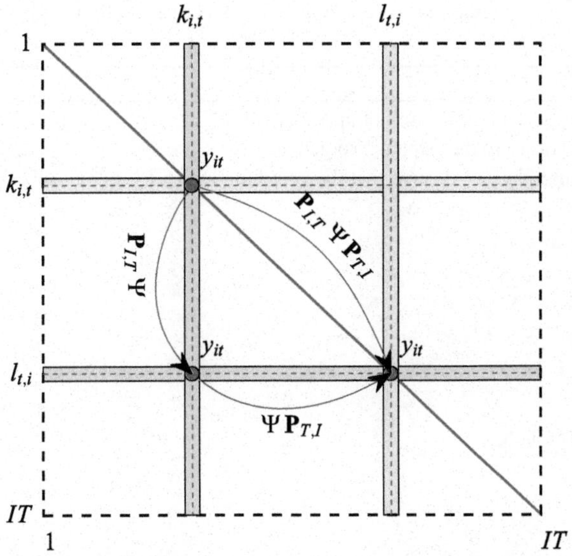

Figure 6.1 Illustration of the permutation of the diagonal matrix $\boldsymbol{\Psi} = \text{diag}\{\text{vec}\,(\mathbf{Y})\}$. Premultiplication permutes row vectors, while postmultiplication permutes column vectors. The result is the permutation of diagonal entries: $\mathbf{P}_{I,T}\,\text{diag}\{\text{vec}\,(\mathbf{Y})\}\mathbf{P}_{T,I} = \text{diag}\{\text{vec}\,(\mathbf{Y}^T)\}$.

With the above result, the Hessian matrix with respect to $\mathbf{A}$ in (6.22) can be expressed as follows

$$\begin{aligned}
\mathbf{H}_{\mathbf{A}}^{(\alpha)} &= (\mathbf{X} \otimes \mathbf{I}_I)\,\text{diag}\left\{\text{vec}\left(\mathbf{Y}^{.[\alpha]} \oslash \mathbf{Q}^{.[\alpha+1]}\right)\right\}\left(\mathbf{X}^T \otimes \mathbf{I}_I\right)\\
&= \mathbf{P}_{J,I}\,(\mathbf{I}_I \otimes \mathbf{X})\,\mathbf{P}_{I,T}\,\text{diag}\left\{\text{vec}\left(\mathbf{Y}^{.[\alpha]} \oslash \mathbf{Q}^{.[\alpha+1]}\right)\right\}\mathbf{P}_{T,I}\left(\mathbf{I}_I \otimes \mathbf{X}^T\right)\mathbf{P}_{I,J}\\
&= \mathbf{P}_{J,I}\,(\mathbf{I}_I \otimes \mathbf{X})\,\text{diag}\left\{\text{vec}\left(\left[\mathbf{Y}^{.[\alpha]} \oslash \mathbf{Q}^{.[\alpha+1]}\right]^T\right)\right\}\left(\mathbf{I}_I \otimes \mathbf{X}^T\right)\mathbf{P}_{I,J}\\
&= \mathbf{P}_{J,I}\begin{bmatrix}\mathbf{H}_{\mathbf{A}^T}^{(1)} & & & 0\\ & \mathbf{H}_{\mathbf{A}^T}^{(2)} & & \\ & & \ddots & \\ 0 & & & \mathbf{H}_{\mathbf{A}^T}^{(I)}\end{bmatrix}\mathbf{P}_{I,J}\,,
\end{aligned} \tag{6.35}$$

where the submatrices $\mathbf{H}_{\mathbf{A}^T}^{(i)}$ are given by

$$\mathbf{H}_{\mathbf{A}^T}^{(i)} = \mathbf{X}\,\text{diag}\left\{\underline{\boldsymbol{y}}_i^{.[\alpha]} \oslash \underline{\boldsymbol{q}}_i^{.[\alpha+1]}\right\}\mathbf{X}^T \qquad \in \mathbb{R}^{J\times J}, \qquad (i = 1, 2, \ldots, I). \tag{6.36}$$

Actually, the partitioned matrix in (6.35) is exactly the Hessian of the cost function with respect to the matrix $\mathbf{A}^T$

$$\mathbf{H}_{\mathbf{A}}^{(\alpha)} = \mathbf{P}_{J,I}\,\mathbf{H}_{\mathbf{A}^T}^{(\alpha)}\,\mathbf{P}_{I,J}. \tag{6.37}$$

The learning rules for $\mathbf{A}$ can be derived from the transposed system: $\mathbf{X}^T\mathbf{A}^T = \mathbf{Y}^T$.

The QNA (Quasi-Newton method based on Alpha-divergence) algorithm is given in Algorithm 6.2.

Algorithm 6.2: QNA-NMF

Input: $\mathbf{Y} \in \mathbb{R}_+^{I \times T}$: input data, J: rank of approximation
α: divergence order in the Alpha-divergence.
Output: $\mathbf{A} \in \mathbb{R}_+^{I \times J}$ and $\mathbf{X} \in \mathbb{R}_+^{J \times T}$ such that the cost function (6.18) is minimized.

begin
  initialization for $\mathbf{A}, \mathbf{X}$
  repeat
    $\mathbf{X} \leftarrow \texttt{QNA}(\mathbf{Y}, \mathbf{A}, \mathbf{X}, \alpha)$ /* Update X */
    $\mathbf{A} \leftarrow \texttt{QNA}(\mathbf{Y}^T, \mathbf{X}^T, \mathbf{A}^T, \alpha)^T$ /* Update A */
  until *a stopping criterion is met* /* convergence condition */
end

function $\mathbf{X} = \texttt{QNA}(\mathbf{Y}, \mathbf{A}, \mathbf{X}, \alpha)$
begin
  repeat
    $\mathbf{Q} = \mathbf{A}\mathbf{X}$
    $\mathbf{G} = \dfrac{1}{\alpha} \mathbf{A}^T \left(\mathbf{1}_{I \times T} - \mathbf{Y}^{.[\alpha]} \oslash \mathbf{Q}^{.[\alpha]}\right)$ /* Gradient with respect to X */
    parfor $t = 1$ to T **do** /* parallel loop for update X */
      $\mathbf{H}_{\mathbf{X}}^{(t)} = \mathbf{A}^T \left(\boldsymbol{y}_t^{.[\alpha]} \oslash \boldsymbol{q}_t^{.[\alpha+1]}\right) \mathbf{A}$
      $\tilde{\boldsymbol{g}}_t = \texttt{pcg}(\mathbf{H}_{\mathbf{X}}^{(t)}, \boldsymbol{g}_t)$ /* solve system $\tilde{\boldsymbol{g}}_t = \mathbf{H}_{\mathbf{X}}^{(t)^{-1}} \boldsymbol{g}_t$ */
      $\boldsymbol{x}_t \leftarrow [\boldsymbol{x}_t - \tilde{\boldsymbol{g}}_t]_+$ /* Update $\boldsymbol{x}_t$ */
    endfor
  until *a stopping criterion is met* /* convergence condition */
end

In the particular case of $\alpha \to 0$, the Alpha-divergence converges to the dual KL I-divergence, that is

$$D_{KL2}(\mathbf{AX}||\mathbf{Y}) = \lim_{\alpha \to 0} D_A^{(\alpha)}(\mathbf{Y}||\mathbf{AX}) = \sum_{it} \left(z_{it} \ln \frac{z_{it}}{y_{it}} + y_{it} - z_{it} \right), \quad z_{it} = [\mathbf{AX}]_{it}, \tag{6.38}$$

and consequently, the gradient and Hessian matrices simplify as

- For $\mathbf{X}$:

$$\mathbf{G}_{\mathbf{X}}^{(0)} = \nabla_{\mathbf{X}} D_{KL2} = \mathbf{A}^T \ln(\mathbf{AX} \oslash \mathbf{Y}) \quad \in \mathbb{R}^{J \times T}, \tag{6.39}$$

and

$$\mathbf{H}_{\mathbf{X}}^{(0)} = \operatorname{diag}\left\{[\mathbf{H}_{\mathbf{X}}^{(t)}]_{t=1,\ldots,T}\right\} \quad \in \mathbb{R}^{JT \times JT}, \tag{6.40}$$

where

$$\mathbf{H}_{\mathbf{X}}^{(t)} = \mathbf{A}^T \operatorname{diag}\left\{\mathbf{1} \oslash \boldsymbol{q}_t\right\} \mathbf{A} \quad \in \mathbb{R}^{J \times J}.$$

- For $\mathbf{A}$:

$$\mathbf{G}_{\mathbf{A}}^{(0)} = \nabla_{\mathbf{A}} D_{KL2} = \ln(\mathbf{AX} \oslash \mathbf{Y})\, \mathbf{X}^T \quad \in \mathbb{R}^{I \times J}, \tag{6.41}$$

and

$$\mathbf{H}_{\mathbf{A}}^{(0)} = \mathbf{P}_{J,I} \operatorname{diag}\left\{[\mathbf{H}_{\mathbf{A}}^{(i)}]_{i=1,\ldots,I}\right\} \mathbf{P}_{I,J} \quad \in \mathbb{R}^{IJ \times IJ}, \tag{6.42}$$

where

$$\mathbf{H}_{\mathbf{A}}^{(i)} = \mathbf{X} \operatorname{diag}\left\{\mathbf{1}^T \oslash \underline{\boldsymbol{q}}_i\right\} \mathbf{X}^T \quad \in \mathbb{R}^{J \times J}.$$

The Alpha-divergence unifies many well-known statistical distances, which makes our NMF algorithm more flexible to various distributions of noise. For $\alpha = 2$ we have the Pearson's distance which can be regarded as a normalized squared Euclidean distance.

6.1.3 Projected Quasi-Newton for Beta-Divergence

The multiplicative algorithms that uses the Beta-divergence were proposed for NMF, for example, in [8,13,22], and also extensively discussed in Chapter 3. The Beta-divergence, given below, can be also minimized with the projected quasi Newton method

$$D_B^{(\beta)}(\mathbf{Y}||\mathbf{AX}) = \sum_{it} \left(y_{it} \frac{y_{it}^{\beta} - q_{it}^{\beta}}{\beta} + \frac{q_{it}^{\beta+1} - y_{it}^{\beta+1}}{\beta+1} \right), \tag{6.43}$$

where $q_{it} = [\mathbf{AX}]_{it}$. Gradients of this cost function with respect to $\mathbf{X}$ and $\mathbf{A}$ are given by

$$\mathbf{G}_{\mathbf{X}}^{(\beta)} = \nabla_{\mathbf{X}} D_B^{(\beta)} = \mathbf{A}^T \left[(\mathbf{Q} - \mathbf{Y}) \circledast \mathbf{Q}^{.[\beta-1]}\right] \quad \in \mathbb{R}^{J \times T}, \tag{6.44}$$

$$\mathbf{G}_{\mathbf{A}}^{(\beta)} = \nabla_{\mathbf{A}} D_B^{(\beta)} = \left[\mathbf{Q}^{.[\beta-1]} \circledast (\mathbf{Q} - \mathbf{Y})\right] \mathbf{X}^T \quad \in \mathbb{R}^{I \times J}. \tag{6.45}$$

This leads to the Hessian for $\mathbf{X}$

$$\mathbf{H}_{\mathbf{X}}^{(\beta)} = \left(\mathbf{I}_T \otimes \mathbf{A}^T\right) \operatorname{diag}\left\{\operatorname{vec}\left(\beta \mathbf{Q}^{.[\beta-1]} - (\beta - 1)\, \mathbf{Y} \circledast \mathbf{Q}^{.[\beta-2]}\right)\right\} (\mathbf{I}_T \otimes \mathbf{A}) \tag{6.46}$$

$$= \operatorname{diag}\left\{[\mathbf{H}_{\mathbf{X}}^{(t)}]_{t=1,\ldots,T}\right\} \in \mathbb{R}^{JT \times JT}, \tag{6.47}$$

where

$$\mathbf{H}_{\mathbf{X}}^{(t)} = \mathbf{A}^T \operatorname{diag}\left\{\left(\beta \boldsymbol{q}_t - (\beta - 1)\, \boldsymbol{y}_t\right) \circledast \boldsymbol{q}_t^{.[\beta-2]}\right\} \mathbf{A} \quad \in \mathbb{R}^{I \times I}. \tag{6.48}$$

The Hessian for $\mathbf{A}$ is given by

$$\mathbf{H}_{\mathbf{A}}^{(\beta)} = (\mathbf{X} \otimes \mathbf{I}_I) \operatorname{diag}\left\{\operatorname{vec}\left(\beta \mathbf{Q}^{.[\beta-1]} - (\beta - 1)\, \mathbf{Y} \circledast \mathbf{Q}^{.[\beta-2]}\right)\right\} \left(\mathbf{X}^T \otimes \mathbf{I}_I\right) \tag{6.49}$$

$$= \mathbf{P}_{I,T} \operatorname{diag}\left\{[\mathbf{H}_{\mathbf{A}}^{(i)}]_{i=1,\ldots,I}\right\} \mathbf{P}_{T,I}, \tag{6.50}$$

where

$$\mathbf{H}_{\mathbf{A}}^{(i)} = \mathbf{X} \operatorname{diag}\left\{\left(\beta \underline{\boldsymbol{q}}_i - (\beta - 1)\, \underline{\boldsymbol{y}}_i\right) \circledast \underline{\boldsymbol{q}}_i^{.[\beta-2]}\right\} \mathbf{X}^T \quad \in \mathbb{R}^{J \times J}. \tag{6.51}$$

Similarly to the learning rule for Alpha-divergence, the learning rule (6.28) can be applied for updating each column vectors $\boldsymbol{x}_t$ as

$$\boxed{\boldsymbol{x}_t \leftarrow [\boldsymbol{x}_t - \tilde{\boldsymbol{g}}_t]_+, \quad (t = 1, 2, \ldots, T),} \tag{6.52}$$

where $\tilde{\boldsymbol{g}}_t$ is the solution of the linear system

$$\boxed{\mathbf{H}_{\mathbf{X}}^{(t)} \, \tilde{\boldsymbol{g}}_t = \boldsymbol{g}_t}. \tag{6.53}$$

In fact, $\mathbf{H}_{\mathbf{X}}^{(t)}$ (6.48) is not sometimes positive definite but symmetric. This problem is discussed in the next section. However, each linear system (6.53) can be solved by the symmetric LQ method [3,27] with the `symmlq` function

$$\tilde{\boldsymbol{g}}_t = \texttt{symmlq}(\mathbf{H}_{\mathbf{X}}^{(t)}, \boldsymbol{g}_t),$$

where $\boldsymbol{g}_t$ is the t-th column of $\mathbf{G}_{\mathbf{X}}$. The learning rules for $\mathbf{A}$ can be derived from the transposed system: $\mathbf{X}^T\mathbf{A}^T = \mathbf{Y}^T$. The QNB (Quasi-Newton method based on Beta-divergence) algorithm is given in Algorithm 6.3.

Algorithm 6.3: QNB-NMF

Input: $\mathbf{Y} \in \mathbb{R}_+^{I\times T}$: input data, J: rank of approximation
β: divergence order in the Beta-divergence.
Output: $\mathbf{A} \in \mathbb{R}_+^{I\times J}$ and $\mathbf{X} \in \mathbb{R}_+^{J\times T}$ such that the cost function (6.43) is minimized.

begin
  initialization for $\mathbf{A}, \mathbf{X}$
  repeat
    $\mathbf{X} \leftarrow \text{QNB}(\mathbf{Y}, \mathbf{A}, \mathbf{X}, \beta)$ /* Update X */
    $\mathbf{A} \leftarrow \text{QNB}(\mathbf{Y}^T, \mathbf{X}^T, \mathbf{A}^T, \beta)^T$ /* Update A */
  until *a stopping criterion is met*   /* convergence condition */
end
function $\mathbf{X} = \text{QNB}(\mathbf{Y}, \mathbf{A}, \mathbf{X}, \beta)$
begin
  repeat
    $\mathbf{Q} = \mathbf{A}\mathbf{X}$
    $\mathbf{G} = \mathbf{A}^T \left[(\mathbf{Q} - \mathbf{Y}) \circledast \mathbf{Q}^{.[\beta-1]}\right]$ /* Gradient with respect to X */
    parfor $t = 1$ **to** T **do**   /* parallel loop for update X */
      $\mathbf{H}_{\mathbf{X}}^{(t)} = \mathbf{A}^T \operatorname{diag}\left\{\left(\beta\,\boldsymbol{q}_t - (\beta - 1)\,\boldsymbol{y}_t\right) \circledast \boldsymbol{q}_t^{.[\beta-2]}\right\}\mathbf{A}$
      $\tilde{\boldsymbol{g}}_t = \texttt{symmlq}(\mathbf{H}_{\mathbf{X}}^{(t)}, \boldsymbol{g}_t)$ /* solve system $\tilde{\boldsymbol{g}}_t = \mathbf{H}_{\mathbf{X}}^{(t)^{-1}} \boldsymbol{g}_t$ */
      $\boldsymbol{x}_t \leftarrow [\boldsymbol{x}_t - \tilde{\boldsymbol{g}}_t]_+$ /* Update $\boldsymbol{x}_t$ */
    endfor
  until *a stopping criterion is met*   /* convergence condition */
end

We recall here that for $\beta = 1$, the QNB algorithm simplifies to the QNE or ALS algorithm; for $\beta = 0$, we obtain the QN algorithm via the generalized Kullback-Leibler I-divergence; and for $\beta = -1$, we receive the QN algorithm via Itakura-Saito distance. Again, the choice of the parameter β depends on the statistics of the data and the Beta-divergence corresponds to the Tweedie models [10,12,13,20,32].

6.1.4 Practical Implementation

In every alternating step of the algorithm, the columns of **A** are normalized to the unit ℓ_1-norm to give:

$$a_{ij} \leftarrow \frac{a_{ij}}{\sum_{i=1}^{I} a_{ij}}.$$

Remark 6.1 *All the Hessians $\mathbf{H_X}$ and $\mathbf{H_A}$ have block-diagonal structures, or can be transformed to block-diagonal structures. We may experience problems with matrix conditioning for very sparse matrices **A** and **X**. For $\alpha > 0$, the Hessian in (6.20) is not positive-definite if $\exists t,\ \forall i : y_{it} = 0$. Similarly, if $\exists i,\ \forall t : y_{it} = 0$, the Hessian (6.22) is also not positive-definite. Conversely, if **Y** has many zero-value entries, the Hessians may be only positive semi-definite. The normalization of the columns in **A** during the alternating minimization is very helpful to keep the Hessians positive-definite, however, they may still be very ill-conditioned, especially in early updates.*

Thus, to avoid the breakdown of Newton iterations, some regularization of the Hessian is essential, which leads to the quasi-Newton iterations. To this end, we employ the Levenberg-Marquardt approach with the exponentially deceasing regularization parameter: $\lambda = \bar{\lambda}_0 + \lambda_0 \exp\{-\tau k\}$. Such regularization substantially reduces the likelihood of ill-conditioning of the Hessian (random initialization) during initial iterations.

We can gain additional control of the convergence by a slight relaxation of the iterative updates. Also, to reduce the computational cost substantially, the inversion of the Hessian can be replaced by the Q-less QR factorization computed with LAPACK, that is

$$\mathbf{H_X} + \lambda \mathbf{I_X} = \mathbf{Q_X R_X}, \qquad \boldsymbol{w}_\mathbf{X} = \mathbf{Q}_\mathbf{X}^T \mathrm{vec}(\mathbf{G_X}),$$

$$\mathbf{H_A} + \lambda \mathbf{I_A} = \mathbf{Q_A R_A}, \qquad \boldsymbol{w}_\mathbf{A} = \mathbf{Q}_\mathbf{A}^T \mathrm{vec}(\mathbf{G_A}),$$

where the matrices $\mathbf{R_X}$ and $\mathbf{R_A}$ are upper triangular, whereas the matrices $\mathbf{Q_X}$ and $\mathbf{Q_A}$ are orthogonal and are not explicitly computed with the Q-less QR factorization. The final form of the updates with the Quasi-Newton (QN) algorithm is

$$\mathrm{vec}\,(\mathbf{X}) \leftarrow \left[\mathrm{vec}\,(\mathbf{X}) - \eta_0\, \mathbf{R}_\mathbf{X}^{-1} \boldsymbol{w}_\mathbf{X}\right]_+, \tag{6.54}$$

$$\mathrm{vec}\,(\mathbf{A}) \leftarrow \left[\mathrm{vec}\,(\mathbf{A}) - \eta_0\, \mathbf{R}_\mathbf{A}^{-1} \boldsymbol{w}_\mathbf{A}\right]_+, \tag{6.55}$$

where the parameter η_0 controls the relaxation, and it is normally set to $\eta_0 \simeq 1$. The matrices $\mathbf{R_X}$ and $\mathbf{R_A}$ are upper-triangular, thus facilitating the Gaussian elimination in terms of computational complexity of the algorithm (6.54) and (6.55). For ill-conditioned problems, however, a direct inversion or a pseudo-inversion may be a better choice.

The MATLAB code for the above quasi-Newton algorithm is listed in Appendix 6.B.

6.2 Gradient Projection Conjugate Gradient

The projection $\mathcal{P}_\Omega[\cdot] = [\cdot]_+$ simply replaces the negative entries with zeros, but may not provide a monotonic convergence to the true solution. Thus, such a projection sometimes may even lead to a local increase in the value of the cost function as illustrated in Figure 6.2 [21].

For monotonic convergence, we can use the Gradient Projection Conjugate Gradient (GPCG) algorithm that only projects the variables that belong to a so-called inactive set. This algorithm was introduced by More and Toraldo [24] to solve large-scale obstacle problems, elastic-plastic torsion problem, and journal bearing problems; Bardsley [1,2] applied it later for solving large-scale problems in image reconstruction and restoration. This algorithm, based on the "reduced" Newton method, only updates positive components

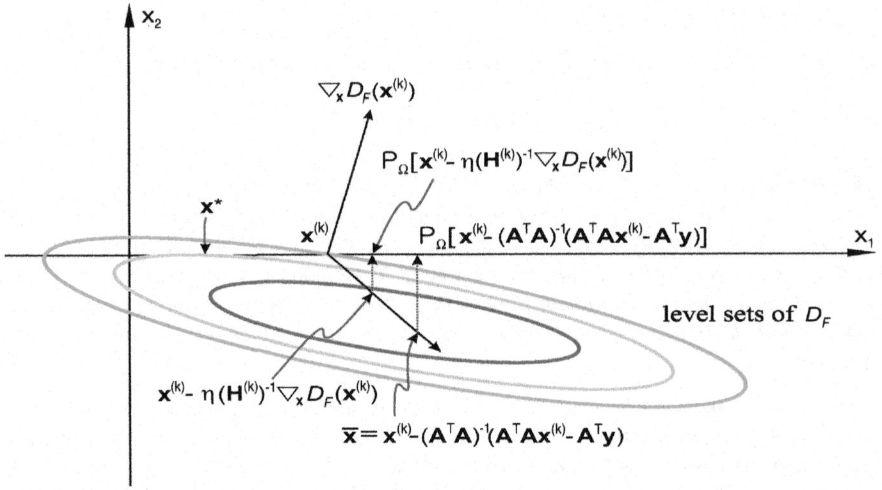

Figure 6.2 Example showing that the projection $\mathcal{P}_{\Omega}[\boldsymbol{x}^{(k)} - \eta \left(\mathbf{H}^{(k)}\right)^{-1} \nabla_x D_F(\boldsymbol{x}^{(k)})]$ (for arbitrary $\eta \in (0, 1)$) moves the current update $\boldsymbol{x}^{(k)}$ from an inner ellipse to an outer one, thus increasing the distance from the minimum. The matrix $\mathbf{H}^{(k)}$ denotes the approximation of the Hessian in the k-th iteration, and $\boldsymbol{x}^*$ is the true solution.

(inactive) of the current approximations of the solution. Due to the inherent sparseness, the set of inactive components is not very large, hence a relatively low computational cost. Such reduced Hessian is not inverted directly, but it is used to compute gradient updates with the Conjugate Gradient (CG) method [18], which converges in a finite number of iterations due to the positive-definiteness of the reduced Hessian. These updates are then used within the Gradient Projection (GP) method to find zero-value components (active) [34].

The GPCG is a hybrid nested iterative method for nonnegative constrained convex optimization. In this section, we shall apply the modified GPCG for updating $\mathbf{A}$ and $\mathbf{X}$, where the cost functions $D(\mathbf{Y}||\mathbf{AX})$ may be defined by the Alpha-divergence (6.18) for $\alpha > 0$, the dual KL I-divergence (6.38) for $\alpha = 0$, and by the Beta-divergence. In NMF, we therefore need to separate optimization tasks:

$$\mathbf{X}^* = \arg\min_{x_{jt} \geq 0} D(\mathbf{Y}||\mathbf{AX}), \tag{6.56}$$

$$\mathbf{A}^* = \arg\min_{a_{ij} \geq 0} D(\mathbf{Y}||\mathbf{AX}). \tag{6.57}$$

We will introduce the GPCG for solving (6.56) for $\mathbf{X}^*$. The solution (6.57) for $\mathbf{A}^*$ can be obtained by applying the GPCG to the transposed system: $\mathbf{X}^T\mathbf{A}^T = \mathbf{Y}^T$.

The solution $\mathbf{X}^*$ should satisfy the KKT conditions, that is

$$\forall j, t: \quad \frac{\partial}{\partial x_{jt}} D(\mathbf{Y}||\mathbf{AX}^*) = 0, \ \text{if } x^*_{jt} > 0, \tag{6.58}$$

$$\forall j, t: \quad \frac{\partial}{\partial x_{jt}} D(\mathbf{Y}||\mathbf{AX}^*) > 0, \ \text{if } x^*_{jt} = 0. \tag{6.59}$$

The GPCG is a two-step algorithm:

1. **Gradient Projection**
   In the first step, the estimate of the solution is updated with Gradient Projection (GP) iterations (see details in Chapter 5),

$$\boxed{\mathbf{X} \leftarrow \mathcal{P}_\Omega[\mathbf{X} - \eta_{\mathbf{X}}^{(p)} \mathbf{G}_{\mathbf{X}}].} \tag{6.60}$$

The step size $\eta_{\mathbf{X}}^{(p)}$ is inexactly estimated with the Armijo rule, that is $\eta_{\mathbf{X}}^{(p)} = \beta^p \eta_{\mathbf{X}}^{(0)}$, for $p = 0, 1, \ldots$, and $\beta \in (0, 1)$. The initial step size $\eta_{\mathbf{X}}^{(0)}$ is defined as the Cauchy point:

$$\boxed{\eta_{\mathbf{X}}^{(0)} = \frac{\boldsymbol{g}_{\mathbf{X}}^T \boldsymbol{g}_{\mathbf{X}}}{\boldsymbol{g}_{\mathbf{X}}^T \mathbf{H}_{\mathbf{X}} \boldsymbol{g}_{\mathbf{X}}},}$$

where $\mathbf{H}_{\mathbf{X}}$ is the Hessian of $D(\mathbf{Y}||\mathbf{AX})$, and the vector $\boldsymbol{g}_{\mathbf{X}}$ is a vectorized version of the gradient matrix $\mathbf{G}_{\mathbf{X}}$: $\boldsymbol{g}_{\mathbf{X}} = \text{vec}\,(\mathbf{G}_{\mathbf{X}})$.

The iterations for $\eta_{\mathbf{X}}^{(p)}$ are terminated at the first p for which

$$D(\mathbf{Y}||\mathbf{AX}^{(p)}) - D(\mathbf{Y}||\mathbf{AX}^{(0)})) \geq \frac{\mu}{\eta_{\mathbf{X}}^{(p)}} ||\mathbf{X}^{(p)} - \mathbf{X}^{(0)})||_F^2. \tag{6.61}$$

2. **Newton Projection**
   In the next step, the estimate $\mathbf{X}$ obtained in the previous step is updated using the Newton iterations (6.6), however, for only positive entries $x_{it} > 0$:

$$\boxed{\text{vec}\left(\mathbf{X}^{\mathcal{Z}}\right) \leftarrow \mathcal{P}_\Omega\left[\text{vec}\left(\mathbf{X}^{\mathcal{Z}}\right) - (\mathbf{H}_{\mathbf{X}}^{\mathcal{Z}})^{-1}\,\text{vec}\left(\mathbf{G}_{\mathbf{X}}^{\mathcal{Z}}\right)\right]} \tag{6.62}$$

where $\mathcal{Z}$ is a binary matrix whose entries are defined as

$$z_{jt} = \begin{cases} 1, & \text{if} \quad x_{jt} > 0, \\ 0, & \text{otherwise} \end{cases}$$

$\mathbf{X}^{\mathcal{Z}}$ is a set of positive entries $x_{it} > 0$, $\mathbf{G}_{\mathbf{X}}^{\mathcal{Z}}$ and $\mathbf{H}_{\mathbf{X}}^{\mathcal{Z}}$ are gradients and Hessian with respect to positive entries x_{it}.

To solve the inverse system in (6.62), we apply the standard CG method [18]

$$\mathbf{H}_{\mathbf{X}}^{\mathcal{Z}} \tilde{\boldsymbol{g}}_{\mathbf{X}} = \text{vec}\left(\mathbf{G}_{\mathbf{X}}^{\mathcal{Z}}\right) = \boldsymbol{g}_{\mathbf{X}}^{\mathcal{Z}}. \tag{6.63}$$

The estimated vector $\tilde{\boldsymbol{g}}_{\mathbf{X}}$ is inserted into the learning rule (6.62) to update $\mathbf{X}^{\mathcal{Z}}$.

The system (6.63) is very sparse, as upon diagonal scaling, the zero-value components of the current GP update are not considered in the evaluation of the gradient used in the next GP update. The subproblem is obtained by removing the components that satisfy (6.59), hence, the name "reduced" Newton optimization.

Remark 6.2 *If $D(\mathbf{Y}||\mathbf{AX})$ is convex and $\mathbf{G}_{\mathbf{X}}$ is a full-rank matrix, the convergence of the CG algorithm is guaranteed in a finite number of iterations due to the positive-definiteness of $\mathbf{H}_{\mathbf{X}}^{\mathcal{Z}}$. The GP with the Armijo rule is always convergent to a unique stationary point of $D(\mathbf{Y}||\mathbf{AX})$ [1,2].*

The criteria for termination of both GP and CG iterations are explained in detail in [1,2,24]. The GPCG algorithm adapted for NMF is given by Algorithm 6.4, and the MATLAB code for updating $\mathbf{X}$ in the GPCG algorithm is given in Appendix 6.B.

Algorithm 6.4: GPCG-NMF

Input: $\mathbf{Y} \in \mathbb{R}_+^{I \times T}$: input data, J: rank of approximation
Output: $\mathbf{A} \in \mathbb{R}_+^{I \times J}$ and $\mathbf{X} \in \mathbb{R}_+^{J \times T}$.

begin
  initialization for $\mathbf{A}, \mathbf{X}$
  repeat
    $\mathbf{X} \leftarrow \text{GPCG}(\mathbf{Y}, \mathbf{A}, \mathbf{X})$ /* Update X */
    $\mathbf{A} \leftarrow \text{GPCG}(\mathbf{Y}^T, \mathbf{X}^T, \mathbf{A}^T)^T$ /* Update A */
  until *a stopping criterion is met* /* convergence condition */
end

function $\mathbf{X} = \text{GPCG}(\mathbf{Y}, \mathbf{A}, \mathbf{X})$
begin
  repeat
    $\mathbf{G_X} = \nabla_{\mathbf{X}} D(\mathbf{Y}||\mathbf{AX}), \quad \mathbf{H_X} = \nabla^2_{\mathbf{X}} D(\mathbf{Y}||\mathbf{AX})$
    $\eta_{\mathbf{X}} = \dfrac{\text{vec}(\mathbf{G_X})^T \text{vec}(\mathbf{G_X})}{\text{vec}(\mathbf{G_X})^T \mathbf{H_X} \text{vec}(\mathbf{G_X})}$
    $\mathbf{X}^{(0)} \leftarrow \mathbf{X}$
    repeat /* Amijo - GP iterations */
      $\mathbf{X} \leftarrow \mathcal{P}_\Omega[\mathbf{X} - \eta_{\mathbf{X}} \mathbf{G_X}]$
      $\eta_{\mathbf{X}} \leftarrow \beta \eta_{\mathbf{X}}$
    until $\left(D(\mathbf{Y}||\mathbf{AX}) - D(\mathbf{Y}||\mathbf{AX}^{(0)})\right) > -\frac{\mu}{\eta_X}||\mathbf{X} - \mathbf{X}^{(0)}||_F^2$
    $\mathcal{Z} = [z_{jt}], z_{jt} = \begin{cases} 1 \text{ if } & x_{jt} > 0, \\ 0 \text{ otherwise}, \end{cases}$ /* select positive entries */
    $\mathbf{G}_{\mathbf{X}}^{\mathcal{Z}} = \nabla_{\mathbf{X}^{\mathcal{Z}}} D(\mathbf{Y}||\mathbf{AX}), \quad \mathbf{H}_{\mathbf{X}}^{\mathcal{Z}} = \nabla^2_{\mathbf{X}^{\mathcal{Z}}} D(\mathbf{Y}||\mathbf{AX})$
    $\tilde{\mathbf{g}}_{\mathbf{X}} = \text{pcg}(\mathbf{H}_{\mathbf{X}}^{\mathcal{Z}}, \text{vec}\left(\mathbf{G}_{\mathbf{X}}^{\mathcal{Z}}\right))$ /* solve with CG algorithm */
    $\eta_{\mathbf{X}} = 1, \mathbf{X}^{(0)} \leftarrow \mathbf{X}$
    repeat /* Newton iterations */
      $\eta_{\mathbf{X}} \leftarrow \beta \eta_{\mathbf{X}}$
      $\text{vec}\left(\mathbf{X}^{\mathcal{Z}}\right) \leftarrow \mathcal{P}_\Omega\left[\text{vec}(\mathbf{X})^{\mathcal{Z}} - \eta_{\mathbf{X}} \tilde{\mathbf{g}}_{\mathbf{X}}\right]$
    until $D(\mathbf{Y}||\mathbf{AX}) \geq D(\mathbf{Y}||\mathbf{AX}^{(0)})$
  until *a stopping criterion is met*
end

6.3 FNMA algorithm

The FNMA algorithm, originally proposed by Kim, Sra, and Dhillon in [21], also separates the updated variables into two sets: active and inactive. In this respect, the FNMA is very similar to the GPCG algorithm; however, it does not involve the CG algorithm to explore inactive variables, that is, the variables that do not satisfy the condition (6.59). In contrast to the GPCG, the active set in the FNMA is defined as:

$$\boxed{\mathcal{A}(\mathbf{X}) = \left\{j, t : x_{jt} = 0, [\nabla_{\mathbf{X}} D(\mathbf{Y}||\mathbf{AX})]_{jt} > 0\right\},} \tag{6.64}$$

which is motivated by the KKT conditions (see Chapter 5). The projection of $\mathbf{X}$ onto the active set (6.64) is obtained with

$$\left[\mathcal{Z}_{+}[\mathbf{X}]\right]_{jt} = \begin{cases} x_{jt}, & \text{if } (j,t) \notin \mathcal{A}(\mathbf{X}) \\ 0, & \text{otherwise.} \end{cases} \tag{6.65}$$

In each iterative step, the FNMA identifies the active set, and then it uses the Newton method to update only the components that belong to the inactive set. The FNMA was proved to monotonically decrease the cost function to the limit point, which is a stationary point of the nonnegative least squares problem [21].

To apply the Newton method, the inverses of the Hessians $\mathbf{H_X} \in \mathbb{R}^{JT \times JT}$ and $\mathbf{H_A} \in \mathbb{R}^{IJ \times IJ}$ must be computed. Since a Hessian matrix may be very large and severely ill-conditioned, many techniques for approximating its inverse have been developed. Typically, the inverse of the Hessian is updated using only the information about the successive gradient vectors, which leads to the Quasi-Newton (QN) method [5,25,28]. The original QN algorithm was proposed by W.C. Davidon in 1959, and later popularized by Fletcher and Powell in 1963; it is also known as the DFP updating formula. The most commonly used QN algorithms are the SR1 formula (for symmetric rank one) and the widespread BFGS method, suggested independently by Broyden, Fletcher, Goldfarb, and Shanno, in 1970.

The first FNMA algorithm, proposed in [21], uses the BFGS method which in the k-th iteration updates the Hessian matrix $\mathbf{H}_{\boldsymbol{x}_t}^{(k)}$ for $\boldsymbol{x}_t$ as

$$\mathbf{H}_{\boldsymbol{x}_t}^{(k+1)} = \mathbf{H}_{\boldsymbol{x}_t}^{(k)} - \frac{\mathbf{H}_{\boldsymbol{x}_t}^{(k)} \boldsymbol{u}_t \boldsymbol{u}_t^T \mathbf{H}_{\boldsymbol{x}_t}^{(k)}}{\boldsymbol{u}_t^T \mathbf{H}_{\boldsymbol{x}_t}^{(k)} \boldsymbol{u}_t} + \frac{\boldsymbol{w}_t \boldsymbol{w}_t^T}{\boldsymbol{u}_t^T \boldsymbol{w}_t}, \tag{6.66}$$

where $\boldsymbol{w}_t = \nabla_{\boldsymbol{x}_t} D(\boldsymbol{y}_t || \mathbf{A}\boldsymbol{x}_t^{(k+1)}) - \nabla_{\boldsymbol{x}_t} D(\boldsymbol{y}_t || \mathbf{A}\boldsymbol{x}_t^{(k)})$, and $\boldsymbol{u}_t = \boldsymbol{x}_t^{(k+1)} - \boldsymbol{x}_t^{(k)}$. Let $\mathbf{S}_{\boldsymbol{x}_t}{}^{(k)}$ denote the inverse of $\mathbf{H}_{\boldsymbol{x}_t}^{(k)}$, then the application of the Sherman-Morrison-Woodbury formula to (6.66) leads to

$$\mathbf{S}_{\boldsymbol{x}_t}{}^{(k+1)} = \mathbf{S}_{\boldsymbol{x}_t}{}^{(k)} - \frac{(\mathbf{S}_{\boldsymbol{x}_t}{}^{(k)} \boldsymbol{w}_t \boldsymbol{u}_t^T + \boldsymbol{u}_t \boldsymbol{w}_t^T \mathbf{S}_{\boldsymbol{x}_t}{}^{(k)})}{\boldsymbol{u}_t^T \boldsymbol{w}_t} + \left(1 + \frac{\boldsymbol{w}_t^T \mathbf{S}_{\boldsymbol{x}_t}{}^{(k)} \boldsymbol{w}_t}{\boldsymbol{u}_t^T \boldsymbol{w}_t}\right) \frac{\boldsymbol{u}_t \boldsymbol{u}_t^T}{\boldsymbol{u}_t^T \boldsymbol{w}_t}. \tag{6.67}$$

For the squared Euclidean cost function $\nabla_{\boldsymbol{x}_t} D_F(\boldsymbol{y}_t || \mathbf{A}\boldsymbol{x}_t^{(k)}) = \mathbf{A}^T \mathbf{A} \boldsymbol{x}_t^{(k)} - \mathbf{A}^T \boldsymbol{y}_t$, (6.67) can be rewritten as

$$\begin{aligned} \mathbf{S}_{\boldsymbol{x}_t}{}^{(k+1)} &= \mathbf{S}_{\boldsymbol{x}_t}{}^{(k)} - \frac{(\mathbf{S}_{\boldsymbol{x}_t}{}^{(k)} \mathbf{A}^T \mathbf{A} \boldsymbol{u}_t \boldsymbol{u}_t^T + \boldsymbol{u}_t \boldsymbol{u}_t^T \mathbf{A}^T \mathbf{A} \mathbf{S}_{\boldsymbol{x}_t}{}^{(k)})}{\boldsymbol{u}_t^T \mathbf{A}^T \mathbf{A} \boldsymbol{u}_t} \\ &+ \left(1 + \frac{\boldsymbol{u}_t^T \mathbf{A}^T \mathbf{A} \mathbf{S}_{\boldsymbol{x}_t}{}^{(k)} \mathbf{A}^T \mathbf{A} \boldsymbol{u}_t}{\boldsymbol{u}_t^T \mathbf{A}^T \mathbf{A} \boldsymbol{u}_t}\right) \frac{\boldsymbol{u}_t \boldsymbol{u}_t^T}{\boldsymbol{u}_t^T \mathbf{A}^T \mathbf{A} \boldsymbol{u}_t}. \end{aligned} \tag{6.68}$$

The pseudo-code for the FNMA-NMF algorithm [21] that uses the QN method (e.g. the update given by (6.68)) for calculating $\mathbf{S}_{\boldsymbol{x}_t}$ (the inverse of the Hessian for $\boldsymbol{x}_t$) is given in Algorithm 6.5:

Algorithm 6.5: FNMA-NMF

Input: $\mathbf{Y} \in \mathbb{R}_+^{I \times T}$: input data, J: rank of approximation
Output: $\mathbf{A} \in \mathbb{R}_+^{I \times J}$ and $\mathbf{X} \in \mathbb{R}_+^{J \times T}$.

begin
  initialization for $\mathbf{A}$, $\mathbf{X}$
  repeat
    $\mathbf{X} \leftarrow \text{FNMA}(\mathbf{Y}, \mathbf{A}, \mathbf{X})$ /* Update $\mathbf{X}$ */
    $\mathbf{A} \leftarrow \text{FNMA}(\mathbf{Y}^T, \mathbf{X}^T, \mathbf{A}^T)^T$ /* Update $\mathbf{A}$ */
  until *a stopping criterion is met* /* convergence condition */
end
function $\mathbf{X} = \text{FNMA}(\mathbf{Y}, \mathbf{A}, \mathbf{X})$
begin
  $\mathbf{S_X} = \mathbf{I}_J$
  repeat
    $\mathbf{G_X} = \nabla_{\mathbf{X}} D(\mathbf{Y}||\mathbf{AX})$
    $\mathcal{A}(\mathbf{X}) = \left\{ j, t : x_{jt} = 0, [\mathbf{G_X}]_{jt} > 0 \right\}$
    $\mathbf{P_X} = [\mathbf{p}_1, \ldots, \mathbf{p}_T] = \mathcal{Z}_+[\mathbf{G_X}]$ /* according to (6.65) */
    foreach $t = 1, \ldots, T$ **do**
      Compute $\mathbf{S}_{\mathbf{x}_t}$ with (6.67)
      $\tilde{\mathbf{p}}_t \leftarrow \mathcal{Z}_+[\mathbf{S}_{\mathbf{x}_t} \mathbf{p}_t]$
      $\tilde{\mathbf{P}}_{\mathbf{X}} = [\tilde{\mathbf{p}}_1, \ldots, \tilde{\mathbf{p}}_T]$
      Compute $\eta_{\mathbf{X}}$ using any line-search method
      $\mathbf{X} \leftarrow \mathcal{P}_\Omega \left[\mathbf{X} - \eta_{\mathbf{X}} \tilde{\mathbf{P}}_{\mathbf{X}} \right]$
    end
  until *a stopping criterion is met*
end

The learning rate η can be computed with any line search method, e.g. with the Armijo rule. Since the Hessian matrices for the Euclidean cost function with respect to $\mathbf{X}$ and $\mathbf{A}$ are block diagonal matrices with the same blocks, the above-presented FNMA algorithm can be considerably simplified, as the blocks for $\mathbf{X}$ and $\mathbf{A}$ can be expressed by small matrices $\mathbf{A}^T\mathbf{A} \in \mathbb{R}^{J \times J}$ and $\mathbf{X}\mathbf{X}^T \in \mathbb{R}^{J \times J}$, and thus the inversion of the blocks can be easily performed. The pseudo-code for the Reduced FNMA (R-FNMA) algorithm that uses such properties of the Hessians is listed in Algorithm 6.6.

Both FNMA algorithms can be easily implemented. The MATLAB code for the function $R-FNMA(\cdot,\cdot,\cdot)$ of the R-FNMA-NMF algorithm, where the learning rate η is estimated with the Armijo rule is listed in Appendix 6.B.

6.4 NMF with Quadratic Programming

In this section, we discuss the NMF method [35] that uses Quadratic Programming (QP) optimization. The motivation for the approach described in [35] was the Interior-Point Trust-Region (IPTR) method which was proposed in [31], and later, modified by Calvetti *et al.* [7]. Rojas and Steihaug [31] considered a least-squares problem with nonnegativity and ℓ_2-norm constraints. They showed that such a problem can be transformed into a QP problem with a logarithmic barrier function. They also used the LSTRS method [30] to solve a trust-region subproblem within the QP problem. The LSTRS boils a QP problem down to a parameterized

Algorithm 6.6: R-FNMA-NMF

Input: $\mathbf{Y} \in \mathbb{R}_{+}^{I \times T}$: input data, J: rank of approximation
Output: $\mathbf{A} \in \mathbb{R}_{+}^{I \times J}$ and $\mathbf{X} \in \mathbb{R}_{+}^{J \times T}$.

begin
  initialization for $\mathbf{A}, \mathbf{X}$
  repeat
    $\mathbf{X} \leftarrow$ `R-FNMA`$(\mathbf{Y}, \mathbf{A}, \mathbf{X})$ `/* Update X */`
    $\mathbf{A} \leftarrow$ `R-FNMA`$(\mathbf{Y}^T, \mathbf{X}^T, \mathbf{A}^T)^T$ `/* Update A */`
  until *a stopping criterion is met* `/* convergence condition */`
end

function $\mathbf{X} =$ `R-FNMA`$(\mathbf{Y}, \mathbf{A}, \mathbf{X})$
begin
  repeat
    $\mathbf{G_X} \leftarrow \nabla_{\mathbf{X}} D(\mathbf{Y}||\mathbf{AX})$
    $\mathcal{A}(\mathbf{X}) = \left\{ j, t : x_{jt} = 0, [\mathbf{G_X}]_{jt} > 0 \right\}$
    $\mathbf{P_X} = \mathcal{Z}_{+}[\mathbf{G_X}]$ `/* according to (6.65) */`
    $\mathbf{H_X} = \mathbf{A}^{-1}\mathbf{A}$
    $\tilde{\mathbf{P}}_{\mathbf{X}} \leftarrow \mathcal{Z}_{+}[\mathbf{H}_{\mathbf{X}}^{-1}\mathbf{P_X}]$
    Compute $\eta_{\mathbf{X}}$ using any line-search method
    $\mathbf{X} \leftarrow \mathcal{P}_{\Omega}\left[\mathbf{X} - \eta_{\mathbf{X}}\tilde{\mathbf{P}}_{\mathbf{X}}\right]$
  until *a stopping criterion is met* `/* convergence condition */`
end

eigenvalue problem that can be solved with the implicitly restarted Arnoldi method. Calvetti *et al.* [7] solved a QP trust-region subproblem with a hybrid method based on the Lanczos bidiagonalization and QR factorization. For a cost function expressed by the squared Euclidean distance, the Hessian is positive definite and has a regular diagonal block structure, hence, the trust-region subproblem can be easily solved using the Q-less QR factorization combined with the Gaussian elimination. The Q-less QR factorization can be obtained with the Givens rotations as implemented in LAPACK. Moreover, to further reduce the computational cost, and to make the Hessian better conditioned for a sparse system matrix, we combine the QR factorization with an active-set method that is often used for convex QP.

NMF with QP has also been discussed by Heiler and Schnörr in [17] in the context of application of sequential quadratic and second-order cone programming to the squared Euclidean cost function subject to the sparsity constraints that were proposed by Hoyer in [19]. They obtained satisfactory results for different datasets: facial expression classification [6,23], Paatero experiments [26], CBCL face dataset[2] and Caltech-101 image dataset [16].

6.4.1 Nonlinear Programming

The minimization problems in NMF are nonlinear due to the nonnegativity constraints. We therefore consider a general nonlinear problem which is then transformed into a QP problem with nonlinear constraints [5,9,25].

[2]CBCL face database no. 1, MIT Central for Biological and Computational Learning, 2000.

Let $F(\boldsymbol{x}) = F(x_1, \ldots, x_J) : \mathbb{R}^J \to \mathbb{R}$, where $\boldsymbol{x} \in \Omega \subseteq \mathbb{R}^J$, be a nonlinear function, and Ω a feasible set. We consider the following nonlinear problem:

$$\min_{\boldsymbol{x} \in \Omega} F(\boldsymbol{x}), \tag{6.69}$$

subject to the constraints:

$$\hat{c}_i(\boldsymbol{x}) \geq \hat{d}_i : \quad i = 1, 2, \ldots, m_1, \tag{6.70}$$

$$\check{c}_i(\boldsymbol{x}) \leq \check{d}_i : \quad i = m_1 + 1, \ldots, m_2, \quad m_1 \leq m_2, \tag{6.71}$$

$$\bar{c}_i(\boldsymbol{x}) = \bar{d}_i : \quad i = m_2 + 1, \ldots, m, \quad m_2 \leq m, \tag{6.72}$$

where $\hat{c}_i(\boldsymbol{x}) : \mathbb{R}^J \to \mathbb{R}$, $\check{c}_i(\boldsymbol{x}) : \mathbb{R}^J \to \mathbb{R}$, and $\bar{c}_i(\boldsymbol{x}) : \mathbb{R}^J \to \mathbb{R}$ are continuously differentiable functions at some point $\boldsymbol{x}^* \in \Omega$. The local minimum $\boldsymbol{x}^*$ to (6.69) can be determined by searching for stationary points of the Lagrangian functional $\mathcal{L}(\boldsymbol{x}, \boldsymbol{\lambda}, \boldsymbol{\theta}, \boldsymbol{\nu})$ associated with the problem (6.69), that is

$$\begin{aligned} \mathcal{L}(\boldsymbol{x}, \boldsymbol{\lambda}, \boldsymbol{\theta}, \boldsymbol{\nu}) &= F(\boldsymbol{x}) + \sum_{i=1}^{m_1} \theta_i (\hat{c}_i(\boldsymbol{x}) - \hat{d}_i) \\ &+ \sum_{i=m_1+1}^{m_2} \lambda_i (\check{c}_i(\boldsymbol{x}) - \check{d}_i) + \sum_{i=m_2+1}^{m} \nu_i (\bar{c}_i(\boldsymbol{x}) - \bar{d}_i), \end{aligned} \tag{6.73}$$

where $\boldsymbol{\lambda} = [\lambda_i] \in \mathbb{R}^{m_1}$, $\boldsymbol{\theta} = [\theta_i] \in \mathbb{R}^{m_2 - m_1}$, and $\boldsymbol{\nu} = [\nu_i] \in \mathbb{R}^{m - m_2}$ are Lagrangian multipliers. If $\boldsymbol{x}^*$ is a local minimum, then the following first-order Karush-Kuhn-Tucker (KKT) conditions are satisfied:

$$\nabla_{\boldsymbol{x}} \mathcal{L}(\boldsymbol{x}^*, \boldsymbol{\lambda}^*, \boldsymbol{\theta}^*, \boldsymbol{\nu}^*) = \boldsymbol{0}, \quad \boldsymbol{\lambda}^* = [\lambda_i^*], \ \boldsymbol{\theta}^* = [\theta_i^*], \ \boldsymbol{\nu}^* = [\nu_i^*], \tag{6.74}$$

$$\theta_i^* (\hat{c}_i(\boldsymbol{x}^*) - \hat{d}_i) = 0, \quad \lambda_i^* (\check{c}_i(\boldsymbol{x}^*) - \check{d}_i) = 0, \quad \nu_i^* (\bar{c}_i(\boldsymbol{x}^*) - \bar{d}_i) = 0, \tag{6.75}$$

$$\theta_i^* \leq 0, \quad \lambda_i^* \geq 0, \tag{6.76}$$

where ν_i^* is of an arbitrary sign, and $\boldsymbol{\lambda}^*$, $\boldsymbol{\theta}^*$, and $\boldsymbol{\nu}^*$ are Lagrange multipliers at point $\boldsymbol{x}^*$. From (6.74), we then have

$$\nabla_{\boldsymbol{x}} F(\boldsymbol{x}^*) + \nabla_{\boldsymbol{x}} \left((\boldsymbol{\theta}^*)^T \hat{\boldsymbol{c}}(\boldsymbol{x}^*) \right) + \nabla_{\boldsymbol{x}} \left((\boldsymbol{\lambda}^*)^T \check{\boldsymbol{c}}(\boldsymbol{x}^*) \right) + \nabla_{\boldsymbol{x}} \left((\boldsymbol{\nu}^*)^T \bar{\boldsymbol{c}}(\boldsymbol{x}^*) \right) = \boldsymbol{0}. \tag{6.77}$$

6.4.2 Quadratic Programming

The update for $\mathbf{X}$ involves solving the multiple linear system: $\mathbf{A}\mathbf{X} = \mathbf{Y}$ with respect to $\mathbf{X}$, where the update for $\mathbf{A}$ is performed by solving the transposed linear multiple system: $\mathbf{X}^T \mathbf{A}^T = \mathbf{Y}^T$ with respect to $\mathbf{A}$. Both systems are subject to the nonnegativity constraints. Apart from these intrinsic constraints, we also admit some additional constraints.

It is well-known that the least-squares solution $\mathbf{X}_{LS} = \mathbf{A}^{\dagger}\mathbf{Y}$ (without any constraints) for a perturbed system of linear equations: $\mathbf{A}\mathbf{X} = \mathbf{Y}$, where $\mathbf{A}^{\dagger}$ is the Moore-Penrose pseudo-inverse of $\mathbf{A}$, satisfies the inequality: $||\mathbf{X}_{LS}||_F \gg ||\mathbf{X}_{exact}||_F$, where $\mathbf{X}_{exact}$ is an exact solution. The stabilization of the solution can be achieved by some regularizations which enforce $||\mathbf{X}_{LS}||_F \leq \delta_{\mathbf{X}}$, where $\delta_{\mathbf{X}}$ is assumed to be known from some prior information about the solution.

We therefore have the following constraints on the factors to be estimated: $||\mathbf{X}||_F \le \delta_{\mathbf{X}}$ and $x_{jt} \ge 0$, and $||\mathbf{A}||_F \le \delta_A$ and $a_{ij} \ge 0$. This allows us to reformulate the QP problem in order to estimate $\mathbf{X}$ and $\mathbf{A}$ separately.

- *Step 1:* Let $F(\boldsymbol{x})$ in (6.69) be defined by a typical quadratic form that appears in the standard QP problem [5, p.152], [25, p.441]:

$$F(\mathbf{X}) = \frac{1}{2}\mathrm{vec}(\mathbf{X})^T \mathbf{Q}_{\mathbf{X}} \mathrm{vec}(\mathbf{X}) + \mathrm{tr}(\mathbf{C}_{\mathbf{X}}^T \mathbf{X}), \quad \text{for } \boldsymbol{x} = \mathrm{vec}\{\mathbf{X}\}. \tag{6.78}$$

Note that $\mathrm{tr}(\mathbf{C}_{\mathbf{X}}^T\mathbf{X}) = \mathrm{vec}(\mathbf{C}_{\mathbf{X}})^T \mathrm{vec}(\mathbf{X})$. Assuming that $\mathbf{Q}_{\mathbf{X}} \in \mathbb{R}^{JT \times JT}$ is symmetric and positive definite, $\mathbf{C}_{\mathbf{X}} \in \mathbb{R}^{J \times T}$, and the constraints (6.70)–(6.72) have the form:

$$\begin{aligned} &\hat{c}_{jt}(\mathbf{X}) = x_{jt}, && \hat{d}_{jt} = 0, \quad j = 1, \dots, J, \ \ t = 1, \dots, T, \\ &\check{c}(\mathbf{X}) = ||\mathbf{X}||_F, && \check{d} = \delta_{\mathbf{X}}, \\ &\bar{c}(\mathbf{X}) = 0, && \bar{d} = 0. \end{aligned}$$

The Lagrangian functional can be written as

$$\mathcal{L}(\mathbf{X}, \lambda_{\mathbf{X}}, \boldsymbol{\Theta}_{\mathbf{X}}) = F(\mathbf{X}) + \frac{\lambda_{\mathbf{X}}}{2}\left(||\mathbf{X}||_F^2 - \delta_{\mathbf{X}}^2\right) + \mathrm{tr}(\boldsymbol{\Theta}_{\mathbf{X}}^T \mathbf{X}), \tag{6.79}$$

with the KKT conditions (6.74)–(6.76):

$$\nabla_{\mathbf{X}} F(\mathbf{X}) + \lambda_{\mathbf{X}}\mathbf{X} + \boldsymbol{\Theta}_{\mathbf{X}} = \mathbf{0}, \ \ \lambda_{\mathbf{X}} \ge 0, \ \ \theta_{jt}^{(x)} \le 0, \ \ \boldsymbol{\Theta}_{\mathbf{X}} = [\theta_{jt}^{(x)}], \tag{6.80}$$

$$\lambda_{\mathbf{X}}\left(||\mathbf{X}||_F^2 - \delta_{\mathbf{X}}^2\right) = 0, \quad \mathrm{tr}(\boldsymbol{\Theta}_{\mathbf{X}}^T\mathbf{X}) = 0. \tag{6.81}$$

Inserting (6.78) into (6.80), $\mathbf{X}$ can be computed from

$$(\mathbf{Q}_{\mathbf{X}} + \lambda_{\mathbf{X}}\mathbf{I}_{JT})\mathrm{vec}(\mathbf{X}) = -\mathrm{vec}(\mathbf{C}_{\mathbf{X}} + \boldsymbol{\Theta}_{\mathbf{X}}). \tag{6.82}$$

- *Step 2:* To solve the transposed system $\mathbf{X}^T\mathbf{A}^T = \mathbf{Y}^T$, the new variable of interest is $\mathrm{vec}(\mathbf{A}^T)$. We also assume that $F(\boldsymbol{x})$ in (6.69) is defined by a similar quadratic form:

$$F(\mathbf{A}^T) = \frac{1}{2}\mathrm{vec}(\mathbf{A}^T)^T \mathbf{Q}_{\mathbf{A}^T} \mathrm{vec}(\mathbf{A}^T) + \mathrm{tr}(\mathbf{C}_{\mathbf{A}^T}^T \mathbf{A}^T), \quad \text{for } \boldsymbol{x} \leftarrow \mathbf{A}^T, \tag{6.83}$$

where $\mathbf{Q}_{\mathbf{A}^T} \in \mathbb{R}^{IJ \times IJ}$ is a symmetric and positive definite matrix, $\mathbf{C}_{\mathbf{A}^T} \in \mathbb{R}^{J \times I}$, and the constraints (6.70)–(6.72) have the form:

$$\begin{aligned} &\hat{c}_{ij}(\mathbf{A}^T) = a_{ji}, && \hat{d}_{ji} = 0, \quad i = 1, \dots, I, \ \ j = 1, \dots, J, \\ &\check{c}(\mathbf{A}^T) = ||\mathbf{A}^T||_F, && \check{d} = \delta_{\mathbf{A}^T}, \\ &\bar{c}(\mathbf{A}^T) = 0, && \bar{d} = 0. \end{aligned}$$

The Lagrangian functional has a similar form given by

$$\mathcal{L}(\mathbf{A}^T, \lambda_{\mathbf{A}^T}, \mathbf{\Theta}_{\mathbf{A}^T}) = F(\mathbf{A}^T) + \frac{\lambda_{\mathbf{A}^T}}{2}\left(||\mathbf{A}^T||_F^2 - \delta_{\mathbf{A}^T}^2\right) + \mathrm{tr}(\mathbf{\Theta}_{\mathbf{A}^T}^T \mathbf{A}^T), \tag{6.84}$$

and the KKT conditions (6.74)–(6.76) are

$$\nabla_{\mathbf{A}^T} F(\mathbf{A}^T) + \lambda_{\mathbf{A}^T}\mathbf{A}^T + \mathbf{\Theta}_{\mathbf{A}^T} = \mathbf{0}, \;\; \lambda_{\mathbf{A}^T} \geq 0, \;\; \theta_{ji}^{(A)} \leq 0, \;\; \mathbf{\Theta}_{\mathbf{A}^T} = [\theta_{ji}^{(A)}], \tag{6.85}$$

$$\lambda_{\mathbf{A}^T}\left(||\mathbf{A}^T||_F^2 - \delta_{\mathbf{A}^T}^2\right) = 0, \quad \mathrm{tr}(\mathbf{\Theta}_{\mathbf{A}^T}^T \mathbf{A}^T) = 0. \tag{6.86}$$

Thus, the matrix $\mathbf{A}$ can be computed from

$$(\mathbf{Q}_{\mathbf{A}^T} + \lambda_{\mathbf{A}^T}\mathbf{I}_{IJ})\mathrm{vec}(\mathbf{A}^T) = -\mathrm{vec}(\mathbf{C}_{\mathbf{A}^T} + \mathbf{\Theta}_{\mathbf{A}^T}). \tag{6.87}$$

Remark 6.3 *The solutions to both equations (6.82) and (6.87) can be regarded as Tikhonov regularized solutions, where the regularization parameters are $\lambda_{\mathbf{X}}$ and $\lambda_{\mathbf{A}^T}$. These parameters can be adaptively estimated, e.g. by the Generalized Cross-Validation (GCV) technique [4]. It is also possible to estimate these parameters based on the Lanczos bidiagonalization [7], assuming $\delta_{\mathbf{X}}$ and $\delta_{\mathbf{A}^T}$ can be roughly estimated using some prior information on the solutions.*

6.4.3 Trust-region Subproblem

Since in NMF problems Gaussian noise occurs very often, in this case it is sufficient to use the squared Euclidean distance:

$$D_F(\mathbf{Y}||\mathbf{AX}) = \frac{1}{2}||\mathbf{Y} - \mathbf{AX}||_F^2, \tag{6.88}$$

for which

$$\begin{aligned} D_F(\mathbf{Y}||\mathbf{AX}) &= \frac{1}{2}||\mathbf{Y} - \mathbf{AX}||_F^2 \\ &= \frac{1}{2}\,\mathrm{tr}(\mathbf{X}^T\mathbf{A}^T\mathbf{AX}) - \mathrm{tr}\left((\mathbf{A}^T\mathbf{Y})^T\mathbf{X}\right) + \frac{1}{2}\,\mathrm{tr}(\mathbf{Y}^T\mathbf{Y}) \\ &= \frac{1}{2}\mathrm{vec}(\mathbf{X})^T(\mathbf{I}_T \otimes \mathbf{A}^T\mathbf{A})\mathrm{vec}(\mathbf{X}) - \mathrm{tr}\left((\mathbf{A}^T\mathbf{Y})^T\mathbf{X}\right) + c, \end{aligned} \tag{6.89}$$

where $c = \frac{1}{2}\,\mathrm{tr}(\mathbf{Y}^T\mathbf{Y})$, and the symbol $\otimes$ denotes the Kronecker product. Upon neglecting the constant c and comparing (6.89) to (6.78), we have $\mathbf{Q}_{\mathbf{X}} = \mathbf{I}_T \otimes \mathbf{A}^T\mathbf{A}$ and $\mathbf{C}_{\mathbf{X}} = -\mathbf{A}^T\mathbf{Y}$ in (6.78), and alternatively $\mathbf{Q}_{\mathbf{A}^T} = \mathbf{I}_I \otimes \mathbf{XX}^T$ and $\mathbf{C}_{\mathbf{A}^T} = -\mathbf{XY}^T$ in (6.83).

After neglecting the terms $\mathbf{\Theta}_{\mathbf{X}}$ and $\mathbf{\Theta}_{\mathbf{A}^T}$, the solutions to (6.82) and (6.87) may be interpreted as standard Tikhonov regularized least-squares solutions. However, note that the nonpositive matrices $\mathbf{\Theta}_{\mathbf{X}}$ and $\mathbf{\Theta}_{\mathbf{A}^T}$ of Lagrange multipliers enforce nonnegative solutions that are no longer least-squares solutions. These terms cannot be readily estimated to satisfy the corresponding equations (6.81) and (6.86), and we may opt to iteratively approximate the solutions to (6.82) and (6.87). In the case considered, we can avoid the complicated duality gaps $\mathrm{tr}(\mathbf{\Theta}_{\mathbf{X}}^T\mathbf{X})$ and $\mathrm{tr}(\mathbf{\Theta}_{\mathbf{A}^T}^T\mathbf{A}^T)$ by adding some penalty terms to the regularized cost function, thus ensuring nonnegative solutions. It is possible to use several such penalty functions (quadratic penalty, logarithmic barrier, or exact penalty) whose properties have been discussed in [25]. Following the proposition by Rojas and Steihaug [31], we adopted the logarithmic barrier functions:

$$\Psi_{\mathbf{X}}(\mathbf{X}) = \theta_{\mathbf{X}} \sum_{j=1}^{J} \sum_{t=1}^{T} \ln x_{jt}, \quad \Psi_{\mathbf{A}^T}(\mathbf{A}^T) = \theta_{\mathbf{A}^T} \sum_{i=1}^{I} \sum_{j=1}^{J} \ln a_{ji}, \tag{6.90}$$

where $\theta_{\mathbf{X}} \leq 0$ and $\theta_{\mathbf{A}^T} \leq 0$. These functions tend to relax the ill-conditioning of Hessians, thus considerably reducing the number of Lagrange multipliers – instead of matrices $\mathbf{\Theta}_{\mathbf{X}}$ and $\mathbf{\Theta}_{\mathbf{A}^T}$ we need to estimate only two parameters $\theta_{\mathbf{X}}$ and $\theta_{\mathbf{A}^T}$. The resulting penalized cost function is given by:

$$\begin{aligned} D_R(\mathbf{Y}||\mathbf{AX}) &= \frac{1}{2}||\mathbf{Y}-\mathbf{AX}||_F^2 + \frac{\lambda_{\mathbf{X}}}{2}\left(||\mathbf{X}||_F^2 - \delta_{\mathbf{X}}^2\right) \\ &+ \frac{\lambda_{\mathbf{A}^T}}{2}\left(||\mathbf{A}^T||_F^2 - \delta_{\mathbf{A}^T}^2\right) + \Psi_{\mathbf{X}}(\mathbf{X}) + \Psi_{\mathbf{A}^T}(\mathbf{A}^T), \end{aligned} \tag{6.91}$$

and can be used for solving both alternating minimization problems in NMF. The penalty terms introduce strong nonlinearity, and to use QP the function $D_R(\mathbf{Y}||\mathbf{AX})$ needs to be approximated around points $\mathbf{X}$ and $\mathbf{A}$ by second-order Taylor expansions. The so modified cost functions become

$$\begin{aligned} D_R(\mathbf{Y}||\mathbf{A}(\mathbf{X}+\mathbf{\Delta X})) &= D_R(\mathbf{Y}||\mathbf{AX}) + \mathrm{tr}(\mathbf{G}_{\mathbf{X}}^T \mathbf{\Delta X}) \\ &+ \frac{1}{2}\mathrm{vec}(\mathbf{\Delta X})^T \mathbf{H}_{\mathbf{X}} \mathrm{vec}(\mathbf{\Delta X}), \end{aligned} \tag{6.92}$$

$$\begin{aligned} D_R(\mathbf{Y}||(\mathbf{A}+\mathbf{\Delta A})\mathbf{X}) &= D_R(\mathbf{Y}||\mathbf{AX}) + \mathrm{tr}(\mathbf{G}_{\mathbf{A}^T}^T \mathbf{\Delta A}^T) \\ &+ \frac{1}{2}\mathrm{vec}(\mathbf{\Delta A}^T)^T \mathbf{H}_{\mathbf{A}^T} \mathrm{vec}(\mathbf{\Delta A}^T), \end{aligned} \tag{6.93}$$

where the corresponding gradients $\mathbf{G}_{\mathbf{X}}$ and $\mathbf{G}_{\mathbf{A}^T}$ with respect to $\mathbf{X}$ and $\mathbf{A}^T$ are given by:

$$\begin{aligned} \mathbf{G}_{\mathbf{X}} &= \nabla_{\mathbf{X}} D_R(\mathbf{Y}||\mathbf{AX}) = \mathbf{A}^T(\mathbf{AX}-\mathbf{Y}) + \lambda_{\mathbf{X}}\mathbf{X} + \theta_{\mathbf{X}}[x_{jt}^{-1}] \in \mathbb{R}^{J\times T}, \\ \mathbf{G}_{\mathbf{A}^T} &= \nabla_{\mathbf{A}^T} D_R(\mathbf{Y}||\mathbf{AX}) = \mathbf{X}(\mathbf{AX}-\mathbf{Y})^T + \lambda_{\mathbf{A}^T}\mathbf{A}^T + \theta_{\mathbf{A}^T}[a_{ji}^{-1}] \in \mathbb{R}^{J\times I}, \end{aligned}$$

and the corresponding Hessians $\mathbf{H}_{\mathbf{X}} \in \mathbb{R}^{JT\times JT}$ and $\mathbf{H}_{\mathbf{A}} \in \mathbb{R}^{IJ\times IJ}$ have forms:

$$\begin{aligned} \mathbf{H}_{\mathbf{X}} &= \nabla_{\mathbf{X}}^2 D_R(\mathbf{Y}||\mathbf{AX}) = \mathbf{I}_T \otimes \mathbf{A}^T\mathbf{A} + \lambda_{\mathbf{X}}\mathbf{I}_{JT} - \theta_{\mathbf{X}}\,\mathrm{diag}(\mathrm{vec}([x_{jt}^{-2}])), \\ \mathbf{H}_{\mathbf{A}^T} &= \nabla_{\mathbf{A}^T}^2 D_R(\mathbf{Y}||\mathbf{AX}) = \mathbf{I}_I \otimes \mathbf{XX}^T + \lambda_{\mathbf{A}^T}\mathbf{I}_{JI} - \theta_{\mathbf{A}^T}\,\mathrm{diag}(\mathrm{vec}([a_{ij}^{-2}]^T)). \end{aligned}$$

The gradient and the Hessian for the matrix $\mathbf{A}$ can be reformulated to

$$\begin{aligned} \mathbf{G}_{\mathbf{A}} &= \nabla_{\mathbf{A}} D_R(\mathbf{Y}||\mathbf{AX}) = (\mathbf{AX}-\mathbf{Y})\mathbf{X}^T + \lambda_{\mathbf{A}}\mathbf{A} + \theta_{\mathbf{A}}[a_{ij}^{-1}] \in \mathbb{R}^{I\times J}, \\ \mathbf{H}_{\mathbf{A}} &= \nabla_{\mathbf{A}}^2 D_R(\mathbf{Y}||\mathbf{AX}) = \mathbf{XX}^T \otimes \mathbf{I}_I + \lambda_{\mathbf{A}}\mathbf{I}_{IJ} - \theta_{\mathbf{A}}\,\mathrm{diag}(\mathrm{vec}([a_{ij}^{-2}])). \end{aligned}$$

We can now formulate the following trust-region subproblems:

$$\min_{\Delta \mathbf{X}} D_R(\mathbf{Y}||\mathbf{A}(\mathbf{X}+\Delta\mathbf{X})), \qquad \text{s.t.} \;\; ||\mathbf{X}+\Delta\mathbf{X}||_F \le \delta_{\mathbf{X}}, \tag{6.94}$$

$$\min_{\Delta \mathbf{A}} D_R(\mathbf{Y}||(\mathbf{A}+\Delta\mathbf{A})\mathbf{X}), \qquad \text{s.t.} \;\; ||\mathbf{A}+\Delta\mathbf{A}||_F \le \delta_{\mathbf{A}}. \tag{6.95}$$

Following the approach by Rojas and Steihaug [31], the trust-region subproblems can be transformed into the following QP problems:

$$\min_{\mathbf{Z_X}} \left\{ \frac{1}{2} \text{vec}(\mathbf{Z_X})^T (\mathbf{Q_X}) \, \text{vec}(\mathbf{Z_X}) + \text{tr}(\mathbf{C}_\mathbf{X}^T \mathbf{Z_X}) \right\}, \qquad \mathbf{Z_X} = \mathbf{X} + \Delta\mathbf{X}, \tag{6.96}$$

$$\min_{\mathbf{Z}_{\mathbf{A}^T}} \left\{ \frac{1}{2} \text{vec}(\mathbf{Z}_{\mathbf{A}^T})^T \left(\mathbf{Q}_{\mathbf{A}^T}\right) \text{vec}(\mathbf{Z}_{\mathbf{A}^T}) + \text{tr}(\mathbf{C}_{\mathbf{A}^T}^T \mathbf{Z}_{\mathbf{A}^T}) \right\}, \qquad \mathbf{Z}_{\mathbf{A}^T} = \mathbf{A}^T + \Delta\mathbf{A}^T, \tag{6.97}$$

where

$$\mathbf{Q_X} = \mathbf{I}_T \otimes \mathbf{A}^T\mathbf{A} + \lambda_\mathbf{X} \mathbf{I}_{JT} - \theta_\mathbf{X} \, \text{diag}(\text{vec}([x_{jt}^{-2}])), \tag{6.98}$$

$$\mathbf{C_X} = 2\theta_\mathbf{X}[x_{jt}^{-1}] - \mathbf{A}^T\mathbf{Y}, \quad \lambda_\mathbf{X} \ge 0, \quad \theta_\mathbf{X} \le 0, \tag{6.99}$$

$$\mathbf{Q}_{\mathbf{A}^T} = \mathbf{I}_I \otimes \mathbf{X}\mathbf{X}^T + \lambda_{\mathbf{A}^T} \mathbf{I}_{IJ} - \theta_{\mathbf{A}^T} \, \text{diag}(\text{vec}([a_{ij}^{-2}]^T)), \tag{6.100}$$

$$\mathbf{C}_{\mathbf{A}^T} = 2\theta_{\mathbf{A}^T}[a_{ij}^{-1}]^T - \mathbf{X}\mathbf{Y}^T, \quad \lambda_{\mathbf{A}^T} \ge 0, \quad \theta_{\mathbf{A}^T} \le 0. \tag{6.101}$$

The QP problems given by (6.96) and (6.97) are therefore equivalent to the following basic QP problems:

$$\min_{\mathbf{X}} \left\{ \frac{1}{2} \text{vec}(\mathbf{X})^T \left(\mathbf{I}_T \otimes \mathbf{A}^T\mathbf{A}\right) \text{vec}(\mathbf{X}) - \text{tr}(\mathbf{Y}^T\mathbf{A}\mathbf{X}) \right\},$$
$$\text{s.t.} \;\; x_{jt} \ge 0, \quad ||\mathbf{X}||_F \le \delta_\mathbf{X} \tag{6.102}$$

$$\min_{\mathbf{A}^T} \left\{ \frac{1}{2} \text{vec}(\mathbf{A}^T)^T \left(\mathbf{I}_I \otimes \mathbf{X}\mathbf{X}^T\right) \text{vec}(\mathbf{A}^T) - \text{tr}(\mathbf{Y}\mathbf{X}^T\mathbf{A}^T) \right\},$$
$$\text{s.t.} \;\; a_{ij} \ge 0, \quad ||\mathbf{A}||_F \le \delta_{\mathbf{A}^T}. \tag{6.103}$$

Remark 6.4 *Assuming* **A** *and* **X** *are full rank matrices, both* $\mathbf{Q}_{\mathbf{A}^T}$ *and* $\mathbf{Q_X}$ *are positive-definite but possibly very ill-conditioned, especially when* **A** *and* **X** *are very sparse. This is due to the fact that* $q_{ss}^{(x)} \to \infty$*, where* $s = j + (t-1)J$, $\mathbf{Q_X} = [q_{mn}^{(x)}]$, $m, n \in \{1, \ldots, JT\}$, if $x_{js} \to 0$ *and* $|\theta_\mathbf{X}| < \infty$. *Also, if* $a_{ij} \to 0$ *and* $|\theta_{\mathbf{A}^T}| < \infty$, *then* $q_{ss}^{(A)} \to \infty$, *where* $s = j + (i-1)J$, $\mathbf{Q}_{\mathbf{A}^T} = [q_{mn}^{(A)}]$, $m, n \in \{1, \ldots, IJ\}$. *For the positive-definite matrices* $\mathbf{Q}_{\mathbf{A}^T}$ *and* $\mathbf{Q_X}$, *the solutions* $\mathbf{X}^*$ *and* $\mathbf{A}^*$ *are global in the separate minimization problems for NMF. However, the NMF alternating minimization does not guarantee the convergence to a global solution.*

6.4.4 Updates for A

The mixing matrix **A** can be updated by solving the QP problem (6.97). Assuming that the objective function in this problem is denoted by $\Phi(\mathbf{A}^T)$, that is $\Phi(\mathbf{A}^T) = \frac{1}{2}\text{vec}(\mathbf{Z}_{\mathbf{A}^T})^T \left(\mathbf{Q}_{\mathbf{A}^T}\right) \text{vec}(\mathbf{Z}_{\mathbf{A}^T}) + \text{tr}(\mathbf{C}_{\mathbf{A}^T}^T \mathbf{Z}_{\mathbf{A}^T})$. The stationary point $\mathbf{A}_*^T$ of $\Phi(\mathbf{A}^T)$ occurs for $\nabla_{\mathbf{A}^T}\Phi(\mathbf{A}_*^T) = \mathbf{0}$, and the global solution can be obtained by solving the system of linear equations:

$$\mathbf{Q}_{\mathbf{A}^T} \mathbf{z} = -\text{vec}(\mathbf{C}_{\mathbf{A}^T}), \tag{6.104}$$

where $\mathbf{z} = \text{vec}(\mathbf{Z}_{\mathbf{A}^T}) \in \mathbb{R}^{IJ}$.

Although the matrix $\mathbf{Q}_{\mathbf{A}^T}$ may be very ill-conditioned (see Remark 6.4), sufficiently small entries a_{ij} do not need to be updated in successive inner iterations during the estimation of the parameter $\theta_{\mathbf{A}^T}$ in (6.100) and (6.101). This way, the size of the system to be solved decreases, and the "reduced" system becomes better conditioned. To perform this operation, we can use some ideas from the well-known active-set methods [25]. Let

$$\mathcal{I}(\mathbf{A}^T) = \left\{t: \; a_t \geq \epsilon_{\mathbf{A}}, \quad \text{where } a_t = [\text{vec}(\mathbf{A}^T)]_t\right\} \tag{6.105}$$

be an inactive-set for matrix $\mathbf{A}^T$, and $\epsilon_{\mathbf{A}}$ be some threshold for entries a_{ij}. Now let us define the matrix $\mathbf{T} = \mathbf{I}_{[*,\mathcal{I}(\mathbf{A}^T)]} \in \mathbb{R}^{IJ \times N}$, where $\mathbf{I} = \mathbf{I}_{IJ}$, the symbol $*$ refers to all the rows, and N is the number of entries in $\mathcal{I}(\mathbf{A}^T)$. Thus, the "reduced" system can be expressed as

$$(\mathbf{T}^T \mathbf{Q}_{\mathbf{A}^T} \mathbf{T}) z_R = -\mathbf{T}^T \text{vec}(\mathbf{C}_{\mathbf{A}^T}), \tag{6.106}$$

where $z_R = \mathbf{T}^T z \in \mathbb{R}^N$. If $\mathbf{A}$ is very sparse, we have $N << IJ$.

The matrix $\mathbf{Q}_{\mathbf{A}^T}^{(R)} = \mathbf{T}^T \mathbf{Q}_{\mathbf{A}^T} \mathbf{T}$ is positive definite, and for the cost function defined by (6.88), it has a regular block diagonal structure with small blocks, that is, the number of entries under the main diagonal is very small. This admits to the factorization of the matrix $\mathbf{Q}_{\mathbf{A}^T}^{(R)}$ using the Q-less QR factorization that is implemented in LAPACK. Applying the Givens rotations to the under-diagonal entries the upper triangular matrix $\mathbf{R}$ can be computed in a relatively small number of arithmetic operations, thus

$$\mathbf{Q}_{\mathbf{A}^T}^{(R)} = \mathbf{QR}, \quad \text{where } \boldsymbol{w} = -\mathbf{Q}^T \mathbf{T}^T \text{vec}(\mathbf{C}_{\mathbf{A}^T}). \tag{6.107}$$

The orthogonal matrix $\mathbf{Q}$ is not computed explicitly in the Q-less QR factorization but only the vector $\boldsymbol{w}$ is returned. The new system: $\mathbf{R} z_R = \boldsymbol{w}$ with the upper triangular matrix $\mathbf{R}$ can be efficiently solved with Gaussian elimination.

The system (6.104) has to be solved iteratively for increasing values of $\theta_{\mathbf{A}^T}$ in (6.100) and (6.101), whereas the iterative schedule for $\theta_{\mathbf{A}^T}$ can be derived from (6.87). Combining (6.104), (6.100) and (6.101) with (6.87), we have

$$\text{vec}(\mathbf{\Theta}_{\mathbf{A}}^T) = 2\theta_{\mathbf{A}^T} \text{vec}([a_{ij}^{-1}]^T) - \theta_{\mathbf{A}^T} \,\text{diag}(\text{vec}([a_{ij}^{-2}]^T)) z. \tag{6.108}$$

When $z = \text{vec}(\mathbf{A}^T)$, the duality gap: $\text{tr}(\mathbf{\Theta}_{\mathbf{A}^T}^T \mathbf{A}^T)$ in (6.84) can be approximated by $\text{vec}(\mathbf{\Theta}_{\mathbf{A}^T})^T \text{vec}(\mathbf{A}^T) = \theta_{\mathbf{A}^T} IJ$. Following the approaches by Rojas and Steihaug [31], and Calvetti *et al.* [7], we use the following iterative rule for this parameter:

$$\begin{aligned}
\tilde{\boldsymbol{\theta}}^{(l)} &= \theta^{(l-1)} \left(2\text{vec}([\tilde{a}_{ij}^{-1}]^T) - \text{diag}(\text{vec}([\tilde{a}_{ij}^{-2}]^T)) z\right), \quad \text{where } \tilde{\mathbf{A}} = \mathbf{A}^{(l-1)}, \\
\theta^{(l)} &= \frac{\rho}{IJ} (\tilde{\boldsymbol{\theta}}^{(l)})^T \text{vec}(\tilde{\mathbf{A}}^T),
\end{aligned} \tag{6.109}$$

where $\mathbf{A}^{(l-1)}$ is the matrix $\mathbf{A}$ estimated in the $(l-1)$-th inner iterative step, and $\rho \in (0, \; 1)$.

The second-order Taylor series, as in (6.92) and (6.93), approximates the function of interest only in some small vicinity of a given point; the solution to the problem (6.97) is more accurate if the vicinity $\Delta\mathbf{A}^T$ is smaller, i.e. $||\Delta\mathbf{A}^T||_F \to 0$. To satisfy this condition, and additionally, to ensure $\forall i, j: \; [\mathbf{A} + \Delta\mathbf{A}]_{ij} > 0$, $\mathbf{A}$ is upgraded by the additive rule: $\mathbf{A} \leftarrow \mathbf{A} + \beta(\mathbf{Z} - \mathbf{A})$, where $\mathbf{Z} = (\text{Matrix}(z))^T$ and z is the solution to (6.104). The parameter β is estimated by line-search rules [7,31] to ensure the upgrade for $\mathbf{A}$ is always positive.

Our algorithm for updating $\mathbf{A}$ combines the trust-region algorithm [25, p.299] with the log-barrier algorithm [25, p.505], and it is also motivated by the $TRUST_\mu$ algorithm in [31] and the Constrained Tikhonov Regularization algorithm in [7]. Algorithm 6.7 is our final QP-NMF algorithm.

Algorithm 6.7: QP-NMF

Input: $\mathbf{Y} \in \mathbb{R}_{+}^{I \times T}$: input data, J: rank of approximation
Output: $\mathbf{A} \in \mathbb{R}_{+}^{I \times J}$ and $\mathbf{X} \in \mathbb{R}_{+}^{J \times T}$.

function $\mathbf{X} = \text{QP}(\mathbf{Y}, \mathbf{A}, \mathbf{X})$
begin
  Randomly initialize: $\mathbf{A}$
  repeat
    $\mathbf{X} \leftarrow \max\{\varepsilon, (\mathbf{A}^T\mathbf{A})^{-1}\mathbf{A}^T\mathbf{Y}\}$
    $\mathbf{I} = \mathbf{I}_{IJ} \in \mathbb{R}^{IJ \times IJ}, \mathbf{A} = \mathbf{E} \in \mathbb{R}^{I \times J}$ – matrix of all ones
    $\bar{\mathbf{Q}} = \mathbf{I}_I \otimes \mathbf{X}\mathbf{X}^T + \lambda_{\mathbf{A}^T}\mathbf{I}_{IJ}, \bar{\boldsymbol{c}} = \text{vec}\left(\mathbf{X}\mathbf{Y}^T\right)$
    $\mathbf{G}_0 = (\mathbf{Y} - \mathbf{A}\mathbf{X})\mathbf{X}^T, \theta = -\frac{\rho}{IJ}\,\text{tr}(\mathbf{G}_0\mathbf{A}^T)$
    $\boldsymbol{a} = \text{vec}(\mathbf{A}^T)$
    repeat
      $\mathcal{I} = \{t : a_t \geq \varepsilon_{\mathbf{A}}\}$ `/* active set */`
      $\mathbf{Q}_{\mathbf{A}^T} \leftarrow \bar{\mathbf{Q}}_{\mathcal{I},\mathcal{I}} - \theta\,\text{diag}\left\{\boldsymbol{a}_{\mathcal{I}}^{.[2]}\right\}$
      $\boldsymbol{c}_{\mathbf{A}} \leftarrow 2\theta\boldsymbol{a}_{\mathcal{I}}^{.[-1]} - \bar{\boldsymbol{c}}_{\mathcal{I}}$
      $[\mathbf{R}, \boldsymbol{w}] \leftarrow lqr(\mathbf{Q}_{\mathbf{A}^T}, \boldsymbol{c}_{\mathbf{A}})$
      $\tilde{\boldsymbol{z}} = \mathbf{R}\backslash\boldsymbol{w}$
      $\tilde{\boldsymbol{h}} = \tilde{\boldsymbol{z}} - \boldsymbol{a}_{\mathcal{I}}$
      $\boldsymbol{a}_{\mathcal{I}} \leftarrow \boldsymbol{a}_{\mathcal{I}} + \min\left\{1, \tau\min(\boldsymbol{a}_{\mathcal{I}} \oslash \tilde{\boldsymbol{h}})\right\}\tilde{\boldsymbol{h}}$
      $\theta \leftarrow \left(\frac{\rho\theta}{IJ}\right)\left(2\texttt{card}(\mathcal{I}) - \mathbf{1}^T(\tilde{\boldsymbol{z}} \oslash \boldsymbol{a}_{\mathcal{I}})\right)$
    until $|\theta| < \frac{\rho\eta}{IJ}||\boldsymbol{a}_{\mathcal{I}}||_2$
    $\mathbf{A} \leftarrow \texttt{reshape}(\boldsymbol{a}, J)^T$
    $a_{ij} \leftarrow \dfrac{a_{ij}}{\sum_{q=1}^{I} a_{qj}}$ `/* ℓ1-norm normalization */`
  until *a stopping criterion is met* `/* convergence condition */`
end

Despite the assumption $\lambda_{\mathbf{A}^T} \leq \delta_{\mathbf{A}^T}$, the column vectors in $\mathbf{A}$ may have different lengths due to intrinsic scale and permutation ambiguities. This may lead to very badly scaled estimations for sources, which in consequence reduces the performance. To overcome this problem, we scale the columns in $\mathbf{A}$ to the unit ℓ_1-norm.

The MATLAB code for Algorithm 6.7 is listed in Appendix 6.B.

6.5 Hybrid Updates

The discussed algorithms can be applied for updating both matrices $\mathbf{A}$ and $\mathbf{X}$. However, when one of the estimated matrices is considerably larger than the other one, we may use the second-order algorithm only for updating the smaller matrix, but the larger one can be estimated with a faster algorithm, e.g. using the ALS or HALS algorithm which are described in Chapter 4. This can be also justified by the fact that the second-order algorithms are more suitable for solving over-determined systems since they exploit the information about the curvature of the cost function.

Assuming that the matrix $\mathbf{X}$ is larger than $\mathbf{A}$, the nonnegative components in $\mathbf{X}$ can be estimated with the projected ALS or HALS algorithms. As a special case, we can also use the Regularized ALS (RALS) algorithm which solves a regularized least-squares problem

$$\mathbf{X}^* = \arg\min_{\mathbf{X}} \left\{\frac{1}{2}||\mathbf{Y} - \mathbf{A}\mathbf{X}||_F^2 + \frac{\lambda_{\mathbf{X}}}{2}J(\mathbf{X})\right\}. \tag{6.110}$$

The regularization term

$$J(\mathbf{X}) = \sum_t (||\boldsymbol{x}_t||_1)^2 = \mathrm{tr}\{\mathbf{X}^T \mathbf{1} \mathbf{X}\},$$

where $\boldsymbol{x}_t$ is the t-th column of $\mathbf{X}$, and $\mathbf{1} \in \mathbb{R}^{J \times J}$ is a matrix of all one entries, which enforces sparsity in the columns of $\mathbf{X}$, is motivated by the diversity measure[3] used in the M-FOCUSS algorithm [14]. Let $\Psi(\mathbf{X})$ be the objective function in (6.110). Thus the stationary point of $\Psi(\mathbf{X})$ is reached when

$$\nabla_{\mathbf{X}} \Psi(\mathbf{X}) = \mathbf{A}^T(\mathbf{A}\mathbf{X} - \mathbf{Y}) + \lambda_{\mathbf{X}} \mathbf{1}\mathbf{X} = 0,$$

which leads to the following regularized least-squares solution

$$\mathbf{X}_{LS} = (\mathbf{A}^T\mathbf{A} + \lambda_{\mathbf{X}}\mathbf{1})^{-1}\mathbf{A}^T\mathbf{Y}.$$

To satisfy the nonnegativity constraints, $\mathbf{X}_{LS}$ is projected onto $\Omega_{\mathbf{X}}$.

The regularization parameter $\lambda_{\mathbf{X}}$ in (6.110) is set according to the exponential rule:

$$\lambda_{\mathbf{X}} = \lambda_{\mathbf{X}}^{(s)} = \lambda_0 \exp\{-\tau s\}, \tag{6.111}$$

where s denotes the s-th alternating step. This rule is motivated by a temperature schedule in the simulated annealing that steers the solution towards a global one. Larger parameter λ_0 and smaller τ should give more stable results but this slows down the convergence, and the higher number of alternating steps is needed.

6.6 Numerical Results

In the experiments, we tested several hybrid NMF algorithms. For example, the QP-ALS algorithm uses the QP method for updating the matrix $\mathbf{A}$ and the ALS algorithm for updating $\mathbf{X}$. The QNE-RALS algorithm combines the Euclidean based Quasi-Newton (QNE) method for updating $\mathbf{A}$ with the RALS algorithm for updating $\mathbf{X}$.

The selected hybrid algorithms have been evaluated based on the same numerical tests as described in Chapters 3, 4, and 5. We used three benchmarks of source signals: benchmark A (Figure 5.1), benchmark B (Figure 3.11), and benchmark C (Figure 4.8). Each NMF algorithm has been initialized with the multi-start initialization as shown in Chapter 1. The tests for the benchmark A have been performed for two cases of inner iterations,[4] that is with $k = 1$ and $k = 5$. For other benchmarks, the number of inner iterations was adjusted to a given algorithm.

In all the simulations, we normalized the rows of matrix $\mathbf{Y}$ to unit variances. This was motivated by practical applications where measured sensor signals have considerable different amplitudes for spatially distributed sensors.

The quality of the estimation is measured with the mean-SIRs that are calculated according to the formulae (5.80)–(5.81) given in Chapter 5.

The selected parameters of the statistics for 100 mean-SIR samples obtained by the analyzed algorithms are shown in Table 6.1 and in Figures 6.3. Table 6.1 lists the mean-SIR values obtained for the best and worst sample (with the highest and lowest SIR values). Note that some algorithms (like QP-ALS) give very stable estimates, with identical the best and worst cases. The number of alternating steps was set to 1000 and 200, respectively, for the benchmark A and B.

In the QP algorithm, we set the following parameters: $\rho = 10^{-3}$, $\varepsilon_{\mathbf{A}} = 10^{-6}$, $\epsilon = 10^{-9}$ (eps), $L = 4$, $\tau = 0.9995$, $\eta = 10^{-4}$. Satisfactory results can be obtained if $\lambda_{\mathbf{A}^T} \in [500, 2000]$. For higher values of $\lambda_{\mathbf{A}^T}$

[3]The diversity measure $J^{(p,q)} = \sum_{j=1}^{J}(||\underline{\mathbf{x}}_j||_q)^p$, $p \geq 0$, $q \geq 1$, was introduced by Cotter [14] to enforce sparsity in the rows of $\mathbf{X}$. Since we are more concerned with a sparsity column profile, we apply the measure to the columns in $\mathbf{X}$ instead of the rows, assuming $q = 1$ and $p = 2$.

[4]The inner iterations means a number of iterative steps that are performed to update only $\mathbf{A}$ (before going to the update of $\mathbf{X}$).

Table 6.1 Mean-SIRs [dB] obtained with 100 samples of Monte Carlo analysis for the estimation of sources and columns of mixing matrix from noise-free mixtures of signals in Figure 5.1 (see Chapter 5). Sources **X** are estimated with the projected pseudo-inverse. The number of inner iterations for updating **A** is denoted by k. The notation *best* or *worst* means that the mean-SIR value is calculated for the best or worst sample from Monte Carlo analysis. In the last column, the elapsed time [in seconds] is given for each algorithm with $k = 1$. The abbreviations QNE and QNA refer to the QN algorithm for minimization of the Euclidean distance and Alpha-divergence, respectively. The number of alternating steps was set to 1000.

Algorithm		Mean-$SIR_{\mathbf{A}}$ [dB]				Mean-$SIR_{\mathbf{X}}$ [dB]				Time [s]
		$k = 1$		$k = 5$		$k = 1$		$k = 5$		
		Mean	Std	Mean	Std	Mean	Std	Mean	Std	
QP–ALS	best	28.4	0	34.2	0	34	0	38.8	0	2.8
$\lambda = 500$	worst	28.4	0	34.2	0	34	0	38.8	0	
QP–ALS	best	23.6	0	26.5	0	29.3	0	31	0	2.8
$\lambda = 1500$	worst	23.6	0	26.5	0	29.3	0	31	0	
QNE–RALS	best	77.8	0	77.8	0	74.4	0	74.4	0	5.5
$\alpha_0 = 1, \tau = 100$	worst	77.8	0	77.8	0	74.4	0	74.4	0	
QNE–HALS	best	48.7	0	48.7	0	35.2	0	35.2	0	4.2
	worst	48.7	0	48.7	0	35.2	0	35.2	0	
QNA–RALS	best	87	0	87	0	77.4	0	77.4	0	26.6
$\alpha_0 = 1, \tau = 100$	worst	87	0	87	0	77.4	0	77.4	0	
QNE–ALS	best	77.2	7.1	100.9	8.2	93.3	2.9	93.2	3.1	2
	worst	58.9	7.1	70.3	8.2	69.4	2.9	78.1	3.1	
GPCG–ALS	best	137.5	33.5	138.1	30.1	134.6	33.2	135.3	29.7	10.3
	worst	32.3	33.5	16.2	30.1	28.8	33.2	29.1	29.7	
GPCG–RALS	best	137.6	28.4	138.6	24.6	134.7	28.3	136.7	24.7	10.6
$\alpha_0 = 1, \tau = 100, \alpha = 0.5$	worst	41.5	28.4	31.2	24.6	25.6	28.3	29.2	24.7	
R-FNMA–RALS	best	75.8	0	75.8	0	74.6	0	74.6	0	3.8
$\alpha_0 = 1, \tau = 100, \alpha = 0.5$	worst	75.8	0	75.8	0	74.6	9	74.6	0	
R-FNMA–HALS	best	48.7	0	48.7	0	35.2	0	35.2	0	6.9
	worst	48.7	0	48.7	0	35.2	0	35.2	0	
R-FNMA–ALS	best	22.1	2.1	23.4	2.4	54.3	4.8	37.4	4	4.6
	worst	14.1	2.1	17.4	2.4	13.6	4.8	14.9	4	
R-FNMA	best	23.4	4.5	32.8	4.4	46.8	9.7	46.8	8.6	36.3
	worst	10.5	4.5	9.5	4.4	9	9.7	9	8.6	

the variance of the distribution of the mean-SIR values is smaller but one should remember that it also decreases the mean of the distribution. More details on the selection of $\lambda_{\mathbf{A}T}$ can be found in [35].

We also tested how the number of observations affects the performance of the algorithms. For these experiments, we used the signals from the benchmark B (`X_spectra` dataset) that were mixed with the uniformly distributed random matrix $\mathbf{A} \in \mathbb{R}^{I \times 5}$. The MC results obtained for $I = 10$ and $I = 1000$ are shown in Figure 6.3(a) and Figure 6.3(b), respectively.

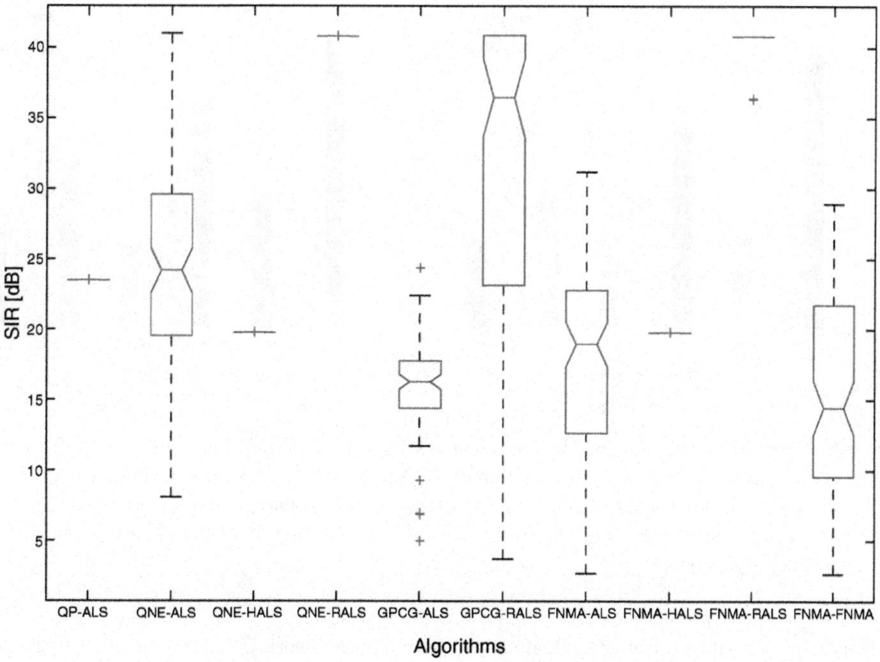

(a) $I = 10$

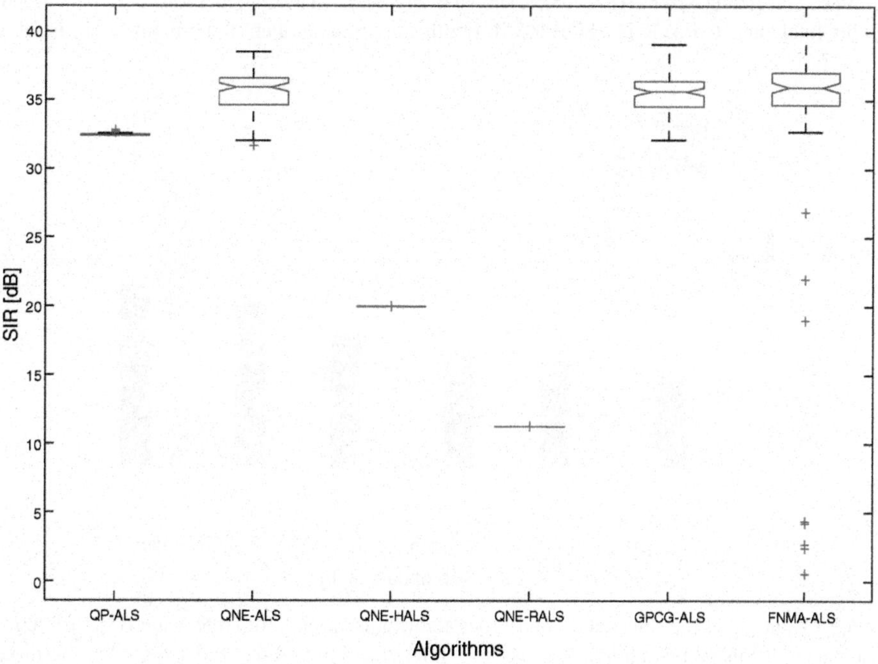

(b) $I = 1000$

Figure 6.3 MC analysis for the selected second-order algorithms, each initialized randomly 100 times. The `X_spectra` dataset of the 5 sources ($J = 5$) was used. For every MC trial, the sources were mixed with the uniformly distributed random matrix $\mathbf{A} \in \mathbb{R}^{I \times J}$, where: (a) $I = 10$, (b) $I = 1000$.

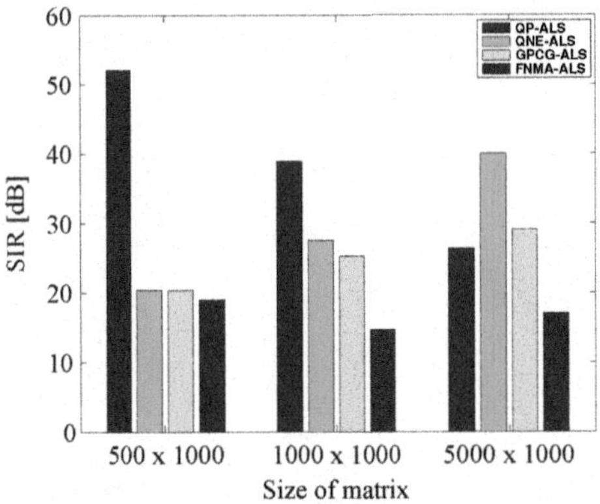

(a) SIR [dB]

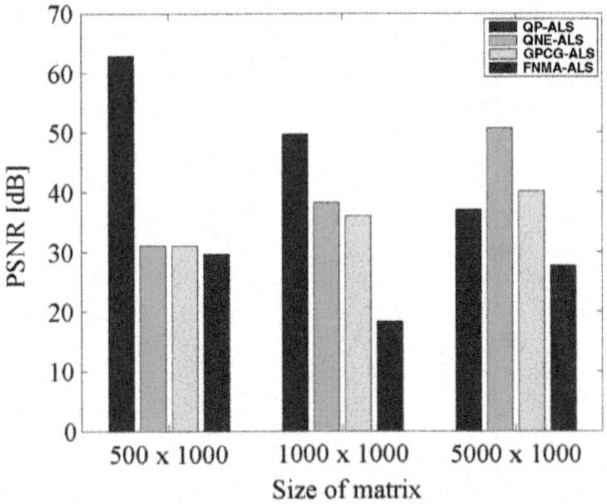

(b) PSNR [dB]

Figure 6.4 Mean-SIRs [dB] obtained by the selected second-order algorithms applied for large-scale BSS problems where the benchmark signals (benchmark C: $\mathbf{X} \in \mathbb{R}^{10\times1000}$) illustrated in Figure 4.8 (Chapter 4) were mixed with the uniformly distributed random mixing matrices of sizes: 500×1000, 1000×1000, and 5000×1000. The number of inner iterations was set to $k = 5$ for the QP-ALS and GPCG algorithms, and to $k = 1$ for other algorithms.

To test the performance for large-scale data, we used the benchmark C. The mean-SIR values for three cases of the observed data ($\mathbf{Y}_1 \in \mathbb{R}^{500\times1000}$, $\mathbf{Y}_2 \in \mathbb{R}^{1000\times1000}$, $\mathbf{Y}_3 \in \mathbb{R}^{5000\times1000}$) are illustrated in Figure 6.4. The runtime measured in MATLAB 2008a is shown in Figure 6.5. The computations were performed on the CPU 9650, Intel Quad Core, 3GHz, 8GB RAM. For this test, the number of alternating steps is set to 200.

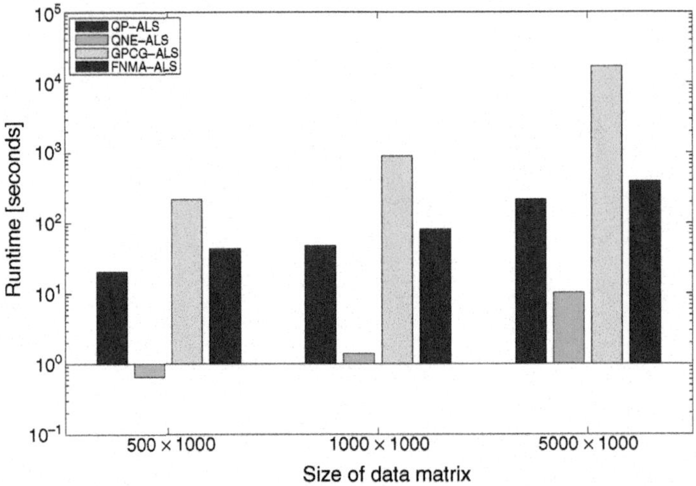

Figure 6.5 Runtime of the selected second-order algorithms applied for large-scale BSS problems where the benchmark signals (benchmark C: $\mathbf{X} \in \mathbb{R}^{10\times1000}$) illustrated in Figure 4.8 (Chapter 4) were mixed with the uniformly distributed random mixing matrices of sizes: 500×1000, 1000×1000, and 5000×1000. The computations have been performed in MATLAB 2008a on CPU 9650, Intel Quad Core, 3GHz, 8GB RAM. The number of inner iterations was set to $k = 5$ for the QP-ALS and GPCG algorithms, and to $k = 1$ for other algorithms.

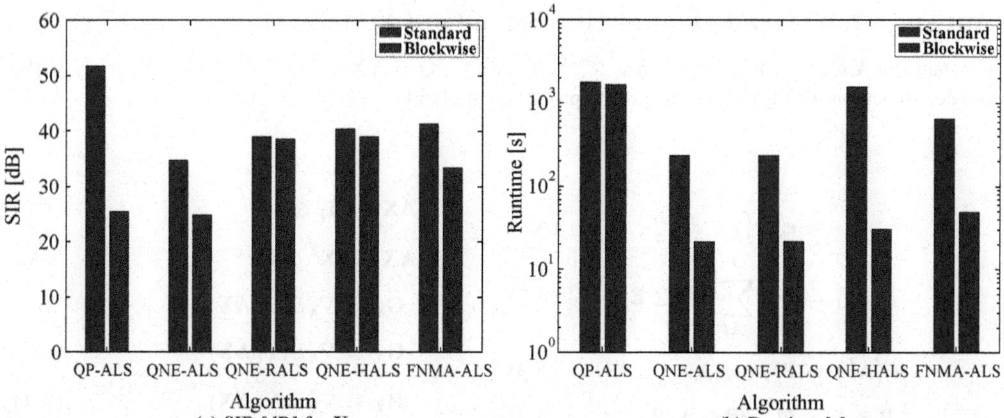

(a) SIR [dB] for **X** (b) Runtime [s]

Figure 6.6 Performance of the selected second-order algorithms applied for estimating a highly redundant large-scale BSS problem where $\mathbf{A} \in \mathbb{R}^{10000\times 10}$ and $\mathbf{X} \in \mathbb{R}^{10\times 1000}$ (benchmark C): (a) mean-SIR [dB] for **X**; (b) Runtime [seconds]. In the `Random blockwise` technique we set $C = 100$ (number of columns), and $R = 50$ (number of rows). The number of inner iterations was set to $k = 5$ for the QP-ALS, and to $k = 1$ for other algorithms.

In addition, we test the performance of the selected algorithms for highly redundant data, where $\mathbf{Y} \in \mathbb{R}^{10000\times 1000}$ was obtained from the benchmark signals (benchmark C: $\mathbf{X} \in \mathbb{R}^{10\times 1000}$) illustrated in Figure 4.8, and the uniformly distributed random mixing matrix $\mathbf{A} \in \mathbb{R}^{10000\times 10}$. For such data, we test the usefulness of the blockwise data selection technique described in Section 1.3.1.

For updating the source matrix **X** we randomly select 50 rows ($R = 50$) from 10000 observations, but for updating the mixing matrix **A**, we randomly select 100 columns ($C = 100$) from the data matrix **Y**. The mean-SIR values and elapsed runtime for 500 alternating steps are shown in Figure 6.6(a) and Figure 6.6(b), respectively.

6.7 Discussions

The results demonstrate that nearly all the analyzed algorithms give the estimations with the mean-SIR higher than 15 dB, which can be considered as the satisfactory results. When the dataset is not very large, the most consistent results can be obtained with the QP-ALS, QNE-RALS, and QNE-HALS algorithms. This is confirmed by the mean-SIR values (best and worst samples) given in Table 6.1 and in Figure 6.3. Unfortunately, the QP-ALS and QNE-RALS algorithms are parameter-dependent. When the dataset is large, the computational cost of the QP-ALS algorithm is significant (see Figure 6.5), and the efficient estimation of the parameters ρ and $\lambda_{\mathbf{A}T}$ is more difficult. Both parameters, and additionally the number of inner iterations k, must be carefully adjusted to obtain the satisfactory results. Hence, the QNE-ALS algorithm seems to be a better choice for large-scale data. Figure 6.3(b) and Figure 6.4 show that the performance of the QNE-ALS algorithm is high, and as demonstrated in Figures 6.5 and 6.6(b) the results were obtained in the shortest time. Moreover, the QNE-ALS algorithm for highly redundant data can be efficiently accelerated using the blockwise data selection technique described in Section 1.3.1. Figure 6.6 demonstrates that this technique only slightly decreases the mean-SIR values but the computational time is shortened about tenfold.

Appendix 6.A: Gradient and Hessian of Cost Functions

For matrices $\mathbf{A} \in \mathbb{R}^{I \times J}$, $\mathbf{B} \in \mathbb{R}^{M \times N}$, $\mathbf{X} \in \mathbb{R}^{J \times T}$, $\mathbf{Y} \in \mathbb{R}^{I \times T}$, $\mathbf{Q} = \mathbf{A}\mathbf{X}$, and vectors basis vector $\boldsymbol{e}_{i,I} = [\mathbf{I}_I]_i$ and cost functions $D(\mathbf{Y}||\mathbf{AX})$, the following properties are hold

$$\mathbf{E}_{i,j} = \boldsymbol{e}_{i,I}\, \boldsymbol{e}_{j,J}^T \in \mathbb{R}^{I \times J} \qquad \text{(A.1)}$$

$$\mathbf{P}_{I,J} = \sum_{i=1}^{I} \sum_{j=1}^{J} \left(\mathbf{E}_{ij} \otimes \mathbf{E}_{ij}^T\right) \qquad \text{(A.2)}$$

$$\operatorname{vec}\left(\mathbf{A}^T\right) = \mathbf{P}_{I,J} \operatorname{vec}(\mathbf{A}) \qquad \text{(A.3)}$$

$$\mathbf{B} \otimes \mathbf{A} = \mathbf{P}_{I,M}^T \left(\mathbf{B} \otimes \mathbf{A}\right) \mathbf{P}_{J,N} \qquad \text{(A.4)}$$

$$\operatorname{diag}\left\{\operatorname{vec}\left(\mathbf{A}^T\right)\right\} = \mathbf{P}_{I,J} \operatorname{diag}\{\operatorname{vec}(\mathbf{A})\} \mathbf{P}_{J,I} \qquad \text{(A.5)}$$

$$\nabla_{\mathbf{X}}(\mathbf{X}^{.[\alpha]}) = \alpha \operatorname{diag}\left\{\operatorname{vec}\left(\mathbf{X}^{.[\alpha-1]}\right)\right\} \qquad \text{(A.6)}$$

$$\nabla_{\mathbf{X}}(\mathbf{AX}) = \mathbf{I}_T \otimes \mathbf{A}, \qquad \text{(A.7)}$$

$$\nabla_{\mathbf{A}}(\mathbf{AX}) = \mathbf{X}^T \otimes \mathbf{I}_I, \qquad \text{(A.8)}$$

$$\mathbf{G}_{\mathbf{X}} = \nabla_{\mathbf{X}} D(\mathbf{Y}||\mathbf{AX}) \qquad \text{(A.9)}$$

$$\mathbf{G}_{\mathbf{A}} = \nabla_{\mathbf{A}} D(\mathbf{Y}||\mathbf{AX}) \qquad \text{(A.10)}$$

$$\mathbf{H}_{\mathbf{X}} = \nabla_{\mathbf{X}}^2 D(\mathbf{Y}||\mathbf{AX}) \qquad \text{(A.11)}$$

$$\mathbf{H}_{\mathbf{A}} = \nabla_{\mathbf{A}}^2 D(\mathbf{Y}||\mathbf{AX}) \qquad \text{(A.12)}$$

$$\mathbf{H}_{\mathbf{A}} = \mathbf{P}_{J,I}\, \mathbf{H}_{\mathbf{A}^T}\, \mathbf{P}_{I,J}. \qquad \text{(A.13)}$$

1. **Frobenius norm**

$$\mathbf{G}_{\mathbf{X}} = \mathbf{A}^T(\mathbf{AX} - \mathbf{Y}) \qquad \text{(A.14)}$$

$$\mathbf{H}_{\mathbf{X}} = \mathbf{I}_T \otimes \mathbf{A}^T\mathbf{A} \qquad \text{(A.15)}$$

$$\mathbf{G}_{\mathbf{A}} = (\mathbf{AX} - \mathbf{Y})\mathbf{X}^T \qquad \text{(A.16)}$$

$$\mathbf{H}_{\mathbf{A}} = \mathbf{X}\mathbf{X}^T \otimes \mathbf{I}_I \qquad \text{(A.17)}$$

2. **Alpha-divergence**

$$\mathbf{G}_{\mathbf{X}} = \frac{1}{\alpha}\mathbf{A}^T\left(\mathbf{1}_{I \times T} - (\mathbf{Y} \oslash \mathbf{AX})^{.[\alpha]}\right) \qquad \text{(A.18)}$$

$$\mathbf{H}_{\mathbf{X}} = \left(\mathbf{I}_T \otimes \mathbf{A}^T\right) \mathbf{D} \left(\mathbf{I}_T \otimes \mathbf{A}\right) \qquad \text{(A.19)}$$

$$\mathbf{D} = \operatorname{diag}\left\{\operatorname{vec}\left(\mathbf{Y}^{.[\alpha]} \oslash \mathbf{Q}^{.[\alpha+1]}\right)\right\} \qquad \text{(A.20)}$$

$$\mathbf{G}_{\mathbf{A}} = \frac{1}{\alpha}\left(\mathbf{1}_{I \times T} - (\mathbf{Y} \oslash \mathbf{AX})^{.[\alpha]}\right)\mathbf{X}^T \qquad \text{(A.21)}$$

$$\mathbf{H}_{\mathbf{A}} = \left(\mathbf{X} \otimes \mathbf{I}_I\right) \mathbf{D} \left(\mathbf{X}^T \otimes \mathbf{I}_I\right) \qquad \text{(A.22)}$$

3. **Beta-divergence**

$$\mathbf{G}_{\mathbf{X}} = \mathbf{A}^T\left[(\mathbf{Q} - \mathbf{Y}) \circledast \mathbf{Q}^{.[\beta-1]}\right] \qquad \text{(A.23)}$$

$$\mathbf{H}_{\mathbf{X}}^{(\beta)} = \left(\mathbf{I}_T \otimes \mathbf{A}^T\right) \mathbf{D} \left(\mathbf{I}_T \otimes \mathbf{A}\right) \qquad \text{(A.24)}$$

$$= \operatorname{diag}\left\{[\mathbf{A}^T \operatorname{diag}\{\boldsymbol{d}_t\}\mathbf{A}]_{t=1,\ldots,T}\right\} \qquad \text{(A.25)}$$

$$\boldsymbol{d}_t = \left(\beta\, \boldsymbol{q}_t - (\beta - 1)\, \boldsymbol{y}_t\right) \circledast \boldsymbol{q}_t^{.[\beta-2]}$$

$$\mathbf{D} = \operatorname{diag}\left\{\operatorname{vec}\left(\beta\, \mathbf{Q}^{.[\beta-1]} - (\beta - 1)\, \mathbf{Y} \circledast \mathbf{Q}^{.[\beta-2]}\right)\right\}$$

$$\mathbf{G}_{\mathbf{A}} = \left[\mathbf{Q}^{.[\beta-1]} \circledast (\mathbf{Q} - \mathbf{Y})\right]\mathbf{X}^T \qquad \text{(A.26)}$$

$$\mathbf{H}_{\mathbf{A}}^{(\beta)} = \left(\mathbf{X} \otimes \mathbf{I}_I\right) \mathbf{D} \left(\mathbf{X}^T \otimes \mathbf{I}_I\right) \qquad \text{(A.27)}$$

$$= \mathbf{P}_{I,T} \operatorname{diag}\left\{[\mathbf{X} \operatorname{diag}\{\underline{\boldsymbol{d}}_i\}\mathbf{X}^T]_{i=1,\ldots,I}\right\} \mathbf{P}_{T,I}. \qquad \text{(A.28)}$$

$$\underline{\boldsymbol{d}}_i = \left(\beta\, \underline{\boldsymbol{q}}_i - (\beta - 1)\, \underline{\boldsymbol{y}}_i\right) \circledast \underline{\boldsymbol{q}}_i^{.[\beta-2]}$$

Appendix 6.B: MATLAB Source Codes

```
% Second-order NMF algorithm
function [A,X,RELRES]=nmf_newton(Y,opts)
%
% Non-negative Matrix Factorization (NMF) with
% standard Regularized Alternating Least-Squares (RALS) algorithm for X,
% and quasi-Newton for A
%
% INPUTS:
% Y:            data,
% opts: structure of optional parameters for algorithm (see defoptions)
%       . J:               number of sources,
%       . MaxIter          maximum number of iterations (default: 1000),
%       . Tol              tolerance (stopping criterion; default: 1E-4),
%       . NoLayer          number of layers (default: 1),
%       . NoRestart        number of restarts (default: 10),
%       . CostFun          cost function (default: Frobenius norm)
%       . Alpha0           initial regularization parameter (default: 100),
%       . Tau              damping factor (default: 50),
%       . Alpha            Free parameter in divergence (default: Alpha = 1),
%       . verbose          if verbose = 1 (default),
%                          the messages (Frobenius norm, No of iterations,
%                          restarts, layers) are displayed,
%                          if verbose = 0, no messages,
%
% OUTPUTS:
%      > A:                   estimated mixing matrix,
%      > X:                   estimated source matrix,
%      > RELRES               matrix [NoLayer by MaxIter] of relative residual
%                             (Frobenius norm) errors,
% Copyright 2008 by A. Cichocki and R. Zdunek and Anh Huy Phan
% =========================================================================
I = size(Y,1);
% Set algorithm parameters from input or by using defaults
defoptions = struct('J',I,'Tol',1e-4,'NoRestart',10,'NoLayer',1,...
    'MaxIter',1000,'CostFun',1,'Alpha',1,'Tau',50,'Alpha0',100,'verbose',1);
if ~exist('opts','var')
    opts = struct;
end

[J,tol,NoRestart,NoLayer,MaxIter,...
    CostFun,Alpha,Tau,Alpha0,verbose] = scanparam(defoptions,opts);
Y(Y< 0) = eps;

% Scaling to unit variances
mx = sqrt(var(Y,[],2) + eps);
Yn = bsxfun(@rdivide,Y,mx);

% Initialization
RELRES = zeros(NoLayer,MaxIter);
Y_norm = norm(Yn,'fro');
No_iter = 10;        % number of initial iterations in "restarts procedure"
Ys = Yn; AH = eye(I); kc = 0;
lambda = 1E-12;          % Levenberg-Marquardt regularization of Hessian
Ie = eye(J);

% Alternatings
```

```
for t = 1:NoLayer
    Fro_outer_temp = 0;
    if verbose, fprintf(1, '\nLayer: %d, \n', t); end

    for  nr = 0: NoRestart
        if (nr == NoRestart)&&(NoRestart> 0)                % initialize
            A = A_best; No_iter = MaxIter;
        else
            while 1, A = rand(size(Ys,1),J); if cond(A) < 50, break; end;
            end                             % Initialization
        end

        for k  = 1: No_iter
            pause(.001)
            Ap = A;
            % ============================================================
            Alpha_reg = Alpha0*exp(-k/Tau); % exp. rule for regular.param.
            if isnan(A),
                disp('Matrix A is too much ill-conditioned. Try again.');
                break;
            end
            if cond(A) > 1E6, alphaX = 1E-6; else alphaX = 0; end
            %Update for X
            X = max(1E6*eps,pinv(A'*A + Alpha_reg + alphaX*Ie)*A'*Ys);

            [GA,HA] = gradhess(Ys',X',A',CostFun);
            [WA,RA] = qr(HA,GA(:)); % Q-less QR factorization
            cond_RA = condest(RA);

            if isinf(cond_RA)
                fprintf(1,['Upper-triangular matrix R is singular'...
                    'in %d iteration(s).\nRestart is needed.\n'],k)
                break;
            end
            A = A - .9*reshape(RA\WA,J,I)'; % Newton iterations
            A(A ≤ 0) = 1E2*eps;
            A = bsxfun(@rdivide,A,sum(A,1));     % L1-norm normalization
            % ============================================================

            if nr ==  NoRestart && nargout > 2 % relative error
                Zw = AH*A*X;
                Zn = bsxfun(@rdivide,Zw,sqrt(var(Zw,[],2) + eps));
                RELRES(t,k) = norm(Yn - Zn,'fro')/Y_norm;
            end

            if (nr ==  NoRestart) && ( (k > 40 && ~mod(k,50)))
                norm_A = norm(A - Ap,'fro');
                if norm_A < tol, break, end             % stopping criterion
            end
        end % Alternatings
        Fro_outer = norm(Ys - A*X,'fro');
        if (nr == 0) || (Fro_outer < Fro_outer_temp)
       A_best = A; Fro_outer_temp = Fro_outer;
        end % multi-conditions

        if verbose % verbose of messages
            fprintf(1, 'Restart %d, Iterations = %d, ',nr,k-1);
            fprintf(1, 'Frobenius norm = %e\n',Fro_outer);
        end
    end % Restarts
```

```

    Ys = X;
    AH = AH*A;
    kc = kc + k-1;
end % Layers

% De-normalization
A = bsxfun(@times,AH,mx);

if verbose
    fprintf(1,'Newton-NMF performed totally %d alternating steps, ',kc);
    fprintf(1,'FIT = %.4f\n',1-norm(Y-A*X,'fro')/norm(Y,'fro'));
end

    function [GX,HX] = gradhess(Y,A,X,CostFun)
        switch CostFun
            case 1 % Frobenius norm
                hA = A'*A + (lambda + 1E8*exp(-k))*eye(J);
                GX = hA*X - A'*Y;  % Gradient
                HX = kron(speye(size(X,2)),hA); % Hessian

            case 2 % Alpha-divergence
                Z = A*X+100*eps;
                if Alpha == 0 % Dual Kullback-Leibler
                    Zx = 1./Z;
                    GX = X*log(Z./Ys)'; % Gradient
                else % Alpha-divergence
                    Zx = (Y./Z).^Alpha;
                    GX = (1/Alpha)*A'*(1-Zx); % Gradient
                    Zx = Zx./Z;
                end
                HX = spalloc(I*J,I*J,I*J^2);
                for i = 1:size(Ys,1)
                    HX(((i-1)*J+1):i*J,((i-1)*J+1):i*J) =...
                        bsxfun(@times,A',Zx(:,i)')*A; % Hessian
                end
                HX = HX + (lambda + 1E15*exp(-2*k))*speye(I*J);

            case 3 % Beta-divergence
                Z = A*X+100*eps;
                Zx = (Alpha*Z - (Alpha-1)*Y).*Z.^(Alpha-2);
                GX = A.'*((Z- Y).*(Z).^(Alpha-1));
                HX = spalloc(I*J,I*J,I*J^2);
                for i = 1:size(Ys,1)
                    HX(((i-1)*J+1):i*J,((i-1)*J+1):i*J) =...
                        bsxfun(@times,A',Zx(:,i)')*A; % Hessian
                end
                HX = HX + (lambda + 1E15*exp(-2*k))*speye(I*J);
        end
    end
end
```

```
function [A,X,RELRES]=nmf_QNA(Y,opts)
%
% Non-negative Matrix Factorization (NMF) with
% RALS algorithm for X, and QNA for A
% Copyright 2008 by A. H. Phan and A. Cichocki and R. Zdunek
% =====================================================================
I = size(Y,1);
```

```
% Set algorithm parameters from input or by using defaults
defoptions = struct('J',I,'tol',1e-8,'nrestart',10,...
    'niter',1000,'CostFun',1,'alpha',1,'tau',50,'alpha0',100,'verbose',0);
if ~exist('opts','var')
    opts = struct;
end

[J,tol,nrestart,MaxIter,...
    CostFun,Alpha,Tau,Alpha0,verbose] = scanparam(defoptions,opts);
Y(Y< 0) = eps;

% Scaling to unit variances
mx = 1./sqrt(var(Y,[],2) + eps);
Yn = bsxfun(@times,Y,mx);

% Initialization
No_iter = 10; % number of initial iterations in "restarts procedure"
kc = 0;

if matlabpool('size') == 0
    matlabpool
end
Fro_outer_temp = 0;
if verbose, fprintf(1, '\nLayer: %d, \n', t); end
Ie = eye(J);
for nr= 0:nrestart
    pause(.0001)
    if (nr == nrestart)&&(nrestart > 0) % initialize
        A = A_best;  No_iter = MaxIter;
    else
        while 1, A = rand(size(Yn,1),J); if cond(A) < 50, break; end;
        end                                   % Initialization
    end

    for k  = 1: No_iter
        pause(.001)
        Ap = A;

        X = ALS(Yn,A);                          % Update X
        A = QNA(Yn',X',A')';                    % Update A
        A = bsxfun(@rdivide,A,sum(A,1)); % normalization to unit L1-norm

        if (nr ==  nrestart) && ( (k > 40 && ~mod(k,50)))
            norm_A = norm(A - Ap,'fro');
            if norm_A < tol, break, end  % stopping criterion
        end
    end % Alternatings

    Fro_outer = norm(Yn - A*X,'fro');
    if (nr == 0) || (Fro_outer < Fro_outer_temp)
        A_best = A; Fro_outer_temp = Fro_outer;
    end % multi-conditions

    if verbose % verbose of messages
        fprintf(1, 'Restart %d, Iterations = %d, ',nr,k-1);
        fprintf(1, 'Frobenius norm = %e\n',Fro_outer);
    end
end % Restarts
```

```
66   % De—normalization
67   A = bsxfun(@rdivide,A,mx);
68
69   if verbose
70       fprintf(1,'Newton—NMF performed totally %d alternating steps, ',kc);
71       fprintf(1,'Rel. residual error = %e\n',norm(Y—A*X,'fro')/norm(Y,'fro'));
72   end
73   matlabpool close
74       function X = QNA(Yn,A,X)
75           Z = A*X+100*eps;
76           Zx = (Yn./Z).^Alpha;
77           GX = (1/Alpha)*A'*(1—Zx);    % Gradient
78           Zx = Zx./Z;
79           G_tilde = zeros(size(X));
80           parfor i=1:size(X,2),
81               HXi = bsxfun(@times,A',Zx(:,i)')*A + ...
82                   (1e—12 + 1E8*exp(—2*k))*eye(J);
83               [G_tilde(:,i),flag] = symmlq(HXi,GX(:,i));
84           end
85           X = X — .9*G_tilde; % Newton iterations
86           X(X ≤ 0) = 1E2*eps;
87       end
88       function X = ALS(Yn,A)
89           Alpha_reg = Alpha0*exp(—k/Tau); % exponential rule for regular.param.
90           if cond(A) > 1E6, alphaX = 1E—6; else alphaX = 0; end
91           X = max(1E6*eps,pinv(A'*A + Alpha_reg + alphaX*Ie)*A'*Yn);
92       end
93   end
```

```
1    function X = gpcg_nmf(Y,A,X,MaxIter)
2    % GPCG algorithm adapted for NMF based on Frobenius norm
3    % Copyright by Andrzej Cichocki and Rafal Zdunek
4    % Optimized by Anh Huy Phan and Andrzej Cichocki, 2008
5
6    T = size(Y,2);
7    beta = 0.5; mu = 0.01; MaxSearchIter = 10;
8    Diff_DF_max = 0; gamma_GP = 0.01;
9
10   Hx = A'*A; AtY = A'*Y;
11   H = kron(speye(T),Hx);                        % Hessian
12   Dcostold = norm(Y — A*X,'fro');               % Cost value
13   for k = 1:MaxIter
14       % Step 1
15       Xp = X;
16       Gx = Hx * X — AtY;                         % Gradient Matrix
17       gx = Gx(:);                                % Vectorization
18
19       eta = (gx'*gx)/(gx'*(H*gx));
20       for i = 0:MaxSearchIter                    % Armijo rule
21           X = max(X + eta*Gx,0);
22           Dcost = norm(Y — A*X,'fro');
23           if (Dcost — Dcostold) ≤ mu/eta*norm(X — Xp,'fro');
24               break;
25           end
```

```
            eta = eta*beta;
        end

        z = X(:) > eps;                                   % Inactive set
        Gx = Hx * X - AtY;                                % Gradient Matrix
        [gtil,flag] = pcg(H(z,z),-Gx(z));                 % Conjugate Gradients
        xz = X(z);
        for i = 0:MaxSearchIter                           % Newton rule
            xz = max(xz + eta*gtil,0);
            Dcost = norm(Y - A*X,'fro');
            if (Dcost - Dcostold) ≤ 0, break, end
            eta = eta*beta;
        end
        X(z) = xz;

        Diff_DF = Dcostold - Dcost;
        Diff_DF_max = max(Diff_DF,Diff_DF_max);
        if (Diff_DF ≤ gamma_GP*Diff_DF_max) & (k > 1)  % Stopping criterion
            break;
        end
        Dcostold = Dcost;
    end
```

```
% Nonnegative Matrix Factorization with Fast Newton-type algorithm
function X = fnma_x(Y,A,X,no_iter)
%
% INPUTS:
% > Y:          observation matrix [I by T] of mixed signals (images)
% > A:          source component matrix [I by J]
% > X:          initial guess [J by T]
% > no_iter:    number of inner iterations
%
% OUTPUT:
% > X:          mixing matrix [J by T]
% ======================================================================
B = A'*A; C = A'*Y; D = inv(B);
sigma = 0.01; beta = 1; kinc = 1;
lower_bound = 1E-6; beta_dec = 0.1;

for l = 1: no_iter                              % inner iterations
    F = .5*norm(Y - A*X,'fro')^2;
    G = B*X - C; % Gradient of D(Y||AX) with respect to X
    I_active = find((X < eps)&(G > 0));
    G(I_active) = 0;
    U = D*G; U(I_active) = 0;

    s = 0;
    while (beta > 1E-15) && (beta < 1E8)     % search step size
        s = s + 1;

        Xn = max(X - beta*U,lower_bound);
        Delta = Xn - X;
        Fn = .5*norm(Y - A*Xn,'fro')^2;
        cond_alpha = ((Fn - F) ≤ sigma*U(:)'*Delta(:));

        if cond_alpha | (Xn == X)
```

```
            % Increase
            if ~kinc && (s > 1), X = Xn; break; end
            beta = beta/beta_dec;
            kinc = 1; Xp = Xn;
        else
            % Decrease
            if kinc && (s > 1), X = Xp; beta = beta*beta_dec; break; end
            beta = beta*beta_dec;
            kinc = 0;
        end
    end % while
end
```

```
function [A,X,RELRES] = nmf_qp(Y,opts)
% Non-negative Matrix Factorization (NMF) with Quadratic Programming
%
% [A,X]=nmf_qp(Y,J,MaxIter,Tol,NoLayer,NoRestart,Lambda,Alpha0,Tau,
%       IterInner,verbose) produces mixing matrix A of dimension [I by J],
%       and source matrix X of dimension [J by T],
%       for the linear mixing model: AX = Y,
%       where Y is an observation matrix [I by T].
%
% Note: > I: number of sensors,
%       > J: number of sources,
%       > T: number of samples,
%
% INPUTS:
%       > Y:            data,
%       > J:            number of sources,
%       > MaxIter       maximum number of iterations (default: 100),
%       > Tol           tolerance (stopping criterion; default: 1E-4),
%       > NoLayer       number of layers (default: 1),
%       > NoRestart     number of restarts (default: 10),
%       > Lambda        Tikhonov regularization parameter (default: 1000)
%       > Alpha0        initial regularization parameter (default: 0),
%       > Tau           damping factor (default: 50),
%       > IterInner     inner iterations for computing A (default: 4),
%       > verbose       if verbose = 1 (default), display
%                       Frobenius norm, No of iterations, restarts,layers
%
% OUTPUTS:
%       > A:            estimated mixing matrix,
%       > X:            estimated source matrix,
%       > RELRES        matrix [NoLayer by MaxIter] of relative residual
%                       (Frobenius norm) errors,
% Copyright 2008 by A. Cichocki and R. Zdunek
% Optimized by Anh Huy Phan and Andrzej Cichocki - 01/2008
% =======================================================================
% Set algorithm parameters from input or by using defaults
I = size(Y,1);
defoptions = struct('J',I,'tol',1e-8,'init',[1 1],'nrestart',10,...
    'NoLayer',1,'niter',1000,'IterInner',3,'tau',50,'alpha0',0,...
    'Lambda',1000,'verbose',0);
if ~exist('opts','var')
    opts = struct;
end

[J,tol,init,NoRestart,NoLayer,MaxIter,IterInner,...
```

```
    Tau,Alpha0,verbose] = scanparam(defoptions,opts);
Y(Y< 0) = eps;

% Normalization to unit variances
[I,T]=size(Y);
mx = sqrt(var(Y,[],2) + eps);
Yn = bsxfun(@rdivide,Y,mx);

% Initialization
A = zeros(I,J); X=zeros(J,T);
RELRES = zeros(NoLayer,MaxIter);
Y_norm = norm(Yn,'fro');
No_iter = 10;              % number of iterations to select the best trial
Ys = Yn; AH = eye(I); kc = 0; Ie = eye(J);

% Parameters for QP-NMF
rho = 1E-3; eta = 1E-4; epsil_A = 1E-6;

for t = 1:NoLayer
    Fro_outer_temp = 0;
    if verbose, fprintf(1, '\nLayer: %d, \n', t); end
    IJ = size(Ys,1)*J;

    for  nr = 0: NoRestart
        if (nr == NoRestart)&&(NoRestart > 0)
            A = A_best; No_iter = MaxIter;
        else
            while 1, A = rand(size(Ys,1),J); if cond(A) < 50, break; end;
            end                                 % Initialization
        end

        for k  = 1: No_iter
            Ap = A;
            % ==============================================================
            Alpha_reg = Alpha0*exp(-k/Tau); % exp. rule for regular.param.
            if isnan(A),
                disp('Matrix A is too much ill-conditioned. Try again.');
                break;
            end
            if cond(A) > 1E6, alphaX = 1E-6; else alphaX = 0; end
            %Update for X
            X = max(1E6*eps,pinv(A'*A + Alpha_reg + alphaX*Ie)*A'*Ys);

            % Settings
            A = ones(size(Ys,1),J);
            At = A';
            B = X*X'; C_bar = X*Ys';
            Q_bar = kron(speye(size(Ys,1)),B); % Hessian with respect to A
            Go = C_bar - B*At;                 % Gradient with respect to A^T
            theta = -(rho/(IJ))*abs(Go(:)'*At(:)); % log barrier parameter
            a = At(:);
            for l = 1: IterInner                    % inner iterations
                inactive = a ≥ epsil_A;            % inactive set
                a_tilde = a(inactive);
                N = length(a_tilde);
                c_tilde = 2*theta./a_tilde - C_bar(inactive);
                Q_tilde = Q_bar(inactive,inactive);
                Q_tilde = Q_tilde - theta*spdiags(1./a_tilde.^2,0,N,N);

                [Q,RA] = qr(Q_tilde,-c_tilde);  % Q-less QR factorization
```

```
                z_tilde = RA\Q;                    % Gaussian elimination
                h_tilde = z_tilde - a_tilde;
                beta = min(1, 0.9995*min(a_tilde./abs(h_tilde)));
                a_tilde = a_tilde + beta*h_tilde;

                theta = theta* (2*N - sum(z_tilde./a_tilde));

                a(~inactive) = 0;
                a(inactive) = a_tilde;

                if abs(theta)<eta*norm(a), break, end % stopping criterion
                theta = (rho/sqrt(IJ))*theta;
            end
            A = reshape(a,J,[])fl;
            A = bsxfun(@rdivide,A,sum(A,1)+eps);      % normalization
            % ============================================================

            if nr ==  NoRestart && nargout > 2        % relative error
                Zw = AH*A*X;
                Zn = bsxfun(@rdivide,Zw,sqrt(var(Zw,[],2) + eps));
                RELRES(t,k) = norm(Yn - Zn,'fro')/Y_norm;
            end
            if (nr ==  NoRestart) && ( (k > 40 && ~mod(k,50)))
                norm_A = norm(A - Ap,'fro');
                if norm_A < tol, break, end           % stopping criterion
            end
        end % Alternatings

        Fro_outer = norm(Ys - A*X,'fro');
        if (nr == 0) || (Fro_outer < Fro_outer_temp)
            A_best = A; Fro_outer_temp = Fro_outer;
        end % multi-conditions

        if verbose % verbose of messages
            fprintf(1,'Restart %d, Iterations = %d, Frob. norm = %e\n',...
                nr,k-1,Fro_outer);
        end
    end % Restarts

    Ys = X; AH = AH*A;
    kc = kc + k-1;
end % Layers

% De-normalization
A = bsxfun(@times,AH,mx);

if verbose
    fprintf(1,'QP-NMF performed totally %d alternating steps\n',kc);
    fprintf(1,'FIT = %.4f\n',1-norm(Y-A*X,'fro')/norm(Y,'fro'));
end
```

References

[1] J. M. Bardsley. A nonnegatively constrained trust region algorithm for the restoration of images with an unknown blur. *Electronic Transactions on Numerical Analysis*, 20:139–153, 2005.

[2] J.M. Bardsley and C.R. Vogel. Nonnegatively constrained convex programming method for image reconstruction. *SIAM Journal on Scientific Computing*, 4:1326–1343, 2004.

[3] R. Barrett, M. Berry, T.F. Chan, J. Demmel, J. Donato, J. Dongarra, V. Eijkhout, R. Pozo, C. Romine, and H. Van der Vorst. *Templates for the Solution of Linear Systems: Building Blocks for Iterative Methods, 2nd Edition*. SIAM, Philadelphia, PA, 1994.
[4] A. Björck. *Numerical Methods for Least-Squares Problems*. SIAM, Philadelphia, 1996.
[5] S. Boyd and L. Vandenberghe. *Convex Optimization*. Cambridge University Press, Cambridge, UK, 2004.
[6] I. Buciu and I. Pitas. Application of non-negative and local non negative matrix factorization to facial expression recognition. pages 288–291. Proc. of Intl. Conf. Pattern Recognition (ICPR), 2004.
[7] D. Calvetti, G. Landi, L. Reichel, and F. Sgallari. Non-negativity and iterative methods for ill-posed problems. *Inverse Problems*, 20:1747–1758, 2004.
[8] A. Cichocki, S. Amari, R. Zdunek, R. Kompass, G. Hori, and Z. He. Extended SMART algorithms for non-negative matrix factorization. *Springer, LNAI-4029*, 4029:548–562, 2006.
[9] A. Cichocki and R. Unbehauen. *Neural Networks for Optimization and Signal Processing*. John Wiley & Sons ltd, New York, 1994.
[10] A. Cichocki and R. Zdunek. Regularized alternating least squares algorithms for non-negative matrix/tensor factorizations. *Springer, LNCS-4493*, 4493:793–802, June 3–7 2007.
[11] A. Cichocki, R. Zdunek, and S. Amari. Csiszar's divergences for non-negative matrix factorization: Family of new algorithms. *Springer, LNCS-3889*, 3889:32–39, 2006.
[12] A. Cichocki, R. Zdunek, S. Choi, R. Plemmons, and S. Amari. Nonnegative tensor factorization using Alpha and Beta divergencies. In *Proc. IEEE International Conference on Acoustics, Speech, and Signal Processing (ICASSP07)*, volume III, pages 1393–1396, Honolulu, Hawaii, USA, April 15–20 2007.
[13] A. Cichocki, R. Zdunek, S. Choi, R. Plemmons, and S.-I. Amari. Novel multi-layer nonnegative tensor factorization with sparsity constraints. *Springer, LNCS-4432*, 4432:271–280, April 11–14 2007.
[14] S.F. Cotter, B.D. Rao, K. Engan, and K. Kreutz-Delgado. Sparse solutions to linear inverse problems with multiple measurement vectors. *IEEE Transactions on Signal Processing*, 53(7):2477–2488, 2005.
[15] I. Dhillon and S. Sra. Generalized nonnegative matrix approximations with Bregman divergences. In *Neural Information Proc. Systems*, pages 283–290, Vancouver, Canada, December 2005.
[16] L. Fei-Fei, R. Fergus, and P. Perona. Learning generative visual models from few training examples: an incremental Bayesian approach tested on 101 object categories. In *IEEE CVPR Workshop of Generative Model Based Vision*, 2004.
[17] M. Heiler and C. Schnoerr. Learning sparse representations by non-negative matrix factorization and sequential cone programming. *Journal of Machine Learning Research*, 7:1385–1407, 2006.
[18] M.R. Hestenes and E. Stiefel. Method of conjugate gradients for solving linear systems. *J. Res. Nat. Bur. Standards*, 49:409–436, 1952.
[19] P.O. Hoyer. Non-negative matrix factorization with sparseness constraints. *Journal of Machine Learning Research*, 5:1457–1469, 2004.
[20] B. Jorgensen. *The Theory of Dispersion Models*. Chapman and Hall, London, 1997.
[21] D. Kim, S. Sra, and I.S. Dhillon. Fast Newton-type methods for the least squares nonnegative matrix approximation problem. In *Proc. 6-th SIAM International Conference on Data Mining*, Minneapolis, Minnesota, USA, April 2007.
[22] R. Kompass. A generalized divergence measure for nonnegative matrix factorization. *Neural Computation*, 19(3):780–791, 2006.
[23] M.J. Lyons, S. Akamatsu, M. Kamachi, and J. Gyoba. Coding facial expressions with Gabor wavelets. In *Proc. of the 3-rd IEEE Intl. Conf. on Auto. Face and Gesture Recog.*, pages 200–205, Nara, Japan, April 14–16 1998.
[24] J.J. More and G. Toraldo. On the solution of large quadratic programming problems with bound constraints. *SIAM Journal on Optimization*, 1(1):93–113, 1991.
[25] J. Nocedal and S.J. Wright. *Numerical Optimization*. Springer Series in Operations Research. Springer, New York, 1999.
[26] P. Paatero. A weighted non-negative least squares algorithm for three-way PARAFAC factor analysis. *Chemometrics Intelligent Laboratory Systems*, 38(2):223–242, 1997.
[27] C.C. Paige, B.N. Parlett, and H.A. van der Vorst. Approximate solutions and eigenvalue bounds from krylov subspaces. *Numerical Linear Algebra with Applications*, 2:115–133, 1993.
[28] A. Peressini, F. Sullivan, and J. Uhl. *The Mathematics of Nonlinear Programming*. Springer-Verlag, Berlin, 1988.
[29] T. Read and N. Cressie. *Goodness-of-Fit Statistics for Discrete Multivariate Data*. Springer Series in Statistics. Springer Verlag, New York, 1988.
[30] M. Rojas, S. A. Santos, and D.C. Sorensen. A new matrix-free algorithm for the large-scale trust-region subproblem. *SIAM Journal on Optimization*, 11(3):611–646, 2000.

[31] M. Rojas and T. Steihaug. An interior-point trust-region-based method for large-scale non-negative regularization. *Inverse Problems*, 18:1291–1307, 2002.

[32] G. K. Smyth. Fitting Tweedies compound Poisson model to insurance claims data: dispersion modelling. *ASTIN Bulletin*, 32:2002.

[33] R. Zdunek and A. Cichocki. Non-negative matrix factorization with quasi-Newton optimization. *Springer LNAI*, 4029:870–879, 2006.

[34] R. Zdunek and A. Cichocki. Nonnegative matrix factorization with constrained second-order optimization. *Signal Processing*, 87:1904–1916, 2007.

[35] R. Zdunek and A. Cichocki. Nonnegative matrix factorization with quadratic programming. *Neurocomputing*, 71(10–12):2309–2320, 2008.

7

Multi-Way Array (Tensor) Factorizations and Decompositions

The problems of nonnegative multi-way array (tensor) factorizations and decompositions arise in a variety of disciplines in the sciences and engineering. They have a wide range of important applications such as in bioinformatics, neuroscience, image understanding, text mining, chemometrics, computer vision and graphics, where tensor factorizations and decompositions can be used to perform factor retrieval, dimensionality reduction, compression, denoising, to mention but a few. For example, in neuroimage processing, images and videos are naturally represented by third-order, or general higher-order tensors. Color video sequences are normally represented by fourth-order tensors, thus requiring three indices for color images and a fourth index for the temporal information.

Almost all NMF algorithms described in the earlier chapters can be extended or generalized to the various nonnegative tensor factorizations and decompositions formulated in Chapter 1. In this chapter we mainly focus on the Nonnegative Tensor Factorization (NTF) (i.e., the PARAFAC with nonnegativity and sparsity constraints), the Nonnegative Tucker Decomposition (NTD) and the Block-Oriented Decomposition (BOD). In order to make this chapter as self-contained as possible, we re-introduce some concepts and derive many efficient heuristic algorithms for nonnegative tensor (multi-way array) factorizations and decompositions. Our particular emphasis is on a detailed treatment of generalized robust cost functions, such as Alpha- and Beta-divergences. Based on these cost functions, several classes of algorithms are introduced, including: (1) multiplicative updating; (2) Alternating Least Squares (ALS); and (3) Hierarchical ALS (HALS). These algorithms are then incorporated into multi-layer networks in order to improve the performance (see also Chapters 3–6), starting from relatively simple third-order nonnegative tensor factorizations through to extensions to arbitrarily high order tensor decompositions.

Practical considerations include the ways to impose nonnegativity or semi-nonnegativity, together with optional constraints such as orthogonality, sparsity and/or smoothness. To follow the material in this chapter it would be helpful to be familiar with Chapters 1, 3 and 4.

7.1 Learning Rules for the Extended Three-way NTF1 Problem

Based on the background given in Chapter 1, we shall now introduce practical learning rules for several extended tensor decompositions.

Nonnegative Matrix and Tensor Factorizations: Applications to Exploratory Multi-way Data Analysis and Blind Source Separation Andrzej Cichocki, Rafal Zdunek, Anh Huy Phan and Shun-ichi Amari

7.1.1 Basic Approaches for the Extended NTF1 Model

Consider the extended NTF1 model with irregular frontal slices, shown in Figure 7.1(a) [24], which can be exploited as follows: "Given a three-way (third-order) tensor formed by a set of matrices $\mathbf{Y}_q \in \mathbb{R}_+^{I \times T_q}$ $(q = 1, 2, \ldots, Q)$, formulate a set of nonnegative and sparse matrices $\mathbf{A} \in \mathbb{R}_+^{I \times J}$, $\mathbf{C} \in \mathbb{R}_+^{Q \times J}$ and $\mathbf{X}_q \in \mathbb{R}_+^{J \times T_q}$ for $q = 1, 2, \ldots, Q$ with reduced dimensions (typically, $J << I < T_q$)".
The extended NTF1 model for a three-way array can be represented in two different mathematical forms, as illustrated in Figures 7.1(b) and 7.1(c). Firstly, it can be described by a set of tri-NMF models:

$$\boxed{\mathbf{Y}_q = \mathbf{A}\mathbf{D}_q\mathbf{X}_q + \mathbf{E}_q, \qquad (q = 1, 2, \ldots, Q),} \tag{7.1}$$

where $\mathbf{D}_q \in \mathbb{R}_+^{J \times J}$ are diagonal matrices (each diagonal matrix contains the q-th row of matrix $\mathbf{C} \in \mathbb{R}_+^{Q \times J}$ in its main diagonal), $\mathbf{X}_q = [x_{jtq}] \in \mathbb{R}_+^{J \times T_q}$ are matrices representing sources (or hidden components), and matrices $\mathbf{E}_q = [e_{itq}] \in \mathbb{R}^{I \times T_q}$ represent errors or noise depending upon the application. The diagonal matrices $\mathbf{D}_q$ can be considered as scaling matrices and can therefore be absorbed into the matrices $\mathbf{X}_q$ upon defining a new set of matrices as $\mathbf{X}_q := \mathbf{D}_q\mathbf{X}_q$ (if no additional constraints on the component matrix $\mathbf{C}$ are imposed), to give

$$\boxed{\mathbf{Y}_q = \mathbf{A}\mathbf{X}_q + \mathbf{E}_q, \qquad (q = 1, 2, \ldots, Q).} \tag{7.2}$$

Thus, only the mixing (bases) matrix $\mathbf{A}$ and the set of scaled source matrices $\mathbf{X}_q$ need to be found whereas due to the scaling ambiguity the matrix $\mathbf{C}$ does not need to be calculated explicitly. This also allows us to use row-wise unfolding to convert the NTF1 problem into the standard NMF problem described by the single matrix equation

$$\boxed{\mathbf{Y}_{(1)} = \mathbf{A}\mathbf{X}_{(1)} + \mathbf{E}_{(1)},} \tag{7.3}$$

where $\mathbf{Y}_{(1)} = \overline{\mathbf{Y}} = [\bar{y}_{i\bar{t}}] = [\mathbf{Y}_1, \mathbf{Y}_2, \ldots, \mathbf{Y}_Q] \in \mathbb{R}^{I \times \overline{T}}$; $\mathbf{X}_{(1)} = \overline{\mathbf{X}} = [\bar{x}_{j\bar{t}}] = [\mathbf{X}_1, \mathbf{X}_2, \ldots, \mathbf{X}_Q] \in \mathbb{R}^{J \times \overline{T}}$; $\mathbf{E}_{(1)} = [\mathbf{E}_1, \mathbf{E}_2, \ldots, \mathbf{E}_Q] \in \mathbb{R}^{I \times \overline{T}}$, and $\overline{T} = \sum_{q=1}^{Q} T_q$, $\bar{t} = 1, 2, \ldots, \overline{T}$.

Based on the above representations, we have several possible approaches to find (identify) the extended NTF1 model:

1. Global strategy, for example, based on an alternating minimization of the cost function:

$$D_F(\overline{\mathbf{Y}} \,||\, \mathbf{A}\overline{\mathbf{X}}) = \frac{1}{2} \left\| \overline{\mathbf{Y}} - \mathbf{A}\overline{\mathbf{X}} \right\|_F^2, \tag{7.4}$$

   which leads to the standard ALS algorithm (see Chapter 4).
2. Local strategy, for example, by sequential alternating minimization of a set of cost functions

$$D_F(\mathbf{Y}_q||\mathbf{A}\mathbf{X}_q) = \frac{1}{2} \left\| \mathbf{Y}_q - \mathbf{A}\mathbf{X}_q \right\|_F^2, \qquad (q = 1, 2, \ldots, Q), \tag{7.5}$$

   where the matrix $\mathbf{A}$ can be estimated as the average of all the Q slices.
3. Hybrid approach, in which $\mathbf{A}$ is found based on the minimization of a global cost function, that is,

$$D_F(\overline{\mathbf{Y}}||\mathbf{A}\overline{\mathbf{X}}) = \frac{1}{2} \left\| \overline{\mathbf{Y}} - \mathbf{A}\overline{\mathbf{X}} \right\|_F^2 \tag{7.6}$$

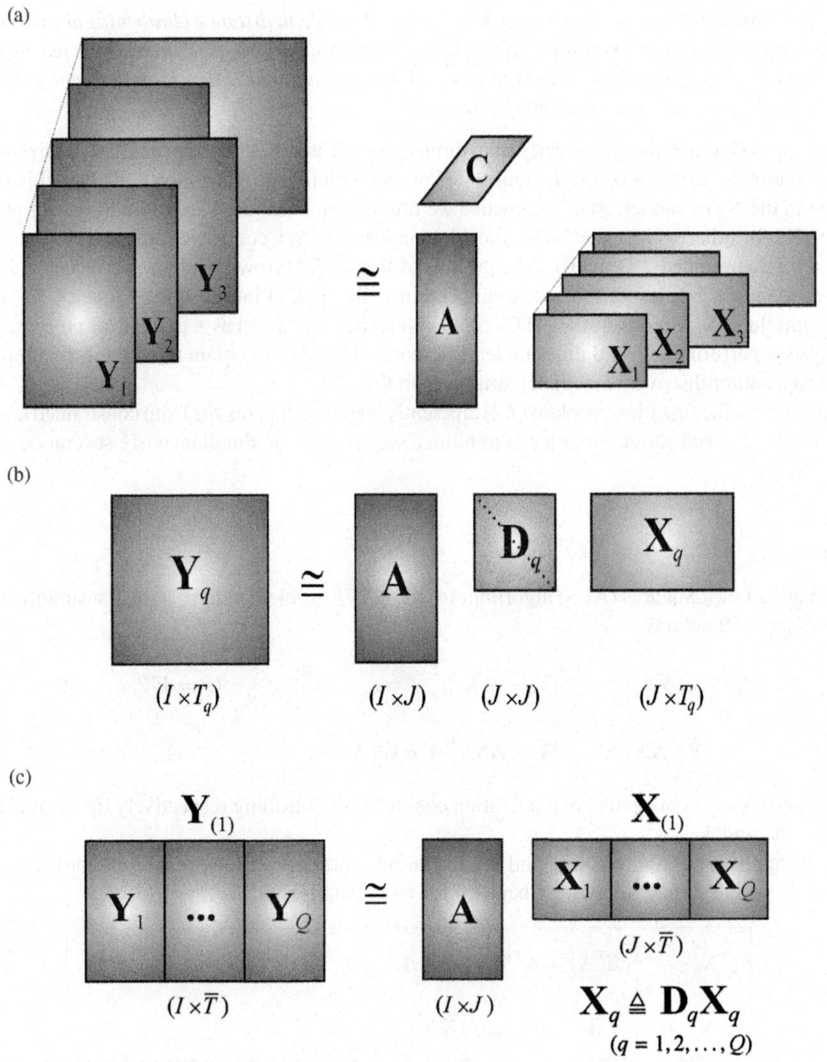

Figure 7.1 (a) NTF1 model that approximately decomposes a three-way array with irregular frontal slices $\mathbf{Y}_q \in \mathbb{R}_+^{I \times T_q}$ into a set of nonnegative matrices $\mathbf{A} = [a_{ij}] \in \mathbb{R}_+^{I \times J}$, $\mathbf{C} = [c_{qj}] \in \mathbb{R}_+^{Q \times J}$ and $\{\mathbf{X}_1, \mathbf{X}_2, \ldots, \mathbf{X}_Q\}$, $\mathbf{X}_q = [x_{jtq}] \in \mathbb{R}_+^{J \times T_q}$; ($\mathbf{E}_q \in \mathbb{R}^{I \times T_q}$ represents errors). (b) An equivalent representation using set of three-factor NMF, where $\mathbf{D}_q = \text{diag}(\boldsymbol{c}_q)$ are diagonal matrices. (c) Global matrix representation using row-wise (mode-1) unfolding of the three-way array; in this case the sub-matrices are defined as $\mathbf{X}_q \stackrel{\Delta}{=} \mathbf{D}_q \mathbf{X}_q$, $(q = 1, 2, \ldots, Q)$.

for fixed matrices $\mathbf{X}_q$, $(q = 1, 2, \ldots, Q)$. For a fixed $\mathbf{A}$, on the other hand, we perform the minimization of set of local cost functions (to find $\mathbf{X}_q$), for instance

$$D_F(\mathbf{Y}_q || \mathbf{A}\mathbf{X}_q) = \frac{1}{2} \left\| \mathbf{Y}_q - \mathbf{A}\mathbf{X}_q \right\|_F^2, \qquad (q = 1, 2, \ldots, Q). \tag{7.7}$$

Remark 7.1 *Instead of cost functions based on squared Euclidean distance (Frobenius norm), we may use a number of alternative cost functions, such as Alpha- or Beta-divergences. Moreover, we may use different cost functions for the estimation of* $\mathbf{A}$ *and the set of matrices* $\mathbf{X}_q$, *assuming they have the same global minima (neglecting scaling and permutation ambiguities).*

The three-way NTF1 model can be clearly transformed into a 2-way nonnegative matrix factorization (NMF) problem by unfolding (matricizing) the tensors. However, strictly speaking such a 2D model is not always equivalent to the NTF1 model, since in practice we often need to impose various additional constraints for each slice $\mathbf{X}_q$. In other words, the NTF1 model cannot be always considered equal to a standard 2-way NMF of a single unfolded 2-D matrix. The profiles of the stacked (row-wise unfolded) $\mathbf{X}_{(1)} = \overline{\mathbf{X}}$ are often not treated as single profiles and the constraints are usually applied independently to each $\mathbf{X}_q$ matrix that forms the unfolded $\overline{\mathbf{X}}$. Moreover, the NTF1 problem can be considered as a dynamical process, where the data analysis is performed several times under different conditions,[1] to obtain the full information about the available data and/or discover some inner structures in the data.

To deal with the factorization problem (7.2) efficiently, we can adopt results from constrained optimization and multi-criteria optimization, in order to minimize sequentially or simultaneously several cost functions [26,44,46,53,54].

7.1.2 ALS Algorithms for NTF1

The Alternating Least Squares (ALS) algorithm for the NTF1 problem with sparsity constraints is based on the following cost functions

$$D_F(\mathbf{Y}_q \,||\, \mathbf{A}\mathbf{X}_q) = \frac{1}{2} \left\|\mathbf{Y}_q - \mathbf{A}\mathbf{X}_q\right\|_F^2 + \alpha_{\mathbf{X}_q} \left\|\mathbf{X}_q\right\|_1, \qquad (q = 1, 2, \ldots, Q) \tag{7.8}$$

$$D_F(\overline{\mathbf{Y}} \,||\, \mathbf{A}\overline{\mathbf{X}}) = \frac{1}{2} \left\|\overline{\mathbf{Y}} - \mathbf{A}\overline{\mathbf{X}}\right\|_F^2 + \alpha_{\mathbf{A}} \left\|\mathbf{A}\right\|_1, \tag{7.9}$$

where $\alpha_{\mathbf{X}_q}$ and $\alpha_{\mathbf{A}}$ are nonnegative regularization coefficients controlling respectively the degree of sparsity for matrices $\mathbf{X}_q$ and $\mathbf{A}$.

Upon the application of alternating and sequential minimization of these cost functions, we can derive the following simple update rules (see Chapter 4 for more details):

$$\mathbf{X}_q \leftarrow \left[(\mathbf{A}^T\mathbf{A})^{-1}(\mathbf{A}^T\mathbf{Y}_q - \alpha_{\mathbf{X}_q}\, \mathbf{1}_{J\times T_q})\right]_+, \qquad (q = 1, 2, \ldots, Q), \tag{7.10}$$

$$\mathbf{A} \leftarrow \left[(\overline{\mathbf{Y}}\,\overline{\mathbf{X}}^T - \alpha_{\mathbf{A}}\mathbf{1}_{I\times J})(\overline{\mathbf{X}}\,\overline{\mathbf{X}}^T)^{-1}\right]_+. \tag{7.11}$$

It is now straightforward to impose additional constraints, for example, the orthogonality of column vectors $\boldsymbol{x}_q$ within matrices $\mathbf{X}_q$, by performing the following transformations

$$\mathbf{X}_q \leftarrow (\mathbf{X}_q\mathbf{X}_q^T)^{-1/2}\mathbf{X}_q, \tag{7.12}$$

together with imposing the orthogonality between the row vectors of the matrix $\mathbf{A}$

$$\mathbf{A} \leftarrow \mathbf{A}(\mathbf{A}^T\mathbf{A})^{-1/2}, \tag{7.13}$$

after every iteration.

[1] With multi-start initializations, multi-layer or recurrent implementation, Monte Carlo analysis, selection of additional natural constraints, etc. For these reasons we shall later illustrate the hybrid approach.

In a special case, when we impose orthogonality constraints on matrices, say $\mathbf{A}$ and $\mathbf{X}_q$, but not on the matrix $\mathbf{C}$, the minimization is based on the following set of local cost functions

$$D_F(\mathbf{Y}_q||\mathbf{A}\mathbf{X}_q) = \frac{1}{2} \left\| \mathbf{Y}_q - \mathbf{A}\mathbf{D}_q\mathbf{X}_q \right\|_F^2, \qquad (q = 1, 2, \ldots, Q), \tag{7.14}$$

with slightly different update rules[2]

$$\mathbf{X}_q \leftarrow \left[(\mathbf{A}\mathbf{D}_q)^{\dagger}\mathbf{Y}_q\right]_+, \qquad \mathbf{X}_q \leftarrow [(\mathbf{X}_q\mathbf{X}_q^T)^{-1/2}\mathbf{X}_q]_+, \tag{7.15}$$

$$\mathbf{A}^{(q)} \leftarrow \left[\mathbf{Y}_q(\mathbf{D}_q\mathbf{X}_q)^{\dagger}\right]_+, \qquad \mathbf{A} \leftarrow \frac{1}{Q}\sum_q \mathbf{A}^{(q)}, \qquad \mathbf{A} \leftarrow [\mathbf{A}(\mathbf{A}^T\mathbf{A})^{-1/2}]_+, \tag{7.16}$$

$$\mathbf{D}_q \leftarrow \left[\mathbf{A}^{\dagger}\mathbf{Y}_q\mathbf{X}_q^{\dagger}\right]_+, \qquad (q = 1, 2, \ldots, Q). \tag{7.17}$$

In practice, it is better to impose the optional orthogonality constraints using a multi-layered model in the second layer of NTF1 rather than directly on the first or single NTF1 layer (see Chapter 1). The MATLAB implementations of the ALS algorithm is given in Appendix 7.G, and a demonstration example is given in Section 7.7.

7.1.3 Multiplicative Alpha and Beta Algorithms for the NTF1 Model

The Alpha- and Beta-divergences are two complementary cost functions [18,26] which are members of a wide class of generalized cost functions suitable for the NTF1 model [18,23]. Let us first consider the Alpha-divergence given by

$$D^{(\alpha)}(\overline{\mathbf{Y}}||\mathbf{A}\overline{\mathbf{X}}) = \frac{1}{\alpha(\alpha-1)} \sum_{i\bar{t}} \left(\bar{y}_{i\bar{t}}^{\alpha} [\mathbf{A}\overline{\mathbf{X}}]_{i\bar{t}}^{1-\alpha} - \alpha \bar{y}_{i\bar{t}} + (\alpha - 1)[\mathbf{A}\overline{\mathbf{X}}]_{i\bar{t}} \right), \tag{7.18}$$

$$D_q^{(\alpha)}(\mathbf{Y}_q||\mathbf{A}\mathbf{X}_q) = \frac{1}{\alpha(\alpha-1)} \sum_{itq} \left(y_{itq}^{\alpha} [\mathbf{A}\mathbf{X}_q]_{it}^{1-\alpha} - \alpha y_{itq} + (\alpha - 1)[\mathbf{A}\mathbf{X}_q]_{it} \right), \tag{7.19}$$

where $\bar{y}_{i\bar{t}} = [\overline{\mathbf{Y}}]_{i\bar{t}}$, $\bar{x}_{j\bar{t}} = [\overline{\mathbf{X}}]_{j\bar{t}}$ and $\bar{t} = 1, 2, \ldots, \overline{T}$.

Apply nonlinearly transformed gradient descent to yield

$$\Phi(x_{jtq}) \leftarrow \Phi(x_{jtq}) - \eta_{jtq} \frac{\partial D_A(\mathbf{Y}_q||\mathbf{A}\mathbf{X}_q)}{\partial \Phi(x_{jtq})}, \qquad \Phi(a_{ij}) \leftarrow \Phi(a_{ij}) - \eta_{ij} \frac{\partial D_A(\overline{\mathbf{Y}}||\mathbf{A}\overline{\mathbf{X}})}{\partial \Phi(a_{ij})}, \tag{7.20}$$

where $\Phi(x) = x^{\alpha}$ is a suitably chosen function.[3] The following learning rates may be used

$$\eta_{jtq} = \alpha^2 \Phi(x_{jtq}) / (x_{jtq}^{1-\alpha} \sum_{i \in S_I} a_{ij}), \qquad \eta_{ij} = \alpha^2 \Phi(a_{ij}) / (a_{ij}^{1-\alpha} \sum_{\bar{t} \in S_{\overline{T}}} \bar{x}_{j\bar{t}}), \tag{7.21}$$

[2]Usually, a sparsity constraint is naturally and intrinsically provided due to the nonlinear projection approach (e.g., half-wave rectifier or adaptive nonnegative shrinkage with gradually decreasing threshold [21]).

[3]For $\alpha = 0$ instead of $\Phi(x) = x^{\alpha}$, we have used $\Phi(x) = \ln(x)$, which leads to the generalized SMART algorithm: $x_{jtq} \leftarrow x_{jtq} \prod_{i \in S_I} (y_{itq}/[\mathbf{A}\mathbf{X}_q]_{it})^{\eta_j a_{ij}}$ and $a_{ij} \leftarrow a_{ij} \prod_{\bar{t} \in S_{\overline{T}}} \left(\bar{y}_{i\bar{t}}/[\mathbf{A}\overline{\mathbf{X}}]_{i\bar{t}}\right)^{\tilde{\eta}_j \bar{x}_{j\bar{t}}}$ [18].

to obtain the following iterative update rules referred to as the multiplicative Alpha NTF1 algorithm

$$x_{jtq} \leftarrow x_{jtq} \left(\frac{\sum_{i \in \mathcal{S}_I} a_{ij} \left(y_{itq}/[\mathbf{A}\mathbf{X}_q]_{it} \right)^{\alpha}}{\sum_{i \in \mathcal{S}_I} a_{ij}} \right)^{1/\alpha}, \qquad x_{jtq} \leftarrow P_{\Omega}^{(X)}(x_{jtq}), \quad \forall j, t, q \tag{7.22}$$

$$a_{ij} \leftarrow a_{ij} \left(\frac{\sum_{\bar{t} \in \mathcal{S}_{\bar{T}}} \left(\bar{y}_{i\bar{t}}/[\mathbf{A}\bar{\mathbf{X}}]_{i\bar{t}} \right)^{\alpha} \bar{x}_{j\bar{t}}}{\sum_{\bar{t} \in \mathcal{S}_{\bar{T}}} \bar{x}_{j\bar{t}}} \right)^{1/\alpha}, \qquad a_{ij} \leftarrow P_{\Omega}^{(A)}(a_{ij}), \quad \forall i, j, \ \alpha \neq 0. \tag{7.23}$$

where $\mathcal{S}_I$ is subset of all indices $1, 2, \ldots, I$, and $\mathcal{S}_{\bar{T}}$ is subset of all indices $1, 2, \ldots, T$. Note that the sparsity constraints can be achieved and controlled via a suitable nonlinear transformation in the form $x_{jtq} \leftarrow P_{\Omega}^{(X)}(x_{jtq})$ and/or $a_{ij} \leftarrow P_{\Omega}^{(A)}(a_{ij})$. For example, $\mathcal{P}_{\Omega}(x)$ can be a suitably chosen shrinkage function with decreasing threshold [19] (see also Chapters 3 and 4 for more details).

Regularized Beta-divergences for the sparse NTF1 problem can be defined as follows [18,26]:

$$D^{(\beta)}(\bar{\mathbf{Y}}||\mathbf{A}\bar{\mathbf{X}}) = \sum_{i\bar{t}} \left(\bar{y}_{i\bar{t}} \frac{\bar{y}_{i\bar{t}}^{\beta} - [\mathbf{A}\bar{\mathbf{X}}]_{i\bar{t}}^{\beta}}{\beta(\beta+1)} + [\mathbf{A}\bar{\mathbf{X}}]_{i\bar{t}}^{\beta} \frac{[\mathbf{A}\bar{\mathbf{X}}]_{i\bar{t}} - \bar{y}_{i\bar{t}}}{\beta+1} \right) + \alpha_{\mathbf{A}} \ \| \mathbf{A} \|_1, \tag{7.24}$$

$$D_q^{(\beta)}(\mathbf{Y}_q||\mathbf{A}\mathbf{X}_q) = \sum_{it} \left(y_{itq} \frac{y_{itq}^{\beta} - [\mathbf{A}\mathbf{X}_q]_{it}^{\beta}}{\beta(\beta+1)} + [\mathbf{A}\mathbf{X}_q]_{it}^{\beta} \frac{[\mathbf{A}\mathbf{X}_q]_{it} - y_{itq}}{\beta+1} \right) + \alpha_{\mathbf{X}_q} \left\| \mathbf{X}_q \right\|_1, \tag{7.25}$$

where the regularization terms $\alpha_{\mathbf{A}} \ \|\mathbf{A}\|_1$ and $\alpha_{\mathbf{X}_q} \ \|\mathbf{X}_q\|_1$ are introduced to impose sparsity on the solution (see Chapter 3). By minimizing the above formulated Beta-divergences sequentially, we can derive various kinds of NTF algorithms: Multiplicative form based on the standard gradient descent algorithm, Exponentiated Gradient (EG), Projected Gradient (PG), Alternating Interior-Point Gradient (AIPG), or ALS algorithms (see Chapters 3 and 4).

For example, based on the standard gradient descent followed by a shrinkage transformation, we obtain the following multiplicative update rules

$$x_{jtq} \leftarrow x_{jtq} \frac{\left[\sum_{i \in \mathcal{S}_I} a_{ij} \, (y_{itq}/[\mathbf{A}\mathbf{X}_q]_{it}^{1-\beta}) - \alpha_{\mathbf{X}_q} \right]_+}{\sum_{i \in \mathcal{S}_I} a_{ij} \, [\mathbf{A}\mathbf{X}_q]_{it}^{\beta}}, \qquad x_{jtq} \leftarrow P_{\Omega}^{(X)}(x_{jtq}), \ \forall j, t, q, \tag{7.26}$$

$$a_{ij} \leftarrow a_{ij} \frac{\left[\sum_{\bar{t} \in \mathcal{S}_{\bar{T}}} (\bar{y}_{i\bar{t}}/[\mathbf{A}\bar{\mathbf{X}}]_{i\bar{t}}^{1-\beta}) \, \bar{x}_{j\bar{t}} - \alpha_{\mathbf{A}} \right]_+}{\sum_{\bar{t} \in \mathcal{S}_{\bar{T}}} [\mathbf{A}\bar{\mathbf{X}}]_{i\bar{t}}^{\beta} \, \bar{x}_{j\bar{t}}}, \qquad a_{ij} \leftarrow P_{\Omega}^{(A)}(a_{ij}), \ \forall i, j. \tag{7.27}$$

7.1.4 Multi-layer NTF1 Strategy

In order to improve the performance of all the developed NTF1 algorithms, especially for ill-conditioned and badly scaled data and also to reduce the risk of converging to local minima in nonconvex alternating minimization computations, we next present a simple multi-stage procedure combined with multi-start initializations, in which we perform sequential decomposition of nonnegative matrices. In the first step, we perform the basic decomposition (factorization) $\mathbf{Y}_q = \mathbf{A}^{(1)}\mathbf{X}_q^{(1)}$ using any available NTF1 algorithm. In the second stage, the results obtained from the first stage are used to perform a similar decomposition as: $\mathbf{X}_q^{(1)} = \mathbf{A}^{(2)}\mathbf{X}_q^{(2)}$ using the same or different update rules, and so on. The process can be repeated based only on the last components until some predefined stopping criteria are satisfied. Thus, the NTF1 model has the form $\mathbf{Y}_q = \mathbf{A}^{(1)}\mathbf{A}^{(2)}\cdots\mathbf{A}^{(L)}\mathbf{X}_q^{(L)}$, with a nonnegative basis matrix defined as $\mathbf{A} = \mathbf{A}^{(1)}\mathbf{A}^{(2)}\cdots\mathbf{A}^{(L)}$. In each step, we usually obtain gradual improvements of the performance. Physically, this means that we build up a system that has many layers or cascaded connections of L mixing subsystems. The most important key issue for the success of this technique is that the estimation in each layer must achieve at least 98% (or maximum) explained variation. This means that the explained variation should be used as a stopping criterion for each processing layer. In statistics, explained variation or explained randomness measures the proportion to which a mathematical model accounts for the variation of a given data set.[4] The complementary part of the total variation is called unexplained or residual. The general Multilayer NTF1 algorithm is described in Algorithm 7.1 [18,22,23].

Algorithm 7.1: Multilayer NTF1

Input: set of Q matrices $\mathbf{Y}_q \in \mathbb{R}_+^{I\times T_q}$: input data, J: number of basis components
Output: factor $\mathbf{A} \in \mathbb{R}_+^{I\times J}$ and set of factors $\mathbf{X}_q \in \mathbb{R}_+^{J\times T_q}$, $q = 1, 2, \ldots, Q$
such that the suitable cost functions are minimized.

begin
  $l = 0$ /* number of layers */
  $\mathbf{X}_q^{(l)} = \mathbf{Y}_q, \forall q, \mathbf{A} = \mathbf{I}$
  repeat
    $l \leftarrow l + 1$
    Initialize randomly $\mathbf{X}_q^{(l)}, \forall q$ and/or $\mathbf{A}^{(l)}$
    /* Minimize alternatively with suitable cost functions */

$$\mathbf{A}^{(l)} \leftarrow \arg\min_{\mathbf{A}^{(l)}\geq 0} \{D^{(A)}(\overline{\mathbf{X}}^{(l-1)} \,||\, \mathbf{A}^{(l)}\overline{\mathbf{X}}^{(l)})\}, \quad a_{ij}^{(l)} \leftarrow a_{ij}^{(l)} / \sum_{i\in S_I} a_{ij}^{(l)} \tag{7.28}$$

$$\mathbf{X}_q^{(l)} \leftarrow \arg\min_{\mathbf{X}_q^{(l)}\geq 0} \{D_q^{(X)}(\mathbf{X}_q^{(l-1)} \,||\, \mathbf{A}^{(l)}\mathbf{X}_q^{(l)})\}, \quad (q = 1, \ldots, Q) \tag{7.29}$$

    $\mathbf{A} \leftarrow \mathbf{A}\,\mathbf{A}^{(l)}$
  until *a stopping criterion is met* /* convergence condition */
  $\mathbf{X}_q = \mathbf{X}_q^{(l)}, \forall q$
end

$\overline{\mathbf{X}}^{(l)} = [\mathbf{X}_1^{(l)}, \mathbf{X}_2^{(l)}, \ldots, \mathbf{X}_Q^{(l)}]$

Similar (dual) algorithms can be derived for the NTF2 model described in Chapter 1. Note that the NTF2 model can be obtained from the NTF1 model by performing matrix transposes.

[4] Often, variation is quantified as variance; then, the more specific term explained variance can be used.

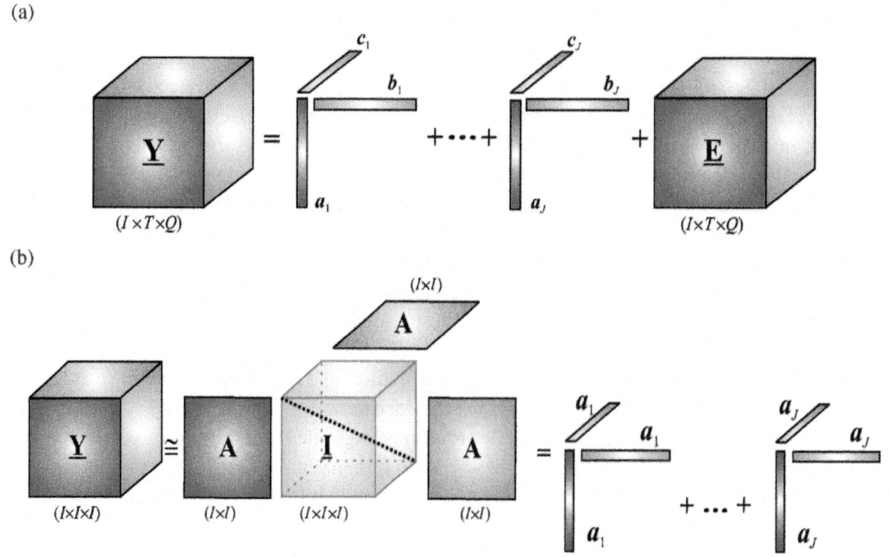

Figure 7.2 (a) Illustration of the standard nonnegative tensor factorization (NTF) and (b) Super-Symmetric Tensor Nonnegative Factorization (SSNTF) for a third-order tensor by sum of rank-one tensors. The SSNTF is a special case of the NTF for $I = T = Q$ and $\mathbf{A} = \mathbf{B} = \mathbf{C}$ (or equivalently $\boldsymbol{a}_j = \boldsymbol{b}_j = \boldsymbol{c}_j \in \mathbb{R}^I, \ \forall j$).

7.2 Algorithms for Three-way Standard and Super Symmetric Nonnegative Tensor Factorization

In this section we derive and discuss a wide class of algorithms for third-order nonnegative tensor factorization (also called nonnegative PARAFAC) and, as a special case, we consider Super-Symmetric Nonnegative Tensor Factorization (SSNTF). NTF and SSNTF have many potential applications including multi-way clustering, feature extraction, multi-sensory or multi-dimensional data analysis, and nonnegative neural sparse coding [73,86]. The main advantage of the algorithms presented in this section is their relatively low complexity for large-scale problems and, in the case of multiplicative algorithms, the possibility for their straightforward extension to higher-order tensor factorizations. We also discuss a wide class of cost functions such as Squared Euclidean, Kullback Leibler I-divergence, Alpha-divergence, and Beta-divergence.

In general, the NTF model for a third-order tensor $\underline{\mathbf{Y}} \in \mathbb{R}^{I \times T \times Q}$ can be factorized into three nonnegative component (factor) matrices: $\mathbf{A} = [\boldsymbol{a}_1, \boldsymbol{a}_2, \ldots, \boldsymbol{a}_J] \in \mathbb{R}^{I \times J}$, $\mathbf{B} = [\boldsymbol{b}_1, \boldsymbol{b}_2, \ldots, \boldsymbol{b}_J] \in \mathbb{R}^{T \times J}$ and $\mathbf{C} = [\boldsymbol{c}_1, \boldsymbol{c}_2, \ldots, \boldsymbol{c}_J] \in \mathbb{R}^{Q \times J}$ (see Figure 7.2 and also Chapter 1)

$$\underline{\mathbf{Y}} = \sum_{j=1}^{J} \left(\boldsymbol{a}_j \circ \boldsymbol{b}_j \circ \boldsymbol{c}_j\right) + \underline{\mathbf{E}} = \underline{\mathbf{I}} \times_1 \mathbf{A} \times_2 \mathbf{B} \times_3 \mathbf{C} + \underline{\mathbf{E}}, \tag{7.30}$$

where $\underline{\mathbf{E}} = \underline{\mathbf{Y}} - \hat{\underline{\mathbf{Y}}} \in \mathbb{R}^{I \times T \times Q}$ is a tensor representing the error.

The goal is to estimate matrices $\mathbf{A}$, $\mathbf{B}$, $\mathbf{C}$ subject to constraints. These include scaling to unit length vectors, nonnegativity, orthogonality, sparseness and/or smoothness of all or some of the columns $\boldsymbol{a}_j$.

A super-symmetric tensor has entries which are invariant under any permutation of the indices. For example, for a third-order super-symmetric tensor $\underline{\mathbf{Y}} \in \mathbb{R}^{I \times T \times Q}$ (with $I = T = Q$) we have $y_{itq} = y_{iqt} = y_{tiq} = y_{tqi} = y_{qit} = y_{qti}$, and its nonnegative factorization (referred to as the SSNTF) simplifies into the form

$$\underline{\mathbf{Y}} = \sum_{j=1}^{J} \left(\boldsymbol{a}_j \circ \boldsymbol{a}_j \circ \boldsymbol{a}_j \right) + \underline{\mathbf{E}} = \underline{\mathbf{I}} \times_1 \mathbf{A} \times_2 \mathbf{A} \times_3 \mathbf{A} + \underline{\mathbf{E}}, \tag{7.31}$$

shown in Figure 7.2(b). The scalar form of (7.31) is given by

$$y_{itq} = \hat{y}_{itq} + e_{itq} = \sum_{j=1}^{J} a_{ij}\, a_{tj}\, a_{qj} + e_{itq}, \tag{7.32}$$

where $\boldsymbol{a}_j \in \mathbb{R}^I$ is the j-th column vector of matrix $\mathbf{A}$.

Comon [30] has shown an intimate relationship between super-symmetric tensors and polynomials, Shashua and Zass addressed application to classification and clustering problems [73], whereas Hazan *et al.* [42] developed multiplicative algorithms. Super-symmetric tensors arise naturally in multi-way clustering where they represent generalized affinity tensors in higher order statistics.

7.2.1 Multiplicative NTF Algorithms Based on Alpha- and Beta-Divergences

The main objective of this section is to derive simple multiplicative algorithms which are suited for both very sparse representations and highly over-determined cases. The main advantage of multiplicative algorithms is their simplicity and relatively straightforward generalization to N-th order tensors ($N > 3$), however, they can be relatively slow (e.g., in comparison to enhanced ALS and HALS algorithms).

7.2.1.1 Multiplicative Alpha NTF Algorithm

Most widely known and often used adaptive algorithms for NTF/NMF and SSNTF are based on alternating minimization of the squared Euclidean distance and the generalized Kullback-Leibler divergence [11,37,54]. In this section, we discuss Alpha- and Beta-divergences (see Chapter 2) as more flexible and general cost functions.

The Alpha-divergence for the NTF model (7.30) can be defined as

$$D_A^{(\alpha)}(\underline{\mathbf{Y}}||\underline{\mathbf{Q}}) = \begin{cases} \displaystyle\sum_{itq} \left(\frac{y_{itq}}{\alpha(\alpha-1)} \left[\left(\frac{y_{itq}}{q_{itq}} \right)^{\alpha-1} - 1 \right] - \frac{y_{itq} - q_{itq}}{\alpha} \right), & \alpha \neq 0{,}1, \quad (7.33a) \\ \displaystyle\sum_{itq} \left(y_{itq} \ln \left(\frac{y_{itq}}{q_{itq}} \right) - y_{itq} + q_{itq} \right), & \alpha = 1, \quad (7.33b) \\ \displaystyle\sum_{itq} \left(q_{itq} \ln \left(\frac{q_{itq}}{y_{itq}} \right) + y_{itq} - q_{itq} \right), & \alpha = 0, \quad (7.33c) \end{cases}$$

where $y_{itq} = [\underline{\mathbf{Y}}]_{itq}$ and $q_{itq} = \hat{y}_{itq} = \sum_j a_{ij}\, b_{tj}\, c_{qj} = [\mathbf{A}\, \mathbf{D}_q(\boldsymbol{c}_q)\, \mathbf{B}^T]_{it}$ for $i = 1, 2, \ldots, I,\ t = 1, 2, \ldots, T,\ q = 1, 2, \ldots, Q$. The gradient of the Alpha-divergence (7.33), for $\alpha \neq 0$, can be expressed in a compact form as

$$\frac{\partial D_A^{(\alpha)}}{\partial a_{ij}} = \frac{1}{\alpha} \sum_{tq} b_{tj}\, c_{qj} \left[1 - \left(\frac{y_{itq}}{q_{itq}} \right)^{\alpha} \right], \qquad \alpha \neq 0. \tag{7.34}$$

Again, instead of applying standard gradient descent, we use the projected (nonlinearly transformed) gradient approach[5] given by

$$\Phi(a_{ij}) \leftarrow \Phi(a_{ij}) - \eta_{ij} \frac{\partial D_A^{(\alpha)}}{\partial a_{ij}}, \tag{7.35}$$

where $\Phi(x)$ is a suitably chosen function. For $\Phi(x) = x^{\alpha}$, we have

$$a_{ij} \leftarrow \Phi^{-1}\left(\Phi(a_{ij}) - \eta_{ij} \frac{\partial D_A^{(\alpha)}}{\partial a_{ij}}\right). \tag{7.36}$$

By choosing the learning rates, $\eta_{ij} = \alpha \Phi(a_{ij}) / \sum_{tq} c_{qj}\, b_{jt}$, we obtain a simple multiplicative update rule (referred to as the Alpha algorithm)

$$a_{ij} \leftarrow a_{ij} \left(\frac{\sum_{tq} c_{qj}\, b_{tj} \left(y_{itq}/q_{itq}\right)^{\alpha}}{\sum_{tq} c_{qj}\, b_{tj}} \right)^{1/\alpha}, \quad a_{ij} \leftarrow P_{\Omega}^{(A)}(a_{ij}), \quad \forall i\, j, \quad \alpha \neq 0. \tag{7.37}$$

and similarly

$$b_{tj} \leftarrow b_{tj} \left(\frac{\sum_{iq} a_{ij}\, c_{qj} \left(y_{itq}/q_{itq}\right)^{\alpha}}{\sum_{iq} a_{ij}\, c_{qj}} \right)^{1/\alpha}, \quad b_{tj} \leftarrow P_{\Omega}^{(B)}(b_{tj}), \quad \forall t\, j, \tag{7.38}$$

$$c_{qj} \leftarrow c_{qj} \left(\frac{\sum_{it} a_{ij}\, b_{tj} \left(y_{itq}/q_{itq}\right)^{\alpha}}{\sum_{it} a_{ij}\, b_{tj}} \right)^{1/\alpha}, \quad c_{qj} \leftarrow P_{\Omega}^{(C)}(c_{qj}), \quad \forall q\, j. \tag{7.39}$$

As desired, the updates preserve the positivity of the estimated components for positive values of the initial estimate of the vectors $\boldsymbol{a}_j$, $\boldsymbol{b}_j$, and $\boldsymbol{c}_j$. For every iteration, the elements of $\boldsymbol{a}_j$ are updated using a Jacobi-type algorithm with respect to the entries $\boldsymbol{a}_j$, and in a Gauss-Seidel manner with respect to the entries of other vectors. These update rules can be considered as a generalization of the EMML algorithm discussed in Chapter 3 and a generalization of algorithms proposed in [42,73].

Since several of the elements of the super-symmetric tensors are highly redundant, it is convenient to perform the summations for only some preselected indices $t, q \in \mathcal{S}_I$, where $\mathcal{S}_I$ is subset of all indices $1, 2, \ldots, I$, and to apply nonlinear projections or filtering using monotonic functions which increase or decrease the sparseness. In the simplest case, it is sufficient to apply the nonlinear transformation $a_{ij} \leftarrow \left(a_{ij}\right)^{1+\alpha_{sp}}$, where α_{sp} is a coefficient whose value ranges typically between 0.001 and 0.005. This parameter is positive if we wish to increase sparseness of an estimated component, and conversely is negative if we wish to decrease the sparseness. The multiplicative Alpha NMF algorithm for the SSNTF model with sparsity

[5] Which can be considered as a generalization of the exponentiated gradient.

control now takes the following form

$$a_{ij} \leftarrow \left[a_{ij} \left(\frac{\sum\limits_{t,\, q \in \mathcal{S}_I} a_{qj}\, a_{tj} \left(y_{itq}/q_{itq} \right)^{\alpha}}{\sum\limits_{t,\, q \in \mathcal{S}_I} a_{qj}\, a_{tj}} \right)^{\omega/\alpha} \right]^{1+\alpha_{sp}}, \tag{7.40}$$

where ω is an over-relaxation parameter (typically in the range (0, 2)) which controls the convergence speed.

7.2.1.2 Alternative Multiplicative Alpha NTF Algorithm

The multiplicative Alpha NTF algorithm can be derived indirectly from the multiplicative Alpha NMF algorithm based on the unfolded matrix representation of the NTF

$$\mathbf{Y}_{(1)} = \mathbf{A}\,(\mathbf{C} \odot \mathbf{B})^T = \mathbf{A}\,\mathbf{Z}_{\mathbf{A}}^T, \tag{7.41}$$

where $\mathbf{Z}_{\mathbf{A}} = \mathbf{C} \odot \mathbf{B}$. Based on the multiplicative Alpha NMF algorithm derived in Chapter 3, the learning rule for factor $\mathbf{A}$ has the following form

$$\mathbf{A} \leftarrow \mathbf{A} \circledast \left\{ \left[\left(\mathbf{Y}_{(1)} \oslash \hat{\mathbf{Y}}_{(1)} \right)^{.[\alpha]} \mathbf{Z}_{\mathbf{A}} \right] \oslash \left(\mathbf{1}\,\mathbf{1}^T \mathbf{Z}_{\mathbf{A}} \right) \right\}^{.[1/\alpha]} \tag{7.42}$$

$$= \mathbf{A} \circledast \left\{ \left[\left(\mathbf{Y}_{(1)} \oslash \hat{\mathbf{Y}}_{(1)} \right)^{.[\alpha]} (\mathbf{C} \odot \mathbf{B}) \right] \oslash \left(\mathbf{1}\,\mathbf{1}^T (\mathbf{C} \odot \mathbf{B}) \right) \right\}^{.[1/\alpha]}, \tag{7.43}$$

where $\hat{\mathbf{Y}}$ denotes the estimated matrix of $\mathbf{Y}$. The learning rules for component matrices $\mathbf{B}$ and $\mathbf{C}$ have the corresponding forms

$$\mathbf{B} \leftarrow \mathbf{B} \circledast \left\{ \left[\left(\mathbf{Y}_{(2)} \oslash \hat{\mathbf{Y}}_{(2)} \right)^{.[\alpha]} (\mathbf{C} \odot \mathbf{A}) \right] \oslash \left(\mathbf{1}\,\mathbf{1}^T (\mathbf{C} \odot \mathbf{A}) \right) \right\}^{.[1/\alpha]}, \tag{7.44}$$

$$\mathbf{C} \leftarrow \mathbf{C} \circledast \left\{ \left[\left(\mathbf{Y}_{(3)} \oslash \hat{\mathbf{Y}}_{(3)} \right)^{.[\alpha]} (\mathbf{B} \odot \mathbf{A}) \right] \oslash \left(\mathbf{1}\,\mathbf{1}^T (\mathbf{B} \odot \mathbf{A}) \right) \right\}^{.[1/\alpha]}. \tag{7.45}$$

The update rules are derived rigorously, but are computationally demanding. For their simplification, we will use the following transformations.

Transformation 7.1 *When the column vectors of component matrices* $\mathbf{A}$ *and* $\mathbf{B}$ *are normalized to unit value* ℓ_1*-norm (i.e.,* $\|\boldsymbol{a}_j\|_1 = \|\boldsymbol{b}_j\|_1 = 1,\ \forall j$*), the factor* $\mathbf{C}$ *can be represented as*

$$\mathbf{C} = \mathbf{C}_\ell\, \boldsymbol{\Lambda}, \tag{7.46}$$

where $\boldsymbol{c}_{\ell j} = \boldsymbol{c}_j / \|\boldsymbol{c}_j\|_1$*, and* $\boldsymbol{\Lambda} = \mathrm{diag}(\|\boldsymbol{c}_1\|_1, \|\boldsymbol{c}_2\|_1, \ldots, \|\boldsymbol{c}_J\|_1)$ *is a scaling diagonal matrix.*

Transformation 7.2 *The Khatri-Rao product of* $\mathbf{C}$ *and* $\mathbf{A}$ *can be expressed as*

$$\begin{aligned}
\mathbf{C} \odot \mathbf{A} &= [\boldsymbol{c}_1 \otimes \boldsymbol{a}_1, \boldsymbol{c}_2 \otimes \boldsymbol{a}_2, \ldots, \boldsymbol{c}_J \otimes \boldsymbol{a}_J] \\
&= \left[\|\boldsymbol{c}_1\|_1 (\boldsymbol{c}_{\ell 1} \otimes \boldsymbol{a}_1), \|\boldsymbol{c}_2\|_1 (\boldsymbol{c}_{\ell 2} \otimes \boldsymbol{a}_2), \ldots, \|\boldsymbol{c}_J\|_1 (\boldsymbol{c}_{\ell J} \otimes \boldsymbol{a}_J) \right] \\
&= \left[\boldsymbol{c}_{\ell 1} \otimes \boldsymbol{a}_1, \boldsymbol{c}_{\ell 2} \otimes \boldsymbol{a}_2, \ldots, \boldsymbol{c}_{\ell J} \otimes \boldsymbol{a}_J \right] \boldsymbol{\Lambda} \\
&= (\mathbf{C}_\ell \odot \mathbf{A})\, \boldsymbol{\Lambda},
\end{aligned} \tag{7.47}$$

yielding

$$\mathbf{1}_{IT}{}^T(\mathbf{A} \odot \mathbf{B}) = \mathbf{1}_J{}^T, \tag{7.48}$$

$$\mathbf{1}_{QI}{}^T(\mathbf{C} \odot \mathbf{A}) = \mathbf{1}_{QI}{}^T(\mathbf{C}_\ell \odot \mathbf{A})\,\mathbf{\Lambda} = \mathbf{1}_J{}^T\,\mathbf{\Lambda}, \tag{7.49}$$

$$\mathbf{1}_{QT}{}^T(\mathbf{C} \odot \mathbf{B}) = \mathbf{1}_{QT}{}^T(\mathbf{C}_\ell \odot \mathbf{B})\,\mathbf{\Lambda} = \mathbf{1}_J{}^T\,\mathbf{\Lambda}. \tag{7.50}$$

From (7.43) and (7.50), we obtain a simplified rule for the update of matrix $\mathbf{A}$, given by

$$\begin{aligned}
\mathbf{A} \leftarrow \mathbf{A} \circledast & \left\{ \left[\left(\mathbf{Y}_{(1)} \oslash \hat{\mathbf{Y}}_{(1)}\right)^{.[\alpha]} (\mathbf{C} \odot \mathbf{B}) \right] \oslash \left(\mathbf{1}\,\mathbf{1}^T (\mathbf{C} \odot \mathbf{B}) \right) \right\}^{.[1/\alpha]} \\
= \mathbf{A} \circledast & \left\{ \left[\left(\mathbf{Y}_{(1)} \oslash \hat{\mathbf{Y}}_{(1)}\right)^{.[\alpha]} (\mathbf{C}_\ell \odot \mathbf{B})\,\mathbf{\Lambda} \right] \oslash \left(\mathbf{1}\,\mathbf{1}_J{}^T\,\mathbf{\Lambda} \right) \right\}^{.[1/\alpha]} \\
= \mathbf{A} \circledast & \left\{ \left[\left(\mathbf{Y}_{(1)} \oslash \hat{\mathbf{Y}}_{(1)}\right)^{.[\alpha]} (\mathbf{C}_\ell \odot \mathbf{B}) \right] \oslash \left(\mathbf{1}\,\mathbf{1}_J{}^T \right) \right\}^{.[1/\alpha]} \\
= \mathbf{A} \circledast & \left[\left(\mathbf{Y}_{(1)} \oslash \hat{\mathbf{Y}}_{(1)}\right)^{.[\alpha]} (\mathbf{C}_\ell \odot \mathbf{B}) \right]^{.[1/\alpha]}.
\end{aligned} \tag{7.51}$$

Since $\mathbf{\Lambda}$ is a diagonal matrix, it gets absorbed in both the numerator and denominator of (7.51). Similarly, we can drive the corresponding learning rules for component matrices $\mathbf{B}$ and $\mathbf{C}$, giving the following simplified form of the multiplicative Alpha NTF algorithm:

$$\mathbf{A} \leftarrow \mathbf{A} \circledast \left[\left(\mathbf{Y}_{(1)} \oslash \hat{\mathbf{Y}}_{(1)}\right)^{.[\alpha]} (\mathbf{C}_\ell \odot \mathbf{B}) \right]^{.[1/\alpha]}, \quad \mathbf{A} \leftarrow \mathbf{A}\,\mathrm{diag}\{\mathbf{1}^T\,\mathbf{A}\}^{-1}, \tag{7.52}$$

$$\mathbf{B} \leftarrow \mathbf{B} \circledast \left[\left(\mathbf{Y}_{(2)} \oslash \hat{\mathbf{Y}}_{(2)}\right)^{.[\alpha]} (\mathbf{C}_\ell \odot \mathbf{A}) \right]^{.[1/\alpha]}, \quad \mathbf{B} \leftarrow \mathbf{B}\,\mathrm{diag}\{\mathbf{1}^T\,\mathbf{B}\}^{-1}, \tag{7.53}$$

$$\mathbf{C} \leftarrow \mathbf{C} \circledast \left[\left(\mathbf{Y}_{(3)} \oslash \hat{\mathbf{Y}}_{(3)}\right)^{.[\alpha]} (\mathbf{B} \odot \mathbf{A}) \right]^{.[1/\alpha]}, \quad \mathbf{C}_\ell \leftarrow \mathbf{C}\,\mathrm{diag}\{\mathbf{1}^T\,\mathbf{C}\}^{-1}. \tag{7.54}$$

The learning rule for a super-symmetric tensor ($\mathbf{A} = \mathbf{B} = \mathbf{C}$) simplifies further into

$$\mathbf{A} \leftarrow \mathbf{A} \circledast \left[\left(\mathbf{Y}_{(1)} \oslash \hat{\mathbf{Y}}_{(1)}\right)^{.[\alpha]} (\mathbf{A}_\ell \odot \mathbf{A}_\ell) \right]^{.[1/\alpha]}, \tag{7.55}$$

where $\mathbf{A}_\ell$ denotes an ℓ_1-norm normalized form of $\mathbf{A}$, i.e., $\boldsymbol{a}_{\ell j} = \boldsymbol{a}_j / \|\boldsymbol{a}_j\|_1$, and $\underline{\hat{\mathbf{Y}}} = [\![\mathbf{A}_\ell, \mathbf{A}_\ell, \mathbf{A}]\!]$.

7.2.1.3 Multiplicative Beta NTF Algorithm

The Beta-divergence for the three-way NTF model (7.30) is defined as follows,

$$D_B^{(\beta)}(\underline{\mathbf{Y}}\|\underline{\mathbf{Q}}) = \begin{cases}
\displaystyle\sum_{itq} \left(y_{itq} \frac{y_{itq}^{\beta} - q_{itq}^{\beta}}{\beta} - \frac{y_{itq}^{\beta+1} - q_{itq}^{\beta+1}}{\beta + 1} \right), & \beta > 0, \quad (7.56a) \\[2ex]
\displaystyle\sum_{itq} \left(y_{itq} \ln(\frac{y_{itq}}{q_{itq}}) + y_{itq} - q_{itq} \right), & \beta = 0, \quad (7.56b) \\[2ex]
\displaystyle\sum_{itq} \left(\ln(\frac{q_{itq}}{y_{itq}}) + \frac{y_{itq}}{q_{itq}} - 1 \right), & \beta = -1. \quad (7.56c)
\end{cases}$$

where $y_{itq} = [\underline{\mathbf{Y}}]_{itq}$ and $q_{itq} = \hat{y}_{itq} = \sum_j a_{ij} b_{tj} c_{qj}$.

This cost function allows us to expand the class of NTF algorithms [11,25,27,29,37]. For instance, to derive a multiplicative NTF learning algorithm, the gradient of the regularized Beta-divergence (7.56) with respect to matrix element of matrices a_{ij} is computed as

$$\frac{\partial D_B^{(\beta)}}{\partial a_{ij}} = \sum_{tq} \left(q_{itq}^{\beta} - y_{itq} q_{itq}^{\beta-1} \right) c_{qj}\, b_{tj}. \tag{7.57}$$

In the first order gradient descent setting

$$a_{ij} \leftarrow a_{ij} - \eta_{ij} \frac{\partial D_B^{(\beta)}}{\partial a_{ij}}, \tag{7.58}$$

and by choosing suitable learning rates, $\eta_{ij} = a_{ij} / \sum_{tq} q_{itq}^{\beta}\, c_{qj}\, b_{tj}$, we obtain the multiplicative Beta NTF update rule

$$a_{ij} \leftarrow a_{ij} \frac{\displaystyle\sum_{tq} b_{tj}\, c_{qj}\, (y_{jtq}/q_{itq}^{1-\beta})}{\sum_{tq} q_{itq}^{\beta}\, b_{tj}\, c_{qj}}, \qquad a_{ij} \leftarrow P_{\Omega}^{(A)}(a_{ij}), \qquad \forall i, j, \tag{7.59}$$

together with

$$b_{tj} \leftarrow b_{tj} \frac{\displaystyle\sum_{iq} a_{ij}\, c_{qj}\, (y_{jtq}/q_{itq}^{1-\beta}),}{\displaystyle\sum_{iq} q_{itq}^{\beta}\, a_{ij}\, c_{qj}}, \qquad b_{tj} \leftarrow P_{\Omega}^{(B)}(b_{tj}), \qquad \forall t, j, \tag{7.60}$$

$$c_{qj} \leftarrow c_{qj} \frac{\displaystyle\sum_{tq} a_{ij}\, b_{tj}\, (y_{jtq}/q_{itq}^{1-\beta})}{\displaystyle\sum_{it} q_{itq}^{\beta}\, a_{ij}\, b_{tj}}, \qquad c_{qj} \leftarrow P_{\Omega}^{(C)}(c_{qj}), \qquad \forall q, j. \tag{7.61}$$

Hence, for a super-symmetric tensor the update rule takes the following form[6]

$$a_{ij} \leftarrow a_{ij} \frac{\displaystyle\sum_{t \le q \in S_I} a_{qj}\, a_{tj}\, (y_{jtq}/q_{itq}^{1-\beta})}{\displaystyle\sum_{t \le q \in S_I} q_{itq}^{\beta}\, a_{qj}\, a_{tj}}, \qquad a_{ij} \leftarrow P_{\Omega}^{(A)}(a_{ij}), \qquad \forall i, j, \tag{7.62}$$

As a special case, for $\beta = 0$, the multiplicative NTF algorithm simplifies to the generalized alternating EMML algorithm, similar to that proposed in [42,72]:

$$a_{ij} \leftarrow a_{ij} \frac{\displaystyle\sum_{t \le q \in S_I} a_{qj}\, a_{tj} (y_{jtq}/q_{itq})}{\displaystyle\sum_{t \le q \in S_I} a_{qj}\, a_{tj}}, \qquad a_{ij} \leftarrow P_{\Omega}^{(A)}(a_{ij}), \qquad \forall i, j. \tag{7.63}$$

[6]For example, if the matrix $\mathbf{A}$ is very sparse and additionally satisfies orthogonality constraints, we may impose the following projection $\mathbf{A} \leftarrow \mathbf{A} \left(\mathbf{A}^T \mathbf{A} \right)^{-1/2}$ after each iteration.

For large arrays, sampling all the possible (ordered) tuples out of I^3 points introduces a considerable computational burden. As the dimension of the tensor grows, most NTF algorithms for processing $\underline{\mathbf{Y}}$ may become impractical, however, for the multiplicative update rules it is sufficient to sample only over a relatively small proportion of the total indices making them suitable for large-scale problems. It is an important advantage of multiplicative algorithms, that they support partial sampling of a tensor. This is a very desirable property, as we can record only entries with indices $1 \leq i \leq t \leq q \leq I$ instead of all possible entries.

7.2.2 Simple Alternative Approaches for NTF and SSNTF

For tensors with large dimensions ($Q >> 1$), the above learning algorithms can be computationally very demanding. To this end, we next present alternative approaches which convert the problem to a simple tri-NMF model

$$\boxed{\overline{\mathbf{Y}} = \mathbf{A}\overline{\mathbf{D}}\mathbf{X} + \overline{\mathbf{E}},} \tag{7.64}$$

where $\overline{\mathbf{Y}} = \sum_{q=1}^{Q} \mathbf{Y}_q \in \mathbb{R}^{I \times T}$; $\overline{\mathbf{D}} = \sum_{q=1}^{Q} \mathbf{D}_q = \operatorname{diag}\{\overline{d}_1, \overline{d}_2, \ldots, \overline{d}_J\}$ and $\overline{\mathbf{E}} = \sum_{q=1}^{Q} \mathbf{E}_q \in \mathbb{R}^{I \times T}$.

7.2.2.1 Averaging Approach

The above system of linear algebraic equations can also be represented in a scalar form: $\overline{y}_{it} = \sum_j a_{ij} x_{jt} \overline{d}_j + \overline{e}_{it}$ and in the vector form: $\overline{\mathbf{Y}} = \sum_j \boldsymbol{a}_j \, \overline{d}_j \, \underline{\boldsymbol{x}}_j + \overline{\mathbf{E}}$, where $\boldsymbol{a}_j$ are the columns of $\mathbf{A}$ and $\underline{\boldsymbol{x}}_j$ are the rows of $\mathbf{X} = \mathbf{B}^T$. This is justified if noise in the frontal slices is uncorrelated. The model can be rewritten as

$$\boxed{\overline{\mathbf{Y}} = \widetilde{\mathbf{A}}\widetilde{\mathbf{X}} + \overline{\mathbf{E}},} \tag{7.65}$$

where $\widetilde{\mathbf{A}} = \mathbf{A}\overline{\mathbf{D}}^{1/2}$ and $\widetilde{\mathbf{X}} = \overline{\mathbf{D}}^{1/2}\mathbf{X}$, assuming that $\overline{\mathbf{D}} \in \mathbb{R}^{J \times J}$ is a diagonal nonsingular matrix. In this way, we now revert to NMF for the evaluation of $\widetilde{\mathbf{A}}$ and $\widetilde{\mathbf{X}}$. For example, by minimizing the following regularized cost function:

$$D(\overline{\mathbf{Y}}||\widetilde{\mathbf{A}}\widetilde{\mathbf{X}}) = \frac{1}{2}\left\|\overline{\mathbf{Y}} - \widetilde{\mathbf{A}}\widetilde{\mathbf{X}}^T\right\|_F^2 + \alpha_{\mathbf{A}}\left\|\widetilde{\mathbf{A}}\right\|_1 + \alpha_{\mathbf{X}}\left\|\widetilde{\mathbf{X}}\right\|_1 \tag{7.66}$$

and applying the ALS approach, we obtain the following update rules

$$\widetilde{\mathbf{A}} \leftarrow \left[(\overline{\mathbf{Y}}\widetilde{\mathbf{X}}^T - \alpha_{\mathbf{A}}\, \mathbf{1}_{I \times J})(\widetilde{\mathbf{X}}\widetilde{\mathbf{X}}^T)^{-1}\right]_+, \tag{7.67}$$

$$\widetilde{\mathbf{X}} \leftarrow \left[(\widetilde{\mathbf{A}}^T\widetilde{\mathbf{A}})^{-1}(\widetilde{\mathbf{A}}^T\overline{\mathbf{Y}} - \alpha_{\mathbf{X}}\, \mathbf{1}_{J \times T})\right]_+. \tag{7.68}$$

7.2.2.2 Row-wise and Column-wise Unfolding Approaches

It is worth noting that the diagonal matrices $\mathbf{D}_q$ can often be considered as scaling matrices that can be absorbed into matrix $\mathbf{A}$ or matrix $\mathbf{X}$. By defining the column-normalized matrices $\mathbf{A}_q = \mathbf{A}\mathbf{D}_q$ or $\mathbf{X}_q = \mathbf{D}_q\mathbf{X}$, we can use the following simplified models:

$$\mathbf{Y}_q = \mathbf{A}_q\mathbf{X} + \mathbf{E}_q, \qquad \text{or} \qquad \mathbf{Y}_q = \mathbf{A}\mathbf{X}_q + \mathbf{E}_q, \qquad (q = 1, 2, \ldots, Q), \tag{7.69}$$

Thus leading to the standard NMF problem.

As a special case, for the SSNTF model, we can have simplified models described by a single compact matrix equation using column-wise or row-wise unfolding as follows

$$\mathbf{Y}_c = \mathbf{A}_c\mathbf{A}^T, \qquad \text{or} \qquad \mathbf{Y}_r = \mathbf{A}\mathbf{A}_r^T, \tag{7.70}$$

where $\mathbf{Y}_c = \mathbf{Y}_r^T = [\mathbf{Y}_1; \mathbf{Y}_2; \dots; \mathbf{Y}_Q] \in \mathbb{R}^{I^2 \times I}$ is the column-wise unfolded matrix of the slices $\mathbf{Y}_q$ and $\mathbf{A}_c = \mathbf{A}_r^T = [\mathbf{A}_1; \mathbf{A}_2; \dots; \mathbf{A}_Q] \in \mathbb{R}^{I^2 \times I}$ is the column-wise unfolded matrix of the matrices $\mathbf{A}_q = \mathbf{A}\mathbf{D}_q$ $(q = 1, 2, \dots, I)$. We can calculate matrix $\mathbf{A}$, using any efficient NMF algorithm (multiplicative, IPN, quasi-Newton, or ALS) [11,25,27–29,37]. For example, by minimizing the following cost function:

$$D(\mathbf{Y}_c || \mathbf{A}_c \mathbf{A}^T) = \frac{1}{2} \left\| \mathbf{Y}_c - \mathbf{A}_c \mathbf{A}^T \right\|_F^2 + \alpha_{\mathbf{A}} \, \|\mathbf{A}\|_1 \tag{7.71}$$

and applying the ALS approach, we obtain the following iterative update rule:

$$\mathbf{A} \leftarrow \left[([\mathbf{Y}_c^T \mathbf{A}_c - \alpha_{\mathbf{A}} \mathbf{1}_{I \times J}]_+)(\mathbf{A}_c^T \mathbf{A}_c)^{-1} \right]_+ \tag{7.72}$$

or equivalently

$$\mathbf{A} \leftarrow \left[([\mathbf{Y}_r \mathbf{A}_r - \alpha_{\mathbf{A}} \, \mathbf{1}_{I \times J}]_+)(\mathbf{A}_r^T \mathbf{A}_r)^{-1} \right]_+, \tag{7.73}$$

where $\mathbf{A}_c = \mathbf{A}_r^T = [\mathbf{A}\mathbf{D}_1; \mathbf{A}\mathbf{D}_2; \dots; \mathbf{A}\mathbf{D}_Q]$, $\mathbf{D}_q = \text{diag}\{\underline{\boldsymbol{a}}_q^{[-1]}\}$ and $\underline{\boldsymbol{a}}_q$ denotes the q-th row of $\mathbf{A}$.

7.3 Nonnegative Tensor Factorizations for Higher-Order Arrays

The three-way NTF model can be readily extended to higher-order NTF (or nonnegative PARAFAC),[7] this can be formulated as follows, "Factorize a given N-th order tensor $\underline{\mathbf{Y}} \in \mathbb{R}^{I_1 \times I_2 \cdots \times I_N}$ into a set of N nonnegative component matrices: $\mathbf{A}^{(n)} = [\boldsymbol{a}_1^{(n)}, \boldsymbol{a}_2^{(n)}, \dots, \boldsymbol{a}_J^{(n)}] \in \mathbb{R}_+^{I_n \times J}$, $(n = 1, 2, \dots, N)$ representing the common (loading) factors", that is,

$$\underline{\mathbf{Y}} = \widehat{\underline{\mathbf{Y}}} + \underline{\mathbf{E}} = \sum_{j=1}^{J} \lambda_j \, (\boldsymbol{a}_j^{(1)} \circ \boldsymbol{a}_j^{(2)} \circ \cdots \circ \boldsymbol{a}_j^{(N)}) \, + \, \underline{\mathbf{E}} \tag{7.74}$$

or in an equivalent form using mode-n multiplications

$$\begin{aligned} \underline{\mathbf{Y}} &= \underline{\mathbf{\Lambda}} \times_1 \mathbf{A}^{(1)} \times_2 \mathbf{A}^{(2)} \cdots \times_N \mathbf{A}^{(N)} + \mathbf{E} \\ &= [\![\boldsymbol{\lambda}; \mathbf{A}^{(1)}, \mathbf{A}^{(2)}, \dots, \mathbf{A}^{(N)}]\!] + \underline{\mathbf{E}} = [\![\boldsymbol{\lambda}; \{\mathbf{A}\}]\!] + \underline{\mathbf{E}}, \end{aligned} \tag{7.75}$$

where all the vectors $\boldsymbol{a}_j^{(n)} \in \mathbb{R}_+^{I_n}$ are normalized unit length column vectors (that is $\|\boldsymbol{a}_j^{(n)}\|_2^2 = \boldsymbol{a}_j^{(n)\,T} \boldsymbol{a}_j^{(n)} = 1$), $\forall j, n, \boldsymbol{\lambda} = [\lambda_1, \lambda_2, \dots, \lambda_J]^T \in \mathbb{R}_+^J$ are scaling factors, and $\underline{\mathbf{\Lambda}} \in \mathbb{R}_+^{J \times J \times \dots \times J}$ is a tensor having nonzero elements $\boldsymbol{\lambda}$ only on the superdiagonal.

We make our usual assumption that the scaling vector $\boldsymbol{\lambda}$ can be absorbed into the (nonnormalized) factor matrix $\mathbf{A}^{(N)}$ to give a simplified form for the NTF model:

$$\begin{aligned} \underline{\mathbf{Y}} &\approx \sum_{j=1}^{J} \boldsymbol{a}_j^{(1)} \circ \boldsymbol{a}_j^{(2)} \circ \dots \circ \boldsymbol{a}_j^{(N)} = \underline{\mathbf{I}} \times_1 \mathbf{A}^{(1)} \times_2 \mathbf{A}^{(2)} \cdots \times_N \mathbf{A}^{(N)} \\ &= [\![\mathbf{A}^{(1)}, \mathbf{A}^{(2)}, \dots, \mathbf{A}^{(N)}]\!] = [\![\{\mathbf{A}\}]\!] = \widehat{\underline{\mathbf{Y}}}, \end{aligned} \tag{7.76}$$

with $\|\boldsymbol{a}_j^{(n)}\|_2 = 1$ for $n = 1, 2, \dots, N-1$ and $j = 1, 2, \dots, J$.

This model is often referred to as the Harshman model [42,50], corresponding to the standard PARAFAC or CANDECOMP with nonnegativity constraints [52]. Our objective here is to estimate the nonnegative

[7] For convenience, Table 7.1 replicates some of the most frequently used notations and symbols and higher-order tensor operations.

Table 7.1 Basic tensor operations and notations.

$\circ$	outer product	$\mathbf{A}^{(n)\dagger}$	pseudo inverse of $\mathbf{A}^{(n)}$
$\otimes$	Kronecker product	$\underline{\mathbf{Y}}$	tensor
$\odot$	Khatri-Rao product	$\mathbf{Y}_{(n)}$	mode-n matricized version of $\underline{\mathbf{Y}}$
$\circledast$	Hadamard product	$\mathbf{A}^{\odot}$	$= \mathbf{A}^{(N)} \odot \mathbf{A}^{(N-1)} \odot \cdots \odot \mathbf{A}^{(1)}$
$\oslash$	element-wise division	$\mathbf{A}^{\odot_{-n}}$	$= \mathbf{A}^{(N)} \odot \cdots \odot \mathbf{A}^{(n+1)} \odot \mathbf{A}^{(n-1)} \odot \cdots \odot \mathbf{A}^{(1)}$
$\times_n$	mode-n product of tensor and matrix	$\mathbf{A}^{\otimes}$	$= \mathbf{A}^{(N)} \otimes \mathbf{A}^{(N-1)} \otimes \cdots \otimes \mathbf{A}^{(1)}$
$\bar{\times}_n$	mode-n product of tensor and vector	$\mathbf{A}^{\otimes_{-n}}$	$= \mathbf{A}^{(N)} \otimes \cdots \otimes \mathbf{A}^{(n+1)} \otimes \mathbf{A}^{(n-1)} \otimes \cdots \otimes \mathbf{A}^{(1)}$
$[\mathbf{AX}]_j$	j-th column vector of matrix $\mathbf{AX}$	$\mathbf{A}^{\circledast}$	$= \mathbf{A}^{(N)} \circledast \mathbf{A}^{(N-1)} \circledast \cdots \circledast \mathbf{A}^{(1)}$
$\mathbf{1}$	column vector or a matrix of all ones	$\mathbf{A}^{\circledast_{-n}}$	$= \mathbf{A}^{(N)} \circledast \cdots \circledast \mathbf{A}^{(n+1)} \circledast \mathbf{A}^{(n-1)} \circledast \cdots \circledast \mathbf{A}^{(1)}$
$\underline{\mathbf{1}}$	tensor of all ones	$\underline{\mathbf{G}} \times \{\mathbf{A}\}$	$= \underline{\mathbf{G}} \times_1 \mathbf{A}^{(1)} \times_2 \mathbf{A}^{(2)} \cdots \times_N \mathbf{A}^{(N)}$
$\mathbf{A}^{(n)}$	the $n-th$ factor	$\underline{\mathbf{G}} \times_{-n} \{\mathbf{A}\}$	$= \underline{\mathbf{G}} \times_1 \mathbf{A}^{(1)} \cdots$
$\boldsymbol{a}_j^{(n)}$	j-th column vector of $\mathbf{A}^{(n)}$		$\times_{n-1} \mathbf{A}^{(n-1)} \times_{n+1} \mathbf{A}^{(n+1)} \cdots \times_N \mathbf{A}^{(N)}$

component matrices $\mathbf{A}^{(n)}$ or equivalently the set of vectors $\boldsymbol{a}_j^{(n)}$, $(n = 1, 2, \ldots, N, \quad j = 1, 2, \ldots, J)$, assuming that the number of factors J is known or can be estimated. In other words, the N-way nonnegative tensor $\underline{\mathbf{Y}} \in \mathbb{R}^{I_1 \times I_2 \times \cdots \times I_N}$, is decomposed into a sum of nonnegative rank-one tensors (see Figure 7.3).

As desired, for $N = 2$ and for $\mathbf{A}^{(1)} = \mathbf{A}$ and $\mathbf{A}^{(2)} = \mathbf{B} = \mathbf{X}^T$ the NTF model (7.76) simplifies to the standard NMF.

Moreover, using the unfolding approach described in Chapter 1, the N-th order NTF model (7.76) can be represented in a matrix form

$$\boxed{\mathbf{Y}_{(n)} = \mathbf{A}^{(n)} \mathbf{Z}_{(-n)} + \mathbf{E}_{(n)}, \qquad (n = 1, 2, \ldots, N),} \tag{7.77}$$

where $\mathbf{Y}_{(n)} \in \mathbb{R}^{I_n \times I_1 \cdots I_{n-1} I_{n+1} \cdots I_N}$ is the mode-n unfolded matrix of the tensor $\underline{\mathbf{Y}} \in \mathbb{R}^{I_1 \times I_2 \times \cdots \times I_N}$ and

$$\mathbf{Z}_{(-n)} = [\mathbf{A}^{(N)} \odot \cdots \odot \mathbf{A}^{(n+1)} \odot \mathbf{A}^{(n-1)} \odot \cdots \odot \mathbf{A}^{(1)}]^T = [\mathbf{A}^{\odot_{-n}}]^T \in \mathbb{R}_+^{J \times I_1 \cdots I_{n-1} I_{n+1} \cdots I_N}. \tag{7.78}$$

In this way, the problem can be converted into a set of standard NMF models with matrices $\mathbf{Y} = \mathbf{Y}_{(n)}$, $\mathbf{A} = \mathbf{A}^{(n)}$ and $\mathbf{X} = \mathbf{B}^T = \mathbf{Z}_{(-n)}$ for $n = 1, 2, \ldots, N$, and basic NMF algorithms can be almost directly adopted as presented in the previous chapters to higher-order NTF problems.

Most working algorithms for the NTF model are based on the ALS minimization of the squared Euclidean distance (Frobenius norm) [11,18,50,51,74]. For example, the ALS algorithm for the unfolded NTF model

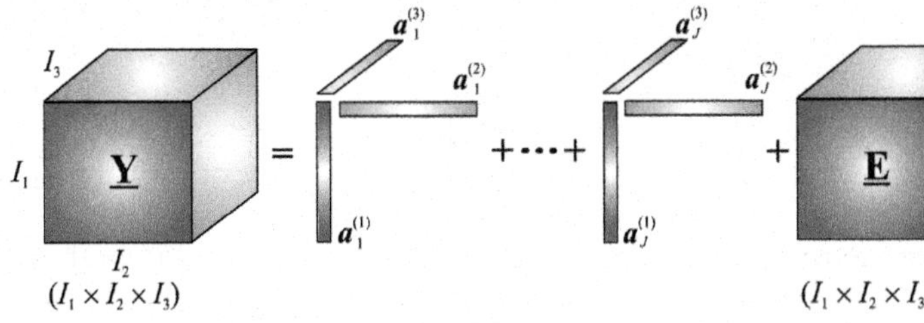

Figure 7.3 Illustration and notations used for a higher-order tensor factorization; the objective is to estimate the nonnegative vectors $\boldsymbol{a}_j^{(n)}$ for $j = 1, 2, \ldots, J$ $(n = 1, 2, 3)$.

takes the following form

$$\mathbf{A}^{(n)} \leftarrow \left[\mathbf{Y}_{(n)}\, \mathbf{Z}_{(-n)}^T \left(\mathbf{Z}_{(-n)} \mathbf{Z}_{(-n)}^T\right)^{-1}\right]_+ = \left[\mathbf{Y}_{(n)}\, \mathbf{A}^{\odot_{-n}} \left(\mathbf{A}^{\odot_{-n}\,T} \mathbf{A}^{\odot_{-n}}\right)^{-1}\right]_+$$

$$= \boxed{\left[\mathbf{Y}_{(n)}\, \mathbf{A}^{\odot_{-n}} \left(\left\{\mathbf{A}^T \mathbf{A}\right\}^{\circledast_{-n}}\right)^{-1}\right]_+, \quad (n = 1, \ldots, N).} \tag{7.79}$$

At present, ALS algorithms for NMF and NTF are considered "workhorse" approaches, however they are not guaranteed to converge to a global minimum or even a stationary point, but only to a solution where the cost functions cease to decrease [11,50,51,74]. However, the ALS method can be considerably improved and the computational complexity reduced as will be shown in the next section [20,66]. For large-scale problems the unfolded matrices $\mathbf{Y}_{(n)}$ can be extremely large, so it would be a challenge to solve such a problem or even to store such matrices. One way to tackle this problem is to use multiplicative algorithms discussed in the previous section, and perform sampling (summations) very sparsely, i.e., only for some preselected entries or tubes of tensors. Alternatively, we may apply the HALS techniques described in Chapter 4. A family of HALS algorithms for higher-order NTF, together with improved ALS methods, is discussed later in this chapter.

7.3.1 Alpha NTF Algorithm

The algorithm for three-way Nonnegative Tensor Factorization based on the Alpha-divergence described by equations (7.43), (7.44), and (7.45) can be naturally generalized to a higher order tensor as follows

$$\mathbf{A}^{(n)} \leftarrow \mathbf{A}^{(n)} \circledast \left\{\left[\left(\mathbf{Y}_{(n)} \oslash \hat{\mathbf{Y}}_{(n)}\right)^{.[\alpha]} \mathbf{A}^{\odot_{-n}}\right] \oslash \left(\mathbf{1}\,\mathbf{1}^T (\mathbf{A}^{\odot_{-n}})\right)\right\}^{.[1/\alpha]}, \tag{7.80}$$

where $\hat{\underline{\mathbf{Y}}} = [\![\mathbf{A}^{(1)}, \mathbf{A}^{(2)}, \ldots, \mathbf{A}^{(N)}]\!]$ is an approximation of data tensor $\underline{\mathbf{Y}}$. Assume that factors $\mathbf{A}_\ell^{(n)}$ denote the ℓ_1-norm normalized version of factors $\mathbf{A}^{(n)}$, that is

$$\mathbf{A}_\ell^{(n)} = \mathbf{A}^{(n)}\ \mathrm{diag}\{\mathbf{1}^T \mathbf{A}^{(n)}\}^{-1}, \quad \forall n. \tag{7.81}$$

Note that for any $n \neq N$, $\mathbf{A}^{(n)}$ is exactly equal to $\mathbf{A}_\ell^{(n)}$ (after being normalized to unit length column vectors in the sense of the ℓ_1-norm) since

$$\mathbf{1}^T \mathbf{A}^{(n)} = \mathbf{1}^T \mathbf{A}_\ell^{(n)} = \mathbf{1}^T, \quad n \neq N. \tag{7.82}$$

The ℓ_1-norms for all column vectors in the product $\mathbf{A}_\ell{}^{\odot_{-n}}$ are calculated as follows

$$\mathbf{1}^T \mathbf{A}_\ell{}^{\odot_{-n}} = \left(\mathbf{1}^{\odot_{-n}}\right)^T \mathbf{A}_\ell{}^{\odot_{-n}} = \left(\mathbf{1}^T \mathbf{A}_\ell^{(n)}\right)^{\circledast_{-n}} = \mathbf{1}^T, \quad \forall n. \tag{7.83}$$

For any $n \neq N$ the Khatri-Rao product $\mathbf{A}^{\odot_{-n}}$ can be expressed as

$$\mathbf{A}^{\odot_{-n}} = \mathbf{A}^{(N)} \odot \mathbf{A}^{\odot_{-\{n,N\}}} = \left(\mathbf{A}_\ell^{(N)}\ \mathrm{diag}\{\mathbf{1}^T \mathbf{A}^{(N)}\}\right) \odot \mathbf{A}_\ell{}^{\odot_{-\{n,N\}}} \tag{7.84}$$

$$= \left(\mathbf{A}_\ell^{(N)} \odot \mathbf{A}_\ell{}^{\odot_{-\{n,N\}}}\right) \mathrm{diag}\{\mathbf{1}^T \mathbf{A}^{(N)}\} = \mathbf{A}_\ell{}^{\odot_{-n}}\ \mathrm{diag}\{\mathbf{1}^T \mathbf{A}^{(N)}\}, \quad n \neq N. \tag{7.85}$$

Furthermore, we have the following relation

$$\mathbf{A}^{\odot_{-n}} = \mathbf{A}_\ell^{\odot_{-n}}\ \mathbf{\Lambda}_n, \quad \forall n, \tag{7.86}$$

where $\mathbf{\Lambda}_n$ is scaling diagonal matrix:

$$\mathbf{\Lambda}_n = \begin{cases} \mathrm{diag}\{\mathbf{1}^T \mathbf{A}^{(N)}\}, & n \neq N \\ \mathbf{I}, & n = N \end{cases}. \tag{7.87}$$

Hence, from equations (7.83), (7.85) and (7.86), the ℓ_1-norms for all column vectors in the product $\mathbf{A}^{\odot_{-n}}$ are given by

$$\mathbf{1}^T \mathbf{A}^{\odot_{-n}} = \mathbf{1}^T \mathbf{A}_\ell^{\odot_{-n}} \boldsymbol{\Lambda}_n = \mathbf{1}^T \boldsymbol{\Lambda}_n, \qquad \forall n. \tag{7.88}$$

By replacing $\mathbf{A}^{\odot_{-n}}$ in (7.80) by $\mathbf{A}_\ell^{\odot_{-n}} \boldsymbol{\Lambda}_n$, we obtain generalized NTF update rule

$$\mathbf{A}^{(n)} \leftarrow \mathbf{A}^{(n)} \circledast \left\{ \left[\left(\mathbf{Y}_{(n)} \oslash \hat{\mathbf{Y}}_{(n)} \right)^{.[\alpha]} (\mathbf{A}_\ell^{\odot_{-n}} \boldsymbol{\Lambda}_n) \right] \oslash \left(\mathbf{1}\, \mathbf{1}^T \boldsymbol{\Lambda}_n \right) \right\}^{.[1/\alpha]}, \tag{7.89}$$

which can be further simplified after neglecting the scaling diagonal matrix $\boldsymbol{\Lambda}_n$ as follows:

$$\boxed{\mathbf{A}^{(n)} \leftarrow \mathbf{A}^{(n)} \circledast \left[\left(\mathbf{Y}_{(n)} \oslash \hat{\mathbf{Y}}_{(n)} \right)^{.[\alpha]} \mathbf{A}_\ell^{\odot_{-n}} \right]^{.[1/\alpha]}.} \tag{7.90}$$

The above update rule is referred to as the Alpha NTF algorithm and its pseudo-code is given in Algorithm 7.2.

For a super-symmetric tensor, factors are identical and the learning rule for the SSNTF model can be simplified as

$$\boxed{\mathbf{A} \leftarrow \mathbf{A} \circledast \left[\left(\mathbf{Y}_{(1)} \oslash \hat{\mathbf{Y}}_{(1)} \right)^{.[\alpha]} \mathbf{A}_\ell^{\odot N-1} \right]^{.[1/\alpha]},} \tag{7.91}$$

where $\mathbf{A}_\ell = \mathbf{A}\, \text{diag}\{\mathbf{1}^T \mathbf{A}\}^{-1}$, $\hat{\underline{\mathbf{Y}}} = [\![\mathbf{A}_\ell, \mathbf{A}_\ell, \ldots, \mathbf{A}_\ell, \mathbf{A}]\!]$ is the SSNTF model with $N-1$ normalized factors $\mathbf{A}_\ell$, and one non-normalized factor $\mathbf{A}$. $\mathbf{A}_\ell^{\odot N-1}$ denotes the Khatri-Rao product of $N-1$ identity factors $\mathbf{A}_\ell$.

Algorithm 7.2: Alpha NTF

Input: $\underline{\mathbf{Y}}$: input data of size $I_1 \times I_2 \times \cdots \times I_N$, J: number of basis components
Output: N component matrices $\mathbf{A}^{(n)} \in \mathbb{R}_+^{I_n \times J}$.

begin
  ALS or random initialization for all factors $\mathbf{A}^{(n)}$
  $\mathbf{A}_\ell^{(n)} = \mathbf{A}^{(n)}\, \text{diag}\{\mathbf{1}^T \mathbf{A}^{(n)}\}^{-1}$ for $\forall n$ `/* normalize to unit length`[a] `*/`
  $\mathbf{A}^{(n)} = \mathbf{A}_\ell^{(n)}$ for $\forall n \neq N$
  repeat
    $\hat{\underline{\mathbf{Y}}} = [\![\mathbf{A}^{(1)}, \mathbf{A}^{(2)}, \ldots, \mathbf{A}^{(N)}]\!]$
    for $n = 1$ *to* N **do**
      $\mathbf{A}^{(n)} \leftarrow \mathbf{A}^{(n)} \circledast \left[\left(\mathbf{Y}_{(n)} \oslash \hat{\mathbf{Y}}_{(n)} \right)^{.[\alpha]} \mathbf{A}_\ell^{\odot_{-n}} \right]^{.[1/\alpha]}$
      $\mathbf{A}_\ell^{(n)} = \mathbf{A}^{(n)}\, \text{diag}\{\mathbf{1}^T \mathbf{A}^{(n)}\}^{-1}$ `/* normalize to unit length */`
      if $n \neq N$ **then** $\mathbf{A}^{(n)} = \mathbf{A}_\ell^{(n)}$
    end
  until *a stopping criterion is met* `/* convergence condition */`
end

[a] In MATLAB, the ℓ_1 normalization of matrix $\mathbf{A}$ can be performed using `Aell = bsxfun(@rdivide, A,sum(A))`.

7.3.2 Beta NTF Algorithm

The following algorithm is an extension of the multiplicative Beta NTF algorithm for three-way factorization given in (7.59), (7.60), (7.61)

$$\mathbf{A}^{(n)} \leftarrow \mathbf{A}^{(n)} \circledast \left[\left(\mathbf{Y}_{(n)} \circledast \hat{\mathbf{Y}}_{(n)}^{.[\beta-1]} \right) \mathbf{A}^{\odot_{-n}} \right] \oslash \left(\hat{\mathbf{Y}}_{(n)}^{.[\beta]} \mathbf{A}^{\odot_{-n}} \right), \tag{7.92}$$

where $\hat{\underline{\mathbf{Y}}} = [\![\mathbf{A}^{(1)}, \mathbf{A}^{(2)}, \ldots, \mathbf{A}^{(N)}]\!]$ is an approximation of data tensor $\underline{\mathbf{Y}}$, and β is the parameter in the Beta-divergence (7.56). The pseudo-code of this algorithm is given in Algorithm 7.3.

Algorithm 7.3: Beta NTF

Input: $\underline{\mathbf{Y}}$: input data of size $I_1 \times I_2 \times \cdots \times I_N$, J: number of basis components
Output: N component matrices $\mathbf{A}^{(n)} \in \mathbb{R}_+^{I_n \times J}$.

begin
  ALS or random initialization for all factors $\mathbf{A}^{(n)}$
  repeat
    $\hat{\underline{\mathbf{Y}}} = [\![\mathbf{A}^{(1)}, \mathbf{A}^{(2)}, \ldots, \mathbf{A}^{(N)}]\!]$
    for $n = 1$ *to* N **do**
      $\mathbf{A}^{(n)} \leftarrow \mathbf{A}^{(n)} \circledast \left[\left(\mathbf{Y}_{(n)} \circledast \hat{\mathbf{Y}}_{(n)}^{.[\beta-1]} \right) \mathbf{A}^{\odot_{-n}} \right] \oslash \left(\hat{\mathbf{Y}}_{(n)}^{.[\beta]} \mathbf{A}^{\odot_{-n}} \right)$
      if $n \neq N$ **then** $\mathbf{A}^{(n)} \leftarrow \mathbf{A}^{(n)} \operatorname{diag}\{\mathbf{1}^T \mathbf{A}^{(n)}\}^{-1}$ `/* normalize */`
    end
  until *a stopping criterion is met* `/* convergence condition */`
end

7.3.3 Fast HALS NTF Algorithm Using Squared Euclidean Distance

Instead of minimizing one or two cost functions we shall now introduce an approach based on the minimization of a set of local cost functions with the same global minima (e.g., squared Euclidean distances and Alpha- or Beta-divergences with a single parameter alpha or beta) which try to approximate rank-one tensors [20,66].

Consider sequential minimization of the following set of local cost functions

$$D_F^{(j)}(\boldsymbol{a}_j^{(1)}, \boldsymbol{a}_j^{(2)}, \ldots, \boldsymbol{a}_j^{(N)}) = \frac{1}{2} \left\| \underline{\mathbf{Y}}^{(j)} - \boldsymbol{a}_j^{(1)} \circ \boldsymbol{a}_j^{(2)} \circ \cdots \circ \boldsymbol{a}_j^{(N)} \right\|_F^2 \tag{7.93}$$

$$= \frac{1}{2} \left\| \mathbf{Y}_{(n)}^{(j)} - \boldsymbol{a}_j^{(n)} \left\{ \boldsymbol{a}_j \right\}^{\odot_{-n} T} \right\|_F^2, \tag{7.94}$$

for $j = 1, 2, \ldots, J$ and $n = 1, 2, \ldots, N$, subject to the nonnegativity constraints, where

$$\left\{ \boldsymbol{a}_j \right\}^{\odot_{-n} T} = [\boldsymbol{a}_j^{(N)}]^T \odot \cdots \odot [\boldsymbol{a}_j^{(n+1)}]^T \odot [\boldsymbol{a}_j^{(n-1)}]^T \odot \cdots \odot [\boldsymbol{a}_j^{(1)}]^T \tag{7.95}$$

and

$$\begin{aligned}\underline{\mathbf{Y}}^{(j)} &= \underline{\mathbf{Y}} - \sum_{p \neq j} \boldsymbol{a}_p^{(1)} \circ \boldsymbol{a}_p^{(2)} \circ \cdots \circ \boldsymbol{a}_p^{(N)} \\ &= \underline{\mathbf{Y}} - \sum_{p=1}^{J} \left(\boldsymbol{a}_p^{(1)} \circ \cdots \circ \boldsymbol{a}_p^{(N)}\right) + \left(\boldsymbol{a}_j^{(1)} \circ \cdots \circ \boldsymbol{a}_j^{(N)}\right) \\ &= \underline{\mathbf{Y}} - \widehat{\underline{\mathbf{Y}}} + \left(\boldsymbol{a}_j^{(1)} \circ \cdots \circ \boldsymbol{a}_j^{(N)}\right) = \underline{\mathbf{Y}} - \widehat{\underline{\mathbf{Y}}} + [\![\{\boldsymbol{a}_j\}]\!]. \end{aligned} \tag{7.96}$$

Note that (7.94) is the mode-n unfolding (matricized) version of (7.93) and that the desired residual tensors $\underline{\mathbf{Y}}^{(j)} \in \mathbb{R}_+^{I_1 \times I_2 \times \cdots \times I_N}$ converge to rank-one tensors during the iteration process. In other words, we update both terms of the cost function (7.93). The gradient of (7.94) with respect to element $\boldsymbol{a}_j^{(n)}$ is given by

$$\begin{aligned}\frac{\partial D_F^{(j)}}{\partial \boldsymbol{a}_j^{(n)}} &= -\mathbf{Y}_{(n)}^{(j)} \left\{\boldsymbol{a}_j\right\}^{\odot_{-n}} + \boldsymbol{a}_j^{(n)} \left\{\boldsymbol{a}_j\right\}^{\odot_{-n} T} \left\{\boldsymbol{a}_j\right\}^{\odot_{-n}} \\ &= -\mathbf{Y}_{(n)}^{(j)} \left\{\boldsymbol{a}_j\right\}^{\odot_{-n}} + \boldsymbol{a}_j^{(n)} \gamma_j^{(n)}, \end{aligned} \tag{7.97}$$

where the scaling coefficients $\gamma_j^{(n)}$ can be computed as follows[8]

$$\begin{aligned}\gamma_j^{(n)} &= \left\{\boldsymbol{a}_j\right\}^{\odot_{-n} T} \left\{\boldsymbol{a}_j\right\}^{\odot_{-n}} = \left\{\boldsymbol{a}_j^T \boldsymbol{a}_j\right\}^{\circledast_{-n}} \\ &= \left\{\boldsymbol{a}_j^T \boldsymbol{a}_j\right\}^{\circledast} / \left(\boldsymbol{a}_j^{(n)\,T} \boldsymbol{a}_j^{(n)}\right) = \left(\boldsymbol{a}_j^{(N)\,T} \boldsymbol{a}_j^{(N)}\right) / \left(\boldsymbol{a}_j^{(n)\,T} \boldsymbol{a}_j^{(n)}\right) \\ &= \begin{cases} \boldsymbol{a}_j^{(N)\,T} \boldsymbol{a}_j^{(N)}, & n \neq N \\ 1, & n = N. \end{cases} \end{aligned} \tag{7.98}$$

Hence, a HALS NTF learning rule for $\boldsymbol{a}_j^{(n)}, \forall j, \forall n$ can be obtained by setting the gradient (7.97) to zero, and applying a nonlinear projection to keep the components nonnegative

$$\boldsymbol{a}_j^{(n)} \leftarrow \left[\mathbf{Y}_{(n)}^{(j)} \left\{\boldsymbol{a}_j\right\}^{\odot_{-n}}\right]_+, \qquad (j = 1, 2, \ldots, J, \quad n = 1, 2, \ldots, N). \tag{7.99}$$

The scaling factors $\gamma_j^{(n)}$ are omitted due to the normalization to unit length of the vectors $\boldsymbol{a}_j^{(n)}$, i.e., $\boldsymbol{a}_j^{(n)} = \boldsymbol{a}_j^{(n)} / \|\boldsymbol{a}_j^{(n)}\|_2$ for $n = 1, 2, \ldots N-1$. The learning rule (7.99) is referred to as the HALS NTF algorithm [66] and can be written in an equivalent form expressed by mode-n multiplication of a tensor by vectors

$$\begin{aligned}\boldsymbol{a}_j^{(n)} &\leftarrow \left[\underline{\mathbf{Y}}^{(j)} \,\bar{\times}_1\, \boldsymbol{a}_j^{(1)} \cdots \bar{\times}_{n-1}\, \boldsymbol{a}_j^{(n-1)} \,\bar{\times}_{n+1}\, \boldsymbol{a}_j^{(n+1)} \cdots \bar{\times}_N\, \boldsymbol{a}_j^{(N)}\right]_+ \\ &= \left[\underline{\mathbf{Y}}^{(j)} \,\bar{\times}_{-n}\, \{\boldsymbol{a}_j\}\right]_+, \qquad (j = 1, 2, \ldots, J, \quad n = 1, 2, \ldots, N). \end{aligned} \tag{7.100}$$

The above updating formula is elegant and relatively simple but involves rather high computational cost for large-scale problems. In order to derive a more efficient (faster) algorithm we exploit the properties of the Khatri-Rao and Kronecker products of two vectors [66]:

$$\left[\mathbf{A}^{(1)} \odot \mathbf{A}^{(2)}\right]_j = \left[\boldsymbol{a}_1^{(1)} \otimes \boldsymbol{a}_1^{(2)} \;\cdots\; \boldsymbol{a}_J^{(1)} \otimes \boldsymbol{a}_J^{(2)}\right]_j = \boldsymbol{a}_j^{(1)} \odot \boldsymbol{a}_j^{(2)}$$

[8] By taking into account $\|\boldsymbol{a}_j^{(n)}\|_2 = 1, n = 1, \ldots, N-1, \forall j$.

or in more general form

$$\left\{\boldsymbol{a}_j\right\}^{\odot_{-n}} = \left[\mathbf{A}^{\odot_{-n}}\right]_j . \tag{7.101}$$

Hence, by replacing $\mathbf{Y}_{(n)}^{(j)}$ terms in (7.99) by those in (7.96), and taking into account (7.101), the update learning rule (7.99) can be expressed as

$$\begin{aligned}
\boldsymbol{a}_j^{(n)} &\leftarrow \mathbf{Y}_{(n)} \left[\mathbf{A}^{\odot_{-n}}\right]_j - \widehat{\mathbf{Y}}_{(n)} \left[\mathbf{A}^{\odot_{-n}}\right]_j + [\![\{\boldsymbol{a}_j\}]\!]_{(n)} \left\{\boldsymbol{a}_j\right\}^{\odot_{-n}} \\
&= \left[\mathbf{Y}_{(n)}\, \mathbf{A}^{\odot_{-n}}\right]_j - \mathbf{A}^{(n)} \mathbf{A}^{\odot_{-n}T} \left[\mathbf{A}^{\odot_{-n}}\right]_j + \boldsymbol{a}_j^{(n)} \left\{\boldsymbol{a}_j\right\}^{\odot_{-n}T} \left\{\boldsymbol{a}_j\right\}^{\odot_{-n}} \\
&= \left[\mathbf{Y}_{(n)}\, \mathbf{A}^{\odot_{-n}}\right]_j - \mathbf{A}^{(n)} \left[\mathbf{A}^{\odot_{-n}T} \mathbf{A}^{\odot_{-n}}\right]_j + \gamma_j^{(n)}\, \boldsymbol{a}_j^{(n)} \\
&= \left[\mathbf{Y}_{(n)}\, \mathbf{A}^{\odot_{-n}}\right]_j - \mathbf{A}^{(n)} \left[\left\{\mathbf{A}^T \mathbf{A}\right\}^{\circledast_{-n}}\right]_j + \gamma_j^{(n)}\, \boldsymbol{a}_j^{(n)} \\
&= \left[\mathbf{Y}_{(n)}\, \mathbf{A}^{\odot_{-n}}\right]_j - \mathbf{A}^{(n)} \left[\left\{\mathbf{A}^T \mathbf{A}\right\}^{\circledast} \oslash \left(\mathbf{A}^{(n)\,T} \mathbf{A}^{(n)}\right)\right]_j + \gamma_j^{(n)}\, \boldsymbol{a}_j^{(n)},
\end{aligned} \tag{7.102}$$

subject to the unit length normalization of vectors $\boldsymbol{a}_j^{(n)}$ for $n = 1, 2, \ldots, N-1$. Algorithm 7.4 gives the pseudo-code for the HALS NTF algorithm.

Bearing in mind that the vectors $\boldsymbol{a}_j^{(n)}$ must be nonnegative we arrive at the update rule referred to as the Fast HALS NTF algorithm [66]

$$\boxed{\boldsymbol{a}_j^{(n)} \leftarrow \left[\gamma_j^{(n)}\, \boldsymbol{a}_j^{(n)} + \left[\mathbf{Y}_{(n)} \mathbf{A}^{\odot_{-n}}\right]_j - \mathbf{A}^{(n)} \left[\left\{\mathbf{A}^T \mathbf{A}\right\}^{\circledast} \oslash \left(\mathbf{A}^{(n)\,T} \mathbf{A}^{(n)}\right)\right]_j\right]_+ ,} \tag{7.103}$$

for $j = 1, 2, \ldots, J$ and $n = 1, 2, \ldots, N$. This new learning rule reduces computation of the Khatri-Rao products in each update step for $\boldsymbol{a}_j^{(n)}$ in (7.99) and (7.100). The detailed pseudo-code of this algorithm is given in Algorithm 7.5. As a special case of $N = 2$, FAST HALS NTF reduces into the FAST HALS NMF algorithm derived in Chapter 4.

Algorithm 7.4: Simple HALS NTF

Input: $\underline{\mathbf{Y}}$: input data of size $I_1 \times I_2 \times \cdots \times I_N$, J: number of basis components

Output: N component matrices $\mathbf{A}^{(n)} \in \mathbb{R}_+^{I_n \times J}$ such that the cost functions (7.94) are minimized.

begin
- ALS or random initialization for all factors $\mathbf{A}^{(n)}$
- $\boldsymbol{a}_j^{(n)} \leftarrow \boldsymbol{a}_j^{(n)} / \|\boldsymbol{a}_j^{(n)}\|_2$ for $\forall j, n = 1, 2, ..., N-1$ `/* normalize to unit length */`
- $\underline{\mathbf{E}} = \underline{\mathbf{Y}} - \widehat{\underline{\mathbf{Y}}} = \underline{\mathbf{Y}} - [\![\{\mathbf{A}\}]\!]$ `/* residual tensor */`
- **repeat**
  - **for** $j = 1$ *to* J **do**
    - $\underline{\mathbf{Y}}^{(j)} = \underline{\mathbf{E}} + [\![\boldsymbol{a}_j^{(1)}, \boldsymbol{a}_j^{(2)}, \ldots, \boldsymbol{a}_j^{(N)}]\!]$
    - **for** $n = 1$ *to* N **do**
      - $\boldsymbol{a}_j^{(n)} \leftarrow \left[\mathbf{Y}_{(n)}^{(j)} \left\{\boldsymbol{a}_j\right\}^{\odot_{-n}}\right]_+$ `/* See Eqs. (7.99) and (7.100) */`
      - **if** $n \neq N$ **then** $\boldsymbol{a}_j^{(n)} \leftarrow \boldsymbol{a}_j^{(n)} / \|\boldsymbol{a}_j^{(n)}\|_2$ `/* normalize to unit length */`
    - **end**
    - $\underline{\mathbf{E}} = \underline{\mathbf{Y}}^{(j)} - [\![\boldsymbol{a}_j^{(1)}, \boldsymbol{a}_j^{(2)}, \ldots, \boldsymbol{a}_j^{(N)}]\!]$
  - **end**
- **until** *a stopping criterion is met* `/* convergence condition */`

end

Algorithm 7.5: FAST HALS NTF

Input: $\underline{\mathbf{Y}}$: input data of size $I_1 \times I_2 \times \cdots \times I_N$, J: number of basis components
Output: N factors $\mathbf{A}^{(n)} \in \mathbb{R}_+^{I_n \times J}$ such that the cost functions (7.94) are minimized.

begin
  Nonnegative random or nonnegative ALS initialization for all factors $\mathbf{A}^{(n)}$ /* [a] */
  $\boldsymbol{a}_j^{(n)} \leftarrow \boldsymbol{a}_j^{(n)} / \|\boldsymbol{a}_j^{(n)}\|_2$ for $\forall j, n = 1, 2, ..., N-1$ /* normalize to unit length */
  $\mathbf{T}^{(1)} = (\mathbf{A}^{(1)\,T}\mathbf{A}^{(1)}) \circledast \cdots \circledast (\mathbf{A}^{(N)\,T}\mathbf{A}^{(N)})$
  repeat
    $\gamma = \operatorname{diag}(\mathbf{A}^{(N)\,T}\mathbf{A}^{(N)})$
    for $n = 1$ *to* N **do**
      if $n = N$ **then** $\gamma = 1$
      $\mathbf{T}^{(2)} = \mathbf{Y}_{(n)}\, \{\mathbf{A}^{\odot_{-n}}\}$
      $\mathbf{T}^{(3)} = \mathbf{T}^{(1)} \oslash (\mathbf{A}^{(n)\,T}\mathbf{A}^{(n)})$
      for $j = 1$ *to* J **do**
        $\boldsymbol{a}_j^{(n)} \leftarrow \left[\gamma_j\, \boldsymbol{a}_j^{(n)} + \boldsymbol{t}_j^{(2)} - \mathbf{A}^{(n)}\, \boldsymbol{t}_j^{(3)}\right]_+$
        if $n \neq N$ **then** $\boldsymbol{a}_j^{(n)} = \boldsymbol{a}_j^{(n)} / \left\|\boldsymbol{a}_j^{(n)}\right\|_2$ /* normalize to unit length */
      end
      $\mathbf{T}^{(1)} = \mathbf{T}^{(3)} \circledast (\mathbf{A}^{(n)\,T}\mathbf{A}^{(n)})$
    end
  until *a stopping criterion is met* /* convergence condition */
end

[a] For a three-way tensor, direct trilinear decomposition can be used for initialization.

7.3.4 Generalized HALS NTF Algorithms Using Alpha- and Beta-Divergences

The HALS algorithms derived in the previous section can be extended to more flexible and generalized algorithms by applying the family of Alpha- and Beta-divergences as cost functions. Using the HALS approach discussed in Chapter 4, we can derive a family of local learning rules for the N-th order NTF problem (7.76). Based on the mode-n matricized (unfolding) version of the tensor $\underline{\mathbf{Y}}$, is given by

$$\mathbf{Y}_{(n)} = \mathbf{A}^{(n)}\, \mathbf{Z}_{(-n)} = \mathbf{A}^{(n)}\, (\mathbf{A}^{\odot_{-n}})^T, \qquad (n = 1, 2, \ldots, N),$$

Thus, NTF model can be considered as a set of regular NMF models with $\mathbf{A} \equiv \mathbf{A}^{(n)}$ and $\mathbf{B} = \mathbf{X}^T = \mathbf{A}^{\odot_{-n}}$. Applying directly the Alpha HALS NMF learning rule derived in Chapter 4 gives

$$\boldsymbol{a}_j^{(n)} \leftarrow \Psi^{-1}\left(\frac{\Psi\left(\left[\mathbf{Y}_{(n)}^{(j)}\right]_+\right)\, \boldsymbol{b}_j}{\boldsymbol{b}_j^T \Psi(\boldsymbol{b}_j)}\right), \tag{7.104}$$

where $\boldsymbol{b}_j$ is the j-th column vector of product matrix $\mathbf{A}^{\odot_{-n}}$ (see also (7.101))

$$\boldsymbol{b}_j = \left[\mathbf{A}^{\odot_{-n}}\right]_j = \left\{\boldsymbol{a}_j\right\}^{\odot_{-n}}, \tag{7.105}$$

Algorithm 7.6: **Alpha HALS NTF**

Input: $\underline{\mathbf{Y}}$: input data of size $I_1 \times I_2 \times \cdots \times I_N$, J: number of basis components
Output: N factors $\mathbf{A}^{(n)} \in \mathbb{R}_+^{I_n \times J}$ such that the cost functions (7.94) are minimized.

begin
  Nonnegative random or nonnegative ALS initialization for all factors $\mathbf{A}^{(n)}$
  $\boldsymbol{a}_j^{(n)} \leftarrow \boldsymbol{a}_j^{(n)} / \|\boldsymbol{a}_j^{(n)}\|_2$ for $\forall j, n = 1, 2, ..., N-1$ /* normalize to unit length */
  $\underline{\mathbf{E}} = \underline{\mathbf{Y}} - \widehat{\underline{\mathbf{Y}}} = \underline{\mathbf{Y}} - [\![\{\mathbf{A}\}]\!]$ /* residual tensor */
  repeat
    for $j = 1$ *to* J **do**
      $\underline{\mathbf{Y}}^{(j)} = \underline{\mathbf{E}} + [\![\boldsymbol{a}_j^{(1)}, \boldsymbol{a}_j^{(2)}, \ldots, \boldsymbol{a}_j^{(N)}]\!]$
      for $n = 1$ *to* N **do**
        $\boldsymbol{a}_j^{(n)} \leftarrow \Psi^{-1}\left(\Psi\left([\underline{\mathbf{Y}}^{(j)}]_+\right) \bar{\times}_{-n} \{\boldsymbol{a}_j\}\right)_+$ /* see (7.109) */
        if $n \neq N$ **then**
          $\boldsymbol{a}_j^{(n)} \leftarrow \boldsymbol{a}_j^{(n)} / \|\boldsymbol{a}_j^{(n)}\|_2$ /* normalize */
        else
          $\boldsymbol{a}_j^{(n)} \leftarrow \boldsymbol{a}_j^{(n)} / \Psi^{-1}\left(\left\{\boldsymbol{a}_j^T \Psi(\boldsymbol{a}_j)\right\}^{\circledast_{-n}}\right)$
        end
      end
      $\underline{\mathbf{E}} = \underline{\mathbf{Y}}^{(j)} - [\![\boldsymbol{a}_j^{(1)}, \boldsymbol{a}_j^{(2)}, \ldots, \boldsymbol{a}_j^{(N)}]\!]$
    end
  until *a stopping criterion is met*   /* convergence condition */
end

and $\mathbf{Y}_{(n)}^{(j)}$ is an mode-n matricized version of $\underline{\mathbf{Y}}^{(j)}$ in (7.96)

$$\begin{aligned}\mathbf{Y}_{(n)}^{(j)} &= \mathbf{Y}_{(n)} - \widehat{\mathbf{Y}}_{(n)} + \boldsymbol{a}_j^{(n)} \boldsymbol{b}_j^T \\ &= \mathbf{Y}_{(n)} - \widehat{\mathbf{Y}}_{(n)} + \boldsymbol{a}_j^{(n)} \left\{\boldsymbol{a}_j\right\}^{\odot_{-n}\,T} \\ &= \mathbf{Y}_{(n)} - \widehat{\mathbf{Y}}_{(n)} + [\![\{\boldsymbol{a}_j\}]\!]_{(n)}.\end{aligned} \tag{7.106}$$

For a specific nonlinear function $\Psi(\cdot)$ for instance, $\Psi(x) = x^{\alpha}$, we have

$$\begin{aligned}\Psi(\boldsymbol{b}_j) &= \Psi(\{\boldsymbol{a}_j\}^{\odot_{-n}}) \\ &= \Psi(\boldsymbol{a}_j^{(N)}) \odot \cdots \odot \Psi(\boldsymbol{a}_j^{(n+1)}) \odot \Psi(\boldsymbol{a}_j^{(n-1)}) \odot \cdots \odot \Psi(\boldsymbol{a}_j^{(1)}) \\ &= \{\Psi(\boldsymbol{a}_j)\}^{\odot_{-n}}\end{aligned} \tag{7.107}$$

and the denominator in (7.104) can be simplified to

$$\boldsymbol{b}_j^T \Psi(\boldsymbol{b}_j) = \{\boldsymbol{a}_j\}^{\odot_{-n} T} \{\Psi(\boldsymbol{a}_j)\}^{\odot_{-n}} = \{\boldsymbol{a}_j^T \Psi(\boldsymbol{a}_j)\}^{\circledast_{-n}}. \tag{7.108}$$

This completes the derivation of a flexible Alpha HALS NTF update rule, which in its tensor form is given by

$$\boxed{\boldsymbol{a}_j^{(n)} \leftarrow \Psi^{-1}\left(\frac{\Psi\left([\underline{\mathbf{Y}}^{(j)}]_+\right) \bar{\times}_{-n} \{\boldsymbol{a}_j\}}{\left\{\boldsymbol{a}_j^T \Psi(\boldsymbol{a}_j)\right\}^{\circledast_{-n}}}\right)_+}, \tag{7.109}$$

for $j = 1, 2, \ldots, J$ and $n = 1, 2, \ldots, N$, where all nonlinear operations (typically, $\Psi(\boldsymbol{a}) = \boldsymbol{a}^{\,.[\alpha]}$) are component-wise.[9] The pseudocode for the Alpha HALS NTF algorithm is given in Algorithm 7.6.

In a similar way, we can derive the Beta HALS algorithm based on Beta-divergence which for the N-th order NTF problem (7.76) can be written as

$$\boldsymbol{a}_j^{(n)} \leftarrow \frac{([\mathbf{Y}_{(n)}^{(j)}]_+)\ \Psi(\boldsymbol{b}_j)}{\Psi(\boldsymbol{b}_j^T)\ \boldsymbol{b}_j}, \qquad (j = 1, 2, \ldots, J), \tag{7.110}$$

where $\boldsymbol{b}_j = \{\boldsymbol{a}_j\}^{\odot_{-n}}$ and $\mathbf{Y}_{(n)}^{(j)}$ are defined by (7.106) and (7.96). Using equation (7.107), the learning rule (7.110) finally gives

$$\boldsymbol{a}_j^{(n)} \leftarrow \frac{([\mathbf{Y}_{(n)}^{(j)}]_+)\ \{\Psi(\boldsymbol{a}_j)\}^{\odot_{-n}}}{\{\Psi(\boldsymbol{a}_j)\}^{\odot_{-n}T}\ \{\boldsymbol{a}_j\}^{\odot_{-n}}} = \frac{[\underline{\mathbf{Y}}^{(j)}]_+ \ \bar{\times}_{-n}\ \{\Psi(\boldsymbol{a}_j)\}}{\{\Psi(\boldsymbol{a}_j)^T\ \boldsymbol{a}_j\}^{\circledast_{-n}}}, \quad (\forall n, \forall j). \tag{7.111}$$

The above Beta HALS NTF algorithm can be further simplified by performing normalization of vectors $\boldsymbol{a}_j^{(n)}$ for $n = 1, 2, \ldots, N-1$ to unit length vectors after each iteration step:

$$\boldsymbol{a}_j^{(n)} \leftarrow \left[\underline{\mathbf{Y}}^{(j)}\ \bar{\times}_{-n}\ \{\Psi(\boldsymbol{a}_j)\}\right]_+ \tag{7.112}$$

$$\boldsymbol{a}_j^{(n)} \leftarrow \boldsymbol{a}_j^{(n)} / \left\| \boldsymbol{a}_j^{(n)} \right\|_2, \qquad (n = 1, 2, \ldots, N-1). \tag{7.113}$$

The detailed pseudo-code for the Beta HALS NTF algorithm is given in Algorithm 7.7. Heuristics show that it is often mathematically intractable or computationally impractical to establish whether an algorithm has converged locally or globally. To this end, local Alpha HALS NTF and Beta HALS NTF algorithms can be combined with multi-start initializations using ALS as follows [64–66]:

1. Perform factorization of a tensor for any value of α or β parameters (preferably, set the value of the parameters to unity due to simplicity and high speed of the algorithm for this value).
2. If the algorithm has converged but has not achieved the desirable FIT value,[10] restart the factorization by keeping the previously estimated component matrices as the initial matrices for the ALS initialization.
3. If the algorithm does not converge, alter the values of α or β parameters incrementally; this may help to avoid local minima.
4. Repeat the procedure until a desirable FIT value is reached or there is a negligible or no change in the FIT value or a negligible or no change in the component matrices, or the value of the cost function is negligible or zero.

7.3.5 Tensor Factorization with Additional Constraints

Natural constraints such as sparsity, smoothness, uncorrelatedness, can be imposed on the factors and the core tensor for NTF and NTD in a similar way as we have done for NMF problems (see Chapters 3 and 4).

[9] In practice, instead of half-wave rectifying we often use alternative nonlinear transformations, e.g., adaptive nonnegative shrinkage function with gradually decreasing threshold till it achieves approximately the value of the variance of noise σ^2_{noise}.

[10] The FIT value is defined as FIT $= (1 - \frac{\|\underline{\mathbf{Y}} - \hat{\underline{\mathbf{Y}}}\|_F}{\|\underline{\mathbf{Y}}\|_F})100\%$.

Algorithm 7.7: Beta HALS NTF

Input: $\underline{\mathbf{Y}}$: input data of size $I_1 \times I_2 \times \cdots \times I_N$, J: number of basis components
Output: N factors $\mathbf{A}^{(n)} \in \mathbb{R}_+^{I_n \times J}$ such that the cost functions (7.94) are minimized.

begin
  Nonnegative random or nonnegative ALS initialization for all factors $\mathbf{A}^{(n)}$
  $\boldsymbol{a}_j^{(n)} \leftarrow \boldsymbol{a}_j^{(n)} / \|\boldsymbol{a}_j^{(n)}\|_2$ for $\forall j, n = 1, 2, ..., N-1$ /* normalize to unit length */
  $\underline{\mathbf{E}} = \underline{\mathbf{Y}} - \widehat{\underline{\mathbf{Y}}} = \underline{\mathbf{Y}} - [\![\{\mathbf{A}\}]\!]$ /* residual tensor */
  repeat
    for $j = 1$ *to* J **do**
      $\underline{\mathbf{Y}}^{(j)} = \underline{\mathbf{E}} + [\![\boldsymbol{a}_j^{(1)}, \boldsymbol{a}_j^{(2)}, \ldots, \boldsymbol{a}_j^{(N)}]\!]$
      for $n = 1$ *to* N **do**
        $\boldsymbol{a}_j^{(n)} \leftarrow \left[\underline{\mathbf{Y}}^{(j)} \,\bar{\times}_{-n}\, \{\Psi(\boldsymbol{a}_j)\}\right]_+$ /* see (7.109) */
        if $n \neq N$ **then**
          $\boldsymbol{a}_j^{(n)} \leftarrow \boldsymbol{a}_j^{(n)} / \|\boldsymbol{a}_j^{(n)}\|_2$ /* normalize */
        else
          $\boldsymbol{a}_j^{(N)} \leftarrow \boldsymbol{a}_j^{(N)} / \{\Psi(\boldsymbol{a}_j)^T\, \boldsymbol{a}_j\}^{\circledast -n}$
        end
      end
      $\underline{\mathbf{E}} = \underline{\mathbf{Y}}^{(j)} - [\![\boldsymbol{a}_j^{(1)}, \boldsymbol{a}_j^{(2)}, \ldots, \boldsymbol{a}_j^{(N)}]\!]$
    end
  until *a stopping criterion is met* /* convergence condition */
end

For example, the HALS NTF algorithm (7.100) with additional constraints can be extended as follows

$$\boldsymbol{a}_j^{(n)} \leftarrow \frac{\left[\mathbf{Y}_{(n)}^{(j)} \left\{\boldsymbol{a}_j\right\}^{\odot -n} - \alpha_{sp}\, \mathbf{1} + \alpha_{sm}\, \mathbf{S}\hat{\boldsymbol{a}}_j^{(n)} - \alpha_{cr}\, \mathbf{A}^{(n)}\, \mathbf{1}\right]_+}{\gamma_j^{(n)} + \alpha_{sm} + \alpha_{cr}}, \qquad \forall j, \forall n, \tag{7.114}$$

where the parameters α_{sp}, α_{sm}, and α_{cr} control respectively the degrees of sparsity, smoothness, and uncorrelatedness, $\gamma_j^{(n)}$ are scaling coefficients defined in (7.98), and $\mathbf{S}$ is a smoothing matrix. These parameters can be different for each factor $\mathbf{A}^{(n)}$.

7.4 Algorithms for Nonnegative and Semi-Nonnegative Tucker Decompositions

The higher-order tensor Tucker decomposition (see Figure 7.4) is described as a "decomposition of a given N-th order tensor $\underline{\mathbf{Y}} \in \mathbb{R}^{I_1 \times I_2 \cdots \times I_N}$ into an unknown core tensor $\underline{\mathbf{G}} \in \mathbb{R}^{J_1 \times J_2 \cdots \times J_N}$ multiplied by a set of N unknown component matrices, $\mathbf{A}^{(n)} = [\boldsymbol{a}_1^{(n)}, \boldsymbol{a}_2^{(n)}, \ldots, \boldsymbol{a}_{J_n}^{(n)}] \in \mathbb{R}^{I_n \times J_n}$ $(n = 1, 2, \ldots, N)$, representing common factors or loadings" [32,49–51,61,74,80]

$$\underline{\mathbf{Y}} = \sum_{j_1=1}^{J_1} \sum_{j_2=1}^{J_2} \cdots \sum_{j_N=1}^{J_N} g_{j_1 j_2 \cdots j_N}\; \boldsymbol{a}_{j_1}^{(1)} \circ \boldsymbol{a}_{j_2}^{(2)} \circ \cdots \circ \boldsymbol{a}_{j_N}^{(N)} + \underline{\mathbf{E}} \tag{7.115}$$

$$= \underline{\mathbf{G}} \times_1 \mathbf{A}^{(1)} \times_2 \mathbf{A}^{(2)} \cdots \times_N \mathbf{A}^{(N)} + \underline{\mathbf{E}} = \underline{\mathbf{G}} \times \{\mathbf{A}\} + \underline{\mathbf{E}} \tag{7.116}$$

$$= \hat{\underline{\mathbf{Y}}} + \underline{\mathbf{E}}, \tag{7.117}$$

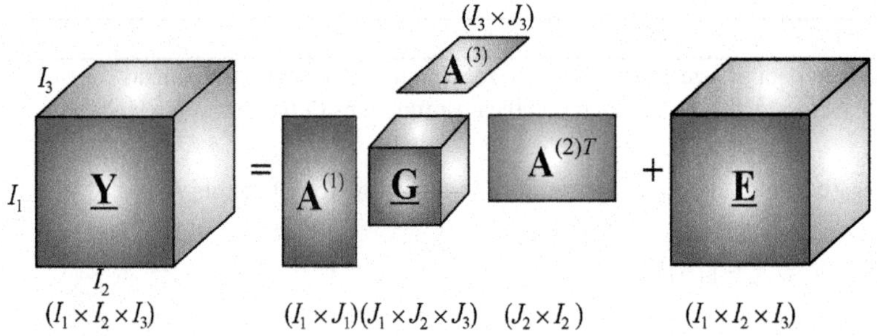

Figure 7.4 Illustration and notations used for a higher-order Tucker decomposition; the objective here is to find optimal component (common factor) matrices $\mathbf{A}^{(n)} \in \mathbb{R}^{I_n \times J_n}$ and a core tensor $\underline{\mathbf{G}} \in \mathbb{R}^{J_1 \times J_2 \times \cdots \times J_n}$. We usually impose additional constraints on the component matrices and/or the core tensor such as nonnegativity and sparsity.

where tensor $\widehat{\underline{\mathbf{Y}}}$ is an approximation of tensor $\underline{\mathbf{Y}}$, and tensor $\underline{\mathbf{E}} = \underline{\mathbf{Y}} - \hat{\underline{\mathbf{Y}}}$ denotes the residual or error tensor. In the next sections, we consider at first a simple Tucker model with orthogonality constraints followed by Tucker models with nonnegativity and sparsity constraints, in which the orthogonality is not necessarily imposed.

7.4.1 Higher Order SVD (HOSVD) and Higher Order Orthogonal Iteration (HOOI) Algorithms

In the general Tucker decomposition, the orthogonality constraints are usually not required. However, if needed, we can use the Higher-Order Singular Value Decomposition (HOSVD) or multi-linear SVD and its low-rank counterpart: Higher Order Orthogonal Iteration (HOOI) [31–34].

The HOSVD is a generalization of SVD to higher-order tensors, and plays an important role in various domains, such as harmonic retrieval, image processing, telecommunications, biomedical applications (magnetic resonance imaging and electrocardiography), world wide web search, face recognition, handwriting analysis and statistical methods involving Independent Component Analysis (ICA). Moreover, it is useful as an initialization tool for Nonnegative Tucker Decomposition discussed later in this chapter.

The HOSVD can be considered as a special form of Tucker decomposition which decomposes an N-th order tensor $\underline{\mathbf{Y}} \in \mathbb{R}^{I_1 \times I_2 \cdots \times I_N}$ as

$$\underline{\mathbf{Y}} = \underline{\mathbf{S}} \times_1 \mathbf{U}^{(1)} \times_2 \mathbf{U}^{(2)} \cdots \times_N \mathbf{U}^{(N)}, \qquad (7.118)$$

where $\mathbf{U}^{(n)} = [\boldsymbol{u}_1^{(n)}, \boldsymbol{u}_2^{(n)}, \ldots, \boldsymbol{u}_{J_N}^{(n)}] \in \mathbb{R}^{I_n \times I_n}$ $(n = 1, 2, \ldots, N)$ are orthogonal matrices and the core tensor $\underline{\mathbf{S}}$ is an all-orthogonal and ordered tensor of the same dimension as the data tensor $\underline{\mathbf{Y}}$. All-orthogonality means that all sub-tensors $\underline{\mathbf{S}}_{i_n=k}$ and $\underline{\mathbf{S}}_{i_n=l}$ obtained by fixing the n-th index to k, l, are mutually orthogonal with respect to inner products for all possible values of n, k, and l, subject to $k \neq l$, whereas ordering means that

$$||\underline{\mathbf{S}}_{i_n=1}||_F \geq ||\underline{\mathbf{S}}_{i_n=2}||_F \geq \cdots \geq ||\underline{\mathbf{S}}_{i_n=I_n}||_F \qquad \forall n. \qquad (7.119)$$

This decomposition is a generalization of the standard SVD for the matrix $\mathbf{Y} \in \mathbb{R}^{I \times T}$

$$\mathbf{Y} = \mathbf{U}\mathbf{S}\mathbf{V}^T = \mathbf{S} \times_1 \mathbf{U} \times_2 \mathbf{V} = \sum_{i=1}^{I} \sigma_i \boldsymbol{u}_i \boldsymbol{v}_i^T. \qquad (7.120)$$

In the matrix form, the HOSVD can be written as [32,33,35]:

$$\mathbf{Y}_{(n)} = \mathbf{U}^{(n)}\mathbf{S}_{(n)} \left(\mathbf{U}^{(N)} \otimes \cdots \otimes \mathbf{U}^{(n+1)} \otimes \mathbf{U}^{(n-1)} \otimes \cdots \otimes \mathbf{U}^{(1)}\right)^T, \tag{7.121}$$

which admits representation in a compact form as

$$\mathbf{Y}_{(n)} = \mathbf{U}^{(n)}\mathbf{S}_{(n)}\mathbf{V}^{(n)\,T}, \tag{7.122}$$

where

$$\mathbf{V}^{(n)} = \left(\mathbf{U}^{(N)} \otimes \cdots \otimes \mathbf{U}^{(n+1)} \otimes \mathbf{U}^{(n-1)} \otimes \cdots \otimes \mathbf{U}^{(1)}\right). \tag{7.123}$$

It then follows that HOSVD can be computed directly in two steps:

1. For $n = 1, 2, \ldots, N$ compute the unfolded matrices $\mathbf{Y}_{(n)}$ from $\underline{\mathbf{Y}}$ and their standard SVD: $\mathbf{Y}_{(n)} = \mathbf{U}^{(n)}\mathbf{S}^{(n)}\mathbf{V}^{(n)\,T}$. The orthogonal matrices $\mathbf{U}^{(n)}$ are leading left singular vectors of $\mathbf{Y}_{(n)}$. Alternatively, compute eigenvalue decomposition (EVD) of the covariance matrices; $\mathbf{Y}_{(n)}^T\mathbf{Y}_{(n)} = \mathbf{U}^{(n)}\mathbf{\Lambda}^{(n)}\mathbf{U}^{(n)\,T}$.
2. Compute the core tensor using the inversion formula.[11]

$$\underline{\mathbf{S}} = \underline{\mathbf{Y}} \times_1 \mathbf{U}^{(1)\,T} \times_2 \mathbf{U}^{(2)\,T} \cdots \times_N \mathbf{U}^{(N)\,T}. \tag{7.124}$$

The HOSVD is computed by means of N standard matrix SVD's, and we can reduce the computational cost of the HOSVD by using fast and efficient SVD algorithms (e.g., without explicitly computing right singular vectors) [17,38,81].

HOSVD results in an ordered orthogonal basis for multi-dimensional representation of input data spanned by each mode of the tensor. In order to achieve dimensionality reduction in each space, we project the data samples onto the principal axis and keep only the components that correspond to the leading (largest) singular values in that subspace. This leads to the concept of the best rank-$R_1, R_2, \ldots, R_N$ approximation [17,32,38,71,81,83], formulated as follows: "Given a real N-th order tensor $\underline{\mathbf{Y}} \in \mathbb{R}^{I_1 \times I_2 \cdots \times I_N}$ find a lower rank tensor $\underline{\hat{\mathbf{Y}}}$ of the same dimension which minimizes the FIT"

$$\underline{\hat{\mathbf{Y}}} = \arg\min_{\underline{\hat{\mathbf{Y}}}} \left\|\underline{\mathbf{Y}} - \underline{\hat{\mathbf{Y}}}\right\|_F^2. \tag{7.125}$$

The approximated (desired) tensor is represented as

$$\underline{\hat{\mathbf{Y}}} = \underline{\tilde{\mathbf{S}}} \times_1 \tilde{\mathbf{U}}^{(1)} \times_2 \tilde{\mathbf{U}}^{(2)} \cdots \times_N \tilde{\mathbf{U}}^{(N)}, \tag{7.126}$$

where $\underline{\tilde{\mathbf{S}}} \in \mathbb{R}^{R_1 \times R_2 \cdots \times R_N}$ is the core tensor and $\tilde{\mathbf{U}}^{(n)} = [\boldsymbol{u}_1^{(n)}, \boldsymbol{u}_2^{(n)}, \ldots, \boldsymbol{u}_{R_n}^{(n)}] \in \mathbb{R}^{I_n \times R_n}$ are reduced orthogonal matrices (with orthonormal columns $\boldsymbol{u}^{(n)}$) with $R_n \le I_n$ (typically, $R_n << I_n$) for $n = 1, 2, \ldots N$.

A simple approach for solving this problem is to apply a truncated HOSVD, whereby the left singular vectors corresponding to the smallest singular values are ignored.[12] In other words, an approximation with prescribed accuracy can be obtained by appropriate truncation of singular vectors of $\mathbf{U}^{(n)}$ component matrices, corresponding to the singular values below a chosen threshold. Unfortunately, the HOSVD does not attempt to minimize the FIT $\|\underline{\mathbf{Y}} - \underline{\hat{\mathbf{Y}}}\|_F$, and does not produce an optimal lower rank $R_1, R_2, \ldots, R_N$ approximation

---

[11] This formula is valid only if matrices $\mathbf{U}^{(n)}$ are orthogonal and consequently the core tensor is also orthogonal in the sense that the slices in any mode are orthogonal. In general, when matrices $\mathbf{U}^{(n)}$ are not orthogonal we can write $\underline{\mathbf{S}} = \underline{\mathbf{Y}} \times_1 \mathbf{U}^{(1)\,\dagger} \times_2 \mathbf{U}^{(2)\,\dagger} \cdots \times_N \mathbf{U}^{(N)\,\dagger}$, where $\mathbf{U}^{(n)\,\dagger}$ is the pseudo inverse of $\mathbf{U}^{(n)\,\dagger}$.

[12] It is natural to ask whether the best rank-$\{R_1, R_2, \ldots, R_N\}$ tensor approximation can be obtained by a truncation of the HOSVD, by analogy to matrix SVD. This is only the case in the absence of noise. For noisy data this is not true, although the largest (leading) mode-n singular values and the corresponding singular vectors yield a tensor $\underline{\hat{\mathbf{Y}}}$ that is a "quite good" but not the best possible approximation under the given rank constraints.

to $\underline{\mathbf{Y}}$, because it optimizes for each mode separately without taking into account interactions among the modes. However, the HOSVD often produces a close to optimal low rank approximation and is relatively fast in comparison with the iterative algorithms discussed below [81]. The computation of the best rank approximation of a tensor requires an iterative ALS algorithm called HOOI [32,33,35].

The HOOI usually uses the HOSVD to initialize the matrices and the whole procedure can be described as follows. Assuming that the basis orthogonal matrices $\tilde{\mathbf{U}}^{(n)}$ are known or estimated, the core tensor can be obtained as [32,33]

$$\underline{\tilde{\mathbf{S}}} = \underline{\hat{\mathbf{Y}}} \times_1 \tilde{\mathbf{U}}^{(1)\,T} \times_2 \tilde{\mathbf{U}}^{(2)\,T} \cdots \times_N \tilde{\mathbf{U}}^{(N)\,T}. \tag{7.127}$$

Therefore, instead of minimizing (7.125) we can equivalently maximize the cost function [32,33,35]

$$J(\tilde{\mathbf{U}}^{(1)}, \tilde{\mathbf{U}}^{(2)}, \ldots \tilde{\mathbf{U}}^{(N)}) = \left\| \underline{\mathbf{Y}} \times_1 \tilde{\mathbf{U}}^{(1)\,T} \times_2 \tilde{\mathbf{U}}^{(2)\,T} \cdots \times_N \tilde{\mathbf{U}}^{(N)\,T} \right\|_F^2, \tag{7.128}$$

where only the basis matrices $\tilde{\mathbf{U}}^{(n)}$ are unknown. For example, with $\tilde{\mathbf{U}}^{(1)}, \ldots, \tilde{\mathbf{U}}^{(n-1)}, \tilde{\mathbf{U}}^{(n+1)}, \ldots, \tilde{\mathbf{U}}^{(N)}$ fixed, we can project tensor $\underline{\mathbf{Y}}$ onto the $\{R_1, \ldots, R_{n-1}, R_{n+1}, \ldots, R_N\}$-dimensional space defined as

$$\underline{\mathbf{W}}^{(-n)} = \underline{\mathbf{Y}} \times_1 \tilde{\mathbf{U}}^{(1)\,T} \cdots \times_{n-1} \tilde{\mathbf{U}}^{(n-1)\,T} \times_{n+1} \tilde{\mathbf{U}}^{(n+1)\,T} \cdots \times_N \tilde{\mathbf{U}}^{(N)\,T} \tag{7.129}$$

and then the orthogonal matrix $\tilde{\mathbf{U}}^{(n)}$ can be estimated as an orthonormal basis for the dominant subspace of the projection by applying the standard matrix SVD for mode-n unfolded matrix $\mathbf{W}_{(n)}^{(-n)}$. The ALS algorithm is used to find the (locally) optimal solutions for (7.125) [32,33,35,49,82]. In each step of the iteration, we optimize only one of the basis matrices, while keeping others fixed. The HOOI algorithm was introduced by De Lathauwer, De Moor and Vandewalle [33] and recently extended and implemented by Kolda and Bader in [50] in their MATLAB Tensor Toolbox [8]. The pseudo-code of the algorithm is described in detail in Algorithm 7.8.

Algorithm 7.8: HOOI Initialization for NTD

Input: $\underline{\mathbf{Y}}$: input data of size $I_1 \times I_2 \times \cdots \times I_N$,
$J_1, J_2, \ldots, J_N$: number of basis components for each factor
Output: N factors $\mathbf{A}^{(n)} \in \mathbb{R}_+^{I_n \times J_n}$ and a core tensor $\underline{\mathbf{G}} \in \mathbb{R}_+^{J_1 \times J_2 \times \cdots \times J_N}$ such that the cost function (7.125) is minimized.

begin
  HOSVD or random initialization for all factors $\mathbf{A}^{(n)}$
  repeat
    for $n = 1$ *to* N **do**
      $\underline{\mathbf{W}}^{(-n)} = \underline{\mathbf{Y}} \times_{-n} \{\mathbf{A}^T\}$
      $[\mathbf{A}^{(n)}, \boldsymbol{\Sigma}^{(n)}, \mathbf{V}^{(n)}] =$ `svds`$(\mathbf{W}_{(n)}^{(-n)}, J_n,$ 'LM') `/*` $\mathbf{W}_{(n)}^{(-n)} \approx \mathbf{A}^{(n)} \boldsymbol{\Sigma}^{(n)} \mathbf{V}^{(n)\,T}$ [a] `*/`
      $\mathbf{A}^{(n)} = [\mathbf{A}^{(n)}]_+$ `/* need to fix signs in advance */`
    end
  until *a stopping criterion is met* `/* convergence condition */`
  $\underline{\mathbf{G}} = \underline{\mathbf{W}}^{(-N)} \times_N \mathbf{A}^{(N)T}$
end

[a] In practice, we use the MATLAB function *eigs*, i.e., $[\mathbf{A}^{(n)}, D_n] = \text{eigs}\,(\mathbf{W}_{(n)}^{(-n)}\, \mathbf{W}_{(n)}^{(-n)\,T}, J_n, \text{'LM'})$, to find the largest magnitude eigenvalues and eigenvectors of a sparse matrix.

To summarize, the idea behind this algorithm is to apply SVD and to find R_n leading left singular vectors of the mode-n matricized version of the product tensor $\underline{\mathbf{W}}^{(-n)} = \underline{\mathbf{Y}} \times_{-n} \{\mathbf{A}^T\}$. By imposing the nonnegative constraints on all factors $\mathbf{A}^{(n)}$ and with only one or two iterations, the ALS procedure becomes a useful initialization tool for NTD.

The HOOI algorithm generally improves the performance of the best rank approximation as compared to the HOSVD, although it does not always guarantee a globally optimal result. Moreover, it is more computationally demanding than HOSVD since HOOI is an iterative algorithm while the HOSVD is not.

7.4.2 ALS Algorithm for Nonnegative Tucker Decomposition

The nonnegative Tucker decomposition (NTD) is another special kind of the Tucker decomposition [80], with nonnegative (and often also sparsity and/or smoothness) constraints, which has already found some applications in neuroscience, bioinformatics and chemometrics [32,48,61]. There are several existing NTD algorithms [41,47,48,61], which process tensors using global learning rules. For large-scale problems, the raw data tensor and its temporary variables stored in memory are very large-scale, and often cause memory overflow error during the decomposition process. One approach to avoid this problem is to process and update the tensor and its factors following block-wise or vector procedures instead of the operation on whole matrices or tensors [64–66]. This approach is referred to as a local decomposition or local learning rule, and will be discussed in the next sections.

For a global ALS algorithm with nonlinear projection, the Tucker decomposition for an N-th order tensor $\underline{\mathbf{Y}} \in \mathbb{R}^{I_1 \times I_2 \times \cdots \times I_N}$ can be written in a matricized (unfolded) form as

$$\mathbf{Y}_{(n)} = \mathbf{A}^{(n)} \mathbf{G}_{(n)} \mathbf{A}^{\otimes_{-n} T} = \mathbf{A}^{(n)} \mathbf{G}_{(n)} \mathbf{Z}_{(-n)}, \qquad (n = 1, 2, \dots, N) \tag{7.130}$$

or equivalently in the vectorized form as

$$\text{vec}(\mathbf{Y}_{(n)}) = \text{vec}\left(\mathbf{A}^{(n)} \mathbf{G}_{(n)} \mathbf{A}^{\otimes_{-n} T}\right) = \left(\mathbf{A}^{\otimes_{-n}} \otimes \mathbf{A}^{(n)}\right) \text{vec}(\mathbf{G}_{(n)}). \tag{7.131}$$

where

$$\mathbf{Z}_{(-n)} = \mathbf{A}^{\otimes_{-n} T} = \left[\mathbf{A}^{(N)} \otimes \cdots \otimes \mathbf{A}^{(n+1)} \otimes \mathbf{A}^{(n-1)} \otimes \cdots \otimes \mathbf{A}^{(1)}\right]^T. \tag{7.132}$$

The above representation forms a basis for the derivation of various updating algorithms for NTD [48,64]. For example, by minimizing alternatively the cost function $||\mathbf{Y}_{(n)} - \mathbf{A}^{(n)}\mathbf{Z}_{(-n)}||$ for $n = 1, 2, \dots, N$, and finding fixed points (using the approach explained in Chapter 4, we obtain the ALS algorithm

$$\mathbf{A}^{(n)} \leftarrow \left[\mathbf{Y}_{(n)} \left(\mathbf{G}_{(n)} \mathbf{Z}_{(-n)}\right)^{\dagger}\right]_+, \qquad (n = 1, 2, \dots, N), \tag{7.133}$$

$$\mathbf{G}_{(n)} \leftarrow \left[\mathbf{A}^{(n)\dagger} \mathbf{Y}_{(n)} \mathbf{Z}^{\dagger}_{(-n)}\right]_+, \tag{7.134}$$

for which the simplified form is given by

$$\mathbf{A}^{(n)} \leftarrow \left[\mathbf{Y}_{(n)} \{\mathbf{A}^{\dagger}\}^{\otimes_{-n}} \mathbf{G}^{\dagger}_{(n)}\right]_+, \qquad (n = 1, 2, \dots, N), \tag{7.135}$$

$$\underline{\mathbf{G}} \leftarrow \left[\underline{\mathbf{Y}} \times \{\mathbf{A}^{\dagger}\}\right]_+ = \underline{\mathbf{Y}} \times_1 \mathbf{A}^{(1)\dagger} \times_2 \mathbf{A}^{(2)\dagger} \cdots \times_N \mathbf{A}^{(N)\dagger}. \tag{7.136}$$

7.4.3 HOSVD, HOOI and ALS Algorithms as Initialization Tools for Nonnegative Tensor Decomposition

The ALS algorithm for Tucker decomposition is very useful (especially for noiseless data), as it can also be used for the initialization of other algorithms for NTD. The HOSVD and HOOI algorithms can also be used for initializations of NTD, especially when loading matrices $\mathbf{A}^{(n)}$ are sparse and orthogonal or close to orthogonal.

The basic idea of the HOSVD and HOOI initialization of NTD is to apply SVD to unfolded matrices $\mathbf{Y}_{(n)}$ or a mode-n matricized version of the product tensor $\underline{\mathbf{W}}^{(-n)} = \underline{\mathbf{Y}} \times_{-n} \{\mathbf{A}^T\}$, find $R_n = J_n$ leading left singular vectors of the unfolded matrices, and finally impose the nonnegativity constraints on all factors $\mathbf{A}^{(n)}$ (typically only one or two iterations is sufficient). See pseudocode listing in Algorithms 7.9 and 7.10.

The advantage of the HOSVD and HOOI approaches over standard ALS is that they can estimate the dimension of the core tensor by analyzing the distribution of singular values.

7.4.4 Multiplicative Alpha Algorithms for Nonnegative Tucker Decomposition

Recently, several versions of multiplicative NTD algorithms have been proposed [48,61,64]. Instead of directly applying Alpha-divergence, we may apply update formulas derived in Chapter 3, that is the multiplicative Alpha NMF algorithm for the standard NMF.

Algorithm 7.9: HOSVD_initialization

Input: $\underline{\mathbf{Y}}$: input data of size $I_1 \times I_2 \times \cdots \times I_N$,
$J_1, J_2, \ldots, J_N$: number of basis components for each factor
Output: N factors $\mathbf{A}^{(n)} \in \mathbb{R}_+^{I_n \times J_n}$ and a core tensor $\underline{\mathbf{G}} \in \mathbb{R}_+^{J_1 \times J_2 \times \cdots \times J_N}$ such that the cost function (7.125) is minimized.

begin
  for $n = 1$ *to* N **do**
    $[\mathbf{A}^{(n)}, \mathbf{S}^{(n)}, \mathbf{V}^{(n)}] = \texttt{svds}(\mathbf{Y}_{(n)}, J_n, \text{'L'})$ /* $[\mathbf{A}^{(n)}, \mathbf{D}_n] = \texttt{eigs}(\mathbf{Y}_{(n)}\mathbf{Y}_{(n)}^T, J_n, \text{'LM'})$ */
    $\mathbf{A}^{(n)} = [\mathbf{A}^{(n)}]_+$ /* need to fix signs in advance */
  end
  $\underline{\mathbf{G}} = \underline{\mathbf{Y}} \times \{\mathbf{A}^T\}$
end

Algorithm 7.10: HOOI_initialization

Input: $\underline{\mathbf{Y}}$: input data of size $I_1 \times I_2 \times \cdots \times I_N$,
$J_1, J_2, \ldots, J_N$: number of basis components for each factor
Output: N factors $\mathbf{A}^{(n)} \in \mathbb{R}_+^{I_n \times J_n}$ and a core tensor $\underline{\mathbf{G}} \in \mathbb{R}_+^{J_1 \times J_2 \times \cdots \times J_N}$ such that the cost function (7.125) is minimized.

begin
  $[\{\mathbf{A}\}, \underline{\mathbf{G}}] = \text{HOSVD}(\underline{\mathbf{Y}}, [J_1, J_2, \ldots, J_N])$ /* or random initialization */
  for $n = 1$ *to* N **do**
    $\underline{\mathbf{W}}^{(-n)} = \underline{\mathbf{Y}} \times_{-n} \{\mathbf{A}^T\}$
    $[\mathbf{A}^{(n)}, \mathbf{D}_n] = \texttt{eigs}(\mathbf{W}_{(n)}^{(-n)} \mathbf{W}_{(n)}^{(-n)T}, J_n, \text{'LM'})$ /* $[\mathbf{A}^{(n)}, \mathbf{\Sigma}^{(n)}, \mathbf{V}^{(n)T}] = \texttt{svds}(\mathbf{W}_{(n)}^{(-n)}, J_n, \text{'L'})$ */
    $\mathbf{A}^{(n)} = [\mathbf{A}^{(n)}]_+$ /* need to fix signs in advance */
  end
  $\underline{\mathbf{G}} = \underline{\mathbf{W}}^{(-N)} \times_N \mathbf{A}^{(N)T}$
end

In order to derive updating rules for NTF, we use the following properties of vector and Kronecker operators:

$$\text{vec}(\mathbf{ABC}^T) = (\mathbf{C} \otimes \mathbf{A})\text{vec}(\mathbf{B}), \tag{7.137}$$

$$(\mathbf{A} \otimes \mathbf{B})^T = \mathbf{A}^T \otimes \mathbf{B}^T, \tag{7.138}$$

$$(\mathbf{A} \otimes \mathbf{B})(\mathbf{C} \otimes \mathbf{D}) = \mathbf{AC} \otimes \mathbf{BD}. \tag{7.139}$$

Recall that mode-n product of the tensor $\underline{\mathbf{1}}$ with the normalized factor $\mathbf{A}^{(n)\,T}$ is also tensor $\underline{\mathbf{1}}$, that is

$$\left[\underline{\mathbf{1}} \times_n \mathbf{A}^{(n)\,T}\right]_{(n)} = \mathbf{A}^{(n)\,T}\mathbf{1}_{(n)} = \mathbf{1}_{(n)}, \tag{7.140}$$

where $\mathbf{1}_{(n)}$ is the mode-n matricization of the tensor with all elements equal to unity, thus giving

$$\underline{\mathbf{1}} \times \{\mathbf{A}^T\} = \underline{\mathbf{1}}. \tag{7.141}$$

We also have $\left(\mathbf{A}^{\otimes_{-n}} \otimes \mathbf{A}^{(n)}\right)\mathbf{1} = \text{vec}(\mathbf{A}^{(n)}\mathbf{1}\mathbf{1}^T\mathbf{A}^{\otimes_{-n}T})$.

7.4.4.1 Learning Rule for Factors $\mathbf{A}^{(n)}$

The multiplicative Alpha-NMF update rules for the regular NMF model $\mathbf{Y} = \mathbf{AX} + \mathbf{E} = \widehat{\mathbf{Y}} + \mathbf{E}$ derived in Chapter 3 [28] are given by

$$\mathbf{A} \leftarrow \mathbf{A} \circledast \left\{\left[\left(\mathbf{Y} \oslash \widehat{\mathbf{Y}}\right)^{.[\alpha]} \mathbf{X}^T\right] \oslash \left(\mathbf{1}\mathbf{1}^T\mathbf{X}^T\right)\right\}^{.[1/\alpha]}, \tag{7.142}$$

$$\mathbf{X} \leftarrow \mathbf{X} \circledast \left\{\left[\mathbf{A}^T\left(\mathbf{Y} \oslash \widehat{\mathbf{Y}}\right)^{.[\alpha]}\right] \oslash \left(\mathbf{A}^T\mathbf{1}\mathbf{1}^T\right)\right\}^{.[1/\alpha]}. \tag{7.143}$$

Taking into account (7.130) and (7.142), the learning rule for estimating the factors $\mathbf{A}^{(n)}$, $(n = 1, 2, \ldots, N)$ can be written as

$$\mathbf{A}^{(n)} \leftarrow \mathbf{A}^{(n)} \circledast \left\{\left[\left(\mathbf{Y}_{(n)} \oslash \widehat{\mathbf{Y}}_{(n)}\right)^{.[\alpha]} \mathbf{A}^{\otimes_{-n}}\mathbf{G}_{(n)}^T\right] \oslash \left(\mathbf{1}\mathbf{1}^T\mathbf{A}^{\otimes_{-n}}\mathbf{G}_{(n)}^T\right)\right\}^{.[1/\alpha]}. \tag{7.144}$$

The term $\mathbf{1}^T\mathbf{A}^{\otimes_{-n}}\mathbf{G}_{(n)}^T$ in the denominator of (7.144) is a $J_n \times 1$ vector whose elements are sums of column vectors of matrix $\mathbf{A}^{\otimes_{-n}}\mathbf{G}_{(n)}^T$. Hence, the denominator in (7.144) is represented by an $I_n \times J_n$ large matrix consisting of an $I_n \times 1$ tiling of copies of vector $\mathbf{1}^T\mathbf{A}^{\otimes_{-n}}\mathbf{G}_{(n)}^T$. In other words, all the elements in each column vector of the denominator matrix of (7.144) are identical. Therefore, the denominator can be omitted if vectors $\boldsymbol{a}_{j_n}^{(n)}$ are normalized to unit ℓ_1-norm [64–66]. Here, we choose the ℓ_1-norm normalization (see the next section for more detail)

$$\boldsymbol{a}_{j_n}^{(n)} \leftarrow \boldsymbol{a}_{j_n}^{(n)} / \left\| \boldsymbol{a}_{j_n}^{(n)} \right\|_1, \quad (n = 1, 2, \ldots, N, \quad j_n = 1, 2, \ldots, J_N). \tag{7.145}$$

The ℓ_1-norm normalization (7.145) enforces all factor coefficients to be in the range of [0,1], thus eliminating the large differences between values of coefficients and allowing us to avoid updating $\underline{\widehat{\mathbf{Y}}}$ at each iteration step.

An alternative way to prevent the coefficients from reaching zero values is to apply a component-wise nonlinear operator to all the factors, that is

$$[\mathbf{A}^{(n)}]_+ = \max\{\varepsilon, \mathbf{A}^{(n)}\}, \qquad (\text{in MATLAB, } \textit{typically} \;\; \varepsilon = 2^{-52}). \tag{7.146}$$

The term $\mathbf{G}_{(n)}\mathbf{A}^{\otimes_{-n}\,T}$ in (7.144) is exactly the mode-n matricized version of the tensor $\underline{\mathbf{G}} \times_{-n} \{\mathbf{A}\}$, and factors $\mathbf{A}^{(n)}$ are tall full-rank matrices, so $\mathbf{A}^{(n)\dagger}\mathbf{A}^{(n)} = \mathbf{I}$ is an identity matrix. We then have:

$$\mathbf{G}_{(n)}\mathbf{A}^{\otimes_{-n}T} = \mathbf{A}^{(n)\dagger}\,\widehat{\mathbf{Y}}_{(n)}. \tag{7.147}$$

After some algebraic manipulation, we obtain a simplified learning rule for the estimation of the factors $\mathbf{A}^{(n)}$, referred to as the multiplicative Alpha NTD algorithm [64]:

$$\begin{aligned} \mathbf{A}^{(n)} &\leftarrow \mathbf{A}^{(n)} \circledast \left\{ \left[\left(\underline{\mathbf{Y}} \oslash \widehat{\underline{\mathbf{Y}}} \right)^{.[\alpha]} \right]_{(n)} \widehat{\mathbf{Y}}_{(n)}^{T}\, \mathbf{A}^{(n)\dagger\, T} \right\}^{.[1/\alpha]}, \\ \mathbf{a}_{j_n}^{(n)} &\leftarrow \mathbf{a}_{j_n}^{(n)} / \left\| \mathbf{a}_{j_n}^{(n)} \right\|_1, \qquad (n = 1, 2, \ldots, N, \qquad j_n = 1, 2, \ldots, J_n).. \end{aligned} \tag{7.148}$$

This above update rule (7.149) can be represented in a compact form using the contracted tensor product (see Chapter 1)

$$\mathbf{A}^{(n)} \leftarrow \mathbf{A}^{(n)} \circledast \left[\left\langle \left\langle \left(\underline{\mathbf{Y}} \oslash \widehat{\underline{\mathbf{Y}}} \right)^{.[\alpha]}, \widehat{\underline{\mathbf{Y}}} \right\rangle_{-n}, \mathbf{A}^{(n)\dagger} \right\rangle_2 \right]^{.[1/\alpha]}. \tag{7.149}$$

In practice, in early iterations, factors $\mathbf{A}^{(n)}$ can be ill-conditioned matrices. To avoid problem with convergence, we can find an approximation $\mathbf{U}^T$ of the expression $\langle(\underline{\mathbf{Y}} \oslash \widehat{\underline{\mathbf{Y}}})^{.[\alpha]}, \widehat{\underline{\mathbf{Y}}}\rangle_{-n}\mathbf{A}^{(n)\dagger\, T}$ by solving the system of linear equations

$$\mathbf{A}^{(n)}\mathbf{U} = \left\langle \widehat{\underline{\mathbf{Y}}}, \left(\underline{\mathbf{Y}} \oslash \widehat{\underline{\mathbf{Y}}} \right)^{.[\alpha]} \right\rangle_{-n} = \mathbf{W}^{(n)}, \tag{7.150}$$

which can be represented in an equivalent form (but much smaller dimension):

$$\mathbf{A}^{(n)T}\mathbf{A}^{(n)}\mathbf{U} = \mathbf{A}^{(n)T}\mathbf{W}^{(n)}, \tag{7.151}$$

where $\mathbf{W}^{(n)}$ is an $I_n \times I_n$ matrix, and $\mathbf{U}$ is $J_n \times I_n$ matrix. The above system of linear equations can be solved, for example, using Levenberg-Marquardt approach with the exponentially decreasing regularization parameter $\lambda = \bar{\lambda}_0 + \lambda_0 \exp\{-\tau k\}$ (see also in Chapters 4 and 6)

$$\left(\mathbf{A}^{(n)T}\mathbf{A}^{(n)} + \lambda \mathbf{I}_{J_n} \right) \mathbf{U} = \mathbf{A}^{(n)T}\mathbf{W}^{(n)}, \tag{7.152}$$

where the least squares solution $\mathbf{U}$ can be computed by Gaussian elimination or by orthogonal-triangular factorization.

Finally, the update rule (7.149) for $\mathbf{A}^{(n)}$ can be formulated as

$$\mathbf{A}^{(n)} \leftarrow \mathbf{A}^{(n)} \circledast \mathbf{U}^{.[1/\alpha]\,T}. \tag{7.153}$$

7.4.4.2 Learning Rule for Core Tensor

From (7.131) and (7.143), the core tensor $\underline{\mathbf{G}}$ can be estimated as

$$\begin{aligned} \mathrm{vec}(\mathbf{G}_{(n)}) \leftarrow \mathrm{vec}(\mathbf{G}_{(n)}) \circledast & \left[\left(\mathbf{A}^{\otimes_{-n}} \otimes \mathbf{A}^{(n)} \right)^T \left(\mathrm{vec}(\mathbf{Y}_{(n)}) \oslash \mathrm{vec}(\widehat{\mathbf{Y}}_{(n)}) \right)^{.[\alpha]} \right]^{.[1/\alpha]} \\ \oslash & \left[\left(\mathbf{A}^{\otimes_{-n}} \otimes \mathbf{A}^{(n)} \right)^T \mathbf{1} \right]^{.[1/\alpha]}, \end{aligned} \tag{7.154}$$

where $[(\mathbf{A}^{\otimes_{-n}} \otimes \mathbf{A}^{(n)})^T \mathbf{1}]$ is employed instead of $((\mathbf{A}^{\otimes_{-n}} \otimes \mathbf{A}^{(n)})^T \mathbf{1}\mathbf{1}^T)$ because $\text{vec}(\mathbf{Y}_{(n)})$ is a column vector. The long vector $\mathbf{1} \in \mathbb{R}^{I_1 I_2 \cdots I_N}$ in this expression can be considered as a Kronecker product of N smaller vectors $\mathbf{1}^{(n)} \in \mathbb{R}^{I_n}$ $(n = 1, 2, \ldots, N)$ whose dimensions are respectively $I_1, I_2, \ldots, I_N$. Taking into account that $\mathbf{A}^{(n)}$ are unit ℓ_1-norm factors, $\mathbf{A}^{(n)\,T}\mathbf{1}^{(n)} = \mathbf{1}$, the denominator of (7.154) can be expressed as

$$\begin{aligned}\left(\mathbf{A}^{\otimes_{-n}} \otimes \mathbf{A}^{(n)}\right)^T \mathbf{1} &= \left(\{\mathbf{A}^T\}^{\otimes_{-n}} \otimes \mathbf{A}^{(n)\,T}\right)\left(\mathbf{1}^{(N)} \otimes \cdots \otimes \mathbf{1}^{(n+1)} \otimes \mathbf{1}^{(n-1)} \otimes \cdots \otimes \mathbf{1}^{(1)} \otimes \mathbf{1}^{(n)}\right) \\ &= \left(\{\mathbf{A}^T\mathbf{1}\}^{\otimes_{-n}}\right) \otimes \left(\mathbf{A}^{(n)\,T}\mathbf{1}^{(n)}\right) = \left(\{\mathbf{1}\}^{\otimes_{-n}}\right) \otimes (\mathbf{1}) = \mathbf{1}. \end{aligned} \tag{7.155}$$

Thus it can be neglected. This justifies the choice of the ℓ_1-norm.

Furthermore, note that the second term in (7.154) can be expressed as

$$\begin{aligned}\left(\mathbf{A}^{\otimes_{-n}} \otimes \mathbf{A}^{(n)}\right)^T \left(\text{vec}(\mathbf{Y}_{(n)}) \oslash \text{vec}(\widehat{\mathbf{Y}}_{(n)})\right)^{.[\alpha]} &= \left(\{\mathbf{A}^T\}^{\otimes_{-n}} \otimes \mathbf{A}^{(n)\,T}\right)\left(\text{vec}(\mathbf{Y}_{(n)} \oslash \widehat{\mathbf{Y}}_{(n)})\right)^{.[\alpha]} \\ = \text{vec}\left(\mathbf{A}^{(n)\,T}\left[\underline{\mathbf{Y}} \oslash \widehat{\underline{\mathbf{Y}}}\right]^{.[\alpha]}_{(n)} \{\mathbf{A}^T\}^{\otimes_{-n}\,T}\right) &= \text{vec}\left(\left[\left(\underline{\mathbf{Y}} \oslash \widehat{\underline{\mathbf{Y}}}\right)^{.[\alpha]} \times \{\mathbf{A}^T\}\right]_{(n)}\right). \end{aligned} \tag{7.156}$$

Hence, the learning rule (7.154) can be finally simplified as [64–66]

$$\boxed{\text{vec}(\mathbf{G}_{(n)}) \leftarrow \text{vec}(\mathbf{G}_{(n)}) \circledast \text{vec}\left(\left[\left(\underline{\mathbf{Y}} \oslash \widehat{\underline{\mathbf{Y}}}\right)^{.[\alpha]} \times \{\mathbf{A}^T\}\right]_{(n)}\right)^{.[1/\alpha]}} \tag{7.157}$$

and can be expressed in a compact tensor form as

$$\boxed{\underline{\mathbf{G}} \leftarrow \underline{\mathbf{G}} \circledast \left\{\left(\underline{\mathbf{Y}} \oslash \widehat{\underline{\mathbf{Y}}}\right)^{.[\alpha]} \times \{\mathbf{A}^T\}\right\}^{.[1/\alpha]}}. \tag{7.158}$$

Learning rules for the Fast Alpha NTD algorithm and a detailed pseudo-code of are given in Algorithm 7.11.

Algorithm 7.11: Fast Alpha NTD Algorithm

Input: $\underline{\mathbf{Y}}$: input data of size $I_1 \times I_2 \times \cdots \times I_N$,
$J_1, J_2, \ldots, J_N$: number of basis components for each factor, α: divergence parameter.
Output: N factors $\mathbf{A}^{(n)} \in \mathbb{R}_+^{I_n \times J_n}$ and a core tensor $\underline{\mathbf{G}} \in \mathbb{R}^{J_1 \times J_2 \times \cdots \times J_N}$

begin
  Nonnegative ALS initialization for all $\mathbf{A}^{(n)}$ and $\underline{\mathbf{G}}$
  repeat
    $\widehat{\underline{\mathbf{Y}}} = \underline{\mathbf{G}} \times \{\mathbf{A}\}$
    for $n = 1$ *to* N **do**
      $\mathbf{A}^{(n)} \leftarrow \mathbf{A}^{(n)} \circledast \left[\left\langle\left\langle\left(\underline{\mathbf{Y}} \oslash \widehat{\underline{\mathbf{Y}}}\right)^{.[\alpha]}, \widehat{\underline{\mathbf{Y}}}\right\rangle_{-n}, \mathbf{A}^{(n)\,\dagger}\right\rangle_2\right]^{.[1/\alpha]}$ /* See (7.149) */
      $\boldsymbol{a}_{j_n}^{(n)} \leftarrow \boldsymbol{a}_{j_n}^{(n)} / \|\boldsymbol{a}_{j_n}^{(n)}\|_1$ /* normalize to unit length */
    end
    $\underline{\mathbf{G}} \leftarrow \underline{\mathbf{G}} \circledast \left\{\left(\underline{\mathbf{Y}} \oslash \widehat{\underline{\mathbf{Y}}}\right)^{.[\alpha]} \times \{\mathbf{A}^T\}\right\}^{.[1/\alpha]}$ /* See (7.158) */
  until *a stopping criterion is met*   /* convergence condition */
end

7.4.5 Beta NTD Algorithm

In this section we extend the multiplicative Beta NMF algorithm proposed in [28] and derived rigorously in Chapter 3 to the NTD algorithm. For convenience, we shall reintroduce the Beta NMF algorithm:

$$\mathbf{A} \leftarrow \mathbf{A} \circledast \left[\left(\mathbf{Y} \circledast \widehat{\mathbf{Y}}^{\cdot [\beta-1]} \right) \mathbf{X}^T \right] \oslash \left(\widehat{\mathbf{Y}}^{\cdot [\beta]} \mathbf{X}^T \right), \tag{7.159}$$

$$\mathbf{X} \leftarrow \mathbf{X} \circledast \left[\mathbf{A}^T \left(\mathbf{Y} \circledast \widehat{\mathbf{Y}}^{\cdot [\beta-1]} \right) \right] \oslash \left(\mathbf{A}^T \widehat{\mathbf{Y}}^{\cdot [\beta]} \right). \tag{7.160}$$

The Beta NTD algorithm can be derived indirectly using the matricized representation of the NTD (7.130) and by applying learning rule (7.159) for factors $\mathbf{A}^{(n)}$, to yield

$$\mathbf{A}^{(n)} \leftarrow \mathbf{A}^{(n)} \circledast \left[\left(\mathbf{Y}_{(n)} \circledast \widehat{\mathbf{Y}}_{(n)}^{\cdot [\beta-1]} \right) \mathbf{A}^{\otimes_{-n}} \mathbf{G}_{(n)}^T \right] \oslash \left(\widehat{\mathbf{Y}}_{(n)}^{\cdot [\beta]} \mathbf{A}^{\otimes_{-n}} \mathbf{G}_{(n)}^T \right), \tag{7.161}$$

whereas the vectorized expression (7.131) and learning rule (7.160) are used to establish update rule for a core tensor $\underline{\mathbf{G}}$ in a vector form

$$\text{vec}(\mathbf{G}_{(n)}) \leftarrow \text{vec}(\mathbf{G}_{(n)}) \circledast \text{vec} \left(\left[\left(\underline{\mathbf{Y}} \circledast \widehat{\underline{\mathbf{Y}}}^{\cdot [\beta-1]} \right) \times \{\mathbf{A}^T\} \right]_{(n)} \right) \oslash \text{vec} \left(\left[\widehat{\mathbf{Y}}_{(n)}^{\cdot [\beta]} \times \{\mathbf{A}^T\} \right]_{(n)} \right) \tag{7.162}$$

or equivalently in a more compact tensor form

$$\underline{\mathbf{G}} \leftarrow \underline{\mathbf{G}} \circledast \left[\left(\underline{\mathbf{Y}} \circledast \widehat{\underline{\mathbf{Y}}}^{\cdot [\beta-1]} \right) \times \{\mathbf{A}^T\} \right] \oslash \left[\widehat{\underline{\mathbf{Y}}}^{\cdot [\beta]} \times \{\mathbf{A}^T\} \right]. \tag{7.163}$$

Note, that we can replace the term $\mathbf{A}^{\otimes_{-n}} \mathbf{G}_{(n)}^T$ by $\widehat{\mathbf{Y}}_{(n)}^T \mathbf{A}^{(n)\dagger T}$ in (7.161), and obtain the alternative update rule for $\mathbf{A}^{(n)}$

$$\mathbf{A}^{(n)} \leftarrow \mathbf{A}^{(n)} \circledast \left\langle \left\langle \underline{\mathbf{Y}} \circledast \widehat{\underline{\mathbf{Y}}}^{\cdot [\beta-1]}, \widehat{\underline{\mathbf{Y}}} \right\rangle_{-n}, \mathbf{A}^{(n)\dagger} \right\rangle_2 \oslash \left\langle \left\langle \widehat{\underline{\mathbf{Y}}}^{\cdot [\beta]}, \widehat{\underline{\mathbf{Y}}} \right\rangle_{-n}, \mathbf{A}^{(n)\dagger} \right\rangle_2. \tag{7.164}$$

The detailed pseudo-code of the Fast Beta HALS NTD algorithm is given in Algorithm 7.12.

7.4.6 Local ALS Algorithms for Nonnegative TUCKER Decompositions

In this section, we illustrate the usefulness of the local ALS approach to NTD, and also to Semi-Nonnegative Tucker Decomposition (SNTD).

Most algorithms for the NTD model are based on ALS minimization of the squared Euclidean distance [11,48,50] used as the global cost function, that is

$$D_F(\underline{\mathbf{Y}} \,||\, \underline{\mathbf{G}}, \{\mathbf{A}\}) = \frac{1}{2} \left\| \underline{\mathbf{Y}} - \widehat{\underline{\mathbf{Y}}} \right\|_F^2, \tag{7.165}$$

subject to nonnegativity constraints. With some adjustments on this cost function, we establish local learning rules for components and core tensor.

Algorithm 7.12: **Fast Beta NTD Algorithm**

Input: $\underline{\mathbf{Y}}$: input data of size $I_1 \times I_2 \times \cdots \times I_N$,
$J_1, J_2, \ldots, J_N$: number of basis components for each factor, β: divergence parameter.
Output: N factors $\mathbf{A}^{(n)} \in \mathbb{R}_+^{I_n \times J_n}$ and a core tensor $\underline{\mathbf{G}} \in \mathbb{R}^{J_1 \times J_2 \times \cdots \times J_N}$

begin
  Nonnegative ALS initialization for all $\mathbf{A}^{(n)}$ and $\underline{\mathbf{G}}$
  repeat
    $\widehat{\underline{\mathbf{Y}}} = \underline{\mathbf{G}} \times \{\mathbf{A}\}$
    for $n = 1$ *to* N **do**
      $\mathbf{A}^{(n)} \leftarrow \mathbf{A}^{(n)} \circledast \left[\left(\mathbf{Y}_{(n)} \circledast \widehat{\mathbf{Y}}_{(n)}^{\,.[\beta-1]}\right) \mathbf{A}^{\otimes -n}\, \mathbf{G}_{(n)}^T\right] \oslash \left(\widehat{\mathbf{Y}}_{(n)}^{\,.[\beta]}\, \mathbf{A}^{\otimes -n}\, \mathbf{G}_{(n)}^T\right)$
      $\boldsymbol{a}_{j_n}^{(n)} \leftarrow \boldsymbol{a}_{j_n}^{(n)} / \|\boldsymbol{a}_{j_n}^{(n)}\|_p$ /* normalize to unit length */
    end
    $\underline{\mathbf{G}} \leftarrow \underline{\mathbf{G}} \circledast \left[\left(\underline{\mathbf{Y}} \circledast \widehat{\underline{\mathbf{Y}}}^{\,.[\beta-1]}\right) \times \{\mathbf{A}^T\}\right] \oslash \left[\widehat{\underline{\mathbf{Y}}}^{\,.[\beta]} \times \{\mathbf{A}^T\}\right]$
  until *a stopping criterion is met*   /* convergence condition */
end

7.4.6.1 Learning Rule for Factors $\mathbf{A}^{(n)}$

We define the residual tensor $\underline{\mathbf{Y}}^{(j_n)}$

$$
\begin{aligned}
\underline{\mathbf{Y}}^{(j_n)} &= \underline{\mathbf{Y}} - \sum_{r_1=1}^{J_1} \cdots \sum_{r_n \neq j_n} \cdots \sum_{r_N=1}^{J_N} g_{r_1 \ldots r_n \ldots r_N}\, \boldsymbol{a}_{j_1}^{(1)} \circ \cdots \circ \boldsymbol{a}_{r_n}^{(n)} \circ \cdots \circ \boldsymbol{a}_{r_N}^{(N)} \\
&= \underline{\mathbf{Y}} - \widehat{\underline{\mathbf{Y}}} + \underline{\mathbf{G}}_{r_n=j_n} \times_{-n} \{\mathbf{A}\} \times_n \boldsymbol{a}_{j_n}^{(n)} \\
&= \underline{\mathbf{E}} + \underline{\mathbf{G}}_{r_n=j_n} \times_{-n} \{\mathbf{A}\} \times_n \boldsymbol{a}_{j_n}^{(n)}, \qquad (j_n = 1, 2, \ldots, J_N),
\end{aligned}
\tag{7.166}
$$

where $\underline{\mathbf{G}}_{r_n=j_n} \in \mathbb{R}^{J_1 \times \cdots \times J_{n-1} \times 1 \times J_{n+1} \times \cdots \times J_N}$ is a subtensor of the tensor $\underline{\mathbf{G}} \in \mathbb{R}^{J_1 \times \cdots \times J_{n-1} \times J_n \times J_{n+1} \times \cdots \times J_N}$ obtained by fixing the n-th index to some value j_n. For example, for the three-way core tensor $\underline{\mathbf{G}} \in \mathbb{R}^{J_1 \times J_2 \times J_3}$, $\underline{\mathbf{G}}_{r_1=1}$ is the first horizontal slice, but of size $1 \times J_2 \times J_3$, $\underline{\mathbf{G}}_{r_2=1}$ is the first lateral slice but of size $J_1 \times 1 \times J_3$, and $\underline{\mathbf{G}}_{r_3=1}$ is the first frontal slice of the tensor $\underline{\mathbf{G}}$, but of size $J_1 \times J_2 \times 1$. The mode-n matricized version of tensor $\underline{\mathbf{G}}_{r_n=j_n}$ is exactly the j_n-th row of the mode-n matricized version of tensor $\underline{\mathbf{G}}$, i.e., $[\underline{\mathbf{G}}_{r_n=j_n}]_{(n)} = [\mathbf{G}_{(n)}]_{\underline{j_n}}$.

To estimate the component $\boldsymbol{a}_{j_n}^{(n)}$, we assume that all the other components in all factors and the core tensor are fixed. Instead of minimizing the cost function (7.165), we can use a more sophisticated approach [65] by minimizing a set of local cost functions, as exemplified by

$$
\begin{aligned}
D_F^{(j_n)}(\underline{\mathbf{Y}}^{(j_n)} \,||\, \boldsymbol{a}_{j_n}^{(n)}) &= \frac{1}{2} \Bigg\| \underline{\mathbf{Y}}^{(j_n)} - \sum_{r_1=1}^{J_1} \cdots \sum_{r_{n-1}=1}^{J_{n-1}} \sum_{r_{n+1}=1}^{J_{n+1}} \cdots \sum_{r_N=1}^{J_N} g_{r_1 \cdots r_{n-1}\, j_n\, r_{n+1} \cdots r_N}\, \boldsymbol{a}_{r_1}^{(1)} \circ \cdots \\
&\qquad \cdots \circ \boldsymbol{a}_{r_{n-1}}^{(n-1)} \circ \boldsymbol{a}_{j_n}^{(n)} \circ \boldsymbol{a}_{r_{n+1}}^{(n+1)} \circ \cdots \circ \boldsymbol{a}_{r_N}^{(N)} \Bigg\|_F^2 \\
&= \frac{1}{2} \left\| \underline{\mathbf{Y}}^{(j_n)} - \underline{\mathbf{G}}_{r_n=j_n} \times_{-n} \{\mathbf{A}\} \times_n \boldsymbol{a}_{j_n}^{(n)} \right\|_F^2 \\
&= \frac{1}{2} \left\| \mathbf{Y}_{(n)}^{(j_n)} - \boldsymbol{a}_{j_n}^{(n)}\, [\underline{\mathbf{G}}_{r_n=j_n}]_{(n)}\, \mathbf{A}^{\otimes -n\, T} \right\|_F^2, \qquad (n = 1, 2, \ldots, N),
\end{aligned}
\tag{7.167}
$$

To derive the learning algorithm, we first calculate the gradient of (7.167) with respect to element $\boldsymbol{a}_{j_n}^{(n)}$

$$\frac{\partial D_F^{(j_n)}}{\partial \boldsymbol{a}_{j_n}^{(n)}} = -\left(\mathbf{Y}_{(n)}^{(j_n)} - \boldsymbol{a}_{j_n}^{(n)} \, [\underline{\mathbf{G}}_{r_n=j_n}]_{(n)} \, \mathbf{A}^{\otimes_{-n}\, T}\right) \mathbf{A}^{\otimes_{-n}} [\underline{\mathbf{G}}_{r_n=j_n}]_{(n)}^T$$
$$= -\left(\mathbf{Y}_{(n)}^{(j_n)} - \boldsymbol{a}_{j_n}^{(n)} \, [\mathbf{G}_{(n)}]_{\underline{j_n}} \, \mathbf{A}^{\otimes_{-n}\, T}\right) \mathbf{A}^{\otimes_{-n}} [\mathbf{G}_{(n)}]_{\underline{j_n}}^T \tag{7.168}$$

and set it to zero to obtain a fixed point learning rule for $\mathbf{A}^{(n)} = [\boldsymbol{a}_1^{(n)}, \boldsymbol{a}_2^{(n)}, \ldots, \boldsymbol{a}_{J_n}^{(n)}]$ given by

$$\boldsymbol{a}_{j_n}^{(n)} \leftarrow \mathbf{Y}_{(n)}^{(j_n)} \, \mathbf{A}^{\otimes_{-n}} \, [\mathbf{G}_{(n)}]_{\underline{j_n}}^T \Big/ \left([\mathbf{G}_{(n)}]_{\underline{j_n}} \, \mathbf{A}^{\otimes_{-n}\, T} \, \mathbf{A}^{\otimes_{-n}} \, [\mathbf{G}_{(n)}]_{\underline{j_n}}^T\right), \tag{7.169}$$

$$\boldsymbol{a}_{j_n}^{(n)} \leftarrow \left[\boldsymbol{a}_{j_n}^{(n)}\right]_+ \tag{7.170}$$

for $n = 1, 2, \ldots, N$ and $j_n = 1, 2, \ldots, J_N$.

In the next step we shall further optimize the derived local learning (update) rules.

7.4.6.2 Update Rules for the Core Tensor

Elements of the core tensor will be sequentially updated with assumption that all components are fixed. In this case the cost function (7.167) can be expressed in the following form [65]

$$D_F^{(j_n)} = \frac{1}{2}\Bigg\| \underline{\mathbf{Y}}^{(j_n)} - \sum_{\substack{(r_1,\ldots,r_{n-1},r_{n+1},\ldots,r_N)\neq \\ (j_1,\ldots,j_{n-1},j_{n+1},\ldots,j_N)}} g_{r_1\cdots r_{n-1}\, j_n\, r_{n+1}\cdots r_N} \boldsymbol{a}_{r_1}^{(1)} \circ \cdots \circ \boldsymbol{a}_{r_{n-1}}^{(n-1)} \circ \boldsymbol{a}_{j_n}^{(n)} \circ \boldsymbol{a}_{r_{n+1}}^{(n+1)} \circ \cdots \circ \boldsymbol{a}_{r_N}^{(N)}$$
$$- g_{j_1\cdots j_n \cdots j_N} \, \boldsymbol{a}_{j_1}^{(1)} \circ \cdots \circ \boldsymbol{a}_{j_n}^{(n)} \circ \cdots \circ \boldsymbol{a}_{j_N}^{(N)} \Bigg\|_F^2$$
$$= \frac{1}{2}\Bigg\| \underline{\mathbf{Y}} - \sum_{\substack{(r_1,\ldots,r_n,\ldots,r_N)\neq \\ (j_1,\ldots,j_n,\ldots,j_N)}} g_{r_1\ldots r_n\ldots r_N} \, \boldsymbol{a}_{r_1}^{(1)} \circ \cdots \circ \boldsymbol{a}_{r_n}^{(n)} \circ \cdots \circ \boldsymbol{a}_{r_N}^{(N)}$$
$$- g_{j_1\cdots j_n \cdots j_N} \, \boldsymbol{a}_{j_1}^{(1)} \circ \cdots \circ \boldsymbol{a}_{j_n}^{(n)} \circ \cdots \circ \boldsymbol{a}_{j_N}^{(N)} \Bigg\|_F^2$$
$$= \frac{1}{2} \left\| \underline{\mathbf{Y}}^{(j_1 j_2 \cdots j_N)} - g_{j_1 j_2 \cdots j_N} \, \boldsymbol{a}_{j_1}^{(1)} \circ \boldsymbol{a}_{j_2}^{(2)} \circ \cdots \circ \boldsymbol{a}_{j_N}^{(N)} \right\|_F^2, \tag{7.171}$$

where the tensor $\underline{\mathbf{Y}}^{(j_1 j_2 \cdots j_N)}$ is defined as

$$\underline{\mathbf{Y}}^{(j_1 j_2 \cdots j_N)} = \underline{\mathbf{Y}} - \sum_{r_1 \neq j_1} \sum_{r_2 \neq j_2} \cdots \sum_{r_N \neq j_N} g_{r_1 r_2 \ldots r_N} \, \boldsymbol{a}_{r_1}^{(1)} \circ \boldsymbol{a}_{r_2}^{(2)} \circ \cdots \circ \boldsymbol{a}_{r_N}^{(N)}$$
$$= \underline{\mathbf{Y}} - \widehat{\underline{\mathbf{Y}}} + g_{j_1 j_2 \cdots j_N} \, \boldsymbol{a}_{j_1}^{(1)} \circ \boldsymbol{a}_{j_2}^{(2)} \circ \cdots \circ \boldsymbol{a}_{j_N}^{(N)}$$
$$= \underline{\mathbf{E}} + g_{j_1 j_2 \cdots j_N} \, \boldsymbol{a}_{j_1}^{(1)} \circ \boldsymbol{a}_{j_2}^{(2)} \circ \cdots \circ \boldsymbol{a}_{j_N}^{(N)}. \tag{7.172}$$

The cost function (7.171) can be expressed in the matricized form and also in its vectorized version by taking into account $\text{vec}(\mathbf{ABC}) = (\mathbf{C}^T \otimes \mathbf{A})\text{vec}(\mathbf{B})$ and that the Khatri-Rao and Kronecker products of two vectors are identical, that is

$$\begin{aligned} D_F^{(j_1 j_2 \cdots j_N)} &= \frac{1}{2} \left\| \mathbf{Y}_{(1)}^{(j_1 j_2 \cdots j_N)} - g_{j_1 j_2 \cdots j_N}\, \boldsymbol{a}_{j_1}^{(1)} \left(\boldsymbol{a}_{j_N}^{(N)} \odot \cdots \odot \boldsymbol{a}_{j_2}^{(2)}\right)^T \right\|_F^2, \\ &= \frac{1}{2} \left\| \operatorname{vec}(\mathbf{Y}_{(1)}^{(j_1 j_2 \cdots j_N)}) - \left(\boldsymbol{a}_{j_N}^{(N)} \odot \cdots \odot \boldsymbol{a}_{j_2}^{(2)} \otimes \boldsymbol{a}_{j_1}^{(1)}\right) g_{j_1 j_2 \cdots j_N} \right\|_F^2 \\ &= \frac{1}{2} \left\| \operatorname{vec}(\mathbf{Y}_{(1)}^{(j_1 j_2 \cdots j_N)}) - \left(\boldsymbol{a}_{j_N}^{(N)} \otimes \cdots \otimes \boldsymbol{a}_{j_2}^{(2)} \otimes \boldsymbol{a}_{j_1}^{(1)}\right) g_{j_1 j_2 \cdots j_N} \right\|_F^2. \end{aligned} \tag{7.173}$$

To obtain the learning rule, calculate the gradient of (7.173) with respect to elements $g_{j_1 j_2 \cdots j_N}$

$$\frac{\partial D_F^{(j_1 j_2 \cdots j_N)}}{\partial g_{j_1 j_2 \cdots j_N}} = -\left(\boldsymbol{a}_{j_N}^{(N)} \otimes \cdots \otimes \boldsymbol{a}_{j_1}^{(1)}\right)^T \left(\operatorname{vec}(\mathbf{Y}_{(1)}^{(j_1 j_2 \cdots j_N)}) - \left(\boldsymbol{a}_{j_N}^{(N)} \otimes \cdots \otimes \boldsymbol{a}_{j_1}^{(1)}\right) g_{j_1 j_2 \cdots j_N}\right) \tag{7.174}$$

and set it to zero, to yield a learning rule for entries of the core tensor $\underline{\mathbf{G}}$, given by

$$g_{j_1 j_2 \cdots j_N} \leftarrow \left[\frac{\left(\boldsymbol{a}_{j_N}^{(N)} \otimes \cdots \otimes \boldsymbol{a}_{j_1}^{(1)}\right)^T \operatorname{vec}(\mathbf{Y}_{(1)}^{(j_1 j_2 \cdots j_N)})}{\left(\boldsymbol{a}_{j_N}^{(N)} \otimes \cdots \otimes \boldsymbol{a}_{j_1}^{(1)}\right)^T \left(\boldsymbol{a}_{j_N}^{(N)} \otimes \cdots \otimes \boldsymbol{a}_{j_1}^{(1)}\right)} \right]_+. \tag{7.175}$$

This update rule can be simplified by taking into account that the Kronecker product of the two unit-length vectors $\boldsymbol{a}$ and $\boldsymbol{b}$, i.e., $\boldsymbol{c} = \boldsymbol{a} \otimes \boldsymbol{b}$, is also a unit-length vector, that is

$$\| \boldsymbol{c} \|_2^2 = \boldsymbol{c}^T \boldsymbol{c} = (\boldsymbol{a} \otimes \boldsymbol{b})^T (\boldsymbol{a} \otimes \boldsymbol{b}) = (\boldsymbol{a}^T \boldsymbol{a}) \otimes (\boldsymbol{b}^T \boldsymbol{b}) = 1 \otimes 1 = 1. \tag{7.176}$$

Hence, if all the factors $\mathbf{A}^{(n)}$ are normalized via projections: $\boldsymbol{a}_{j_n}^{(n)} \leftarrow \boldsymbol{a}_{j_n}^{(n)} / \| \boldsymbol{a}_{j_n}^{(n)} \|_2$, the denominator of (7.175) can be ignored, and by replacing $\mathbf{Y}_{(1)}^{(j_1 j_2 \cdots j_N)}$ in (7.175) by (7.172), we have

$$\begin{aligned} g_{j_1 j_2 \cdots j_N} &\leftarrow \left(\boldsymbol{a}_{j_N}^{(N)} \otimes \cdots \otimes \boldsymbol{a}_{j_1}^{(1)}\right)^T \operatorname{vec}\left(\underline{\mathbf{E}}_{(1)} + \left(\boldsymbol{a}_{j_N}^{(N)} \otimes \cdots \otimes \boldsymbol{a}_{j_1}^{(1)}\right) g_{j_1 j_2 \cdots j_N}\right) \\ &= g_{j_1 j_2 \cdots j_N} + \operatorname{vec}\left(\boldsymbol{a}_{j_1}^{(1)\,T} \underline{\mathbf{E}}_{(1)} \left(\boldsymbol{a}_{j_N}^{(N)\,T} \otimes \cdots \otimes \boldsymbol{a}_{j_2}^{(2)\,T}\right)^T\right) \\ &= g_{j_1 j_2 \cdots j_N} + \underline{\mathbf{E}} \,\bar{\times}_1\, \boldsymbol{a}_{j_1}^{(1)} \,\bar{\times}_2\, \boldsymbol{a}_{j_2}^{(2)} \cdots \bar{\times}_N\, \boldsymbol{a}_{j_N}^{(N)}. \end{aligned} \tag{7.177}$$

Moreover, the denominator of the learning rule (7.169) can also be neglected as it is a scale factor equal to one.

7.4.6.3 Optimized Update Rules for NTD

Finally, the learning rules for NTD factors $\mathbf{A}^{(n)}$ $(n = 1, 2, \ldots, N)$ and core tensor $\underline{\mathbf{G}}$ can be summarized as follows (referred here to as the ℓ_2 HALS-NTD algorithm) [65]

$$\boldsymbol{a}_{j_n}^{(n)} \leftarrow \left[\mathbf{Y}_{(n)}^{(j_n)} \left[\left(\underline{\mathbf{G}} \times_{-n} \{\mathbf{A}\}\right)_{(n)} \right]_{j_n}^T \right]_+, \qquad \boldsymbol{a}_{j_n}^{(n)} \leftarrow \boldsymbol{a}_{j_n}^{(n)} / \left\| \boldsymbol{a}_{j_n}^{(n)} \right\|_2, \tag{7.178}$$

$$g_{j_1 j_2 \cdots j_N} \leftarrow \left[g_{j_1 j_2 \cdots j_N} + \underline{\mathbf{E}} \,\bar{\times}_1\, \boldsymbol{a}_{j_1}^{(1)} \,\bar{\times}_2\, \boldsymbol{a}_{j_2}^{(2)} \cdots \bar{\times}_N\, \boldsymbol{a}_{j_N}^{(N)} \right]_+, \quad (j_n = 1, \ldots, J_N). \tag{7.179}$$

Depending on the application, it may be beneficial to use the ℓ_1-norm normalization [64] instead of the ℓ_2-norm, which leads to the following alternative update rules referred to as the ℓ_1

Algorithm 7.13: Local HALS-NTD Algorithm

Input: $\underline{\mathbf{Y}}$: input data of size $I_1 \times I_2 \times \cdots \times I_N$,
$J_1, J_2, \ldots, J_N$: number of basis components for each factor
Output: N factors $\mathbf{A}^{(n)} \in \mathbb{R}_+^{I_n \times J_n}$ and a core tensor $\underline{\mathbf{G}} \in \mathbb{R}^{J_1 \times J_2 \times \cdots \times J_N}$

begin
  Nonnegative ALS initialization for all $\mathbf{A}^{(n)}$ and $\underline{\mathbf{G}}$
  Normalize all $\boldsymbol{a}_{j_n}^{(n)}$ for $n = 1, 2, ..., N$ to unit length
  $\underline{\mathbf{E}} = \underline{\mathbf{Y}} - \widehat{\underline{\mathbf{Y}}}$ /* residual tensor */
  repeat
    for $n = 1$ *to* N **do**
      for $j_n = 1$ *to* J_n **do**
        $\mathbf{Y}_{(n)}^{(j_n)} = \mathbf{E}_{(n)} + \boldsymbol{a}_{j_n}^{(n)} [\mathbf{G}_{(n)}]_{j_n} \mathbf{A}^{\otimes_{-n}\, T}$
        $\boldsymbol{a}_{j_n}^{(n)} \leftarrow \left[\mathbf{Y}_{(n)}^{(j_n)} \left[\left(\underline{\mathbf{G}} \times_{-n} \{\mathbf{A}\} \right)_{(n)} \right]_{j_n}^{T} \right]_+$ /* see (7.180) */
        $\boldsymbol{a}_{j_n}^{(n)} \leftarrow \boldsymbol{a}_{j_n}^{(n)} / \|\boldsymbol{a}_{j_n}^{(n)}\|_p$ /* $p = 1, 2$ */
        $\mathbf{E}_n \leftarrow \mathbf{Y}_{(n)}^{(j_n)} - \boldsymbol{a}_{j_n}^{(n)} [\mathbf{G}_{(n)}]_{j_n} \mathbf{A}^{\otimes_{-n}\, T}$
      end
    end
    /* $\underline{\mathbf{G}} \leftarrow \left[\underline{\mathbf{G}} \circledast \left(\underline{\mathbf{Y}} \oslash \widehat{\underline{\mathbf{Y}}} \right) \times \{\mathbf{A}^T\} \right]_+$ */ /* see (7.182) and (7.177) */
    foreach $j_1 = 1, \ldots, J_1, j_2 = 1, \ldots, J_2, \ldots, j_N = 1, \ldots, J_N$ **do**
      $g_{j_1 j_2 \cdots, j_N} \leftarrow g_{j_1 j_2 \cdots, j_N} + \underline{\mathbf{E}} \times_1 \boldsymbol{a}_{j_1}^{(1)} \bar{\times}_2 \boldsymbol{a}_{j_2}^{(2)} \cdots \bar{\times}_N \boldsymbol{a}_{j_N}^{(N)}$
      $\underline{\mathbf{E}} \leftarrow \underline{\mathbf{E}} + \Delta_{g_{j_1 j_2 \cdots j_N}} \boldsymbol{a}_{j_1}^{(1)} \circ \cdots \circ \boldsymbol{a}_{j_N}^{(N)}$, /* $\Delta_{g_{j_1 j_2 \cdots j_N}} = g_{j_1 j_2 \cdots, j_N}^{(k)} - g_{j_1 j_2 \cdots, j_N}^{(k-1)}$ */
    end
  until *a stopping criterion is met* /* convergence condition */
end

HALS-NTD algorithm:

$$\boldsymbol{a}_{j_n}^{(n)} \leftarrow \left[\mathbf{Y}_{(n)}^{(j_n)} \left[\left(\underline{\mathbf{G}} \times_{-n} \{\mathbf{A}\} \right)_{(n)} \right]_{j_n}^{T} \right]_+, \tag{7.180}$$

$$\boldsymbol{a}_{j_n}^{(n)} \leftarrow \boldsymbol{a}_{j_n}^{(n)} / \left\| \boldsymbol{a}_{j_n}^{(n)} \right\|_1, \quad (j_n = 1, \ldots, J_N), \tag{7.181}$$

$$\underline{\mathbf{G}} \leftarrow \left[\underline{\mathbf{G}} \circledast \left(\underline{\mathbf{Y}} \oslash \widehat{\underline{\mathbf{Y}}} \right) \times \{\mathbf{A}^T\} \right]_+ . \tag{7.182}$$

7.4.7 Semi-Nonnegative Tucker Decomposition

In the semi-nonnegative Tucker decomposition (SNTD) model, the entries of a core tensor $\underline{\mathbf{G}}$ do not have to be nonnegative. Based on the same local learning rule as that for factors $\mathbf{A}^{(n)}$, we can derive a simple algorithm for the SNTD by modifying the learning rule for the core tensor. Note that the factors $\mathbf{A}^{(n)}$ are tall matrices, so, $\mathbf{A}^{(n)\dagger} \mathbf{A}^{(n)} = \mathbf{I}$ is an identity matrix. Hence, we have the following relation

$$\text{vec}(\underline{\mathbf{Y}}_{(1)}) = \mathbf{A}^{\otimes} \, \text{vec}(\underline{\mathbf{G}}_{(1)}) = [\mathbf{A}^{(N)} \otimes \mathbf{A}^{(N-1)} \otimes \cdots \otimes \mathbf{A}^{(1)}] \, \text{vec}(\underline{\mathbf{G}}_{(1)}). \tag{7.183}$$

Therefore, we obtain

$$\text{vec}(\underline{\mathbf{G}}_{(1)}) = \left(\mathbf{A}^{\otimes} \right)^{\dagger} \text{vec}(\underline{\mathbf{Y}}_{(1)}) = \left[\left(\mathbf{A}^{\dagger} \right)^{\otimes_{-1}} \otimes \mathbf{A}^{(1)\dagger} \right] \text{vec}(\underline{\mathbf{Y}}_{(1)}) = \text{vec}\left(\left(\underline{\mathbf{Y}} \times \{\mathbf{A}^{\dagger}\} \right)_{(1)} \right)$$

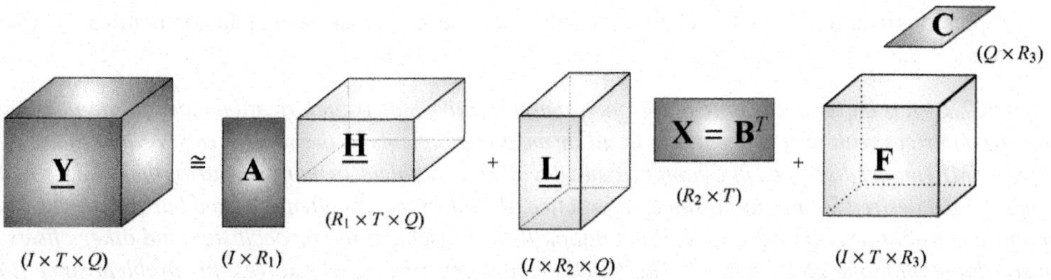

Figure 7.5 Illustration of Block Oriented Decomposition (BOD1) for a third-order tensor. The core tensors and factor (component) matrices are linked via mode-n multiplications which construct basic subtensors of the same size as a data tensor: $\underline{\mathbf{Y}} \cong \underline{\mathbf{H}} \times_1 \mathbf{A} + \underline{\mathbf{L}} \times_2 \mathbf{B} + \underline{\mathbf{F}} \times_3 \mathbf{C}$, where core tensors $\underline{\mathbf{H}} \in \mathbb{R}^{R_1 \times T \times Q}$, $\underline{\mathbf{L}} \in \mathbb{R}^{I \times R_2 \times Q}$, $\underline{\mathbf{F}} \in \mathbb{R}^{I \times T \times R_3}$ have usually much smaller dimensions than a data tensor $\underline{\mathbf{Y}} \in \mathbb{R}^{I \times T \times Q}$, i.e., $R_1 << I$, $R_2 << T$, and $R_3 << Q$.

or equivalently in a compact tensor notation

$$\boxed{\underline{\mathbf{G}} = \underline{\mathbf{Y}} \times \{\mathbf{A}^{\dagger}\} = \underline{\mathbf{Y}} \times_1 \mathbf{A}^{(1)\,\dagger} \times_2 \mathbf{A}^{(2)\,\dagger} \cdots \times_N \mathbf{A}^{(N)\,\dagger}.} \tag{7.184}$$

7.5 Nonnegative Block-Oriented Decomposition

In this section we derive algorithms for the Nonnegative Block-Oriented Decomposition (NBOD1) based on the Tucker1 model (Figure 7.5). In the simplest scenario, for a three-way data tensor $\underline{\mathbf{Y}} \in \mathbb{R}^{I \times T \times Q}$, NBOD1 is basically a sum of three subtensors $\underline{\mathbf{Y}}^{(n)}$, $(n = 1, 2, 3)$ factorized by Tucker1 in each mode (that is, using horizontal, lateral and frontal slices). Since each subtensor itself is a Tucker1 model with only one core tensor and one factor (component matrix), the mathematical description of the NBOD1 model is given by

$$\boxed{\underline{\mathbf{Y}} \cong \underline{\mathbf{H}} \times_1 \mathbf{A} + \underline{\mathbf{L}} \times_2 \mathbf{B} + \underline{\mathbf{F}} \times_3 \mathbf{C},} \tag{7.185}$$

where tensors $\underline{\mathbf{H}} \in \mathbb{R}^{R_1 \times T \times Q}$, $\underline{\mathbf{L}} \in \mathbb{R}^{I \times R_2 \times Q}$, $\underline{\mathbf{F}} \in \mathbb{R}^{I \times T \times R_3}$ are core tensors in the Tucker1 models in three different modes, and $\mathbf{A} \in \mathbb{R}^{I \times R_1}$, $\mathbf{B} \in \mathbb{R}^{T \times R_2}$ and $\mathbf{C} \in \mathbb{R}^{Q \times R_3}$ are corresponding factors. The objective is to find all three core tensors $\underline{\mathbf{H}}$, $\underline{\mathbf{L}}$, $\underline{\mathbf{F}}$ and the corresponding factor matrices $\mathbf{A}$, $\mathbf{B}$ and $\mathbf{C}$ with nonnegativity and sparsity constraints for their entries. We assume that dimensions R_1, R_2, R_3 are known in advance or can be estimated, typically, $R_1 << I$, $R_2 << T$ and $R_3 << Q$ [16].

Using the matricization approach the NBOD1 model can be described in several equivalent matrix factorization forms:

$$\mathbf{Y}_{(1)} \cong \mathbf{A}\,\mathbf{H}_{(1)} + \mathbf{L}_{(1)} \left(\mathbf{I}_Q \otimes \mathbf{B}\right)^T + \mathbf{F}_{(1)} \left(\mathbf{C} \otimes \mathbf{I}_T\right)^T$$

$$= \boxed{\begin{bmatrix} \mathbf{A} & \mathbf{L}_{(1)} & \mathbf{F}_{(1)} \end{bmatrix} \begin{bmatrix} \mathbf{H}_{(1)} & \mathbf{I}_Q \otimes \mathbf{B}^T & \mathbf{C}^T \otimes \mathbf{I}_T \end{bmatrix},} \tag{7.186}$$

$$\mathbf{Y}_{(2)} \cong \mathbf{H}_{(2)} \left(\mathbf{I}_Q \otimes \mathbf{A}\right)^T + \mathbf{B}\,\mathbf{L}_{(2)} + \mathbf{F}_{(2)} \left(\mathbf{C} \otimes \mathbf{I}_I\right)^T$$

$$= \boxed{\begin{bmatrix} \mathbf{H}_{(2)} & \mathbf{B} & \mathbf{F}_{(2)} \end{bmatrix} \begin{bmatrix} \mathbf{I}_Q \otimes \mathbf{A}^T & \mathbf{L}_{(2)} & \mathbf{C}^T \otimes \mathbf{I}_I \end{bmatrix},} \tag{7.187}$$

$$\mathbf{Y}_{(3)} \cong \mathbf{H}_{(3)} \left(\mathbf{I}_T \otimes \mathbf{A}\right)^T + \mathbf{L}_{(3)} \left(\mathbf{B} \otimes \mathbf{I}_I\right)^T + \mathbf{C}\,\mathbf{F}_{(3)}$$

$$= \boxed{\begin{bmatrix} \mathbf{H}_{(3)} & \mathbf{L}_{(3)} & \mathbf{C} \end{bmatrix} \begin{bmatrix} \mathbf{I}_T \otimes \mathbf{A}^T & \mathbf{B}^T \otimes \mathbf{I}_I & \mathbf{F}_{(3)} \end{bmatrix}.} \tag{7.188}$$

Such representations allow us to relatively easily estimate core tensors and factor matrices via NMF or SCA.

Remark 7.2 *It is important to note that standard nonnegative matrix factorization models cannot uniquely factorize matrices with offsets occurring in all or most components at the same levels. Although the affine NMF (aNMF) model discussed in Chapter 3 can alleviate this problem, in many scenarios the aNMF still fails to achieve the desired factorization since, in practice, the offset level is often not fixed but slightly fluctuates around a constant level. For example, illumination flicker, discontinuity or occlusion and other offsets are causes that violate the aNMF model. The NBOD model described above solves this problem since it can extract or pick up fluctuating offsets in each of the modes. This was our main motivation for investigating and developing such models. NBOD model can be considered as an extension and generalization of the affine NMF model. We illustrate performance of the NBOD model in the computer simulation section for difficult benchmarks.*

7.5.1 Multiplicative Algorithms for NBOD

Multiplicative algorithms for the NBOD model can be derived from the multiplicative NMF algorithms presented in Chapter 3. We illustrate briefly this approach by minimizing the Frobenius cost function

$$D(\underline{\mathbf{Y}}||\hat{\underline{\mathbf{Y}}}) = \frac{1}{2}||\underline{\mathbf{Y}} - \hat{\underline{\mathbf{Y}}}||_F^2, \tag{7.189}$$

where $\hat{\underline{\mathbf{Y}}}$ is the current estimation of data tensor $\underline{\mathbf{Y}}$. In this case, by applying the mode-1 matricization (7.186), the multiplicative learning rules for $\mathbf{H}_{(1)}$ and $\mathbf{A}$ can be written as follows (see the ISRA NMF algorithm in Chapter 3 for comparison)

$$\mathbf{A} \leftarrow \mathbf{A} \circledast \left(\mathbf{Y}_{(1)}\mathbf{H}_{(1)}^T\right) \oslash \left(\hat{\mathbf{Y}}_{(1)}\mathbf{H}_{(1)}^T\right), \tag{7.190}$$

$$\mathbf{H}_{(1)} \leftarrow \mathbf{H}_{(1)} \circledast \left(\mathbf{A}^T\mathbf{Y}_{(1)}\right) \oslash \left(\mathbf{A}^T\hat{\mathbf{Y}}_{(1)}\right), \tag{7.191}$$

where $\hat{\mathbf{Y}}_{(1)}$ is the mode-1 matricization of the current estimation $\hat{\underline{\mathbf{Y}}}$. These learning rules can be formulated in the tensor form by using the contracted tensor product

$$\mathbf{A} \leftarrow \mathbf{A} \circledast \langle \underline{\mathbf{Y}}, \underline{\mathbf{H}} \rangle_{-1} \oslash \langle \hat{\underline{\mathbf{Y}}}, \underline{\mathbf{H}} \rangle_{-1}, \tag{7.192}$$

$$\underline{\mathbf{H}} \leftarrow \underline{\mathbf{H}} \circledast \left(\underline{\mathbf{Y}} \times_1 \mathbf{A}^T\right) \oslash \left(\hat{\underline{\mathbf{Y}}} \times_1 \mathbf{A}^T\right). \tag{7.193}$$

Similarly, using the mode-2 matricization of the tensors gives us the learning rules for $\mathbf{B}$ and $\mathbf{L}_{(2)}$

$$\mathbf{B} \leftarrow \mathbf{B} \circledast \left(\mathbf{Y}_{(2)}\mathbf{L}_{(2)}^T\right) \oslash \left(\hat{\mathbf{Y}}_{(2)}\mathbf{L}_{(2)}^T\right), \tag{7.194}$$

$$\mathbf{L}_{(2)} \leftarrow \mathbf{L}_{(2)} \circledast \left(\mathbf{B}^T\mathbf{Y}_{(2)}\right) \oslash \left(\mathbf{B}^T\hat{\mathbf{Y}}_{(2)}\right) \tag{7.195}$$

and in the tensor form

$$\mathbf{B} \leftarrow \mathbf{B} \circledast \langle \underline{\mathbf{Y}}, \underline{\mathbf{L}} \rangle_{-2} \oslash \langle \hat{\underline{\mathbf{Y}}}, \underline{\mathbf{L}} \rangle_{-2}, \tag{7.196}$$

$$\underline{\mathbf{L}} \leftarrow \underline{\mathbf{L}} \circledast \left(\underline{\mathbf{Y}} \times_2 \mathbf{B}^T\right) \oslash \left(\hat{\underline{\mathbf{Y}}} \times_2 \mathbf{B}^T\right). \tag{7.197}$$

Finally, applying the mode-3 matricization to the tensors gives us the learning rules for $\mathbf{C}$ and $\mathbf{F}_{(3)}$

$$\mathbf{C} \leftarrow \mathbf{C} \circledast \langle \underline{\mathbf{Y}}, \underline{\mathbf{F}} \rangle_{-3} \oslash \langle \hat{\underline{\mathbf{Y}}}, \underline{\mathbf{F}} \rangle_{-3}, \tag{7.198}$$

$$\underline{\mathbf{F}} \leftarrow \underline{\mathbf{F}} \circledast \left(\underline{\mathbf{Y}} \times_3 \mathbf{C}^T\right) \oslash \left(\hat{\underline{\mathbf{Y}}} \times_3 \mathbf{C}^T\right). \tag{7.199}$$

In Appendix 7.H, we provide MATLAB code for the multiplicative NBOD1 algorithm based on the Frobenius cost function. In similar way, we can adapt a wide class of NMF algorithms discussed in previous chapters to the NBOD models.

7.6 Multi-level Nonnegative Tensor Decomposition - High Accuracy Compression and Approximation

Performance of NTD can be further improved by multi-stage hierarchical tensor decomposition. Wu *et al.* [85] recently proposed a hierarchical tensor approximation for multi-dimensional images using the standard Tucker decomposition based on approximations of all sub-blocks which are partitioned from the redundancies at each level. We here present a novel hierarchical multi-level scheme for NTD, which includes the following steps:

1. Approximate the given data tensor $\underline{\mathbf{Y}}$ by the level-1 tensor $\widehat{\underline{\mathbf{Y}}}_1 = \underline{\mathbf{G}} \times \{\mathbf{A}\}$.
2. Compute the residual error tensor $\underline{\mathbf{R}}_1 = \underline{\mathbf{Y}}$- $\widehat{\underline{\mathbf{Y}}}_1$, and divide it into two parts by threshold values set up by its most frequent values (defined by the mode function): $\underline{\mathbf{R}}_{1up} = \max(\underline{\mathbf{R}}_1, \text{mode}(\underline{\mathbf{R}}_1))$, $\underline{\mathbf{R}}_{1low} = \min(\underline{\mathbf{R}}_1, \text{mode}(\underline{\mathbf{R}}_1))$. Then, we normalize these two tensors $\underline{\mathbf{R}}_{1up}$ and $\underline{\mathbf{R}}_{1low}$ to unit scale [0, 1], and also invert $\underline{\mathbf{R}}_{1low} = \underline{\mathbf{1}} - \underline{\mathbf{R}}_{1low}$.
3. Decompose these two nonnegative residue tensors to get two new approximation tensors $\widehat{\underline{\mathbf{Y}}}_{1up}$ and $\widehat{\underline{\mathbf{Y}}}_{1low}$. Invert and scale these two tensors to the original ranges of their corresponding tensors $\underline{\mathbf{R}}_{1up}$ and $\underline{\mathbf{R}}_{1low}$.
4. Obtain the level-2 approximation tensor $\widehat{\underline{\mathbf{Y}}}_2$ and return to step 2 for the next level.

The residual tensor $\underline{\mathbf{R}}$ does not need to be split if we use the standard or semi-nonnegative Tucker decomposition. Multi-level decomposition allows much smaller errors and higher performance to be achieved. Figure 7.18 illustrates the approximated slices in the multi-level scheme illustrated in Figure 7.6. The accuracy of approximation increases gradually with the number of decomposition levels (Figures 7.18(b)–(h)). It should be noted that the approximated tensor obtained in the first layer is similar to the low-resolution data of its raw data. In the case of noisy data, to receive the high-resolution details, we must make a tradeoff between the level of detail and noise. Upon applying denoising to the residue tensors, NTD may become an

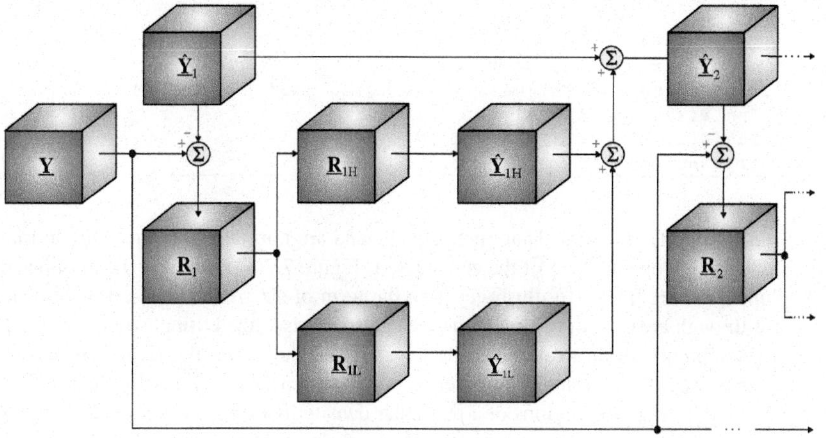

Figure 7.6 Hierarchical multi-level nonnegative tensor decomposition.

efficient tool for multi-way restoration and compression. In reconstruction/denoising applications, we take into account an approximation tensor $\underline{\widehat{\mathbf{Y}}}$, but for feature extraction applications, factors $\mathbf{A}^{(n)}$ and core tensor $\underline{\mathbf{G}}$ are analyzed. Another advantage of this scheme is that we can avoid decomposition using a large core tensor, since in hierarchical decomposition the dimension of the core tensor can be much smaller.

7.7 Simulations and Illustrative Examples

The performance of the algorithms introduced in this chapter are now illustrated with several case studies for various benchmarks and real-world data (see also [19,65,66]). For convenience, the case studies are explained through examples which reveal performance and convergence properties of the algorithms for both synthetic benchmarks and real-world data. Source code for the derived algorithms were written in MATLAB 7.5b and can be found in the Appendix. They require the use of the MATLAB Tensor Toolbox Version 2.2 (or higher) developed by Bader and Kolda [6–8].

7.7.1 Experiments for Nonnegative Tensor Factorizations

We next present illustrative examples for well controlled benchmarks, in order of increasing complexity, starting from a rank-one tensor decomposition to higher rank tensor factorization. These are accompanied with examples for basic sparse and smooth signals such as half-wave rectified sine, cosine, chirp and saw-tooth waveforms. Some real-world datasets are then discussed in detail and their potential applications are introduced.

Example 7.1 *Basic example for tensor construction and visualization*

This example provides several ways of visualizing a third-order data tensor, and its representation by component matrices. In this example the input third-order data tensor is rank-one, i.e., it is composed of three basic waveforms defined in the code listing.

Listing 7.1 Example 7.1 for basic sparse components.

```
freq = {10 3 2};
time2 = -.51:1/50:.51;
time1 = 0.01:1/50:.99;
time = { time2   time1   time1};
width = {.5 .6 .5};
signals = { 'chirp'  'square'  'cos' };
A = cellfun(@signalsuit,signals ,time ,freq ,width,'uni',0);
A = cellfun(@(x) x-min(x),A,'uni',0);
A = cellfun(@(x) x/norm(x),A,'uni',0);
Y = tensor(ktensor(A));
```

Factors $\mathbf{A}^{(n)} = \boldsymbol{a}^{(n)}$ $(J = 1,\ n = 1, 2, 3)$ are nonnegative and are normalized to unit-length. In MATLAB, factors $\mathbf{A}^{(n)}$ are the n-th element A{n} of the cell array A. Figure 7.7(a) visualizes the rank-one data tensor $\underline{\mathbf{Y}}$ as a 3D volume, and Figure 7.7(b) illustrates $\underline{\mathbf{Y}}$ in the form of six frontal slices. The zero elements are displayed as see-through areas. A third-order tensor can also be visualized using iso-surface rendering. The iso-surface is a set of points within a volume of space which have the same value (e.g., velocity, temperature, density); in other words, it represents a level set of continuous 3-D functions. In medical imaging, iso-surfaces may be used to represent regions of a particular density in a three-dimensional CT scan, allowing the visualization of internal organs, bones, or other structures. Iso-surfaces also tend to be a popular form of visualization for 3D datasets since they can be rendered by a simple polygonal model. Figure 7.7(c) shows

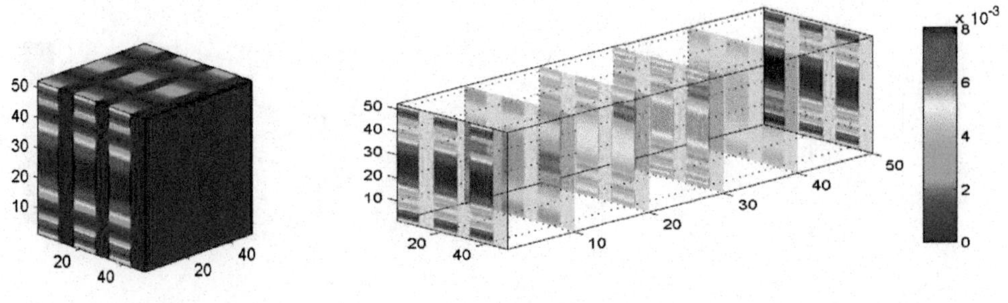

(a) 3D volume visualization of data tensor

(b) Six frontal slices of data tensor

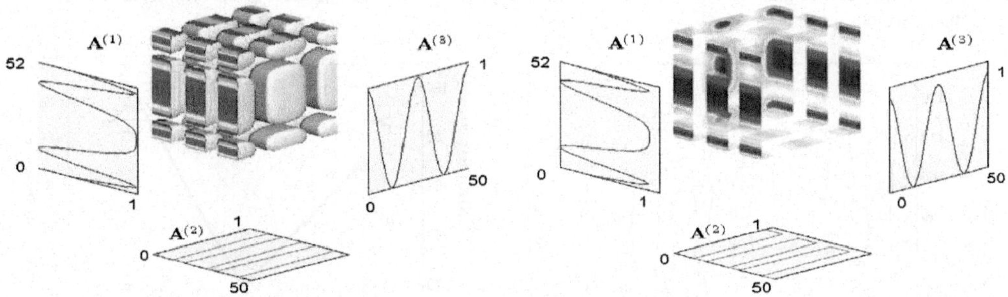

(c) Iso-surface visualization of the reconstructed data tensor with estimated factors

(d) Transparent visualization of the data reconstructed tensor with the estimated factors

Figure 7.7 Nonnegative Tensor Factorization of a rank-one third-order tensor from Example 7.1 employing different visualization modes.

the iso-surface of a data tensor with the iso-value of 0.001 and Figure 7.7(d) displays boundary slices of a tensor in a transparent (see-through) mode.

Example 7.2 *Factorization of a three-way symmetric tensor*

This example illustrates nonnegative tensor factorization for a third-order data tensor constructed from three identical factors composed of five components: three chirp waveforms (instantaneous frequencies: 0–5 Hz, 0–10 Hz, 0–7 Hz) and half-wave rectified cosine and sine waveforms (3Hz). A visualization of the original data tensor and its factors is presented in the transparent form in Figure 7.8(a), and in the frontal slice mode in Figure 7.8(b). Note that this is a symmetric tensor due to identical component matrices. Factorization of this data tensor by the Alpha and Beta NTF algorithms gives high performance. For instance, by choosing appropriate values of parameter $\alpha \in [0.2 - 2]$ we obtained almost perfect factorization (FIT of the model). Figure 7.8(c) provides a comparison of the performance index (SIR in dB) for different values of α for SIR > 70 dB. Figure 7.8(d) illustrates the dependence of the performance index (SIR) on three initialization techniques: HOOI, ALS and random mode, for the Fast HALS NTF algorithm. The MC (Monte Carlo) analysis was performed using 100 trials and the stopping criterion was the difference between the explained variations of two consecutive estimations with threshold value of 10^{-6}. For the noiseless tensor data, the HALS NTF algorithm with the ALS initialization usually exhibits the best performance (SIR $= 90.2$ dB) and requires the lowest number of iterations (168). The outliers (red cross-points) in Figure 7.8(d) show that NTF algorithms sometimes deliver much lower performance due to a poor initialization state.

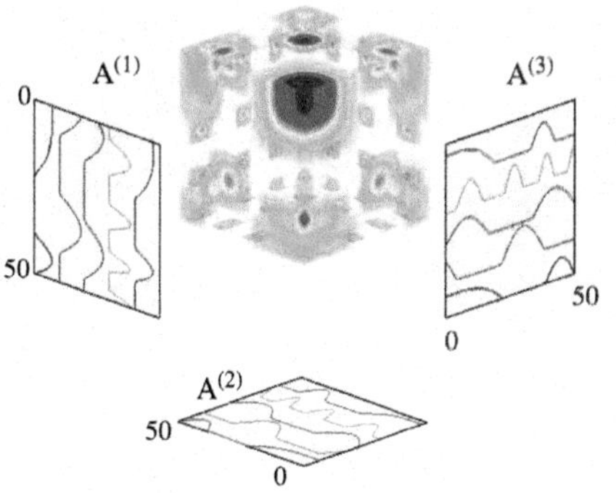

(a) 3D volume of original data tensor with corresponding input factors

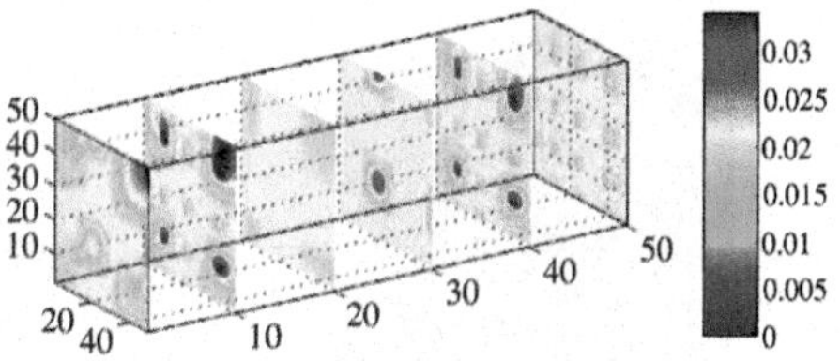

(b) Frontal slices of the original data tensor

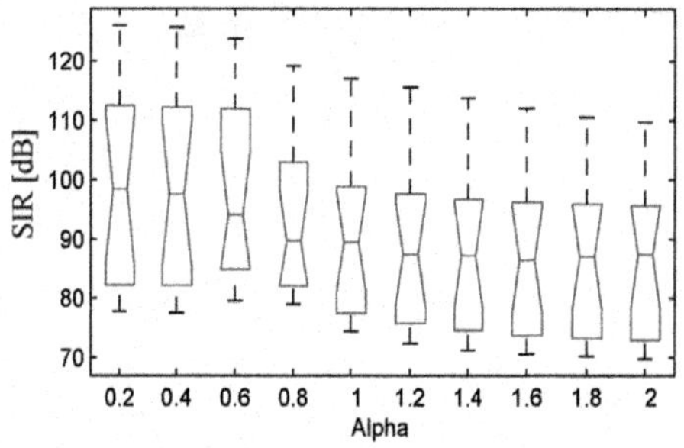

(c) Performance of the Alpha NTF Algorithm

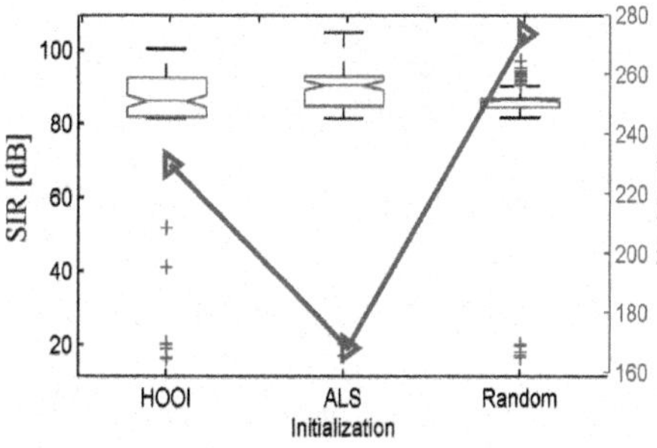

(d) Comparison of initialization techniques (SIR on the left vertical axis, No. iterations on the right vertical axis)

Figure 7.8 Example 7.2: The data tensor was constructed using five basis waveforms (all three factors are identical) (a) 3D visualization of the original tensor with the corresponding factors; (b) Frontal slices of the original tensor (c) SIR distributions obtained by Alpha NTF algorithm with $\alpha = [0.2 - 2]$; (reconstruction is almost perfect and estimated factors are almost identical to original factors, ignoring scaling and permutation ambiguities); (d) Performance comparison of Alpha NTF algorithm for three initialization techniques: HOOI, ALS and Random mode.

Listing 7.2 Example 7.2.

```
% Alpha NTF with alpha = 0.2:.2:2 using ALS initialization
randn('state',7256157);
Ainit = mat2cell(rand(sum(In),R),In,R);
alpha = .2:.2:2;
R = 5;
SIR = zeros(3*R,numel(alpha));

for k = 1:numel(alpha)
    options = struct('verbose',1,'tol',1e-6,'maxiters',500,'init',3,...
        'nonlinearproj',1,'alpha',alpha(k),'fixsign',0,'Ainit',{Ainit});
    [Y_alpha,Atemp,Ahat] = parafac_alpha(Y,R,options);
    SIR(:,k) = cell2mat(cellfun(@CalcSIR,A,Ahat,'uni',0));
end

figure
labels = mat2cell(sprintf('%.1f\n',alpha),1,4*ones(1,numel(alpha)))
boxplot(SIR,'label',labels,'notch','on')
xlabel('Alpha');ylabel('SIR [dB]');
```

Listing 7.3 Example 7.3 generates membrane tensor of 40 slices.

```
L0 = membrane(1,25); L0 = L0 - min(L0(:));
nlayers = 40;
Y = L0(:,:,ones(1,nlayers));
sc = reshape(1:nlayers,[1,1,nlayers]);
Y = bsxfun(@times,Y,sc);
```

Example 7.3 *Reconstruction of a noisy L-shaped membrane tensor*

The NTF was used to reduce noise in a 3D image. The third-order tensor $\underline{\mathbf{Y}} \in \mathbb{R}_+^{51\times 51\times 40}$ for which each slice was generated by the first Eigenfunction of the L-shaped membrane (which generates the MATLAB logo) $\underline{\mathbf{Y}}[:,:,q] = q*$`membrane(1, 25)`, $q = 1, \ldots, 40$ was corrupted by additive Gaussian noise with SNR = 10 dB (see Figure 7.9(a)). For the data tensor without additive noise (Figure 7.9(b)) the NTF model was able to explain 98.59% of the variation with $J = 4$ components, and 99.20% of the variation with $J = 5$ components. Figure 7.9(c) presents the estimated factors and the corresponding reconstructed data tensor obtained by the Fast HALS NTF algorithm for the data tensor without additive noise. In the next experiment the noisy data tensor (see Figure 7.9(a)) was approximated by the NTF model based on the Alpha and Beta HALS NTF algorithms with smoothness constraints, which gave the FIT value of 96.1% using only $J = 4$ components. Figure 7.9(d) displays the reconstructed data tensor for the estimated components by applying the Beta HALS NTF ($\beta = 2$) algorithm. The performance is also illustrated by the visualization of the reconstructed exemplary horizontal 40-th slice (see Figure 7.9(f)) where the original noisy slice is shown as a reference in Figure 7.9(e). In addition, the performance for different values of parameters α and β for the Alpha and Beta HALS NTF algorithms are illustrated in Figures 7.9(g) and 7.9(h) with PSNR in the left (blue) axis and number of iterations in the right (red) axis.

Example 7.4 *Large-scale tensor*

We constructed a large-scale data tensor (of size $500 \times 500 \times 500$) corrupted by additive white Gaussian noise with SNR = 0 dB by employing three basic original factors of 5 components (benchmarks: `X_spectra_sparse`, `ACPos24sparse10` and `X_spectra` [23], see Figure 7.10(a)). Figure 7.10 (b) presents the reconstructed data tensor with FIT = 99.99% (with respect to the original data tensor). The performance is illustrated via a 3D volume, iso-surface, and estimated factors, as shown in Figures 7.10(b)–7.10(f). The 10-th noisy slice of the raw noisy data tensor and its corresponding reconstructed slice are displayed in Figures 7.10(d) and 7.10(e), respectively.

Example 7.5 *Decomposition of amino acids fluorescence data*

We tested the Fast HALS NTF algorithm on real-world data: decomposition of amino acids fluorescence data (Figure 7.11(a)) composed from five samples containing tryptophan, phenylalanine and tyrosine (`claus.mat`) [4,15]. The data tensor was additionally corrupted by Gaussian noise with SNR = 0 dB (Figure 7.11(b)), and the factors were estimated with $J = 3$. The Beta HALS NTF algorithm had $\beta = 1.2$, and the Alpha HALS NTF algorithm had $\alpha = 0.9$, and both algorithms were run with the same number of iterations (100 times). The performances and running times are compared in Figure 7.11(g), and also in Table 7.3. Additionally, smoothness constraints were used for the Fast NTF, Alpha and Beta HALS NTF algorithms. Based on the FIT ratio and the PSNR index observe that the HALS algorithms usually exhibited better performance than standard NTF algorithms. For example, the first recovered slice (Figure 7.11(c)) was almost identical to the corresponding slice of the clean original data tensor (with 99.51% Fit value). In comparison, the NMWF, lsNTF, ALS_K, ALS_B algorithms produced artifacts as illustrated in Figure 7.11(d). Some other aspects of the performance are illustrated more intuitively in Figures 7.11(e) and 7.11(f). Additional examples for 4-D fluorescence data are given in Chapter 8.

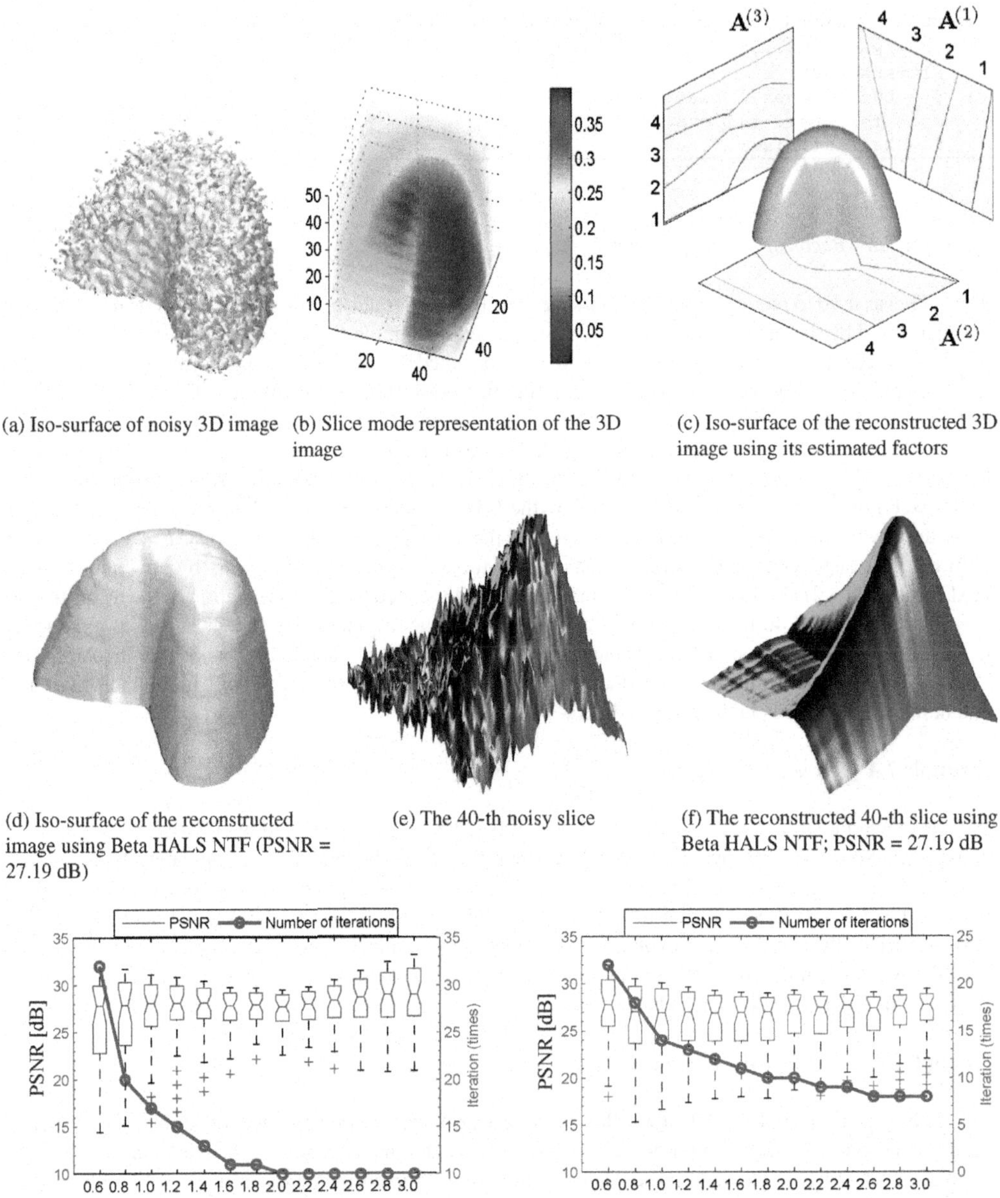

(a) Iso-surface of noisy 3D image (b) Slice mode representation of the 3D image (c) Iso-surface of the reconstructed 3D image using its estimated factors

(d) Iso-surface of the reconstructed image using Beta HALS NTF (PSNR = 27.19 dB) (e) The 40-th noisy slice (f) The reconstructed 40-th slice using Beta HALS NTF; PSNR = 27.19 dB

(g) Alpha HALS sNTF with smoothness constraints (h) Beta HALS sNTF with smoothness constraints

Figure 7.9 Reconstruction of noisy 3D image for the membrane tensor $\underline{\mathbf{Y}} \in \mathbb{R}_+^{51\times 51\times 40}$ from Example 7.3: (a)–(b) iso-surface and slice mode visualizations of the noisy and "clean" data tensors, (c) estimated factors and the data tensor by the Fast HALS-NTF for the "clean" data tensor (without additive noise), (d) iso-surface visualization of the reconstructed tensor by using Beta-HALS-NTF algorithm; (e)–(f) surface visualizations of the 40-th slices of the noisy data tensor and reconstructed tensor by Beta ($\beta = 2$) HALS NTF algorithm, respectively; (g)–(h) Performance of multiplicative Alpha NTF and Beta NTF algorithms with smoothness constraints.

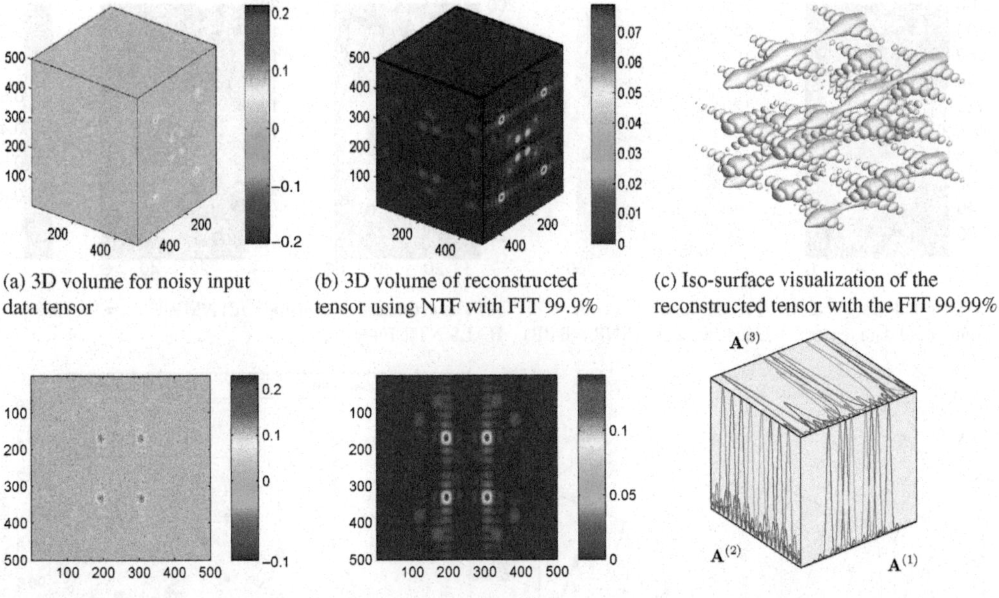

(a) 3D volume for noisy input data tensor

(b) 3D volume of reconstructed tensor using NTF with FIT 99.9%

(c) Iso-surface visualization of the reconstructed tensor with the FIT 99.99%

(d) Original 10-th slice of the noisy data tensor

(e) 10-th reconstructed slice using HALS NTF algorithm

(f) Estimated factors using HALS NTF algorithm

Figure 7.10 NTF for a large-scale data tensor $\underline{\mathbf{Y}} \in \mathbb{R}_+^{500\times 500\times 500}$ degraded by additive Gaussian noise with SNR $= 0$ dB in Example 7.4.

Example 7.6 *Feature extraction for the CBCL face database*

We performed feature extraction for the CBCL face database. The tensor was formed using the first 100 face images of dimension 19×19 and then factorized in two ways, by using 49 components and 100 components. The Beta HALS NTF algorithm was selected with $\beta = 1$ and was compared with the NMWF and the lsNTF algorithms. In the case with 100 components, the reconstructed tensors explained 98.24 %, 97.83 % and 74.47% of the variation of the original data tensor for the Beta HALS NTF, NMWF and lsNTF algorithms, respectively (see Table 7.3). Note that the estimated components obtained by using the Beta HALS NTF (Figure 7.12(d)) are relatively sparse and the reconstructed face images are very similar to the original images (see Figure 7.12(c)). The montage plot of 49 basis components is given in Figure 7.12(d).

Example 7.7 *NTF-1 with orthogonal components*

We consider the NTF1 model with eight nonnegative sources $\mathbf{X}_q$, $(q = 1, 2, \ldots, 8)$ obtained via the ALS algorithm. Each source $\mathbf{X}_q$ has $J = 5$ orthogonal components generated by the function `nngorth.m` (see Chapter 4). Matrices $\mathbf{A} \in \mathbb{R}_+^{20\times 5}$ and $\mathbf{C} \in \mathbb{R}_+^{8\times 5}$ are randomly generated nonnegative matrices. The following MATLAB code (Listing 7.4) illustrates how to generate the data tensor $\underline{\mathbf{Y}}$ and decompose it under two layers. In Figure 7.13, we compare two orthogonal matrices $\mathbf{G}$ formed by eight separated matrices $\mathbf{G}_q$ for two processes: decomposition under one single layer with 1000 iterations; decomposition under two layers: 300 iterations for layer one, and 200 iterations for layer two in which we apply orthogonal projection for $\mathbf{X}_q$. Each blob at a specified position (i, j) in the Hinton diagrams (Figures 7.13(a) and 7.13(b)) has its size proportional to the orthogonal index $\mathbf{G}_{ij}$. It is obvious and intuitive to see that the process with two layers and additional an orthogonality constraint returns much better performance.

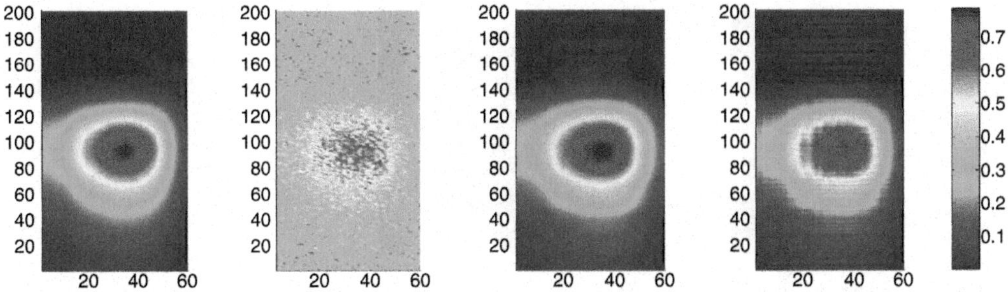

(a) Original first slice of amino acid data tensor (b) The same slice with Gaussian noise (SNR = 0 dB) (c) The estimated slice using HALS NTF, Fit = 99.51% (d) NMWF, Fit = 98.76%

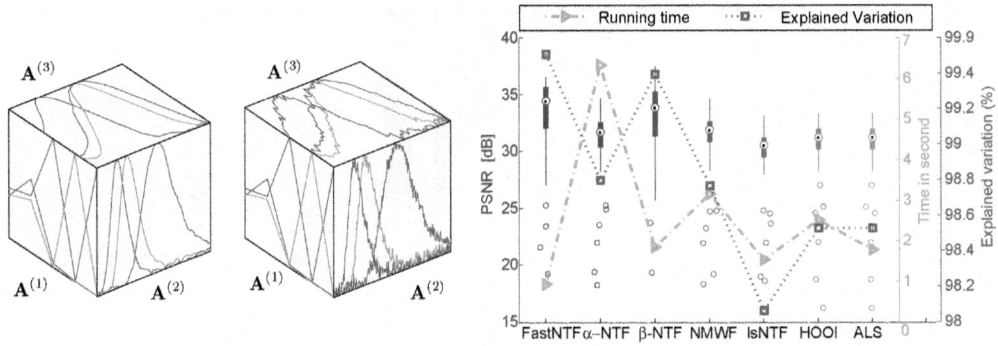

(e) Estimated factors using HALS NTF algorithm (f) The estimated factors using NMWF algorithm (g) PSNR, running time and Fit for various algorithms

Figure 7.11 Comparison of performance for various algorithms in Example 7.5. Illustration of estimated factors by the FAST HALS NTF in comparison with the multiplicative NMWF algorithm for three-way tensor factorization of amino acid data. (a)–(b) the first slices of the original amino acid tensor and the same tensor corrupted by large Gaussian noise, (c)–(d) the reconstructed slice using the HALS NTF and NMWF algorithms, (e)–(f) three estimated factors using the HALS NTF and NMWF algorithms (the estimated factors should be as smooth as possible), (g) comparison of distributions of PSNR (dB) for all slices, CPU time (seconds) and the explained variation (FIT %).

In summary, our extensive simulations confirmed that the NTF algorithms presented in this chapter gave consistently high performance results. In particular, we have confirmed that the FAST HALS NTF, Alpha HALS NTF and Beta HALS NTF algorithms are more robust to noise and generally exhibit better performance and provide faster convergence than many other recently developed NMF/NTF algorithms.

Although we have tested NTF algorithms mostly for approximations and on the denoising of benchmarks, they can also be applied in quite different scenarios where underlying hidden components and their features and structures are of interest, as illustrated in the following section.

7.7.2 Experiments for Nonnegative Tucker Decomposition

We shall next consider a simple example which involves sparse nonnegative sources, in order to demonstrate a successful and unique decomposition based on the NTD model for sparse and non-overlapping nonnegative components.

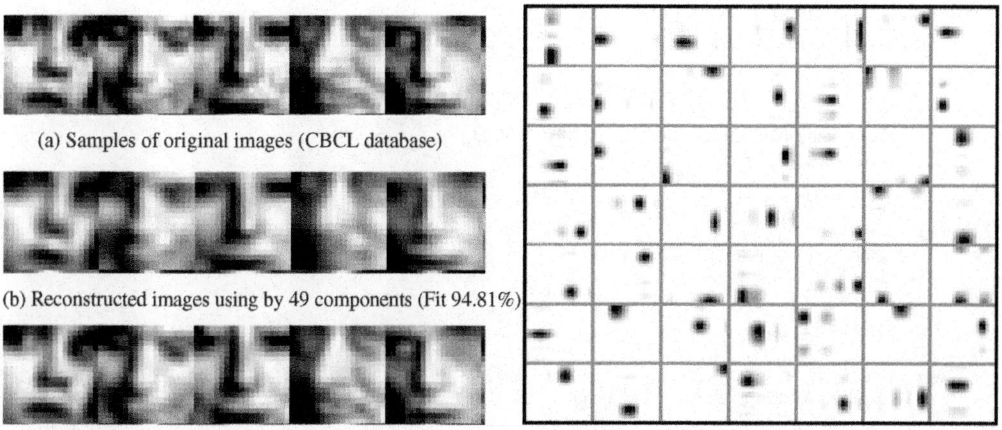

(a) Samples of original images (CBCL database)

(b) Reconstructed images using by 49 components (Fit 94.81%)

(c) Reconstructed images by 100 components (Fit 98.24%)

(d) 49 basis components estimated by Beta HALS NTF algorithm, (FIT 94.95%)

Figure 7.12 Illustration of NTF for 100 CBCL face images in Example 7.6 for 49 and 100 basis components based on the Beta HALS NTF algorithm.

Example 7.8 *NTD with the benchmark* `ACPos24sparse10`

This example considers a third-order data tensor $\underline{\mathbf{Y}}$ built up from nonnegative smooth components (the benchmark `ACPos24sparse10` used also in the previous examples). However, in this case we used a core tensor $\underline{\mathbf{G}} \in \mathbb{R}^{4\times3\times4}$ drawn from a uniform random distribution (see Figure 7.14(a) for visualization of factors and core tensor). The NTD algorithms presented in this chapter (Alpha NTD, Beta NTD and HALS NTD) successfully returned the factors which explained nearly 100 percent of the variation (FIT) of the original tensor, with high SIR performance. Listing 7.5 gives the MATLAB code illustrating how to construct this data tensor.

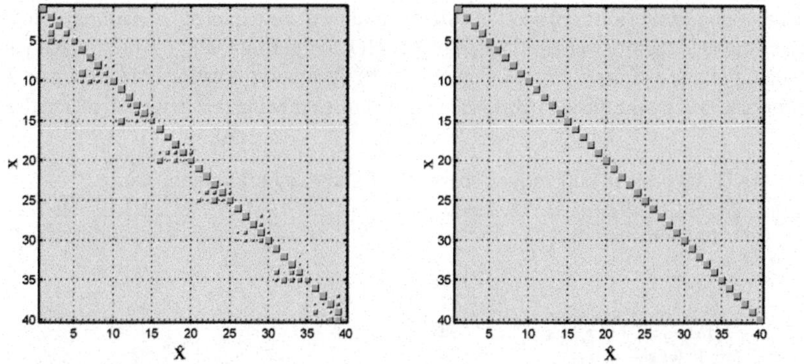

(a) 1000 iterations

(b) Layer 1: 300 iterations + Layer 2: 200 iterations

Figure 7.13 Illustration for performance of ALS NTF1. Eight matrices $\mathbf{X}_q$ contain 40 components, and form eight orthogonal matrices $\mathbf{G}_q$. Hinton diagrams of two common orthogonal matrices $\mathbf{G}$ for two cases of decomposition: (a) ALS NTF1 under 1000 iterations; (b) ALS NTF1 with orthogonal transform for layer 2 under 300 iterations (layer 1) + 200 iterations (layer 2). The blob size is proportional to the orthogonal value.

Listing 7.4 Example 7.7 for NTF-1.

```
% Example 7.7 for NTF-1 model
% with 8 orthogonal factors X
% A and C are random nonnegative matrices
rand('twister',210052.2)
T = [100,200,400,500, 300 200 500 1000];
Q = numel(T); I = 20; J = 5;

X = cell(1,Q);
for k = 1:Q
    X{k} = nngorth(J,T(k),0.5);
end
A = rand(I,J);
C = rand(Q,J);

for q = 1:Q
    Y{q} = A*diag(C(q,:))*X{q};
end
%% Layer-1: ALS NTF-1 with output variable Fact = {AH CH XH}
opts = struct('maxiters',300);
Fact = ntf1_als(Y,J,opts);
%% Layer-2: ALS NTF-1 with orthogonal constraints
% and initialized by the previous result
opts = struct('orthoforce',1,'initialize',0,'maxiters',200);
Fact = ntf1_als(Y,J,opts,Fact);
%% Get XH and compute performance
[G,mapind] = cellfun(@nmf_G,Fact{3},X,'uni',0);
[c,r] = meshgrid(1:J*Q,1:J);
r = bsxfun(@plus,r,reshape(repmat(0:J:J*Q-1,J,1),1,[]));
idx2 = sub2ind([J*Q J*Q],r,c);
Gm = zeros(J*Q); Gm(idx2) = [G{:}];
figure;hinton(Gm)
xlabel('$\bf \hat X$','interpreter','latex')
ylabel('$\bf X$','interpreter','latex')
```

The components (columns of factor matrices) were estimated by imposing additional orthogonality constraints and the stopping criterion used was the difference value of the explained variations (with the threshold of 10^{-6}). The estimated factors were initialized by the HOOI algorithm, and the orthogonal parameter λ_{ort} was set to 0.05 for all the estimated factors. Figure 7.14(b) illustrates the estimated component matrices and the core tensor obtained by the basic HALS NTD algorithm after being rearranged to match the component

Listing 7.5 MATLAB code to build a 3-rd sparse Tucker tensor for Example 7.8.

```
filename = 'ACPos24sparse10_norm';
data = load(filename,'S');
R = [ 4 3 4];
% Generate factors
for k = 1: 3
    temp = data.S(2:1+R(k),1:120)fl;
    for l = 1:R(k)
        A{k}(:,l) = decimate(temp(:,l),3);
    end
    A{k} = bsxfun(@rdivide,max(eps,A{k}),sqrt(sum(A{k}.^2))); %normalize
end
% Generate core tensor G
randn('state',7256267);
G = tensor(rand(R));
Y = ttm(G,A); %build tensor Y
```

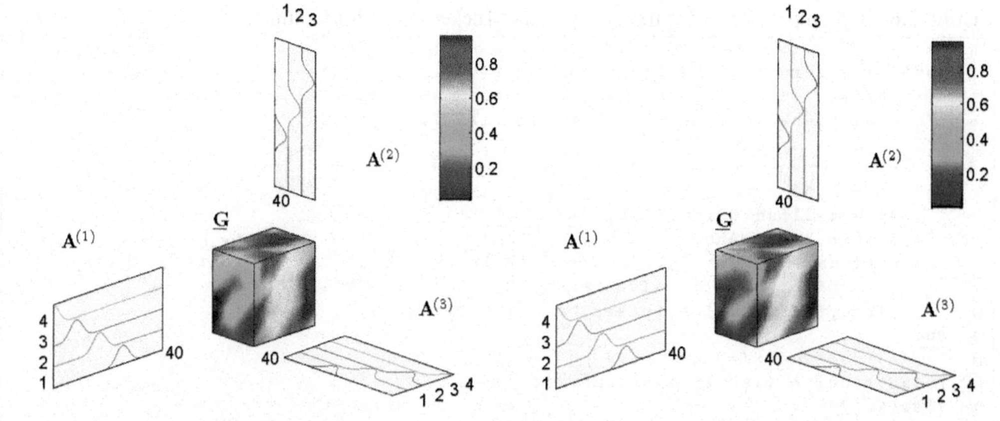

(a) Original core tensor and factors

(b) Estimated factors with core tensor using the ℓ_2 HALS NTF algorithm for the data tensor without noise

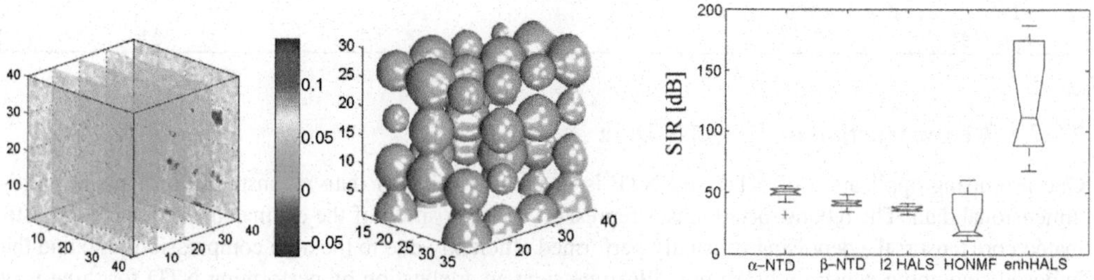

(c) Frontal slices of the noisy data tensor with SNR = 0 dB

(d) The reconstructed tensor for noisy data with PSNR = 85.49 dB

(e) SIR distributions for the reconstructed tensors for various Tucker decomposition algorithms

Figure 7.14 Illustration of NTD for a data tensor with sparse factors (with and without noise) for Examples 7.8 and 7.9: (a)–(b) original and estimated factors and core tensors; (c) frontal slices of the noisy tensor with SNR $= 0\,$dB; (d) reconstructed tensor by ℓ_2 HALS NTF algorithm (scaled to the unit ℓ_2-norm) with PSNR $= 85.49\,$dB; (e) SIR distributions obtained by Alpha, Beta, ℓ_2 HALS and enhanced HALS NTD algorithms for the data tensor without noise.

order of the original factors. Next, we performed nonnegative Tucker decomposition using the multi-layer model (referred to as enhanced HALS or briefly enhHALS). The estimated core tensor $\underline{\mathbf{G}}$ at each layer was adjusted by the product of the observed tensor and the pseudo-inverse factors $\mathbf{A}^{(n)\dagger}$ as

$$\underline{\mathbf{G}} = \underline{\mathbf{Y}} \times_1 \mathbf{A}^{(1)\dagger} \times_2 \mathbf{A}^{(2)\dagger} \cdots \times_N \mathbf{A}^{(N)\dagger} . \tag{7.200}$$

In each layer, the new core tensor and the estimated factors were used for the initialization of the succeeding layer. Using this model, we were able to improve the performance of nonnegative Tucker decomposition significantly (with SIR $\approx 111.23\,$dB), whereas using the standard NTD approach, we achieved the performance of $50.21\,$dB, $41.19\,$dB, $35.98\,$dB and $17.99\,$dB for the Alpha NTD, Beta NTD, ℓ_2 HALS NTD and the HONMF [61], respectively. Figure 7.14(e) illustrates this comparison.

The MATLAB code for this multi-layer model is given in Listing 7.6.

Listing 7.6 Decomposition of the three-way sparse Tucker tensor for Example 7.8.

```
%% Decompose tensor in Example 7.8 by ℓ-2 HALS NTD
options = struct('tol',1e-6,'maxiters',1000,...
    'init','eigs','orthoforce',0,'lda_ortho',0.05,'ellnorm',2);
[Yhat,Ahat,Ghat,fit,iter] = tucker_hals(Y,R,options);

%% Enhance results by hierarchical scheme (enhHALS)
for ii = 1:10
    Ap = cellfun(@pinv,Ahat,'uni',0);
    Ghat = ttm(Y,Ahat); %equation 7.200
    options = struct('tol',1e-6,'maxiters',1000,'init',{{Ahat{:} Ghat}},...
        'orthoforce',0,'lda_ortho',0.001,'ellnorm',2);
    [Yhat,Ahat,Ghat] = tucker_hals(Y,R,options);
end
%% Iso-surface visualization
[hcap,hiso] = visualize3D(That,.1,[]);
view(45,20)
set(hiso,'AmbientStrength',.5,'DiffuseStrength',.6)

%% Visualization of factors and core tensor
figure
visualize3D_core_fac(G,A)
view(45,20)
```

7.7.2.1 Reconstruction of 3D Noisy Data

One promising application of NTD and NTF is in the denoising or data reconstruction of noisy multi-dimensional data. The reconstructed tensor represents a composition of the estimated factors and the estimated core tensor; the denoising is usually performed efficiently due to the high compression ratio and the imposed smoothing constraints. We now illustrate such an application by performing NTD for three-way synthetic and real-world data tensors degraded by large additive Gaussian noise (with SNR = [-10-0] dB). The simulation results will be compared with the HOOI algorithm [8,33], the ALS algorithm of Anderson and Bro [4] and HONMF [61].

Example 7.9 *Noisy tensor degraded by additive Gaussian noise with SNR = 0 dB*

We used the same real-world data tensor as in **Example 7.8** but degraded by additive Gaussian noise with SNR = 0 dB. The frontal slices of this noisy data tensor are displayed in Figure 7.14(c). We performed NTD for the data tensor by employing the ℓ_2 HALS NTD algorithm (scaled to the unit length of ℓ_2-norm). The reconstructed tensor is obtained by composing the estimated factors and core tensor, almost perfect denoising was achieved with PSNR = 85.49 dB, as illustrated in Figure 7.14(d) (iso-surface visualization of the reconstructed tensor). The MATLAB code is given in Listing 7.7.

Example 7.10 *Noisy tensor degraded by additive Gaussian noise with SNR = −10 dB*

Listing 7.7 Example 7.9 for decomposition of noisy tensor with Gaussian noise SNR = 0 dB.

```
Yori = Y;
Y = Y + randn(size(Y))*sqrt(var(Y(:)))*1/10^(0/20);

% Reconstruct - Denoising
options = struct('tol',1e-6,'maxiters',500,...
    'init','eigs','orthoforce',0,'lda_ortho',[0.3 .1 .1],'ellnorm',2);
[Yhat,Ahat,Ghat] = tucker_hals(Y,R,options);
PSNR = PSNRtensor(Yori,Yhat);
```

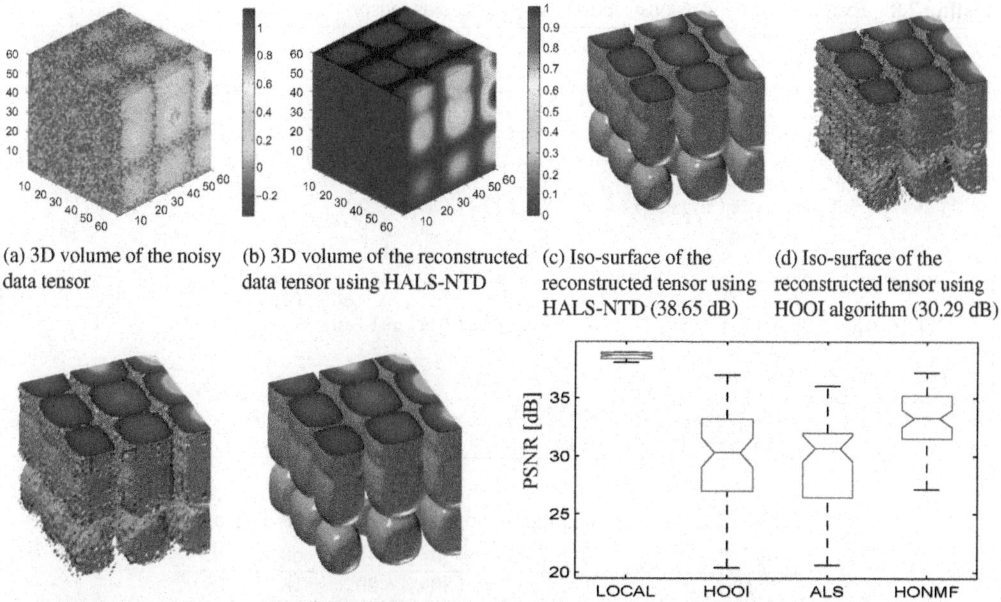

(a) 3D volume of the noisy data tensor

(b) 3D volume of the reconstructed data tensor using HALS-NTD

(c) Iso-surface of the reconstructed tensor using HALS-NTD (38.65 dB)

(d) Iso-surface of the reconstructed tensor using HOOI algorithm (30.29 dB)

(e) Iso-surface of the reconstructed tensor using ALS (30.66 dB)

(f) Iso-surface of the reconstructed tensor using the HONMF (33.24 dB)

(g) PSNR performance of various Tucker decomposition algorithms

Figure 7.15 Visualization of performance of various Tucker decomposition algorithms in Example 7.10 for data tensor $\underline{\mathbf{Y}} \in \mathbb{R}_+^{60\times60\times60}$ corrupted by large additive Gaussian noise with SNR $= -10$ dB in both volume and the iso-surface visualizations.

A third-order tensor $\underline{\mathbf{Y}} \in \mathbb{R}^{60\times60\times60}$, corrupted by large Gaussian noise to give SNR $= -10$ dB, was constructed using three benchmarks: `X_spectra_sparse.mat`, `ACPos24sparse10` and `X_spectra` [24], and a randomly generated core tensor $\underline{\mathbf{G}} \in \mathbb{R}^{4\times5\times4}$ (see Figure 7.15). The local HALS NTD algorithm provided the best reconstruction of the original data tensor.

Example 7.11 *Amino acids fluorescence data degraded by additive Gaussian noise with SNR = 0 dB*

In this example we considered the same real-world data as in **Example 7.5**. Instead of NTF, the NTD model with a core tensor size of $4 \times 3 \times 3$ was applied to the observed data tensor $\underline{\mathbf{Y}} \in \mathbb{R}^{5\times201\times61}$ corrupted by Gaussian noise with SNR = 0 dB. In order to reconstruct the original data tensor we imposed the smoothness constraint on the estimated factors $\mathbf{A}^{(2)}$ and $\mathbf{A}^{(3)}$. The simulation results presented in Figure 7.16(c) were obtained using the ℓ_2 HALS NTD algorithm. The estimated factors and the core tensor are visualized in Figure 7.16(d). Next, tensor decompositions using the HONMF and standard ALS algorithms are illustrated in Figures 7.16(e) and 7.16(f). Again, the additional constraints, such as orthogonality or smoothness, help to improve the performance considerably.

7.7.2.2 Multi-Way Data Compression

As has been shown earlier, the Tucker decomposition allows us to dramatically reduce the number of the required data samples to describe real world data, and has potential application in multi-dimensional data compression and sparse representation, especially for 3D medical images. Instead of storing the full data tensor, it can be approximately represented by the factors (component matrices) and the core tensor; these

Listing 7.8 Example 7.11 for reconstruction the amino acids tensor.

```
clear
load('claus.mat')
R = [4 3 3];
In = size(Y);
Y = Y- min(Y(:)); Y(isnan(Y)) = 0; Y = tensor(Y);

% Degraded raw data by Gaussian noise with SNR = 0dB
Y = Y + randn(size(Y))*sqrt(var(Y(:)))*1/10^(0/20);

% Decompose and reconstruct data by ell_2 HALS
options = struct('tol',1e-6,'maxiters',500,'init','eigs',...
    'orthoforce',1,'ellnorm',2,'lda_smooth',[0 4000 1000]);
[Yhat,Ahat,Ghat,fit,iter] = tucker_hals(Y,R,options);

% Visualize tensor in slice mode
visopt.trans = 1;
visualize3D_slice(Yhat,1:5,[],[],visopt);
view(45,20); daspect([1,.7,.05])
```

Table 7.2 Viewing and illumination angles of the BTF database.

$\theta[^\circ]$	0	15	30	45	60	75
$\Delta\phi[^\circ]$		60	30	20	18	15
No. of images	1	6	12	18	20	24

can then be stored. In practice, the NTD decomposition is not always perfect (lossy compression), however the resulting compact approximate representation is still efficient and very promising. In other words, for real-world data, 100% FIT is hard to achieve due to noise and a limited number of components. However, we can improve this ratio gradually by using a multi-layer scheme. In this approach we decompose the residual tensor until the desired FIT value is satisfied [65]. This is illustrated by the following example.

Example 7.12 *The* `Ceiling` *textures*

The input data tensor consists of 16 selected `Ceiling` textures[13] [70,78] of size 128 × 128 (see Figures 7.17(a) and 7.17(b)). The texture images were captured at different viewing and illumination angles (θ_v, ϕ_v) and (θ_i, ϕ_i). For a fixed viewing angle, there are total 81 images taken at light positions as shown in Table 7.2. For this experiment, we selected 16 images [5, 6, 7, 13, 14, 15, 24, 25, 26, 45, 46, 47, 65, 66, 67, 68] taken at the view angle ($\theta_v = 0, \phi = 0$). Each frontal slice of the data tensor represents a top-left sub-image of the original image, scaled to be half-size. We decomposed this data tensor by applying a core tensor of dimension $\underline{\mathbf{G}} \in \mathbb{R}^{20\times20\times16}$, and the HALS NTD algorithm reached the FIT value of 97.2738% and the median PSNR = 30.44 dB. Figure 7.17(c) illustrates the reconstruction of the original data (see Figure 7.17(b)). Note that the total number of entries in estimated factors and core tensor was: $128 \times 20 + 128 \times 20 + 16 \times 16 + 20 \times 20 \times 16 = 11776$ with a FIT value of 97.2738%, while the original data tensor consists of 262144 entries. In this case, we achieved a compression ratio of 22.2609:1 (262.144/11.776) or a space savings of 95.51% (= 1 − 11.776/262.144).

For the noisy data in **Example 7.11** we were able to perform reconstruction (denoising and compression simultaneously in a single processing step). Only $5 \times 4 + 201 \times 3 + 61 \times 3 + 4 \times 3 \times 3 = 842$ samples were needed to represent the original tensor, instead of $5 \times 201 \times 61 = 61305$ samples, giving a compression ratio of 72.81 : 1, and space savings of 98.63%. Note that the factors estimated by the NTD model are often

[13] The `Ceiling` database contains 6561 images with 800 × 800 pixels; our example is simplified.

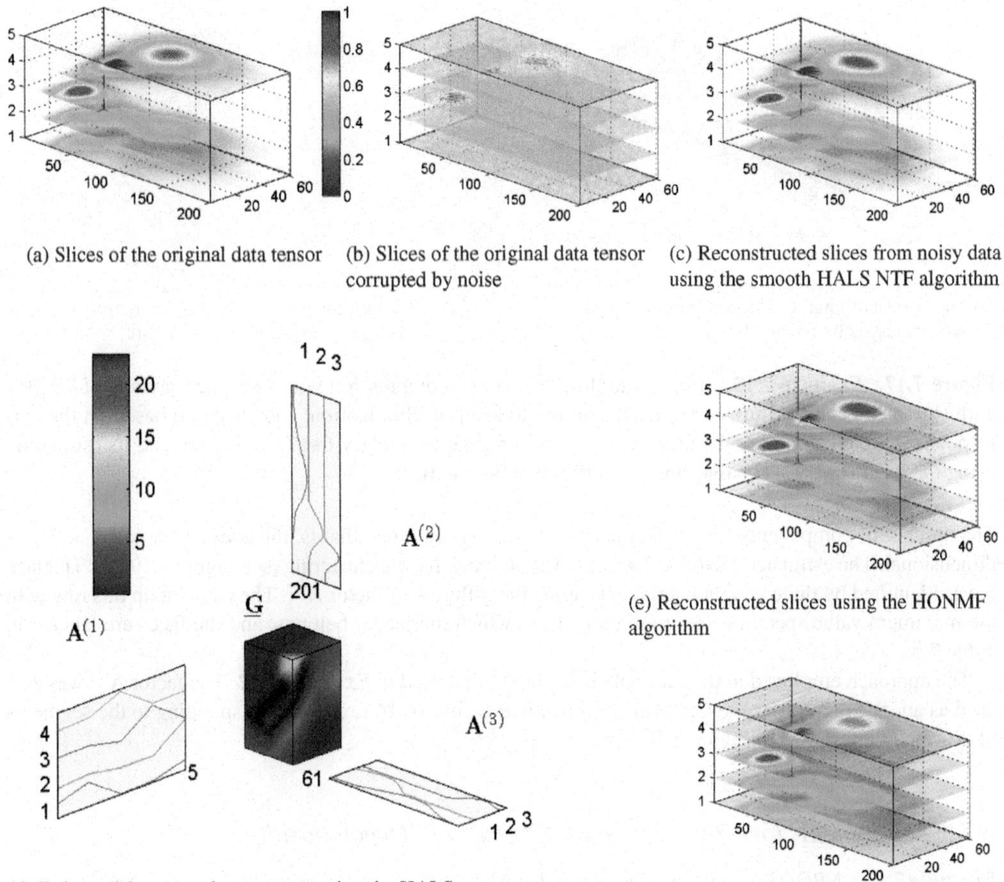

(a) Slices of the original data tensor

(b) Slices of the original data tensor corrupted by noise

(c) Reconstructed slices from noisy data using the smooth HALS NTF algorithm

(d) Estimated factors and core tensor using the HALS NTD algorithm

(e) Reconstructed slices using the HONMF algorithm

(f) Reconstructed slices using the ALS algorithm

Figure 7.16 Example 7.11: Tensor reconstruction for the real-world data `clauss.mat` represented by a $5 \times 201 \times 61$ dimensional data tensor corrupted by Gaussian noise with SNR = 0 dB. The reconstructed tensor is enforced to be smooth due to the smoothness constraints imposed on the estimated factors. Slice mode visualizations use the normalized color-bar given in (a).

sparse, and the actual number of samples to be stored can be even smaller than the total number of all the entries in the factors and the core tensor. For example, for the estimated factors and the core tensor, the number of near-zero elements (of value less than 10^{-4}) is six for $\mathbf{A}^{(1)}$, 283 for $\mathbf{A}^{(2)}$, 75 for $\mathbf{A}^{(3)}$ and 51 for the core $\underline{\mathbf{G}}$. Therefore, the total number of nonzero elements is only $842 - 415 = 427$, giving a revised space savings of 99.30%.

7.7.2.3 Multi-Way Data Clustering

Example 7.13 *The ORL face database*

This example considers a dataset with 30 face images of three subjects: 22, 33 and 34 selected from the ORL face database [68], organized as a third-order data tensor for which the frontal slices are faces (Figure 7.19(a)).

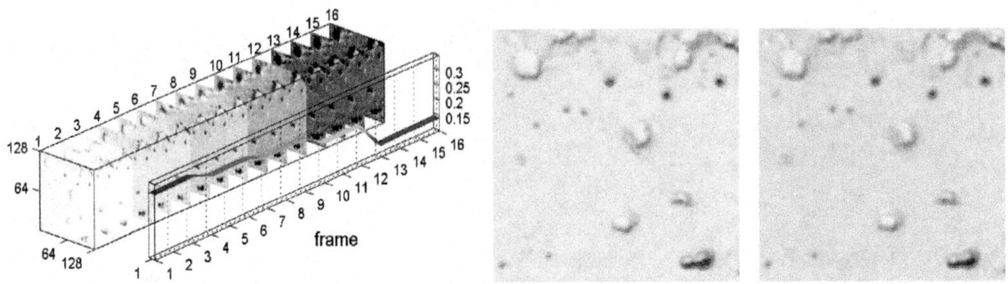

(a) The sixteen original `Ceiling` textures organised in five clusters by factor $\mathbf{A}^{(3)}$ (b) `Ceiling` texture (c) Approximation at 30.44 dB

Figure 7.17 Example 7.12 illustrates multi-dimensional compression with the space savings of 95.51% with (a) 16 **`Ceiling`** textures arranged according to level of illumination, and clustered based on the corresponding weights on the 3-rd factor $\mathbf{A}^{(3)}$; (b)-(c) original exemplary **`Ceiling`** texture and its estimation using the local HALS NTD algorithm with PSNR = 30.44 dB.

The number of components for all the factors was set to only three, that is, the core tensor was $3 \times 3 \times 3$ dimensional. The estimated factor $\mathbf{A}^{(3)}$ was used as an index for the clustering (see Figure 7.19(b)). The face q was identified by three weight coefficients along the q-th row of factor $\mathbf{A}^{(3)}$. The element on this row with the maximum value specifies the group (person) to which the face q belongs, and the faces are labeled in Table 7.5.

The approach employed in this example is similar to that used in **Example 7.12**. The factor $\mathbf{A}^{(3)}$ was also used as an index to cluster the `Ceiling` textures; five groups of 16 textures corresponding to the segments of this factor are shown in Figure 7.17(a).

7.7.3 Experiments for Nonnegative Block-Oriented Decomposition

Example 7.14 *NBOD1 with the* `swimmer` *benchmark*

Using the NBOD model the `swimmer` dataset was decomposed into 15 frontal components and 1 lateral component. It means that in this case $R_1 = 0$, $R_2 = 1$ and $R_3 = 15$. We imposed additional sparsity and

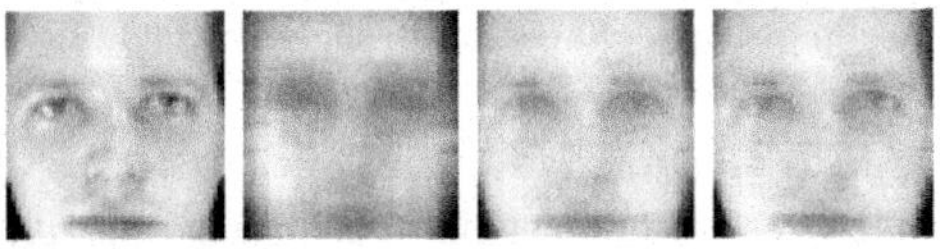

(a) Raw data (b) level-1 21.63 dB (c) level-3 27.48 dB (d) level-4 29.45 dB

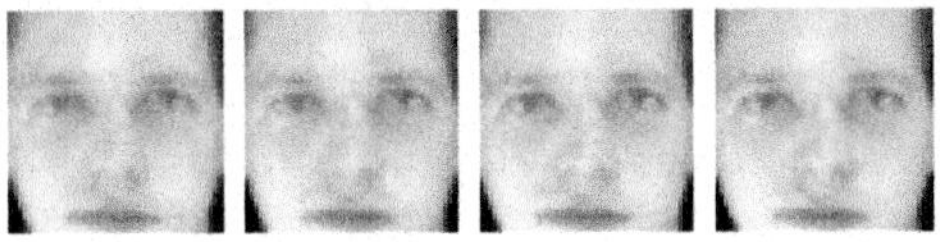

(e) level-5 31.63 dB (f) level-6 33.23 dB (g) level-7 34.57 dB (h) level-8 35.97 dB

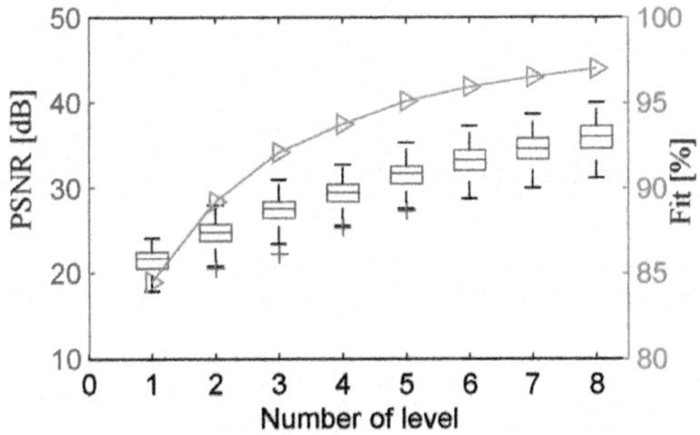

(i) Increase of PSNR and Fit indices over layer

Figure 7.18 Gradual improvement of the reconstruction of face images by applying a multi-level hierarchical NTD decomposition based on the local HALS-NTD algorithm.

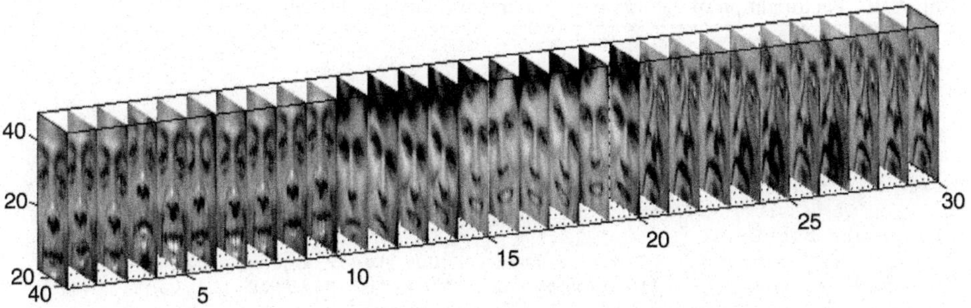

(a) Selected 30 face images for ORL database of three different persons

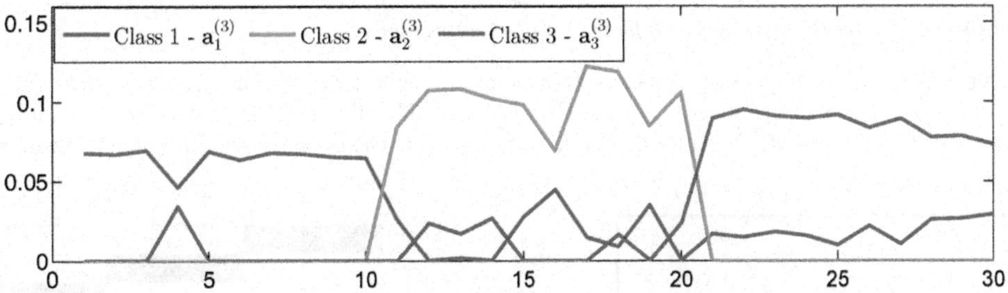

(b) Plot of three components of the factor $\mathbf{A}^{(3)}$

Figure 7.19 Clustering of 30 face images for three persons (ORL database) in Example 7.13 using the Beta HALS NTD algorithm. The estimated factor $\mathbf{A}^{(3)}$ is used as an index for clustering (discrimination of) the faces.

orthogonality constraints on basis frontal slices $\underline{\mathbf{F}}$, and basis lateral slices $\underline{\mathbf{L}}$. MATLAB code for this example is given in Listing 7.10, whereas the decomposition results are illustrated in Figure 7.20. The core tensor $\underline{\mathbf{F}}$ explains 15 basis pictures including the "torso" part. Two "limb" components are explained by one basis lateral picture shown in Figure 7.20(c). Figure 7.20(d) indicates the coding values of this basis picture in all 32 lateral slices of the data. Note that the basis lateral picture appears in lateral slices 6,7, ... ,10.

Listing 7.9 Example 7.13: classification for the ORL faces.

```
data = load('orl_48_48_non_scale.mat');
ind = [ 22 33 34 ];
indx = bsxfun(@plus,10*(ind-1)',1:10);
Y= reshape(data.V(:,indx),48,48,[]);
Y = Y- min(Y(:))+eps; Y = Y/max(Y(:)); Y = tensor(Y);
% Decompose with Beta NTD
R = [3 3 3];
options = struct('tol',1e-6,'maxiters',1000,'alpha',1,...
    'init','eigs','nonlinearproj',1,'projector','max',...
    'orthoforce',1,'updateYt',0,'ellnorm',1,'Galpha',1);
[Yhat,A,G]  = tucker_beta_fb(Y,R,options);
% Display factor A^{(3)}
figure
plot(A{3})
```

Listing 7.10 Factorization of the `swimmer` benchmark with the NBOD model.

```
load('swimmer.mat')
In = size(Y);
R = [0 1 15];              % number of components in each modes.
A = rand(In(1),R(1)); H = rand(R(1),In(2),In(3));
B = rand(In(2),R(2)); L = rand(In(1),R(2),In(3));
C = rand(In(3),R(3)); F = rand(In(2),In(2),R(3));
err = inf
% NBOD with the multiplicative algorithm
while err >1e-2
    options = struct('R',R,'maxIter',2000,'lspar',[0 .5 .5],...
        'lortho',[0 0 1],'verbose',1,'A',A,'H',H,'B',B,'L',L,'C',C,'F',F);
    [A,H,B,L,C,F,err] = nbod_ISRA(Y,options);
end
```

Example 7.15 *NBOD for images with flickering (fluctuating) offset*

In order to illustrate a challenging problem in this example, we generated a variation of the `swimmer` dataset. The coding component of the offset part ("torso") was not fixed but fluctuated around dc value as displayed in Figure 7.21(a). The aNMF failed for this difficult benchmark. The offset component cannot be correctly

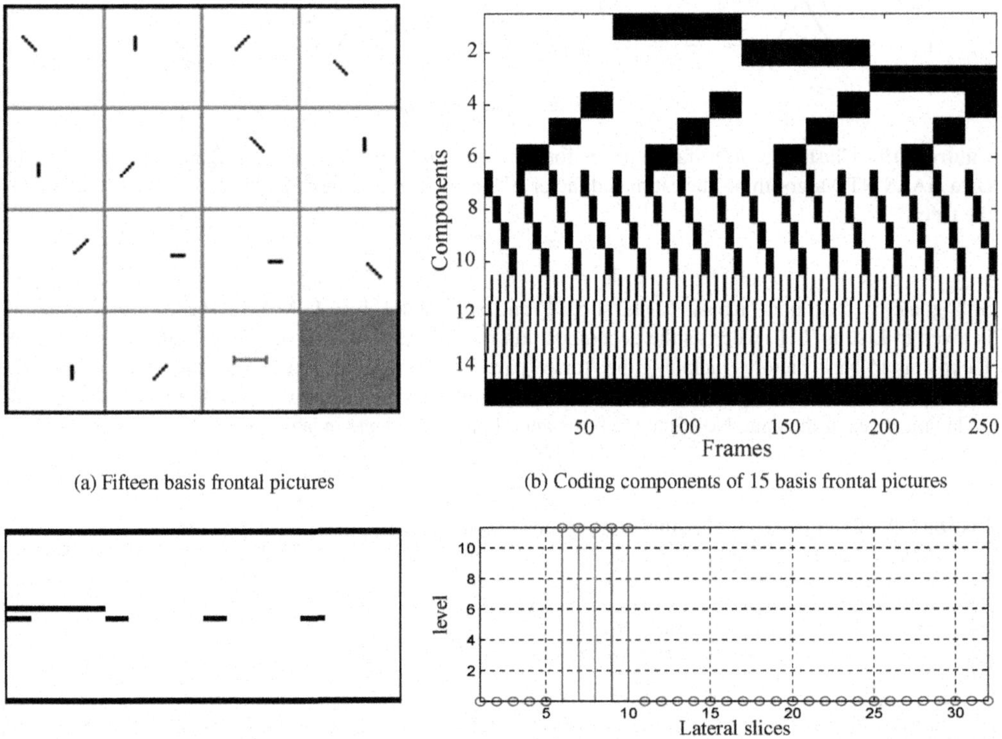

(a) Fifteen basis frontal pictures (b) Coding components of 15 basis frontal pictures

(c) Basis lateral picture (d) Coding component of the basis lateral picture

Figure 7.20 Factorization of the `swimmer` benchmark by using the multiplicative NBOD algorithm. (a)–(b) 15 basis frontal pictures $\underline{\mathbf{F}}$ and their coding components $\mathbf{C}$; (c)–(d) 1 basis lateral picture and its coding component. This basis picture exists in lateral slices 6, 7,..., 10 of the data tensor.

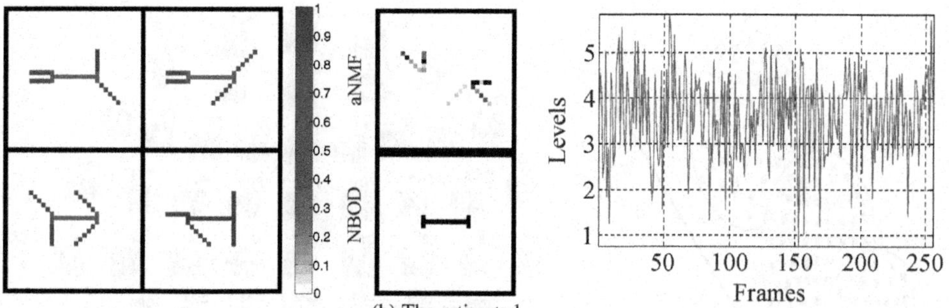

(a) The `swimmer` with the "torso" flickered (b) The estimated offset pictures (c) The coding component estimated by using NBOD

Figure 7.21 The variation of the `swimmer` benchmark with the "torso" degraded by flickering. Its coding values fluctuated around 1: (a) four `swimmer` samples from 256 pictures. (b) two offset pictures estimated by using aNMF (top) and NBOD (bottom). The aNMF failed for this data while the NBOD1 perfectly extracted all desired components: (c) the coding component of the "torso" picture estimated by using NBOD1 model.

estimated via aNMF model (see Figure 7.21(b)). However, the NBOD1 model with the multiplicative algorithm perfectly extracted all basis components. In Figures 7.21(b) and 7.21(c), we provide the "torso" picture and its coding values in 256 frames estimated by using the NBOD1 model.

7.7.4 Multi-Way Analysis of High Density Array EEG – Classification of Event Related Potentials

Example 7.16 *Analysis of the EEG dataset* `tutorialdataset2.zip`

This example illustrates the analysis of real-world EEG data which can be found in the file `tutorial-dataset2.zip` [60]. The dataset consists of 28 files containing inter-trial phase coherence (ITPC) measurements [77] of EEG signals of 14 subjects during a proprioceptive stimulation provided by a pull of the left and right hands. For every subject the ITPC measurements provide a third-order tensor of 64 channels $\times$ 72 times $\times$ 61 frequencies, giving a total of 28 replicated measurement arrays from different trials as visualized in Figure 7.22(a). Figure 7.22(b) illustrates the topological montage plot of 64 time-frequency (horizontal) slices for the 64 channels of subject 1 during the left hand stimulation. One slice from channel C3 is zoomed in and displayed in Figure 7.22(c). The whole ITPC data set can be seen as a 4-way tensor of 64 channels $\times$ 72 times $\times$ 61 frequencies $\times$ 28 measurements. The main objective in this experiment was to extract factors which could explain the stimulus activities along these 28 measurements. The number of components in each factor is specified based on the number of activities: one for left hand stimulus, one for right hand stimulus, and one more for both stimuli. To reduce tensor dimension, the 2-D time-frequency data were vectorized. As a result, the preprocessed ITPC tensor is formed as channel $\times$ (time-frequency) $\times$ measurement ($64 \times 4392 \times 28$). Figure 7.22(d) shows a 3D visualization of the observed tensor, which can be found in the file `EEG_multitask.mat`.

The following results are obtained by the NTF with Fast HALS algorithm. Exemplary illustrative results are shown in Figure 7.23 with scalp topographic maps and their corresponding IPTC time-frequency measurements, whereas the performance comparisons are given in Table 7.3. The components of the first factor $\mathbf{A}^{(1)}$ are relative to the location of electrodes, and are used to illustrate the scalp topographic maps (the first row in Figure 7.23); whereas the 2-nd factor $\mathbf{A}^{(2)}$ represents the time-frequency spectral maps which were vectorized, and presented in the second row. Each component of these factors corresponds to a specific stimulus (left, right and both hand actions).

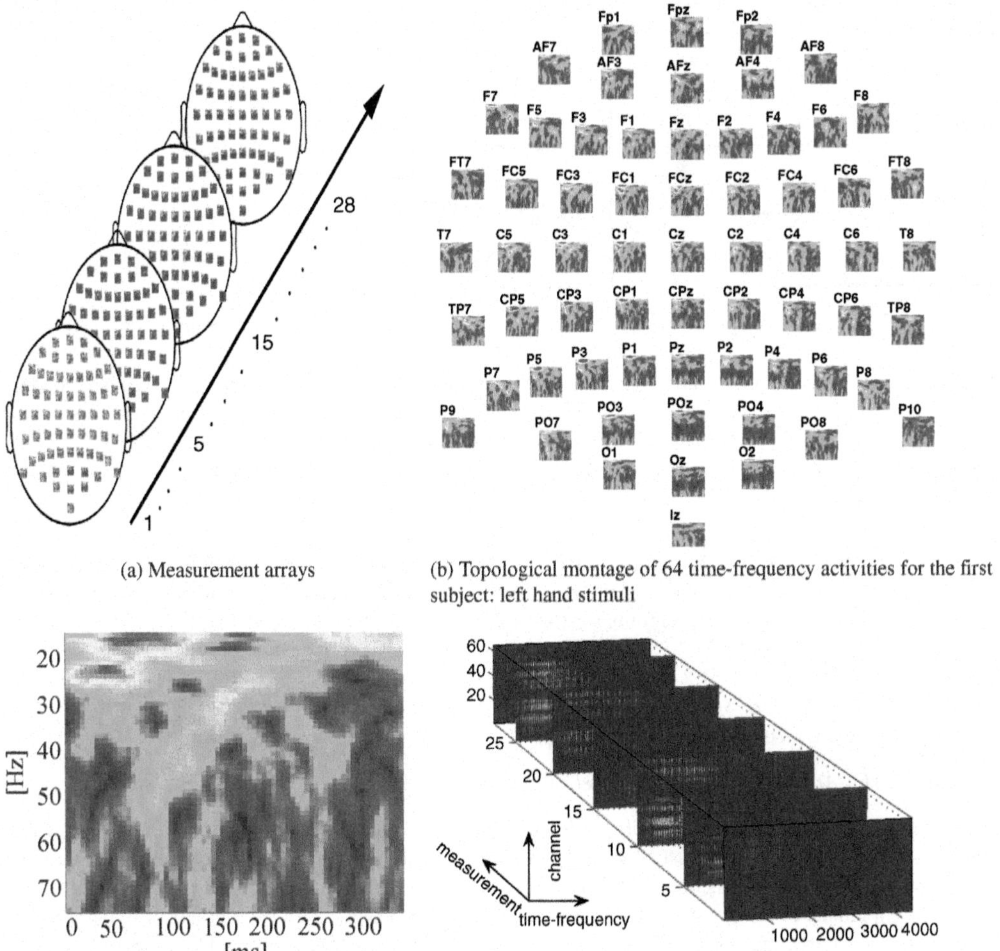

(a) Measurement arrays

(b) Topological montage of 64 time-frequency activities for the first subject: left hand stimuli

(c) A time-frequency plot zoomed in on channel C3

(d) A 3D ITPC tensor of 64 channels× 4392 (time-frequency) × 28 measurements

Figure 7.22 Visualization of EEG data `tutorialdataset2.zip` [60] in Example 7.16.

Example 7.17 *Decomposition and classification of visual and auditory EEG signals*

We illustrate decomposition and classification of EEG signals according to the nature of the stimulus: visual and auditory for the benchmark `EEG_AV_stimuli` [3,9,59,84]. The stimuli were

1. Auditory stimulus with a single tone of 2000 Hz of 30 ms duration.
2. Visual stimulus in the form of a 5×5 checkerboard (600×600 pixels) displayed on a LCD screen (32×25cm). The stimulus duration was also 30 ms.
3. Both the auditory and the visual stimuli simultaneously.

A single class (stimulus type) consisted $N = 25$ trials, and was stored in a separate file. In each trial, EEG signals were recorded from 61 channels (except channels VEOG, HEOG, FP1) during 1.5 seconds after

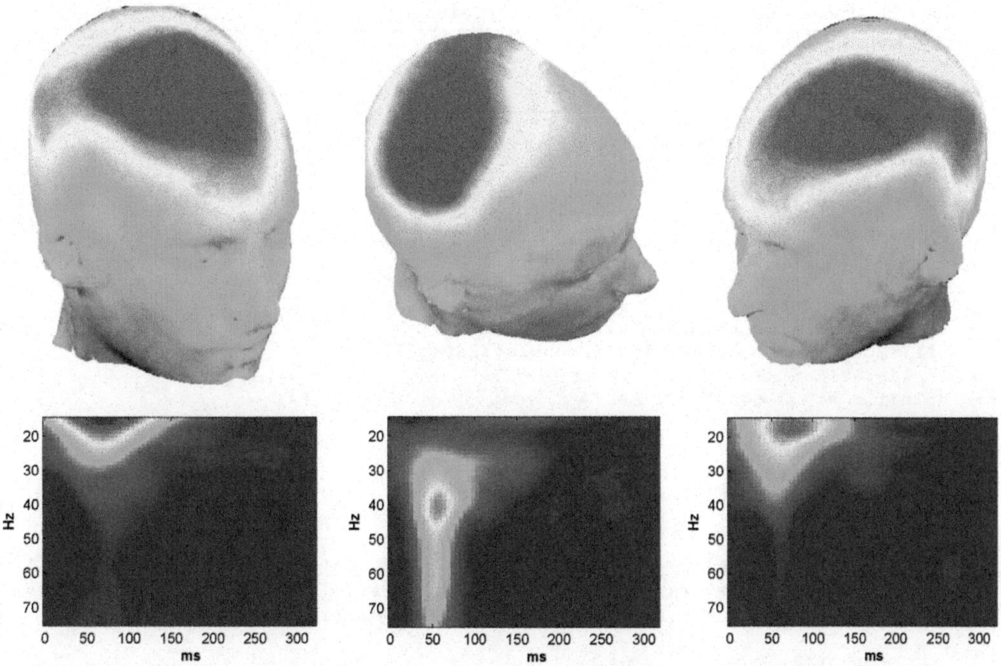

(a) Left hand stimuli (b) Gamma activity of both stimuli (c) Right hand stimuli

Figure 7.23 EEG analysis based on the FAST HALS NTF for Example 7.16, the component matrices are $\mathbf{A}^{(1)}$ for a scalp topographic map (first row), and factor $\mathbf{A}^{(2)}$ for spectral (time-frequency) map (second row) (see [60] for details). Results are consistent with the previous analysis [60] but convergence properties of the FAST HALS NTF algorithm are different.

stimulus presentation at a sampling rate of 1kHz. All the EEG signals in one class formed a third-order tensor of 25 trials $\times$ 1500 samples $\times$ 61 channels. Hence, the full dataset is a 4-way tensor of size 25 trials $\times$ 1500 samples $\times$ 61 channels $\times$ 3 classes.

Table 7.3 Comparison of performances of NTF algorithms.

	FIT (%)				**Time** (seconds)		
Example No.	7.4	7.5	7.6	7.16	7.4	7.5	7.16
FastNTF	**99.9955**	**99.51**		**52.41**	**51.73**	**0.93**	**7.08**
α-NTF		98.77				6.33	
β-NTF	99.9947	99.39	**98.24**		470.53	1.85	
NMWF [a]	99.9918	98.76	97.83	52.38	513.37	3.16	58.19
lsNTF [b]		98.06	74.47	51.33		3.30	4029.84
ALS	99.9953	98.53		**53.17**	145.73	2.52	67.24
HOOI	99.9953	98.53		53.13	965.76	1.78	66.39

[a] In fact, the NMWF failed for very noisy data due to the large negative entries. We enforced the estimated components to have nonnegative values by half-wave rectifying.

[b] The lsNTF algorithm failed for the large-scale example with the tensor of 500 $\times$ 500 $\times$ 500. However, for the same problem when the dimension of tensor is reduced to 300 $\times$ 300 $\times$ 300, the lsNTF needed 2829.98 seconds and achieved 99.9866% of FIT value.

Listing 7.11 Factorization of the EEG dataset `tutorialdataset2` using the HALS NTF algorithm.

```
% Factorize the EEG dataset: tutorialdataset2 (the ERPwavelab toolbox)
% into 3 components with the HALS NTF algorithm.
% Use the headplot function of the EEGLAB toolbox for visualization.
% Copyright 2008 by Anh Huy Phan and Andrzej Cichocki
clear
load EEG_ch_x_fr_tim_x_subj_2
R = 3; % number of components
In = size(X);
X = tensor(X);
%% Processing using HALS NTF algorithm
options = struct('verbose',1,'tol',1e-6,'maxiters',200,'init','eigs',...
     'nonlinearproj',1,'fixsign',0);
[Xhat,Uinit,A] = parafac_hals(X,R,options);
FIT = CalcExpVar(double(X(:)),double(Xhat(:)));
%% Visualization
splnfile = '3D-64CHANNEL.SPL';
for k = 1:R
    subplot(3,3,k)
    T=A{1}(:,k);
    headplot(T-mean(T),splnfile,'electrodes','off');
    title(sprintf('${\\bf a}^{(1)}_{%d}$',k),'interpreter','latex')

    subplot(3,3,3+k)
    imagesc(tim,fre,reshape(A{2}(:,k),numel(fre),[]));
    title(sprintf('${\\bf a}^{(2)}_{%d}$',k),'interpreter','latex')

    subplot(3,3,6+k)
    plot(A{3}(:,k));
    title(sprintf('${\\bf a}^{(3)}_{%d}$',k),'interpreter','latex')
end
```

These three types of event related potentials can be classified according to the latency at which their components occur after stimulus presentation. The latency features of the three classes (corresponding to three different conditions) in the time domain can be found in Figures 7.24(d), 7.24(e) and 7.24(f). For instance, there is a N100 peak appearing around at 100 ms following the stimulus for auditory stimuli as shown in Figure 7.24(d). We applied the nonnegative Tucker decomposition in analyzing the dataset in order to find the complex interactions and relationships between components expressing three modes: channels (space), spectra (time frequency representation), and classes (corresponding to three stimuli).

Table 7.4 Performance of NTD algorithms for various datasets (Examples 7.8–7.13).

Example	Data Tensor Size	Core Tensor Size	PSNR [dB]				Ratio
			LOCAL	HOOI	ALS	HONMF	
7.8	$40 \times 40 \times 40$	$4 \times 3 \times 4$	**111.23**	0	0	17.99	131.15
7.9 (0 dB)	$40 \times 40 \times 40$	$4 \times 3 \times 4$	**85.49**	0	0	0	131.15
7.11 (-10 dB)	$60 \times 60 \times 60$	$4 \times 5 \times 4$	**38.65**	30.29	30.66	33.24	251.16
7.13	$128 \times 128 \times 16$	$20 \times 20 \times 16$	30.44				22.26

Table 7.5 Labeling of ORL faces based on the factor $\mathbf{A}^{(3)}$.

Row	$\mathbf{A}^{(3)}$			Label
1	**0.0676**	0.0000	0.0000	1
15	0.0812	**0.2383**	0.0000	2
30	0.0295	0.0000	**0.0729**	3

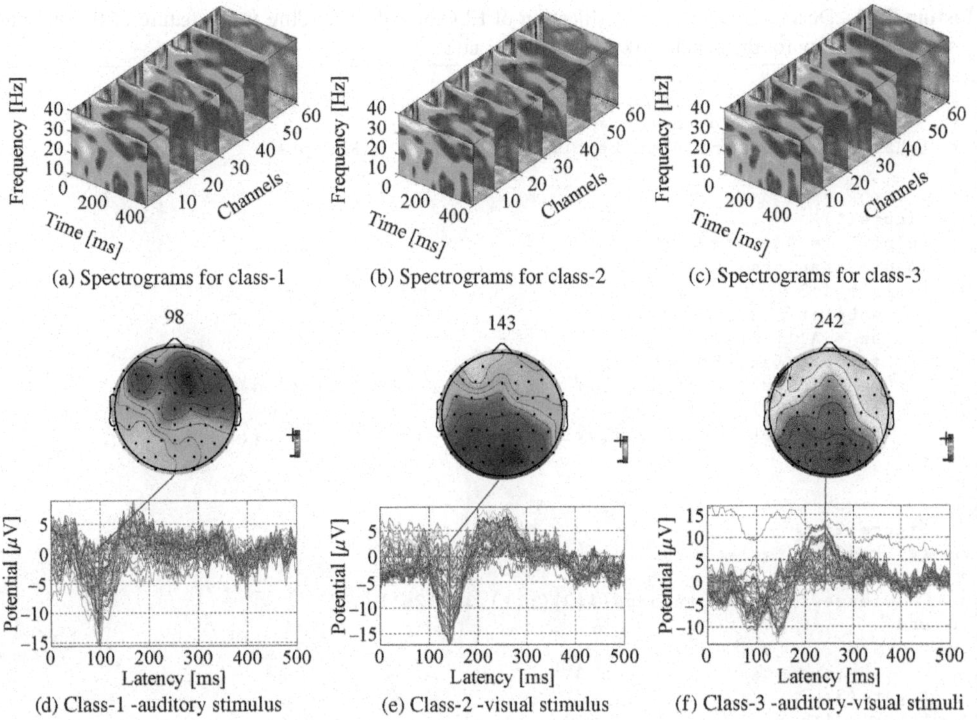

(a) Spectrograms for class-1 (b) Spectrograms for class-2 (c) Spectrograms for class-3

(d) Class-1 -auditory stimulus (e) Class-2 -visual stimulus (f) Class-3 -auditory-visual stimuli

Figure 7.24 Spectrograms and corresponding ERP for three types of stimuli in Example 7.17: (a)–(d) auditory, (b)–(e) visual, (c)–(f) auditory-visual stimulus.

First, the EEG signals were transformed by the complex Morlet wavelet into the time-frequency spectrograms of 31 frequency bins (10-40Hz) $\times$ 126 time frames (0-500ms) to form a raw data tensor $\underline{\mathbf{W}}$ of 61 channels $\times$ 31 frequency bins $\times$ 126 time frames $\times$ 25 trials $\times$ 3 classes. To analyze this spectral tensor, we averaged the wavelet coefficient magnitudes along all the trials as

$$y_{c,f,t,l} = \frac{1}{N} \sum_{n}^{N} \left| w_{c,f,t,n,l} \right|. \tag{7.201}$$

This data tensor corresponds to the time-frequency transformed Event Related Potentials (ERP) [36]. In practice, to reduce the high computational cost, we computed averages of the tensor along 25 trials, to obtain 183 (61 channels $\times$ 3 classes) EEG signals for all three classes (each with 1500 time samples); then transformed them to the time-frequency domain. In this way, we avoided wavelet transformation for all 4575 (61 channels $\times$ 25 trials $\times$ 3 classes) EEG signals. The preprocessed data tensor $\underline{\mathbf{Y}}$ has size of 61 channels $\times$ 31 frequency bins $\times$ 126 time frames $\times$ 3 classes. Figures 7.24(a), 7.24(b), and 7.24(c) show some selected spectra for three stimulus classes.

Finally, we reshaped data tensor $\underline{\mathbf{Y}}$ into a third-order nonnegative tensor of the size $61 \times 3906 \times 3$. The Nonnegative Tucker Decomposition model was chosen to decompose the preprocessed tensor data, and the ℓ_2 HALS NTD was selected to extract underlying components. We set the number of components to three, that is, the size of core tensor was $3 \times 3 \times 3$. The listing code 7.12 shows how the data tensor was processed and results were visualized. The three estimated components of the factor $\mathbf{A}^{(1)}$ express spatial activations

Listing 7.12 Decomposition and classification of EEG signals according to the nature of the stimulus: visual and auditory for the benchmark `EEG_AV_stimuli`.

```
% Copyright 2008 by Anh Huy Phan and Andrzej Cichocki
opts = struct('tol',1e-6,'maxiters',500,'init','nvecs',...
     'ellnorm',2,'orthoforce',1,'fixsign',0);
[Yhat,A,G,fit,iter] = tucker_localhals(Y,[3 3 3],opts);

% Topographic and Spherically-splined EEG field maps - factor A^(1)
figure(1)
splnfile = 'AudVis4_6Nov2001.spl';
load AVchanlocs;
for k = 1:3
     subplot(3,2,2*(k-1)+1)
     ak = A{1}(:,k);
     topoplot(ak,chanlocs,...
     'maplimits',[min(A{1}(:)) max(A{1}(:))],'electrodes','off');
     subplot(3,2,2*(k-1)+2)
     headplot(ak - mean(ak),splnfile,'electrodes','off');
end

%% Spectra components - factor A^(2)
figure(2)
for k = 1:3
     subplot(3,1,k)
     imagesc(tim,Fa,reshape(A{2}(:,k),31,126))
end

% Class components - factor A^(3)
figure(3)
for k = 1:3
     subplot(3,1,k)
     bar(A{3}(:,k))
end

% Hinton diagrams for frontal slices of the core tensor
figure(4)
Gd = double(G);
for k = 1:3
     subplot(3,1,k)
     hinton(Gd(:,:,k),max(Gd(:)))
end

%% Joint rate of components a_i^(3) to three classes
figure(5)
JR = squeeze(sum(Gd.^2,2));
JR = bsxfun(@rdivide,Jrate,sum(Jrate));
hinton(Jrate,1)
```

(channels) distributed over 61 channels. The topographic maps and the corresponding spherically-splined field maps are displayed in Figure 7.25(a). The three basis spectral components $\boldsymbol{a}_{j_2}^{(2)}$, $j_2 = 1, 2, 3$ were reshaped and displayed in Figure 7.25(b). The three components $\boldsymbol{a}_{j_3}^{(3)}$, $j_3 = 1, 2, 3$ indicating the category of stimuli are plotted in Figure 7.25(c). Using such multi-way analysis, the three classes of stimuli were clearly classified, as illustrated by Table 7.6.

In the next step we attempt to specify which spectral components and spatial components mainly affect each class and how they interact with each other. Note that for the NTD model, each component in a factor may have multiple interactions with some or all the components in the other factors. In other words, we cannot say that the category component $\boldsymbol{a}_1^{(3)}$ is characterized by spatial and spectral components $\boldsymbol{a}_1^{(1)}$ and $\boldsymbol{a}_1^{(2)}$.

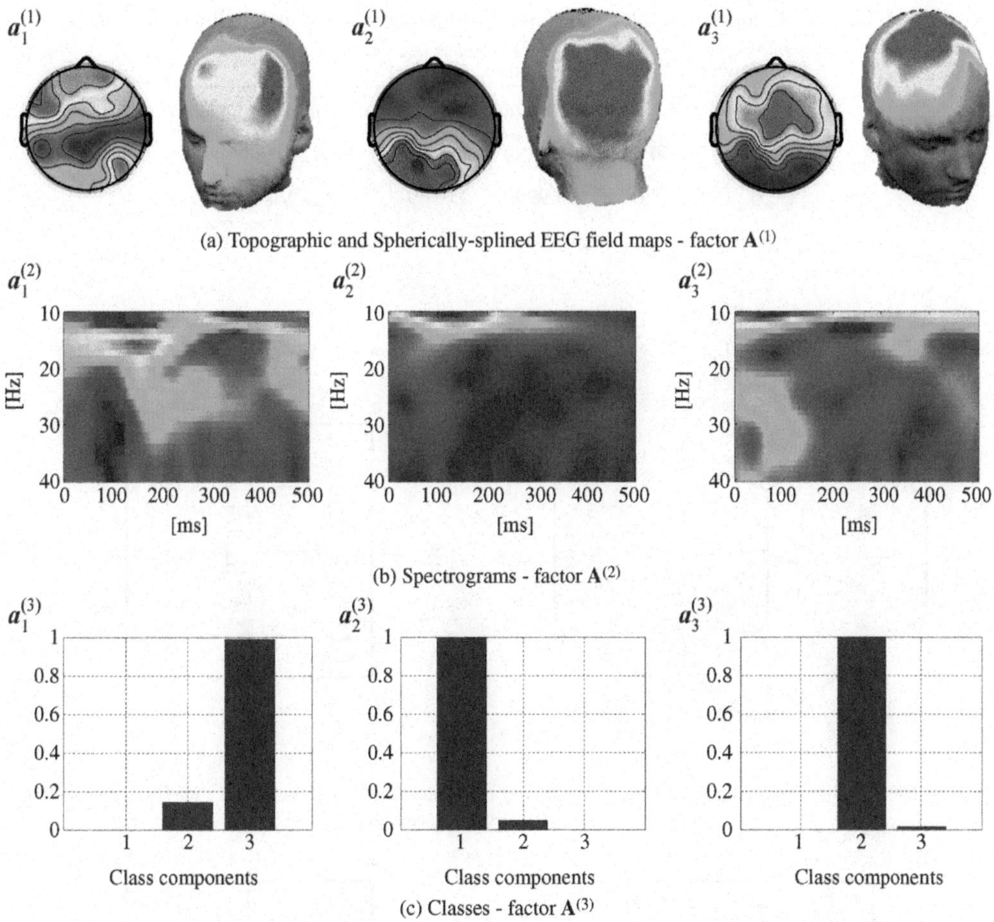

Figure 7.25 Visualization of components of the NTD model in Example 7.17. (a) factor 1 $\mathbf{A}^{(1)}$ characterizes spatial components displayed in topographic maps and sherically-spline EEG field maps; (b) spectral components expressed by factor $\mathbf{A}^{(2)}$; (c) expression of factor $\mathbf{A}^{(3)}$ for 3 classes: component 1 $\boldsymbol{a}_1^{(3)}$ - auditory-visual class, component 2 $\boldsymbol{a}_2^{(3)}$ - auditory class, and component 3 $\boldsymbol{a}_3^{(3)}$ - visual class in Example 7.17.

However, we can identify the dominant or most significant interactions via the analysis of entries of the core tensor $\underline{\mathbf{G}}$. We assumed that all the component vectors are normalized to unit length in the sense of the ℓ_2-norm, hence the coefficients of the core tensor $\underline{\mathbf{G}}$ express the energy of rank-one tensors[14] which are built up from the basis components $\boldsymbol{a}_j^{(n)}$, $(n = 1, \ldots, N,\ j = 1, \ldots, J_n)$. By using the ℓ_2 HALS NTD, the estimated components in all factors $\boldsymbol{a}_{j_n}^{(n)}$ have been already normalized to unit length vectors. In general, we can apply the fast fixing method described in Appendix 7.B.

For example, the energy coefficient g_{311}^2 represents links or relationship among components $\boldsymbol{a}_3^{(1)}$, $\boldsymbol{a}_1^{(2)}$ and $\boldsymbol{a}_1^{(3)}$. Note that the category component $\boldsymbol{a}_j^{(3)}$, $(j = 1, 2, 3)$ only locates on the j-th frontal slice $\mathbf{G}_j = \underline{\mathbf{G}}_{:,:,j}$. Figure 7.26 shows Hinton diagrams visualizing all the three frontal slices $\mathbf{G}_j$ and the corresponding category

[14]Justification is given in Appendix 7.1.

Table 7.6 Components of factor $\mathbf{A}^{(3)}$ for the three classes of stimuli in Example 7.17.

Component	Coefficients			Class
$\boldsymbol{a}_1^{(3)}$	0.0	0.1426	**0.9898**	3 - Auditory+Visual
$\boldsymbol{a}_2^{(3)}$	**0.9988**	0.0489	0.0	1 - Auditory
$\boldsymbol{a}_3^{(3)}$	0.0	**0.9999**	0.0169	2 - Visual

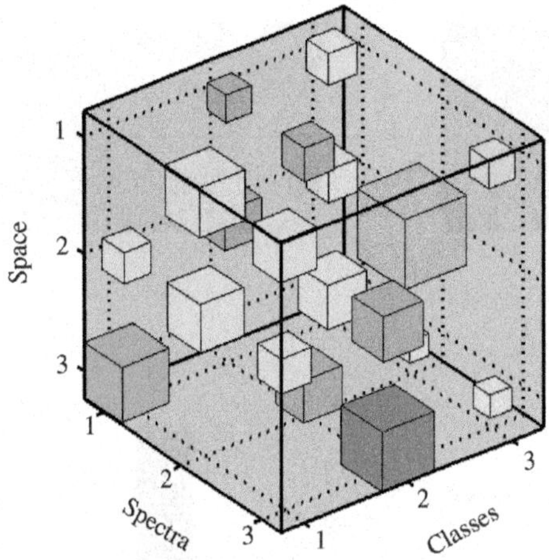

(a) The 3-D Hinton diagram of the core tensor of $\underline{\mathbf{G}}$

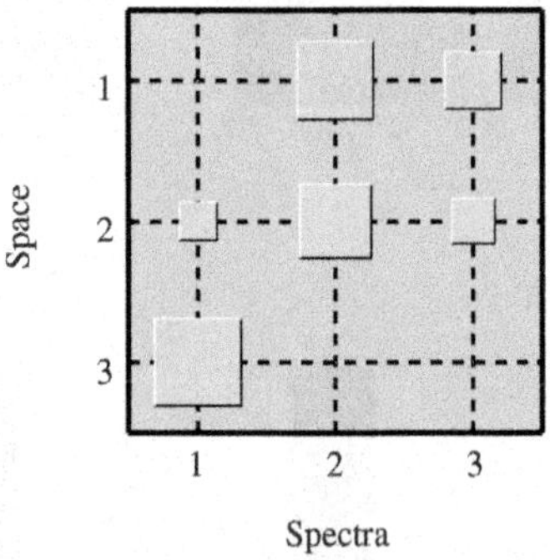

(b) Hinton diagram of the frontal slice $\mathbf{G}_1$ (corresponding to class-3, auditory+visual stimuli)

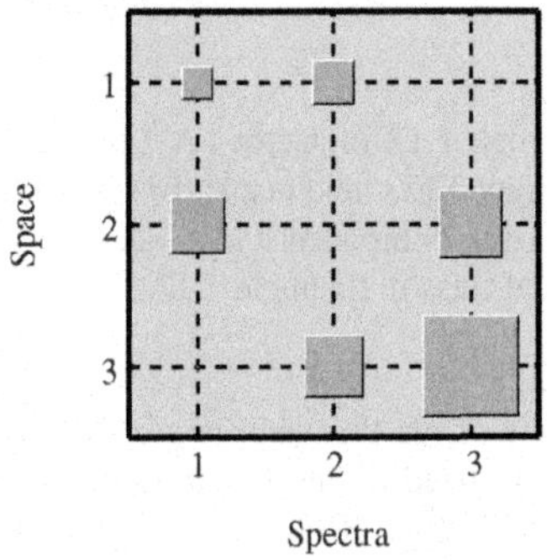

(c) Hinton diagram of the frontal $\mathbf{G}_2$

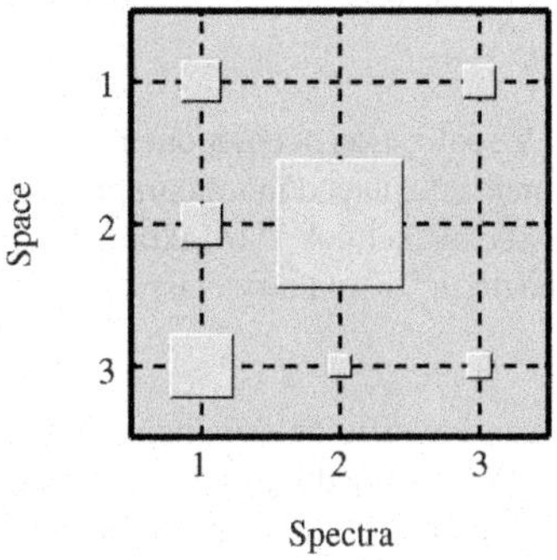

(d) Hinton diagram of the frontal slice $\mathbf{G}_3$

Figure 7.26 Illustration of the core tensor via Hinton diagrams: (a) full core tensor $\underline{\mathbf{G}}$ in 3-D mode; (b)–(d) Frontal slices $\mathbf{G}_j = \underline{\mathbf{G}}_{:,:,j}$, $j = 1, 2, 3$ express the interaction of the j-th component $\boldsymbol{a}_j^{(3)}$ with components expressing channel and spectrogram in factors $\mathbf{A}^{(1)}$ and $\mathbf{A}^{(2)}$: Auditory class - $\boldsymbol{a}_2^{(3)}$ concentrates mainly on coefficient g_{332}, or spreads on $\boldsymbol{a}_3^{(1)}$ by spectrogram $\boldsymbol{a}_3^{(2)}$; Visual class - $\boldsymbol{a}_3^{(3)}$ spreads on $\boldsymbol{a}_2^{(1)}$ by spectrogram $\boldsymbol{a}_2^{(2)}$, auditory + visual class - $\boldsymbol{a}_1^{(3)}$ spreads on $\boldsymbol{a}_3^{(1)}$, $\boldsymbol{a}_2^{(1)}$, and $\boldsymbol{a}_1^{(1)}$ by spectrograms $\boldsymbol{a}_1^{(2)}$, $\boldsymbol{a}_2^{(2)}$ and $\boldsymbol{a}_3^{(2)}$. Darker colors in (a) indicate dominant components for each of the three factors.

Table 7.7 Frontal slices of the estimated core tensor $\underline{\mathbf{G}}$ in Example 7.17.

$\mathbf{G}_1$			$\mathbf{G}_2$			$\mathbf{G}_3$		
0.00	316.66	175.46	50.80	91.92	0.00	79.53	0.00	54.39
74.51	275.45	104.64	157.36	0.00	222.92	85.07	839.18	0.00
404.20	0.00	0.00	0.00	188.03	511.55	205.80	26.08	32.18

components $\boldsymbol{a}_j^{(3)}$. The volume of box is proportional to the intensity of a corresponding coefficient. The dominant coefficients in each slice $\mathbf{G}_j$ indicate the most significant interactions of spatial and spectral components for each j-th category. The coefficients of the estimated core tensor $\underline{\mathbf{G}}$ provided in Table 7.7 illustrate links or interactions between components. Using this visualization, $g_{332} = 511.55$ is the major coefficient on the 2-nd frontal slice, the auditory class (corresponding to the component $\boldsymbol{a}_2^{(3)}$) is therefore mainly characterized by the third spectral component $\boldsymbol{a}_3^{(2)}$ (see in Figure 7.25(c)) and spreads on the spatial component 3 ($\boldsymbol{a}_3^{(1)}$).

For the class-2 (visual stimulus) (represented by the component $\boldsymbol{a}_3^{(3)}$), the element g_{223} on the third frontal slice is the dominant coefficient. This can be interpreted as evidence that the second spectral component $\boldsymbol{a}_2^{(2)}$ (see Figure 7.25(b)) and the second spatial component $\boldsymbol{a}_2^{(1)}$ is dominant for this class. For the class-3 (auditory-visual stimuli, represented by the component $\boldsymbol{a}_1^{(3)}$), energy was more widely distributed amongst all the spatial components and all the spectral components (see Figure 7.25(b)).

In order to qualitatively evaluate the interactions among components we defined the Joint Rate (JR) index (see Appendix 7.A). For example, the Joint Rate index $JR_{j_3}^{j_1}$ which reflects how strongly the spatial component $\boldsymbol{a}_{j_1}^{(1)}$ affects the category component $\boldsymbol{a}_{j_3}^{(3)}$ is expressed as

$$JR_{j_3}^{j_1} = \frac{\sum_{j_2=1}^{J_2} g_{j_1 j_2 j_3}^2}{\sum_{j_1=1}^{J_1} \sum_{j_2=1}^{J_2} g_{j_1 j_2 j_3}^2}. \tag{7.202}$$

The maximum value of this index is 1, meaning that component $\boldsymbol{a}_{j_3}^{(3)}$ is completely expressed by spatial component $\boldsymbol{a}_{j_1}^{(1)}$, and does not depend on other components of the spatial factor $\mathbf{A}^{(1)}$. Otherwise, the minimum value of this index is 0, meaning that $\boldsymbol{a}_{j_3}^{(3)}$ it is not related to $\boldsymbol{a}_{j_1}^{(1)}$. In Figure 7.29, we illustrate the interactions among components by using three Hinton diagrams of the Joint Rate indices for various pairs of components, especially between spatial and category components (Figure 7.29(a)) and spectral and category components (Figure 7.29(b)). From these diagrams (see also Figures 7.27 and 7.28), it can be seen that the auditory class interacts predominantly with the third spatial component, and the third spectral component, whereas the visual class links with the second spatial component, and the second spectral component, and the auditory+visual class links with all the spatial and spectral components.

In Figure 7.27, we present the three bar plots of the computed JR indices among three spatial components (see Figure 7.25(a)) and the three class components (see Figure 7.25(c)) for the auditory (row 2), visual (row 3) and auditory+visual (row 1) stimuli. The significant interactions among components for each class can be evaluated directly from these plots; for example, the $JR_{2/3}^{3/1}$ has the largest value. This means that the 2nd category component affects primarily the third spatial component, the 3rd category component correlates mainly with the second spatial component ($JR_{3/3}^{2/1}$), whereas the 1st one spreads over all the three components.

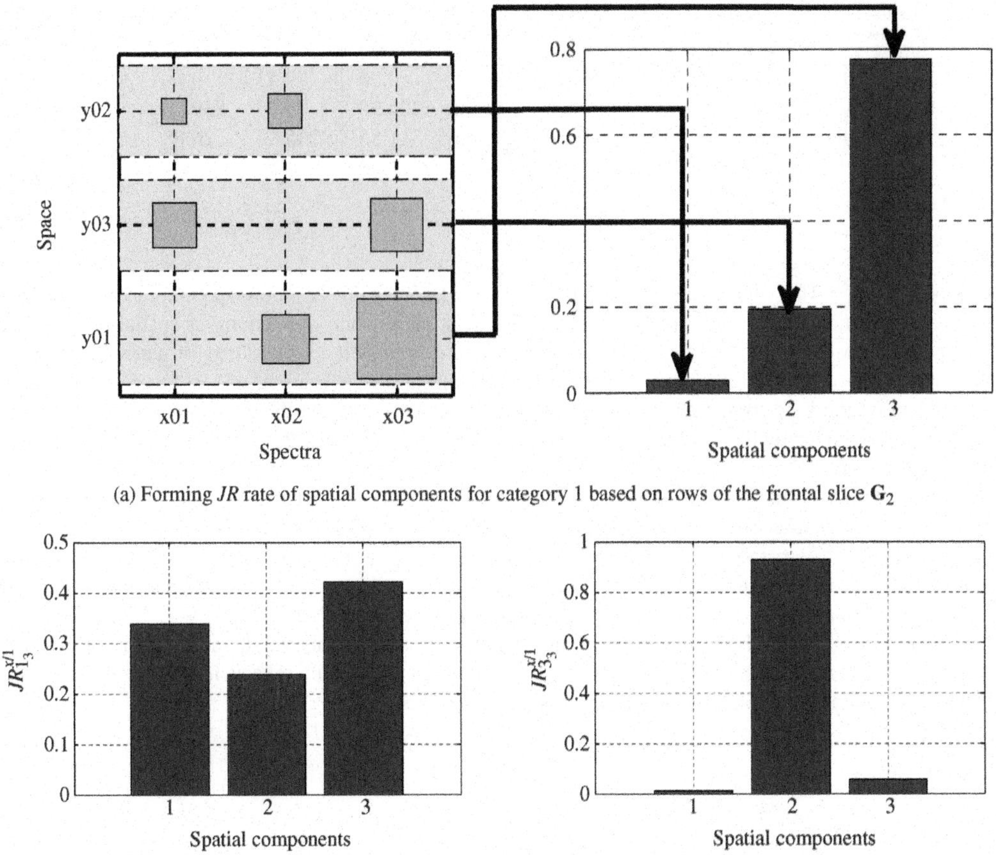

(a) Forming *JR* rate of spatial components for category 1 based on rows of the frontal slice $\mathbf{G}_2$

(b) Joint rate of spatial components for category 1

(c) Joint rate of spatial components for category 2

Figure 7.27 (a) Illustration of analysis using the *JR* index of spatial components for the category component 2; (b)–(c) The joint rates (JR) show clearly the concentration of category components 1 and 3 on spatial components.

7.7.5 Application of Tensor Decomposition in Brain Computer Interface – Classification of Motor Imagery Tasks

In comprehensive Brain Computer Interface (BCI) studies, the brain data structures often contain higher-order ways (modes) such as trials, tasks conditions, subjects, and groups despite the intrinsic dimensions of space, time and frequency. In fact, specific mental tasks or stimuli are often presented repeatedly in a sequence of trials leading to a large volume stream of data encompassing many dimensions: Channels (space), time-frequencies, trials, and conditions.

As Figure 7.30 shows, most existing BCI systems use three basic signal-processing blocks. The system applies a preprocessing step to remove noise and artifacts (mostly related to ocular, muscular, and cardiac activities) to enhance the SNR. In the next step, the system performs feature extraction and selection to detect the specific target patterns in brain activity that encode the user's mental tasks, detect an event-related response, or reflect the subject's motor intentions. The last step is aimed at translating or associating these specific features into useful control (command) signals to be sent to an external device. Ideally, the translator block supports the noncontrol state, because without NC support, all classifier output states are considered

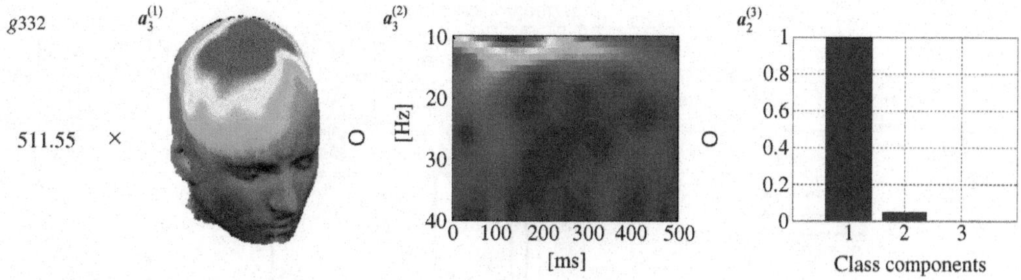

(a) Channel map (b) Spectrogram (c) Class expression

Figure 7.28 Visualization of the three significant components $a_3^{(1)}, a_3^{(2)}, a_2^{(3)}$ for a dominant rank-one tensor in the NTD for the auditory class.

intentional. With NC support, the user can control whether or not the output is considered intentional. In the latter case, a self-paced NC state paradigm is monitored continuously, where users can perform specific mental tasks whenever they want.

In the preprocessing step, the system can decompose the recorded brain signals into useful signal and noise subspaces using standard techniques (like ICA or nonlinear adaptive filtering). One promising approach to enhance signals, extract significant features, and perform some model reduction is to apply blind source separation techniques, especially multiway blind source separation and multiway array (tensor) decomposition.

A promising and popular approach based on the passive endogenous paradigm is to exploit temporal/spatial changes or spectral characteristics of the sensorimotor rhythm (SMR) oscillations, or mu-rhythm (8-12 Hz) and beta rhythm (18-25 Hz). These oscillations typically decrease during, or immediately before a movement event related desynchronization (ERD). External stimuli-visual, auditory, or somatosensory – drive exogenous BCI tasks, which usually do not require special training. Two often used paradigms are P300 and steady-state visually evoked potentials (SSVEP). P300 is an event-related potential that appears approximately 300 ms after a relevant and rare event. SSVEP uses a flicker stimulus at relatively low frequency (typically, 5-45 Hz).

Another promising and related extension of BCI is to incorporate real-time neuro-feedback capabilities to train subjects to modulate EEG brain patterns and parameters such as ERPs, ERD, SMR, and P300 to meet a specific criterion or learn self-regulation skills where users change their EEG patterns in response to feedback. Such integration of neuro-feedback in BCI is an emerging technology for rehabilitation, but it is also a new paradigm in neuroscience that might reveal previously unknown brain activities associated with behavior or self-regulated mental states (see Figure 7.31). In a neuro-feedback-modulated response (active

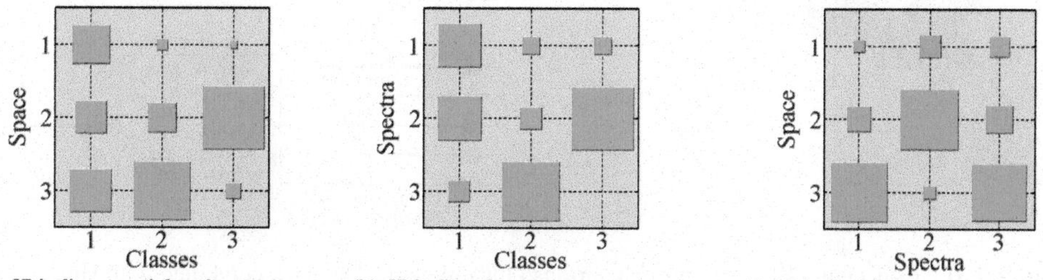

(a) *JR* indices spatial and category components (b) *JR* indices between spectral and category components (c) *JR* indices between spatial and spectral components

Figure 7.29 Hinton diagrams of the Joint Rate indices between (a) spatial and category components, (b) spectral and category components, and (c) spatial and spectral components.

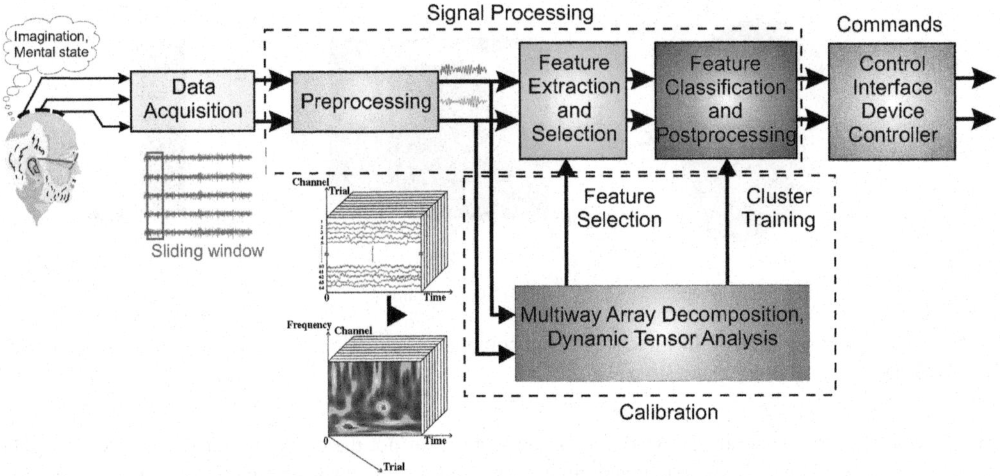

Figure 7.30 Multistage procedure for online BCI. Preprocessing and feature extraction play a key role in real-time, high-performance BCI systems. In the calibration step, most BCI ERP studies are based on multi-subject and multi-condition analysis. For such scenarios, the tensor decompositions naturally encompass extra modalities such as trials, subjects, conditions, and so on and allow the system to find the dominant sources of activity differences without supervision.

endogenous) BCI paradigm, users learn to generate specific brain waves through various mental strategies while monitoring the outcome of their efforts in near real time. Typically, the user visualizes the preprocessed and translated target brain signal to increase motivation and improve recognition accuracy. However, the successful control of the interface in this way usually requires a quite long process and up to several weeks

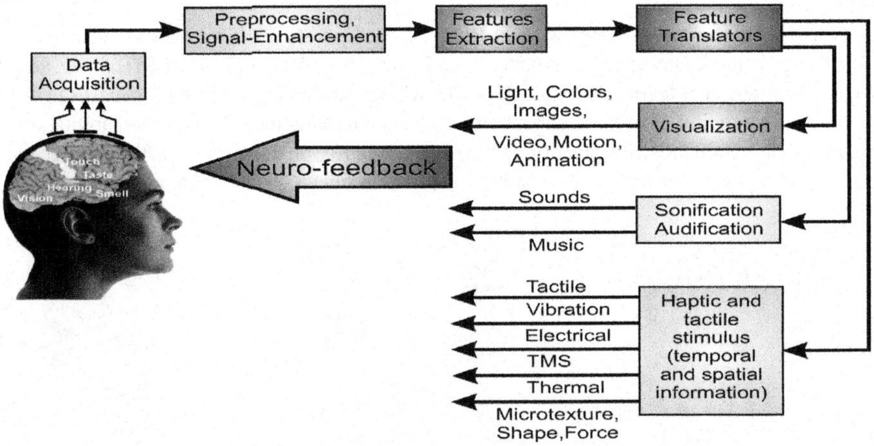

Figure 7.31 Conceptual BCI system with various kinds of neuro-feedback combined with Human Computer Interactions (HCI). The development of a BCI must handle two learning systems: The computer should learn to discriminate between different complex patterns of brain activity as accurately as possible, and BCI users should learn via different neuro-feedback configurations to modulate and self-regulate or control BCI activity.

of training. BCI neuro-feedback in any of these paradigms should be as speedy as possible, which requires fast real-time signal processing algorithms. Recent neurofeedback experiments confirm that performance increases with richer feedback. For example, a simple bar gives lower accuracies than a full immersive 3D dynamic visualization or sonification.

Standard matrix factorizations, such as PCA, SVD, ICA and NMF and their variants, are invaluable tools for BCI feature selection, dimensionality reduction, noise reduction, and mining [5,12,45,55–58,62,67]. However, they have only two modes or 2-way representations (e.g., channels and time) and therefore have severe intrinsic limitations. For such kind of data 2-way matrix factorizations (ICA, NMF) or "flat-world view" may be insufficient for future BCI systems. In order to obtain more natural representations of the original multi-dimensional data structure, it is necessary to use tensor decomposition approaches, since additional dimensions or modes can be retained only in multi-linear models to produce structures that are unique and which admit interpretations that are neurophysiologically meaningful [1,61].

Recent advances in developing high-spatial density array EEG have called for multi-dimensional signal processing techniques (referred to here as the multi-way analysis (MWA), multi-way array (tensor) factorization/decomposition or dynamic tensor analysis (DTA) or window-based tensor analysis (WTA)) to be employed in order to increase performance of BCI systems [1,2,20,29,48,50,61,64,75,76].

Note that, in general, tensor decompositions allow multi-channel and multi-subject time-frequency-space sparse representation, artifacts rejection in the time-frequency domain, feature extraction, multi-way clustering and coherence tracking. Our main objective is to decompose multichannel time-varying EEG into multiple components with distinct modalities in the space, time and frequency domains to identify among them common components across these different domains and, at the same time, discriminate across different conditions.

Further operations performed on these components can remove redundancy and achieve compact sparse representation. There are at least two possible operations we can perform. First, extracted factors or hidden latent components can be grouped (clustered) together and represented collectively in a lower dimensional space to extract features and remove redundancy. Second, a component can be simply pruned if it is uncorrelated with a specific mental task. Note that by adding extra dimensions, it is possible to investigate topography and time and frequency pattern (e.g., Morlet wavelets) in one analysis.

The resulting components can be described, not only by the topography and the time-frequency signature but also by the relative contribution from the different users or conditions (see Figures 7.32 and 7.33 which can be considered as examples). In practice, various oscillatory activities might overlap, but the sparse and nonnegative representations of tensor data given, e.g., by the time-frequency-space transformation, enables the decompositions to isolate each oscillatory behavior even when these activities are not well separated in the space-time domain alone.

The core of our BCI system consists of a family of fast unsupervised algorithms for tensor decompositions [20,64]. Thanks to nonnegativity and sparsity constraints, NTF and NTD decompose a tensor in additive (not subtractive) factors or components. Moreover, results can be given a probabilistic interpretation.

However, it should be emphasized that the standard off-line algorithms for PARAFAC or TUCKER models are usually limited due to memory/space and computational constraints, and they assume static input data, while for BCI applications we are interested in multi-way analysis and decomposition of dynamic streams (sequence) of tensors. In other words, brain patterns change over time, so the tensor stream should be decomposed in order to perform feature extraction and to capture the structured dynamics in such collections of streams. For this purpose we may use new concepts and algorithms for Dynamic Tensor Analysis (DTA) and Window-based Tensor Analysis (WTA) [75,76]. Recently, we have proposed new local, fast online algorithms which are promising for such models and future BCI applications [20,29,64].

In summary, it appears that higher order tensor decomposition is a promising approach for BCI due to the following reasons:

1. The dimensions of factors are physically meaningful (e.g., scalp plots, temporal patterns, and trial-to-trial variability) and can be interpreted relatively easily from a neurophysiological context.

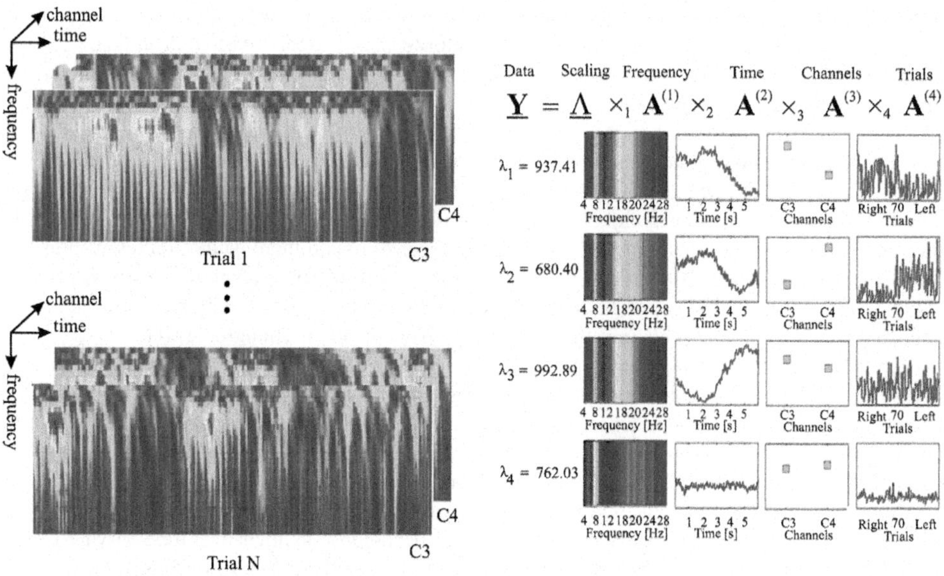

(a) 4-way tensor based on $N = 140$ three-way EEG spectral tensors

(b) 4 components explain the 4-way EEG tensor

Figure 7.32 Decomposition of the 4-way time-frequency-spectral EEG data into basic components during motor imaginary tasks.

2. In BCI, we are typically dealing with very large and high dimensional data matrices in which we need to reduce the number of parameters to be evaluated. Multi-way array decomposition is an efficient way to do this.
3. Tensor decompositions can impose many objectives such as discrimination, statistical independence, smoothness, sparse representation and multi-way clustering.

Figure 7.32 illustrates off-line analysis of a 4-way spectral (Morlet wavelets) EEG tensor (frequency × time × channel × trial) factorization including 140 trials recorded from the C3 and C4 electrodes during right and left hand motor imagery (70 right-hand trials and 70 left-hand ones). Each trial is represented by the three-way tensor (frequency × time × channel) in Figure 7.32(a). A spectral tensor was factorized into four components displayed in Figure 7.32(b). Component 1 corresponds to left-hand imagery (due to the significantly greater C3 weight than the C4 one), component 2 represents the right-hand imagery, and component 3 reflects both left and right hand imagery stimuli. The theta rhythm (4-8 Hz), which is related to concentration is represented by component 4 [56].

Figure 7.33 illustrates the experimental results using the 4-way tensor decomposition of multi-channel (62 electrodes) EEG data (channel, frequency, time, conditions) into four component (factor) matrices in the space (topographic map), frequency, time and class domain shown from left-to-right on this figure. In order to find the most discriminative components for different classes (i.e., left hand and right hand motor imagery), we imposed a sparseness constraint on the spatial and category components. Each row of Figure 7.32(b) represents one component of the factor matrices. From these plots components 4 and 5 are recognized as discriminative components corresponding to the motor imaginary tasks due to their scalp maps covering sensorimotor areas. Component 4 illustrates the ERD/ERS phenomena that indicates in the spatial distribution a larger amplitude on the left hemisphere and lower amplitude for the right hemisphere (see column 1), and the energy of oscillations dominated by the mu rhythm frequency range mainly at 8-12Hz (see column 2),

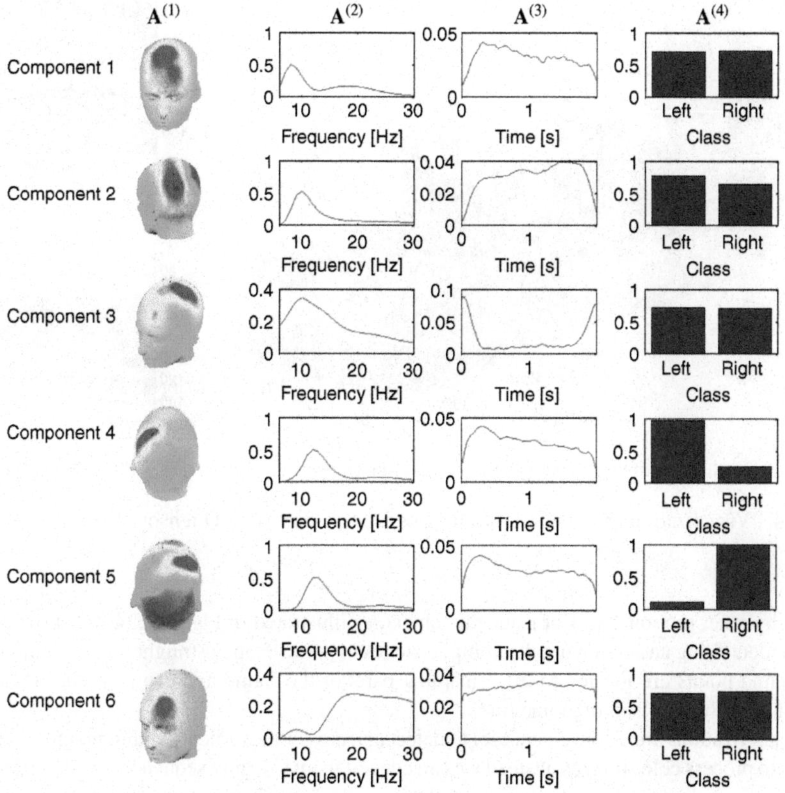

Figure 7.33 Decomposition of 62 channels EEG signals into basis components. Four columns represent the four factors in the analysis.

and observation of quasi-stationary oscillations through the whole trial duration (see column 3). Hence the larger amplitude is shown for class-1 and lower amplitude on class-2 (column 4). Similarly, component 5 shows ERD on the left hemisphere and ERS on the right hemisphere. Other components represent the existing artifacts (EOG, EMG) and other brain activities uncorrelated to event related potentials.

The multi-way analysis approach and the related concepts (tensor decompositions, especially their extensions to dynamic tensor analysis) presented in this chapter are only a subset of a number of promising and emerging signal processing and data mining tools with potential applications to future BCI systems. The main advantage of the presented approach is its flexibility when dealing with multi-dimensional data and the possibility to enforce various physical constraints.

7.7.6 Image and Video Applications

Both NTF and NTD have enormous potential for applications dealing with the analysis of sequence of images. To illustrate tensor representation of images, consider first the `Hall-monitor` image sequence which has 330 frames of size 144×176 pixels. Assume that objects move through frames (frontal slices) of the image sequence, as illustrated in Figure 7.34(a). The motion orbit of an object is expressed by the horizontal slices as in Figure 7.34(b) for the horizontal slice at row 80. A fixed pixel without any movement or a background point (pixel), is represented by a straight line through all the frames. The orbit line can be

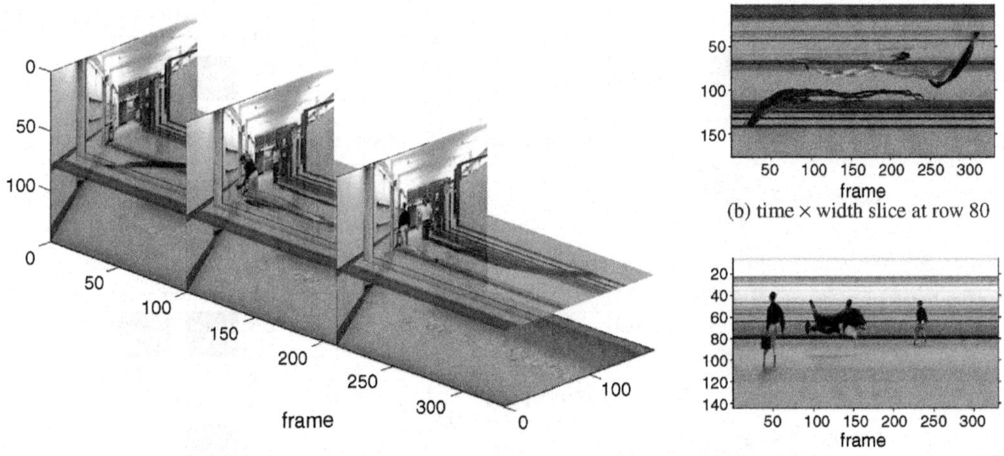

(a) 3D visualization of the image sequence as a 3D data tensor

(b) time × width slice at row 80

(c) time × height slice at column 70

Figure 7.34 Visualization of the Hall Monitor image sequence as a 3D tensor of height × width × time (frame).

interrupted by other motion orbits of a specific object, as illustrated in Figure 7.34(b) for the orbits of two objects. By identifying and removing the abrupt changes in points from a straight line in a horizontal frame, the background points are located in the remaining parts of the frame and can be reconstructed by using median values of the pixels in the same line.

Tensor factorizations are a convenient tool to detect and cluster movements within images. In practice, we do not need to process color images, instead we can convert them to a gray sequence to reduce tensor size and computational cost. In order to reduce the complexity and tensor size, it is convenient to form sub-tensors based on blocks of 8×8 pixels as shown in Figure 7.35(a). We have a set of sub-tensors of size $8 \times 8 \times 3 \times 330$ for color images and $8 \times 8 \times 330$ for gray images. In Figure 7.35(b), the sub-tensor at block (49,65) is visualized in 3D mode. We can easily recognize which (8×8) frontal slices belong to frontal objects (in dark color) and which ones are background blocks. Factorization of this sub-tensor with only one $J = 1$ component by any NTF algorithm (in our case, we used the Fast HALS NTF) returns the last factor $\mathbf{A}^{(4)}$ (or $\mathbf{A}^{(3)}$ for gray tensor), which expresses the variation of the time dimension. Figure 7.35(b) shows this factor with stationary points in red color which correspond to background frame indices. A simple approach to locate the stationary points is to find the maximum peak of the histogram of this factor. Figures 7.35(c), and 7.35(d) show other sub-tensors and their corresponding estimated factors.

The final step is to compute the median block of all blocks at stationary indices. Then, we can combine all median blocks to form the background of the observed sequence as shown in Figure 7.37(a).

The MATLAB code of this example is given in Listing 7.13.

In practice, due to some unexpected interferences such as camera movement (pan, roll, zoom or dolly) due to hand shaking or car movement, the horizontal slices will incorrectly express the motion orbit of pixels. In such case, besides the local movement across the pixels, there is global motion as well (motion of frames) [10,63]. Figure 7.36(a) illustrates an example of an unstable image sequence (the `Panorama` sequence [79]) which has 48 frames of size 240×360 pixels; three motion orbits of three corresponding pixels at (44,44) , (211,75) and (164,215) are shown.

This problem can be alleviated as follows:

1. Assume the global motion obeys an affine motion model with six parameters (two for rotation, two for translation and two for scaling). Estimate the global motion between two consecutive frames k and $(k + 1)$,

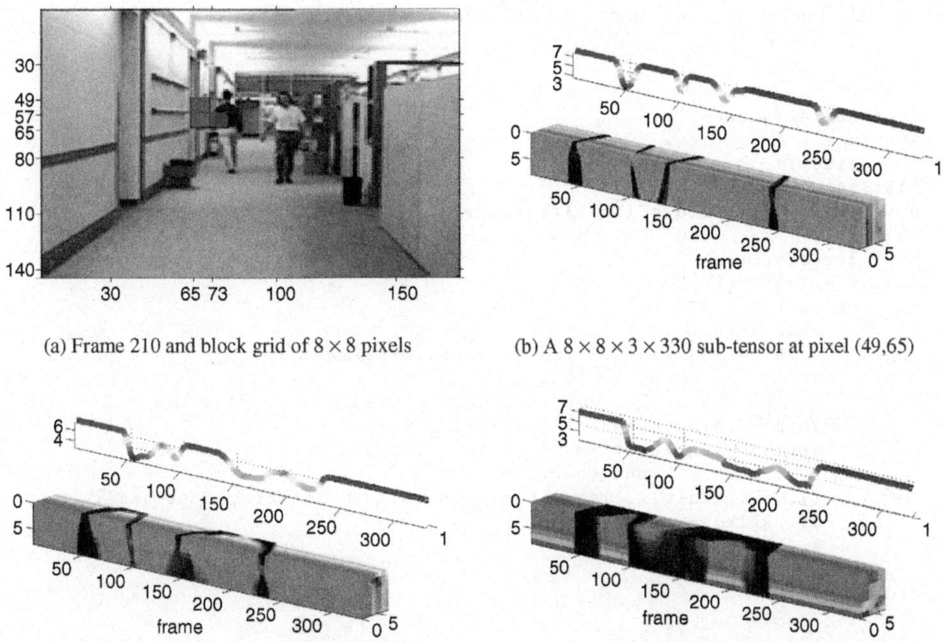

(a) Frame 210 and block grid of 8 × 8 pixels

(b) A 8 × 8 × 3 × 330 sub-tensor at pixel (49,65)

(c) A 8 × 8 × 3 × 330 sub-tensor at pixel (49,73)

(d) A 8 × 8 × 3 × 330 sub-tensor at pixel (57,73)

Figure 7.35 Illustration of three sub-tensors built from 8 × 8 blocks and extending through time along all frames. The tensor size is 8 × 8 × 3 × 330 for a color tensor, or 8 × 8 × 330 for a gray-scale tensor. The three red squares in (a) indicate the locations of the three subtensors in (b), (c), and (d). Factorizations of these sub-tensors with only component $J = 1$ return the last factors $\mathbf{A}^{(4)}$ (or $\mathbf{A}^{(3)}$ for gray-scale tensor) which express the stationary or dynamic blocks (frontal slices). Detection of the stationary points on $\mathbf{A}^{(4)}$ (in red) gives a set of frame indices of background blocks at the observed points (pixels).

to obtain the affine transformation matrix $\mathbf{M}_k \in \mathbb{R}^{3\times 3}$, $k = 1, 2, \ldots, K-1$, where K is the number of frames and

$$\mathbf{I}_{k+1} = \mathbf{M}_k \, \mathbf{I}_k = \begin{pmatrix} a_1 & a_2 & a_3 \\ a_4 & a_5 & a_6 \\ 0 & 0 & 1 \end{pmatrix} \mathbf{I}_k, \tag{7.203}$$

where $\mathbf{I}_k$ expresses the N pixels of the frame k

$$\mathbf{I}_k = \begin{pmatrix} x_1 & x_2 & \cdots & x_N \\ y_1 & y_2 & \cdots & y_N \\ 1 & 1 & \cdots & 1 \end{pmatrix}. \tag{7.204}$$

2. Compute the accumulated global motion between frame k and frame 1.

$$\mathbf{M}_k^a = \mathbf{M}_{k-1} \, \mathbf{M}_{k-1}^a, \quad k = 2, \ldots, K-1. \tag{7.205}$$

3. Transform all the $(K-1)$ frames by affine matrices $\mathbf{M}_k^a$ to obtain a new sequence which compensates for the global motion.

Listing 7.13 Background estimation for the `HallMonitor` sequence.

```
% Copyright 2008 by Anh Huy Phan and Andrzej Cichocki
% Load Hall sequence
clear
load Hall4Dtensor;
%% Process full block
sT = size(T);
blksize = 8;
d = zeros(sT(4),prod(sT(1:2))/blksize^2);
kblk = 0;
xy = zeros(1,2);
for xu = 1:8:sT(1)-7
    for yu = 1:8:sT(2)-7
        kblk = kblk + 1;
        Tblk  = T(xu:xu+blksize-1, yu:yu+blksize-1,:,:);

        %% Factorize subtensor with Parafac algorithms R = 1
        Yblk = permute(tensor(Tblk),[4 1 2 3]);
        options = struct('verbose',1,'tol',1e-6,'maxiters',500,...
            'init',2,'nonlinearproj',1);
        [X_hals,Uinit,A_,ldam,iter] = parafac_hals(Yblk,1,options);
        d(:,kblk) = double(A_{1});
        xy(kblk,:) = [xu yu];
    end
end

%% Find stationary blocks and build background image
maxd = max(d); mind = min(d);
thresh = 0.005;
Imbgr = zeros(sT(1:3));
for k = 1:size(d,2);
    edges = [mind(k):thresh:maxd(k) maxd(k)+eps] ;
    [n,bin] = histc(d(:,k),edges);
    m = mode(bin);
    indbgr = find((d(:,k)≥edges(m)) & (d(:,k)≤ edges(m+1)));
    bgrblk  = median(T(xy(k,1):xy(k,1)+blksize-1,...
        xy(k,2):xy(k,2)+blksize-1,:,indbgr),4);
    Imbgr(xy(k,1):xy(k,1)+blksize-1, xy(k,2):xy(k,2)+blksize-1,:) = bgrblk;
end

%% Display the estimated background image
imshow(Imbgr)
```

Figure 7.36(b) shows that after global motion compensation using IMAGE REGISTRATION software [43] the sequence of images is now aligned. All background points have orbits represented by straight lines. The same procedure can be used to reconstruct the background, as shown in Figure 7.36.

The MATLAB source code for this example is given in Listing 7.14.

7.8 Discussion and Conclusions

The focus of this chapter has been on fast and efficient algorithms for tensor factorizations and decompositions using nonnegativity and sparsity constraints. We have introduced a wide class of heuristic NTF and NTD algorithms and confirmed their validity and practicality by extensive simulations. In particular, the class of ALS, local HALS NTF and HALS NTD algorithms has been discussed and their robustness to noise

Listing 7.14 Background estimation for the `Panorama` sequence.

```
% Copyright 2008 by Anh Huy Phan and Andrzej Cichocki
% load parnorama files
list = dir('*.png');
T = cell(1,numel(list));
for k = 1:numel(list)
    T{k} = imread(list(k).name);
end
sT = size(T{1});
T = im2double(reshape(cell2mat(T),sT(1),sT(2),[]));
In = size(T); % 3-D tensor
%%  Estimate affine global motion with 6-parameters
iters = [5,3,3];
M = zeros(3,3,In(3));
for k = 1:In(3)
    M(:,:,k) = estMotionMulti2(T(:,:,1),T(:,:,k),iters,[],1,1);
end
%% Inverse affine transform and build new tensor
Tnew = T;
for k = 2:In(3)
    Tnew(:,:,k) = warpAffine2(T(:,:,k),M(:,:,k));
end
%% Process all sub-blocks
sT = size(T);blksz = 8;
d = zeros(sT(end),prod(sT(1:2))/blksz^2);
kblk = 0; thresh = 0.005;
xy = zeros(1,2);
Imbgr = zeros(sT(1:2));
for xu = 1:8:sT(1)-7
    for yu = 1:8:sT(2)-7
        kblk = kblk + 1;
        Tblk  = Tnew(xu:xu+blksz-1, yu:yu+blksz-1,:);
        nanind = find(isnan(Tblk));
        [r,c,nanfr] = ind2sub(size(Tblk),nanind);
        Tblknew = Tblk(:,:,setdiff(1:sT(end),unique(nanfr)));
        %% Factorize subtensor with Parafac algorithms R = 1
        Yblk = permute(tensor(Tblknew),[3 1 2 ]);
        options = struct('verbose',1,'tol',1e-6,'maxiters',500,...
          'nonlinearproj',1,'fixsign',0,'fitmax',.99);
        [X_hals,Uinit,A_,ldam,iter] = parafac_hals2_acc(Yblk,1,options);
        d = full(A_{1});

        edges = [min(d):thresh:max(d) max(d)+eps] ;
        [n,bin] = histc(d ,edges);
        m = mode(bin);
        indbgr  = find((d ≥edges(m)) & (d ≤ edges(m+1)));
        Imbgr(xu:xu+blksz-1,yu:yu+blksz-1)=median(Tblknew(:,:,indbgr),3);
    end
end
%% Display the estimated background image
imshow(Imbgr)
```

and excellent convergence properties for synthetic and real-world data sets have been illustrated. It has been shown that for large-scale problems, the developed algorithms help to reduce computational costs and avoid memory overflow error. The algorithms considered can be extended to semi-NTF and to sparse PARAFAC using suitable nonlinear projections and regularization terms [21]. They can be further extended by imposing additional, natural physical constraints, such as smoothness, continuity, closure, unimodality, local

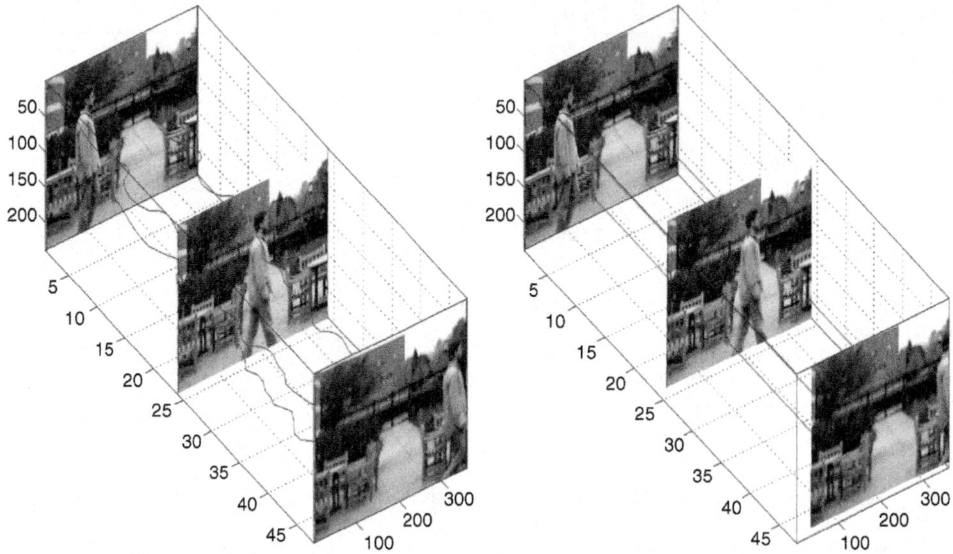

(a) Motion orbits of some background pixels are unstable due to camera motion

(b) Fixing of the unstable alignment of frames by compensation of global motion

Figure 7.36 Visualization of the Panorama sequence recorded with an unstable camera motion. (a) background pixels move on rough motion orbits. (b) stabilizing background pixels is the first step prior to sub-tensor factorization. Based on global motion compensation the frames are aligned with the first frame and a new image sequence is generated for the next processing steps.

rank - selectivity, and/or by taking into account *a prior* knowledge about some specific multi-dimensional models and multi-way data. The algorithms presented in this chapter can likely find applications in three areas of data analysis (especially, EEG and fMRI) and signal/image processing: (i) multi-way blind source separation, (ii) model reductions and selection and (iii) sparse image coding. The MATLAB code for the algorithms and experiments can be found in the toolboxes NMFLAB/NTFLAB/MULTIWAYLAB.

(a) Background of the Hall Monitor sequence

(b) Background of the panorama sequence

Figure 7.37 Background of the Hall Monitor and Panorama sequences.

Appendix 7.A: Evaluation of Interactions and Relationships Among Hidden Components for NTD Model

The Tucker model allows for the extraction of different numbers of factors in each of the modes. The main difference between PARAFAC and Tucker models is that the Tucker model permits interactions within each modality while the PARAFAC model does not. In other words, the Tucker model encompasses all possible linear interactions between the components (vectors) pertaining to the various modalities of the data. These interactions can be explicitly expressed by a core tensor, if the components $\boldsymbol{a}_{j_n}^{(n)}$ are normalized to unit length; they are assumed to be additive if nonnegativity constraints are imposed.

In general, each component in the Nonnegative Tucker Decomposition model may have J_n different interactions with all the components for all the other factors $\mathbf{A}^{(n)}$. For each pair (j_p, j_q), we define the Joint Rate (JR) index as follows

$$JR_{j_q}^{j_p} = \frac{\displaystyle\sum_{\substack{j_1,\ldots,j_{p-1},j_{p+1},\ldots,\\ j_{q-1},j_{q+1},\ldots,j_N}} \left\| g_{j_1,\ldots,j_p,\ldots,j_q,\ldots,j_N}\, \boldsymbol{a}_{j_1}^{(1)} \circ \cdots \circ \boldsymbol{a}_{j_p}^{(p)} \circ \cdots \circ \boldsymbol{a}_{j_q}^{(q)} \circ \cdots \circ \boldsymbol{a}_{j_N}^{(N)} \right\|_F^2}{\displaystyle\sum_{j_1,\ldots,j_{q-1},j_{q+1},\ldots,j_N} \left\| g_{j_1,\ldots,j_q,\ldots,j_N}\, \boldsymbol{a}_{j_1}^{(1)} \circ \cdots \circ \boldsymbol{a}_{j_q}^{(q)} \circ \cdots \circ \boldsymbol{a}_{j_N}^{(N)} \right\|_F^2}, \tag{A.1}$$

which allows us to evaluate the strength of interaction among various components.

The $JR_{j_q}^{j_p}$ index between component $\boldsymbol{a}_{j_p}^{(p)}$, $(j_p = 1, \ldots, J_p)$ and the component $\boldsymbol{a}_{j_q}^{(q)}$, $(j_q = 1, \ldots, J_q)$ can be interpreted as the ratio of the total energy of all the rank-one tensors, to which both components $\boldsymbol{a}_{j_p}^{(p)}$ and $\boldsymbol{a}_{j_q}^{(q)}$ contribute, and the total energy of all the rank-one tensors in which the component $\boldsymbol{a}_{j_q}^{(q)}$ interacts.

Our objective is to identify the dominant relationships between components. Let $\underline{\mathbf{Y}}_j$ denote the rank-one tensor built up from N unit length components $\boldsymbol{a}_{j_1}^{(1)}, \boldsymbol{a}_{j_2}^{(2)}, \ldots, \boldsymbol{a}_{j_N}^{(N)}$ in the ℓ_2-norm sense: $\left\|\boldsymbol{a}_{j_n}^{(n)}\right\|_2^2 = \boldsymbol{a}_{j_n}^{(n)T} \boldsymbol{a}_{j_n}^{(n)} = 1$ for $n = 1, 2, \ldots, N$, that is

$$\underline{\mathbf{Y}}_j = \boldsymbol{a}_{j_1}^{(1)} \circ \boldsymbol{a}_{j_2}^{(1)} \circ \cdots \circ \boldsymbol{a}_{j_N}^{(N)}. \tag{A.2}$$

Note that the ℓ_2-norm of this tensor in its vectorized form is given by

$$\begin{aligned} \left\|\mathrm{vec}\left(\underline{\mathbf{Y}}_j\right)\right\|_2^2 &= \left\|\boldsymbol{a}_{j_1}^{(1)} \otimes \boldsymbol{a}_{j_2}^{(2)} \otimes \cdots \otimes \boldsymbol{a}_{j_N}^{(N)}\right\|_2^2 \\ &= \left(\boldsymbol{a}_{j_1}^{(1)} \otimes \boldsymbol{a}_{j_2}^{(2)} \otimes \cdots \otimes \boldsymbol{a}_{j_N}^{(N)}\right)^T \left(\boldsymbol{a}_{j_1}^{(1)} \otimes \boldsymbol{a}_{j_2}^{(2)} \otimes \cdots \otimes \boldsymbol{a}_{j_N}^{(N)}\right) \\ &= \left(\boldsymbol{a}_{j_1}^{(1)T} \boldsymbol{a}_{j_1}^{(1)}\right) \left(\boldsymbol{a}_{j_2}^{(2)T} \boldsymbol{a}_{j_2}^{(2)}\right) \cdots \left(\boldsymbol{a}_{j_N}^{(N)T} \boldsymbol{a}_{j_N}^{(N)}\right) = 1, \end{aligned} \tag{A.3}$$

and a rank-one tensor of ℓ_2-norm unit-length components has unit energy. Therefore, for NTD, if all components are ℓ_2-norm unit length vectors, the energy of the original data will be expressed only by the coefficients of the core tensor. This leads to a simplified expression of the JR index as follows

$$JR_{j_q}^{j_p} = \frac{\displaystyle\sum_{\substack{j_1,\ldots,j_{p-1},j_{p+1},\ldots,\\ j_{q-1},j_{q+1},\ldots,j_N}} g^2_{j_1,\ldots,j_p,\ldots,j_q,\ldots,j_N}}{\displaystyle\sum_{j_1,\ldots,j_{q-1},j_{q+1},\ldots,j_N} g^2_{j_1,\ldots,j_q,\ldots,j_N}} = \frac{\left\|\underline{\mathbf{G}}_{(r_p,r_q)=(j_p,j_q)}\right\|_F^2}{\left\|\underline{\mathbf{G}}_{r_q=j_q}\right\|_F^2}, \tag{A.4}$$

where $\underline{\mathbf{G}}_{r_q=j_q}$ is an $(N-1)$-th order subtensor of size $J_1 \times \cdots \times J_{q-1} \times J_{q+1} \times \cdots \times J_N$ obtained by fixing the q-th index of the core tensor $\underline{\mathbf{G}}$ to value j_q, and $\underline{\mathbf{G}}_{(r_p,r_q)=(j_p,j_q)}$ is an $(N-2)$-th order subtensor of size

$J_1 \times \cdots \times J_{p-1} \times J_{p+1} \cdots \times J_{q-1} \times J_{q+1} \times \cdots \times J_N$ by fixing the p-th index to value j_p, and the q-th index to value j_q.

For example, for a third-order core tensor, the JR index between the first component $\boldsymbol{a}_1^{(1)}$ of the factor $\mathbf{A}^{(1)}$ and the second component $\boldsymbol{a}_2^{(3)}$ of the factor $\mathbf{A}^{(3)}$ is given by

$$JR_{2_3}^{1_1} = \frac{\sum_{j_2=1}^{J_2} g_{1,j_2,2}^2}{\sum_{j_1=1}^{J_1}\sum_{j_2=1}^{J_2} g_{j_1,j_2,2}^2} = \frac{\left\|\boldsymbol{g}_{2_1}\right\|_2^2}{\|\mathbf{G}_2\|_2^2}, \tag{A.5}$$

where $\mathbf{G}_2$ is the second frontal slice of core tensor $\underline{\mathbf{G}}$, and $\boldsymbol{g}_{2_1}$ is the first row vector of this slice. The following MATLAB code illustrates how to compute all the JR indices between pair of components from the factor $\mathbf{A}^{(1)}$ and the factor $\mathbf{A}^{(3)}$; it returns the JR in the form $JR(j_1, j_3) = JR_{j_3/3}^{j_1/1}$.

Listing A.15 Compute the JR index of the factor 1 for the factor 3 in a three-way core tensor.

```
% Copyright 2008 by Anh Huy Phan, Andrzej Cichocki
JR = squeeze(sum(double(G).^2,2));
JR = bsxfun(@rdivide,Jrate,sum(Jrate));
```

The index $JR_{j_q}^{j_p}$, $(j_p = 1, 2, \ldots, J_p)$ of a J_p between component $\boldsymbol{a}_{j_p}^{(p)}$ and component $\boldsymbol{a}_{j_q}^{(q)}$ gives the percent rate of the interaction for each pair.

Appendix 7.B: Computation of a Reference Tensor

In order to compute JR indices we need to evaluate a reference (normalized) core tensor. This section describes an efficient method for computing the reference core tensor $\underline{\mathbf{G}}$ by normalizing factors $\mathbf{A}^{(n)}$ to unit length vectors. For a general case, a reference core tensor $\underline{\mathbf{G}}$ can be defined as follows

$$\underline{\mathbf{G}}_f = \underline{\mathbf{G}} \times_1 \mathbf{D}_1 \times_2 \mathbf{D}_2 \cdots \times_N \mathbf{D}_N = \underline{\mathbf{G}} \times \{\mathbf{D}\}, \tag{B.1}$$

where $\underline{\mathbf{G}}_f$ is the core tensor after fixing, $\mathbf{D}_n$ is the diagonal matrix of ℓ_p-norms[15] of the column vectors in factor $\mathbf{A}^{(n)}$, that is,

$$\boldsymbol{d}_n = \left[\|\boldsymbol{a}_1^{(n)}\|_2, \|\boldsymbol{a}_2^{(n)}\|_2, \ldots, \|\boldsymbol{a}_N^{(n)}\|_2\right], \tag{B.2}$$

$$\mathbf{D}_n = \operatorname{diag}\{\boldsymbol{d}_n\}, \quad \forall n. \tag{B.3}$$

[15] Any ℓ_p-norm with $p \in [1, \infty]$ can be used; typically, for calculating JR index, we use the ℓ_2-norm.

A considerably simpler and faster method for the reference core tensor as it is given below. Let $\mathbf{C}$ denote a Kronecker product of two diagonal matrices $\text{diag}\{\boldsymbol{d}_p\}$ and $\text{diag}\{\boldsymbol{d}_q\}$

$$\mathbf{C} = \text{diag}\{\boldsymbol{d}_p\} \otimes \text{diag}\{\boldsymbol{d}_q\} = \tag{B.4}$$

$$= \begin{bmatrix} d_{p_1}d_{q_1} & 0 & \cdots & 0 & \cdots & \cdots & 0 & 0 & \cdots & 0 \\ 0 & d_{p_1}d_{q_2} & \cdots & 0 & \cdots & \cdots & 0 & 0 & \cdots & 0 \\ \vdots & \vdots & \ddots & \vdots & & & \vdots & \vdots & \ddots & \vdots \\ 0 & 0 & \cdots & d_{p_1}d_{q_{J_q}} & \cdots & \cdots & 0 & 0 & \cdots & 0 \\ \vdots & \vdots & & \vdots & \ddots & & \vdots & \vdots & & \vdots \\ \vdots & \vdots & & \vdots & & \ddots & \vdots & \vdots & & \vdots \\ 0 & 0 & \vdots & 0 & \cdots & \cdots & d_{p_{J_p}}d_{q_1} & 0 & \cdots & 0 \\ 0 & 0 & \vdots & 0 & \cdots & \cdots & 0 & d_{p_{J_p}}d_{q_2} & \cdots & 0 \\ \vdots & \vdots & \ddots & \vdots & \vdots & \vdots & \vdots & \vdots & \ddots & \vdots \\ 0 & 0 & 0 & 0 & 0 & 0 & 0 & 0 & \cdots & d_{p_{J_p}}d_{q_{J_q}} \end{bmatrix}. \tag{B.5}$$

From the definition of the Kronecker product, a coefficient c_{tt} at the position $t = (j_p - 1)\, J_q + j_q$ ($j_p = 1, \ldots, J_p$, $j_q = 1, \ldots, J_q$) on the diagonal of matrix $\mathbf{C}$ is calculated as

$$c_{tt} = d_{p_{j_p}}\, d_{q_{j_q}}. \tag{B.6}$$

Thus, the diagonal of matrix $\mathbf{C}$ is exactly the Kronecker product of two vectors $\boldsymbol{d}_p$ and $\boldsymbol{d}_q$

$$\text{diag}\{\mathbf{C}\} = \boldsymbol{d}_p \otimes \boldsymbol{d}_q \tag{B.7}$$

or

$$\text{diag}\{\boldsymbol{d}_p\} \otimes \text{diag}\{\boldsymbol{d}_q\} = \text{diag}\{\boldsymbol{d}_p \otimes \boldsymbol{d}_q\}. \tag{B.8}$$

This result can be readily extended to the Kronecker product of N vectors $\boldsymbol{d}_1, \boldsymbol{d}_2, \ldots, \boldsymbol{d}_N$

$$\text{diag}\{\boldsymbol{d}_1\} \otimes \text{diag}\{\boldsymbol{d}_2\} \otimes \cdots \otimes \text{diag}\{\boldsymbol{d}_N\} = \text{diag}\{\boldsymbol{d}_1 \otimes \boldsymbol{d}_2 \cdots \otimes \boldsymbol{d}_N\}. \tag{B.9}$$

Equation (B.1) can be represented in vectorized form as

$$\begin{aligned} \text{vec}\left(\underline{\mathbf{G}}_f\right) &= (\mathbf{D}_N \otimes \cdots \otimes \mathbf{D}_2 \otimes \mathbf{D}_1)\ \text{vec}\left(\underline{\mathbf{G}}\right) \\ &= (\text{diag}\{\boldsymbol{d}_N\} \otimes \cdots \otimes \text{diag}\{\boldsymbol{d}_2\} \otimes \text{diag}\{\boldsymbol{d}_1\})\ \text{vec}\left(\underline{\mathbf{G}}\right) \\ &= \text{diag}\{\boldsymbol{d}_N \otimes \cdots \otimes \boldsymbol{d}_2 \otimes \boldsymbol{d}_1\}\ \text{vec}\left(\underline{\mathbf{G}}\right) \\ &= (\boldsymbol{d}_N \otimes \cdots \otimes \boldsymbol{d}_2 \otimes \boldsymbol{d}_1) \circledast \text{vec}\left(\underline{\mathbf{G}}\right) \\ &= \text{vec}\,(\boldsymbol{d}_1 \circ \boldsymbol{d}_2 \circ \cdots \circ \boldsymbol{d}_N) \circledast \text{vec}\left(\underline{\mathbf{G}}\right) \end{aligned} \tag{B.10}$$

and its matricization yields the final expression

$$\begin{aligned}\underline{\mathbf{G}}_f &= (\boldsymbol{d}_1 \circ \boldsymbol{d}_2 \circ \cdots \circ \boldsymbol{d}_N) \circledast \underline{\mathbf{G}} \\ &= \underline{\mathbf{D}} \circledast \underline{\mathbf{G}}.\end{aligned} \tag{B.11}$$

Finally, computation of the reference core tensor $\underline{\mathbf{G}}_f$ can be achieved conveniently via the Hadamard product of this core tensor and the rank-one tensor built up from N ℓ_2-norm vectors.

Appendix 7.C: Trilinear and Direct Trilinear Decompositions for Efficient Initialization

The Direct Tri-Linear Decomposition (DTLD) was developed by Sanchez and Kowalski [69] and can be used as a very efficient tool for the initialization of our NTF algorithms.

Let $\underline{\mathbf{Y}}$ denote a three-way tensor of size $J \times J \times 2$ with only two frontal slices: $\mathbf{Y}_1 = \underline{\mathbf{Y}}_{j_3=1}$ and $\mathbf{Y}_2 = \underline{\mathbf{Y}}_{j_3=2}$. We can factorize this tensor into three factors $\mathbf{A} \in \mathbb{R}^{J\times J}$, $\mathbf{B} \in \mathbb{R}^{J\times J}$ and $\mathbf{C} = [\underline{\boldsymbol{c}}_1^T, \underline{\boldsymbol{c}}_2^T]^T \in \mathbb{R}^{2\times J}$, according to the PARAFAC model by unfolding the data tensor in the mode-1, that is

$$\mathbf{Y}_{(1)} = [\mathbf{Y}_1, \mathbf{Y}_2] = \mathbf{A}\,(\mathbf{C} \odot \mathbf{B})^T \tag{C.1}$$

$$\mathbf{Y}_1 = \mathbf{A}\,\mathbf{D}_1\,\mathbf{B}^T, \qquad \mathbf{Y}_2 = \mathbf{A}\,\mathbf{D}_2\,\mathbf{B}^T, \tag{C.2}$$

where $\mathbf{D}_1 = \operatorname{diag}\left\{\underline{\boldsymbol{c}}_1\right\}$, $\mathbf{D}_2 = \operatorname{diag}\left\{\underline{\boldsymbol{c}}_2\right\}$. This yields to the following expressions:

$$\begin{aligned}\mathbf{Y}_1\,(\mathbf{B}^T)^{-1} &= \mathbf{A}\,\mathbf{D}_1\,\mathbf{B}^T\,(\mathbf{B}^T)^{-1} = \mathbf{A}\,\mathbf{D}_2\,\mathbf{D}_2^{-1}\mathbf{D}_1 = \mathbf{A}\,\mathbf{D}_2\mathbf{B}^T(\mathbf{B}^T)^{-1}\mathbf{D} \\ &= \mathbf{Y}_2\,(\mathbf{B}^T)^{-1}\mathbf{D},\end{aligned} \tag{C.3}$$

where $\mathbf{D} = \mathbf{D}_2^{-1}\mathbf{D}_1 = \operatorname{diag}\{\underline{\boldsymbol{c}}_1 \oslash \underline{\boldsymbol{c}}_2\}$. Eq. (C.3) corresponds to a generalized eigenvalue problem of two square matrices $\mathbf{Y}_1$ and $\mathbf{Y}_2$, in which $\mathbf{D}$ is the diagonal matrix of generalized eigenvalues, and columns of a full matrix $(\mathbf{B}^T)^{-1}$ are the corresponding eigenvectors. This problem can be solved by the MATLAB function `eig` as $[(\mathbf{B}^T)^{-1}, \mathbf{D}] = \texttt{eig}(\mathbf{Y}_1, \mathbf{Y}_2)$. Due to the scaling ambiguity, we can set the second row vector $\underline{\boldsymbol{c}}_2 = \mathbf{1}^T$, hence, the first row vector $\underline{\boldsymbol{c}}_1$ represents generalized eigenvalues:

$$\underline{\boldsymbol{c}}_1 = \operatorname{diag}\{\mathbf{D}\}^T. \tag{C.4}$$

The above derivation is a special case of the generalized rank annihilation method (GRAM) [13,14,40]. This is a simple yet an efficient initialization technique for the PARAFAC models for which the pseudo-code is listed in Algorithm C.14.

Algorithm C.14: GRAM

Input: $\underline{\mathbf{Y}}$: input data of size $J \times J \times 2$, J: number of basis components
Output: $\mathbf{A} \in \mathbb{R}^{J\times J}$, $\mathbf{B} \in \mathbb{R}^{J\times J}$ and $\mathbf{C} \in \mathbb{R}^{2\times J}$ such that $\underline{\mathbf{Y}} = [\![\mathbf{A}, \mathbf{B}, \mathbf{C}]\!]$

begin
  $[\boldsymbol{\Phi}, \mathbf{D}] = \texttt{eig}(\mathbf{Y}_1, \mathbf{Y}_2)$
  $\mathbf{B} = (\boldsymbol{\Phi}^T)^{-1}$
  $\mathbf{A} = \mathbf{Y}_2\boldsymbol{\Phi}$
  $\mathbf{C} = \left[\operatorname{diag}\{\mathbf{D}\}\,,\ \mathbf{1}\right]^T$
end

Algorithm C.15: DTLD–Initialization for three-way PARAFAC model

Input: $\underline{\mathbf{Y}}$: input data of size $I \times T \times Q$, J: number of basis components
Output: $\mathbf{A} \in \mathbb{R}^{I \times J}$, $\mathbf{B} \in \mathbb{R}^{T \times J}$ and $\mathbf{C} \in \mathbb{R}^{Q \times J}$ such that $\underline{\mathbf{Y}} = [\![\mathbf{A}, \mathbf{B}, \mathbf{C}]\!]$

begin
  $\underline{\mathbf{Y}} = [\![\underline{\mathbf{G}}; \mathbf{A}_t, \mathbf{B}_t, \mathbf{C}_t]\!]$ /* Tucker ALS */
  $[\boldsymbol{\Phi}, \mathbf{D}] = \texttt{eig}(\mathbf{G}_1, \mathbf{G}_2)$
  $\mathbf{A} = \mathbf{A}_t \, \mathbf{G}_2 \, \boldsymbol{\Phi}$
  $\mathbf{B} = \mathbf{B}_t \, (\boldsymbol{\Phi}^T)^{-1}$
  $\mathbf{C} = \mathbf{C}_t \, \left[\mathrm{diag}\{\mathbf{D}\}, \mathbf{1}\right]^T$ /* or $\mathbf{C} = \mathbf{Y}_{(3)} \, (\mathbf{B} \odot \mathbf{A})$ */
end

If the frontal slices $\mathbf{Y}_q$, $(q = 1, 2)$ are nonsquare matrices of size $I \times T$, $I, T \geq J$, then they can be transformed to matrices $\mathbf{T}_q$ of size $J \times J$ by projecting onto the two orthonormal matrices $\mathbf{U} \in \mathbb{R}^{I \times J}$ and $\mathbf{V} \in \mathbb{R}^{T \times J}$, usually via a singular value decomposition $[\mathbf{U}, \mathbf{S}, \mathbf{V}] = \texttt{svd}(\mathbf{Y}_1 + \mathbf{Y}_2, J)$

$$\mathbf{T}_q = \mathbf{U}^T \, \mathbf{Y}_q \, \mathbf{V}, \qquad (q = 1, 2). \tag{C.5}$$

Applying the GRAM algorithm to the tensor $\underline{\mathbf{T}}$, factors $\mathbf{A}$, $\mathbf{B}$ and $\mathbf{C}$ can be found as

$$[\boldsymbol{\Phi}, \mathbf{D}] = \texttt{eig}(\mathbf{T}_1, \mathbf{T}_2) \tag{C.6}$$

$$\mathbf{A} = \mathbf{U} \, \mathbf{T}_2 \boldsymbol{\Phi} \tag{C.7}$$

$$\mathbf{B} = \mathbf{V} (\boldsymbol{\Phi}^T)^{-1} \tag{C.8}$$

$$\mathbf{C} = \left[\mathrm{diag}\{\mathbf{D}\}, \mathbf{1}\right]^T. \tag{C.9}$$

If matrix $\mathbf{D}$ has complex eigenvalues, we can use the MATLAB function `cdf2rdf` to transform the system so $\mathbf{D}$ is in real diagonal form, with 2-by-2 real blocks along the diagonal replacing the original complex pairs. The eigenvectors are also transformed so that $\mathbf{T}_1 \, \mathbf{T}_2^{-1} = \boldsymbol{\Phi} \, \mathbf{D} \, \boldsymbol{\Phi}^{-1}$ continues to hold. The individual columns of $\boldsymbol{\Phi}$ are no longer eigenvectors, but each pair of vectors associated with a 2-by-2 block in $\mathbf{D}$ spans the corresponding invariant vectors, see [39] for further discussion.

The GRAM method is restricted to tensors having only two slices. However, it can be applied to decompose a general three-way tensor with more than two slices. The following algorithm describes this approach and is called the direct trilinear decomposition (DTLD) (see Algorithm C.15). The idea is to decompose a three-way data tensor $\underline{\mathbf{Y}} \in \mathbb{R}^{I \times T \times Q}$ using the Tucker model with core tensor $\underline{\mathbf{G}} \in \mathbb{R}^{J \times J \times 2}$

$$\underline{\mathbf{Y}} = [\![\underline{\mathbf{G}}; \mathbf{A}_t, \mathbf{B}_t, \mathbf{C}_t]\!], \tag{C.10}$$

where $\mathbf{A}_t \in \mathbb{R}^{I \times J}$, $\mathbf{B}_t \in \mathbb{R}^{T \times J}$, and $\mathbf{C}_t \in \mathbb{R}^{Q \times 2}$ are factors obtained by using the ALS Tucker algorithm. Then we factorize the core tensor $\underline{\mathbf{G}}$ using the GRAM algorithm

$$\underline{\mathbf{G}} = [\![\mathbf{A}_f, \mathbf{B}_f, \mathbf{C}_f]\!]. \tag{C.11}$$

The final factors $\mathbf{A}$, $\mathbf{B}$ and $\mathbf{C}$ of the factorizing model: $\underline{\mathbf{Y}} = [\![\mathbf{A}, \mathbf{B}, \mathbf{C}]\!]$ are determined as the products

$$\mathbf{A} = \mathbf{A}_t \, \mathbf{A}_f \tag{C.12}$$

$$\mathbf{B} = \mathbf{B}_t \, \mathbf{B}_f \tag{C.13}$$

$$\mathbf{C} = \mathbf{C}_t \, \mathbf{C}_f. \tag{C.14}$$

Since **A** and **B** are orthogonal loading matrices, the factor **C** can be derived directly from the ALS algorithm

$$\mathbf{C} = \mathbf{Y}_{(3)} \left((\mathbf{B} \odot \mathbf{A})^T \right)^{\dagger} \tag{C.15}$$

$$= \mathbf{Y}_{(3)} (\mathbf{B} \odot \mathbf{A}) \left((\mathbf{B} \odot \mathbf{A})^T (\mathbf{B} \odot \mathbf{A}) \right)^{-1} \tag{C.16}$$

$$= \mathbf{Y}_{(3)} (\mathbf{B} \odot \mathbf{A}) \left(\left(\mathbf{B}^T\mathbf{B}\right) \circledast \left(\mathbf{A}^T\mathbf{A}\right) \right)^{-1} \tag{C.17}$$

$$= \mathbf{Y}_{(3)} (\mathbf{B} \odot \mathbf{A}). \tag{C.18}$$

This algorithm was originally developed in [69], and implemented in the N-way toolbox [4] as a useful initialization algorithm for third-order tensor factorizations. Note that decomposition of the tensor $\underline{\mathbf{Y}}$ following the Tucker model requires very few iterations, typically less than 10. For nonnegative constraints, the half-wave rectifier can be applied to the estimated factors.

Appendix 7.D: MATLAB Source Code for Alpha NTD Algorithm

```
function [T,A,G,fit,iter] = tucker_alpha(Y,R,opts)
% Alpha NTD algorithm using the Alpha divergence
% INPUT
% Y       :    tensor with size of I1 x I2 x ... x IN
% R       :    size of core tensor R1 x R2 x ... x RN: [R1, R2, ..., RN]
% opts    :    structure of optional parameters for algorithm (see defoptions)
%    .tol:     tolerance of stopping criteria (explained variation)     (1e-6)
%    .maxiters: maximum number of iteration                              (50)
%    .init:    initialization type: 'random', 'eigs', 'nvecs' (HOSVD) (random)
%    .alpha: alpha parameter                                              (1)
%    .nonlinearproj: apply half-wave rectifying or not                    (1)
%    .orthoforce:  orthogonal constraint to initialization using ALS
%    .updateYhat:  update Yhat or not, using in Fast Alpha NTD with ell-1
%                  normalization for factors                              (0)
%    .ellnorm:    normalization type                                      (1)
%    .projector: projection type for reconstructed tensor Yhat: max,(real)
%    .fixsign:   fix sign for components of factors.                      (1)
%
% Copyright of Andrzej Cichocki, Anh Huy Phan
% Ver 1.0 12/2008, Anh Huy Phan

% Set algorithm parameters from input or by using defaults
defoptions = struct('tol',1e-6,'maxiters',50,'init','random','alpha',1,...
     'nonlinearproj',1,'orthoforce',1,'updateYhat',0,'ellnorm',1,...
     'projector','real','fixsign',1);
if ~exist('opts','var')
     opts = struct;
end
opts = scanparam(defoptions,opts);

% Extract number of dimensions and norm of Y.
N = ndims(Y); mY = max(Y(:)) ; Y = Y/mY; normY = norm(Y);
if numel(R) == 1,      R = R(ones(1,N)); end
%% Set up and error checking on initial guess for U.
[A,G] = ntd_initialize(Y,opts.init,opts.orthoforce,R);

fprintf('\nAlpha NTD:\n');
UpdateYhat
normresidual = norm(Y(:) - Yhat(:)/max(Yhat(:)));
fit = 1 - (normresidual / normY); %fraction explained by model
```

```
fprintf(' Iter %2d: fit = %e \n', 0, fit);
%% Main Loop: Iterate until convergence
for iter = 1:opts.maxiters
    pause(.0001) % force to interrupt
    fitold = fit;
     % Iterate over all N modes of the tensor
    for n = 1:N
        if opts.updateYhat,  UpdateYhat, end
        A{n} = A{n}.* real((double(tenmat(ttm(Ytilde,A,-n,'t'),n)...
            *tenmat(G,n,'t'))).^(1/opts.alpha));
        if opts.fixsign
            A{n} = fixsigns(A{n});
        end
        if opts.nonlinearproj
            A{n} = max(eps,A{n});
        end
        if opts.ellnorm>0
            A{n} = bsxfun(@rdivide,A{n},(sum(A{n}...
                .^opts.ellnorm)).^(1/opts.ellnorm));
        end
    end
    UpdateYhat
    G = G.*tenfun(@real,ttm(Ytilde,A,'t').^(1/opts.alpha));
    UpdateYhat
    if (mod(iter,10) ==1) || (iter == opts.maxiters)
        % Compute fit
        normresidual = sqrt(normY^2 + norm(Yhat/max(Yhat(:)))^2 - ...
            2 * innerprod(Y,Yhat/max(Yhat(:))));
        fit = 1 - (normresidual/normY);          %fraction explained by model
        fitchange = abs(fitold - fit);
        fprintf('Iter %2d: fit = %e Δfit = %7.1e\n',iter, fit, fitchange);
        if (fitchange < opts.tol) && (fit>0)          % Check for convergence
            break;
        end
    end
end
%% Compute the final result
T = ttm(G,A)*mY;

    function  UpdateYhat
        Yhat = ttm(G,A);
        Ytilde = ((Y+eps)./(Yhat+eps)).^opts.alpha;
        switch opts.projector
            case 'max'
                Ytilde = Ytilde.*(Ytilde>eps);
            case 'real'
                Ytilde = tenfun(@real,Ytilde);
        end
    end
end
```

Appendix 7.E: MATLAB Source Code for Beta NTD Algorithm

```
function [T,A,G,fit,iter] = tucker_beta(Y,R,opts)
% Beta NTD algorithm using beta divergence
% INPUT
% Y    :   tensor with size of I1 x I2 x ... x IN
```

```
% R       :   size of core tensor R1 x R2 x ... x RN: [R1, R2, ..., RN]
% opts    :   structure of optional parameters for algorithm (see defoptions)
%    .tol:    tolerance of stopping criteria (explained variation)    (1e-6)
%    .maxiters: maximum number of iteration                             (50)
%    .init:   initialization type: 'random', 'eigs', 'nvecs' (HOSVD)(random)
%    .beta: beta parameter                                               (1)
%    .nonlinearproj: apply half-wave rectifying or not                   (1)
%    .orthoforce:  orthogonal constraint to initialization using ALS
%    .updateYhat:  update Yhat or not, using in Fast Alpha NTD with ell-1
%                  normalization for factors                            (0)
%    .ellnorm:   normalization type                                     (1)
%    .projector: projection type for reconstructed tensor Yhat: max,(real)
%    .fixsign:  fix sign for components of factors.                     (1)
% Copyright 2008 of Andrzej Cichocki and Anh Huy Phan

% Set algorithm parameters from input or by using defaults
defoptions = struct('tol',1e-6,'maxiters',50,'init','random','Gbeta',1,...
    'beta',1,'ellnorm',1,'nonlinearproj',1,'orthoforce',1,'lda_ortho',0,...
    'updateYhat',0,'projector','real','fixsign',1,'getbestfit',0);
if ~exist('opts','var')
    opts = struct;
end
opts = scanparam(defoptions,opts);

% Extract number of dimensions and norm of Y.
N = ndims(Y); mY = max(Y(:)) ; Y = Y/mY; normY = norm(Y);
if numel(R) == 1, R = R(ones(1,N)); end
if numel(opts.lda_ortho) == 1
    opts.lda_ortho = opts.lda_ortho(ones(1,N));
end
%% Set up and error checking on initial guess for U.
[A,G] = ntd_initialize(Y,opts.init,opts.orthoforce,R);
%%
fprintf('\nBeta NTD:\n');
Yhat = ttm(G,A);
normresidual = norm(Y(:) - Yhat(:)/max(Yhat(:)));
fit = 1 - (normresidual / normY); %fraction explained by model
fprintf(' Iter %2d: fit = %e \n', 0, fit);
fitbest = -inf;
%% Main Loop: Iterate until convergence
for iter = 1:opts.maxiters
    pause(.0001)
    fitold = fit;
    Yalpha = Yhat.^opts.beta;
    % Iterate over all N modes of the tensor
    for n = 1:N
        Z = ttm(G,A,-n);
        A{n} = A{n}.* double(tenmat((Y.* Yalpha./(Yhat+eps)),n) *...
            tenmat(Z,n,'t'))./(double(tenmat(Yalpha,n)* tenmat(Z,n,'t'))...
            + eps - opts.lda_ortho(n) * bsxfun(@minus,A{n},sum(A{n},2)));
        if opts.fixsign
            A{n} = fixsigns(A{n});
        end
        A{n} = max(eps,A{n});
        if opts.ellnorm>0
            A{n} = bsxfun(@rdivide,A{n},(sum(A{n}.^opts.ellnorm))...
                .^(1/opts.ellnorm));
        end
        if opts.updateYhat
```

```
                Yhat = ttm(Z,A,n);
                Yalpha = Yhat.^opts.beta;
            end
        end
        Yhat = ttm(G,A);
        switch opts.Gbeta
            case 1
                G = G.*ttm(Y./Yhat,A,'t');
            otherwise
                G = G.*((ttm(Y.*(Yhat.^(opts.beta-1)),A,'t')+eps)./...
                    (ttm((Yhat.^(opts.beta)),A,'t'))+eps);
        end

        Yhat = ttm(G,A);
        if (mod(iter,10) ==1) || (iter == opts.maxiters)
            % Compute fit
            normresidual = sqrt(normY^2 + norm(Yhat/max(Yhat(:)))^2 - ...
                2 * innerprod(Y,Yhat/max(Yhat(:))));
            fit = 1 - (normresidual/normY);        %fraction explained by model
            fitchange = abs(fitold - fit);
            fprintf('Iter %2d: fit = %e Δfit = %7.1e\n',iter, fit, fitchange);
            if fit > fitbest                      % Check for convergence
                Ap = A; Gp= G;
                fitbest = fit;
            end
            if (fitchange < opts.tol) && (fit>0)
                break;
            end
        end
    end % iter
    %% Compute the final result
    if opts.getbestfit
        A = Ap; G = Gp;
    end
    T = ttm(G,A)*mY;
    end
```

Appendix 7.F: MATLAB Source Code for HALS NTD Algorithm

```
function [T,A,G,fit,iter] = tucker_localhals(Y,R,opts)
% HALS NTD algorithm
% INPUT
% Y      :   tensor with size of I1 x I2 x ... x IN
% R      :   size of core tensor R1 x R2 x ... x RN: [R1, R2, ..., RN]
% opts   :   structure of optional parameters for algorithm (see defoptions)
%    .tol:   tolerance of stopping criteria (explained variation)     (1e-6)
%    .maxiters: maximum number of iteration                             (50)
%    .init:  initialization type: 'random', 'eigs', 'nvecs' (HOSVD) (random)
%    .orthoforce:  orthogonal constraint to initialization using ALS
%    .ellnorm:   normalization type                                      (1)
%    .fixsign:  fix sign for components of factors.                      (0)
%
% Copyright 2008 of Anh Huy Phan and Andrzej Cichocki
```

```
% Set algorithm parameters from input or by using defaults
defoptions = struct('tol',1e-6,'maxiters',50,'init','random',...
    'ellnorm',1,'orthoforce',1,'lda_ortho',0,'lda_smooth',0,'fixsign',0);
if ~exist('opts','var')
    opts = struct;
end
opts = scanparam(defoptions,opts);

% Extract number of dimensions and norm of Y.
N = ndims(Y); normY = norm(Y);

if numel(R) == 1,  R = R(ones(1,N)); end
if numel(opts.lda_ortho) == 1
    opts.lda_ortho = opts.lda_ortho(ones(1,N));
end
if numel(opts.lda_smooth) == 1
    opts.lda_smooth = opts.lda_smooth(ones(1,N));
end
%% Set up and error checking on initial guess for U.
[A,G] = ntd_initialize(Y,opts.init,opts.orthoforce,R);
%%
fprintf('\nLocal NTD:\n');
% Compute approximate of Y
Yhat = ttm(G,A);
normresidual = norm(Y(:) - Yhat(:));
fit = 1 - (normresidual / normY); %fraction explained by model
fprintf(' Iter %2d: fit = %e \n', 0, fit);
Yr = Y- Yhat;
%% For smooth constraint
Uf = A;
for n =1:N
    Uf{n}(:) = 0;
end
%% Main Loop: Iterate until convergence
for iter = 1:opts.maxiters
    pause(0.001)
    fitold = fit;

    % Iterate over all N modes of the tensor
    % for smoothness constraint
     for ksm = 1:N
        Uf{ksm} = opts.lda_smooth(ksm) * [A{ksm}(2,:);
                 (A{ksm}(1:end-2,:)+A{ksm}(3:end,:))/2; A{ksm}(end-1,:)];
     end

    for n = 1:N
        Sn = double(tenmat(ttm(G,A,-n),n));
        Yrn = double(tenmat(Yr,n));
        As = sum(A{n},2);
        for ri = 1:R(n)
            Ani = A{n}(:,ri) + (Yrn * Sn(ri,:)fl- ...
                opts.lda_ortho(n)*(As -A{n}(:,ri))  + Uf{n}(:,ri))...
                /(opts.lda_smooth(n)+Sn(ri,:)*Sn(ri,:)fl);

            if opts.fixsign
                Ani = fixsigns(Ani);
            end
            Ani = max(eps,Ani);
            Yrn = Yrn -(Ani - A{n}(:,ri)) * Sn(ri,:);
            A{n}(:,ri) = Ani;
```

```
        end
        A{n} = bsxfun(@rdivide,A{n},sum(A{n}.^opts.ellnorm)...
            .^(1/opts.ellnorm));
        Yhat = ttm(G,A);
        Yr = Y- Yhat;
    end
    switch opts.ellnorm
        case 1
            G = G.*ttm(((Y+eps)./(Yhat+eps)),A,'t');
        case 2
            Yr1 = double(tenmat(Yr,1));
            for jgind = 1:prod(R)
                jgsub = ind2sub_full(R,jgind);
                va = arrayfun(@(x) A{x}(:,jgsub(x)),2:N,'uni',0);
                ka = khatrirao(va(end:-1:1));
                gjnew = max(eps,G(jgsub) + A{1}(:,jgsub(1))'*Yr1 * ka);
                Yr1 = Yr1 + (G(jgind) - gjnew)* A{1}(:,jgsub(1))*ka';
                G(jgind) = gjnew;
            end
    end
    Yhat = ttm(G,A);
    if (mod(iter,10) ==1) || (iter == opts.maxiters)
        % Compute fit
        normresidual = sqrt(normY^2 + norm(Yhat)^2 -2*innerprod(Y,Yhat));
        fit = 1 - (normresidual/normY);        %fraction explained by model
        fitchange = abs(fitold - fit);
        fprintf('Iter %2d: fit = %e Δfit = %7.1e\n',iter, fit, fitchange);
        if (fitchange < opts.tol) && (fit>0)    % Check for convergence
            break;
        end
    end
end
%% Compute the final result
T = ttm(G, A);
end
```

Appendix 7.G: MATLAB Source Code for ALS NTF1 Algorithm

```
function Fact = ntf1_als(Y,J,opts,Fact)
%
% ALS algorithm for NTF-1 model
% Copyright 2008 of Andrzej Cichocki and Anh Huy Phan
% Set algorithm parameters from input or by using defaults
defoptions = struct('tol',1e-6,'maxiters',500,'ortho',0,'initialize',1);
if ~exist('opts','var'),     opts = struct; end
opts = scanparam(defoptions,opts);
% Extract dimensions of Y.
Q = numel(Y); I = size(Y{1},1);
T = cell2mat(cellfun(@(x) size(x,2),Y,'uni',0));
Y_1 = cell2mat(Y); %unfold tensor
%% Initialization for $\bA$, $\bX$, $\bC$
if opts.initialize
    AH = rand(I,J); CH = rand(Q,J);
    XH = mat2cell(rand(J,sum(T)),J,T);
else
    AH = Fact{1}; CH = Fact{2}; XH = Fact{3};
end
```

```
%% Parameters
fitold= inf; AHtemp = cell(1,Q); YpXH = cell(1,Q); cT = [0 cumsum(T)];
for iter = 1:opts.maxiters
    X_ = max(eps,pinv(AH)*Y_1);
    for q = 1:Q
        XH{q} = max(eps,bsxfun(@rdivide,X_(:,cT(q)+1:cT(q+1)),CH(q,:)fl));
        if opts.ortho
            XH{q} = max(eps,pinv(sqrtm(XH{q}*XH{q}'))*XH{q});
        end
        XH{q} = bsxfun(@rdivide,XH{q},sum(XH{q},2));
        YpXH{q} = Y{q}*pinv(XH{q});
    end

    for q = 1:Q
        AHtemp{q} = max(eps,bsxfun(@rdivide,YpXH{q},CH(q,:)));
        AHtemp{q} = bsxfun(@rdivide,AHtemp{q},sum(AHtemp{q}));
    end
    AH = mean(reshape(cell2mat(AHtemp),I,[],Q),3);
    pAH = pinv(AH)';
    for q = 1:Q
        CH(q,:)= max(eps,sum(pAH.*YpXH{q}));
    end

    for q = 1:Q
        X_(:,cT(q)+1:cT(q+1)) = bsxfun(@times,XH{q},CH(q,:)fl);
    end
    Yhat = AH*X_;

    if (mod(iter,10)==0) || (iter == opts.maxiters)
        fit = CalcExpVar(Yhat,Y_1);
        if abs(fit-fitold) < opts.tol
            break
        end
        fitold = fit;
    end
end
Fact = {AH CH XH};
```

Appendix 7.H: MATLAB Source Code for ISRA BOD Algorithm

```
function [A,H,B,L,C,F] = nbod_ISRA(Y,opts)
% Multiplicative algorithm for Nonnegative BOD based on the Frobenius cost
% function || Y - Yhat||_F
%% INPUT
% Y      :   3 way tensor I1 x I2 x I3
% opts   :   structure of optional parameters for algorithm (see defoptions)
%% OUTPUT
% Factors A, B, C, and core tensors H, L, F
%
% Copyright 2009 by Anh Huy Phan and Andrzej Cichocki
%% =====================================================================
Y(Y< 0) = eps;  In = size(Y);
% Set algorithm parameters from input or by using defaults
defoptions = struct('R',[1 1 1],'maxIter',1000,'lspar',[0 0 0],...
    'lortho',[0 0 0],'verbose',0,'A',[],'H',[],'B',[],'L',[],'C',[],'F',[]);
if ~exist('opts','var'),  opts = struct;    end
[R,maxiter,lspar,lortho,verbose,A,H,B,L,C,F] = scanparam(defoptions,opts);
```

```
% Initialization
if ~all([size(A) == [In(1),R(1)] size(H) == [R(1),In(2),In(3)]])
    A = rand(In(1),R(1));H = rand(R(1),In(2),In(3));
end
if ~all([size(B) == [In(2),R(2)] size(L) == [In(1), R(2),In(3)]])
    B = rand(In(2),R(2));L = rand(In(1),R(2),In(3));
end
if ~all([size(C) == [In(3),R(3)] size(F) == [In(1), In(2),R(3)]])
    C = rand(In(3),R(3));F = rand(In(1),In(2),R(3));
end

Y1 = double(tenmat(Y,1)); Y2 = double(tenmat(Y,2));
Y3 = double(tenmat(Y,3));

errprev = inf;
for k = 1:maxiter
    Yhat = SODbuild(A,B,C,H,L,F);
    Yhat1 = double(tenmat(Yhat,1)); Yhat2 = double(tenmat(Yhat,2));
    Yhat3 = double(tenmat(Yhat,3));

    [A,H] = tucker1(Y1,Yhat1,1, A,H,lspar(1),lortho(1));
    [B,L] = tucker1(Y2,Yhat2,2, B,L,lspar(2),lortho(2));
    [C,F] = tucker1(Y3,Yhat3,3, C,F,lspar(3),lortho(3));

    if (mod(k,20)==1 ) || (k == maxiter)
        err = norm(Y(:) - Yhat(:)); derr = abs(err - errprev);
        fprintf('Iter %f , error %.5d, del: %.5d\n',k,err,derr);
        if (err < 1e-5) %|| abs(err - errprev) < 1e-5
            break
        end
        errprev = err;
    end
end

function [A,G] = tucker1(Ym,Yhatm,mode, A,G,lspar,lortho)
Gm = double(tenmat(G,mode))+eps;
Gm = Gm.*(A'* Ym )./(A'* Yhatm + lspar - lortho*bsxfun(@minus,Gm,sum(Gm)));
Gm = bsxfun(@rdivide,Gm,sqrt(sum(Gm.^2,2)));
G = itenmat(Gm,mode,size(G));
A = A.* (Ym * Gm') ./ (Yhatm * Gm' +eps);
```

Appendix 7.I: Additional MATLAB functions

```
function [A,G] = ntd_initialize(Y,init,orthoforce,R)
% Initialization for NTD algorithms
% Output:   factors A and core tensor G
% Copyright 2008 of Andrzej Cichocki and Anh Huy Phan
N = ndims(Y);In = size(Y);
if iscell(init)
    if numel(init) ≠ N+1
        error('OPTS.init does not have %d cells',N+1);
    end
    for n = 1:N;
        if ~isequal(size(init{n}),[In(n) R(n)])
            error('Factor{%d} is the wrong size',n);
        end
    end
```

```
    if ~isequal(size(init{end}),R)
        error('Core is the wrong size');
    end
    A = init(1:end-1); G = init{end};
else
    switch init
        case 'random'
            A = arrayfun(@rand,In,R,'uni',0); G = tensor(rand(R));
        case {'nvecs' 'eigs'}
            A = cell(N,1);
            for n = 1:N
                A{n} = nvecs(Y,n,R(n));
            end
            G = ttm(Y, A, 't');
        otherwise
            error('Undefined initialization type');
    end
end
%% Powerful initialization
if orthoforce
    for n = 1:N
        Atilde = ttm(Y, A, -n, 't');
        A{n} = max(eps,nvecs(Atilde,n,R(n)));
    end
    A = cellfun(@(x) bsxfun(@rdivide,x,sum(x)),A,'uni',0);
    G = ttm(Y, A, 't');
end
```

```
function sdx = ind2sub_full(siz,ndx)
%IND2SUB Multiple subscripts from linear index.
% Modify the Matlab function IND2SUB
siz = double(siz);
sdx = zeros(1,numel(siz));

n = length(siz);
k = [1 cumprod(siz(1:end-1))];
for i = n:-1:1,
    vi = rem(ndx-1, k(i)) + 1;
    vj = (ndx - vi)/k(i) + 1;
    sdx(i) = vj;
    ndx = vi;
end
```

```
function y = signalsuit(k,t,f,w)
% Copyright 2008 of Andrzej Cichocki and Anh Huy Phan
sigs = {@cos,@sin,@sawtooth,@square,@chirp,@sinc,@rectpuls,...
    @tripuls,@gauspuls,@pulstran};
typesigs = {'cos','sin','sawtooth','square','chirp','sinc','rectpuls',...
    'tripuls','gauspuls','pulstran'};
if ischar(k),    k = find(strcmp(typesigs,k)); end
switch k
    case {1, 2,6,'cos','sin','sinc'}
        y = feval(sigs{k},t*2*pi*f);
    case {3,'sawtooth'}
        y = feval(sigs{k},t*2*pi*f,w);
    case {4,'square'}
        y = feval(sigs{k},t*2*pi*f,w*100);
```

```
        case {5,'chirp'}
            y = feval(sigs{k},t,0,1,f);
        otherwise
            y = feval(@pulstran,t,d,sigs{k},w);
    end
    y = y(:);
```

```
function At = itenmat(A,mode,I)
% Inverse of the tenmat function
% Am = tenmat(A,mode); I = size(A);
% A = itenmat(Am,mode,I);
% Copyright of Anh Huy Phan, Andrzej Cichocki

N = numel(I);
ix = [mode setdiff(1:N,mode)];
At = reshape(A,I(ix));
At = permute(At,[2:mode 1 mode+1:N]);
```

References

[1] E. Acar, C.A. Bingol, H. Bingol, R. Bro, and B. Yener. Multiway analysis of epilepsy tensors. *Bioinformatics*, 23:10–18, 2007.

[2] E. Acar and B. Yener. Unsupervised multiway data analysis: A literature survey. *IEEE Transactions on Knowledge and Data Engineering*, 21:6–20, 2008.

[3] A.H. Andersen and W.S. Rayens. Structure-seeking multilinear methods for the analysis of fMRI data. *NeuroImage*, 22:728–739, 2004.

[4] C.A. Andersson and R. Bro. The N-way toolbox for MATLAB. *Chemometrics Intell. Lab. Systems*, 52(1):1–4, 2000.

[5] F. Babiloni, A. Cichocki, and S. Gao. Brain computer interfaces: Towards practical implementations and potential applications. *Computational Intelligence and Neuroscience*, 2007.

[6] B.W. Bader and T.G. Kolda. Algorithm 862: MATLAB tensor classes for fast algorithm prototyping. *ACM Transactions on Mathematical Software*, 32(4):635–653, 2006.

[7] B.W. Bader and T.G. Kolda. Efficient MATLAB computations with sparse and factored tensors. Technical Report SAND2006-7592, Sandia National Laboratories, Albuquerque, NM and Livermore, CA, December 2006.

[8] B.W. Bader and T.G. Kolda. *MATLAB Tensor Toolbox Version 2.2*, January 2007.

[9] H. Bakardjian and A. Cichocki. Extraction and classification of common independent components in single-trial crossmodal cortical responses. In *Proceedings of the 5th Annual Meeting of the International Multisensory Research Forum*, pages 26–27, Barcelona, Spain, June 2004.

[10] J.R. Bergen, P. An, Th. J. Hanna, and R. Hingorani. Hierarchical model-based motion estimation. pages 237–252. Springer-Verlag, 1992.

[11] M. Berry, M. Browne, A. Langville, P. Pauca, and R. Plemmons. Algorithms and applications for approximate nonnegative matrix factorization. *Computational Statistics and Data Analysis*, 52(1):155–173, 2007.

[12] B. Blankertz, M. Kawanabe, R. Tomioka, V. Nikulin, and K.-R. Müller. Invariant common spatial patterns: Alleviating nonstationarities in brain-computer interfacing. In J.C. Platt, D. Koller, Y. Singer, and S. Roweis, editors, *Advances in Neural Information Processing Systems 20*, pages 113–120. MIT Press, Cambridge, MA, 2008.

[13] K.S. Booksh and B.R. Kowalski. Error analysis of the generalized rank annihilation method. *Journal of Chemometrics*, 8:45–63, 1994.

[14] K.S. Booksh, Z. Lin, Z. Wang, and B.R. Kowalski. Extension of trilinear decomposition method with an application to the flow probe sensor. *Analytical Chemistry*, 66:2561–2569, 1994.

[15] R. Bro. PARAFAC. Tutorial and applications. In *Special Issue 2nd Internet Conf. in Chemometrics (INCINC'96)*, volume 38, pages 149–171. Chemom. Intell. Lab. Syst, 1997.

[16] C. Caiafa and A. Cichocki. Slice Oriented Decomposition: A new tensor representation for 3-way data. *(submitted to Journal of Signal Processing)*, 2009.

[17] J. Chen and Y. Saad. On the tensor SVD and the optimal low rank orthogonal approximation of tensors. *SIAM Journal on Matrix Analysis and Applications (SIMAX)*, 30:1709–1734, 2009.

[18] A. Cichocki, S. Amari, R. Zdunek, R. Kompass, G. Hori, and Z. He. Extended SMART algorithms for non-negative matrix factorization. *Springer, LNAI-4029*, 4029:548–562, 2006.

[19] A. Cichocki and A.H. Phan. Fast local algorithms for large scale nonnegative matrix and tensor factorizations. *IEICE (invited paper)*, March 2009.

[20] A. Cichocki, A.H. Phan, and C. Caiafa. Flexible HALS algorithms for sparse non-negative matrix/tensor factorization. In *Proc. of 18-th IEEE workshops on Machine Learning for Signal Processing*, Cancun, Mexico, 16–19, October 2008.

[21] A. Cichocki, A.H. Phan, R. Zdunek, and L.-Q. Zhang. Flexible component analysis for sparse, smooth, nonnegative coding or representation. In *Lecture Notes in Computer Science, LNCS-4984*, volume 4984, pages 811–820. Springer, 2008.

[22] A. Cichocki and R. Zdunek. Multilayer nonnegative matrix factorization. *Electronics Letters*, 42(16):947–948, 2006.

[23] A. Cichocki and R. Zdunek. NMFLAB for Signal and Image Processing. Technical report, Laboratory for Advanced Brain Signal Processing, BSI, RIKEN, Saitama, Japan, 2006.

[24] A. Cichocki and R. Zdunek. NTFLAB for Signal Processing. Technical report, Laboratory for Advanced Brain Signal Processing, BSI, RIKEN, Saitama, Japan, 2006.

[25] A. Cichocki and R. Zdunek. Regularized alternating least squares algorithms for non-negative matrix/tensor factorizations. *Springer, LNCS-4493*, 4493:793–802, June 3–7 2007.

[26] A. Cichocki, R. Zdunek, and S. Amari. Csiszar's divergences for non-negative matrix factorization: Family of new algorithms. *Springer, LNCS-3889*, 3889:32–39, 2006.

[27] A. Cichocki, R. Zdunek, and S. Amari. New algorithms for non-negative matrix factorization in applications to blind source separation. In *Proc. IEEE International Conference on Acoustics, Speech, and Signal Processing, ICASSP2006*, volume 5, pages 621–624, Toulouse, France, May 14–19 2006.

[28] A. Cichocki, R. Zdunek, S. Choi, R. Plemmons, and S. Amari. Nonnegative tensor factorization using Alpha and Beta divergencies. In *Proc. IEEE International Conference on Acoustics, Speech, and Signal Processing (ICASSP07)*, volume III, pages 1393–1396, Honolulu, Hawaii, USA, April 15–20 2007.

[29] A. Cichocki, R. Zdunek, S. Choi, R. Plemmons, and S.-I. Amari. Novel multi-layer nonnegative tensor factorization with sparsity constraints. *Springer, LNCS-4432*, 4432:271–280, April 11–14 2007.

[30] P. Comon. Tensor decompositions: State of the art and applications. In J.G. McWhirter and I.K. Proudler, editors, *Institute of Mathematics and its Applications Conference on Mathematics in Signal Processing*, pages 1–26. Oxford University Press, UK, 2001.

[31] L. De Lathauwer. Decompositions of a higher-order tensor in block terms – Part I: Lemmas for partitioned matrices. *SIAM Journal on Matrix Analysis and Applications (SIMAX)*, 30(3):1022–1032, 2008. Special Issue on Tensor Decompositions and Applications.

[32] L. De Lathauwer, B. de Moor, and J. Vandewalle. A multilinear singular value decomposition. *SIAM Journal of Matrix Analysis and Applications*, 21(4):1253–1278, 2000.

[33] L. De Lathauwer, B. De Moor, and J. Vandewalle. On the best rank-1 and rank-(R1,R2,...,RN) approximation of higher-order tensors. *SIAM Journal of Matrix Analysis and Applications*, 21(4):1324–1342, 2000.

[34] L. De Lathauwer and D. Nion. Decompositions of a higher-order tensor in block terms – Part III: Alternating least squares algorithms. *SIAM Journal on Matrix Analysis and Applications (SIMAX)*, 30(3):1067–1083, 2008. Special Issue Tensor Decompositions and Applications.

[35] L. De Lathauwer and J. Vandewalle. Dimensionality reduction in higher-order signal processing and rank-$(R_1, R_2, \ldots, R_n)$ reduction in multilinear algebra. *Linear Algebra Applications*, 391:31–55, November 2004.

[36] A. Delorme and S. Makeig. EEGLAB: an open source toolbox for analysis of single-trial EEG dynamics. *Journal of Neuroscience Methods*, 134:9–21, 2004.

[37] I. Dhillon and S. Sra. Generalized nonnegative matrix approximations with Bregman divergences. In *Neural Information Proc. Systems*, pages 283–290, Vancouver, Canada, December 2005.

[38] L. Eldén and B. Savas. A Newton–Grassmann method for computing the best multi-linear rank-(r_1, r_2, r_3) approximation of a tensor. *SIAM Journal on Matrix Analysis and Applications*, 31:248–271, 2009.

[39] N.M. Faber. On solving generalized eigenvalue problems using matlab. *Journal of Chemometrics*, 11:87–91, 1997.

[40] N.M. Faber, L.M.C. Buydens, and G. Kateman. Generalized rank annihilation. iii: practical implementation. *Journal of Chemometrics*, 8:273–285, 1994.

[41] M.P. Friedlander and K. Hatz. Computing nonnegative tensor factorizations. *Computational Optimization and Applications*, 23(4):631–647, March 2008.

[42] T. Hazan, S. Polak, and A. Shashua. Sparse image coding using a 3D non-negative tensor factorization. In *Proc. Int. Conference on Computer Vision (ICCV)*, pages 50–57, 2005.
[43] D. Heeger. Image registration software. Copyright 1997, 2000 by Stanford University, 2000.
[44] P.O. Hoyer. Non-negative matrix factorization with sparseness constraints. *Journal of Machine Learning Research*, 5:1457–1469, 2004.
[45] P.R. Kennedy, R.A.E. Bakay, M.M. Moore, K. Adams, and J. Goldwaithe. Direct control of a computer from the human central nervous system. *IEEE Transactions on Rehabilitation Engineering*, 8(2):198–202, June 2000.
[46] M. Kim and S. Choi. Monaural music source separation: Nonnegativity, sparseness, and shift-invariance. *Springer LNCS*, 3889:617–624, 2006.
[47] Y.-D. Kim and S. Choi. Nonnegative Tucker Decomposition. In *Proc. of Conf. Computer Vision and Pattern Recognition (CVPR-2007)*, Minneapolis, Minnesota, June 2007.
[48] Y.-D. Kim, A. Cichocki, and S. Choi. Nonnegative Tucker Decomposition with Alpha Divergence. In *Proceedings of the IEEE International Conference on Acoustics, Speech, and Signal Processing, ICASSP2008*, Nevada, USA, 2008.
[49] T.G. Kolda. Multilinear operators for higher-order decompositions. Technical report, Sandia National Laboratories, 2006.
[50] T.G. Kolda and B.W. Bader. Tensor decompositions and applications. *SIAM Review*, 51(3):(in print), September 2009.
[51] P.M. Kroonenberg. *Applied Multiway Data Analysis*. John Wiley & Sons Ltd, New York, 2008.
[52] J.B. Kruskal, R.A. Harshman, and M.E. Lundy. How 3-MFA data can cause degenerate parafac solutions, among other relationships. In *Multiway Data Analysis.*, pages 115–122. North-Holland Publishing Co., Amsterdam, The Netherlands, 1989.
[53] A.N. Langville, C.D. Meyer, and R. Albright. Initializations for the nonnegative matrix factorization. In *Proc. of the Twelfth ACM SIGKDD International Conference on Knowledge Discovery and Data Mining*, Philadelphia, USA, August 20–23 2006.
[54] D.D. Lee and H.S. Seung. *Algorithms for Nonnegative Matrix Factorization*, volume 13. MIT Press, 2001.
[55] H. Lee, A. Cichocki, and S. Choi. Nonnegative matrix factorization for motor imagery EEG classification. *LNCS*, 4132:250–259, 2006.
[56] H. Lee, Y.-D. Kim, A. Cichocki, and S. Choi. Nonnegative tensor factorization for continuous EEG classification. *International Journal of Neural Systems*, 17(4):1–13, August 2007.
[57] Y.-Q. Li, A. Cichocki, and S. Amari. Blind estimation of channel parameters and source components for EEG signals: a sparse factorization approach. *IEEE Transactions on Neural Networks*, 17(2):419–431, 2006.
[58] P. Martinez, H. Bakardjian, and A. Cichocki. Fully online, multi-command brain computer interface with visual neurofeedback using SSVEP paradigm. *Journal of Computational Intelligence and Neuroscience*, 2007, 2007.
[59] E. Martínez-Montes, J.M. Sánchez-Bornot, and P.A. Valdés-Sosa. Penalized PARAFAC analysis of spontaneous EEG recordings. *Statistica Sinica*, 18:1449–1464, 2008.
[60] M. Mørup, L.K. Hansen, and S.M. Arnfred. ERPWAVELAB a toolbox for multi-channel analysis of time-frequency transformed event related potentials. *Journal of Neuroscience Methods*, 161:361–368, 2007.
[61] M. Mørup, L.K. Hansen, and S.M. Arnfred. Algorithms for Sparse Nonnegative Tucker Decompositions. *Neural Computation*, 20:2112–2131, 2008.
[62] K.-R. Müller, M. Krauledat, G. Dornhege, G. Curio, and B. Blankertz. Machine learning and applications for brain-computer interfacing. 4557, 2007.
[63] O. Nestares and D.J. Heeger. Robust multiresolution alignment of MRI brain volumes. In *Magnetic Resonance in Medicine, 43:705715*, pages 705–715, 2000.
[64] A.H. Phan and A. Cichocki. Fast and efficient algorithms for nonnegative Tucker decomposition. In *Proc. of The Fifth International Symposium on Neural Networks, Springer LNCS-5264*, pages 772–782, Beijing, China, 24–28, September 2008.
[65] A.H. Phan and A. Cichocki. Local learning rules for nonnegative Tucker decomposition. *(submitted)*, 2009.
[66] A.H. Phan and A. Cichocki. Multi-way nonnegative tensor factorization using fast hierarchical alternating least squares algorithm (HALS). In *Proc. of The 2008 International Symposium on Nonlinear Theory and its Applications*, Budapest, Hungary, 2008.
[67] P. Sajda, K-R. Müller, and K.V. Shenoy. Brain computer interfaces. *IEEE Signal Processing Magazine*, January 2008.
[68] F. Samaria and A.C. Harter. Parameterisation of a stochastic model for human face identification. In *Proceedings of the Second IEEE Workshop on Applications of Computer Vision*, 1994.

[69] E. Sanchez and B.R. Kowalski. Tensorial resolution: a direct trilinear decomposition. *J. Chemometrics*, 4:29–45, 1990.

[70] M. Sattler, R. Sarlette, and R. Klein. Efficient and realistic visualization of cloth. In *EGSR '03: Proceedings of the 14th Eurographics Symposium on Rendering*, pages 167–177, Aire-la-Ville, Switzerland, Switzerland, 2003. Eurographics Association.

[71] B. Savas and L. Eldén. Handwritten digit classification using higher order singular value decomposition. *Pattern Recognition*, 40:993–1003, 2007.

[72] A. Shashua and T. Hazan. Non-negative tensor factorization with applications to statistics and computer vision. In *Proc. of the 22-th International Conference on Machine Learning*, Bonn, Germany, 2005.

[73] A. Shashua, R. Zass, and T. Hazan. Multi-way clustering using super-symmetric non-negative tensor factorization. In *European Conference on Computer Vision (ECCV)*, Graz, Austria, May 2006.

[74] A. Smilde, R. Bro, and P. Geladi. *Multi-way Analysis: Applications in the Chemical Sciences*. John Wiley & Sons Ltd, New York, 2004.

[75] J. Sun. *Incremental Pattern Discovery on Streams, Graphs and Tensors*. PhD thesis, CMU-CS-07-149, 2007.

[76] J. Sun, D. Tao, and C. Faloutsos. Beyond streams and graphs: dynamic tensor analysis. In *Proc.of the 12th ACM SIGKDD International Conference on Knowledge Discovery and Data Mining*, pages 374–383, 2006.

[77] C. Tallon-Baudry, O. Bertrand, C. Delpuech, and J. Pernier. Stimulus specificity of phase-locked and non-phase-locked 40 Hz visual responses in human. *Journal of Neuroscience*, 16 (13):4240–4249, 1996.

[78] The BTF Database Bonn. *CEILING Sample*. http://btf.cs.uni-bonn.de/download.html.

[79] M.K. Titsias. *Unsupervised Learning of Multiple Objects in Images*. PhD thesis, School of Informatics, University of Edinburgh, June 2005.

[80] L.R. Tucker. Some mathematical notes on three-mode factor analysis. *Psychometrika*, 31:279–311, 1966.

[81] P.D. Turney. Empirical evaluation of four tensor decomposition algorithms. Technical report, National Research Council, Institute for Information Technology, 2007, Technical Report ERB-1152. (NRC 49877).

[82] J.H. Wang, P.K. Hopke, T.M. Hancewicz, and S.-L. Zhang. Application of modified alternating least squares regression to spectroscopic image analysis. *Analytica Chimica Acta*, 476:93–109, 2003.

[83] W. Wang. Squared Euclidean distance based convolutive non-negative matrix factorization with multiplicative learning rules for audio pattern separation. In *Proc. 7th IEEE International Symposium on Signal Processing and Information Technology (ISSPIT 2007)*, Cairo, Egypt, December 15–18, 2007.

[84] Z. Wang, A. Maier, N.K. Logothetis, and H. Liang. Single-trial decoding of bistable perception based on sparse nonnegative tensor decomposition. *Journal of Computational Intelligence and Neuroscience*, 30:1–10, 2008.

[85] Q. Wu, T. Xia, and Y. Yu. Hierarchical tensor approximation of multi-dimensional images. In *14th IEEE International Conference on Image Processing*, volume IV, pages 49–52, San Antonio, 2007.

[86] R. Zass and A. Shashua. Doubly stochastic normalization for spectral clustering. In *Neural Information Processing Systems (NIPS)*, Vancuver, Canada, Dec. 2006.

8

Selected Applications

Early applications of the concept inherited by NMF appeared in the middle 1990s under the name Positive Matrix Factorization (PMF). This kind of factorization was applied by Paatero *et al.* [82] to process environmental data. However, the popularity of NMF significantly increased since Lee and Seung published simple multiplicative NMF algorithms which they applied to image data [62,63]. At present, NMF and its variants have already found a wide spectrum of applications.

In this chapter, we briefly discuss some selected applications of NMF and multi-dimensional array decompositions, with special emphasis on those applications to which the algorithms described in the previous chapters are applicable. We review the following applications: data clustering [5,14,16,24,27,67,100,126,127], text mining [67,86,100,119], email surveillance [9], musical instrument classification [6–8], face recognition [42,43,45,113,120,121,128], handwritten digit recognition [61], texture classification [87,89], Raman spectroscopy [65,75,95], fluorescence spectroscopy [37,38,46], hyperspectral imaging [28,40,50,76,85,95,106], chemical shift imaging [11,95,96], and gene expression classification [13,14,33,53,55,73,83,84,109].

8.1 Clustering

Data clustering can be regarded as an unsupervised classification of patterns into groups (clusters) that have similar features, as illustrated in Figure 8.1. The data points reside in a 2D space and can be classified into three disjoint groups. The grouping can basically be obtained with hierarchical or partitioning techniques [49]. In hierarchical clustering, a nested series of partitions is performed with varying dissimilarity level whereas partitioned clustering techniques yield a single partition of data for which a given clustering criterion is optimized (usually locally). The former technique is usually more robust but due to its high computational complexity it is not suitable for large datasets. In such cases, partitioning techniques are more favorable, and are of particular interest.

In partitional clustering a whole set of data is processed simultaneously in one updating cycle. There are many approaches to partitional clustering: K-means, [72] and its variants [2] (ISODATA [4], dynamic clustering [23,102], Mahalanobis distance based clustering, [71], spectral K-means [22]), graph-theoretic clustering [81,125], mixture-resolving and mode-seeking algorithms [47,74], nearest neighbor clustering [70], fuzzy clustering [10,93,124], neural networks based clustering [48,99], Kohonen's learning vector quantization (LVQ), self-organizing map (SOM) [56], evolutionary clustering [51,90], branch-and-bound technique [15,57], and simulated annealing clustering [3,92].

Recently, the use of NMF for partitional clustering of nonnegative data has attracted much interest. Some examples can be found in [5,14,16,27,67,79,100,117,126]. Ding *et al.* [24] showed the equivalence between NMF, spectral clustering and K-means clustering.

Nonnegative Matrix and Tensor Factorizations: Applications to Exploratory Multi-way Data Analysis and Blind Source Separation Andrzej Cichocki, Rafal Zdunek, Anh Huy Phan and Shun-ichi Amari

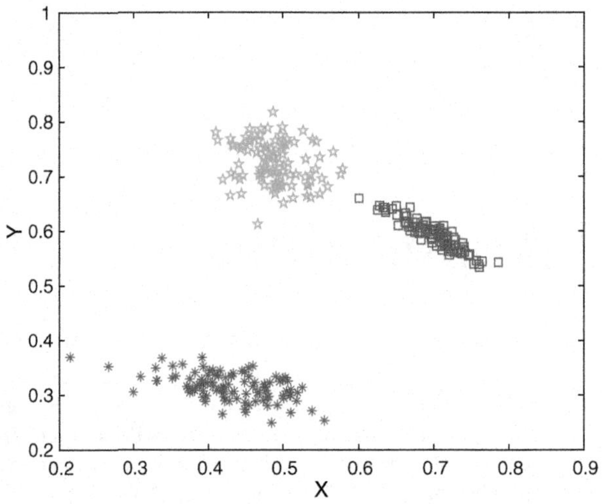

Figure 8.1 Exemplary distribution of clustered data points in 2D space.

8.1.1 Semi-Binary NMF

The basic idea behind the use of NMF for partitional clustering is very simple. Assuming that one of the NMF factors contains the vectors of centroids (central points of clusters), the other one should contain binary numbers that assign cluster indicators to the observation (data) vectors. Assuming non-overlapping clusters, each column vector of this factor should contain exactly one nonzero entry, thus leading to semi-binary NMF.

NMF aims at finding lower-rank nonnegative matrices $\mathbf{A} \in \mathbb{R}^{I \times J}$ and $\mathbf{X} \in \mathbb{R}^{J \times T}$ such that $\mathbf{Y} \cong \mathbf{AX} \in \mathbb{R}^{I \times T}$, given the data matrix $\mathbf{Y}$, the lower rank J, and possibly some prior knowledge about the matrices $\mathbf{A}$ or $\mathbf{X}$. Assuming the clustered data are stored in the matrix $\mathbf{Y}$ in such a way that each column vector of $\mathbf{Y} = [\boldsymbol{y}_1, \ldots, \boldsymbol{y}_T]$ represents a single observation (a datum point in $\mathbb{R}^I$), the clusters are disjoint (non-overlapping), and J is an *a priori* known number of clusters, we can interpret the column vectors of $\mathbf{A} = [\boldsymbol{a}_1, \ldots, \boldsymbol{a}_J]$ as the centroids (indicating the directions of central points of clusters in $\mathbf{R}^I$) and binary values in $\mathbf{X} = [x_{jt}]$ as indicators of the clusters. If $x_{jt} = 1$, then $\boldsymbol{y}_t$ belongs to the j-th cluster; $x_{jt} = 0$, otherwise.

Example 8.1 *Let us consider the following example:*

$$\mathbf{Y} = \begin{bmatrix} 0.5847 & 0.5714 & 0.5867 & 0.7301 & 0.7309 & 0.7124 & 0.7246 \\ 0.1936 & 0.1757 & 0.1920 & 0.4193 & 0.3955 & 0.3954 & 0.4147 \end{bmatrix}. \tag{8.1}$$

It is apparent that the first three columns should be in one cluster, and the last four columns in the other, and $\mathbf{X}$ *takes the form*

$$\mathbf{X} = \begin{bmatrix} 1 & 1 & 1 & 0 & 0 & 0 & 0 \\ 0 & 0 & 0 & 1 & 1 & 1 & 1 \end{bmatrix}. \tag{8.2}$$

The matrix $\mathbf{A}$ *can be computed as* $\mathbf{A} = \mathbf{Y}\mathbf{X}^{\dagger}$, *where* $\mathbf{X}^{\dagger}$ *is a pseudo-inverse to* $\mathbf{X}$, *yielding*

$$\mathbf{A} = \begin{bmatrix} 0.5809 & 0.7245 \\ 0.1871 & 0.4062 \end{bmatrix}. \tag{8.3}$$

The column vector $\mathbf{a}_1$ represents the algebraic mean of the first three columns of $\mathbf{Y}$, and $\mathbf{a}_2$ the algebraic mean of the last four columns of $\mathbf{Y}$. Therefore, the column vectors in $\mathbf{A}$ can be interpreted as the centroids of the clusters.

Upon inspection of (8.2) it is (even visually) clear that $\mathbf{X}$ is an orthogonal matrix, i.e. $\mathbf{X}\mathbf{X}^T = \mathbf{I}_J$, where $\mathbf{I}_J$ is an identity matrix, thus the orthogonality constraints are met here intrinsically.

The algorithm for performing semi-binary NMF was introduced by Zdunek in [127]. To update a binary matrix $\mathbf{X}$, the logistic function which measures the degree of misfitting between the observations (matrix $\mathbf{Y}$) and the forward projected data (matrix $\mathbf{AX}$) was associated with Gibbs-Boltzmann statistics and was maximized using simulated annealing. For updating the matrix $\mathbf{A}$, the FNMA-like [52] algorithm (described in Chapter 6) was used. As a result, the algorithm in [127] outperforms the algorithm proposed in [27] for uni-orthogonal NMF.

8.1.2 NMF vs. Spectral Clustering

In the context of clustering, Ding *et al.* [24] introduced a symmetric NMF, expressed as:

$$\mathbf{Y} \cong \mathbf{A}\mathbf{A}^T, \tag{8.4}$$

where $\mathbf{A} \in \mathbb{R}^{I \times J}$ ($I >> J$) is a matrix with nonnegative entries. They also demonstrated that this kind of NMF is equivalent to Kernel K-means clustering and Laplacian-based clustering. The equivalence follows from the similarity in the objective functions. The K-means approach clusters the data based on the minimization of the least-squares function

$$\Psi = \sum_{k=1}^{K} \sum_{t \in C_k} ||\mathbf{z}_t - \mathbf{m}_k||_2^2, \tag{8.5}$$

where $\mathbf{Z} = [\mathbf{z}_1, \ldots, \mathbf{z}_t, \ldots, \mathbf{z}_T]$ is the data matrix, and $\mathbf{m}_k = \sum_{t \in C_k} \frac{1}{n_k} \mathbf{z}_t$ is the centroid of the cluster C_k of n_k points. The function Ψ in (8.5) can be rewritten as $\Psi = \text{tr}\{\mathbf{Z}^T\mathbf{Z}\} - \text{tr}\{\mathbf{A}^T\mathbf{Z}^T\mathbf{Z}\mathbf{A}\}$ with the matrix $\mathbf{A}$ containing the cluster indicators. As a consequence, the K-means clustering can be obtained by solving the following minimization problem:

$$\min_{\mathbf{A}^T\mathbf{A}=\mathbf{I}_J, \mathbf{A} \geq 0} \left\{ -\text{tr}\{\mathbf{A}^T\mathbf{Z}^T\mathbf{Z}\mathbf{A}\} \right\}. \tag{8.6}$$

The minimization problem (8.6) can also be expressed as

$$\begin{aligned} P &= \min_{\mathbf{A}^T\mathbf{A}=\mathbf{I}_J, \mathbf{A} \geq 0} \left\{ -2\,\text{tr}\{\mathbf{A}^T\mathbf{Y}\mathbf{A}\} \right\} \\ &= \min_{\mathbf{A}^T\mathbf{A}=\mathbf{I}_J, \mathbf{A} \geq 0} \left\{ ||\mathbf{Y}||_F^2 - 2\,\text{tr}\{\mathbf{A}^T\mathbf{Y}\mathbf{A}\} + ||\mathbf{A}^T\mathbf{A}||_F^2 \right\} \\ &= \min_{\mathbf{A}^T\mathbf{A}=\mathbf{I}_J, \mathbf{A} \geq 0} \left\{ ||\mathbf{Y} - \mathbf{A}\mathbf{A}^T||_F^2 \right\}, \end{aligned} \tag{8.7}$$

with $\mathbf{Y} \cong \mathbf{Z}^T\mathbf{Z}$. The matrix $\mathbf{A}$ is very sparse and by keeping its nonnegativity the factorization $\mathbf{Y} \cong \mathbf{A}\mathbf{A}^T$ with $\mathbf{A} \geq \mathbf{0}$ retains the approximately orthogonal columns in $\mathbf{A}$ [24]. Hence, factorizing the square matrix $\mathbf{Y}$ obtained from the data matrix $\mathbf{Z}$, we get the matrix $\mathbf{A}$ that contains the indicators of the clusters.

Ding *et al.* [24] also introduced the weighted symmetric (three-factor) NMF, expressed as $\mathbf{Y} = \mathbf{A}\mathbf{S}\mathbf{A}^T$, where the role of the weighting matrix $\mathbf{S}$ is to improve the factorization when $\mathbf{Z}^T\mathbf{Z}$ is poorly conditioned, and to allow more freedom in the clustering of overlapping data.

Zass and Shashua in [126] demonstrated that spectral clustering, normalized cuts, and Kernel K-means are particular cases of the clustering with nonnegative matrix factorization under a doubly stochastic constraint. They also considered the symmetric matrix decomposition under nonnegativity constraints – similar to that as formulated by Ding *et al.* [24] – however, their optimization strategy leads to a different multiplicative algorithm.

The analysis of NMF versus K-means clustering has been recently discussed by Kim and Park [54], who proposed the Sparse NMF (SNMF) algorithm for data clustering. Their algorithm outperforms the K-means and ordinary NMF in terms of the consistency of the results.

Example 8.2 *An example of data clustering with the selected clustering methods is shown in Figure 8.2. The data contains 1500 data points that constitute three non-overlapping clusters in a 3D space. As demonstrated in Figure 8.2(d), only the semi-binary NMF provides the clusters that fit exactly the original clusters shown in Figure 8.2(a). The uni-orthogonal NMF gives a small number of misclassified vectors, which is visible in the middle cluster in Figure 8.2(c). The worst clustering is obtained with the K-means method, however, it should be noted that K-means method is also based on nonconvex optimization (similarly as NMF) and the result is initialization-dependent. Clustering of such data with K-means is not always wrong, but it is likely to occur more frequently than with uni-orthogonal NMF.*

A comparison of NMF and spectral clustering leads to the following conclusions:

- Advantages of spectral clustering
  - basis vectors that are obtained with the SVD are better separated than with K-means,
  - noise reduction,
  - uniqueness,
  - no need for selection of a distance measure.
- Disadvantages of spectral clustering
  - negative values in basis vectors make interpretation of lateral components more difficult,
  - orthogonality constraints for overlapping clusters are inappropriate,
  - difficulty in identification of subspace clusters,
  - optimal number of basis vectors can be difficult to estimate.
- Advantages of NMF
  - produces factors that have physical meaning and are easy to interpret,
  - deals with overlapping clusters,
  - basis vectors are sparse,
    - including sparsity constraints, subspace clusters can be relatively easily identified,
    - irrelevant features can be removed.
- Disadvantages of NMF
  - non-unique solution – optimization is nonconvex,
  - nonnegativity constraints can restrict correct clustering to only nonnegative data,
  - optimal number of basis vectors J can be difficult to identify.

8.1.3 Clustering with Convex NMF

To improve clustering with NMF various constraints can be adopted. As already mentioned, at least one of the estimated factors may be very sparse, and for such a case it is reasonable to assume additional sparsity constraints to enforce the factorization process to be triggered to the desired solution. For non-overlapping clusters orthogonality constraints can be imposed on the indicator matrix, i.e. $\mathbf{X}\mathbf{X}^T = \mathbf{I}_J$. Such a version of NMF has already been proposed by Ding *et al.* [27], and is known as uni-orthogonal

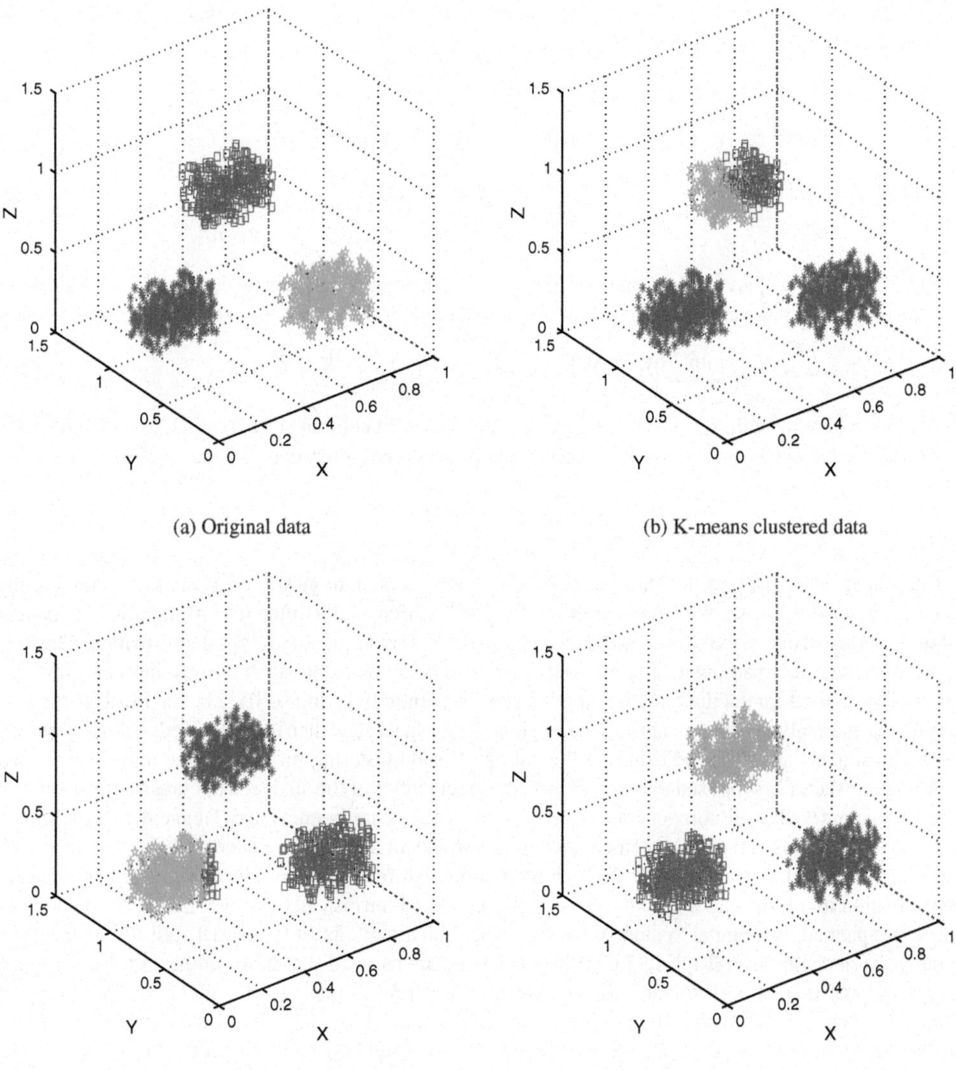

(a) Original data

(b) K-means clustered data

(c) Uni-orthogonal NMF clustered data

(d) Semi-binary NMF clustered data

Figure 8.2 Examples of data clustering techniques.

NMF. A particular case of uni-orthogonal NMF for clustering is semi-binary NMF that is described in Section 8.1.1.

When the data to be clustered are not constrained to be nonnegative, we may scale the data accordingly or another version of NMF can be used, that is, convex NMF [26]. This version assumes that the column vectors of the basis matrix should lie in the column space of $\mathbf{Y}$, and the linear combination of the column vectors should be convex. Each column of $\mathbf{A}$ can be rewritten as

$$\boldsymbol{a}_j = w_{1j}\boldsymbol{y}_1 + \ldots + w_{tj}\boldsymbol{y}_t + \ldots + w_{Tj}\boldsymbol{y}_T, \quad \text{where} \quad \forall t, j : w_{tj} \geq 0. \tag{8.8}$$

Thus we have $\mathbf{A} = \mathbf{YW}$, where $\mathbf{W} \in \mathbb{R}_+^{T\times J}$ is a nonnegative matrix, and the Euclidean objective function in the convex NMF takes the following forms:

$$\begin{aligned} D(\mathbf{Y}||\hat{\mathbf{Y}}) &= ||\mathbf{Y} - \mathbf{YWX}||_F^2 \\ &= \mathrm{tr}\{(\mathbf{I}_T - \mathbf{X}^T\mathbf{W}^T)\mathbf{Y}^T\mathbf{Y}(\mathbf{I}_T - \mathbf{WX})\} \\ &= \sum_{t=1}^{T} \sigma_t^2 ||\mathbf{v}_t^T(\mathbf{I}_T - \mathbf{WX})||_2^2, \end{aligned} \tag{8.9}$$

where $\mathbf{Y}^T\mathbf{Y} = \sum_{t=1}^{T} \sigma_t^2 \mathbf{v}_t \mathbf{v}_t^T$. Note that we assume $\mathbf{Y}$ is unsigned, and consequently $\mathbf{A}$ is also unsigned, whereas only $\mathbf{W}$ and $\mathbf{X}$ are nonnegative matrices. As stated in [26], the solution to the associated problem

$$\min_{\mathbf{W}\geq 0, \mathbf{X}\geq 0} ||\mathbf{I}_T - \mathbf{WX}||_F^2, \quad \text{s.t.} \quad \mathbf{W} \in \mathbb{R}_+^{T\times J}, \mathbf{X} \in \mathbb{R}_+^{J\times T}, \tag{8.10}$$

is $\mathbf{W} = \mathbf{X}^T$, where the columns of $\mathbf{W}$ are obtained from a certain column permutation of the identity matrix $\mathbf{I}_T$. Hence, the factors $\mathbf{W}$ and $\mathbf{X}$ are very sparse, which additionally improves the factorization.

8.1.4 Application of NMF to Text Mining

Text mining usually involves the classification of text documents into groups or clusters according to their similarity in semantic characteristics. For example, a web search engine often returns thousands of pages in response to a broad query, making it difficult for users to browse or to identify relevant information. Clustering methods can be used to automatically group the retrieved documents into a list of meaningful topics.

A typical method for document clustering is Latent Semantic Indexing (LSI) [21] that involves the SVD. LSI projects the analyzed documents into the singular vector space, and then the model reduction is performed and the documents are clustered using traditional partitional clustering methods in the transformed space. Each singular vector is supposed to capture basic latent semantics of the analyzed document. However, SVD does not guarantee that all components of the singular vectors are nonnegative. Hence the assumption on additive combinations of the basis vectors is generally not satisfied, which often does not give any meaningful interpretation. For this reason, NMF is much more attractive for document clustering, and exhibits better discrimination for clustering of partially overlapping data. LSI intrinsically assumes that the latent semantic directions spanned by singular vectors are orthogonal, which may not fit the model well if the clusters are partially overlapped. Several results [86,100,119] have demonstrated that NMF outperforms LSI in terms of accuracy and that its performance is comparable to hierarchical methods. Ding *et al.* [25] explored the relationship between NMF and Probabilistic LSI (PLSI), concluding that the hybrid connections of NMF and PLSI give the best results. Gaussier and Goutte [35] analyzed NMF with respect to Probabilistic Latent Semantic Analysis (PLSA), and they claimed that PLSA solves NMF with KL I-divergence, and for this cost function PLSA provides a better consistency.

Preprocessing strategies for document clustering with NMF are very similar to those for LSI. First, the documents of interest are subjected to stop-words removal and word streaming operations. Then, for each document a weighted term-frequency vector is constructed that assigns to each entry the occurrence frequency of the corresponding term. Assuming I dictionary terms and T documents, the sparse term-document matrix $\mathbf{Y} = [y_{it}] \in \mathbb{R}^{I\times T}$ is constructed from weighted term-frequency vectors, that is

$$y_{it} = F_{it} \log\left(\frac{T}{T_i}\right), \tag{8.11}$$

where F_{it} is the frequency of occurring the i-th term in the t-th document, and T_i is the number of documents containing the i-th term. The entries of $\mathbf{Y}$ are always nonnegative and equal to zero when either the i-th term does not appear in the t-th document or appears in all the documents.

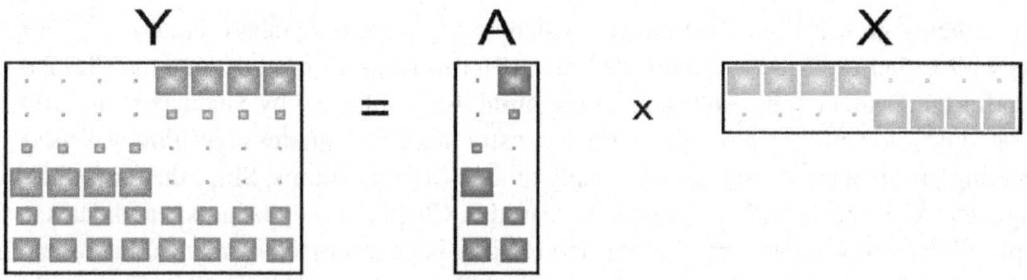

(a) NMF approach

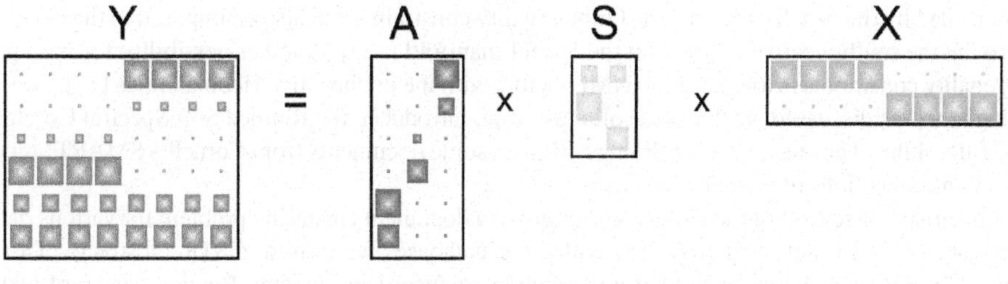

(b) tri-NMF approach

Figure 8.3 Document clustering based on the non-negative factorization of the term-document matrix **Y**: (a) NMF approach with the columns of the matrix **A** represent cluster centers, and the columns in **X** refer to cluster indicator. (b) tri-NMF approach with the rows of the matrix **S** represent term-clustering, and the columns in **S** refer to document-clustering.

The aim is to factorize the matrix **Y** into the nonnegative basis matrix $\mathbf{A} \in \mathbb{R}_+^{I \times J}$ and the nonnegative topics-document matrix $\mathbf{X} \in \mathbb{R}_+^{J \times T}$, where J denotes the number of topics. The position of the maximum value in each column-vector in **X** informs us to which topic a given document can be classified. An illustrative example of the term-document clustering is shown in Figure 8.3(a), where the term-document matrix $\mathbf{Y} \in \mathbb{R}_+^{6 \times 8}$ is factorized into the basis matrix $\mathbf{A} \in \mathbb{R}_+^{6 \times 2}$ and the topics-document matrix $\mathbf{X} \in \mathbb{R}_+^{2 \times 8}$. The columns of **A** refer to the cluster centers, and the columns in **X** are associated with the cluster indicators. Thus, the first four documents can be grouped into one cluster, and the other documents into the other cluster.

A more general scheme for simultaneous clustering both with respect to terms and documents can be modeled by the 3-factor NMF, sometimes referred to as tri-NMF. This model illustrated in Figure 8.3(b) can be expressed as

$$\mathbf{Y} = \mathbf{ASX}, \tag{8.12}$$

where $\mathbf{A} \in \mathbb{R}_+^{I \times K}$, $\mathbf{S} \in \mathbb{R}_+^{K \times J}$, and $\mathbf{X} \in \mathbb{R}_+^{J \times T}$. The matrix **S** is a block matrix. Its rows reveal term-clustering whereas the columns refer to document-clustering. The number of rows and columns in **S** denotes the number of term- and document-clusters, respectively. Normalizing the columns in **A** and the rows in **X** to the unit ℓ_1-norm, a_{ik} denotes the probability that the i-th term belongs to the k-th term-cluster ($k = 1, \ldots, K$), and x_{jt} represents the probability that the t-th document is assigned to the j-th document-cluster. The algorithms for tri-NMF were given in [27].

Typically, the redundancy of a document clustering problem is very high, i.e. $J, K << \min\{I, T\}$, and the topics-document matrix is very sparse, which is a perfect setting for the multiplicative algorithms. Indeed, the well-known Lee-Seung algorithm that minimizes the Euclidean distance was used by W. Xu *et al.* in [119] for clustering the TDT2[1] and the Reuters documents.[2] The TDT2 database contains 64,527 specifically topics-

[1] http://www.nist.gov/speech/tests/tdt/tdt98/index.htm

[2] http://kdd.ics.uci.edu/databases/reuters21578/reuters21578.html

oriented documents grouped into 100 clusters whereas the Reuters database counts 21,578 broad topics documents with 135 clusters. As demonstrated in [119], the accuracy of clustering the TDT2 documents is higher than for the Reuters ones. The same observation was confirmed by Shahnaz *et al.* [100], where for updating the topics-document matrix the additive constrained least-squares algorithm was used. Obviously, a text clustering problem can be solved with many other NMF algorithms. Since the term-document matrix is very large, the ALS and HALS algorithms discussed in Chapter 4 may be good candidates.

The application of NMF to document clustering has also been discussed by many other researchers. B. Xu *et al.* [118] proposed to use orthogonality constraints in their Constrained NMF (CNMF) algorithm, where the orthogonality of lateral components is enforced by the additional penalty terms added to the KL I-divergence and controlled by the penalty parameters. Orthogonality constraints can also be imposed to the NMF factors by deriving the multiplicative updates on the Stiefel manifold [123]. Another possibility for incorporating orthogonality constraints is to update the NMF factors with the nonnegative Hebbian rules [17]. Greene and Cunningham [41] discussed NMF based soft clustering, introduced the Refined Soft Spectral Co-clustering (RSSC) algorithm. They tested the RSSC algorithm on some documents from Cornell's SMART[3] repository that contains collections of technical abstracts.

A comparison of several of the NMF algorithms for a document clustering problem for various databases was performed by Li and Ding [67]. The following databases were used: CSTR (abstracts of technical reports), WebKB (webpages gathered from computer science departments), Reuters (standard text documents), WebACE (news articles from Reuters), and Log (log data collected from several different computers with different operating systems). The clustering has been performed with the following methods: K-means, standard NMF, semi-NMF (nonnegativity constraint imposed only on one factor), convex NMF, tri-NMF, and PLSI. The performance of clustering was estimated with the accuracy measure:

$$Accuracy = \max \left\{ \frac{1}{N} \sum_{\mathcal{C}_j, \mathcal{L}_m} T(\mathcal{C}_j, \mathcal{L}_m) \right\}, \tag{8.13}$$

where $\mathcal{C}_j$ stands for the j-th cluster, $\mathcal{L}_m$ is the m-th class, N is the number of clusters, and $T(\mathcal{C}_j, \mathcal{L}_m)$ is the number of entries in the m-th class which are assigned to the j-th cluster. The clustering results obtained by Li and Ding [67] are shown in Figure 8.4. They concluded that the NMF algorithms generally give better performance than K-means. In fact, the NMF approach is rather equivalent to soft K-means, and also PLSI usually gives the same results as NMF. The tri-NMF provides a good framework for simultaneous clustering of the rows and columns. However, the observed difference in the accuracy between the various NMF algorithms is not very considerable.

8.1.5 Email Surveillance

NMF has also been successfully applied to the extraction and detection of topics from electronic mail messages [9]. The underlying idea of clustering semantic features (topics) from email messages is similar to that exploited in document clustering described above. In the preprocessing mode, the term-message matrix is constructed from the analyzed mail messages. A collection of T messages indexed by I terms (or keywords) can be represented by a $I \times T$ term-message matrix $\mathbf{Y}$. Each entry y_{it} of $\mathbf{Y}$ denotes a weighted frequency at which the term i occurs in the message t. The weighting is determined with some degree of importance of the i-th term in the t-th message. Several weighting strategies can be used [9], for instance, the columns in $\mathbf{Y}$ may be normalized to a unit length. After applying NMF to the term-message matrix, the interpretation of the estimated factors is also similar to document clustering. The column vectors of the matrix $\mathbf{A}$ denote semantic features, and the rows in $\mathbf{X}$ contain the coefficients of linear combinations of the

[3] ftp://ftp.cs.cornell.edu/pub/smart

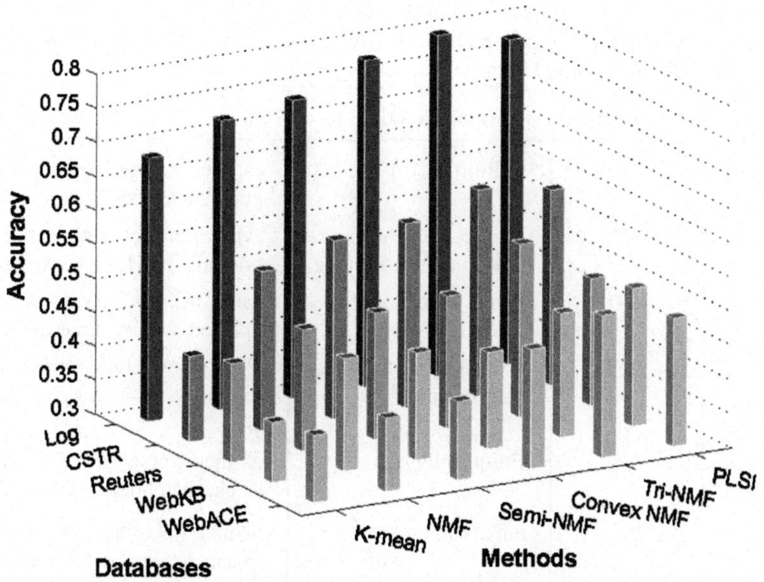

Figure 8.4 Performance of several clustering methods for a document clustering problem.

semantic features. The rank of the factorization determines the number of topics, and it can be estimated from a distribution of singular values of the covariance matrix related to the term-message matrix. The t-th message is classified to the j-th topic if the element x_{jt} takes the maximum value in the t-th column of $\mathbf{X}$. Setting all other entries in the t-th column of $\mathbf{X}$ (besides the entry x_{jt}) to zero, all the nonzero entries in the j-th row of $\mathbf{X}$ determine the j-th cluster. This clustering technique refers to so-called hard clustering. Another strategy for selecting the documents that can be assigned to the same topic is to impose a threshold to the entries in $\mathbf{X}$, and consequently to select only those terms which have their scores (values in $\mathbf{X}$) greater than the threshold, however, this might attribute multiple clusters to one message. Clustering can also be performed with respect to these terms by analyzing dominant values in the column vectors of $\mathbf{A}$.

Berry and Browne in [9] applied NMF to classify email messages from the PRIVATE subcollection of the Enron corpus.[4] The set contains more than 517,000 email messages sent by 150 employees over several years. The aim is to identify relevant underlying topics of the messages without the need to read individual messages, and to classify the messages according to the semantic features (topics). Statistical information on the messages gathered from each month of 2001 is given in Table 8.1 [9].

After removing stop-words and applying word streaming operations, the messages were parsed into the basic terms, the weighted term-message matrix was constructed, and the hybrid NMF algorithm called the GD-CLS was applied. The algorithm combines the Lee-Seung multiplicative algorithm for updating $\mathbf{A}$ and

Table 8.1 Statistics of email messages from the PRIVATE subcollection of the Enron corpus, for each month of 2001. The corresponding number of terms is parsed from each monthly subset [9].

Month	Jan	Feb	Mar	Apr	May	Jun	Jul	Aug	Sep	Oct	Nov	Dec
Messages	3,621	2,804	3,525	4,273	4,261	4,324	3,077	2,828	2,330	2,821	2,204	1,489
Terms	17,888	16,958	20,305	24,010	24,335	18,599	17,617	16,417	15,405	20,995	18,693	8,097

[4] http://www-2.cs.cmu.edu/ enron

Table 8.2 Example of clusters (topics) obtained with the NMF GD-CLS algorithm from the PRIVATE subcollection [9]. A total number of clusters is 50. The ten dominant (with the largest values in **X**) terms for the selected feature vectors are listed in the last column.

Feature Index (j)	Cluster Size	Topic Description	Dominant Terms
10	497	California	ca, cpuc, gov, socalgas, sempra, org, sce, gmssr, aelaw, ci
23	43	Louise Kitchen named top woman by Fortune	evp, fortune, britain, woman, ceo, avon, fiorinai, cfo, Hewlett, Packard
26	231	Fantasy football	game, wr, qb, play, rb, season, injury, updated, fantasy, image
33	233	Texas longhorn football newsletter	UT, orange, longhorn[s], texas, true, truorange, recruiting, oklahoma, defensive
34	65	Enron collapse	partnership[s], fastow, shares, sec, stock, shareholder, investors, equity, lay
39	235	emails about India	dahhol, dpc, india, mseb, maharashtra, indian, lenders, delhi, foreign, minister

the Tikhonov regularized constrained least squares algorithm for updating **X**. The selected results are shown in Table 8.2. The ten dominant terms for each selected feature vector (topic) are given in the fourth column. The terms have been selected because they have the largest values in the corresponding rows of **X**. The topics were inferred from the analysis of the terms that are spanned by the semantic feature vectors. A detailed analysis can be found in [9].

8.2 Classification

A classification problem for objects is related to the separation of those objects into smaller classes, and assigning a given object to a particular class. Classification can be characterized in terms of both supervised and unsupervised learnings. However, NMF is usually applied in the context of unsupervised learning in which the model automatically assigns the objects from a given dataset into the classes, without prior knowledge of the features of the objects. Several studies successfully applied NMF to the classification of the objects such as sounds generated by various instruments, faces, images, textures, genes, neural signals, email messages, or text documents. The last two applications of NMF have been discussed in the previous section in the context of clustering.

8.2.1 Musical Instrument Classification

Classification of musical instruments from audio recordings has many practical applications such as automatic music transcription, internet search, or music information retrieval from multimedia databases.

Benetos *et al.* in [6–8] applied various NMF algorithms for the classification of musical instruments in several sound recordings from the UIOWA database.[5] Their approach uses a supervised learning strategy, where at the training stage the basis matrix **A** and the encoding matrix **X** are obtained with NMF applied to the feature matrix **Y** containing a certain number of sound recordings, separately for each class of instruments.

[5]Publicly available from http://theremin.music.uiowa.edu/index.html

Table 8.3 Classification results of different musical instruments [6].

Instrument	Piano	Bassoon	Cello	Flute	Sax	Violin
Piano	18	0	0	0	0	0
Bassoon	0	9	0	0	0	0
Cello	0	0	16	0	0	0
Flute	1	0	0	8	0	0
Sax	0	0	0	0	9	0
Violin	0	0	0	0	0	29

They carried out the training using a training set of 210 audio files having a duration of about 20 sec and sampled at 44.1 kHz sampling rate. The recordings should be classified into six different instrument classes: piano, violin, cello, flute, bassoon, and soprano saxophone. For the n-th class, several sound recordings of the same instrument are selected, and then transformed into feature vectors according a feature extraction algorithm, and $\mathbf{Y}^{(n)}$ is constructed from the feature vectors. For N classes, this gives a set of N feature matrices $\mathbf{Y}^{(n)}$ with $n = 1, \ldots, N$. The feature extraction algorithm extracts the most relevant features from sound recordings which can discriminate between musical instruments. For instance, mean and variance of the first mel-frequency cepstral coefficients, and several other parameters (such as mean and variance of AudioSpectrumCentroid, AudioSpectrumEnvelope, AudioSpectrumSpread, or AudioSpectrumFlatness) that are part of the MPEG-7 spectral basis descriptors. Each feature matrix is decomposed with NMF as follows: $\mathbf{Y}^{(n)} = \mathbf{A}^{(n)}\mathbf{X}^{(n)}$. The rank of the factorization is set as

$$J_n = \left\lfloor \frac{T_n I_n}{T_n + I_n} \right\rfloor, \tag{8.14}$$

with $\mathbf{Y}^{(n)} \in \mathbb{R}^{I_n \times T_n}$. After obtaining a pair of matrices $\mathbf{A}^{(n)}$ and $\mathbf{X}^{(n)}$ for each instrument class, the training process is completed. For classification a test recording is transformed into the corresponding feature vector $\boldsymbol{y}_{test}$ which is then projected onto the space spanned by the basis matrix $\mathbf{A}^{(n)}$, separately for each class. Thus we get N encoding vectors $\boldsymbol{x}_{test}^{(n)}$ for $n = 1, \ldots, N$. For each class, the Cosine Similarity Measure (CSM) between the vector $\boldsymbol{x}_{test}^{(n)}$ and each column vector of $\mathbf{X}^{(n)}$ is calculated according to the formula

$$CSM_n = \max_{t=1,\ldots,T_n} \left\{ \frac{(\boldsymbol{x}_{test}^{(n)})^T \boldsymbol{x}_t^{(n)}}{||\boldsymbol{x}_{test}^{(n)}||_2 ||\boldsymbol{x}_t^{(n)}||_2} \right\}. \tag{8.15}$$

Then, the highest value of CSM determines the class label of the test recording, i.e. $n_{class} = \arg\max_{n=1,\ldots,N}\{CSM_n\}$.

Table 8.3 shows the classification results given, as obtained by Benetos *et al.* [6]. The columns of Table 8.3 refer to the predicted musical instrument and the rows to the true ones. Note that only misclassification takes place for the flute, which is wrongly classified as piano. This is because both instruments have similar dynamics and spectral shape.

8.2.2 Image Classification

Image classification is an essential part of a pattern or object recognition system [29,32,114], and it is usually based on supervised or unsupervised learning. In our approach that uses NMF the learning is supervised. The classification system first extracts a set of feature vectors that capture the structure of the training data. Once the subspace spanned by the feature vectors is determined, the classification of a new feature vector is accomplished by projecting the test image vector onto the feature subspace and finding the nearest training neighbor. This strategy is similar to that for the musical instrument classification described in Section 8.2.1.

8.2.2.1 Face Recognition

One of the most important and challenging applications of image classification is face recognition. Several recently published works are focused on the application of NMF to face recognition [42,43,45,113,120,121,128]. The feature vectors are usually learned from a set of frontal faces distinguished by subjects, facial expressions, illuminations conditions, viewing conditions, ageing, and partial occlusions such as individuals wearing sunglasses or scarfs. Typically, all frontal pose images are scaled and aligned by the center of eyes and mouths.

A selected number of such different faces forms a training set that is represented by the data matrix $\mathbf{Y}^{(t)} \in \mathbb{R}^{I_t \times T_t}$, where T_t is the number of images in the training set, and I_t is the number of pixels in each training image. The training images are usually scanned in lexicographic order to form the column vectors in $\mathbf{Y}^{(t)}$. Then the matrix $\mathbf{Y}^{(t)}$ is decomposed with NMF into the basis matrix $\mathbf{A}^{(t)} \in \mathbb{R}^{I_t \times J_t}$ and the weighting matrix $\mathbf{X}^{(t)} \in \mathbb{R}^{J_t \times T_t}$, where the rank J_t should satisfy the criterion (8.14).

The test images to be classified should be stored in the matrix $\mathbf{Y}^{(c)} \in \mathbb{R}^{I_c \times T_c}$, where $I_c = I_t$ and T_c is the number of test images. The matrix $\mathbf{Y}^{(c)}$ is then projected onto the subspace spanned by the feature vectors in $\mathbf{X}^{(t)}$, and as a result, the test feature matrix $\mathbf{X}^{(c)}$ is obtained. This operation can be accomplished by $\mathbf{X}^{(c)} = \max\{\mathbf{0}, (\mathbf{A}^{(t)})^{\dagger}\mathbf{Y}^{(c)}\}$, where $(\mathbf{A}^{(t)})^{\dagger}$ is the pseudo-inverse of $\mathbf{A}^{(t)}$, or given the test image matrix $\mathbf{Y}^{(c)}$ it can be obtained by updating the weighting matrix $\mathbf{X}$ and keeping the basis matrix $\mathbf{A}^{(t)}$ fixed. Then for each test image $\boldsymbol{y}_w^{(c)}$ $(w = 1, \ldots, T_c)$, some distance measure between $\boldsymbol{x}_w^{(c)}$ and any vector $\boldsymbol{x}_t^{(t)}$ $(t = 1, \ldots, T_t)$ is calculated. The index of the vector $\boldsymbol{x}_t^{(t)}$ minimizing the distance determines the class to which $\boldsymbol{y}_w^{(c)}$ should be assigned. A study on various distance measures used in this application can be found in [44,121,122].

NMF differs from other decomposition methods such as Principal Component Analysis (PCA), as PCA provides global image representations. By avoiding subtractive combinations of the basis vectors, NMF gives instead sparse and part-based (local) representations of the faces, as was first demonstrated by Lee and Seung in [62,63]. Due to these properties, NMF often outperforms PCA-based methods for face recognition. For illustration, we perform decomposition on the ORL face database [97], where various facial expressions for one subject were selected. The selected training faces from the ORL face database are shown in Figure 8.5. The basis vectors transformed to their equivalent matrix representations are presented in Figure 8.6(a) for PCA and Figure 8.6(b) for NMF. The PCA-based representations have negative values and reflect the nature of global representations, whereas the images obtained with NMF can be regarded as sparse and local part-based representations, that is, eyes, eyebrows, ears, and mouths. The results of face recognition under various facial expressions, under different illumination conditions, and partial occlusion can be found in [43]. Normally, NMF gives better characteristics of recognition than PCA.

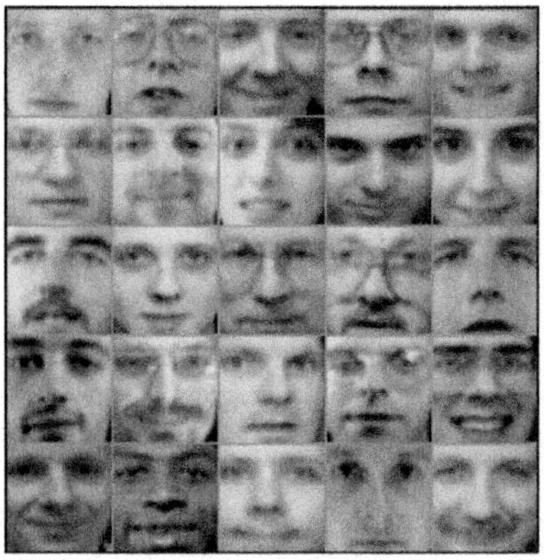

Figure 8.5 Example of training faces from the ORL face database [97].

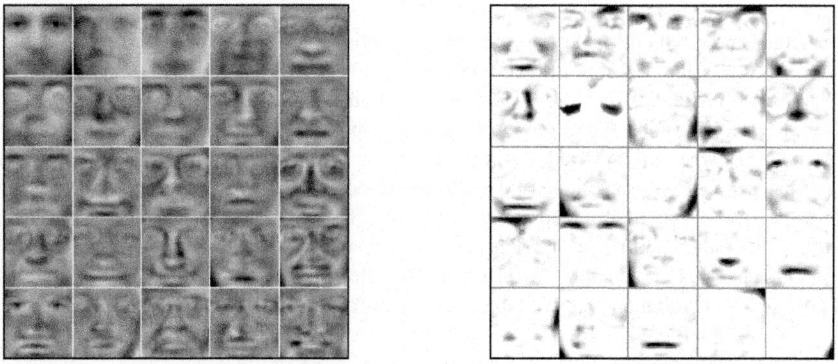

(a) PCA (b) NMF

Figure 8.6 Basis images obtained with: (a) PCA, (b) NMF.

Face recognition with NMF has also been discussed by Okun in [79]. The experiments have been performed for the images taken from the JAFFE database,[6] which contains facial images with different facial expressions of 10 Japanese females. The classification results obtained with the standard NMF were not very satisfactory. However, Local NMF [31,66,112] and Weighted NMF [45] can achieve better results.

8.2.2.2 Handwritten Digit Recognition

Another example comparing the basis vectors obtained with PCA and NMF is shown in Figure 8.7. Handwritten digit recognition has been extensively studied in the literature. A survey of the methods for such

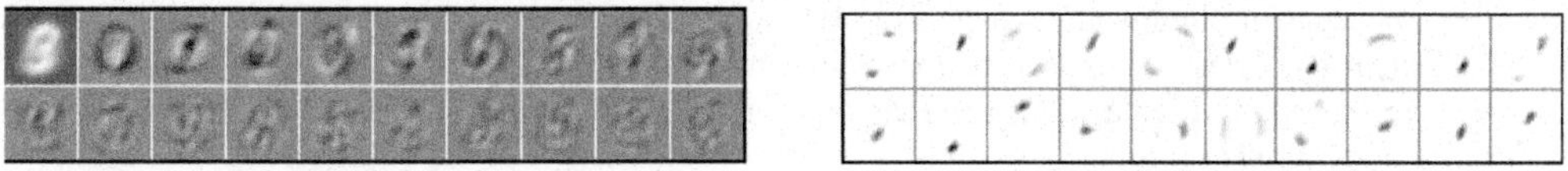

(a) PCA - $J = 20$ (b) NMF - $J = 20$

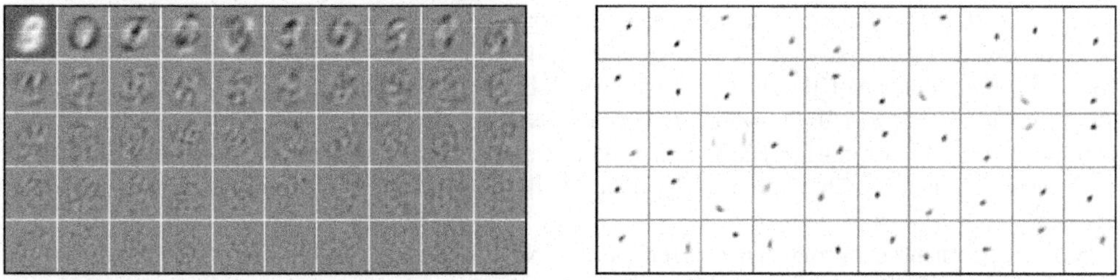

(c) PCA - $J = 50$ (d) NMF - $J = 50$

Figure 8.7 Basis images obtained with different methods (and the rank J) applied to handwritten digits: (a) PCA, $J = 20$; (b) NMF, $J = 20$; (c) PCA, $J = 50$; (d) NMF, $J = 50$.

[6] http://www.mis.atr.co.jp/~mlyons/jaffe.html

classification is given in [61]. Some of them involve a tensor decomposition, which is more attractive with respect to memory requirements and performance. An example of such an approach is the method that uses higher order singular value decomposition (HOSVD) of a tensor [98].

The basis vectors are learnt from a set of handwritten digits that are extracted from the MNIST database[7] used in [60]. Figure 8.7(a) and Figure 8.7(b) illustrate the basis (image) vectors obtained correspondingly with PCA and NMF when the low rank of the factorization is equal to 20. The basis vectors learnt for the same images but with the low rank increased to 50 are presented in Figure 8.7(c) and Figure 8.7(d) for PCA and NMF, respectively. Note that an increase in the low rank for NMF is followed by an increase in a number and sparsity of the part-based representations. A whole image is hence composed from smaller but more numerous elements. This is a completely different effect than provided by PCA, where an increase in the low rank changes only global chessboard effects of the basis images.

8.2.2.3 Texture Classification

Texture classification is an important topic in computer vision and graphics. Many texture classification methods have been reported in the literature [20,36,58,59,64,108], and the application of NMF to texture classification has been proposed, e.g. in [87,89]. In comparison to the above-mentioned supervised classification methods, the approach discussed in [89] uses unsupervised learning, and is similar to document clustering discussed in Section 8.1.4. The main difference appears to be in the preprocessing algorithm that creates the data matrix $\mathbf{Y}$. In text classification this matrix is referred to as a term-document matrix, whereas in the email surveillance application $\mathbf{Y}$ is named the term-message matrix. In text classification we have the "bag-of-keypoint" matrix. Texture can be characterized by simple patterns that are repeated throughout. Each kind of texture has various numbers of different patterns recognized as so-called keypoints [19] which are generated from local regions invariant under viewpoints and illumination effects. There are a few methods for detecting such regions, e.g. Harris-Affine, Hessian-Affine, or Laplacian-Affine detectors [89]. All these-mentioned methods extract ellipsoidal regions. The Harris-Affine detector localizes the regions with significant changes in illumination whereas the Hessian-Affine detector finds blobs of uniform intensity. The Laplacian-Affine detector extracts regions with similar properties as with the Hessian-Affine one. Each region is represented by a scale, affine and rotation invariant SIFT descriptor. A set of the SIFT descriptors for each image is used to form the keypoints.

Figure 8.8(b) illustrates an example of the keypoints extracted from the textures shown in Figure 8.8(a) [89]. The textures are taken from the UIUC database[8] that contains 1000 images grouped in 25 classes. The images are very challenging for texture classification because they have irregular texture patterns, nonplanar surface, nongrid deformations, scale variations and considerable viewpoint changes. Each image to be classified is encoded with the histogram of keypoint occurrence as a column vector in $\mathbf{Y} \in \mathbb{R}^{I \times T}$, where T is the number of the images. Applying NMF to $\mathbf{Y}$ we obtain the basis matrix $\mathbf{A} \in \mathbb{R}^{I \times J}$ and the weighting matrix $\mathbf{X} \in \mathbb{R}^{J \times T}$, where J is an *a priori* known number of classes. The weighting matrix with the column vectors normalized to unit l_1 norm can be interpreted as a probability matrix. The index of the largest value in each column vector indicates the class to which a given texture image should be assigned.

Many experiments confirm that the performance of NMF in texture classification is comparable or sometimes better than for the PLSI method [89]. The following example [89] presents the classification results of 320 texture images that belong to eight classes (fabric textures, wall paper, fur, carpets). The results are illustrated in the form of a confusion matrix given in Table 8.4. Note that only two images are misclassified, and this is mostly due to the high similarity of patterns in the misclassified images.

[7] http://www.cs.cmu.edu/~15781/web/digits.html

[8] http://www-cvr.ai.uiuc.edu/ponce_grp/

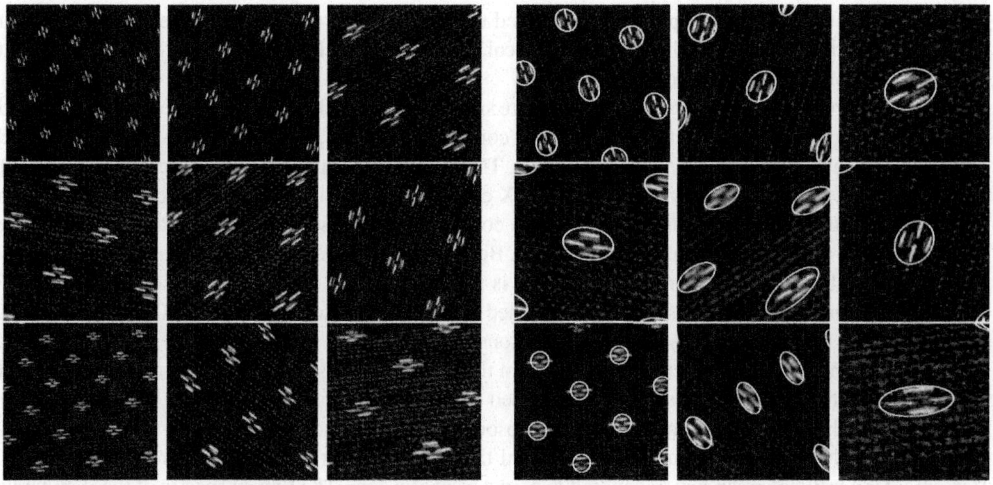

(a) Textures (b) Keynotes

Figure 8.8 Example of (a) one sample of textures, (b) corresponding keynotes.

Table 8.4 Classification results of eight different texture images. [89].

True classes	Class 1	Class 2	Class 3	Class 4	Class 5	Class 6	Class 7	Class 8
Class 1	40	0	0	0	0	0	0	0
Class 2	0	40	0	0	0	0	0	0
Class 3	0	0	40	0	0	0	0	0
Class 4	0	0	0	39	0	0	0	0
Class 5	0	0	0	0	40	1	0	0
Class 6	0	0	0	0	0	39	0	0
Class 7	0	0	0	1	0	0	40	0
Class 8	0	0	0	0	0	0	0	40

8.3 Spectroscopy

NMF is also successfully applied to the selected modalities of spectrographic imaging. They particularly include Raman spectroscopy [65,75,95], Fluorescence Spectroscopy (FS) [37,38,46], Hyperspectral Imaging (HSI) [28,40,50,76,85,95,106], and Chemical Shift Imaging (CSI) [11,95,96].

8.3.1 Raman Spectroscopy

Raman spectroscopy can be used to identify a wide range of chemical compounds in the examined specimen. Each chemical compound has unique Raman spectra (fingerprint) that can be measured due to the Raman scattering effect [34,68]. A monochromatic light generated by a laser interacts with molecular structure of the compound, resulting in different scattering effects. A majority of scattered photons is emitted due to Rayleigh scattering, and such photons have the same energy and wavelength as the photons emitted from a laser light beam. However, a small amount of scattered photons occurring due to inelastic scattering (Raman scattering) have the energy and spectral lines shifted up or down with reference to the energy of incident photons. The shifting is very specific for each molecular structure, which can be considered as a fingerprint by which the molecule is identified.

Due to many reasons (measurement errors, impure specimen, and background fluorescence) the observed spectra represent a mixture of spectra for underlying consistent materials, and the aim is to extract the desired spectra to uniquely identify the examined specimen. The mixture is assumed to be represented by the linear model: $\mathbf{Y} \cong \mathbf{AX}$ where $\mathbf{A}$ is the mixing matrix and $\mathbf{X}$ contains the constituent Raman spectra. Each column of $\mathbf{A}$ represents the concentration/abundence of the corresponding constituent material, and the respective row vector of $\mathbf{X}$ represents the unknown spectrum. Both mixing coefficients and spectra are nonnegative, and hence, the choice of NMF for this application is natural. Because of nonnegativity constraints, NMF should outperform ICA that was successfully applied to Raman spectroscopy in [75,110,111]. Moreover, the performance may even increase assuming only one target spectrum and the others being considered as disturbing spectra that are completely different from the target one. Assuming the spectra are independent an additional orthogonality constraint can be imposed to the row vectors in $\mathbf{X}$ [65].

An example that shows the Raman spectra of Epsomite and Demantoid is taken from [95]. Epsomite is a water soluble hydrous magnesium sulfate mineral that is well-known as artificially created epsom salt. Demantoid is a brilliant green mineral andradite garnet used as a gem. The Raman spectra of these minerals observed in a bandwidth of 201 to 1200 Raman shift (cm^{-1}) are illustrated in Figure 8.9(a) and Figure 8.9(b).

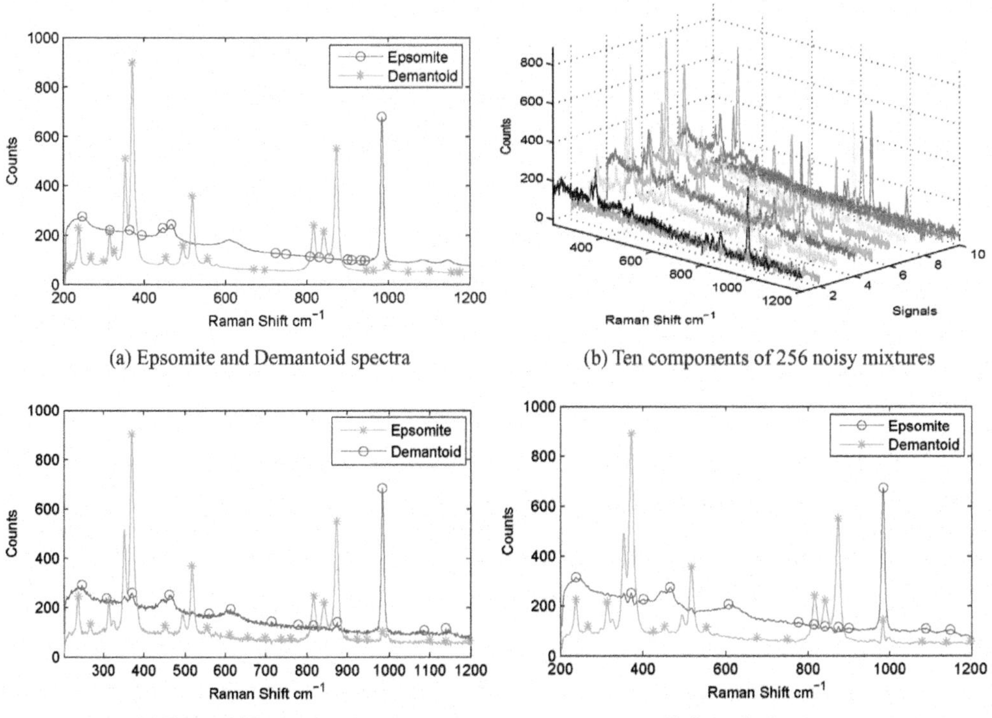

(a) Epsomite and Demantoid spectra

(b) Ten components of 256 noisy mixtures

(c) Estimated Raman spectra

(d) Smoothed spectra

Figure 8.9 Raman spectra: (a) target spectra of Epsomite and Demantoid, (b) ten sample components of 256 mixtures, (c) estimated spectra of Epsomite and Demantoid, (d) NMF with smoothness constraints.

They can be found in the database of the Division of Geological and Planetary Sciences at CalTech.[9] Then the spectra were artificially mixed by a strictly positive random matrix $\mathbf{A} \in \mathbb{R}^{256\times 2}$ and perturbed with an additive Gaussian noise. The estimated spectra with NMF can be found in Figure 8.9(c) and Figure 8.9(d) for Epsomite and Demantoid, respectively. Note that the estimated spectra fit the target ones very well. The results are very good but this is probably due to a strong redundancy – only two signals are estimated from 256 observations.

8.3.2 Fluorescence Spectroscopy

Fluorescence spectroscopy uses a fluorescence effect to analyze concentration of chemical species in the surveyed object. The species being exited by a strong monochromatic light absorb photons from the incident light, changing an electronic energy state of molecules from a ground state (a low energy state) to a vibrational state of higher energy. As a result of molecular collisions, the exited molecule gradually loses its energy, and finally returns to the ground electronic state. Due to the change in electronic state, the photons that have different energy and frequency than the incident ones are emitted. The molecular structure of the species can thereby be determined, by analyzing the spectrum of the emitted light with respect to the frequency of monochromatic light.

Nevertheless, the spectral analysis is not straightforward since the observed spectra are instantaneous mixtures of pure species spectra and other intermediate species spectra. The problem of extracting pure spectra from the mixtures can be formulated in terms of a blind source separation problem, and solved with many algorithms for ICA [18]. Furthermore, considering intrinsic nonnegativity constraints on spectra and their concentrations/abundances, the problem can be solved with NMF, which is more profitable since the separated spectra could be partially statistically dependent.

Assuming the mixed spectra are observed by I sensors and each spectrum has T samples, all the observations can be stored in the observation matrix $\mathbf{Y} \in \mathbb{R}^{I\times T}$. Applying NMF to $\mathbf{Y}$ under the assumption that J is the number of constituent spectra, we obtain the abundance matrix $\mathbf{A} \in \mathbb{R}^{I\times J}$ and the nonnegative matrix $\mathbf{X} \in \mathbb{R}^{J\times T}$ of the pure spectra.

Applications of NMF to fluorescence spectroscopy can be found, e.g., in [37,38,78]. Gobinet *et al.* [37] applied NMF to analyze a distribution of some organic compounds such as bound ferulic acid, free ferulic acid, and p-coumaric acid in durum wheat and barley grains. They used a laser scanning microspectrofluorometer to acquire fluorescence signals. A transversal section of a wheat grain was scanned with a 365 nm laser at a spatial resolution of approximately 1 μm, and the fluorescence signal spectra were measured by a CCD detector in the range of 350 to 670 nm. The observed area was discretized into 20×20 pixels, and each observed spectrum was sampled to yield 128 points. Hence, the observations were stored in the matrix $\mathbf{Y} \in \mathbb{R}^{400\times 128}$. Applying NMF under the assumption that the number of the constituent compounds is three, they obtained the pure species spectra and the pure species concentration/abundance maps of the corresponding pure species. Each abundance map was obtained by matricization of the corresponding column vector of $\mathbf{A} \in \mathbb{R}^{400\times 3}$.

We present an extension of fluorescence spectra using tensor factorization algorithms. The benchmark [12,91] contains 405 recorded measurements of five replicated fluorescence spectra for a total of six different fluorophores in the dataset: catechol, hydroquinone, indole, resorcinol, tryptohpane and tyrosine. Each spectra was expressed by two factors: Emission (136 wavelengths) and Excitation (19 wavelengths). In total 405×5 Emission-Excitation spectra were recorded. If we vectorize all spectra, the NMF model will be applied to find the six basis components. However, tensor factorization helps to return a much more accurate result. For example, we can factorize the tensor with size of 405 samples $\times$ 136 emission wavelengths $\times$ 19

[9] http://minerals.gps.caltech.edu/FILES/raman/Caltech data/

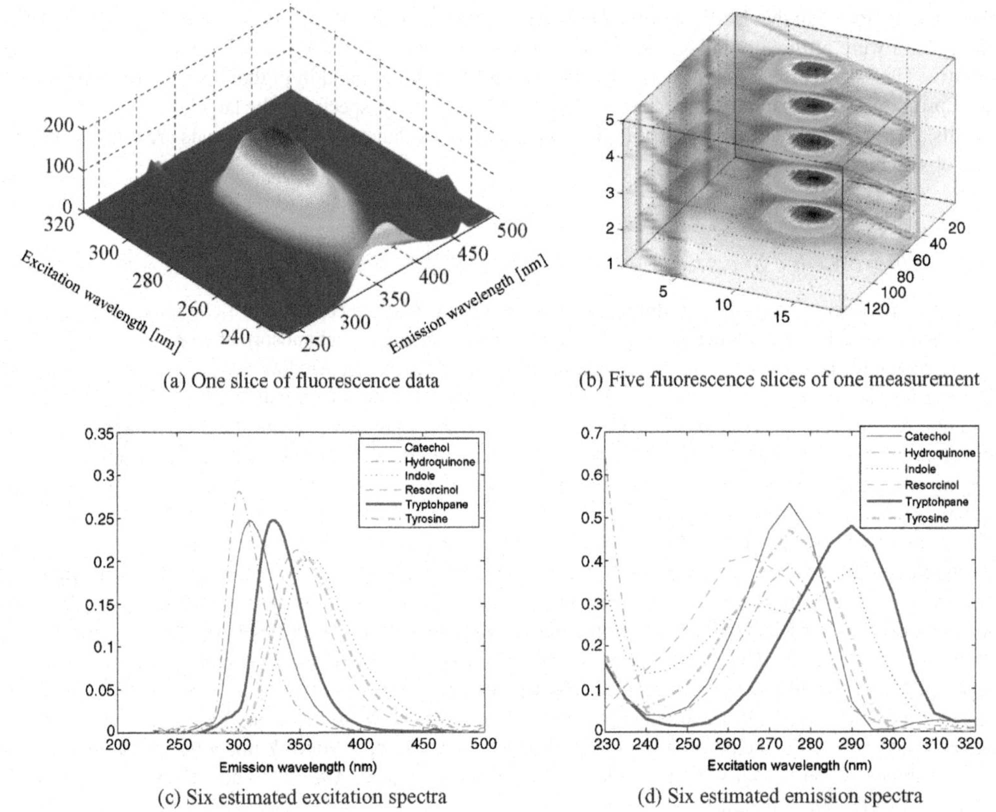

(a) One slice of fluorescence data

(b) Five fluorescence slices of one measurement

(c) Six estimated excitation spectra

(d) Six estimated emission spectra

Figure 8.10 Illustration of factorization for fluorescence data of size 405 samples × 136 emission wavelengths × 19 excitation wavelengths × 5 replicates : (a) one spectra slice of the tensor , (b) one 3-D sample with five fluorescence replicates of size 36 emission wavelengths × 19 excitation wavelengths, (c)–(d) estimated excitation and emission spectra of six different fluorophores in the dataset: catechol, hydroquinone, indole, resorcinol, tryptohpane and tyrosine.

excitation wavelengths × 5 replicates into six components. The second and third factors $\mathbf{A}^{(2)}$ and $\mathbf{A}^{(3)}$ are the Emission and Excitation spectral components, respectively. Figures 8.10(b) and 8.10(c) depict these basis components which correspond to six fluorophores: catechol, hydroquinone, indole, resorcinol, tryptohpane and tyrosine. Another factorization is presented in Figure 8.11 with six basis spectra with size of 136 Emission wavelengths × 19 Excitation wavelengths. Each 2-D spectrum slice is a composition of the six basis spectra. Figure 8.10(a) is an example of one Emission × Excitation slice.

8.3.3 Hyperspectral Imaging

Hyperspectral imaging has found many real-life applications [40]. In the mining and oil industries it is mostly used for identifying various minerals or for searching ore or oil fields. In agriculture it is useful for monitoring the development and health control of crops, detection of the chemical composition of plans, and water quality control. Physicists use this technique in electron microscopy, and soldiers for military surveillance.

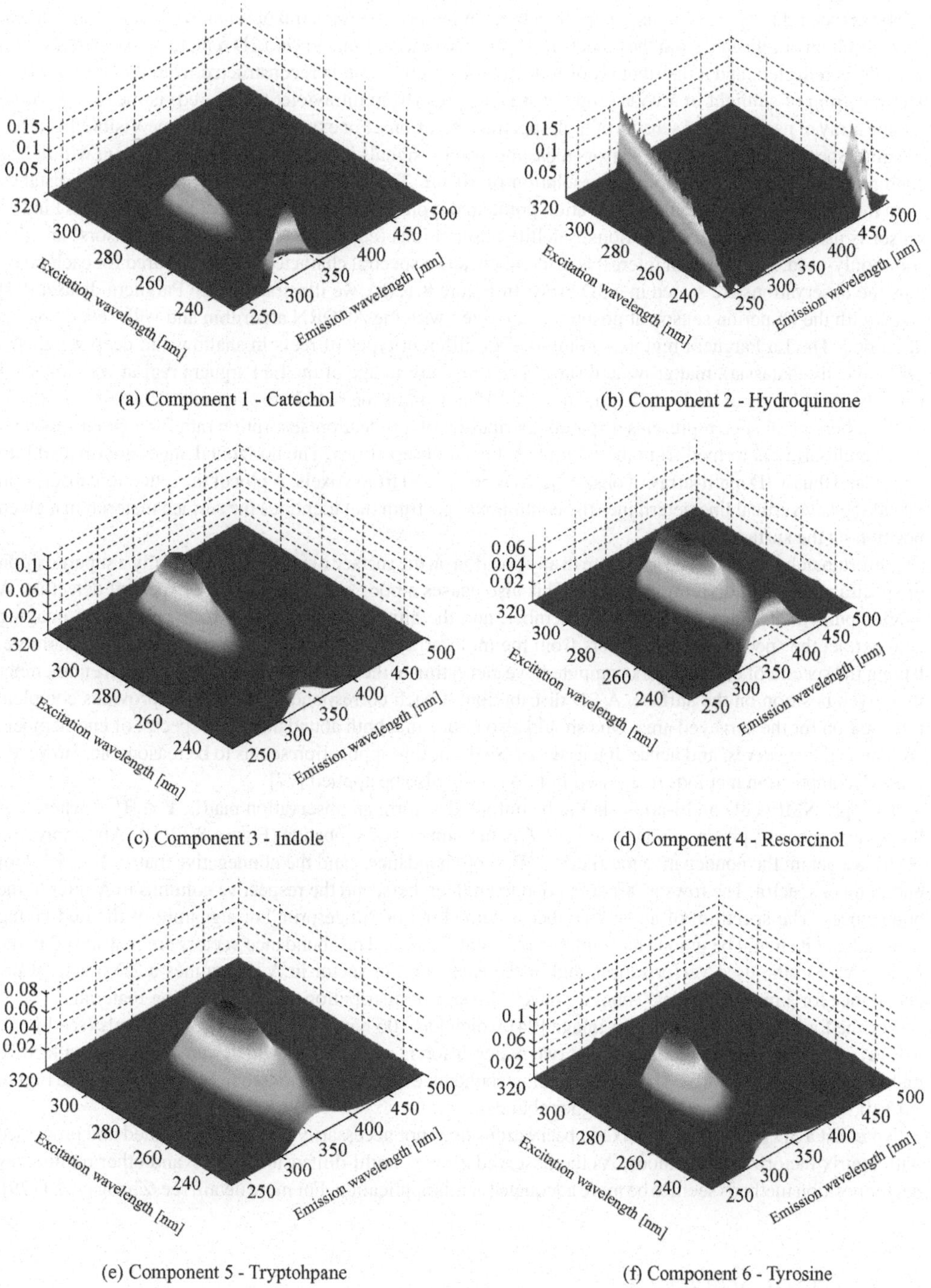

Figure 8.11 Six spectra components were estimated from the tensor 405 samples × 2584 (emission-excitation) wavelengths × 5 replicates. Slice in Figure 8.10(a) was expressed by addition of these six basis spectra.

Hyperspectral imaging remotely maps the object of interest with spectral observations of electromagnetic waves emitted or reflected from the object. Typically, the object of interest is a 2D remote surface from which sunlight is reflected, and a distribution of reflection/absorption rate is reconstructed from the observations. The spectrum of sunlight in a wide range of wavelengths (from ultraviolet to infrared) is measured with a remote array of narrow-band sensors of high spectral resolutions. For example, the Airborne Visible/InfraRed Imaging Spectrometer (AVIRIS) sensors measure spectral signals in the range of $0.4 - 2.45\ \mu m$ within 224 contiguous subbands with a spectral resolution of $10\,nm$. The spatial resolution depends on the distance from the object of interest to the observation point, and typically, it varies from 4 to 20 meters. This kind of sensor is used in airborne observations; satellite observations (e.g., by NASA's Hyperion sensors) are also commonly-used. The object of interest is discretized, and a spectral characteristic is measured for each pixel, thus the observations are stored in a 3D array. In Figure 8.12(b), we illustrate the La Parguera dataset [94] taken with the Hyperion sensor, atmospheric corrected with the ACORN algorithm and with several bands discarded. The La Parguera region is composed of different types of reefs in shallow and deep water, sea grass (mostly thalassia), mangrove and sand. The true color image of the La Parguera region was collected from the three bands of Red = 634 nm, Green = 545 nm, and Blue = 463 nm (Figure 8.12(a)).

Each horizontal slice represents a spatial distribution of a reflection/absorption rate for a given subband. A 3D multi-array is formed from the multiple subband observations. The horizontal slices are divided into pixels, and thus a 3D multi-array of observations is composed from voxels. A plot of the reflection/absorption rate along any vertical line determines the continuous spectrum that identifies the surface material in a given position on the surface.

Unfortunately, the observed spectrum in any position on the surface of interest is practically a superposition of spectra of many underlying materials. This also causes a poor spatial resolution of spectral detectors. A single "pure" material is called the endmember, and the aim of applying NMF to hyperspectral imaging is to extract the spectra of endmembers from the multi-array of mixed spectral observations. Furthermore, having the spectra of endmembers computed, we can estimate the maximum abundance of each endmember in a given position on the surface. A 2D distribution of each corresponding abundance provides complete information for the surveyed area. One should also notice that both abundances and spectra of endmembers are nonnegative curves, and hence, the usage of NMF for this application seems to be reasonable. However, other decomposition methods (e.g., used in ICA) could also be applied [77].

To apply NMF a 3D multi-array should be unfolded to form an observation matrix $\mathbf{Y} \in \mathbb{R}^{I \times T}$ where I is the total number of pixels in one slice, and T is the number of subbands (Figure 8.13(a)). After applying NMF, we obtain the nonnegative matrix $\mathbf{A} \in \mathbb{R}^{I \times J}$ of abundances, and the nonnegative matrix $\mathbf{X} \in \mathbb{R}^{J \times T}$ of endmember spectra. The rows in $\mathbf{X}$ correspond to endmembers, and the respective columns in $\mathbf{A}$ refer to the abundances. The spectrum of an endmember is some kind of "fingerprint" or signature of the underlying material and it should be unique and suitable for identification. The abundance vectors are matricized in the reverse way to the used vectorization, and hence each column vector in $\mathbf{A}$ determines a 2D image of the abundance for a given material. Thus, we have J images of distribution of the underlying materials.

For the La Parguera hyperspectra image [94] which has 106 bands of size 250×239 pixels (see Figure 8.13(a)), the mixed region is composed of two major different endmembers which were recovered with NMF and shown in Figure 8.13(b). The corresponding abundance maps are illustrated in Figure 8.13(c) and Figure 8.13(d), where lighter pixels denote higher abundance.

Note that a 3D multi-array of mixed observations does not necessarily need to be unfolded and processed with matrix factorization methods. As the observed array is multi-dimensional, NTF and other multi-array decomposition methods seem to be more adequate for this application. For more detail, see, Zhang *et al.* [129].

8.3.4 Chemical Shift Imaging

Chemical Shift Imaging (CSI) [11] uses Nuclear Magnetic Resonance (NMR) spectra to identify biochemical properties of the surveyed object. NMR has already found many real-life applications such as medical

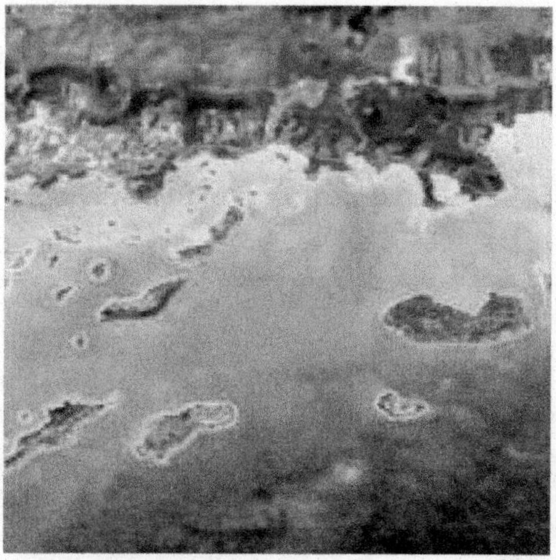

(a) True color of the La Parguera image

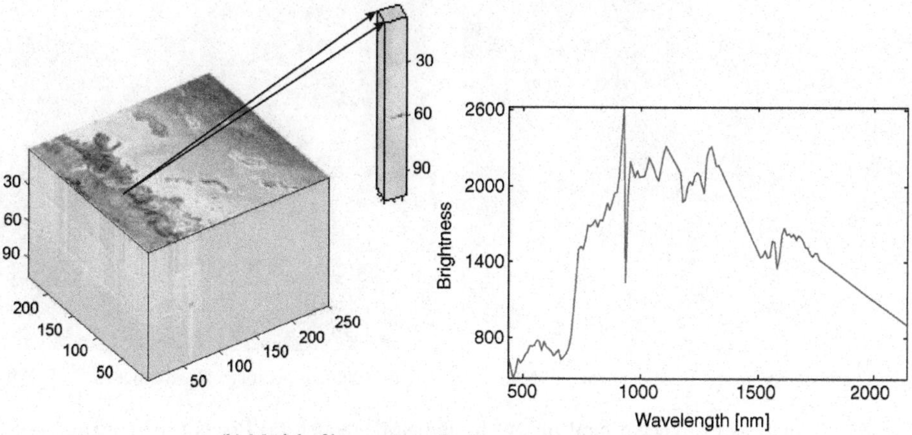

(b) Model of hyperspectra La Parguera image

Figure 8.12 Hyperspectral imagery illustration for the La Parguera hyperspectra image [94], captured with the EO-1 (Hyperion sensor) over La Parguera area Lajas, and Puerto Rico. The dataset processed using ACORN has 106 bands of size 250 $\times$ 239 pixels, where multiple subbands form a 3D multi-array. A vertical profile through a single pixel position shows the continuous spectrum assigned to this position. The spectrum identifies the surface material: (a) true color visualization includes bands at Red = 634 nm, Green = 545 nm, and Blue = 463 nm, (b) relative brightness at pixel (45,102) through 106 bands.

diagnosis (especially for brain tumors), non-destructive testing, chemistry (identification of chemical compounds), petroleum industry (analysis of sedimentary rocks), process control, and magnetometers. CSI is mostly applied for noninvasive detection, identification, and treatment of brain tumors. This imaging modality allows us to precisely characterize a biological tissue by a nuclear resonance frequency or chemical shift of nuclei from a given isotope that is introduced to the tissue as a molecular marker. The typical isotopes used for CSI are as follows: ^{1}H, ^{13}C, ^{113}Cd, ^{15}N, ^{14}N, ^{19}F, ^{31}P, ^{17}O, ^{29}Si, ^{10}B, ^{11}B, ^{23}Na, ^{35}Cl, ^{195}Pt. Nuclei from the isotopes resonate in a given magnetic field at the particular frequency that depends also on its chemical

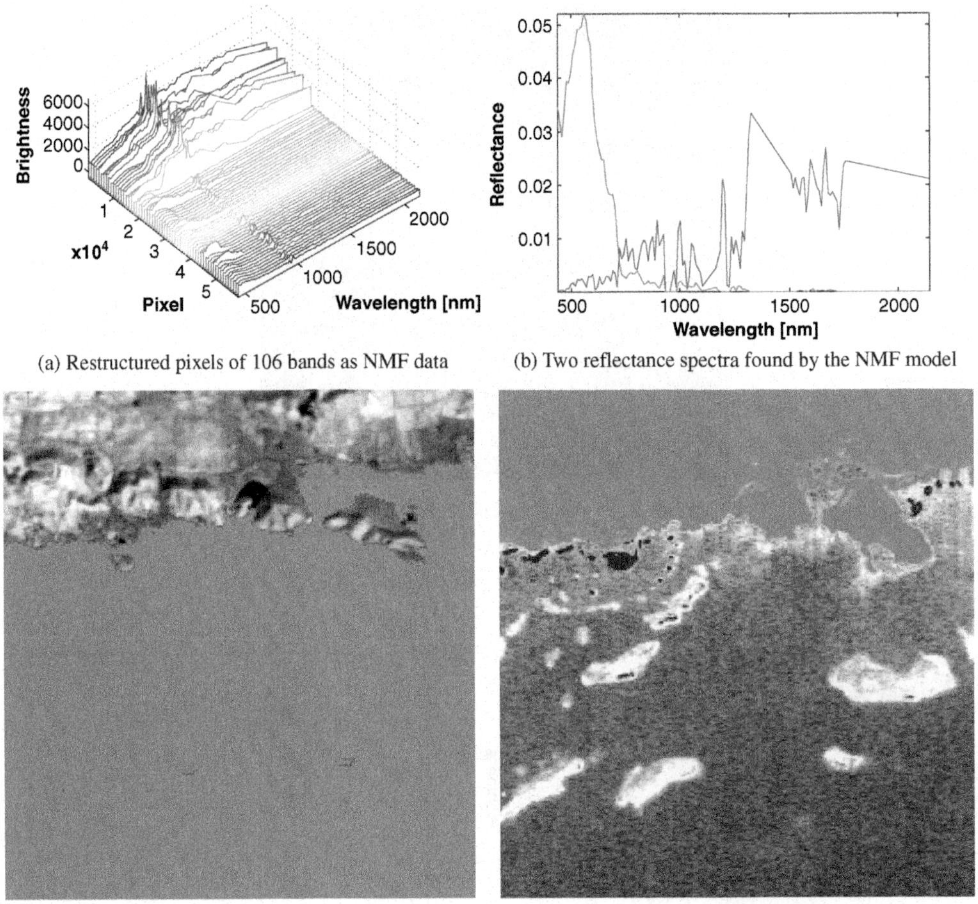

(a) Restructured pixels of 106 bands as NMF data

(b) Two reflectance spectra found by the NMF model

(c) Endmember image 1 - component 1 of NMF

(d) Endmember image 2 - component 2 of NMF

Figure 8.13 Analysis of hyperspectral images by the NMF model: (a) Pixels was reordered according to wavelengths for analysis, (b) two first reflectance components were estimated by NMF, (c)–(d) two endmember spectra correspond to the reflectance components.

environment, i.e. biochemical properties of a tissue. CSI maps the spatial distribution of nuclei associated with a particular chemical shift. Each tissue has a characteristic spectral profile which discriminates healthy brain tissues from tumors. However, the surveyed object might be composed of many types of a tissue with inhomogeneous chemical composition, which means that the observed spectra are a mixture of different constituent spectra. Dividing the surveyed object into I voxels and the observed spectrum in each voxel into T subbands, then all the observations can be stored in the observation matrix $\mathbf{Y} \in \mathbb{R}^{I \times T}$. The observed spectra represent a linear mixture of J constituent spectra (endmembers), which can be modeled as follows: $\mathbf{Y} \cong \mathbf{AX}$, where $\mathbf{A} \in \mathbb{R}^{I \times J}$ is a mixing matrix, and the rows of $\mathbf{X} \in \mathbb{R}^{J \times T}$ contain the spectra of endmembers. Each column vector in $\mathbf{A}$ represents the concentration/abundance of the corresponding endmember. Note that both matrices $\mathbf{A}$ and $\mathbf{X}$ are nonnegative, which justifies the usage of NMF within this decomposition. An example of an 8×8 voxel axial slice of the observed spectra selected from the 3D ^{31}P CSI database [130] of the human brain is shown in Figure 8.14. The spectra near the edges are seriously perturbed with noise. The spectra are composed of 512 subbands with the chemical shift range from 38.66 to -38.51 ppm.

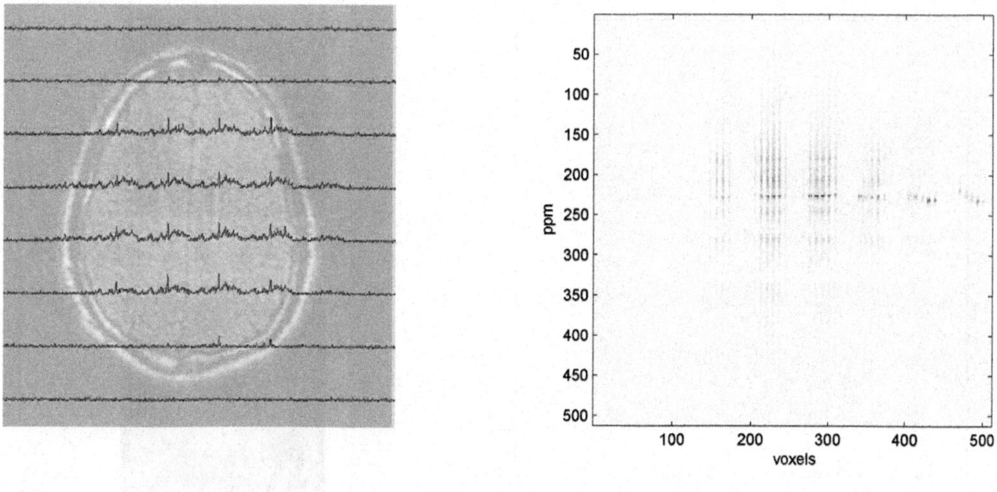

(a) (b)

Figure 8.14 Illustration of a ^{31}P CSI database [130] (a) 64 (8×8) selected observed spectra overlaid as axial slices, (b) imaging of the complete dataset of 512 voxels (8 × 8 × 8) with spectra of 512 points; most voxels are contaminated with noise, while few contain signals.

Figure 8.15 presents the results obtained with NMF applied to the 3D ^{31}P CSI database [130]. The plots in Figure 8.15(a) represent the spectra of endmembers, and the abundance maps are illustrated in Figure 8.15(b). As stated in Sajda *et al.* [96], the upper spectrum in Figure 8.15(a) represents muscle tissue whereas the bottom one refers to brain tissue. Consequently, the corresponding abundance maps present distributions of muscular and brain tissues. Indeed, the muscle tissue is distributed near the skull border, which is visible in the top image in Figure 8.15(b), whereas the brain tissue is centered in the interior of the skull.

8.4 Application of NMF for Analyzing Microarray Data

Matrix factorization and decomposition methods have also found many relevant applications in biomedical data processing and analysis. Several works are concerned with application of NMF to gene expression classification [13,14,33,53,55,73,83,84,109], mostly in order to classify different types of cancers. Other exemplary applications include muscle identifications in the nervous system [107], classification of PET images [1], and protein fold recognition [80].

8.4.1 Gene Expression Classification

The aim of applying NMF to the analysis of DNA microarrays is to group genes and experiments according to their similarity in gene expression patterns. The groups of genes that are referred to as *metagenes* (see [13]) capture latent structures in the observed data and may provide biological insight into underlying biological processes and the mechanisms of diseases. Metagenes also provide meaningful information for clustering the genes as well as the related experiments. Nested and partially overlapped clusters can also be identified with the NMF approach. Nested clusters reflect local properties of expression patterns, and overlapping is due to global properties of multiple biological processes (selected genes can participate in many processes). Typically, there are a few metagenes in the observed DNA microarray that may monitor several thousands of genes. Thus, the redundancy in this application is extremely high, which is very profitable for NMF.

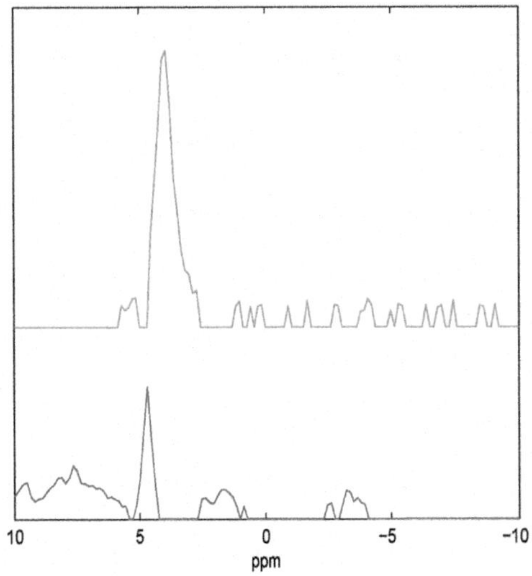

(a) Basis spectral components

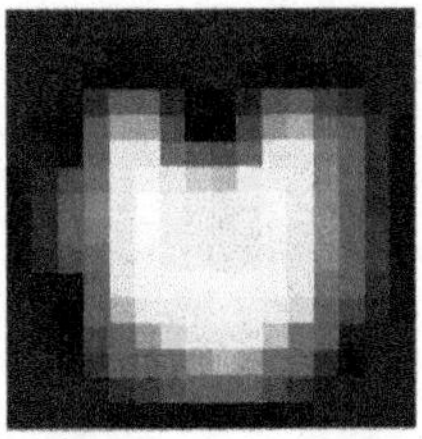

(b) Fifth axial slice of component 1

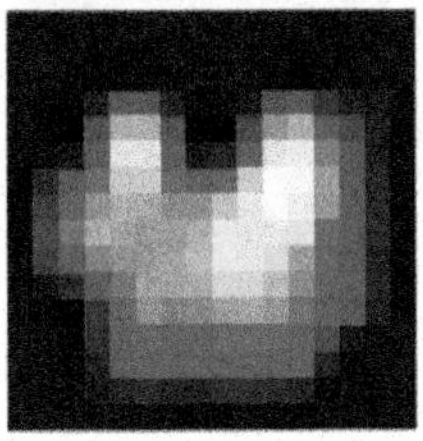

(c) Fifth axial slice of component 2

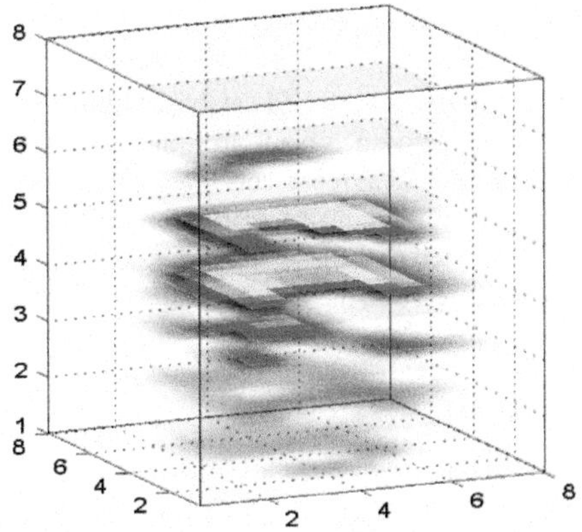

(d) Eight axial slices of components 1

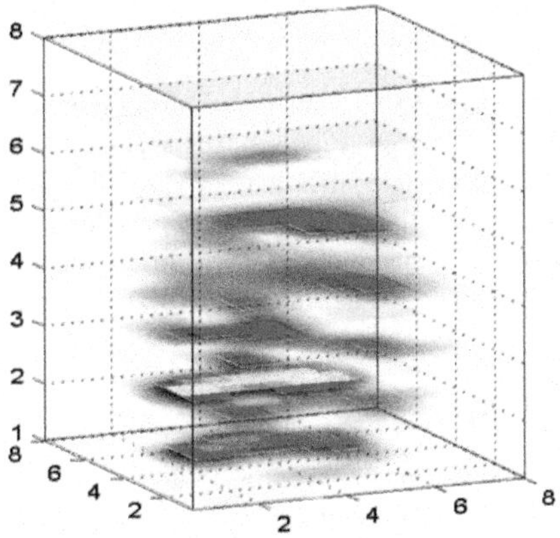

(e) Eight axial slices of components 2

Figure 8.15 Separation results with NMF for a CSI dataset [130] (a) two basis spectral components **A**; (b)–(c) fifth axial slices of two spectral components **A** which are parts of the matrix **X**; (d)–(e) eight axial slices of two basis spectral components or 3-D visualizations of **X**.

Furthermore, metagenes and gene expression patterns can often be described by sparse representations, and imposing sparsity constraints on the recovered factors may considerably improve the performance [14,33].

DNA microarrays simultaneously monitor expression levels of all the genes in a genome. The output from a DNA microarray is stored in the gene-expression matrix $\mathbf{Y} \in \mathbb{R}^{I \times T}$, where I is the number of monitored genes, and T is the number of samples. The t-th column vector of **Y** represents the expression levels of all the genes in the t-th sample that may represent distinct tissue, experiment, or time point. The i-th row vector of **Y** expresses levels of the i-th gene across all the samples. Assuming gene expressions can be grouped into J metagenes, **Y** can be factorized with NMF into the product of two nonnegative matrices $\mathbf{A} \in \mathbb{R}^{I \times J}$ and $\mathbf{X} \in \mathbb{R}^{J \times T}$. Each column vector of **A** represents a metagene, thus a_{ij} denotes the contribution of the i-th

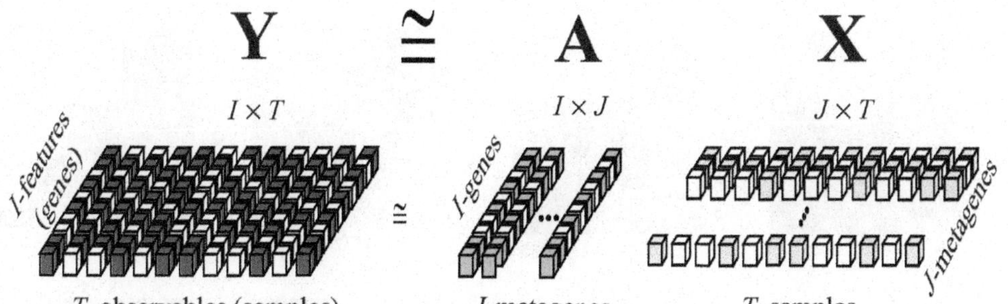

Figure 8.16 Schematic representation of the NMF model applied to gene-expression matrix **Y**.

gene into the j-th metagene, and x_{jt} is the expression level of the j-th metagene in the t-th sample. This factorization is illustrated in Figure 8.16.

Gene clustering can have a dual aspect – in terms of grouping samples (experiments) or genes. The position of the maximum value in each column vector of **X** indicates the index of the cluster to which the sample is assigned. Thus there will be J clusters of the samples. Looking for the maximum value in each row vector of **A**, we can group the genes according to their coherent expression patterns. Double sense clustering which can be applied in this context is referred to as *biclustering* [14]. For such clustering several techniques such as K-means [104], self-organizing maps (SOR), [103,105], or hierarchical clustering [30] have been proposed. However, as reported in [13,14,33] the NMF approach despite its stochastic nature is the most promising and gives the most stable results.

NMF is obtained with the alternating optimization process which is intrinsically nonconvex. Thus, different initializations may not give consistent results. Stability (consistency) of the decomposition can be tested, e.g. with a so-called connectivity matrix [13] that reflects cluster relationships between samples. For T samples, the connectivity matrix $C \in \mathbb{R}^{T \times T}$ is defined as a binary matrix, where $c_{st} = 1$ if the s-th sample and t-th sample belong to the same cluster, and $c_{st} = 0$, otherwise. The connectivity matrix from each run of NMF is reordered to form a block diagonal matrix. After performing several runs, a consensus matrix is calculated by averaging all the connectivity matrices. The entries of a consensus matrix should be ranged from 0 to 1, and they can be interpreted in terms of probabilities that determine the reproducibility of the class assignment.

When the number of metagenes is not known, NMF can be applied several times to each case with a different value of J. If the number of metagenes is well selected, NMF should provide consistent results, which should be observed in a consensus matrix (it should tend to a binary matrix).

To illustrate the use of NMF for classification of different types of cancers, we took the example from [13]. Brunet *et al.* use three different datasets to analyze the performance of NMF. The leukemia dataset is the reduced version of the original data used in [39]. The data contains 38 bone marrow samples of the 5000 most highly varying genes of acute myelogenous leukemia (AML) and acute lymphoblastic leukemia (ALL). The ALL cells can also be classified as the B or T lymphocytes, which in terms of a clustering viewpoint can be considered as hierarchical (nested) clusters. Indeed, the consensus matrix obtained by averaging 50 reordered connectivity matrices for this dataset confirms the hierarchical nature of the clusters as depicted in Figure 8.17(a). The consistent clusters can be observed for $J = 2$ when the classification is between the AML and ALL samples, and for $J = 3$ when the ALL samples are classified as the B or T subclasses. An increase in J results in stronger dispersion in the consensus matrix.

NMF also gave very promising results for the classification of childhood brain tumors known as medulloblastomas. The related gene expressions were taken from the dataset [88] that consists of the following samples: 10 classic medulloblastomas, 10 malignant gliomas, 10 rhabdoids, and four normal samples. The goal of using NMF for analyzing the medulloblastoma samples from this dataset is to find genes that are

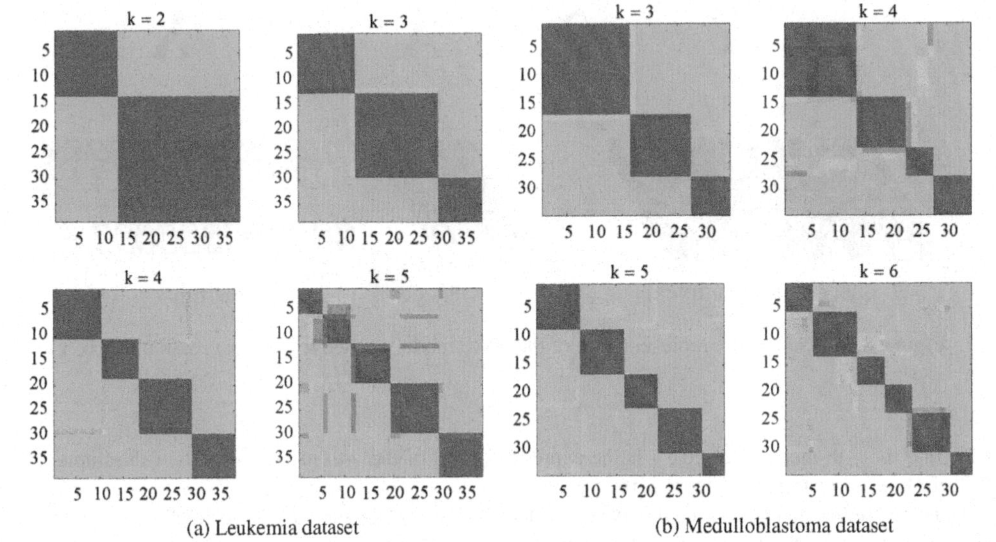

Figure 8.17 Patch plots of consensus matrices obtained with NMF (a) NMF applied for leukemia dataset, (b) NMF applied for medulloblastoma dataset.

statistically correlated with two basic classes of medulloblastomas: classic and desmoplastic. As reported in [13], only clustering with NMF provides easily interpretable clusters that are shown in Figure 8.17(b). For $J = 5$, one nested cluster is almost entirely related to the samples of the desmoplastic class.

Brunet *et al.* [13] successfully applied NMF to the classification of four types of central nervous system embryonal tumors [88]. Moreover, NMF gave much more accurate results than the hierarchical clustering and SOM. The SOR evidently identifies only three classes, merging the malignant glioma and normal samples. Also, the hierarchical clustering misclassifies the data. NMF for $J = 4$ gives a nearly binary consensus matrix that gives strong evidence for a four-class split of the data.

The NMF approach to gene classification given by Brunet *et al.* [13] has been extended by Carmona-Saez *et al.* [14] to dual classification, where the classification is not only with respect to the samples but also with respect to the genes. The aim of this approach is to identify gene expression clusters, i.e. sets of genes that have local coherent expression patterns, as well as to identify the samples highly associated with these clusters. One way to tackle this problem is to identify a gene expression bicluster separately for each metagene. The bicluster contains the genes that correspond to the largest values of entries in a given column vector of $\mathbf{A}$, and consequently, these samples form the same bicluster which corresponds to the largest values of entries in the respective row vector of $\mathbf{X}$. A bicluster can also be readily identified from a graphical representation of the reordered gene-expression matrix. For a given metagene, all the entries in the corresponding column vector of $\mathbf{X}$ and the row vector of $\mathbf{A}$ are sorted in decreasing order, and then a product of the matrices $\mathbf{A}$ and $\mathbf{X}$ adequately sorted gives the reordered gene-expression matrix whose entries with large values in the upper left corner determine a bicluster. This technique is illustrated in Figure 8.18.

Lighter entries in the upper left corner of the reordered gene-expression matrix indicate the genes and the samples that form the bicluster. Associating the gene expressions in a single bicluster with phenotypic variation of cells or tissues can provide meaningful information about tissue-specific functions or the molecular organization of diverse cells.

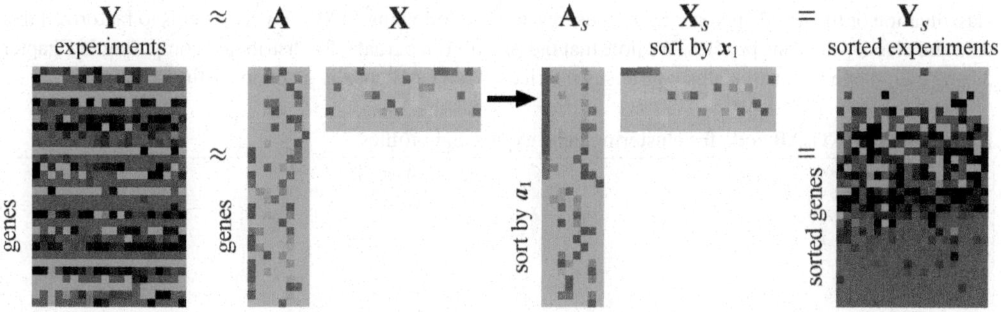

Figure 8.18 Biclustering using NMF.

8.4.2 Analysis of Time Course Microarray Data

From an additional viewpoint, gene expression data can be modeled as time course experiments. Factorization of such data with a small number of components $\mathbf{Y} = \mathbf{A}\,\mathbf{X}$ yields

- the matrix of basis expression profiles $\mathbf{X}$ which have a strong correlation with cell cycle-associated genes $\mathbf{Y}$,
- the contribution matrix $\mathbf{A}$ in which each row determines how a gene responds to the individual basis profiles $\mathbf{X}$.

Wentzell *et al.* [116] addressed this application with their method MCR-WALS (Multivariate Curve Resolution - Weighted Alternating Least Squares) for the yeast cell cycle data of Saccharomyces cerivisiae. An illustrative example for this section will replicate their results through the same dataset, with explanation in detail and by applying NMF algorithms discussed in this book.

In particular, we explain how to find the basis expression profiles of gene expression data and to classify and rearrange the gene profiles. The used datasets were first described by Spellman *et al.* [101]. They consisted of microarray measurements for 6178 open reading frames at seven minute intervals from 0 to 119 minutes, for a total of 18 experiments. However, after excluding genes for which there were more than four missing measurements, there remain 6044 genes which will be referred to as "Alpha-full" (in the spreadsheet "Alpha-Full" of the Excel file `2105-7-343-S3.xls`.[10]) A smaller set of 696 pre-selected genes from this group [69,116] was also used, and will be referred to as "Alpha-696" (in the spreadsheet "Alpha-696"). The missing measurements (2042 in Alpha-full, and zero in Alpha-696) were set to zero and their corresponding error standard deviations were set to a value much greater than the largest proportional error value in the data (a value of 100) [115].

To extract a small number of basis functions from nonlog-transformed microarray data with nonnegativity constraints, we apply the Fast HALS NMF algorithm to factorize the Alpha-696 dataset $\mathbf{Y} \in \mathbb{R}_+^{696\times 18}$ (instead of the full data) into $4, 5, \ldots, 9$ components under 1000 trials for each case. As mentioned above, each row of factor $\mathbf{A}$ expresses the levels that expression profiles $\mathbf{X}$ affected the gene data (see Figure 8.20 for illustration of expression profiles). Therefore, identification of genes expressed by the same expression levels is equivalent to classifying the set of row vectors of matrix $\mathbf{A}$ into distinct groups. The number of groups depends on the number of basis components in $\mathbf{X}$. This can be manipulated similarly as in the previous section based on the consensus matrices for each estimated result (see Brunet *et al.* [13]), however, instead of columns of matrix $\mathbf{X}$, we process rows of matrix $\mathbf{A}$. The best number of components can be defined based on the most consistent

[10]This additional file is associated with Wentzell's paper [116]: `http://www.biomedcentral.com/content/supplementary/1471-2105-7-343-S3.xls`

classification or the less dispersive consensus matrix. The following MATLAB code helps to factorized this data and classify the gene profiles. We note that the function "mc_nmf" for distributed computing in Chapter 3 can be employed to analyze the consensus matrices. MATLAB code is given in Listing 8.1.

Listing 8.1 MATLAB code for clustering gene expression profiles

```
% Factorize the yeast cell cycle data with NMF algorithm, and
% Classify genes based on the matrix of expression level A
% See also "Multivariate curve resolution of time course microarray data"
% Peter D Wentzell et al.
%
%% Load data
clear
load Alpha696
Yn = Y; In = size(Y);

%% Interpolate missing values
[r,c] = find((Yn(:,2:end-1) ==0)& (Ystdf(:,2:end-1)==100));
c = c+1;
for k = 1:numel(c)
    x = setdiff(1:size(Yn,2),c(k));
    Yn(r(k),c(k)) = interp1(x,Yn(r(k),x),c(k));
end
Yn = srcnorm(Yn,2);

%% Factorize the Alpha-696 into 4-9 components
Jarr = 4:9;                                            % number of components
k = 0;
consensus=zeros(numel(Jarr),In(1),In(1));
Ntrials = 1000;
AH = cell(1,numel(Jarr));
XH = cell(1,numel(Jarr));
for J = Jarr
    k = k+1;
    connac=zeros(In(1),In(1));
    for iter = 1:Ntrials
        options = struct('J',J,'ellnorm',1,'init',[4 4],'algtype',[6 6],...
            'nrestart',10,'niter',500,'algorithm','nmf_ghals','Nlayer',3);
        [AH{k},XH{k}]= nmf_multi_layer(Yn,options,[],1);
        connac = connac + nmfconnectivity(AH{k}');
    end
    consensus(k, :, :) = connac/Ntrials ;              % average
end
%% Cluster and reorder consensus matrices
[ordcons,clustid,ordindex,coph] = nmforderconsensus_sgl(consensus,Jarr);
%% Plot Consensus matrices
load gencolormap
for k = 1:size(ordcons)
    figure; imagesc(squeeze(ordcons(k,:,:))); axis square
    colormap(map); title(sprintf('k = %d',Jarr(k)))
end
%% Save results
save Alpha696_results
```

In Figure 8.19, we illustrate the six basis expression profiles estimated by NMF for 4, 5, ..., 9 components from the dataset Alpha-696. In each plot, the components (row vectors of $\mathbf{X}$) were normalized to ℓ_2-norm unit-length and arranged in order of appearance of the first major peak. Whereas the corresponding consensus clustering (similarity) matrices were computed based on 1000 matrices $\mathbf{A}$. Each case for a different number of components is displayed in Figure 8.20. Gene expression profiles in the same cluster are displayed in

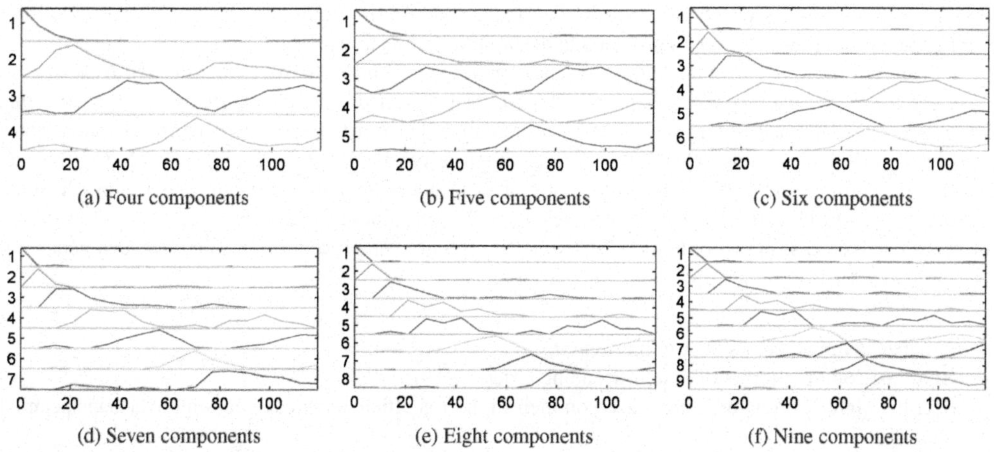

Figure 8.19 NMF results for the Alpha-696 dataset ranging from four to nine components shown in 119 minute intervals of the total study. The components are arranged in order of appearance of the first major peak in the time expression profile.

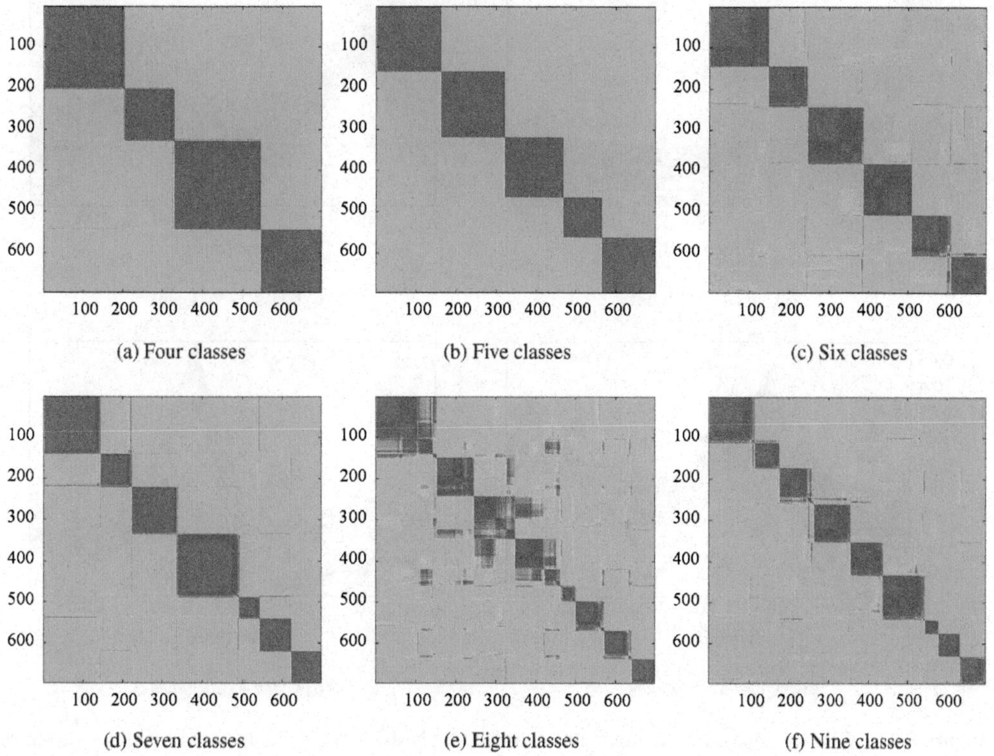

Figure 8.20 Visualization of consensus matrices for clustering gene expression profiles with 4, 5, ..., 9 classes from the Alpha-696 dataset. The similarity matrices are measured and reordered from 1000 trials for each case.

dark gray, while profiles which are never in the same cluster are in bright gray. Partitioning of the subset Alpha-696 into four or five classes provided high stabilities with strong consensus.

In the next step, we identify the highly correlated genes in the entire dataset "Alphafull" with each of the basis profile $\boldsymbol{x}_j$ extracted from "Alpha-696". This can be performed based on their correlation coefficients which are greater than a threshold of 0.8 [116]. An alternative approach to finding such highly correlated genes is to decompose the full dataset ($\mathbf{Y} \in \mathbb{R}_+^{6044\times 18}$) but with the given and fixed basis components $\mathbf{X}$. Coefficients in the same column vector of $\mathbf{A}$ reflect the contribution levels of the corresponding basis profiles $\mathbf{X}$ in the analyzed genes $\mathbf{Y}$. Therefore, a group of coefficients with high values in each column of $\mathbf{A}$ will show the set of genes that have a strong correlation with extracted profiles. The procedure for selecting such genes can be summarized as:

- normalize column vectors of $\mathbf{A}$ to unit-length,
- sort all the column vectors of $\mathbf{A}$ in descending order,
- select the highly correlated genes based on their highest coefficients in each column over one specified threshold.

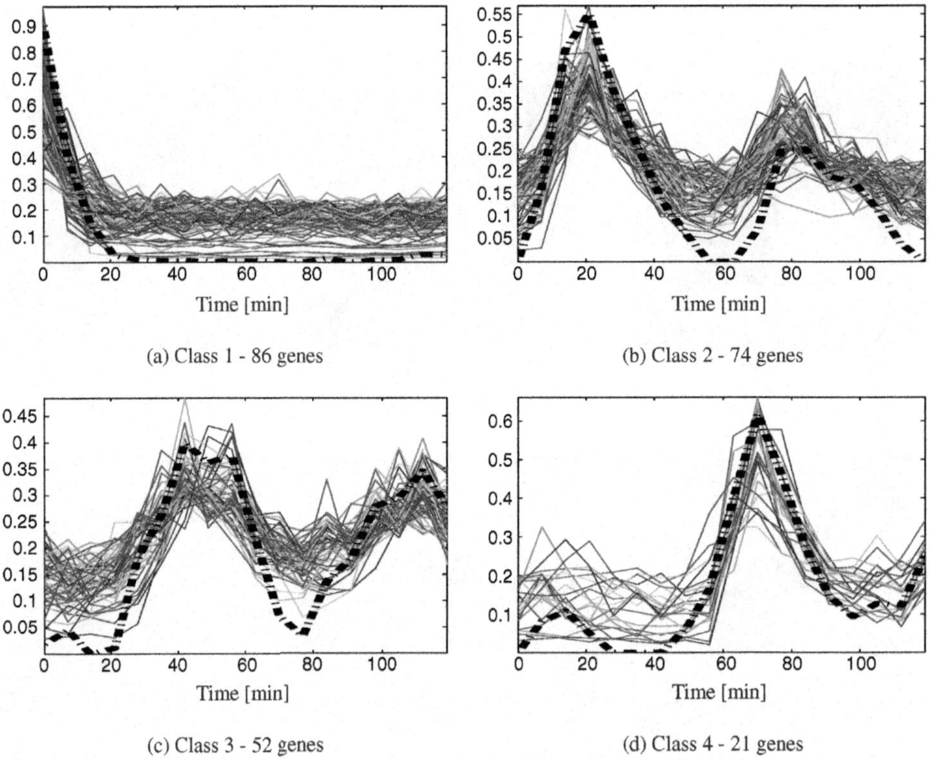

Figure 8.21 Comparison of four components extracted by NMF (thick dash lines) with highly correlated gene expression profiles from the Alpha_full dataset. The basis expression components are extracted from the small dataset: Alpha-696. All selected profiles have correlation coefficients $r > 0.8$, and are normalized to unit-length. The 20 most highly correlated gene expression profiles of each class are listed with their correlation coefficients.

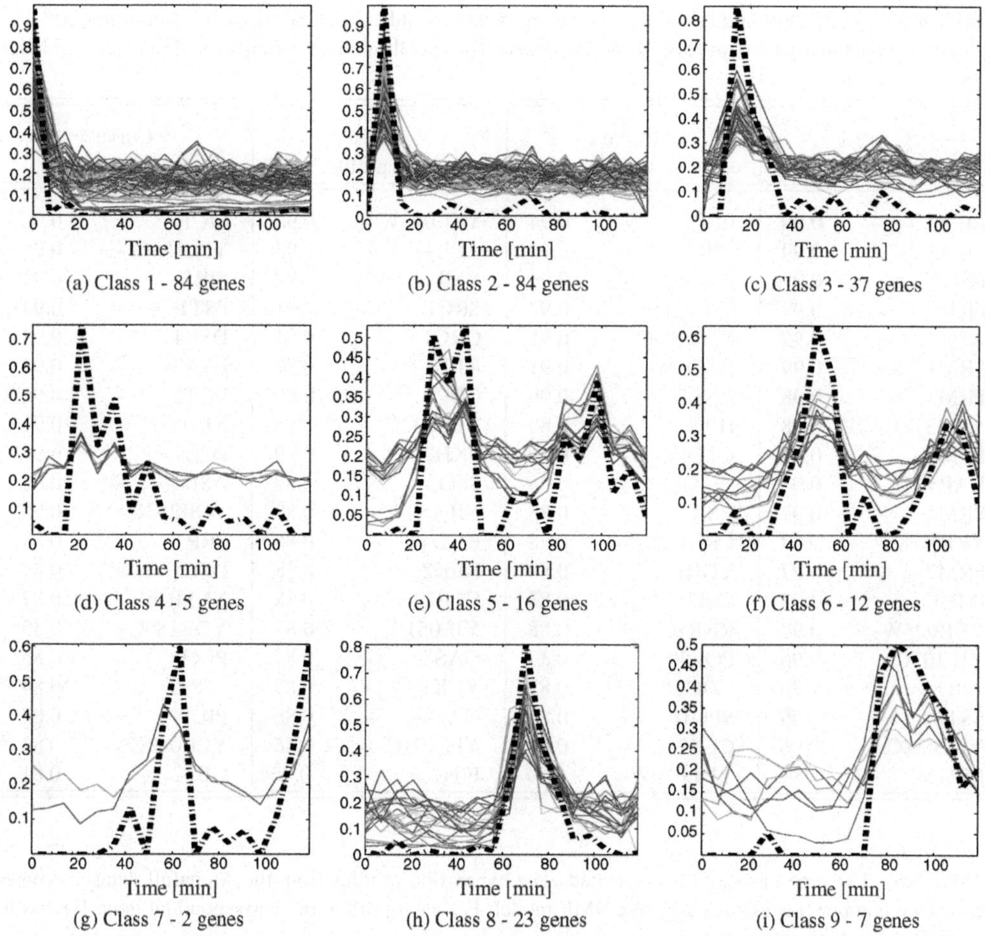

Figure 8.22 Comparison of nine components extracted by NMF (thick dash lines) with highly correlated genes expression profiles from the Alpha_full dataset. The basis expression components are extracted from the small dataset: Alpha-696. All selected profiles have correlation coefficients $r > 0.8$, and are normalized to unit-length.

The results of the selected gene profiles for the cases of four and nine component decompositions are shown in Figure 8.21 and 8.22, respectively. The most highly correlated profiles with their gene ID are also listed in Tables 8.5 and 8.6.

The final application will illustrate how to sort gene profiles according to the basis expression profiles. The procedure is performed as follows:

1. The basis gene expression profiles **X** are rearranged in order of appearance of the first major peaks (events). The columns of the contribution matrix **A** are then rearranged according to this order.
2. The coefficients in the columns of **A** are sorted in descending order.
3. The gene profiles are sorted according to both profiles in **X** and weighting in **A**.

Table 8.5 The 20 most highly correlated gene expression profiles from the Alphafull dataset mapped onto four components extracted by the NMF model. For each profile, the corresponding gene ID and its correlation coefficient are also given.

Curve 1 (86 profiles)			Curve 2 (74 profiles)			Curve 3 (52 profiles)			Curve 4 (21 profiles)		
AGA1	-	0.99	ERP3	-	0.94	YCL013W	-	0.93	SIC1	-	0.96
AGA2	-	0.99	ERP5	-	0.93	ALK1	-	0.93	YPL158C	-	0.95
BSC1	-	0.99	RAD51	-	0.93	YML033W	-	0.92	PIR1	-	0.94
FIG1	-	0.99	YIL141W	-	0.92	SRC1	-	0.91	PST1	-	0.94
ASG7	-	0.99	YDL009C	-	0.91	ORC1	-	0.91	DSE4	-	0.93
PRM1	-	0.99	RAD53	-	0.91	KIP2	-	0.90	FAA3	-	0.92
PRM3	-	0.98	RNR3	-	0.90	CDC5	-	0.89	EGT2	-	0.91
YIL037C	-	0.98	HIF1	-	0.89	YLL032C	-	0.89	NIS1	-	0.91
PRM10	-	0.98	CTF4	-	0.89	FKH2	-	0.89	PCL9	-	0.91
HAP4	-	0.97	JEM1	-	0.88	SSO2	-	0.89	ASH1	-	0.91
PRM5	-	0.97	RSR1	-	0.88	YIL158W	-	0.89	YOR263C	-	0.90
YCRX18C	-	0.97	CRH1	-	0.88	BUD4	-	0.88	PRR1	-	0.89
PRM7	-	0.97	YGR151C	-	0.88	HEK2	-	0.88	DSE3	-	0.89
IME4	-	0.97	CSI2	-	0.88	CLB2	-	0.88	YNL046W	-	0.87
YBR005W	-	0.97	RNR1	-	0.88	YJL051W	-	0.87	YDR186C	-	0.85
YJL107C	-	0.96	POL30	-	0.87	GAS5	-	0.87	PIR3	-	0.85
AIR1	-	0.96	CDC21	-	0.87	YCK1	-	0.87	YRO2	-	0.84
SST2	-	0.96	SPO16	-	0.87	GAS3	-	0.86	PIL1	-	0.83
YDL038C	-	0.96	CLN2	-	0.87	YPL141C	-	0.86	YLR049C	-	0.83
ERG24	-	0.96	YDL010W	-	0.87	FIN1	-	0.86	GBP2	-	0.81

Table 8.6 The nine most highly correlated gene expression profiles from the Alphafull dataset mapped onto nine components extracted by the NMF model. For each profile, the corresponding gene ID and its correlation coefficient are also given.

Curve 1 (84 profiles)			Curve 2 (84 profiles)			Curve 3 (37 profiles)			Curve 4 (5 profiles)			Curve 5 (16 profiles)		
FUS2	-	0.99	ECM9	-	0.99	SNO2	-	0.95	MDJ1	-	0.88	SUR4	-	0.88
YDR124W	-	0.99	MSF1'	-	0.99	YJR149W	-	0.92	YLR385C	-	0.86	HTA1	-	0.87
YOR385W	-	0.99	YDR193W	-	0.99	APJ1	-	0.91	HXK2	-	0.84	HTB1	-	0.87
YCL074W	-	0.99	SLF1	-	0.99	YOR111W	-	0.88	DAK1	-	0.82	SVL3	-	0.86
YCL076W	-	0.98	YIL172C	-	0.98	UBP2	-	0.87	HSP12	-	0.80	SEC61	-	0.86
KAR4	-	0.98	IFH1	-	0.98	YOX1	-	0.87		-		PMT4	-	0.84
PGU1	-	0.98	PHD1	-	0.98	CYT1	-	0.86		-		YNL116W	-	0.84
YCRX18C	-	0.98	RBL2	-	0.97	YIR003W	-	0.86		-		ERG8	-	0.84
GLG1	-	0.98	PBP1	-	0.97	ESC8	-	0.85		-		CWH41	-	0.82
PRM4	-	0.98	YJR136C	-	0.97	YMR196W	-	0.85		-		HHF2	-	0.82

(continued)

Table 8.6 (*Continued*)

Curve 6 (12 profiles)	Curve 7 (2 profiles)	Curve 8 (23 profiles)	Curve 9 (7 profiles)
NUF2 - 0.87	YNL057W- 0.82	YOR263C - 0.96	YOL155C - 0.89
GAS5 - 0.86	GAS4 - 0.81	PIR1 - 0.95	MSB2 - 0.83
SML1 - 0.86	-	PST1 - 0.95	YAR068W- 0.82
CLB1 - 0.86	-	YNL046W- 0.94	SUN4 - 0.82
YML033W- 0.85	-	PCL9 - 0.94	PRY3 - 0.82
YJL051W - 0.82	-	SIC1 - 0.94	YER152C - 0.81
YAR1 - 0.83	-	DSE4 - 0.93	CTS1 - 0.80
SCS2 - 0.82	-	EGT2 - 0.90	-
YCL012W - 0.81	-	NIS1 - 0.90	-
ALK1 - 0.81	-	PIR3 - 0.90	-

Figure 8.23 demonstrates this result for the Alpha-696 dataset and its five basis components. Only 80 gene profiles on the top and bottom of the dataset are displayed. The MATLAB source code is given in Listing 8.2.

Listing 8.2 MATLAB code for sorting gene profiles according to the basis expression profiles.

```
% load gene profile data
clear
load('Alpha696_results','AH','XH')
load('Alpha696','Yn') % load normalized gene profiles

Yn = bsxfun(@minus, Yn,min(Yn,[],2));
%% Sort gene profiles to Ys
As = AH{2};
Xs = XH{2};
% sort by events of X
[Xs,idc] = sortpeak(Xs);
% sort by weigthts in A
[As,idxr] = sortrows(As(:,idc),-1:-1:-5);
% sort Yn
Ys = Yn(idxr,:);
%% Read GeneId
[Num,GeneId]= xlsread('1471-2105-7-343-S3.xls.xls','Alpha-696','A7:A702');
GeneId = deblank(GeneId);

%% Display and comparison
   C = exp_colormap('green-red',64);

figure
   imagesc(Yn([1:40 end-40:end],: ))
    set(gca,'Ytick',1:size(Yn([1:40 end-40:end],:),1),'Yticklabel',...
      GeneId([1:40 end-40:end]))
    colormap(C)

    axis image
    ylim([.5 80])
    figure
    imagesc(Yn(idxr([1:40 end-40:end]),:))
    set(gca,'Ytick',1:size(Yn(idxr([1:40 end-40:end]),:),1),'Yticklabel',...
      GeneId(idxr([1:40 end-40:end])))
    colormap(C)
    axis image
    ylim([.5 80])
```

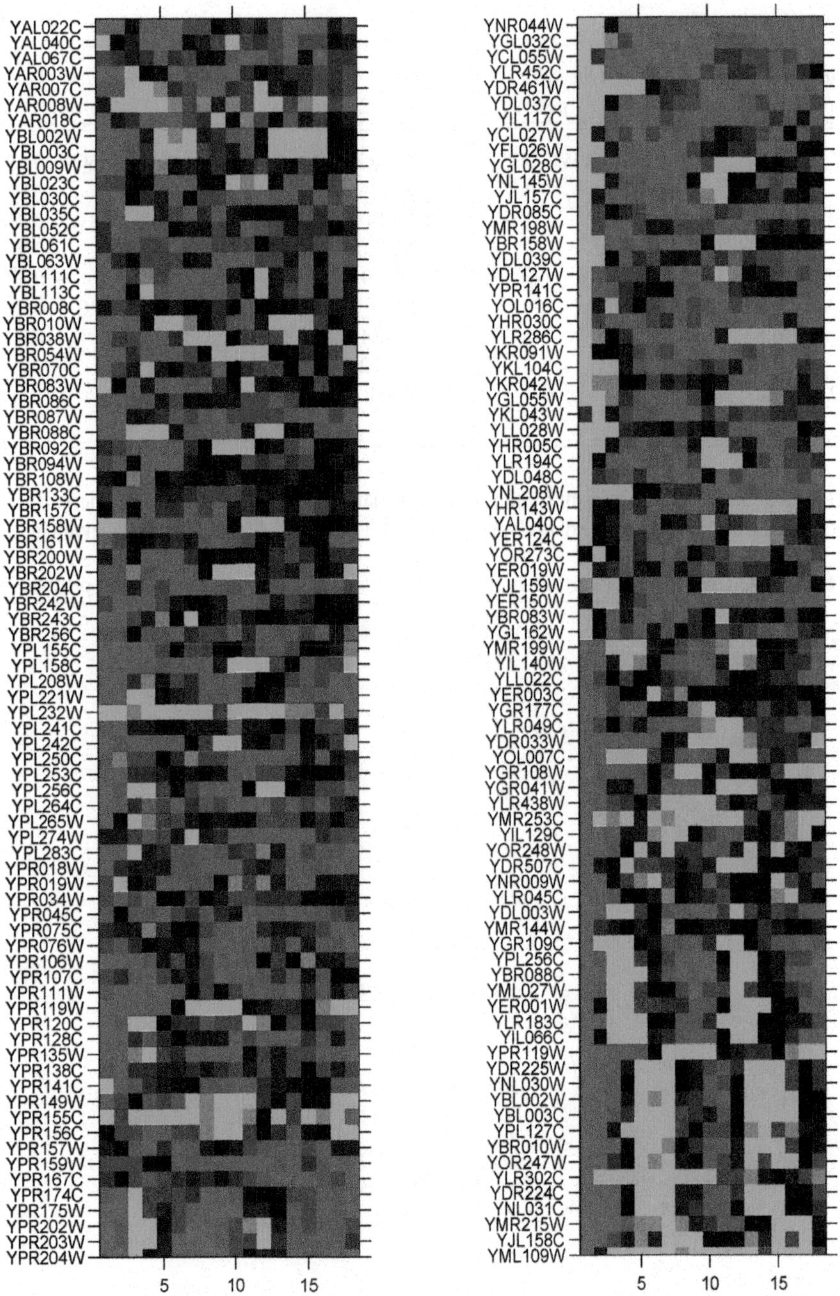

(a) Original Alpha-696 (b) Sorted Alpha-696

Figure 8.23 Sorting the dataset Alpha-696 according to the five extracted basis components. Basis gene expression profiles **X** are rearranged in order of appearance of the first major peaks (events), then the columns of the mixing matrix **A** are rearranged accordingly, and the coefficients in columns of **A** are sorted in descending order. Finally, the genes of the original data are sorted according to both events in **X** and weighting in **A**.

References

[1] J-H. Ahn, S. Kim, J-H. Oh, and S. Choi. Multiple nonnegative matrix factorization of dynamic PET images. In *ACCV*, page 5, 2004.

[2] M.R. Anderberg. *Cluster Analysis for Applications*. Monographs and Textbooks on Probability and Mathematical Statistics. Academic Press, Inc., New York, 1973.

[3] R.A. Baeza-Yates. Introduction to data structures and algorithms related to information retrieval. In *Information Retrieval: Data Structures and Algorithms*, pages 13–27. Prentice-Hall, Inc., Upper Saddle River, NJ, USA, 1992.

[4] G.H. Ball and D.J. Hall. ISODATA, a novel method of data analysis and classification. Technical report, Stanford University, Stanford, CA, April 1965.

[5] A. Banerjee, S. Merugu, I.S. Dhillon, and J. Ghosh. Clustering with Bregman divergences. In *SIAM International Conf. on Data Mining*, Lake Buena Vista, Florida, April 2004. SIAM.

[6] E. Benetos, M. Kotti, and C. Kotropoulos. Applying supervised classifiers based on non-negative matrix factorization to musical instrument classification. In *Proc. 2006 IEEE International Conference on Multimedia and Expo*, pages 2105–2108, Toronto, Canada, July 2006.

[7] E. Benetos, M. Kotti, and C. Kotropoulos. Musical instrument classification using non-negative matrix factorization algorithms. In *Proc. International Symposium on Circuits and Systems (ISCAS 2006)*, Island of Kos, Greece, May 21–24 2006.

[8] E. Benetos, M. Kotti, C. Kotropoulos, J.J. Burred, G. Eisenberg, M. Haller, and T. Sikora. Comparison of subspace analysis-based and statistical model-based algorithms for musical instrument classification. In *Proc. 2nd Workshop on Immersive Communication and Broadcast Systems (ICOB)*, Berlin, Germany, October 27–28 2005.

[9] M. Berry and M. Browne. Email surveillance using non-negative matrix factorization. *Computational and Mathematical Organization Theory*, 11(3):249–264, 2005.

[10] J.C. Bezdek. *Pattern Recognition With Fuzzy Objective Function Algorithms*. Plenum Press, New York, 1981.

[11] L. Brateman. Chemical shift imaging: A review. *American Journal of Roentgenology*, 146(5):971–980, 1986.

[12] R. Bro, N.M. Faber, and A. Rinnan. Standard error of prediction for multilinear PLS. 2. Practical implementation in fluorescence spectroscopy. *Chemometrics and Intelligent Laboratory Systems*, 75:69–76, 2005.

[13] J.P. Brunet, P. Tamayo, T.R. Golub, and J.P. Mesirov. Metagenes and molecular pattern discovery using matrix factorization. *Proceedings National Academy of Science U S A*, 101(12):4164–4169, March 2004.

[14] P. Carmona-Saez, R.D. Pascual-Marqui, F. Tirado, J.M. Carazo, and A. Pascual-Montano. Biclustering of gene expression data by non-smooth non-negative matrix factorization. *BMC Bioinformatics*, 7(78), 2006.

[15] C.H. Cheng. A branch-and-bound clustering algorithm. *IEEE Transactions on Systems Man Cyberntics*, 25(5):895–898, 1995.

[16] H. Cho, I.S. Dhillon, Y. Guan, and S. Sra. Minimum sum squared residue based co-clustering of gene expression data. In *Proc. 4th SIAM International Conference on Data Mining (SDM)*, pages 114–125, Florida, 2004.

[17] S. Choi. Algorithms for orthogonal nonnegative matrix factorization. In *Proc. of the International Joint Conference on Neural Networks (IJCNN)*, pages 1828–1832, Hong Kong, June 1–8 2008.

[18] A. Cichocki and S. Amari. *Adaptive Blind Signal and Image Processing*. John Wiley & Sons Ltd, New York, 2003.

[19] G. Csurka, C. Bray, C. Dance, and L. Fan. Visual categorization with bags of keypoints. In *Proc. Workshop Stat. Learn. Comput. Vis., ECCV*, pages 1–22, 2004.

[20] O.G. Cula and K.J. Dana. Compact representation of bidirectional texture functions. In *Proc. Comput. Vis. Pattern Recognit.*, volume 1, pages 1041–1047, Hawaii, USA, December 11–13 2001.

[21] S.C. Deerwester, S.T. Dumais, T.K. Landauer, G.W. Furnas, and R. A. Harshman. Indexing by latent semantic analysis. *Journal of the American Society of Information Science*, 41(6):391–407, 1990.

[22] I.S. Dhillon and D.M. Modha. Concept decompositions for large sparse text data using clustering. *Machine Learning Journal*, 42:143–175, 2001.

[23] E. Diday. The dynamic cluster method in non-hierarchical clustering. *Journal of Computing and Information Sciences*, 2:61–88, 1973.

[24] C. Ding, X. He, and H.D. Simon. On the equivalence of nonnegative matrix factorization and spectral clustering. In *Proc. SIAM International Conference on Data Mining (SDM'05)*, pages 606–610, 2005.

[25] C. Ding, T. Li, and W. Peng. Nonnegative matrix factorization and probabilistic latent semantic indexing: Equivalence, chi-square statistic, and a hybrid method. In *Proc. of AAAI National Conf. on Artificial Intelligence (AAAI-06)*, 2006.

[26] C. Ding, T. Li, W. Peng, and M.I. Jordan. Convex and semi-nonnegative matrix factorizations. Technical Report 60428, Lawrence Berkeley National Laboratory, November 2006.

[27] C. Ding, T. Li, W. Peng, and H. Park. Orthogonal nonnegative matrix tri-factorizations for clustering. In *KDD06: Proceedings of the 12th ACM SIGKDD international conference on Knowledge Discovery and Data Mining*, pages 126–135, New York, NY, USA, 2006. ACM Press.

[28] Q. Du, I. Kopriva, and H. Szu. Independent-component analysis for hyperspectral remote sensing imagery classification. *Opt. Eng*, 45, 2006.

[29] R.O. Duda, P.E. Hart, and D.G. Stork. *Pattern Classification*. John Wiley & Sons Ltd, New York, 2001.

[30] M.B. Eisen, P.T. Spellman, P.O. Brown, and D. Botstein. Cluster analysis and display of genome-wide expression patterns. *PNAS*, 95:14863–14868, 1998.

[31] T. Feng, S.Z. Li, H-Y. Shum, and H. Zhang. Local nonnegative matrix factorization as a visual representation. In *Proceedings of the 2nd International Conference on Development and Learning*, pages 178–193, Cambridge, MA, June 2002.

[32] K. Fukunaga. *Statistical Pattern Recognition*. Morgan Kaufmann, 1990.

[33] Y. Gao and G. Church. Improving molecular cancer class discovery through sparse non-negative matrix factorization. *Bioinformatics*, 21(21):3970–3975, 2005.

[34] D.J. Gardiner and P.R. Graves, editors. *Practical Raman Spectroscopy*. Springer, Berlin, 1989.

[35] E. Gaussier and C. Goutte. Relation between PLSA and NMF and implications. In *Proceedings of the 28th Annual International ACM SIGIR, Conference on Research and Development in Information Retrieval*, pages 15–19, Salvador, Brazil, August 15–19 2005.

[36] B. Georgescu, I. Shimshoni, and P. Meer. Mean shift based clustering in high dimensions: a texture classification example. In *Proc. 9th IEEE International Conference on Computer Vision (ICCV 2003)*, volume 1, pages 456–463, Nice, France, October 13–16 2003.

[37] C. Gobinet, A. Elhafid, V. Vrabie, R. Huez, and D. Nuzillard. About importance of positivity constraint for source separation in fluorescence spectroscopy. In *Proc. European Signal Processing Conference (EUSIPCO 2005)*, Antalya, Turkey, September 4–8 2005.

[38] C. Gobinet, E. Perrin, and R. Huez. Application of nonnegative matrix factorization to fluorescence spectroscopy. In *Proc. European Signal Processing Conference (EUSIPCO 2004)*, Vienna, Austria, September 6–10 2004.

[39] T.R. Golub, D.K. Slonim, P. Tamayo, C. Huard, M. Gaasenbeek, J.P. Mesirov, H. Coller, M.L. Loh, J.R. Downing, and M.A. Caligiuri. Molecular classification of cancer: Class discovery and class prediction by gene expression monitoring. *Science*, 286:531–537, 1999.

[40] H.F. Grahn and P. Geladi, editors. *Techniques and Applications of Hyperspectral Image Analysis*. John Wiley & Sons Ltd, Chichester, England, 2007.

[41] D. Greene and P. Cunningham. Producing accurate interpretable clusters from high-dimensional data. *Lecture Notes in Computer Science*, 3721:486–494, 2005.

[42] D. Guillamet, M. Bressan, and J. Vitrià. A weighted nonnegative matrix factorization for local representations. In *CVPR*, 2001.

[43] D. Guillamet and J. Vitrià. Classifying faces with nonnegative matrix factorization. In *Proc. 5th Catalan Conference for Artificial Intelligence*, Castello de la Plana, Spain, 2002.

[44] D. Guillamet and J. Vitrià. Determining a suitable metric when using non-negative matrix factorization. In *16th International Conference on Pattern Recognition (ICPR'02)*, volume 2, Quebec City, Canada, 2002.

[45] D. Guillamet, J. Vitrià, and B. Schiele. Introducing a weighted nonnegative matrix factorization for image classification. *Pattern Recognition Letters*, 24(14):2447–2454, 2003.

[46] F. Guimet, R. Boque, and J. Ferre. Application of non-negative matrix factorization combined with Fisher's linear discriminant analysis for classification of olive oil excitation emission fluorescence spectra. *Chemometrics and Intelligent Laboratory Systems*, 81(1):94–106, 2006.

[47] A.K. Jain and R.C. Dubes. *Algorithms for Clustering Data*. Prentice-Hall, Inc., Upper Saddle River, NJ, USA, 1988.

[48] A.K. Jain and J. Mao. Neural networks and pattern recognition. *Computational Intell. Imitating Life*, pages 194–212, 1994.

[49] A.K. Jain, M.N. Murty, and P.J. Flynn. Data clustering: a review. *ACM Comput. Surv.*, 31(3):264–323, 1999.

[50] S. Jia and Y. Qian. A complexity constrained nonnegative matrix factorization for hyperspectral unmixing. *Lecture Notes in Computer Science*, 4666:268–276, 2007.

[51] D. Jones and M.A. Beltramo. Solving partitioning problems with genetic algorithms. In *Proc. of the Fourth International Conference on Genetic Algorithms*, pages 442–449. Morgan Kaufmann Publishers, 1991.

[52] D. Kim, S. Sra, and I.S. Dhillon. Fast Newton-type methods for the least squares nonnegative matrix approximation problem. In *Proc. 6-th SIAM International Conference on Data Mining*, Minneapolis, Minnesota, USA, April 2007.

[53] H. Kim and H. Park. Sparse non-negative matrix factorizations via alternating non-negativity-constrained least squares for microarray data analysis. *Bioinformatics*, 23(12):1495–1502, 2007.

[54] J. Kim and H. Park. Sparse nonnegative matrix factorization for clustering. Technical report, Georgia Institute of Technology, 2008. http://smartech.gatech.edu/handle/1853/20058.

[55] P.M. Kim and B. Tidor. Subsystem identification through dimensionality reduction of large-scale gene expression data. *Genome Research*, 13:1706–1718, 2003.

[56] T. Kohonen. *Self-organization and associative memory: 3rd edition*. Springer-Verlag New York, Inc., New York, NY, USA, 1989.

[57] W.L.G. Koontz, K. Fukunaga, and P.M. Narendra. A branch and bound clustering algorithm. *IEEE Transactions on Computer*, 23:908–914, 1975.

[58] S. Lazebnik, C. Schmid, and J. Ponce. A maximum entropy framework for part-based texture and object recognition. In *Proc. 10th IEEE International Conference on Computer Vision (ICCV 2005)*, volume 1, pages 832–838, Beijing, China, October 17–20 2005.

[59] S. Lazebnik, C. Schmid, and J. Ponce. A sparse texture representation using local affine regions. *IEEE Transactions on Pattern Analysis and Machine Intelligence*, 27(8):1265–1278, 2005.

[60] Y. LeCun, L. Bottou, Y. Bengio, and P. Haffner. Gradient-based learning applied to document recognition. *Proceedings of the IEEE*, 86(11):2278–2324, 1998.

[61] Y. LeCun, L. Jackel, L. Bottou, C. Cortes, J. S. Denker, H. Drucker, I. Guyon, U.A. Muller, E. Sackinger, P. Simard, and V. Vapnik. Learning algorithms for classification: A comparison on handwritten digit recognition. In J.H. Oh, C. Kwon, and S. Cho, editors, *Neural Networks: The Statistical Mechanics Perspective*, pages 261–276. World Scientific, 1995.

[62] D.D. Lee and H.S. Seung. Learning of the parts of objects by non-negative matrix factorization. *Nature*, 401:788–791, 1999.

[63] D.D. Lee and H.S. Seung. *Algorithms for Nonnegative Matrix Factorization*, volume 13. MIT Press, 2001.

[64] T. Leung and J. Malik. Representing and recognizing the visual appearance of materials using three-dimensional textons. *International Journal of Computer Vision*, 43(1):29–44, 2001.

[65] H. Li, T. Adali, W. Wang, D. Emge, and A. Cichocki. Non-negative matrix factorization with orthogonality constraints and its application to Raman spectroscopy. *The Journal of VLSI Signal Processing*, 48(1–2):83–97, 2007.

[66] S.Z. Li, X.W. Hou, H. J. Zhang, and Q.S. Cheng. Learning spatially localized, parts-based representation. In *Proceedings of the IEEE Computer Society Conference on Computer Vision and Pattern Recognition (CVPR 2001)*, volume 1, pages I–207–I–212, 2001.

[67] T. Li and Ch. Ding. The relationships among various nonnegative matrix factorization methods for clustering. In *Proc. 6th International Conference on Data Mining (ICDM06)*, pages 362–371, Washington, DC, USA, 2006. IEEE Computer Society.

[68] J. Loader. *Basic Laser Raman Spectroscopy*. Heyden and Sons, London, 1970.

[69] P. Lu, A. Nakorchevskiy, and E.M. Marcotte. Expression deconvolution: A reinterpretation of DNA microarray data reveals dynamic changes in cell populations. In *Proceedings of the National Academy of Sciences of the United States of America*, 2003.

[70] S.Y. Lu and K.S. Fu. A sentence-to-sentence clustering procedure for pattern analysis. *IEEE Transactions on Systems Man Cyberntics*, 8:381–389, 1978.

[71] J. Mao and A.K. Jain. A self-organizing network for hyperellipsoidal clustering (HEC). *IEEE Transactions on Neural Networks*, 7(1):16–29, January 1996.

[72] J. Mcqueen. Some methods for classification and analysis of multivariate observations. In *Proceedings of the Fifth Berkeley Symposium on Mathematical Statistics and Probability*, pages 281–297, 1967.

[73] E. Mejia-Roa, P. Carmona-Saez, R. Nogales, C. Vicente, M. Vazquez, X. Y. Yang, C. Garcia, F. Tirado, and A. Pascual-Montano. bioNMF: a web-based tool for nonnegative matrix factorization in biology. *Nucleic Acids Research*, 36:W523–W528, 2008.

[74] T. Mitchell, editor. *Machine Learning*. McGraw Hill, Inc., New York, NY, 1997.

[75] S. Moussaoui. *Separation de sources non-negatives. Application au traitement des signaux de spectroscopie*. PhD thesis, Universite Henri Poincare, Nancy, France, December 2005.

[76] S. Moussaoui, D. Brie, C. Carteret, and A. Mohammad-Djafari. Application of Bayesian non-negative source separation to mixture analysis in spectroscopy. In *Proc. 24th Int. Workshop on Bayesian Inference and Maximum*

Entropy Methods in Science and Engineering, pages 25–30, Max-Planck Institute, Garching, Munich, Germany, July 2004.

[77] S. Moussaoui, H. Hauksdóttir, F. Schmidt, C. Jutten, J. Chanussot, D. Brie, S. Douté, and J. A. Benediktsson. On the decomposition of Mars hyperspectral data by ICA and Bayesian positive source separation. *Neurocomputing*, 71(10-12):2194–2208, 2008.

[78] D. Nuzillard and C. Lazar. Partitional clustering techniques for multi-spectral image segmentation. *Journal of Computers*, 2(10):1–8, 2007.

[79] O.G. Okun. Non-negative matrix factorization and classifiers: experimental study. In *Proc. of the Fourth IASTED International Conference on Visualization, Imaging, and Image Processing (VIIP04)*, pages 550–555, Marbella, Spain, 2004.

[80] O.G. Okun and H. Priisalu. Fast nonnegative matrix factorization and its application for protein fold recognition. *EURASIP Journal on Applied Signal Processing*, pages Article ID 71817, 8 pages, 2006.

[81] K. Ozawa. A stratificational overlapping cluster scheme. *Pattern Recognintion*, 18:279–286, 1985.

[82] P. Paatero and U. Tapper. Positive matrix factorization: A nonnegative factor model with optimal utilization of error estimates of data values. *Environmetrics*, 5:111–126, 1994.

[83] A. Pascual-Montano, J.M. Carazo, K. Kochi, D. Lehmean, and R. Pacual-Marqui. Nonsmooth nonnegative matrix factorization (nsNMF). *IEEE Transactions Pattern Analysis and Machine Intelligence*, 28(3):403–415, 2006.

[84] A. Pascual-Montano, P. Carmona-Saez, M. Chagoyen, F. Tirado, J.M. Carazo, and R. Pacual-Marqui. bioNMF: A versatile tool for non-negative matrix factorization in biology. *BMC Bioinformatics*, 7(366), 2006.

[85] V.P. Pauca, J. Pipera, and R.J. Plemmons. Nonnegative matrix factorization for spectral data analysis. *Linear Algebra and its Applications*, 416(1):29–47, 2006.

[86] V.P. Pauca, F. Shahnaz, M.W. Berry, and R.J. Plemmons. Text mining using non-negative matrix factorizations. In *Proc. SIAM Inter. Conf. on Data Mining*, Orlando, FL, April 2004.

[87] G. Peyre. Non-negative sparse modeling of textures. *Lecture Notes in Computer Science*, 4485:628–639, 2007.

[88] S.L. Pomeroy, P. Tamayo, M. Gaasenbeek, L.M. Sturla, M. Angelo, M.E. McLaughlin, J.Y.H. Kim, L.C. Goumnerovak, P.M. Black, C. Lau, J.C. Allen, D. Zagzag, J.M. Olson, T. Curran, C. Wetmore, J.A. Biegel, T. Poggio, S. Mukherjee, R. Rifkin, A. Califanok, G. Stolovitzkyk, D. N. Louis, J.P. Mesirov, E.S. Lander, and T.R. Golub. Prediction of central nervous system embryonal tumour outcome based on gene expression. *Nature*, 415:436–442, 2002.

[89] L. Qin, Q. Zheng, S. Jiang, Q. Huang, and W. Gao. Unsupervised texture classification: Automatically discover and classify texture patterns. *Image and Vision Computing*, 26(5):647–656, 2008.

[90] V.V. Raghavan and K. Birchard. A clustering strategy based on a formalism of the reproductive process in natural systems. *SIGIR Forum*, 14(2):10–22, 1979.

[91] Å. Rinnan. *Application of PARAFAC on Spectral Data*. PhD thesis, Royal Veterinary and Agricultural University (DK), 2004.

[92] M. Rojas, S.A. Santos, and D.C. Sorensen. Deterministic annealing approach to constrained clustering. *IEEE Transactions Pattern Analyssis and Machine Intelligence*, 15:785–794, 1993.

[93] E. H. Ruspini. A new approach to clustering. *Inf. Control*, 15:22–32, 1969.

[94] S.D. Hunt S. Rosario-Torres, M. Vélez-Reyes and L.O. Jiménez. New developments and application of the UPRM MATLAB hyperspectral image analysis toolbox. In *Proceedings of SPIE: Algorithms and Technologies for Multispectral, Hyperspectral, and Ultraspectral Imagery*, volume 6565 of *XII*, May 2007.

[95] P. Sajda, S. Du, T. Brown, L.C. Parra, and R. Stoyanova. Recovery of constituent spectra in 3D chemical shift imaging using nonnegative matrix factorization. In *Proc. of 4th International Symposium on Independent Component Analysis and Blind Signal Separation*, pages 71–76, Nara, Japan, April 2003.

[96] P. Sajda, S. Du, T.R. Brown, R. Stoyanova, D.C. Shungu, X. Mao, and L.C. Parra. Nonnegative matrix factorization for rapid recovery of constituent spectra in magnetic resonance chemical shift imaging of the brain. *IEEE Transactions on Medical Imaging*, 23(12):1453–1465, 2004.

[97] F. Samaria and A.C. Harter. Parameterisation of a stochastic model for human face identification. In *Proceedings of the Second IEEE Workshop on Applications of Computer Vision*, 1994.

[98] B. Savas and L. Eldén. Handwritten digit classification using higher order singular value decomposition. *Pattern Recognition*, 40(3):993–1003, 2007.

[99] I. Sethi and A.K. Jain, editors. *Artificial Neural Networks and Pattern Recognition: Old and New Connections*. Elsevier Science Inc., New York, NY, 1991.

[100] F. Shahnaz, M. Berry, P. Pauca, and R. Plemmons. Document clustering using non-negative matrix factorization. *Journal on Information Processing and Management*, 42:373–386, 2006.

[101] P.T. Spellman, G. Sherlock, M.Q. Zhang, V.R. Iyer, K. Anders, M.B. Eisen, P.O. Brown, D. Botstein, and B. Futcher. Comprehensive identification of cell cycle-regulated genes of the yeast saccharomyces cerevisiae by microarray hybridization. *Mol Biol Cell*, 9(12):3273–3297, December 1998.

[102] M.J. Symon. Clustering criterion and multi-variate normal mixture. *Biometrics*, 77:35–43, 1977.

[103] P. Tamayo, D. Slonim, J. Mesirov, Q. Zhu, S. Kitareewan, E. Dmitrovsky, E.S. Lander, and T.R. Golub. Interpreting patterns of gene expression with self-organizing maps: methods and application to hematopoietic differentiation. *PNAS*, 96:2907–2912, 1999.

[104] S. Tavazoie, J.D. Hughes, M. J. Campbell, R.J. Cho, and G.M. Church. Systematic determination of genetic network architecture. *Nature Genetics*, 22:281–285, 1999.

[105] P. Toronen, M. Kolehmainen, G. Wong, and E. Castren. Analysis of gene expression data using self-organizing maps. *FEBS Letters*, 451(2):142–146, 1999.

[106] S.R. Torres. Iterative Algorithms for Abundance Estimation on Unmixing of Hyperspectral Imagery. Master's thesis, University of Puerto Rico, Mayaguez Campus, 2004.

[107] M.C. Tresch, V.C.K. Cheung, and A. d'Avella. Matrix factorization algorithms for the identification of muscle synergies: Evaluation on simulated and experimental data sets. *Journal of Neurophysiology*, 95:2199–2212, 2006.

[108] M. Varma and A. Zisserman. Classifying images of materials: achieving viewpoint and illumination independence. In *Proc. Eur. Conf. Comput. Vis.*, volume 3, pages 255–271, Copenhagen, Denmark, May 2002.

[109] G. Wang, A.V. Kossenkov, and M.F. Ochs. LS-NMF: A modified non-negative matrix factorization algorithm utilizing uncertainty estimates. *BMC Bioinformatics*, 7(175), 2006.

[110] W. Wang and T. Adali. Constrained ICA and its application to Raman spectroscopy. In *2005 IEEE Antennas and Propagation Society International Symposium*, volume 4B, pages 109–112, Washington, DC, USA, July 3–8 2005.

[111] W. Wang, T. Adali, H. Li, and D. Emge. Detection using correlation bound and its application to Raman spectroscopy. In *2005 IEEE Workshop on Machine Learning for Signal Processing*, pages 259–264, Mystic, CT, September 2005.

[112] Y. Wang, Y. Jia, C. Hu, and M. Turk. Fisher nonnegative matrix factorization for learning local features. In *Proc. 6th Asian Conf. on Computer Vision*, Jeju Island, Korea, 2004.

[113] Y. Wang, Y. Jia, C. Hu, and M. Turk. Non-negative matrix factorization framework for face recognition. *International Journal of Pattern Recognition and Artificial Intelligence*, 19(4):495–511, 2005.

[114] A.R. Webb. *Statistical Pattern Recognition*. John Wiley & Sons Ltd, 2002.

[115] P.D. Wentzell. Alpha-696 data set. http://www.biomedcentral.com/content/supplementary/1471- 2105-7-343-S3.xls, 2006.

[116] P.D. Wentzell, T.K. Karakach, S. Roy, J. Martinez, C.P. Allen, and M. Werner-Washburne. Multivariate curve resolution of time course microarray data. *BMC Bioinformatics*, 7:343+, July 2006.

[117] S. Wild. *Seeding Non-negative Matrix Factorization with the Spherical K-means Clustering*. MSc. Thesis, University of Colorado, 2000.

[118] B. Xu, J. Lu, and G. Huang. A constrained non-negative matrix factorization in information retrieval. In *Proc. IEEE International Conference on Information Reuse and Integration (IRI 2003)*, pages 273–277, Las Vegas, NV, USA, October 27–29 2003.

[119] W. Xu, X. Liu, and Y. Gong. Document clustering based on non-negative matrix factorization. In *SIGIR '03: Proceedings of the 26th annual international ACM SIGIR conference on Research and development in informaion retrieval*, pages 267–273, New York, NY, USA, 2003. ACM Press.

[120] Y. Xue, C. S. Tong, W.-S. Chen, and W. Zhang. A modified non-negative matrix factorization algorithm for face recognition. In *Proc. 18th International Conference on Pattern Recognition (ICPR 2006)*, volume 3, pages 495–498, Hong Kong, August 20–24 2006.

[121] Y. Xue, C.S. Tong, and W. Zhang. Evaluation of distance measures for nmf-based face recognition. In *International Conference on Computational Intelligence and Security*, volume 1, pages 651–656, Guangzhou, China, 2006.

[122] Y. Xue, C.S. Tong, and W. Zhang. Survey of distance measures for NMF-based face recognition. *Lecture Notes in COmputer Science*, 4456:1039–1049, 2007.

[123] J. Yoo and S. Choi. Orthogonal nonnegative matrix factorization: Multiplicative updates on Stiefel manifolds. In *Intelligent Data Engineering and Automated Learning IDEAL 2008*, volume 5326 of *Lecture Notes in Computer Science*, pages 140–147. Springer-Verlag, Berlin, 2008.

[124] L.A. Zadeh. Fuzzy sets. *Inf. Control*, 8:338–353, 1965.

[125] C.T. Zahn. Graph-theoretical methods for detecting and describing gestalt clusters. *IEEE Transactions on Computers*, C-20:68–86, April 1971.

[126] R. Zass and A. Shashua. A unifying approach to hard and probabilistic clustering. In *International Conference on Computer Vision (ICCV)*, Beijing, China, October 2005.

[127] R. Zdunek. Data clustering with semi-binary nonnegative matrix factorization. *Springer, LNAI-5097*, 5097:705–716, 2008.

[128] D. Zhang, S. Chen, and Z.-H. Zhou. Two-dimensional non-negative matrix factorization for face representation and recognition. *Lecture Notes in Computer Science*, 3723:350–363, 2005.

[129] Q. Zhang, H. Wang, R. Plemmons, and P. Pauca. Tensor methods for hyperspectral data processing: A space object identification study. *J. Opt. Soc. Am. A*, 25:3001–3012, Dec 2008. http://www.wfu.edu/~plemmons/papers.htm.

[130] Q. Zhao, P. Patriotis, F. Arias-Mendoza, R. Stoyanova, and T.R. Brown. 3D interactive chemical shift imaging: A comprehensive software program for data analysis and quantification. In *Proceedings of the 47th Experimental Nuclear Magnetic Resonance Conference*, California, April 2006.

Index

Nonnegative Matrix and Tensor Factorizations: Applications to Exploratory Multi-way Data Analysis and Blind Source Separation Andrzej Cichocki, Rafal Zdunek, Anh Huy Phan and Shun-ichi Amari

CPSIA information can be obtained
at www.ICGtesting.com
Printed in the USA
BVOW07*0443071117
499708BV00003B/5/P

9 780470 746660